T0180384

Lecture Notes in Computer Science 13113

More information about this subseries at http://www.springer.com/series/7409

Hujun Yin · David Camacho · Peter Tino ·
Richard Allmendinger ·
Antonio J. Tallón-Ballesteros · Ke Tang ·
Sung-Bae Cho · Paulo Novais ·
Susana Nascimento (Eds.)

Intelligent Data Engineering and Automated Learning – IDEAL 2021

22nd International Conference, IDEAL 2021
Manchester, UK, November 25–27, 2021
Proceedings

Editors
Hujun Yin
University of Manchester
Manchester, UK

Peter Tino
University of Birmingham
Birmingham, UK

Antonio J. Tallón-Ballesteros
University of Huelva
Huelva, Spain

Sung-Bae Cho
Yonsei University
Seoul, Korea (Republic of)

Susana Nascimento
NOVA University of Lisbon
Lisbon, Portugal

David Camacho
Universidad Politecnica de Madrid
Madrid, Spain

Richard Allmendinger
University of Manchester
Manchester, UK

Ke Tang
Southern University of Science
and Technology
Shenzhen, China

Paulo Novais
University of Minho
Braga, Portugal

ISSN 0302-9743 ISSN 1611-3349 (electronic)
Lecture Notes in Computer Science
ISBN 978-3-030-91607-7 ISBN 978-3-030-91608-4 (eBook)
https://doi.org/10.1007/978-3-030-91608-4

LNCS Sublibrary: SL3 – Information Systems and Applications, incl. Internet/Web, and HCI

This Springer imprint is published by the registered company Springer Nature Switzerland AG
The registered company address is: Gewerbestrasse 11, 6330 Cham, Switzerland

Preface

The International Conference on Intelligent Data Engineering and Automated Learning (IDEAL) is an annual international conference dedicated to emerging and challenging topics in intelligent data analytics and associated machine learning paradigms and systems. The conference provides a unique opportunity and stimulating forum for presenting and discussing the latest theoretical advances and real-world applications in computational intelligence and intelligent data analysis.

After two decades of successful and memorable events, the 22nd edition, IDEAL 2021, was back in Manchester, UK, during November 25–27, 2021, for the third time. Although the organisers had intended to hold it as a traditional physical event, given the uncertainty of the COVID-19 situation and many travel restrictions and difficulties, it was decided in July that it would not be practical. Like the previous year and many other similar conferences, IDEAL 2021 was held as a virtual conference. It was technically co-sponsored by the IEEE Computational Intelligence Society UK and Ireland Chapter as well as the World Federation on Soft Computing.

For the past two decades, IDEAL has served an important role in the data analytics, machine learning, and AI communities. The conference aims to bring together researchers and practitioners to exchange the latest findings, disseminate state-of-the-art results, and share experiences and forge alliances on tackling many real-world challenging problems. Despite the unprecedented turbulence in the last eighteen months around the world, the IDEAL conference has continued to play its role in these communities during challenging times. The core themes of IDEAL 2021, as usual, included big data challenges, machine learning, deep learning, data mining, information retrieval and management, bio-/neuro-informatics, bio-inspired models, agents and hybrid intelligent systems, real-world applications of intelligence techniques, and AI.

In total, IDEAL 2021 received 85 submissions, which subsequently underwent rigorous peer reviews by the Program Committee members and experts. Only the papers judged to be of the highest quality and novelty were accepted and included in the proceedings. These proceedings contain 52 accepted papers for the main track and 9 accepted papers for special sessions, which were presented at IDEAL 2021.

In addition to the IDEAL 2021 main track, there had been a few special session proposals. After submission and peer-review, the following special sessions had sufficient numbers of accepted papers:

Special Session 1: Clustering for Interpretable Machine Learning
Special Session 2: Machine Learning Towards Smarter Multimodal Systems
Special Session 3: Computational Intelligence for Computer Vision and Image Processing

We deeply appreciate the efforts of our distinguished keynote speakers: Francisco Herrera of the University of Granada, Spain, Joaquin Vanschoren of Eindhoven University of Technology, The Netherlands, and Deming Chen of the University of Illinois at Urbana-Champaign, USA. We cannot thank them enough for their stimulating lectures.

We would like to thank our sponsors for their technical support. We would also like to thank all the people who devoted so much time and effort to the successful running of the conference, in particular the members of the Program Committee and reviewers, the organisers of the special sessions, and the authors who contributed to the conference.

A special thank you to the special sessions and workshop chairs, Antonio J. Tallón-Ballesteros and Susana Nascimento, and the publicity chairs, Bing Li, Guilherme Barreto, Jose A. Costa and Yimin Wen, for their fantastic work.

Finally, we are very grateful for the hard work of the local organising team at the University of Manchester and our event management collaborators, Magnifisence, in particular Lisa Carpenter and Gail Crowe. We also greatly appreciate the continued support, collaboration, and sponsorship for the best paper awards from Springer LNCS.

October 2021

Hujun Yin
David Camacho
Peter Tino
Richard Allmendinger
Antonio J. Tallón-Ballesteros
Ke Tang
Sung-Bae Cho
Paulo Novais
Susana Nascimento

Organisation

General Chairs

Hujun Yin	The University of Manchester, UK
David Camacho	Technical University of Madrid, Spain
Peter Tino	University of Birmingham, UK

Program Chairs

Richard Allmendinger	The University of Manchester, UK
Antonio J. Tallón-Ballesteros	University of Huelva, Spain
Ke Tang	Southern University of Science and Technology, China
Sung-Bae Cho	Yonsei University, South Korea
Paulo Novais	University of Minho, Portugal
Susana Nascimento	Universidade Nova de Lisboa, Portugal

Steering Committee

Hujun Yin	The University of Manchester, UK
Colin Fyfe	University of the West of Scotland, UK
Guilherme Barreto	Federal University of Ceará, Brazil
Jimmy Lee	Chinese University of Hong Kong, Hong Kong, China
John Keane	The University of Manchester, UK
Jose A. Costa	Federal University of Rio Grande do Norte, Brazil
Juan Manuel Corchado	University of Salamanca, Spain
Laiwan Chan	Chinese University of Hong Kong, Hong Kong, China
Malik Magdon-Ismail	Rensselaer Polytechnic Institute, USA
Marc van Hulle	KU Leuven, Belgium
Ning Zhong	Maebashi Institute of Technology, Japan
Peter Tino	University of Birmingham, UK
Samuel Kaski	Aalto University, Finland
Vic Rayward-Smith	University of East Anglia, UK
Yiu-ming Cheung	Hong Kong Baptist University, Hong Kong, China
Zheng Rong Yang	University of Exeter, UK

Publicity and Liaisons Chairs

Bin Li	University of Science and Technology of China, China
Guilherme Barreto	Federal University of Ceará, Brazil
Jose A. Costa	Federal University of Rio Grande do Norte, Brazil
Yimin Wen	Guilin University of Electronic Technology, China

Program Committee

Hector Alaiz Moreton	University of León, Spain
Jesus Alcala-Fdez	University of Granada, Spain
Richardo Aler	Universidad Carlos III de Madrid, Spain
Cesar Analide	University of Minho, Portugal
Romis Attux	University of Campinas, Brazil
Bruno Baruque	University of Burgos, Spain
Carmelo Bastos Filho	University of Pernambuco, Brazil
Gema Bello Orgaz	Universidad Politécnica de Madrid, Spain
José Alberto Benítez-Andrades	University of León, Spain
Lordes Borrajo	University of Vigo, Spain
Vicent Botti	Universitat Politècnica de València, Spain
Federico Bueno de Mata	Universidad de Salamanca, Spain
Robert Burduk	Wroclaw University of Science and Technology, Poland
Jose Luis Calvo-Rolle	University of A Coruña, Spain
Roberto Carballedo	University of Deusto, Spain
Joao Carneiro	Instituto Superior de Engenharia do Porto, Portugal
Mercedes Carnero	Universidad Nacional de Rio Cuarto, Argentina
Carlos Carrascosa	Universidad Politecnica de Valencia, Spain
José Luis Casteleiro-Roca	University of A Coruña, Spain
Pedro Castillo	University of Granada, Spain
Luís Cavique	University of Aberta, Potugal
Richard Chbeir	Université de Pau et des Pays de l'Adour, France
Songcan Chen	Nanjing University of Aeronautics and Astronautics, China
Xiaohong Chen	Nanjing University of Aeronautics and Astronautics, China
Stelvio Cimato	Università degli Studi di Milano, Italy
Manuel Jesus Cobo Martin	University of Cádiz, Spain
Leandro Coelho	Pontifícia Universidade Católica do Parana, Brazil
Carlos Coello Coello	CINVESTAV-IPN, Mexico
Roberto Confalonieri	Free University of Bozen-Bolzano, Italy
Rafael Corchuelo	University of Seville, Spain
Francesco Corona	Aalto University, Finland
Luís Correia	Universidade de Lisboa, Portugal
Paulo Cortez	University of Minho, Portugal

Jose Alfredo F. Costa	Federal University of Rio Grande do Norte, Brazil
Ângelo Costa	University of Minho, Portugal
Raúl Cruz-Barbosa	Universidad Tecnológica de la Mixteca, Mexico
Alfredo Cuzzocrea	University of Calabria, Italy
Ireneusz Czarnowski	Gdynia Maritime University, Poland
Ernesto Damiani	University of Milan, Italy
Andre de Carvalho	University of São Paulo, Brazil
Fernando De La Prieta	University of Salamanca, Spain
Pedro José de Oliveira	University of Minho, Portugal
Javier Del Ser	University of the Basque Country, Spain
Juan Manuel Dodero	Universidad de Cádiz, Spain
Dinu Dragan	University of Novi Sad, Serbia
Dalila Alves Durães	Universidade do Minho, Portugal
Florentino Fdez-Riverola	University of Vigo, Spain
Bruno Fernandes	University of Minho, Portugal
Antonio Fernández-Caballero	Universidad de Castilla-La Mancha, Spain
João Ferreira	ISCTE, Portugal
Lino Figueiredo	Instituto Superior de Engenharia do Porto, Portugal
Joaquim Filipe	EST-Setubal/IPS, Portugal
Giancarlo Fortino	Università della Calabria, Italy
Felipe M. G. França	COPPE-UFRJ, Brazil
Pedro Freitas	Universidade Católica Portuguesa, Portugal
Dariusz Frejlichowski	West Pomeranian University of Technology, Poland
Hamido Fujita	Iwate Prefectural University, Japan
Marcus Gallagher	University of Queensland, Australia
Isaias Garcia	University of León, Spain
Luis Javier Garcia Villalba	Universidad Complutense de Madrid, Spain
María José Ginzo Villamayor	Universidad de Santiago de Compostela, Spain
Antonio Gonzalez-Pardo	Universidad Rey Juan Carlos, Spain
Anna Gorawska	Silesian University of Technology, Poland
Marcin Gorawski	Silesian University of Technology, Poland
Manuel Graña	University of the Basque Country, Spain
Maciej Grzenda	Warsaw University of Technology, Poland
Pedro Antonio Gutierrez	Universidad de Cordoba, Spain
Ioannis Hatzilygeroudis	University of Patras, Greece
J. Michael Herrmann	University of Edinburgh, UK
Ignacio Hidalgo	Universidad Complutense de Madrid, Spain
Wei-Chiang Hong	Jiangsu Normal University, China
Jean-Michel Ilie	Université Pierre et Marie Curie, France
Dariusz Jankowski	Wrocław University of Technology, Poland
Esteban Jove	Universidade da Coruña, Spain
Vicente Julian	Universitat Politècnica de València, Spain
Jason Jung	Chung-Ang University, South Korea
Rushed Kanawati	Université Paris 13, France
Raul Lara-Cabrera	Universidad Politécnica de Madrid, Spain
Bin Li	University of Science and Technology of China, China

Victor Lobo	Universidade Nova de Lisboa, Portugal
Wenjian Luo	Harbin Institute of Technology, China
Jesús López	Tecnalia Research & Innovation, Spain
José Machado	University of Minho, Portugal
Rui Neves Madeira	Instituto Politécnico de Setúbal, Portugal
Ana Madureira	Instituto Superior de Engenharia do Porto, Portugal
Diana Manjarres	Tecnalia Research & Innovation, Spain
Goreti Marreiros	Instituto Superior de Engenharia do Porto, Portugal
Alejando Martín	Universidad Politécnica de Madrid, Spain
José F. Martínez-Trinidad	INAOE, Mexico
Armando Mendes	University of Azores, Portugal
Cristian Mihaescu	University of Craiova, Romania
Boris Mirkin	Higher School of Economics Moscow, Russia
José M. Molina	Universidad Carlos III de Madrid, Spain
Paulo Moura Oliveira	University of Trás-os-Montes and Alto Douro, Portugal
Tatsuo Nakajima	Waseda University, Japan
Grzegorz J. Nalepa	AGH University of Science and Technology, Poland
Antonio Neme	Universidad Nacional Autonoma de Mexico, Mexico
Jose Neves	University of Minho, Portugal
Ngoc-Thanh Nguyen	Wroclaw University of Science and Technology, Poland
Fernando Nuñez	University of Seville, Spain
Eva Onaindia	Universitat Politècnica de València, Spain
Eneko Osaba	Tecnalia Research & Innovation, Spain
Vasile Palade	Coventry University, UK
Jose Palma	University of Murcia, Spain
Juan Pavón	Universidad Complutense de Madrid, Spain
Carlos Pereira	Instituto Superior de Engenharia de Coimbra, Portugal
Radu-Emil Precup	Politehnica University of Timisoara, Romania
Paulo Quaresma	Universidade de Evora, Portugal
Héctor Quintián	University of A Coruña, Spain
Cristian Ramírez-Atencia	Universidad Politécnica de Madrid, Spain
Izabela Rejer	University of Szczecin, Poland
Matilde Santos	Universidad Complutense de Madrid, Spain
Richardo Santos	Polytechnic Institute of Porto, Portugal
Jose Santos	University of A Coruña, Spain
Fábio Silva	University of Minho, Portugal
Ivan Silva	University of São Paulo, Brazil
Dragan Simic	University of Novi Sad, Serbia
Anabela Simões	Coimbra Institute of Engineering, Portugal
Marcin Szpyrka	AGH University of Science and Technology, Poland
Jesús Sánchez-Oro	Universidad Rey Juan Carlos, Spain
Murat Caner Testik	Hacettepe University, Turkey

Qing Tian	Nanjing University of Information Science and Technology, China
Stefania Tomasiello	University of Salerno, Italy
Carlos M. Travieso-González	University of Las Palmas de Gran Canaria, Spain
Alexandros Tzanetos	University of the Aegean, Greece
Eiji Uchino	Yamaguchi University, Japan
Paulo Urbano	Universidade de Lisboa, Portugal
José Valente de Oliveira	Universidade do Algarve, Portugal
Alfredo Vellido	Universitat Politècnica de Catalunya, Spain
Gianni Vercelli	University of Genoa, Italy
José R. Villar	University of Oviedo, Spain
Tzai-Der Wang	Cheng Shiu University, Taiwan
Dongqing Wei	Shanghai Jiao Tong University, China
Michal Wozniak	Wroclaw University of Technology, Poland
Xin-She Yang	Middlesex University, UK

Additional Reviewers

Julio Palacio
Manuel António Martins
Jon Timmis
Víctor Flores
Luis Rus Pegalajar
Beatriz Ruiz Reina
Barbara Pes
Ying Bi
Victor Flores
Wenbin Pei
Martín Molina
Ji-Yoon Kim
Alexandre Nery
Alvaro Huertas
Rui Neves Madeira
Tri-Hai Nguyen
Carlos Camacho
Gen Li
Gunju Lee
Boyan Xu

Special Session on Clustering for Interpretable Machine Learning

Organisers

Susana Nascimento	NOVA University Lisbon, Portugal
José Valente de Oliveira	University of Algarve, Portugal
Victor Sousa Lobo	NOVA University Lisbon, Portugal
Boris Mirkin	National Research University Higher School of Economics, Russia

Special Session on Machine Learning Towards Smarter Multimodal Systems

Organisers

Nuno Correia	NOVA University Lisbon, Portugal
Rui Neves Madeira	NOVA University Lisbon and Polytechnic Institute of Setúbal, Portugal
Susana Nascimento	NOVA University Lisbon, Portugal

Special Session on Computational Intelligence for Computer Vision and Image Processing

Organisers

Ying Bi	Victoria University of Wellington, New Zealand
Bing Xue	Victoria University of Wellington, New Zealand
Antonio J. Tallón-Ballesteros	University of Huelva, Spain

Contents

Special Session on Clustering for Interpretable Machine Learning

Special Session on Machine Learning towards Smarter Multimodal Systems

Special Session on Computational Intelligence for Computer Vision and Image Processing

Main Track

A Comparison of Machine Learning Approaches for Predicting In-Car Display Production Quality

Luís Miguel Matos[1], André Domingues[1], Guilherme Moreira[2], Paulo Cortez[1](✉), and André Pilastri[3]

[1] ALGORITMI Centre, Department of Information Systems, University of Minho, Guimarães, Portugal
{luis.matos,pcortez}@dsi.uminho.pt, a76953@alunos.uminho.pt

[2] Bosch Car Multimedia, Braga, Portugal
Guilherme.Moreira2@pt.bosch.com

[3] EPMQ, CCG ZGDV Institute, Guimarães, Portugal
andre.pilastri@ccg.pt

Abstract. In this paper, we explore eight Machine Learning (ML) approaches (binary and one-class) to predict the quality of in-car displays, measured using Black Uniformity (BU) tests. During production, the industrial manufacturer routinely executes intermediate assembly (screwing and gluing) and functional tests that can signal potential causes for abnormal display units. By using these intermediate tests as inputs, the ML model can be used to identify the unknown relationships between intermediate and BU tests, helping to detect failure causes. In particular, we compare two sets of input variables (A and B) with hundreds of intermediate quality measures related with assembly and functional tests. Using recently collected industrial data, regarding around 147 thousand in-car display records, we performed two evaluation procedures, using first a time ordered train-test split and then a more robust rolling windows. Overall, the best predictive results (92%) were obtained using the full set of inputs (B) and an Automated ML (AutoML) Stacked Ensemble (ASE). We further demonstrate the value of the selected ASE model, by selecting distinct decision threshold scenarios and by using a Sensitivity Analysis (SA) eXplainable Artificial Intelligence (XAI) method.

Keywords: Anomaly detection · Automated machine learning · Deep learning · Explainable artificial intelligence · Supervised learning · One-class learning.

1 Introduction

The Industry 4.0 generates big data that can be used by Machine Learning (ML) algorithms to provide value [12]. In this work, we predict in-car display quality based on hundreds of assembly and functional tests from Bosch Car Multimedia. Currently, the manufacturer uses Black Uniformity (BU) tests to measure the display of solid black over the entire device screen. By adopting predictive ML models, the goal is to model

H. Yin et al. (Eds.): IDEAL 2021, LNCS 13113, pp. 3–11, 2021.
https://doi.org/10.1007/978-3-030-91608-4_1

the currently unknown relationships between the assembly and functional tests (the inputs) with the in-car display final quality (measured using BU tests), allowing to identify failure causes (e.g., screwing defect).

Most related works using ML to predict production faults consider a binary classification approach [1]. A less adopted approach is to use a one-class learning, such as Isolation Forest (iForest) and deep dense Autoencoder (AE), which only uses normal examples during the training phase [9, 11]. Moreover, the best ML algorithm is often selected by using trial-and-error experiments that consume time. The Automated ML (AutoML) and Automated Deep Learning (ADL) concepts were proposed to reduce the ML analyst effort [4]. Within our knowledge, there are no studies that use AutoML or ADL to predict production failures. In addition, there are no studies that model BU quality using ML based on assembly and functional tests. Thus, the novelty of this paper comes from a practical application point of view, where we compare several ML algorithms (e.g., one-class, AutoML and ADL) and estimate their potential value for the analyzed industrial use case. In particular, we explore eight ML methods: one-class - iForest and deep AE; and binary classification - Decision Tree (DT), Logistic Regression (LR), Random Forest (RF), Deep Feedforward Neural Network (DFFN), AutoML Stacked Ensemble (ASE) and an ADL. To evaluate the methods, we collected a dataset with around 147 thousand in-car display quality tests and adopt two evaluation schemes: an initial time ordered holdout train and test split and then a more robust rolling windows, which simulates several training and test iterations through time. Finally, we demonstrate how the best predictive ML model can provide a value for the analyzed industrial domain by selecting diverse decision thresholds and by using a Sensitivity Analysis (SA) eXplainable Artificial Intelligence (XAI) method [2].

2 Materials and Methods

2.1 Industrial Dataset

For this study, we collected 146,536 records related with in-car displays produced from a Bosch Car Multimedia in the year of 2020. The left of Fig. 1 exemplifies a produced in-car display. For the analyzed product, there are three main intermediate quality tests that are executed during the production process: gluing and screwing process (both executed during assembly), and functional tests (performed after assembly and prior to BU testing). The final quality is assessed by using a BU test that returns a percentage (the higher the value, the better is the display screen). The right of Fig. 1 shows the result of a failed BU test (due to a large detected red region).

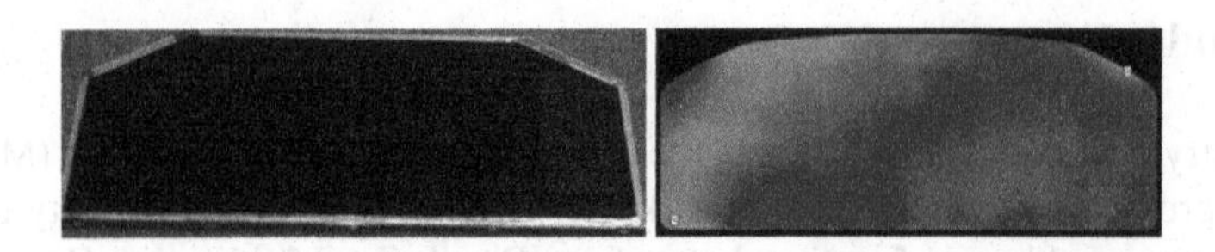

Fig. 1. Example of a produced display (left) and a BU failure test (right). (Color figure online)

In an initial phase of the ML project, the manufacturer provided the raw BU output values and an initial set of input features (a subset of the screwing process tests) that were considered by the manufacturer as more relevant to influence the BU quality, resulting in dataset A. Since the obtained ML results were not considered sufficiently good (as shown in Sect. 3), in a second ML project stage, the full intermediate quality tests (including the gluing process and functional tests) were also requested, leading to the creation of dataset B. All inputs are numeric the Carrier Cavity (CC) and System (SCC) attributes, which are categorical with 8 and 7 distinct levels (used in both A and B sets). In total, dataset A includes 110 screwing input variables (108 numeric and 2 categorical), while dataset B has 1032 input assembly and functional attributes (1030 numeric and 2 categorical).

The data preprocessing involved the transformation of the BU values into a binary target, by using the manufacturer quality rule: y ="Fail" if BU<40% else y ="Pass". The output target (y) is unbalanced, including only 2,138 abnormal instances. When the ML algorithm requires a numeric output (e.g., LR, DFFN), we assume the "Fail" class as the positive concept, thus transformed into $y=1$ (if true) or $y=0$ (if false). Moreover, the CC and SCC categorical variables were encoded into binary numeric ones (within $\{0,1\}$) by adopting the popular one-hot encoding. Finally, the numeric variables were rescaled into the $[0,1]$ range by adopting the popular min-max normalization.

2.2 Anomaly Detection Methods

All ML methods were implemented by using the Python language and the following modules: `scikit-learn` – for iForest, DT, LR and RF (https://scikit-learn.org/stable/); `TensorFlow` – for AE and DFFN (https://www.tensorflow.org/); and `H2O` – ASE and ADL (https://docs.h2o.ai/).

The iForest is a recently proposed one-class ML algorithm [7]. The `scikit-learn` iForest implementation provides a decision score that ranges from $\hat{y}_i$ =-1 (highest abnormal score) to $\hat{y}_i$ =1 (highest normal score). In order to obtain an anomaly probability score ($d_i \in [0,1]$, for an input example i), we rescale the iForest scores by computing $d_i = (1-\hat{y}_i)/2$.

Autoencoders (AE) compresses and encode data into a lower-dimensional representation by assuming a bottleneck layer (with L_b hidden units) [5]. Let $(L_I, L_1, ..., L_H, L_O)$ denote the structure of a dense (fully connected) DFFN architecture with the layer node sizes, where L_I and L_O represent the input and output layer sizes and H is the number of hidden layers. The proposed AE assumes $L_I = L_O$, a symmetrical encoder and decoder structure (e.g., $L_1 = L_{O-1}$) and the popular ReLu activation function is used by all neural units. In the encoder component, the number of hidden layer units decreases by half in each subsequent hidden layer until the bottleneck size (L_b) is reached: $L_1 = L_I/2$, $L_2 = L_1/2$, and so on. Each hidden layer is also attached with a Batch Normalization layer. When adapted to anomaly detection, the AE training algorithm is only fed with standard (normal) instances, aiming to generate output values that are identical to its inputs. In this work, the AE is trained with the Adam optimizer using a batch size of 1024, 100 epochs and early stopping (using 10% of the training data as the validation set). The Mean Absolute Error (MAE) is used as the loss function and reconstruction error: $MAE_i = \sum_{k=1}^{n} \frac{|x_{i,k} - \hat{x}_{i,k}|}{n}$, where $x_{i,k}$ and $\hat{x}_{i,k}$ denote the AE input and output value for the i-th data instance and k-th input or output node. The reconstruction MAE error is used as the decision score $d_i = MAE_i$, where higher reconstruction errors should correspond to a higher anomaly probability. In preliminary experiments, using the training data from the first partition (P1, Sect. 2.3), a grid search was used to search for the best $L_b \in \{2,4,8,16\}$ value that provided the lowest reconstruction error. The best result was achieved using $L_b = 8$, which was kept fixed in the remaining AE experiments.

Focusing on the supervised learning models used, the LR, DT and RF algorithms were set to output an anomaly class probability (d_i for instance i). Turning to the more complex deep learning DFFN model, it assumes the base dense architecture presented in [8] with no additional tuning. We note that this DFFN is adopted as a default supervised deep learning model since the ADL (described below) already performs a substantial DFFN hyperparameter selection. The default DFFN has H=9 hidden layers, under the structure (L_I, 1024, 512, 256, 128, 64, 32, 16, 8, 2, 1). The ReLu activation function is used in the hidden layers, while the logistic function is adopted in the output node, in order to output an anomaly class probability (d_i). To avoid overfitting, we network includes a Dropout with the values 0.5 and 0.2 in the fourth and sixth hidden layers. The DFFN assumes the same AE Adam training, with the difference that a different loss function is used (binary cross entropy).

The ASE and ADL methods were implemented by using the AutoML H2O tool. The ASE first trains 5 distinct regression algorithms: RF, Generalized Linear Model (GLM), XGBoost, Gradient Boosting Machine and a default DFFN network. Then, it employs a Stacking Ensemble (SE), which uses all previously trained models to generate inputs for another GLM model. As for ADL, it uses a dense DFFN, with the H2O automatically tuning 7 of its hyperparameters (e.g., number of hidden units per layer, learning rate). During the AutoML and ADL search, the H2O tool was configured to maximize the Area Under the Curve (AUC) of the Receiver Operating Characteristic (ROC) analysis [3], using a 5-fold cross-validation. Preliminary experiments using the training data from the first partition (P1, Sect. 2.3), allowed to set the stopping criterion for AutoML and ADL, which was fixed to stop after a time limit of 300 (dataset A) and 1,200 (dataset B) seconds.

Descriptive knowledge can be directly extracted from any trained ML model (e.g., ensemble, deep learning) by applying a SA XAI method. In this paper, we adopt the computationally efficient one-dimensional SA (1D-SA) [2], which holds all ML inputs at their average values, except one target input, which is changed with L=7 distinct levels. The ML output responses are stored, allowing to compute an input relevance, which is proportional to the Average Absolute Deviation (AAD) measure applied to the responses, and Variable Effect Characteristic (VEC) curves, which plot the SA ML responses for each input. This SA XAI was implemented by using the `rminer` package of the R tool (https://CRAN.R-project.org/package=rminer) with the following parameters: SA `method=1D-SA`; default values for other parameters (e.g., number of levels `L=7, measure = "AAD"`).

2.3 Evaluation

We adopt two evaluation schemes. Assuming time ordered records, the data (146,536 instances) is divided into two main partitions (Fig. 2): **P1** - with the oldest 70% elements

(102,575 records); and **P2** - with the more recent 40% examples (58,613 records). The first partition (P1) was used to explore an initial single train and test Holdout Split (HS). The second partition (P2) was used to execute a more robust Rolling Window (RW) procedure [13], which simulates a real classifier usage through time by adopting a fixed training window (W), which is rolled in different iterations (with step size of S), generating U training and testing updates. As shown in Fig. 2, the test data from the P1 (HS) and P2 (RW) partitions do not overlap. The HS uses the oldest 70% P1 data to fit a model (71,802 instances, includes model selection and training) and the remaining 30% (30773 records) to test the predictive capability of the ML methods. The first partition test results for dataset A (Sect. 3) were shown to the industrial experts, which identified a higher predictive performance need. This triggered the collection of further intermediate quality inspection tests (e.g., functional tests), that resulted in dataset B. Then, dataset B was also evaluated on P1 test data, obtaining improved results. The second RW evaluation scheme was applied to the best dataset B ML algorithms. In the first RW iteration ($u = 1$), the more recent W examples from P1 were used as the ML fit data, allowing to predict the next test T examples. In the second iteration ($u = 2$), the training data is updated with S newer examples allowing to fit a new ML model and perform T predictions, and so on. In total, the RW results in $U = \frac{P2-(W+T)}{S}$ model updates, where $P2$ is the second data partition length. We use $U = 20$ RW iterations by fixing the values: W =25,000, T =5,000 and $S = 1,400$.

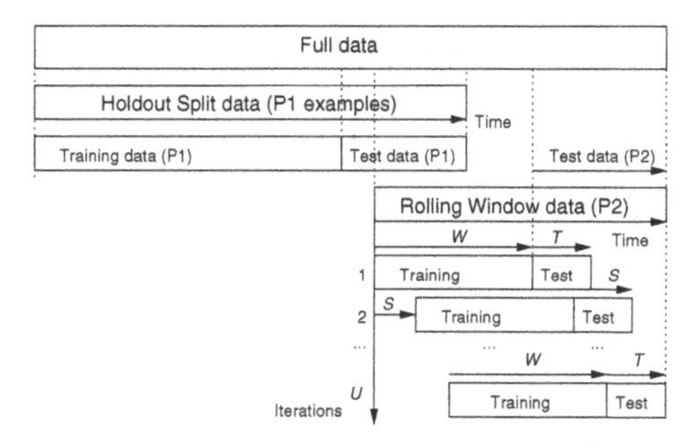

Fig. 2. Schematic of Holdout Split (HS) and Rolling Window (RW) evaluations.

The predictive classification performance is based on the ROC curve [3]. When a classifier outputs a decision score d_i, the class can be interpreted as positive if $d_i > K$, where K is a fixed decision threshold, otherwise it is considered negative. With the class predictions there will be True Positives (TP), True Negatives (TP), False Positives (FP) and False Negatives (FN). The ROC curve shows the performance of a two class classifier across all $K \in [0, 1]$ values, plotting one minus the specificity (x-axis), or False Positive Rate (FPR), versus the sensitivity (y-axis), or True Positive Rate (TPR). The discrimination performance is given by the $AUC = \int_0^1 ROCdK$. The AUC metric has two main advantages [10]. Firstly, when the data is unbalanced (which is our case), the interpretation of the goodness of the metric values does not change. Secondly, the AUC values can be interpreted as: 50% - performance of a random classifier; 60% -

reasonable; 70% - good; 80% - very good; 90% - excellent; and 100% - perfect. If needed, the best K threshold can be selected by using the ROC curve of a validation set (Sect. 3). Since the RW produces several test sets, for each RW iteration we store the AUC value on test data. We also record the computational effort, in terms of the total training time (in s) and prediction response time for one instance (in μs) when using an 2.4 GHz i9 Intel processor. To aggregate all $u \in \{1, ..., U\}$ execution results, we compute the median values, it is less sensitive to outliers when compared with the average. The Wilcoxon non parametric test is used to check if paired differences are significant [6].

3 Results

Table 1 presents HS predictive test results. For all compared ML algorithms, the usage of the first dataset (A) results in a lower class discrimination capability, with the AUC values ranging from 51% (almost random classifier for DFFN) to 69% (LR). When using more inputs (dataset B), there is a substantial improvement in the BU anomaly detection, with all ML algorithms presenting a much higher AUC value.

Table 1. Anomaly detection results (AUC values) for P1 and the HS evaluation (values higher than 0.75 are in **bold**).

	Unsupervised		Supervised					
	iForest	AE	DT	LR	RF	DFFN	ASE	ADL
Dataset A	0.61	0.60	0.55	0.69	0.64	0.51	0.52	0.64
Dataset B	0.71	**0.76**	0.60	**0.87**	**0.84**	**0.87**	**0.84**	0.60

The ML algorithms that obtained an AUC>75% using Dataset B in Table 1 were selected for the second stage evaluation: AE, LR, RF, DFFN and ASE. The respective RW results are shown in Table 2 and left of Table 3. For comparing purposes, we also present the dataset A RW results in Table 2. Similarly to the previous HS evaluation, the best RW results were achieved when using dataset B. In effect, for all tested ML, the AUC differences between dataset B and A are statistically significant (p-value$<$ 0.05). Using dataset B, the best overall anomaly detection performance was obtained by ASE (AUC of 92%), followed by RF (91%), LR and DFFN (89%) and AE (87%). The individual AUC results for each RW iteration are plotted in the left of Fig. 3. Both ASE and RF (purple and green curves) present a consistent excellent discrimination ($\geq$90%) after iteration $u = 6$. Regarding the computational effort, LR and RF provide the fastest training and predict response times. Yet, even the more demanding ML algorithm (ASE) requires a computational effort that is acceptable for the analyzed domain. Thus, we select the ASE model for the remainder demonstration results.

To demonstrate the ASE value, we adopt the last RW iteration ($u = 20$) results, using the most recent non overlapping RW test set ($u = 17$) as a validation set for selecting a K threshold. The ROC curve is shown in the right of Fig. 3, allowing to define several

Table 2. Anomaly detection results for P2 and the RW evaluation (best AUC values per dataset in **bold**; best global AUC is underlined).

	AUC					Training (s)					Predict (μs)				
	AE	LR	RF	DFFN	ASE	AE	LR	RF	DFFN	ASE	AE	LR	RF	DFFN	ASE
Dataset A	0.66	0.72	0.66	0.71	0.66	48.6	0.4	1.3	8.1	300.0	30	1	10	30	50
Dataset B	**0.87**	**0.89**	**0.91**	**0.89**	**<u>0.92</u>**	88.9	3.6	2.0	13.7	1200.0	150	4	10	100	60

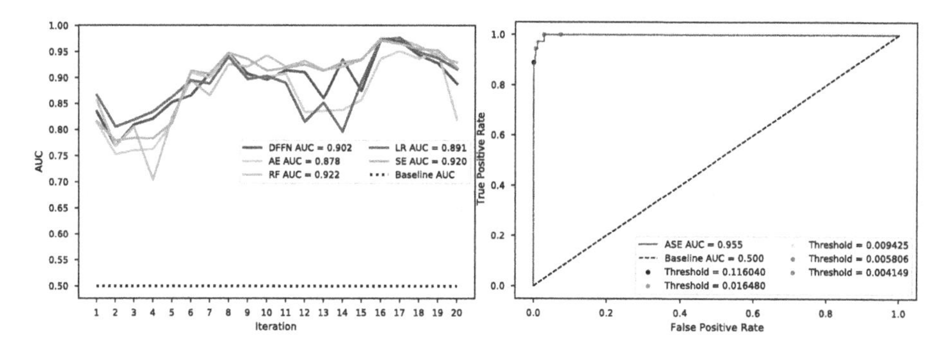

Fig. 3. Evolution of the AUC measure for the distinct RW iterations (left) and the ROC curve for ASE and iteration u=17 (right).

Table 3. Class label ASE prediction results for five threshold values ($u = 20$).

K	TP	TN	FP	FN	TPR	FPR
0.004149	56	4035	239	41	90.3%	18.3%
0.005806	49	4534	403	13	79.0%	8.2%
0.009425	41	4698	239	21	66.1%	4.8%
0.016480	30	4803	134	32	48.4%	2.7%
0.116040	13	4910	27	49	30.0%	0.5%

FPR and TPR trade-offs. Five thresholds (K) were fixed and applied to the last RW iteration test results (Table 3), confirming that the selected thresholds (from $u = 17$) correlate highly with similar sensitive and specific test (from $u = 20$) trade-offs results. Secondly, we applied the SA XAI approach to the ASE model that was fit using the last RW training data ($u = 20$). The left of Fig. 4 plots the top 20 relevant input variables from dataset B (total of 1032 inputs). For instance, the most influential input is related with a functional test (total relevance of 6%). The top 20 inputs account for 51% of the influence in the ASE model. The right of Fig. 4 shows the VEC curves for the top 5 input variables. The plot clearly reveals that the most influential input (Func02_T4105.02) produces the largest ASE output response change, where an increase in the numeric test value results in a nonlinear BU failure probability decrease.

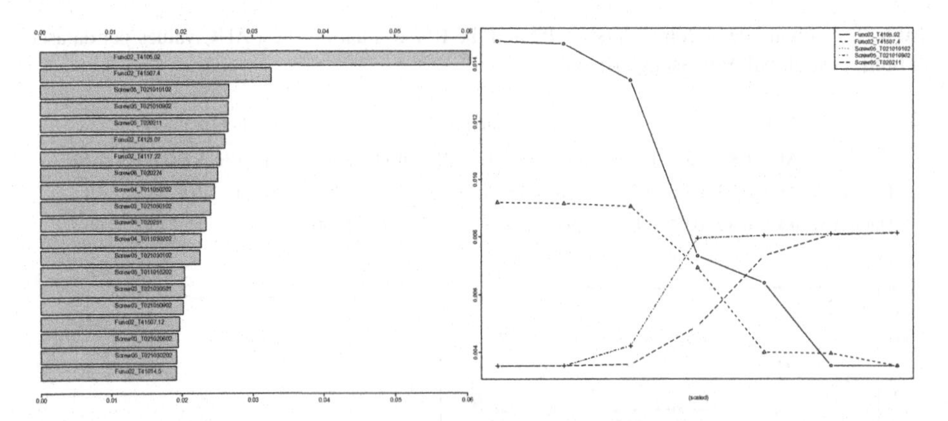

Fig. 4. Extracted knowledge from the ASE model ($u = 20$) using the SA XAI method: top 20 relevant input variables (left) and top 5 VEC curves (right).

4 Conclusions

In this paper, we use ML to model BU quality tests of in-car displays based on assembly and functional intermediate tests. A large set of comparative experiments was held, involving 147 thousand in-car display records. Overall, an excellent discrimination level was obtained by the ASE model (92%) when using both assembly and functional inputs (input set B), followed by the RF (91%), LR and DFFN (89%) and AE (87%). The AE results are particularly appealing for initial product production stages, since the model does not require labeled data and anomaly instances are rare. As for the best predictive model (ASE), it requires a computational effort that is affordable in this domain. The ASE model is particularly appealing for the analyzed industrial domain because it can be automatically adapted to new dynamic data changes, without needing an ML expert to tune or select the model. Moreover, we have shown that more specific or sensitive decision thresholds can be selected by using a validation set, producing a similar classification performance on unseen data. Furthermore, by using a SA XAI procedure, we have shown how descriptive knowledge can be extracted from a trained ASE model, which is valuable for BU failure cause identification. The obtained results were discussed with the manufacturer experts, which returned a very positive feedback.

In future work, we aim to deploy the proposed ASE in a real industrial setting. In addition, we plan to define intermediate production checkpoints before reaching the final BU test. This would allow the setting of quality assessment ML models at early production stages, thus saving production time and costs, and ultimately reducing the number of BU faulty tests.

Acknowledgments. This work is supported by: European Structural and Investment Funds in the FEDER component, through the Operational Competitiveness and Internationalization Programme (COMPETE 2020) [Project n° 39479; Funding Reference: POCI-01-0247-FEDER-39479].

References

1. Angelopoulos, A., et al.: Tackling faults in the industry 4.0 era-a survey of machine-learning solutions and key aspects. Sensors **20**(1), 109 (2020)
2. Cortez, P., Embrechts, M.J.: Using sensitivity analysis and visualization techniques to open black box data mining models. Inf. Sci. **225**, 1–17 (2013)
3. Fawcett, T.: An introduction to ROC analysis. Patt. Recogn. Lett **27**, 861–874 (2006)
4. Ferreira, L., Pires, P.M., Pilastri, A., Martins, C.M., Cortez, P.: A comparison of AutoML tools for machine learning, deep learning and XGBoost. In: International Joint Conference on Neural Networks, IJCNN 2021. IEEE (2021)
5. Hinton, G.E., Salakhutdinov, R.R.: Reducing the dimensionality of data with neural networks. Science **313**(5786), 504–507 (2006)
6. Hollander, M., Wolfe, D.A., Chicken, E.: Nonparametric Statistical Methods. John Wiley & Sons, Hoboken (2013)
7. Liu, F.T., Ting, K.M., Zhou, Z.H.: Isolation forest. In: Proceedings of the 8th IEEE International Conference on Data Mining (ICDM), Pisa, Italy, pp. 413–422. IEEE (2008)
8. Matos, L.M., Cortez, P., Mendes, R., Moreau, A.: Using deep learning for mobile marketing user conversion prediction. In: International Joint Conference on Neural Networks, IJCNN 2019 Budapest, Hungary, 14–19 July 2019, pp. 1–8. IEEE (2019)
9. Pandya, D., et al.: Increasing production efficiency via compressor failure predictive analytics using machine learning. In: Offshore Technology Conference, OTC-28990-MS (2018)
10. Pereira, P.J., Cortez, P., Mendes, R.: Multi-objective grammatical evolution of decision trees for mobile marketing user conversion prediction. Expert Syst. Appl. **168**, 114287 (2021)
11. Ribeiro, D., Matos, L.M., Cortez, P., Moreira, G., Pilastri, A., et al.: A comparison of anomaly detection methods for industrial screw tightening. In: Gervasi, O. (ed.) ICCSA 2021. LNCS, vol. 12950, pp. 485–500. Springer, Cham (2021). https://doi.org/10.1007/978-3-030-86960-1_34
12. Silva, A.J., Cortez, P., Pereira, C., Pilastri, A.: Business analytics in industry 4.0: a systematic review. Exp. Syst. J. Knowl. Eng. **38**, 12741 (2021)
13. Tashman, L.J.: Out-of-sample tests of forecasting accuracy: an analysis and review. Int. Forecast. J. **16**(4), 437–450 (2000)

A Parallel Variable Neighborhood Search for Solving Real-World Production-Scheduling Problems

Eneko Osaba[1(✉)], Erlantz Loizaga[1], Xabier Goenaga[2], and Valentin Sanchez[1]

[1] TECNALIA, Basque Research and Technology Alliance (BRTA), 48160 Derio, Spain
eneko.osaba@tecnalia.com
[2] UKABI GROUP, 48200 Durango, Spain

Abstract. In recent years, industry has evolved towards the efficient digitalization and optimization of products and processes. This situation is the consequence of the huge amount of information available in industrial environments and its efficient management for reaching unprecedented productivity levels. The momentum that enjoys this application field has led to the proposal of advanced methods for the dealing of robotic processes in industrial plants, optimal packaging of goods and the efficient scheduling of production plans, among many others. This paper is focused on the last of these categories. More concretely, we present a Parallel Variable Neighborhood Search for solving an industrial problem in which a fixed amount of materials should be constructed into a limited number of production lines. The construction of these materials has several particularities, such as the need of some specific tools to be correctly produced. It is also relevant to underscore that the problem solved in this research corresponds to a real-world situation, and that it is currently deployed in a production plant in the Basque Country (Spain).

Keywords: Combinatorial optimization · Metaheuristics · Variable neighborhood search · Production-scheduling problems

1 Introduction

Optimization is a widely studied field, which is the main focus of hundreds of studies published every year. The efficient solving of optimization problems usually demands high amounts of computational resources. This fact is especially frequent for problems formulated for dealing with complex real-world situations, with many variables and very specific constraints. This is the main reason why many heterogeneous solving schemes have been proposed by the related community in last years [1]. Arguably, the main reasons for the importance of optimization problems are twofold: the social interest they generate, and their inherent scientific interest.

H. Yin et al. (Eds.): IDEAL 2021, LNCS 13113, pp. 12–20, 2021.
https://doi.org/10.1007/978-3-030-91608-4_2

All these affirmations have been supported along last decades through the proposal of new problem modeling paradigms (such as multi-objective problems, large scale optimization or transfer optimization), and the popularity gained by knowledge streams such as Evolutionary Computation (EC) and Swarm Intelligence (SI) [2]. A myriad of approaches has been introduced for being applied to a wide variety of use cases, being exact methods, heuristics and metaheuristics the ones most frequently used in the literature. In this paper, we will focus our attention on the design and implementation of a metaheuristic, which is the most popular choice among the three mentioned categories, mainly because of their efficiency and adaptability [3].

In recent years, many different metaheuristics have been successfully implemented for giving a response for a wider variety of real-world situations. To cite a few, the well-known Genetic Algorithm has been applied to topics ranging from medicine to food distribution [4]. Ant colony Optimization has also been adapted for solving problems related to logistics or energy [5]. Bat Algorithm has been used for dealing with sport planning or routing problems [6]. Many other examples can be found with reputed methods drawn from EC and SI research branches, such as Particle Swarm Optimization, Cuckoo Search, Water Cycle Algorithm or Firefly Algorithm. We refer interested readers to interesting recent works such as [7,8].

More related with the topic dealt in this paper, a growing amount of papers are being published year by year fully focused on solving problems arising in industrial settings. In this aspect, also EC and SI methods have been demonstrated a great efficiency. Specifically, some branches particularly prolific in this context are related with robotic applications, optimal packaging of goods and the efficient scheduling of production plans. In this paper, we focus on the last of these categories: scheduling. In a nutshell, scheduling can be defined as the efficient assigning of resources to tasks in order to minimize the duration of these tasks. Along the years, these problems have been demonstrated to be adequate for areas such as production planning, computer design, logistic or flexible manufacturing system. Within scheduling problems, the well-known Job-Shop Scheduling Problem (JSP, [9]) is the one that has attracted most of the attention, being also one of the most studied problems in Operation Research field. For readers interested on works revolving around problems arisen in industrial environments, we recommend the following studies [10,11].

Bearing this in mind, the research work presented in what follows revolves on the design and implementation of an EC metaheuristic solver for solving a real-world production scheduling problem. More concretely, we present a Parallel Variable Neighborhood Search (PVNS) for solving an industrial problem in which a fixed amount of materials should be constructed into a limited number of production lines. The building of these materials has several restrictions, such as the need of some specific tools to be produced, among others. It is also relevant to underscore here that the problem solved in this paper has been directly drawn from the real-world, and that it is currently being used in a production plant in the Basque Country (Spain). Furthermore, the development of this system has been conducted among UKABI Group and TECNALIA as a result of a project funded by the Basque Government, precisely to incentive the implementation of technological solutions in the Basque industry, a sector of great importance in this territory.

The rest of the paper is organized as follows. Section 2 is devoted to the description of the problem solved in this research. This section also includes a deep description of the data employed for this research, which is directly extracted from the real use case. Next, Sect. 3 exposes in detail the main features of the proposed PVNS. The experimentation setup, analysis and discussion of the results are given in Sect. 4. This paper finishes in Sect. 5 with conclusions and future research lines.

2 Description of the Problem

As mentioned in the introduction, the problem tackled in this paper is directly drawn from a real-world situation, meaning that the solution developed in this research is currently working on an industrial plant. More concretely, the complete optimization system is being used in an industrial enterprise in Biscay, in the Basque Country.

The problem itself consists on a group $O = \{o_i\}_{i=1}^{n}$ of n independent orders to produce and a set of k production lines $L = \{l^j\}_{j=1}^{k}$. Each $o_i \in O$ order can only be produced on a subset of production lines $L_i \subset L$. Furthermore, each order consists on the construction of a predefined m amount of materials, which should be subsequently and uninterruptedly produced. This means that for each $o_i \in O$ order and each feasible $l_i^j \in L_i^j$ line a fixed construction time t_{ij} is needed.

Also, for every pair of $o_i \in O$ order and feasible $l_i^j \in L_i^j$ line, a set of x tools $U_{ij} = \{u_z\}_{z=1}^{x}$ are required for the production. In this specific problem, every order would need up to 5 different tools, which specifically are $U = \{$`row`,`form`,`head`,`plug`,`material`$\}$. Thus, each U_{ij} must be a subset of U. For each tool type multiple options are available, this situation means that, for example, the `row` tool needed for producing two specific orders would not be the same. In this case, the previous tool should be changed for the correct one, involving a fixed preparation time pt_u. Furthermore, this procedure is conducted in a sequential way, that is, if two orders o_i and o_{ii} will be consecutively produced in the same l_j production line, involving the changing of 3 different tools (e.g. `form`,`head` and `plug`), the total preparation time will be $pt_{total} = pt_{form} + pt_{head} + pt_{plug}$. Furthermore, it should be also mentioned that the tool preparation shall be made also for the conduction of first orders. In other words, all the production lines L in a specific use case have a default set of tools assigned, which are the ones used for producing the last order in the previous planning.

With all this, the main objective of the problem is to find a scheduling for efficiently produce all the planned orders and related materials, minimizing not only the overall production time, but the occupation of all the lines L. In this sense, and because of the requirement related with the tools, the main challenge

```
"ORDERS":[
  {
    "ID":561.0000,
    "QUANTITY":2660.0,
    "LINES":
    [
      {
        "IDHRCAB":1950.0000,
        "IDLINE":3.0000,
        "PRODUCTIONTIME":1.3333333333333333,
        "TOOLSSET":[
          {
            "CODPHASE":1.0000,
            "TOOLS":[
              {
                "IDTOOL":4.0000,
                "CODTOOL":"TCAIRFIL",
                "DESCTOOL":"AIRFIL HEAR"
              },
              {
                "IDTOOL":3.0000,
                "CODTOOL":"TMAIRFIL",
                "DESCTOOL":"AIRFIL PLUG"
              }
```


Fig. 1. A brief except of a JSON automatically generated by the company, and used as input in the optimization system.

is both to assign each order o_i to the most adequate line l_i^j, and to decide the best sequence of the production, aiming at minimize the preparation time of each order.

2.1 Description of the Data

We describe in this subsection the main data used in this research. This information is automatically generated by the industrial company in JSON format. After that, it is directly introduced as input for the developed optimization system, which starts the execution with a data preprocessing phase. We depict in Fig. 1a brief excerpt of a real data file. As can be seen in this extract, the main category `ORDERS` is composed by each order o_i. Every order is identified by its `ID`, depicting also the amount of materials to produce. That is, `QUANTITY` $= m$. The variable `LINES` represent the set of lines in which the order can be produced, that is, L_i^j.

Every line l_i^j is identified by its `IDHRCAB`, and characterized by the ID of the line (`IDLINE`) and the time needed for producing one single material on that line (`PRODUCTIONTIME`, that is, t_{ij}). It is interesting here that this production time varies not only by line but also by order. That is, specific orders need more time than others for producing one single material in the same line. Lastly, a line l_i^j has associated the set of required tools (`TOOLS`, that is, U_{ij}), describing each tool u_z using its ID, code and description. On the other hand, the company also introduces to the system a brief input file pointing the time needed pt_u for changing a tool on each of the available tool categories in U.

3 Parallel Variable Neighborhood Search

As mentioned in the introduction of this paper, a PVNS has been developed to efficiently tackle the problem posed above. For this implementation, we have been inspired by previously published VNS methods, such as the ones that can be found in [12]. Before a deep detail on the method, the first issue to be tackled is the encoding of solutions or individuals. In this work we adopt a label-based representation. In other words, each solution is represented as a permutation $\mathbf{x} = [l_0^a, l_1^b, \ldots, l_n^c]$ of n integers, where we recall that n depicts the number of orders to be produced. Furthermore, the value of each l_i^j denotes the line in which order o_i is produced. For example, assuming a $n = 8$ use case, one feasible solution could be $\mathbf{x} = [1, 2, 1, 1, 2, 2, 3, 3]$, meaning that orders o_1, o_3 and o_4 are made in l_1; while o_2, o_5 and o_6 are produced in l_2, and o_7 and o_8 in l_3.

Furthermore, and because several orders can be produced in the same line, an additional permutation of elements $\mathbf{y} = [po_0, po_1, \ldots, po_n]$ of size n is needed, in which po_i represents the sequential order in which each o_i ir produced in its corresponding line. Following the same example depicted above, a feasible order solution could be $\mathbf{y} = [1, 3, 3, 2, 2, 1, 2, 1]$, meaning that the works in line l_1 are produced in the following order: o_1, o_4 and o_3. Analogously, tasks in line l_2 are made following the sequence o_6, o_5 and o_2. Finally, o_8 is produced first and o_7 secondly in the line l_3.

Additionally, three different movement operators have been developed for evolving individuals along the search process. These operators are called CL_{first}, CL_{random}, and CL_{last}. For each of these functions, the acronym CL means *Change Line*. Briefly explained, these operators select a random order along the whole individual, and they randomly modify the production line in which they are made. Once this modification is made, the sequential order of the works conducted in the production line is re-calculated. In this context, the subscript of each operator represents the sequential order in which the works in placed.

As example, let us to assume CL_{first} operator and the feasible solution above mentioned: $\mathbf{x} = [1, 2, 1, 1, 2, 2, 3, 3]$ and $\mathbf{y} = [1, 3, 3, 2, 2, 1, 2, 1]$. Supposing that the order modified is o_2 and the new assigned line is l_2^1, the new value of $\mathbf{x}$ would be $\mathbf{x}' = [1, 1, 1, 1, 2, 2, 3, 3]$. After that, and considering that CL_{first} introduced the work in the first position of the sequential order, the new $\mathbf{y}$ would be $\mathbf{y}' = [2, 1, 4, 3, 2, 1, 2, 1]$.

Regarding the developed PVNS, we have designed it as a population-based approach of the naïve VNS. Furthermore, following the same philosophy considered as in other recent works, each individual of the population has its own main movement operator, randomly selected among CL_{first}, CL_{random}, and CL_{last}. After that, at each iteration, every individual of the population performs a movement using its main operator, but it may choose a different one with probability 0.33. In case the algorithm choses to use a different movement function, it is selected among the remaining ones in an equiprobable way.

Table 1. Characteristics of the 10 datasets used in the experimentation

ID	Orders o_i	Production lines L	Avg. lines per order L_i	Avg. tools per $o_i - l_i$
SMARTLAN_1	20	10	10	2
SMARTLAN_2	20	10	5	3
SMARTLAN_3	20	5	2.5	5
SMARTLAN_4	30	10	10	2
SMARTLAN_5	30	10	5	3
SMARTLAN_6	30	5	2.5	5
SMARTLAN_7	40	10	10	2
SMARTLAN_8	40	10	5	3
SMARTLAN_9	40	5	2.5	5
SMARTLAN_10	50	10	5	5

4 Experimentation and Results

To measure the performance of the designed PVNS, a dedicated experimentation has been conducted, which is deeply described in this section.

4.1 Benchmark Problems

For assessing the quality and the adequacy, not only of the developed algorithm, but also of the modeled mathematical approach to deal with the real-world problem, an experimentation has been conducted comprising 10 different real use cases. All these instances of the problem faithfully represent the daily working scheduling of the industrial company. In order to obtain representative and insightful outcomes, heterogeneous instances have been employed in these tests, considering diverse number of orders and lines. We depict in Table 1 the main characteristics of the whole dataset. As can be seen in that table, each instance can be described using four different concepts: i) the number of orders o_i, implying a greater complexity as long as the amount of works increases; ii) the number of available production lines L, for which a higher number of lines entails a wider flexibility for the L_i planning, iii) the average amount of feasible lines L_i per order, in which the lower the number, the more restrictive the problem; and iv) the average number of tools needed for each combination of order and feasible line $o_i - l_i$. This last value can range from 2 to 5, and a higher number involves a more costly transition (in terms of preparation time) between two different orders produced on the same line, making the optimization of the sequential order a crucial aspect for the problem.

4.2 Experimental Setup

Once the PVNS was firstly developed and taking into account that the method was to be deployed in a real scenario, different fine-tuning procedures were conducted with the main intention of improving the performance of the method as much as possible. This process of tuning was divided into two different phases: to optimize the algorithmic code in order to avoid inefficient operations, and

Table 2. Results of the conducted experimentation. We depict the average time need for producing all the orders, the standard deviation, runtime and the time converted into days

ID	Avg. Time (in min.)	Std. Deviation	Avg. Runtime (in sec.)	Real Time (in days)
SMARTLAN_1	4523.0	50.5	20.4	4 days and 17 h
SMARTLAN_2	4880.7	73.7	25.0	5 days and 2 h
SMARTLAN_3	5280.3	45.9	27.5	5 days and 12 h
SMARTLAN_4	5040.0	60.2	35.0	5 days and 6 h
SMARTLAN_5	5440.6	80.1	32.5	5 days and 16 h
SMARTLAN_6	6750.3	85.2	39.5	7 days and 1 h
SMARTLAN_7	5147.4	75.3	35.1	5 days and 9 h
SMARTLAN_8	5600.0	84.2	35.7	5 days and 20 h
SMARTLAN_9	8670.1	93.6	32.9	9 days and 1 h
SMARTLAN_10	7570.2	49.3	40.1	7 days and 21 hs

to properly adjust the parameter setting of the solver. For the second of these phases, this fine adjustment can be performed using ad-hoc pilot tests or automatic configuration tools. In this case, we opted for the first of these alternatives, which is the most recommended one when having a certain degree of expertise.

Additionally, with the intention of obtaining statistically reliable insights on the performance of the algorithm, 20 independent runs have been executed for each real use case. In relation to parameterization of the method, a population of 50 individuals has been employed, using the above mentioned CL_{first}, CL_{random}, and CL_{last} operators. For the ending criterion, each run ends when there are 1K iterations without improvements in the best solution found.

4.3 Results and Discussion

All the tests conducted in this paper have been performed on an Intel Core i7-7900U laptop, with 2.80 GHz and a RAM of 16 GB, using PYTHON as programming language. As mentioned in previous Sect. 4.2, 20 independent executions have been run for each use case. The results of the conducted experimentation are shown in Table 2, in which we depict the average time need for producing all the orders in the use case, along with the standard deviation, runtime and the time converted into days. This last information is really important for the real use case, since the industrial company is organized in work shifts of 16 h, needing the supervision of a worker in all the process. For this reason, the result should consider this work shift, since this information is the most important for the company to be used. Another interesting value to be observed is the standard deviation shown by the algorithm. Since the solving system is deployed in a real environment, the less this value the more robust the method. Furthermore, the optimizing system not only returns the time needed to produce all orders, but offers the complete scheduling for being automatically used by the company.

We show in Fig. 2 an example of the outcome provided by the system. As can be seen, the document that the algorithm gives as outcome is self-contained and fully usable for the company as it is, which undoubtedly increases the value

```
"orders":"[327.0, 330.0, 333.0, 340.0, 343.0, 346.0, 349.0, 356.0, 359.0, 362.0]",
"solution":"[4.0, 8.0, 1952.0, 5036.0, 5050.0, 5890.0, 10166.0, 2.0, 12.0, 14.0]",
"productionLines":"[2.0, 4.0, 4.0, 3.0, 5.0, 2.0, 3.0, 2.0, 4.0, 2.0]",
"sequentialOrder":"[1, 1, 2, 1, 1, 2, 2, 3, 3, 4]",
"totalTime":"4000.0",
"TimeByLine":"{2.0: 607.83, 4.0: 3905.49, 3.0: 3244.249, 5.0: 4000.0}",
"StartDate":"2021-03-08 09:30:34.929569",
"FinishDate":"2021-03-11 04:10:34.929569",
"DatesByLine":"{2.0: '2021-03-08 19:38', 4.0: '2021-03-11 02:36', 3.0: ...}",
"unitsToBuild":"[200.0, 1000.0, 2660.0, 2000.0, 8000.0, 2000.0, 3000.0, 50.0 ...]",
"timesPerMaterial":"[0.26, 0.2, 1.33, 0.54, 0.5, 0.2, 0.72, 0.24, 0.3, 0.27]",
"tools":[
    [
        [
            {
                "IDTOOL":4.0,
                "CODTOOL":"TCNL",
                "DESCTOOL":"HEAD NL"
            },
            {
                "IDTOOL":2.0,
                "CODTOOL":"TF12-LP",
                "DESCTOOL":"FORM 15-LP"
            },
        ...
```


Fig. 2. An example of a JSON automatically generated by the algorithm as outcome. This file is self-contained, being able to be directly used by the company as it is.

of the optimization system. On the one hand, we have descriptives fields such as `orders`, `unitsToBuild`, and `StartDate`, which summarize, respectively, the id of the set of orders $o_i \in O$ to produce, the units that should be built on each of these orders, and the starting date of the scheduling. On the other hand, fields as `solution`, `productionLines` and `sequentialOrder` represent the planning proposed by the algorithm for carrying out the production of the orders. Firstly, `solution` represent the list of the above described `IDHRCAB` identifiers; secondly, `productionLines` is the list of lines l^j in which each order has been assigned; finally, `sequentialOrder` depicts the sequential order in which the orders should be produced. Furthermore, additional data regarding the time needed for completing the scheduling is also provided, with fields such as `totalTime`, `timeByLine`, `FinishDate` and `DatesByLine`, which describe, respectively, the total time needed for completing the whole scheduling (in minutes), the time needed by line l^j, the ending date for the complete production and the ending date per line l^j. Finally, the solution also provides the tools employed in the final order conducted on each production line. This last information is used as the starting status for the next scheduling.

5 Conclusions and Future Work

In this paper, a real-world industrial production-scheduling problem has been presented and solved by means of a Parallel Variable Neighborhood Search. The main goal of this study has been the effective tackling of the problem modeled in order to help the industrial company to be more efficient in its daily scheduling procedures. One of the strengths of this work is its real-world nature, which has been highlighted along this paper. For this reason, we have described the input data as it is in the real environment, as well as the real output format of our optimization system, which is automatically used by the industrial company in their production duties. Several future work lines have been planned for the short term. The first one is the development of further tests using different use cases and solving approaches. Moreover, further research will be done regarding this industrial problem, applying additional realistic restrictions and conditions.

Acknowledgments. We would like to thank the Basque Government for its support through SMARTLAN project, ZL-2020/00735, and ELKARTEK program.

References

1. Desale, S., Rasool, A., Andhale, S., Rane, P.: Heuristic and meta-heuristic algorithms and their relevance to the real world: a survey. Int. J. Comput. Eng. Res. Trends **351**(5), 296–304 (2015)
2. Del Ser, J., et al.: Bio-inspired computation: where we stand and what's next. Swarm Evol. Comput. **48**, 220–250 (2019)
3. Hussain, K., Mohd Salleh, M.N., Cheng, S., Shi, Y.: Metaheuristic research: a comprehensive survey. Artif. Intell. Rev. **52**(4), 2191–2233 (2018). https://doi.org/10.1007/s10462-017-9605-z
4. Srinivas, M., Patnaik, L.M.: Genetic algorithms: a survey. Computer **27**(6), 17–26 (1994)
5. Mohan, B.C., Baskaran, R.: A survey: ant colony optimization based recent research and implementation on several engineering domain. Expert Syst. Appl. **39**(4), 4618–4627 (2012)
6. Osaba, E., Yang, X.-S., Fister, I., Jr., Del Ser, J., Lopez-Garcia, P., Vazquez-Pardavila, A.J.: A discrete and improved bat algorithm for solving a medical goods distribution problem with pharmacological waste collection. Swarm Evol. Comput. **44**, 273–286 (2019)
7. Osaba, E., Del Ser, J., Sadollah, A., Bilbao, M.N., Camacho, D.: A discrete water cycle algorithm for solving the symmetric and asymmetric traveling salesman problem. Appl. Soft Comput. **71**, 277–290 (2018)
8. Ouaarab, A., Ahiod, B., Yang, X.-S.: Discrete cuckoo search algorithm for the travelling salesman problem. Neural Comput. Appl. **24**(7), 1659–1669 (2014)
9. Manne, A.S.: On the job-shop scheduling problem. Oper. Res. **8**(2), 219–223 (1960)
10. Precup, R.-E., David, R.-C.: Nature-inspired optimization algorithms for fuzzy controlled servo systems. Butterworth-Heinemann (2019)
11. Precup, R.-E., David, R.-C., Petriu, E.M.: Grey wolf optimizer algorithm-based tuning of fuzzy control systems with reduced parametric sensitivity. IEEE Trans. Industr. Electron. **64**(1), 527–534 (2016)
12. Marinakis, Y., Migdalas, A., Sifaleras, A.: A hybrid particle swarm optimization-variable neighborhood search algorithm for constrained shortest path problems. Eur. J. Oper. Res. **261**(3), 819–834 (2017)

Inheritances of Orthogonality in the Bio-inspired Layered Networks

Naohiro Ishii[1](✉), Toshinori Deguchi[2], Masashi Kawaguchi[3], Hiroshi Sasaki[4], and Tokuro Matsuo[1]

[1] Advanced Institute of Industrial Technology, Tokyo, Japan
nishii@acm.org, matsuo@aiit.ac.jp
[2] National Institute of Technology, Gifu College, Gifu, Japan
deguchi@gifunct.ac.jp
[3] National Institute of Technology, Suzuka College, Mie, Japan
masashi@elec.suzukact.ac.jp
[4] Fukui University of Technology, Fukui, Japan
hsasaki@fukui-ut.ac.jp

Abstract. Layered neural networks are extensively studied for the machine learning, AI and deep learning. Adaptive mechanisms are prominent characteristics in the biological visual networks. In this paper, adaptive orthogonal properties are studied in the layered networks. This paper proposes a model of the bio-inspired asymmetric neural networks. The important features are the nonlinear characteristics as the squaring and rectification functions in the retinal and visual cortex networks. It is shown that the proposed asymmetric network with Gabor filters has adaptive orthogonality under stimulus conditions. In the experiments, the asymmetric networks are superior to the symmetric networks in the classification. The adaptive orthogonality is inherited in the layered asymmetric network from the asymmetric network with Gabor filters. Thus, it is shown that the bio-inspired asymmetric network is effective for generating the basis of orthogonality function and independent subspaces, which will be useful for the creation of features spaces and efficient computations in the learning.

Keywords: Asymmetric neural network · Gabor filter · Correlation and orthogonality analysis · Energy model · Inheritances of orthogonality

1 Introduction

Neural networks play an important role in the processing complex tasks for the visual perception and the deep learning. For the visual motion, adaptation to stimulus and other complex functions, biological models have been studied [1–3, 15]. For the learning efficiently, an independent projection from the inputs is studied in the convolutional neural networks [4] and neurons nonlinear characteristics are shown to generate independent outputs [5]. For the efficient deep learning, the orthogonalization in the weight matrix of neural networks are studied using optimization methods [6]. Further, the feature vectors from different classes are expected to be as orthogonal as possible for efficient learning

H. Yin et al. (Eds.): IDEAL 2021, LNCS 13113, pp. 21–32, 2021.
https://doi.org/10.1007/978-3-030-91608-4_3

[7, 8, 11]. To remove redundancy in the sensory input, neural responses are studied to be statistically independent [9]. Thus, it is important to make clear the network structures how to generate the independence and orthogonality relations in the networks [16]. In this paper, it is shown that the proposed asymmetric network with Gabor filters have adaptive orthogonal properties strongly under stimulus conditions which is followed by the layered asymmetric ones, while the conventional energy symmetric network is weak in the orthogonality. In the experiments, the classification is superior in the asymmetric networks. Further, the orthogonalities are inherited in cases of the odd-odd, odd-even and even-even orders nonlinearities. Thus, the nonlinear characteristics in the bio-inspired layered networks generate the orthogonality functions and independent subspaces, which will be useful for the creation of features spaces and efficient computations in the learning. Then, the inheritances of the orthogonality from the network can derive the orthogonal weights for the next network which will be useful in the learning and features creation.

2 Background of Asymmetric Neural Networks

In the biological neural networks, the structure of the network, is closely related to the functions of the network. Naka, Sakai and Ishii [14] presented a simplified, but essential networks of catfish inner retina as shown in Fig. 1.

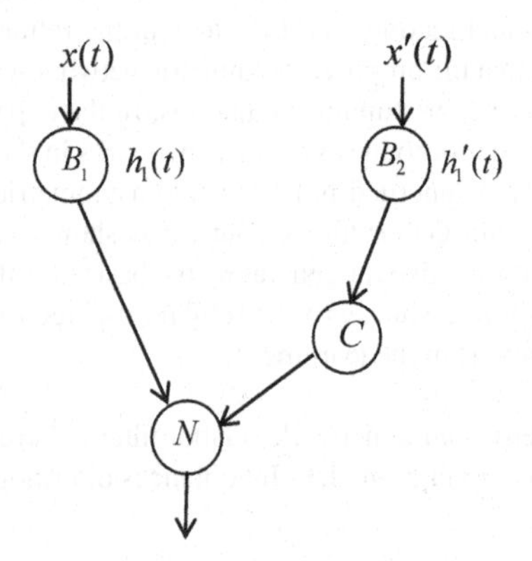

Fig. 1. Asymmetric network with linear and squaring nonlinear pathways [14]

The following asymmetric neural network is extracted from the catfish retinal network [14]. The asymmetric structure network with a quadratic nonlinearity composes of the pathway from the bipolar cell B to the amacrine cell N and that from the bipolar cell B, via the amacrine cell C to the N [14, 15]. That N cell response is realized by a linear filter, which is composed of a differentiation filter followed by a low-pass filter. Thus, the asymmetric network is composed of a linear pathway and a nonlinear pathway with the cell C, which works as a squaring function.

2.1 Orthogonality in the Asymmetric Network Under the Stimulus Condition

The inner orthogonality under the constant value stimulus is computed in the asymmetric networks as shown in Fig. 2. The vatiable t in the Gabor filters is changed to t', where by setting $\xi = 2\pi\omega$ in the Eq. (1), $t' = 2\pi\omega t = \xi t$ and $dt = dt/\xi$ hold. Then, Gabor filters are shown as

$$G_s(t') = \frac{1}{\sqrt{2\pi}\sigma} e^{-\frac{t'^2}{2\sigma^2\xi^2}} sin(t') \quad \text{and} \quad G_c(t') = \frac{1}{\sqrt{2\pi}\sigma} e^{-\frac{t'^2}{2\sigma^2\xi^2}} cos(t'). \tag{1}$$

The impulse response functions $h_1(t)$ and $h_1'(t)$ are replaced by $G_s(t')$ and $G_c(t')$ or vice versa. The outputs of these linear filters are given as follows,

$$y_{11}(t) = \int_0^{\infty} h_1(t')x(t-t')dt' \tag{2}$$

$$y_{21}(t) = \int_0^{\infty} h_1'(t')x(t-t')dt' \tag{3}$$

Adelson and Bergen [1] proposed an energy model with Gabor filters and their squaring function as a functional unit. To verify the orthogonality among their units, a parallel units are located as shown in Fig. 2, in which the asymmetric networks units are proposed here. In Fig. 2, inputs and output of the left asymmetrical network unit are shown in $x(t)$, $x'(t)$ and $y(t)$, while those of the right asymmetrical unit are shown in $v(t')$, $v'(t')$ and $z(t')$. Static conditions imply the brightness of input images do not change to be constant to time, while the dynamic conditions show they change to time. In Fig. 3, the first row (a) shows the same stimuli is inputted in $x(t)$, $x'(t)$, $v(t')$ and $v'(t')$, which are shown in white circle. In the second row (b) in Fig. 3, the rightmost stimulus $v'(t')$ is changed to new different one. The (c), (d) and (e) in Fig. 3 show the stimulus changes are moved.

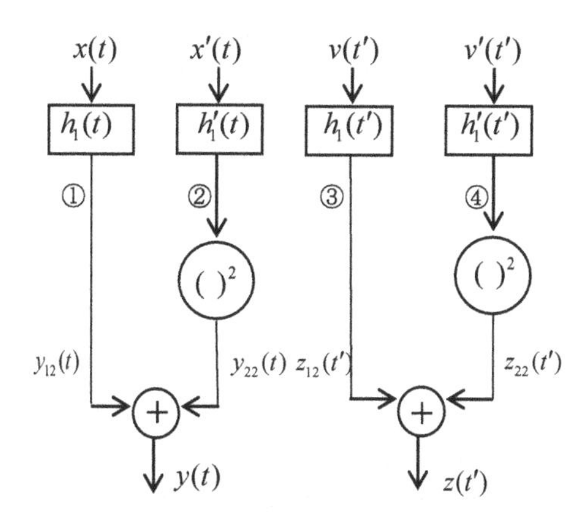

Fig. 2. Orthogonality computations between asymmetric networks units

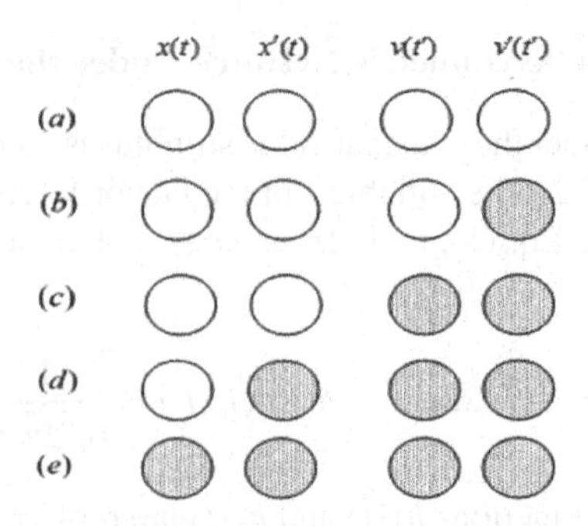

Fig. 3. Stimulus conditions on the two units of the asymmetric networks

2.2 Orthogonality Under Dynamic Condition

In the first row, (a) in Fig. 3, one white noise [13] is a low pass filtered one with zero mean and its power p, which is shown in the circles only under the input variables $x(t)$, $x'(t)$, $v(t)$ and $v'(t')$. Similarly, in the second row, (b) in Fig. 3, the other white noise is a high pass filtered one with zero mean and its power p', which is shown in the gray circles under input variable $v'(t')$. The impulse response functions $h_1(t)$ and $h_1'(t)$ are replaced by the Gabor filters, $G_s(t')$ and $G_c(t')$ as shown in the Eq. (1). Under the stimulus condition (a) in Fig. 3, the correlation between outputs $y(t)$ and $z(t)$ between the asymmetrical networks with Gabor filters in Fig. 2, is as follows,

$$\begin{aligned}
&\int_{-\infty}^{\infty} y(t)z(t)\mathrm{dt} = \int_{-\infty}^{\infty} dt\{\int_0^{\infty} h_1(\tau)x(t-\tau)d\tau \\
&+\int_0^{\infty}\int_0^{\infty} h_1'(\tau_1)h_1'(\tau_2)x'(t-\tau_1)x'(t-\tau_2)d\tau_1 d\tau_2\} \\
&\times\{\int_0^{\infty} h_1(\sigma)v(t-\sigma)d\sigma + \int_0^{\infty}\int_0^{\infty} h_1'(\sigma_1)h_1'(\sigma_2)v'(t-\sigma_1)v'(t-\sigma_2)d\sigma_1 d\sigma_2\} \\
&= \int_0^{\infty}\int_0^{\infty} h_1(\tau)h_1'(\sigma)d\tau d\sigma E[x(t-\tau)v(t-\sigma)] \\
&+\int_0^{\infty}\int_0^{\infty}\int_0^{\infty} h_1(\tau)h_1'(\sigma_1)h_1'(\sigma_2)E[x(t-\tau)v'(t-\sigma_1)v'(t-\sigma_2)]d\tau d\sigma_1 d\sigma_2 \\
&+\int_0^{\infty}\int_0^{\infty}\int_0^{\infty} h_1(\sigma)h_1'(\tau_1)h_1'(\tau_2)E[v(t-\sigma)x'(t-\tau_1)x'(t-\tau_2)]d\sigma d\tau_1 d\tau_2 \\
&+\int_0^{\infty}\int_0^{\infty}\int_0^{\infty}\int_0^{\infty} h_1'(\tau_1)h_1'(\tau_2)h_1'(\sigma_1)h_1'(\sigma_2)E[x'(t-\tau_1)x'(t-\tau_2)v'(t-\sigma_1)v'(t-\sigma_2)]d\tau_1 d\tau_2 d\sigma_1 d\sigma_2 \\
&= \left\{\int_0^{\infty} h_1(\tau)d\tau \int_0^{\infty} h_1'(\sigma)d\sigma\right\}\cdot p + 0 + 0 + 3p^2\left\{\int_0^{\infty} h_1'(\tau)d\tau\right\}^4
\end{aligned} \tag{4}$$

where the first term of the Eq. (4) shows value of the pathways ①and ③. The second and third terms by ①④ and ②③, are 0, respectively. The fourth term is by ②④. The terms ①③ and ②④ are not zero, respectively because the following equations hold,

$$\int_0^{\infty} h_1(\tau)d\tau = \frac{1}{\sqrt{2\pi}\sigma}\int_0^{\infty} e^{-\frac{\tau^2}{2\sigma^2\xi^2}}\sin(\tau)d\tau = \frac{\xi}{\sqrt{\pi}}e^{\frac{1}{2}\sigma^2\xi^2}\int_0^{\frac{1}{\sqrt{2}}\sigma\xi} e^{\tau^2}d\tau > 0 \tag{5}$$

and

$$\int_0^\infty h_1'(\tau)d\tau = \frac{1}{\sqrt{2\pi}\sigma}\int_0^\infty e^{-\frac{\tau^2}{2\sigma^2\xi^2}}\cos(\tau)d\tau = \frac{\xi}{2}e^{-\frac{1}{2}\sigma^2\xi^2} > 0, \tag{6}$$

where ξ is the center frequency of the Gabor filter. Thus, since two pathways are zero in the correlation, while other two pathways(25% and 25%) are non-zero, the orthogonality becomes 50% for the stimuli (a) in Fig. 3. Under the stimulus condition (b), the correlation between outputs $y(t)$ and $z(t)$ in Fig. 2, is computed as (7),

$$\int_{-\infty}^\infty y(t)z(t)dt = p\left\{\int_0^\infty h_1(\tau)d\tau\int_0^\infty h_1'(\sigma)d\sigma\right\} + 0 + 0 + 0 \tag{7}$$

Since three pathways are zero in the correlation, while the first pathway is non-zero, (25%), the orthogonality becomes 75% for the stimuli (b). Similarly, under (c), the orthogonality becomes 92%. Under (d), the orthogonality becomes 75%. Under (e), the orthogonality becomes 75%. The orthogonalities in the symmetric network called energy model [1] are shown in Fig. 4, which are lower ratios.

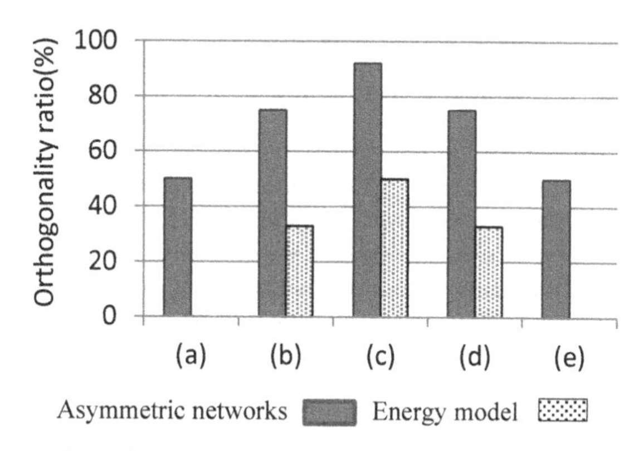

Fig. 4. Comparison of orthogonality under dynamic stimulus conditions in Fig. 3.

3 Experiments for Classification by Random Input Patterns

To investigate the ability for the arbitrary classification to the given random input patterns, 4 random patterns are set as the test one set. This arbitrary classification is evaluated by the determinant of input patterns to be not zero, which means input patterns to be independent. Here, the 4-dimensional square matrix is made as the one set. Then, the 50 sets of input patterns with the 4-dimensional data matrix, are generated using random numbers. In Fig. 5, a symmetric network with Gabor filters $\{G_s, G_c\}$ is followed by the squaring and a threshold cell with weights $\{w_1, w_2, w_3, w_4\}$ and threshold θ.

In Fig. 5, an asymmetric network with Gabor filters $\{G_s, G_c\}$ is followed by squaring for G_c and a threshold cell. For arbitrary classification by the networks in Fig. 5 and Fig. 6, the determinants of the matrix of random input patterns are computed. The classification ratios among networks in Fig. 5, Fig. 6 and a threshold cell are shown in Fig. 7.

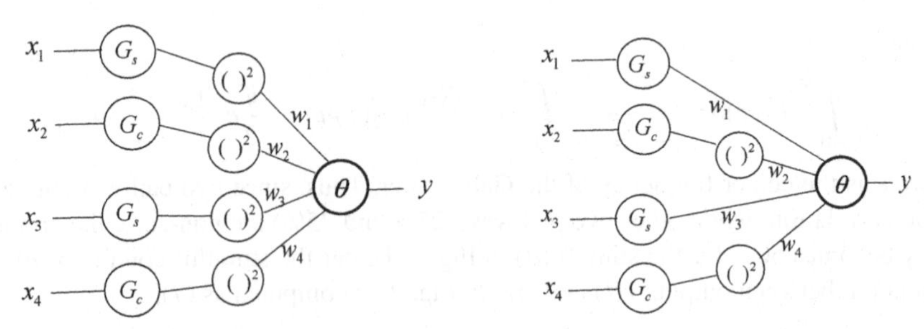

Fig. 5. Symmetric network with Gabor filters

Fig. 6. Asymmetric network with Gabor filters

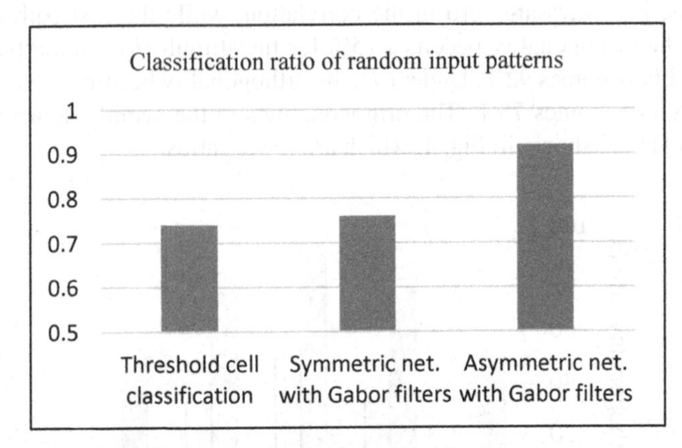

Fig. 7. Comparison of classification ratios using random input patterns among a threshold cell, symmetric network with Gabor filters and asymmetric network with Gabor filters

4 Generation of Extended Orthogonal Units and Bases

We show an example of layered neural network in Fig. 8, which is developed from the neural network in the brain cortex [10, 12]. Figure 8 is a network model of V1 followed by MT, where V1 is the front part of the total network, while MT is the rear part of it. Then, Fig. 8 is transformed to the approximated one in Fig. 9 by the rectification nonlinearity.

4.1 Nonlinearity to Generate Orthogonal Units and Bases

Figure 9 is an approximated network by Tailor expansion of the nonlinear rectification in Fig. 8, which works as extended asymmetric networks. The half-wave rectification in Fig. 8 is approximated in the following equation.

$$f(x) = \frac{1}{1 + e^{-\eta(x-\theta)}} \tag{8}$$

By Taylor expansion of the Eq. (8) at $x = \theta$, the Eq. (9) is derived as follows,

$$\begin{aligned} f(x)_{x=\theta} &= f(\theta) + f'(\theta)(x-\theta) + \frac{1}{2!}f''(\theta)(x-\theta)^2 + \cdots \\ &= \frac{1}{2} + \frac{\eta}{4}(x-\theta) + \frac{1}{2!}\left(-\frac{\eta^2}{4} + \frac{\eta^2 e^{-\eta\theta}}{2}\right)(x-\theta)^2 + \cdots \end{aligned} \tag{9}$$

In Fig. 9, the nonlinear terms, $x^2, x^3, x^4, \ldots$ are generated in the Eq. (9). Thus, the combinations of asymmetric networks with Gabor function pairs $(G_{ab\,\sin}, G_{ab\,\cos})$ are generated as shown in Fig. 9. Since the model in Fig. 8 consists of two pathways(left and right), two layered networks of Fig. 16 are made.

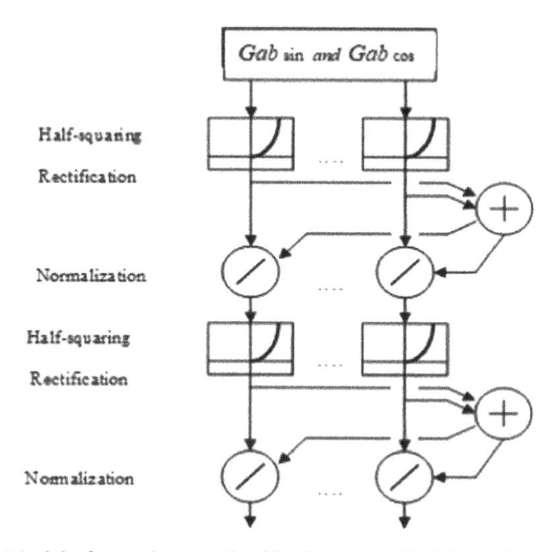

Fig. 8. Model of neural network of brain cortex V1 followed by MT [10]

Figure 8 shows a model of the V1 followed by the MT area in the cortex[10] after the retinal network, which is proposed here as the asymmetric neural network in Fig. 1 and the units of the asymmetric neural networks are shown in Fig. 2. Figure 9 shows the decomposed model of the V1 and the MT in Fig. 8.

4.2 Normalization Network for the Inheritance of Orthogonality in the First Layer

In Fig. 10, (A) shows a symmetric unit after the asymmetric unit in Fig. 2, while (B) shows an asymmetric net after the asymmetric unit in Fig. 2. For the simplified analysis, the impulse response functions $h_{v1}(t) = h'_{v1}(t) = 1$. The correlation between $\xi(t)$ and $\eta(t)$ in (A) in Fig. 10 becomes

$$\int_{-\infty}^{\infty} \xi(t) \cdot \eta(t)\, dt = \int_{-\infty}^{\infty} y(t) \cdot z(t)\, dt \tag{10}$$

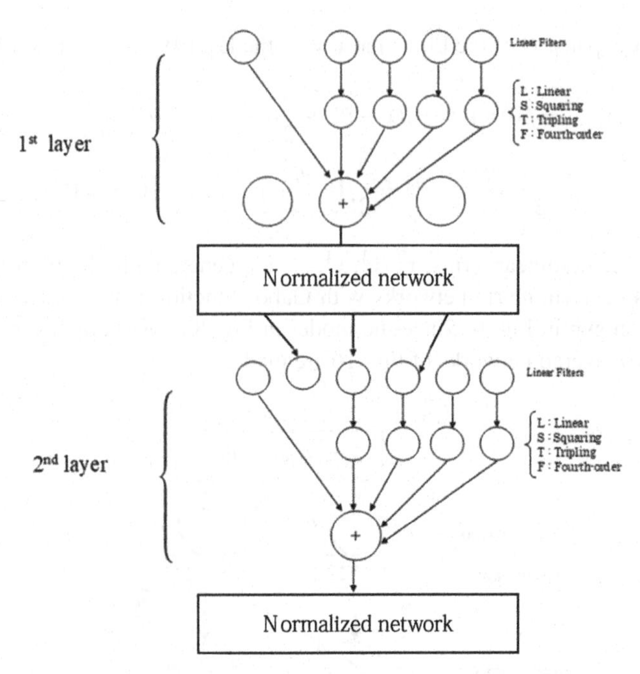

Fig. 9. A decomposed network for the layered network in Fig. 8

The Eq. (10) is the same as (20) under the stimulus condition in Fig. 2. Then the orthogonality values are sama as shown in Fig. 2. Under the dynamic stimulus conditions in Fig. 3, orthogonalirity ratio between $\xi(t)$ and $\eta(t)$ in the Eq. (10) is shown in first black bar in Fig. 11, in which the product $\xi(t) \cdot \eta(t)$ is same to that of $y(t) \cdot z(t)$. Similarly, the orthogonality ratio of the asymmetric network (B) in Fig. 10 is shown in the second gray bar in Fig. 11, in which the product $\xi(t) \cdot \eta(t)$ is same to that of $y(t) \cdot z(t)^2$.

4.3 Orthogonality in the Second Layer

In Fig. 8, the first layer of the V1 area is followed by the second layer MTarea, which shows the same structure of the first layer[9]. Thus, the decomposed model of the V1 and MT in Fig. 9 shows also the same structures for the first and the second layers. As the example, since the out put of the first layer is $\xi(t) + \eta(t)$ in (A) of Fig. 10, it is expressed as follows,

$$\xi(t) + \eta(t) = \int h_1(\tau)x(t-\tau)d\tau + \iint h_1'(\tau_1)h_1'(\tau_2)x'(t-\tau_1)x'(t-\tau_2)d\tau_1 d\tau_2$$
$$+ \int h_1(\sigma)x(t-\sigma)d\tau + \iint h_1'(\sigma_1)h_1'(\sigma_2)x'(t-\sigma_1)x'(t-\sigma_2)d\sigma_1 d\sigma_2 \tag{11}$$

The nonlinear rectification in the second layer in Fig. 8 generates polinomial terms with the quadratic, the cubic, the fourth order...... as

$$(\xi(t) + \eta(t)),\ (\xi(t) + \eta(t))^2,\ (\xi(t) + \eta(t))^3 \ldots, \tag{12}$$

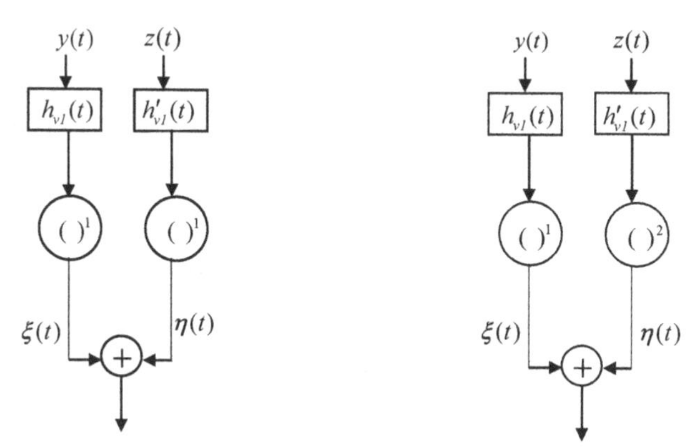

Fig. 10. Inheritance of orthogonality in the layered network after asymetric net in Fig. 2. In (A), the unit is odd-odd nonlinearity, while in (B,.the unit is odd-even nonlinearity

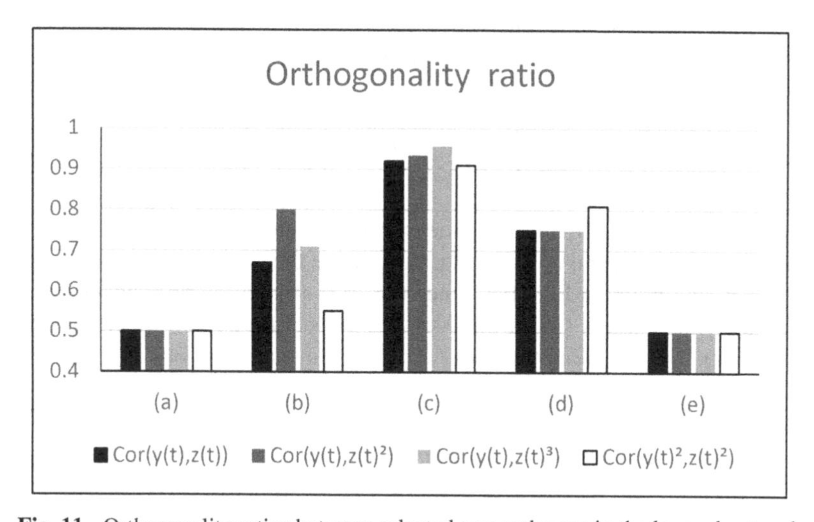

Fig. 11. Orthogonality ratios between selected two pathways in the layered network

in which each term is decomposed as shown in the 2nd layer in Fig. 9.

5 Generation of Wavelet Orthogonal Bases

The layered network proposed here consists of three layers. The first layer is in Fig. 2, in which the impuse funtions $h_1(t)$ and $h'_1(t)$ are replaced with Gabor filters $G_s(t)$ and $G_c(t)$ in the Eq. (1). Then, the wavelets of Gabor filters are generated. The wavelets are inherited to the first layer of the decomposed network in Fig. 9.

5.1 Wavelet Orthogonal Bases Generated in the 1st Layer

Orthogonal bases are generated in the 1st layer in Fig. 9. The combination pairs of the Gabor filters in the1st layer are generated in the output, which are shown in the following Eq. (13).

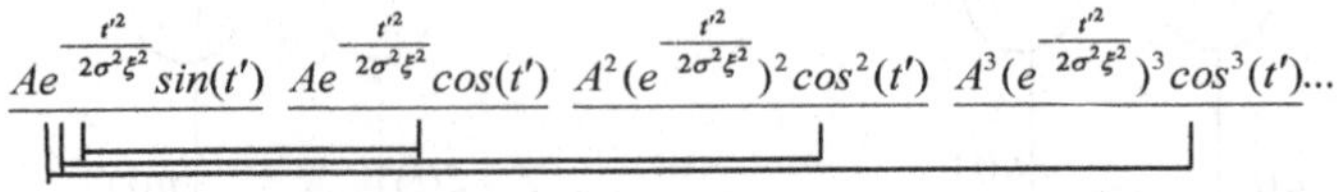

$$Ae^{-\frac{t'^2}{2\sigma^2\xi^2}} sin(t') \quad Ae^{-\frac{t'^2}{2\sigma^2\xi^2}} cos(t') \quad A^2(e^{-\frac{t'^2}{2\sigma^2\xi^2}})^2 cos^2(t') \quad A^3(e^{-\frac{t'^2}{2\sigma^2\xi^2}})^3 cos^3(t')...$$

Fig. 12. Wavelet orthogonal combination pairs among Gabor sine and cosines functions

Under the same stimulus condition in Fig. 3, the orthogonality is computed between pathways in the connecting lines in Fig. 2. Here, the connecting line shows over 50% orthogonality ratio in Fig. 2. Since $Ae^{-\frac{t'^2}{2\sigma^2\xi^2}} \sin(t')$ and $Ae^{\frac{t'^2}{2\sigma^2\xi^2}} \cos(t')$ are orthogonal, thus they become orthogonal bases. This orthogonal connection in bold lines in Fig. 12. In Fig. 13, the Gaussian function is included between the approximated by triangle functions (a) and (b). The triangle function (a) in Fig. 13 is represented by the linear equations as

$$\begin{cases} at + b \text{ for } t < 0 \\ -at + b \text{ for } t \geq 0 \end{cases}, \tag{14}$$

where $a > 0$ and $b > 0$ hold. Similarly, the triangle function (b) in Fig. 13 is represented using the coefficients a' and b' in the Eq. (14). Since the cross trigo-nometric product coefficients a' and b' in the Eq. (14). Since the cross trigonometric product,sin t and $\cos^2 t$ of the Gabor filters (1) on the $[-\pi, +\pi]$, becomes an odd function, which is shown in $f(-t) = -f(t)$ using function $f(t)$, the following equations are developed.

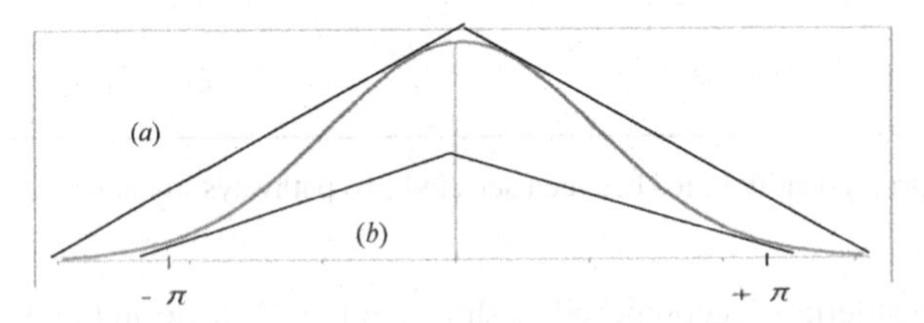

Fig. 13. Gaussian function is included between two triangle functions (a) and (b)

$$\int_{-\pi}^{+\pi} |a't + b'| \cdot f(t)dt \leq \int_{-\pi}^{+\pi} \left(\frac{1}{\sqrt{2\pi}\sigma} e^{-\frac{t^2}{2\sigma^2\xi}} \right)^3 \sin t \cdot \cos^2 t dt \leq \int_{-\pi}^{+\pi} |at + b| \cdot f(t)dt \tag{15}$$

$$\begin{aligned} &\int_{-\pi}^{+\pi} \left(\frac{1}{\sqrt{2\pi}\sigma} e^{-\frac{t^2}{2\sigma^2\xi}} \right)^3 \sin t \cdot \cos^2 t dt \leq \int_{-\pi}^{+\pi} |at + b| \cdot f(t)dt \\ &= \int_{-\pi}^{0} (at + b)f(-t)dt + \int_{0}^{+\pi} (-at + b)f(t)dt = 0 \end{aligned} \tag{16}$$

Similar independent relations hold between the left $cos(t')$ and the right $sin(t')$, $sin^3(t')$ terms, respectively in the Eq. (17) in Fig. 14, which are shown in the solid line.

$$\underline{Ae^{-\frac{t'^2}{2\sigma^2\xi^2}}cos(t')}\ \underline{Ae^{-\frac{t'^2}{2\sigma^2\xi^2}}sin(t')}\ \underline{A^3(e^{-\frac{t'^2}{2\sigma^2\xi^2}})^3 sin^3(t')}\ \underline{A^5(e^{-\frac{t'^2}{2\sigma^2\xi^2}})^5 sin^5(t')}\ldots$$

Fig. 14. Wavelet orthogonal combination pairs among Gabor cosine and sine functions

6 Conclusion

This paper proposes a model of the bio-inspired asymmetric neural networks. That network is based on the retinal and visual cortex model. The important features of these networks among them are the nonlinear characteristics as the squaring and rectification functions on the ways of the networks. In this paper, it is shown that the proposed asymmetric network with Gabor filters has adaptive orthogonality under stimulus conditions. In the experiments, the asymmetric networks are superior to the symmetric ones in the classification. The asymmetric network with Gabor filters is followed by the bio-inspired layered networks, in which the nonlinear characteristics play important roles. Then, the adaptive orthogonality in the asymmetric networl is inherited and extended in the layered networks, which derives the weight orthogonality for the effective learning and features creation.

References

1. Adelson, E.H., Bergen, J.R.: Spatiotemporal energy models for the perception of motion. J. Optical Soc. Am. A **2**(2), 284–298 (1985)
2. Reichard, W.: Autocorrelation, A Principle for the Evaluation of Sensory Information by the Central Nervous System. Rosenblith Edition, Wiley, NY(1961)
3. Beyeler, M., Rounds, E. L. Carlson, K., Dutt, N., Krichmar, J. L.: Neural correlates of sparse coding and dimensionality reduction, PLoS Comput. Biol. **5**(6), 1–33 (2019)
4. Pan, H., Jiang, H.: Learning Convolutional Neural Networks using Hybrid Orthogonal Projection and Estimation, ACML 2017, Proc. Mach. Learn. Res. **77**, 1–16 (2017)
5. Widrow, B., Greenblatt, A., Kim, Y., Park, D.: The *No-Prop* algorithm: a new learning algorithm for multilayer neural networks. Neural Netw. **37**, 182–188 (2013)
6. Huang, l., Liu, X., Lang, B., Yu, A.W., Wang, Y., Li, B.: Orthogonal weight normalization: solution to optimization over multiple dependent stiefel manifolds in deep neural networks. In: 32nd AAAI Conference on AI, AAAI-18, pp. 3271–3278 (2018)
7. Momma, M., Bennett, K.P.: Constructing orthogonal latent features for arbitrary loss, Chapt. 28. In: Guyon I., Nikravesh M., Gunn S., Zadeh L.A. (eds.) Feature Extraction. Studies in Fuzziness and Soft Computing, vol. 207. Springer, Heidelberg. https://doi.org/10.1007/978-3-540-35488-8_29
8. Shi, W., Gong, Y., Cheng, D., Tao, X., Zheng, N.: Entropy orthogonality based deep discriminative feature learning for object recognition. Pattern Recogn. **81**, 71–80 (2018)

9. Simoncelli, E.P., Olhausen, B.A.: Natiral image statistics and neural representation. Ann. Rev. Neurosci. **24**, 1193–1216 (2001)
10. Simonceli, E.P., Heeger, D.J.: A model of neuronal responses in visual area MT. Vision. Res. **38**, 743–761 (1996)
11. Schtze, H., Barth, E., Martinetz, T.: Learning efficient data representations with orthogonal sparse coding. IEEE Trans. Computat. Imaging **2**(3),177–189 (2016)
12. Heeger, D.J.: Models of motion perception, University of Pennsylvania, Department of Computer and Information Science, Technical Report No.MS-CIS-87-91, Sept 1987
13. Marmarelis, P.Z., Marmarelis, V.Z.: Analysis of Physiological Systems – The White Noise Approach. Plenum Press, New York (1978)
14. Naka, K.-I., Sakai, H.M., Ishii, N.: Generation of transformation of second order nonlinearity in catfish retina. Ann. Biomed. Eng. **16**, 53–64 (1988)
15. Ishii, N., Deguchi, T., Kawaguchi, M., Sasaki, H.: Motion detection in asymmetric neural networks. In: Cheng, L., Liu, Q., Ronzhin, A. (eds.) ISNN 2016. LNCS, vol. 9719, pp. 409–417. Springer, Cham (2016). https://doi.org/10.1007/978-3-319-40663-3_47
16. Ishii, N., Deguchi, T., Kawaguchi, M., Sasaki, H.: Distinctive features of asymmetric neural networks with gabor filters. In: de Cos Juez, F.J., et al. (eds.) HAIS 2018. LNAI, vol. 10870, pp. 185–196. Springer, Cham (2018). https://doi.org/10.1007/978-3-319-92639-1_16

A Neural Architecture for Detecting Identifier Renaming from Diff

Qiqi Gu(✉) and Wei Ke

School of Applied Sciences, Macao Polytechnic Institute, Macao SAR, China
{qiqi.gu,wke}@ipm.edu.mo

Abstract. In software engineering, code review controls code quality and prevents bugs. Although many commits to a codebase add features, some commits are code refactoring, including renaming of identifiers. Reviewing code refactoring requires a bit of different efforts than that of reviewing functional changes. For instance, renaming an identifier has to make sure that the new name not only is more descriptive and follows the naming convention of the institution, but also does not collide with any other identifiers. We propose in this paper a machine learning model to automatically identify commits consisting of pure identifier renaming, from only the diff files. This technique helps code review enforce naming and coding conventions of the institution, and let quality assurance testers focus more on functional changes. In contrast to the traditional way of detecting such changes by parsing the full source code before and after the commit, which is less efficient and requires rigorous syntactical completeness and correctness, our novel approach based on neural networks is able to read only the diff and gives a confidence value of whether it is a renaming or not. Since there had been no existing labeled dataset on repository commits, we labeled a dataset with more than 1,000 repos from GitHub by Java syntax analysis. Then we trained a neural network to classify these commits as whether they are renaming, obtaining the test accuracy of 85.65% and the false positive rate of 2.03%. The methods in our experiment also have significance for general static analysis with neural network approaches.

Keywords: Neural networks · Program analysis · Diff classification · Code review · Supervised learning

1 Introduction

Commit messages are a key component of software maintenance—they can help developers validate changes, locate and triage defects, and understand modifications [25]. However, commit messages can be incomplete or inaccurate [5]. For example, when writing commit messages, a developer may say only identifiers were changed if he or she didn't check the diff very carefully, while the developer actually changed some white spaces or fixed a few minor bugs. In order to enforce software quality and the quality of commit messages, code review is

H. Yin et al. (Eds.): IDEAL 2021, LNCS 13113, pp. 33–44, 2021.
https://doi.org/10.1007/978-3-030-91608-4_4

introduced as a process in software engineering, where another developer reviews the committed code and the commit messages made by the former developer [23].

We believe that commits can be classified based on their diff. Proper classification of commits can remedy incomplete or inaccurate commit messages written manually, and guide code reviewers to pick those proper commits to review.

There are existing works in the field of program analysis, but research is struggling to answer questions like "Why does this method crash?", "Is this code thread-safe?" [3]. To enrich the experience towards solving the broader problem of code classification, we carry out this experiment to tackle one specific type of code modification recognition—renaming of identifiers. In this work, we propose a probabilistic machine-learning classification model, per taxonomy of [4], to automatically identify the commits to a version-control system that only consist of identifier renaming. This technique is novel in that the neural network is able to read merely the diff and give a confidence value of whether it belongs to the category. On the contrary, the traditional way of detecting such changes is done by analyzing the full source code before and after the commit, which is less efficient and often complicated with partial source code. For typical software projects nowadays each with a lot of imported libraries, the ability to carry out analysis based on local partial source code is certainly an advantage.

Our model can be used as a component in the Continuous Integration (CI) with a Pull-Request (PR) workflow, for instance, on GitHub Actions. When a developer submits a pull request, CI runs to build and test the code and can add labels to the PR. Our model plays a role in CI that we take the unified diff of the PR as input, and classify whether the diff is identifier renaming, "yes" or "no". Then, CI can label the PR or assign it to the corresponding code reviewers. For our model to work, if one commit changes multiple files, the changes to these files should be concatenated to a single diff file, as the `git/diff` program does.

Our work focuses on the classification of diff in a typical programming language—Java diff files, because Java is a popular language and the syntax is relatively simple. We specifically target Java 7, since Java 8 and beyond have introduced many new syntactical improvements, which could cause complication in the parsing and needlessly shift our focus away.

There is no off-the-shelf dataset of commits and labels of whether the commit is identifier renaming. Jiang *et al.* [11] collected an unlabeled dataset of 1,006 repositories and over 2 million commits from GitHub. Although Jiang *et al.* claimed their dataset contained only Java source code, we found it actually containing many non-Java contents. We filtered out non-Java code and 62K commits remain. To further expand the dataset, we employed the crawling GitHub tool provided by [3] to collect additional commits from top Java repos on GitHub in January 2021. The final dataset obtained has 73,080 examples.

We contribute in this work a probabilistic neural network classification model with diff files as input, telling whether the input is renaming or not. The model was specifically trained against the Java programming language, and we believe minor tweaks can allow it to work on other programming languages. Besides the neural network model, we also present a dataset that labels 73K Java diff files

```
1   diff --git a/src/.../TldPatterns.java b/src/.../TldPatterns.java
2   old mode 100644
3   new mode 100755
4   diff --git a/src/.../Chars.java b/src/.../Chars.java
5   index a0cf5bd..e26dca8 100644
6   --- a/src/.../Chars.java
7   +++ b/src/.../Chars.java
8   @@ -43,7 +43,7 @@ import java.util.RandomAccess;
9    * @author Kevin Bourrillion
10   * @since 1
11   */
12  - public class Chars {
13  + public final class Chars {
14         private Chars() {}
15   }
16  \ No newline at end of file
```

Fig. 1. Example diff file in dataset. Line 8 to 15 are valid input to our model.

with whether they are renaming commits, together with a syntax analyzer for Java diff files based on ANTLR.

The structure of this paper is outlined as follows. Section 2 briefly reviews the related work in the field of code summarization and neural network approaches. Section 3 gives the details of our methods for creating the dataset and building the neural network model. Section 4 evaluates the performance of our model. Section 5 discusses our approach, followed by the conclusion in Sect. 6.

2 Related Work

A method can be characterized by its input, techniques, and output. The input of our work is diff files, the techniques are neural networks, and the output is classification. Thus, in this section we first describe the format of diff files, and then introduce the related work in the aspects of input, techniques, and output.

Both GNU/diff and git/diff are able to show the difference character by character between two text files or commits. We concatenate the output so that changes to multiple files are recorded in a single diff file. In terms of the diff format, a line starting with double at-symbol (@@) signals the starting of a hunk where the files differ. Diff files generated by git/diff have optional hunk headers appended at the end of the @@ lines. Figure 1 shows an example diff file.

Code summarization is a sub-field of natural language processing (NLP), here the input is the program source code and the output is the summary. A large body of work has been carried out in this field with assorted techniques. Despite the fact that code summarization does not match our work in all the three aspects, some observations inspired us.

Some techniques are shared from NLP to code summarization after Allamanis *et al.* [4] proposed the naturalness hypothesis which expounded the similarities between programming languages and natural languages. Researchers [13,27] working in commit message generation gradually found neural machine translation (NMT), which was originally used for translating a natural language to another, excelled at translating diff to commit messages. Such an NMT approach has two major components, the encoder and the decoder. The encoder reads the diff and encodes it to some internal matrix representation, then the decoder transforms the internal representation to human readable text, i.e., the commit messages. Commit classification can use a similar technique as in NMT, but replacing the decoder to other neural layers.

When the input is source code rather than diff, the field becomes program analysis, where more and more work has employed neural network techniques. Alexandru *et al.* [3] used NMT to annotate source code tokens with typing, and they also implemented an ANTLR-based parser. JSNice [21] is a neural network model that predicts names of JavaScript identifiers and type annotations of variables. Mou *et al.* [19] used a convolutional neural network to classify programs by functionalities, such as string manipulations and matrix operations. Other uses of neural networks in program analysis include detection of variable misuse [18], bug localization [10], API suggestion [6], and code completion [22].

Taking diff as input, program analysis is specialized to diff analysis. Moreno *et al.* [17] generated release notes by reading changes to program source code and documentations, as well as taking into consideration issues from software repositories. ChangeScribe [12], ChangeDoc [9], and DeltaDoc [5] read diff and followed a set of rules to generate commit messages. None of them used machine learning, and howbeit they all processed diff, their goal was not to classify the diff. RefDiff [24] and RMiner [26] read the complete content of the changed files before and after a commit and construct a diff of an internal format, from which they detect refactoring types. They do not accept the git/diff format which shows barely changed lines of changed files.

Later, machine learning techniques emerged in diff analysis. Loyola *et al.* [14] developed a neural network model to generate text description from diff files. Their work relies on a lexer that divides source code into tokens, thus it is not end-to-end machine learning. Macho *et al.* [15] employed a random forest classifier and categorized commits into forward engineering, re-engineering, corrective engineering and management. By contrast, our work labels commits into their refactoring types.

3 Approach

The overall process of our approach is illustrated in Fig. 2. Rhomboids indicate input and/or output data, and rectangles are processes. The two processes on the left column are the way we label the raw dataset, and the ones on the right column are model layers.

We define the identifier renaming in a commit as follows, with certain restrictions.

1. Renaming of only one identifier is allowed. If a commit changes multiple identifiers, the commit is classified as "no".
2. Renaming of method overloads is treated as a single renaming. Renaming of a class leading to renaming of its constructors is a single renaming.

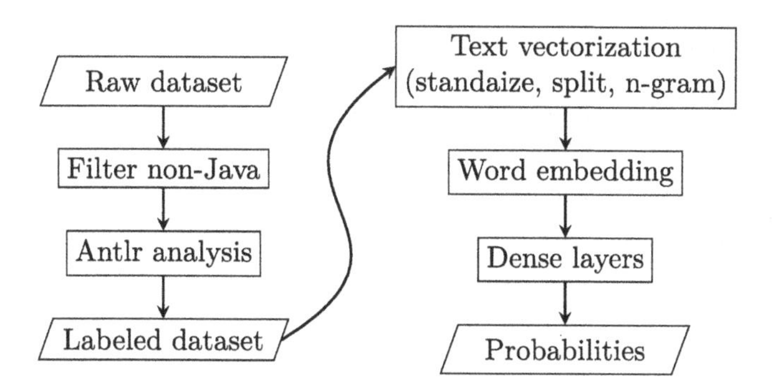

Fig. 2. Overall process of our approach. Rhomboids indicate input and/or output data, and rectangles are operations. The two operations on the left are for creating labeled dataset, and the three on the right are model layers.

3. Package renaming is labeled as "no". Java organizes packages in folders. If a file is moved from one folder to another, Java views the classes in the file as having changed their package. Since the code does not reveal folder information, we treat `package` or `import` changes as "no".
4. If comments or doc comments are changed, the diff is "no" for simplicity because our model is designed to analyze code in a programming language while comments are written in a natural language.
5. An empty diff file, resulted from file renaming or permission change, is "no".

3.1 Creating Labeled Dataset

The dataset of the diff files is pre-processed, as the left column of Fig. 2 shows. The algorithm of creating a labeled dataset is depicted in Algorithm 1.

Pre-processing. Since our focus is to detect identifier renaming in the Java language, changes to non-Java parts in the diff files are removed from the dataset. Line endings and the new line symbols at the end of the file are ignored. To reduce complexity, file paths and modes are ignored as well. Figure 1 expounds an exemplary snippet, and only line 8 to 15 are forwarded to the machine learning model. This step corresponds to the `getJavaFiles()` function in Algorithm 1.

Parsing and Labeling. While we are developing a supervised machine learning model, we must have a labeled dataset. Jiang's dataset has abundant examples but is unlabeled. We would like to make use of Jiang's dataset, therefore we used traditional formal syntax parsing for the labeling. Specifically we implemented a diff syntax analyzer in ANTLR 4 with Java 7 grammar retrieved from ANTLR's GitHub repository[1], which labeled an individual commit as "yes" or "no".

[1] https://github.com/antlr/grammars-v4.

Algorithm 1. Labeling dataset

```
function LABELDATASET(files)
    javaFiles ← getJavaFiles(files)
    unlabeledFiles ← interlace(javaFiles)
    for all file ∈ unlabeledFiles do
        if checkIdentifierChanges(file) then
            label file as "yes"
        else
            label file as "no"
        end if
    end for
end function
function CHECKIDENTIFIERCHANGES(diff)
    (before, after) ← recoverBeforeAfter(diff)
    beforeIds ← findIdentifiers(before)
    afterIds ← findIdentifiers(after)
    ub ← beforeIds \ afterIds
    ua ← afterIds \ beforeIds
    beforeWithoutId ← removeIdentifiers(before, beforeIds)
    afterWithoutId ← removeIdentifiers(after, afterIds)
    if ||ua|| = 1 ∧ ||ub|| = 1 ∧ beforeWithoutId = afterWithoutId then
        return true
    else
        return false
    end if
end function
```

The code in the beginning and at the end of a hunk may not be of complete syntax units. Fortunately, ANTLR 4 comes with error recovery heuristics summarized as single-token insertion and single-token deletion [20], so that it can figure out the type of a unit by the longest match. Once line 8 to 15 in Fig. 1 are fed into our diff syntax analyzer, ANTLR knows line 9, 10, and 11 is a comment block by implicitly inserting `/*` at the beginning of the hunk. Overall, this hunk has one identifier, `Chars`. Moreover, `git/diff` originally shows changes to consecutive lines as a group of deletion followed by a group of addition. The `interlace()` function in Algorithm 1 interlaces added and deleted lines, so that the comparison is easier. The `interlace()` function also does another optimization that if the numbers of added lines and deleted lines in a diff file are not equal, the file is marked as "no".

Our diff-based syntax analyzer is not 100% accurate comparing to full source code analysis. However we argue that when key information cannot be revealed by diff-based syntax analysis, neither can it be done by diff-based neural networks.

3.2 Model

Our model is implemented in TensorFlow 2 with three main parts, illustrated on the right column of Fig. 2. The first is the text vectorization layer, the second is the word embedding [16], and the last is the dense layers.

Algorithm 2. Text vectorization layer

```
function TEXTVECTORIZE(text)
    text ← replace numbers in text to special token
    words ← split text by spaces and punctuation
    tokens ← create 2-gram tokens from words
    voc ← create token-integer mappings
    sort voc by token frequency and keep top 10K tokens
    voc ← voc ∪ {UNK}
    return voc
end function
```

The text vectorization layer is an API in Tensorflow which allows developers to provide custom normalization and split functions. Algorithm 2 is an overview of our text vectorization layer.

Our text vectorization layer first standardizes numbers, including floating points, to a special token in Unicode private use areas, thus safe for internal representation. The layer then splits the text by white spaces and punctuation characters, i.e., !"#$%&\'()*+,-./:;<=>?@[]^`{|}~. We discard white spaces but keep the punctuation. Although splitting multi-word identifiers written in `CamelCase` or `snake_case` reduces vocabulary size, keeping them intact can achieve higher accuracy, as discussed in [7]. Next, tokens are combined into 2-grams, i.e., each token has maximum two words. Finally, the layer limits maximum vocabulary to 10K and the infrequent ones are discarded. When the text vectorization layer sees a word not in the capped vocabulary, the word will be represented by a special `UNK` token.

The word embedding layer transforms each token to a vector of length 16. We did not use pre-trained word embeddings like Word2Vec and GloVe because they were trained from natural language corpus. Instead, the word embedding layer was trained from our diff dataset. Next the input is flattened and goes through the dense layers of size 100, 10, and 1, respectively with RELU [1] as the activation function for each. The total number of parameters of our neural network model is 6,428,332.

4 Evaluation

Splitting Dataset. Our dataset has 73,080 examples but during evaluation we ignore examples bigger than 10 KB, getting a filtered dataset of 72,079 files. According to [8] and [2], small changes are likely to be corrective while large

Table 1. Statistics of the dataset

Name	# Yes	# No	Total
Training and validation	6,602	6,602	13,204
Test	738	6,469	7,207
Unused	0	51,668	51,668
Filtered dataset	7,340	64,739	72,079

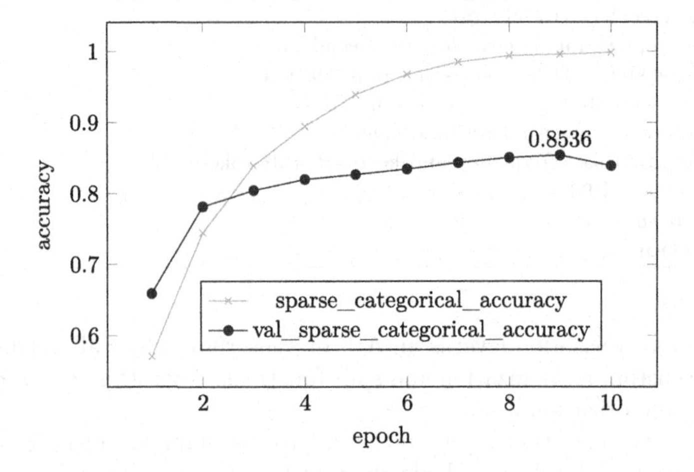

Fig. 3. Training and test accuracy for each epoch. The `sparse_categorical_accuracy` and `val_sparse_categorical_accuracy` refer to the training and test accuracy respectively in the Tensorflow notation.

ones are forward engineering. From the filtered dataset, we randomly selected 10% examples as the test set, consisting of 738 "yes" examples and 6,469 "no" examples. The test set was not used for hyper-parameter tuning. The remaining 90% data items were used for model training and hyper-parameter tuning. As the "no" examples are much more than the "yes" ones, we sub-sampled the "no" examples so that the "yes" class and the "no" class both had 6,602 examples. Our program shuffled the dataset and used four-fifth examples for training and the others for validation. Table 1 shows the statistics of the dataset.

Tuning Hyper-Parameters. The hyper-parameters of the model are tuned by hand. We set the max sequence length to 3,800 because most diff files with size less than 10 KB have less than 3,800 tokens. If max sequence length is too small, our model does not read all the words from a diff file, and the accuracy is lowered. If the value is too large, the model trains and runs too slowly.

The batch size is 200. For each epoch, the training set is reshuffled per [11]. Early stopping kicks in if validation accuracy does not improve after 5 epochs. When training stops, best weights are restored.

Table 2. Confusion matrix

(a) Confusion matrix of our model, from which we calculate false positive rate, precision, and recall

	Pred. "yes"	Pred. "no"
Actual "yes"	5581	888
Actual "no"	146	592

(b) Normalized confusion matrix

	Pred. "yes"	Pred. "no"
Actual "yes"	0.863	0.137
Actual "no"	0.198	0.802

Table 3. Comparison of precision and recall among our neural network approach, RefDiff, and RMiner

	NN		RefDiff 2.0		RMiner	
Refactoring types	Precision	Recall	Precision	Recall	Precision	Recall
Other renaming	0.975	0.863				
Class renaming			0.922	0.874	0.983	0.621
Method renaming			0.946	0.694	0.978	0.771

During training, the text vectorization layer only generates tokens from the training set. The best model took roughly 10 epochs to train. Figure 3 plots the growth of training and validation accuracy. The training accuracy lay at 0.5707 in the first epoch and eventually rose to 0.9984. The validation accuracy went from 0.6588 to 0.8536, then dropped.

Evaluating Performance. After we found the best model from the training and validation set, we saved the model and ran it against the unseen test set. The test accuracy is 85.65%, similar to the validation accuracy. The confusion matrix of the test set is shown in Table 2a. The false-positive rate (FPR), representing the case that a diff is "no" but incorrectly classified as "yes", is 0.0203. Besides, the precision is 0.975 and the recall is 0.863. The normalized confusion matrix is presented in Table 2b.

RefDiff and RMiner take two revisions of a system, build two models, and compare the two models, then detect refactoring types. They consume different datasets and use a dissimilar definition for identifier renaming that as long as a commit involves an identifier renaming, the label is "yes" even if the commit has functional changes. RefDiff and RMiner distinguish method renaming and class renaming, and ignore variable and field renaming. Despite the different datasets, Table 3 tells that our precision is higher than that of RefDiff and close to RMiner, and our recall is close to that of RefDiff and higher than RMiner.

In terms of classification speed, on the same machine, our neural network is able to load and classify the 7,207 test examples in 15 s with the aforementioned accuracy, i.e., 480.47 examples per second (eps), while the speed of the ANTLR syntax analyzer is 3.00 eps with 4-core parallelism. RefDiff 2.0 requires the SHA1 of a git commit for computing its CST diff. Hence we transformed our filtered dataset to this format and fed to RefDiff 2.0, and the speed is 14.26 eps when only renaming detection is turned on. In a nutshell, the speed of our neural network is skyrocketing.

5 Discussion

The advantages of our method over formal syntax analysis are that our model uses only the output of git/diff for classification, and the speed is much faster and still gives a good accuracy. When deployed on a continuous integration system, it's best for our method to classify pull requests in batch, because loading neural network parameters and running other setup take time.

The hyper-parameters of our model are set through experience and experiments. Hyper-parameter search may improve the accuracy of the model. The text vectorization layer of our model makes some assumptions on the programming language, and this layer may need to be changed for our model to adapt to other languages. Firstly, our text vectorization layer accepts numbers without 0 before the decimal point and with underscores as thousands separator as the Java specification permits. If another language does not allow these variations, the layer can be revised for more accurate word splitting. Secondly, our text vectorization assumes identifiers may contain underscores. If another language is good with other symbols in identifiers, the split function should be adjusted too. Thirdly, we removed spaces but spaces are important for parsing layout sensitive languages like Python.

The word embeddings were trained from the Java diffs. To use the model for another programming language, the word embedding layer must be retrained, as a keyword in Java is something different in another language.

If we need to classify other refactoring types, for instance if the diff is about a method moving, we can keep the text vectorization layer in the same way but use more advanced neural layers, such as long short-term memory (LSTM) and gated recurrent units (GRU) because the model would have to capture all tokens in a method, and check if the method body is unchanged and if the method is moved to another place.

6 Conclusion

Code review is a vital step in modern software engineering to control code quality and prevent bugs. We present a novel neural network approach to classify code commits for identifier renaming, only from the changed part, i.e., the diff, without the full source code, to assist the the code review process. Correct classification of pull requests of commits allows faster and easier code review. Thus, we propose a binary classification model with diff as input and a confidence value of whether the diff is pure renaming as output, along with a dataset that labels 73,080 Java diff files with whether they are renaming, and an ANTLR syntax analyzer for Java diff files. The accuracy of our model is 85.65% against unseen data, false positive rate is 2.03%. Our model runs faster comparing to traditional syntax analyses, thus suitable on continuous integration servers.

There are at least three directions in our future work. First, our current hyper-parameters are tuned by hand, and we would like to try hyper-parameter search. Second, we hope to employ LSTM or Transformers as they are best choices for NLP problems. Third, we are eager to detect other software engineering activities, for instance method moving and bug fixing.

Acknowledgement. This work is part of the research project (RP/ESCA-03/2020) funded by Macao Polytechnic Institute, Macao SAR.

References

1. Agarap, A.F.: Deep learning using rectified linear units (relu) (2018). arXiv preprint arXiv:1803.08375
2. Alali, A., Kagdi, H., Maletic, J.I.: What's a typical commit? a characterization of open source software repositories. In: 2008 16th IEEE International Conference on Program Comprehension, pp. 182–191. IEEE (2008)
3. Alexandru, C.V., Panichella, S., Gall, H.C.: Replicating parser behavior using neural machine translation. In: 2017 IEEE/ACM 25th International Conference on Program Comprehension (ICPC), pp. 316–319. IEEE (2017)
4. Allamanis, M., Barr, E.T., Devanbu, P., Sutton, C.: A survey of machine learning for big code and naturalness. ACM Comput. Surv. (CSUR) **51**(4), 1–37 (2018)
5. Buse, R.P., Weimer, W.R.: Automatically documenting program changes. In: Proceedings of the IEEE/ACM Iternational Conference on Automated Software Engineering, pp. 33–42 (2010)
6. Gu, X., Zhang, H., Zhang, D., Kim, S.: Deep API learning. In: Proceedings of the 2016 24th ACM SIGSOFT International Symposium on Foundations of Software Engineering, pp. 631–642 (2016)
7. Haiduc, S., Aponte, J., Moreno, L., Marcus, A.: On the use of automated text summarization techniques for summarizing source code. In: 2010 17th Working Conference on Reverse Engineering, pp. 35–44. IEEE (2010)
8. Hattori, L.P., Lanza, M.: On the nature of commits. In: 2008 23rd IEEE/ACM International Conference on Automated Software Engineering-Workshops, pp. 63–71. IEEE (2008)
9. Huang, Y., Jia, N., Zhou, H.J., Chen, X.P., Zheng, Z.B., Tang, M.D.: Learning human-written commit messages to document code changes. J. Comput. Sci. Technol. **35**(6), 1258–1277 (2020)
10. Huo, X., Li, M., Zhou, Z.H., et al.: Learning unified features from natural and programming languages for locating buggy source code. In: IJCAI, vol. 16, pp. 1606–1612 (2016)
11. Jiang, S., Armaly, A., McMillan, C.: Automatically generating commit messages from diffs using neural machine translation. In: 2017 32nd IEEE/ACM International Conference on Automated Software Engineering (ASE), pp. 135–146. IEEE (2017)
12. Linares-Vásquez, M., Cortés-Coy, L.F., Aponte, J., Poshyvanyk, D.: Changescribe: a tool for automatically generating commit messages. In: 2015 IEEE/ACM 37th IEEE International Conference on Software Engineering, vol. 2, pp. 709–712. IEEE (2015)

13. Liu, Q., Liu, Z., Zhu, H., Fan, H., Du, B., Qian, Y.: Generating commit messages from diffs using pointer-generator network. In: 2019 IEEE/ACM 16th International Conference on Mining Software Repositories (MSR), pp. 299–309. IEEE (2019)
14. Loyola, P., Marrese-Taylor, E., Matsuo, Y.: A neural architecture for generating natural language descriptions from source code changes. In: Proceedings of the 55th Annual Meeting of the Association for Computational Linguistics, vol. 2: Short Papers, pp. 287–292 (2017)
15. Macho, C., McIntosh, S., Pinzger, M.: Predicting build co-changes with source code change and commit categories. In: 2016 IEEE 23rd International Conference on Software Analysis, Evolution, and Reengineering (SANER), vol. 1, pp. 541–551. IEEE (2016)
16. Mikolov, T., Chen, K., Corrado, G., Dean, J.: Efficient estimation of word representations in vector space (2013). arXiv preprint arXiv:1301.3781
17. Moreno, L., Bavota, G., Di Penta, M., Oliveto, R., Marcus, A., Canfora, G.: Arena: an approach for the automated generation of release notes. IEEE Trans. Softw. Eng. **43**(2), 106–127 (2016)
18. Morgachev, G., Ignatyev, V., Belevantsev, A.: Detection of variable misuse using static analysis combined with machine learning. In: 2019 Ivannikov Ispras Open Conference (ISPRAS), pp. 16–24. IEEE (2019)
19. Mou, L., Li, G., Zhang, L., Wang, T., Jin, Z.: Convolutional neural networks over tree structures for programming language processing. In: Proceedings of the AAAI Conference on Artificial Intelligence, vol. 30 (2016)
20. Parr, T.: The Definitive ANTLR 4 Reference. Pragmatic Bookshelf, Raleigh (2013)
21. Raychev, V., Vechev, M., Krause, A.: Predicting program properties from "big code.". ACM SIGPLAN Notices **50**(1), 111–124 (2015)
22. Raychev, V., Vechev, M., Yahav, E.: Code completion with statistical language models. In: Proceedings of the 35th ACM SIGPLAN Conference on Programming Language Design and Implementation, pp. 419–428 (2014)
23. Shimagaki, J., Kamei, Y., McIntosh, S., Hassan, A.E., Ubayashi, N.: A study of the quality-impacting practices of modern code review at sony mobile. In: Proceedings of the 38th International Conference on Software Engineering Companion, pp. 212–221 (2016)
24. Silva, D., Silva, J., Santos, G.J.D.S., Terra, R., Valente, M.T.O.: Refdiff 2.0: a multi-language refactoring detection tool. IEEE Trans. Softw. Eng (2020)
25. Tao, Y., Dang, Y., Xie, T., Zhang, D., Kim, S.: How do software engineers understand code changes? an exploratory study in industry. In: Proceedings of the ACM SIGSOFT 20th International Symposium on the Foundations of Software Engineering, pp. 1–11 (2012)
26. Tsantalis, N., Mansouri, M., Eshkevari, L., Mazinanian, D., Dig, D.: Accurate and efficient refactoring detection in commit history. In: 2018 IEEE/ACM 40th International Conference on Software Engineering (ICSE), pp. 483–494. IEEE (2018)
27. Xu, S., Yao, Y., Xu, F., Gu, T., Tong, H., Lu, J.: Commit message generation for source code changes. In: IJCAI (2019)

Spell Checker Application Based on Levenshtein Automaton

Alexandru Bușe-Dragomir, Paul Ștefan Popescu, and Marian Cristian Mihăescu(✉)

University of Craiova, Craiova, Romania
{stefan.popescu,cristian.mihaescu}@edu.ucv.ro

Abstract. This paper presents a spell checker project based on Levenshtein distance and evaluates the system's performance on both parallel and sequential implementations. The Levenshtein algorithm approaches are presented in this paper: Levenshtein Matrix Distance, Levenshtein Vector Distance, Levenshtein automaton (along with an optimised version), Levenshtein trie and the performance evaluation is performed using three edit distances. Each edit distance is evaluated based on a set of misspelt words, so the results are relevant for various cases. For this scenario, the Levenshtein trie, along with the Levenshtein automaton, performed the best in both sequential and parallel versions for a large amount of misspelt words.

Keywords: Spell checker · Levenshtein automaton · Parallel programming

1 Introduction

This project aimed to create a spell-checking application based on the concept of the Levenshtein distance concept. The Levenshtein distance between two words is the minimum number of single-character edits (insertions, deletions or substitutions) required to change one word into the other. The main idea behind the project was to have both a more technical part represented by the algorithms and experiments and a practical side represented by the REST API. The final application will choose the best performing algorithm (based on the benchmarks generated using the experiments) and use it as an engine for the spell-checking process. In addition to providing suggestions for the misspelt words, the application will also communicate with the Merriam Webster API to obtain and return complete definitions when requested by the client. This is useful in the contexts in which, being undecided about the suggestion that should be picked, a user can request this additional information. Spell-checking and the algorithms and techniques behind this process are continuously evolving as more and more tools and technologies use them.

The motivation behind this project was raised by the possibility of conjunction between 3 different fields of work: algorithms and data structures, testing and experimental work and the practical side represented by the restful API.

H. Yin et al. (Eds.): IDEAL 2021, LNCS 13113, pp. 45–53, 2021.
https://doi.org/10.1007/978-3-030-91608-4_5

The system is constructed on top of several larger modules, and two of the main modules are the algorithm implementations, including the experiments, and the actual REST API built on the Spring framework, representing the Web Development component. There are two main flows, one external initiated by the application client (there are three actions that can be performed by the API user) in the REST API and one internal, triggered through the testing framework, to collect the experimental data. The algorithms, API and experiments have interdependencies, while the Merriam Webster API only communicates with the basic REST service.

The spell checking problem is essential and has been addressed using different approaches, some of them implying the usage of machine learning and deep learning algorithms and some using specifically designed methods which is also the focus of this paper.

2 Related Work

Regarding the automatic correction, several papers address this problem, one of them using deep learning for Spanish [3], some for context-awareness [6] other using Prediction by Partial Matching text compression for Arabic dyslexic text [2] or others as related in [13]. In the case of [3] they trained a Seq2seq Neural Machine Translation Model on two corpora: the Wikicorpus [12] and a collection of three clinical datasets and used GloVe [11], and Word2Vec [7] pre-trained word embeddings to study the model performance. The paper used two strategies for dataset compilation (generating multiple erroneous sentences from one correct sentence and only one erroneous sentence from the correct sentence by applying only one rule) and for validation presented both an intuition on specific phrases and evaluated the model's performance using Recall, Precision and F0.5 measure. Regarding the Arabic dyslexic text, [2] the language model is based on the Prediction by Partial Matching, a text compression scheme that generates several alternatives for every misspelt word. The generated candidate list for each word presented in the paper is based on the edit operations (e.g., insertion, deletion, substitution or transposition), and the correct alternative for each misspelt word is selected on the basis of the compression code length of the trigram [5]. For validation purposes, the system is compared with the Farasa tool [1]. The system provided better results than the other tools, with a recall of 43%, precision 89%, F1 58% and accuracy 81%.

Even if the Levenshtein method is old [4] it is still used in recent research like [10] which proposes a hybrid approach focusing on OCR generated Hindi text. The paper presented Vartani Spellcheck, which is a context-sensitive approach for spelling correction of Hindi text using a state-of-the-art transformer – BERT [9] in conjunction with the Levenshtein distance algorithm. Their proposed technique was validated on a large corpus of text generated by Tesseract OCR [14] on the Hindi epic Ramayana. They obtained an accuracy of 81%, and the results present significant improvements over several previously established context-sensitive error correction techniques for Hindi. They also explain how

Vartani Spellcheck can be used for on-the-fly autocorrect suggestions during continuous typing in a text editor. Another recent research [15] used in Levenshtein method for autocorrect in the context of autocomplete for a search engine. Their research aims to apply the autocomplete and spell-checking features based on the Levenshtein distance algorithm to get text suggestions for error data searching in the library and evaluate based on accuracy. For validation purposes, they used data obtained from UNNES Library, and the accuracy of the Levenshtein algorithm was 86% based on 1055 source cases and 100 target cases.

3 System Design

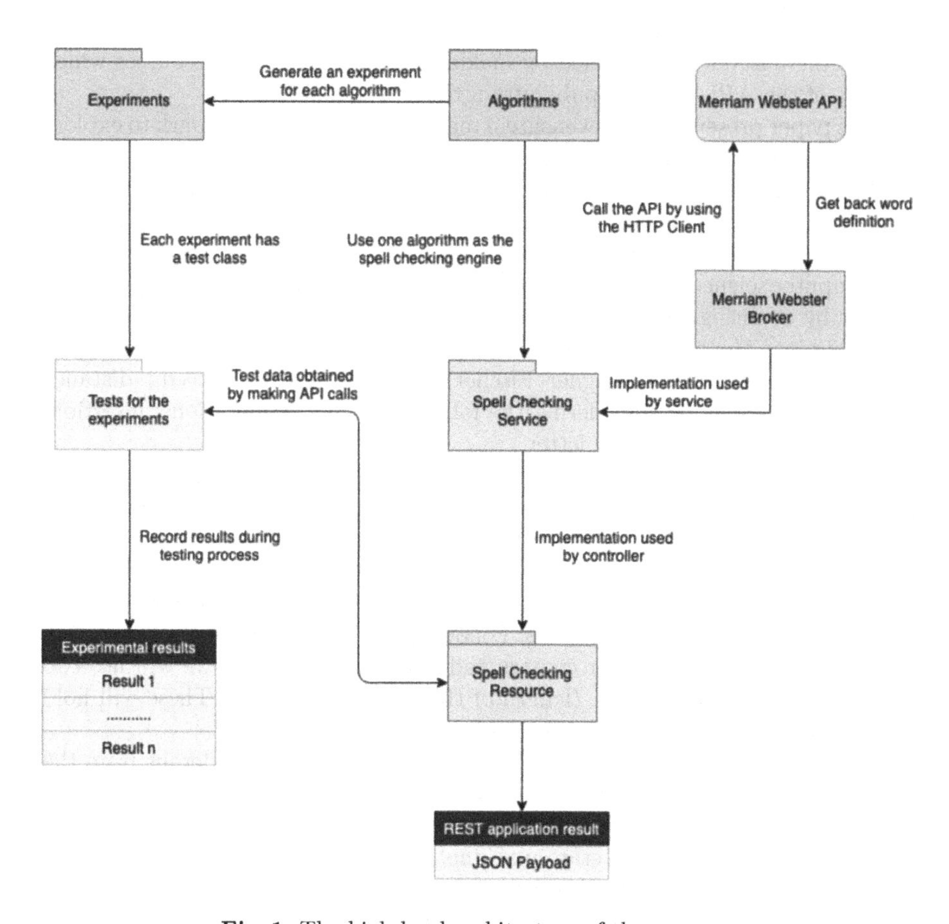

Fig. 1. The high-level architecture of the system

Figure 1 presents an overview of the whole system. The system's central component is represented by the algorithms that play a role in both the data structures side of the application and the REST API. Regarding the experiments module, each algorithm is used to check its performance and correctness during the spell-checking process. The REST API requires only one algorithm in order to work and, due to how the spell checker service is designed, any of the existing ones can be used as the engine. On top of this, the REST API is dependent on the database and so on the repository for the security checks, which are in turn performed by specific classes like the authentication manager. Other packages like the *utils*, *mapper*, *validation*, *dtos*, *entities* and the *broker* are also dependencies for the restful API.

On the other hand, the testing framework is also dependent on specific files which are constructed by the API for convenience by using the application restricted resources like the data set retriever or the test input builder. So, there is an interdependence between the experimental part of the application and the practical side through both the algorithms which are used by both, but also because the test data is obtained by making calls to the helper resources which are part of the API (but not public from the user perspective).

This paper presents four Levenshtein methods used as a benchmark to explore which performs best for our validation context.

Levenshtein Matrix Distance was the first implementation used for the spell-checker as it is the most straightforward and most well-known technique for calculating the edit distance[8] between two strings. In computational linguistics and computer science, edit distance quantifies how different two strings are to one another by counting the minimum number of operations required to transform one string into the other. The matrix approach represents the primary form of the dynamic algorithms (Wagner–Fischer), which computes the edit distance between two strings by considering the following possible operations: insertion, deletion and replacement of a letter.

Levenshtein Vector Distance implementation is, in fact, a space optimization of the matrix version in which, instead of constructing the whole matrix, two arrays are used. This is based on the apparent fact that, in order to fill a matrix row, only values from the current and previous rows are needed. So, there is no point in storing all results when one array can hold the previous results and the other one the new computations. Following the rule from the matrix version in which we had a row header constructed from the first word, we will use two arrays of size $|firstword| + 1$ (length of the first word plus 1). These will hold the previous and current computations for the edit distance.

Levenshtein automaton combines some implementation details from the vector distance algorithm with the standard structure of a classic NFA (non-deterministic finite automaton). The methods offered by the automaton algorithm are typical for a finite state machine.

Levenshtein trie implementation represents another approach for the spell-checking problem by using this well-known data structure and a recursive algorithm for performing the actual suggestion search. The implementation is based on an article from Steve Hanov's blog and the implementation done by Umberto Griffo and hosted on GitHub. The implementation is slightly faster than its version due to the Stream API usage, small amount of parallelization, and removal of redundant operations; it also provides a different API and a more straightforward and concise format.

4 Experimental Results

First, we started our experiments considering a list of 20000 most common English words ordered by frequency as determined by n-gram frequency analysis of Google's Trillion Word Corpus. After a few runs, we found that the input was too extensive for our setup, and the sweet-point was 10000 words for having both relevant and fast experiments. The experimental context is the following: we have an input text with a determined size (in our case, all experiments will have 14 scenarios, each with 10000 input words), we set an edit distance which can have values of 1, 2 or 3 and we use a predefined dictionary (for this experiment, we use 10000 words) and we obtain a set of results. Excluding the case of the optimized parallel automaton, all other experiments had the same overall process of parsing the input and looking up suggestions.

Each of the experiments will be presented with its results, except for the optimized parallel automaton. The Levenshtein Matrix Distance Experiments implementation is straightforward: it iterates over the dictionary using the Stream API, it filters only the words at an edit distance of less than the maximum allowed from the misspelt word and collects them into a list. The edit distance is computed by using the matrix Levenshtein distance algorithm. Since the maximum edit distance does not influence the performance of the matrix and vector algorithms, the three different contexts which are based on it will not lead to different execution times. The experiments have been run for three contexts (edit distance 1, 2 and 3) and with 14 scenarios each (0, 5, 10, 25, 50, 75, 100, 150, 200, 250, 350, 500, 750, 1000 misspelt words, all in input files with a size of 10.000 words). The results have been recorded for all 14 scenarios and all three edit distances and presented for each experiment. In the end, comparations between the best implementations will be graphically presented.

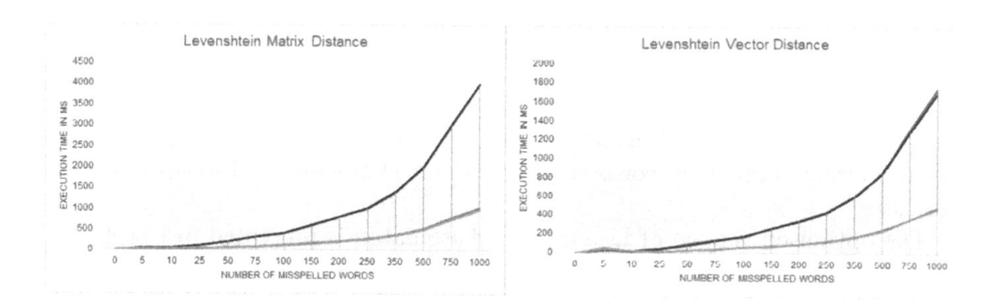

Fig. 2. Results for Levenstein maxtrix and vector distance. (Color figure online)

Figure 2 presents the results for Levenstein Maxtrix and Vector distance; the green line represents the sequential implementation and the red one the parallel version. The performance gain is significant for the parallel version and becomes more prominent as the number of misspelt words rises. Since the maximum edit distance does not influence the performance of the matrix and vector algorithms, the three different contexts which are based on it will not lead to different execution times.

On the right side of Fig. 2, there are the results for the same setup but in Levenstein Vector distance, and we can see a significant improvement on both approaches over the Matrix distance implementation. In the same way, since the edit distance value does not influence the vector Levenshtein algorithm, the experiment results will be presented only for one of the contexts (the chart for the edit distance with a value of 1 is presented). The results are identical for all three edit distances, and there is no point in cluttering up the graph by presenting all three overlapping lines. Comparing these approaches, we can see that the time for vector distance is almost half the time for matrix distance.

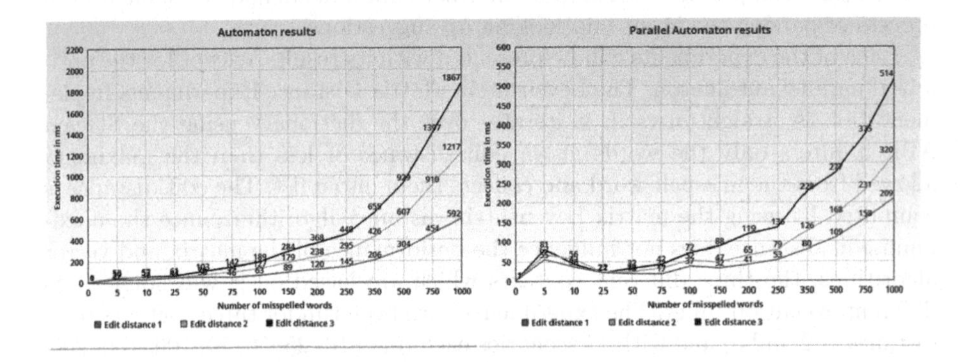

Fig. 3. Results for Levenshtein automaton

Regarding the Levenshtein Automaton Experiments presented in Fig. 3, we can observe that, in this case, the edit distance influences the execution time of the experiment. This is because of the automaton algorithm, where the performance decreases as the edit distance increases. In the case of the parallel implementation, again, we can notice the spike in the execution time when the first scenario with misspelt words is executed. For the parallel automaton, the scenario with ten misspelt words is also relatively slow compared to the following test cases.

The optimised version of Levenshtein Automaton is presented in Fig. 4, and the optimisation comes from reusing the same automaton object, by resetting some fields, instead of creating a new object for each combination misspelt word, dictionary word and edit distance. The cost of creating new objects is higher than that of setting some fields, but the overall performance gain is minimal. In the case of parallel optimised representation, we can see from the graph that manually doing the parallelisation is not paying off due to the overhead that comes with this process. Only for edit distances of 2 or 3 does this approach

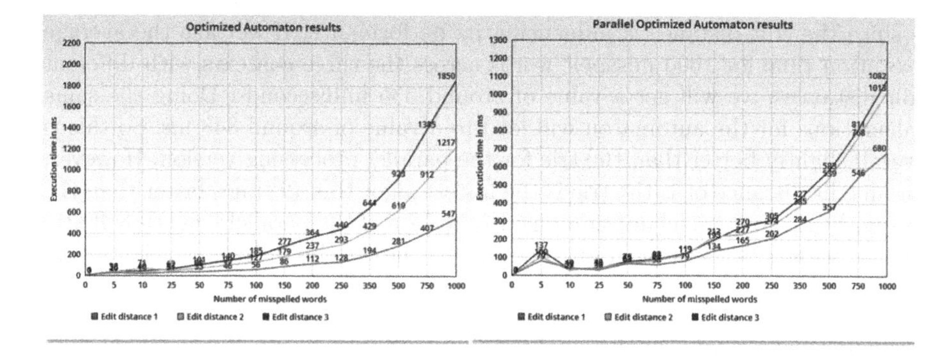

Fig. 4. Results for the optimised version of Levenshtein automaton

perform better than the serial automaton without the memory optimisation. Nevertheless, for the most common edit distance of 1, it is, in fact, slower than both the simple and memory optimised serial versions of the automaton (Fig. 4).

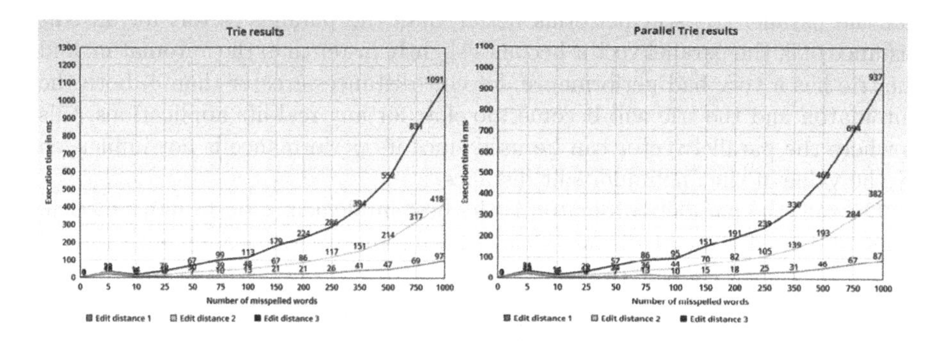

Fig. 5. Results for the Trie Levenshtein

In the case of Fig. 5, which presents the Trie Levenshtein distance, there is a small spike in the execution time for the case with five misspelt words. This is caused by the fact that, for that scenario, the method for finding suggestions is called for the first time, and the trie is initialized with the whole dictionary. The performance is better than that of the parallel automaton but decreases rapidly for more considerable edit distances. In contrast, the automaton keeps performing decently for edit distances of 3 and more (even though it is no match for the vector Levenshtein distance for large edit distances). So, from a practical point of view, the trie and the automaton perform very well, with an edge in the case of the trie for an edit distance of 1, which is very common in real-life applications. Therefore, the trie is presented as a good alternative for automaton implementation. As presented for the algorithm, and is also clear from the chart is that the trie is not performing well for edit distances greater than 1. This is since the edit distance is influencing its performance. If we take the average execution time for 1000 misspelt words across the three contexts with different edit distances, we will get a value of around 535 milliseconds. Doing the same calculations for the automaton will lead to a value of around 348 ms. So this is overall slightly better than the trie for the parallel processing version. However, for an edit distance of 1, the trie is the fastest algorithm (2 times faster than the second-best, the parallel automaton). In the case of the parallel trie, we will also have a slight spike in execution time due to the trie initialization. The overall times are almost identical to those from the simple trie, and this is because the parallelization is very limited. Nevertheless, we can notice a slightly better performance for each edit distance (87 ms vs 97 ms for edit distance 1, 382 ms vs 418 ms for edit distance 2, 937 ms vs 1091 ms for edit distance 3).

5 Conclusions and Future Work

In conclusion, based on the results presented in the previous section for an edit distance of 1, the parallel trie implementation is by far the fastest, followed by the parallel automaton; for an edit distance of 2, the parallel automaton is faster, but the parallel trie still performs better than the parallel vector; for an edit distance of 3, the parallel vector becomes slightly faster than the automaton, and the trie has a very bad performance; for edit distances greater than 4, both the automaton and the trie will become too slow for any real-life applications; this is where the parallel vector can be used since its performance is not influenced by the value of the maximum edit distance.

Overall, the parallel automaton is the best implementation from a practical point of view if the edit distances can variate, and we are dealing with typical values for a spell checker (edit distances of 1 or 2, rarely 3). As shown in the experimental results chapter, its average performance is better than that of the trie for these cases.

If we know that the edit distance is 1, we can pick the trie as it is much faster in this context (about two times faster than the second-best, the parallel automaton). If the edit distances go over a value of 3, we should pick the parallel vector algorithm, but this is not a real-life scenario.

For results replication, the code used for the experiments presented in this paper can be found at: https://bitbucket.org/AlexandruDragomir/spell-checker-application.

Acknowledgements. This work was partially supported by the grant 135C/ 2021 "Development of software applications that integrate machine learning algorithms", financed by the University of Craiova.

References

1. Abdelali, A., Darwish, K., Durrani, N., Mubarak, H.: Farasa: a fast and furious segmenter for arabic. In: Proceedings of the 2016 Conference of the North American Chapter of the Association for Computational Linguistics: Demonstrations, pp. 11–16 (2016)
2. Alamri, M.M., Teahan, W.J.: Automatic correction of arabic dyslexic text. Computers **8**(1), 19 (2019)
3. Bravo-Candel, D., López-Hernández, J., García-Díaz, J.A., Molina-Molina, F., García-Sánchez, F.: Automatic correction of real-word errors in spanish clinical texts. Sensors **21**(9), 2893 (2021)
4. Damerau, F.J.: A technique for computer detection and correction of spelling errors. Commun. ACM **7**(3), 171–176 (1964)
5. Islam, A., Milios, E., Kešelj, V.: Text similarity using google tri-grams. In: Kosseim, L., Inkpen, D. (eds.) AI 2012. LNCS (LNAI), vol. 7310, pp. 312–317. Springer, Heidelberg (2012). https://doi.org/10.1007/978-3-642-30353-1_29
6. Lhoussain, A.S., Hicham, G., Abdellah, Y.: Adaptating the levenshtein distance to contextual spelling correction. Int. J. Comput. Sci. Appl **12**(1), 127–133 (2015)
7. Ma, L., Zhang, Y.: Using word2vec to process big text data. In: 2015 IEEE International Conference on Big Data (Big Data), pp. 2895–2897. IEEE (2015)
8. Marzal, A., Vidal, E.: Computation of normalized edit distance and applications. IEEE Trans. Pattern Anal. Mach. Intell. **15**(9), 926–932 (1993)
9. Naseer, M., Asvial, M., Sari, R.F.: An empirical comparison of bert, roberta, and electra for fact verification. In: 2021 International Conference on Artificial Intelligence in Information and Communication (ICAIIC), pp. 241–246. IEEE (2021)
10. Pal, A., Mustafi, A.: Vartani spellcheck-automatic context-sensitive spelling correction of ocr-generated hindi text using bert and levenshtein distance (2020). arXiv preprint arXiv:2012.07652
11. Pennington, J., Socher, R., Manning, C.D.: Glove: global vectors for word representation. In: Proceedings of the 2014 Conference on Empirical Methods in Natural Language Processing (EMNLP), pp. 1532–1543 (2014)
12. Reese, S., Boleda, G., Cuadros, M., Rigau, G.: Wikicorpus: a word-sense disambiguated multilingual wikipedia corpus (2010)
13. Singh, S., Singh, S.: Systematic review of spell-checkers for highly inflectional languages. Artif. Intell. Rev **53**(6), 4051–4092 (2019). https://doi.org/10.1007/s10462-019-09787-4
14. Smith, R.: An overview of the tesseract OCR engine. In: Ninth International Conference on Document Analysis and Recognition (ICDAR 2007), vol. 2, pp. 629–633. IEEE (2007)
15. Yulianto, M.M., Arifudin, R., Alamsyah, A.: Autocomplete and spell checking levenshtein distance algorithm to getting text suggest error data searching in library. Sci. J. Inf. **5**(1), 75 (2018)

Ensemble Synthetic Oversampling with Manhattan Distance for Unbalanced Hyperspectral Data

Tajul Miftahushudur[1,2](✉), Bruce Grieve[1], and Hujun Yin[1]

[1] Department of Electrical and Electronic Engineering, The University of Manchester, Manchester M13 9PL, UK
mtaj001@lipi.go.id

[2] Research Center for Electronics and Telecommunication, Indonesian Institute of Sciences (LIPI), Bandung, Indonesia

Abstract. Hyperspectral imaging is a spectroscopic imaging technique that can cover a broad range of electromagnetic wavelengths and subdivide those into spectral bands. As a consequence, it may distinguish specific features more effectively than conventional colour cameras. This technology has been increasingly used in agriculture for various applications such as crop leaf area index, plant classification and disease monitoring. However, the abundance of information in hyperspectral imagery may cause high dimensionality problem, leading to computational complexity and storage issues. Furthermore, data availability is another major issue. In agriculture application, typically, it is difficult to collect equal number of samples as some classes or diseases are rare while others are abundant and easy to collect. This may give rise to an imbalanced data problem that can severely reduce machine learning performance and introduce bias in performance measurement. In this paper, an oversampling method is proposed based on Safe-Level synthetic minority oversampling technique (Safe-Level SMOTE), which is modified in terms of its k-nearest neighbours (KNN) function to make it fit better with high dimensional data. Using convolutional neural networks (CNN) as the classifier combined with ensemble bagging with differentiated sampling rate (DSR), the approach demonstrates better performances than the other state-of-the-art methods in handling imbalance situations.

Keyword: Imbalanced data · Hyperspectral imaging · Plant analysis · Safe-level SMOTE · CNN · Ensemble

1 Introduction

In the past decades, the use of hyperspectral imaging (HSI) for analyzing object characteristics has attracted increased attention in many fields due to its various advantages over conventional imaging. Hyperspectral images comprise of numerous spectral band making the technology richer in information than colour (RGB) imaging. The ability of hyperspectral technology to cover spectral bands beyond the visible spectral region

H. Yin et al. (Eds.): IDEAL 2021, LNCS 13113, pp. 54–64, 2021.
https://doi.org/10.1007/978-3-030-91608-4_6

and into the near-infrared (circa 700–2,500 nm wavelengths) or mid-infrared regions (circa 2,500–30,000 nm wavelengths) can be used to characterise chemical or physical characteristics of a specimen, such as a plant leaf.

Furthermore, hyperspectral imaging, when combined with machine learning techniques, can be used to handle a range of agriculture challenges such as plant classification [1], early disease detection and other biotic or abiotic stresses monitoring [2]. To maximise machine learning performance, a large data source is essential in the development of machine learning methods. Unfortunately, the availability of large amount of data can become an issue in many areas including agriculture. In practice, it can be difficult to obtain certain samples due to resources and sample scarcity. Conversely for some categories samples may be easier to obtain due to abundance sources. For example, it is easier to collect healthy leaf samples rather than to collect a leaf that has been stressed by a rare disease. The significant differences in size of the collected samples in various categories may lead to a serious imbalance problem that can severely hamper the performance of machine learning. Furthermore, in performance measurement, differences in sample size or test may introduce bias if the performance only measured with a simple matrix accuracy. Therefore, handling with imbalanced data in machine learning is a complex challenge.

Some oversampling techniques to overcome imbalance data problems in agriculture have been developed in previous studies [3, 4] and [5]. However, all these techniques were implemented on low dimensional data such as 2D images with RGB channels. Several studies have also been conducted for hyperspectral imagery. A combination of ensemble technique, synthetic minority oversampling technique (SMOTE) and rotation forest was discussed in [6], in which, data was iteratively balanced with a different number of rotation decision trees. Each classification result was then combined with a bagging approach. The result showed that the approach could surpass other hybrid techniques. Another method is called support vector sampling [7]. This technique has lower complexity compared to the standard sampling method. Finally, the latest technique is called composite clustering sampling [8]. It can reduce complexity in 3D convoulutional neural networks (CNN), which is commonly used to extract spatial and spectral feature of the HSI simultaneously. Most of the existing techniques have been tested on benchmark datasets such as the Indian pines dataset [9], a benchmark for classifying crop variety from HSI. To sum up, most of the previous studies perform a hybrid of methods to improve the accuracy. However, there is still room for improvement without sacrificing the complexity of implementation. In this paper we aim to tackle the imbalanced data problem with a simple approach and evaluate its performance on multiple indicators.

In this work, we present a classification framework for hyperspectral data using a modified SMOTE algorithm and ensemble learning. The SMOTE algorithm is modified by replacing its distance function to make the algorithm more suited for hyperspectral data. A differentiable sampling rate (DSR) is adopted in the ensemble learning to generate various training sets with different numbers of synthesised samples. Then, the training sets were used to train with a classifier based on a deep learning network.

2 Synthetic Minority Oversampling Technique (SMOTE)

In general, there are three approaches to tackling an imbalance problem: algorithm level, data level and hybrid (combination of algorithm and data level). The algorithm level approach aims to modify the existing algorithm by compensating majority class domination on the training process to reduce the bias of the majority class. The most discussed approach in the literature are threshold moving (adjust the threshold decision from normally 0.5 to a particular value) and cost-sensitive learning (modify the weight of minority sample). The most common technique used at the data level is resampling, which aims to balance sample distribution among classes. Resampling can be divided into two classes: undersampling and oversampling. The difference is that in oversampling methods new synthetics samples of the minority class are create whilst in undersampling some data in the majority class are removed.

One of the most popular oversampling technique is SMOTE [10]. This technique works by synthesizing samples in the minority class by finding k-nearest neighbour (KNN) from the sample data. After the neighbouring samples are found, new synthetic data is then generated between the sample data and the nearest neighbours. The main limitation of the SMOTE is that this method does not consider data distribution of the majority class; as a result, the new synthesis data could spread between the minority class and the majority class clusters. This problem may increase the number of false-negative data and reduce machine learning performance in recall of the majority class. This problem may occur more frequently with high dimensionality data, making it ineffective on such data types as hyper or multispectral images [11]. To solve this problem, modification of the distance calculation in KNN can be considered.

Some research has been done to tackle the SMOTE issue, for instance, by combining the SMOTE and undersampling approach like Tomeks-Link to remove synthetic data in majority cluster [12], Borderline SMOTE [13] and Safe-Level SMOTE [14] to categorize the synthetic data candidate based the number of the majority samples around the nearest neighbours:

- Noisy area: If the number of the majority samples in the neighbourhood is larger than the minority sample.
- Borderline area: If the number of the majority samples is equal to the minority sample.
- Safe area: If most of minority samples are around the neighbour.

The difference between the Borderline SMOTE and Safe-Level SMOTE is in the location of generated synthesis data. The Borderline SMOTE focuses on the borderline area in generating synthetics samples, while Safe-Level SMOTE focuses on a safe area. Furthermore, the Safe-Level SMOTE categorizes five criteria to determine the safe level; therefore, it is expected that new synthetic data will not overlap with the data in the majority class.

3 Approach

3.1 Proposed Method

The Euclidean distance function is typically used to measure the distance between minority sample and a neighbouring data point on SMOTE algorithm. With low dimensional data, say less than ten features, this function can effectively distinguish distances between data points. However, for high dimensional data like hyperspectral imagery which has numerous spectral features, the Euclidean distance is less effective as data distributes more sparsely than the low dimensional data. Measuring distances using the Euclidean distance in sparse data makes the distance ratios close to 1. Hence, in [15] it was suggested the use of the Manhattan distance function for high dimensional data. The Manhattan formula is given as,

$$d(x, y) = \sum_{i=1}^{m} |x_i - y_i| \tag{1}$$

where m is the number of dimensions.

The proposed method further incorporates an ensemble by implementing a differentiated sampling ratio (DSR) [16] to increase the base learner's diversity in the ensemble bagging framework. To illustrate how DSR works, say the sampling ratio is of B% of the population. N1 is the number of samples in the majority class. For the first classifier, the sampling ratio B is set to 10%. Sample data in minority class is divided into two groups. In the first group, 10% of N1 data is resampled as a replacement of the original minority class data. To balance with the majority samples (100%–10%), N1 of minority class data is generated by Safe-Level SMOTE and stored in the second group. Both groups then combined to construct the first balanced data set to be trained in the first classifier. In the second classifier, the DSR ratio is updated to 20%, up to the 10th classifier with 100%. Flowchart of the proposed method is shown in Fig. 1. In this paper, each classifier is trained with a CNN. The CNN architecture for the Indian Pines and University of Manchester (UoM) datasets are illustrated in Tables 1 and 2, respectively.

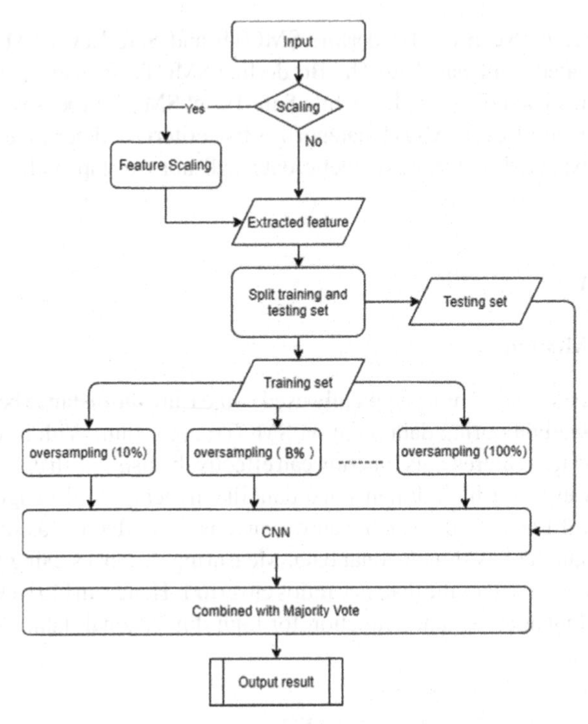

Fig. 1. Flow diagram of the proposed resampling and classification method.

Table 1. CNN architecture for Indian Pines

Input 1 × 200 spectral vector
1 × 3 conv. 64 ReLU
1 × 3 conv. 64 ReLU
1 × 2 max-pooling
1 × 3 conv. 128 ReLU
1 × 2 max-pooling
1 × 3 conv. 256 ReLU
1 × 2 max-pooling
1 × 3 conv. 512 ReLU
1 × 2 max-pooling
Dropout, prob = 0.5
16-class softmax

Table 2. CNN architecture for UoM dataset

Input 1 × 33 spectral vector
1 × 3 conv. 32 ReLU
1 × 2 max-pooling
1 × 3 conv. 64 ReLU
1 × 2 max-pooling
1 × 3 conv. 128 ReLU
1 × 2 max-pooling
1024 dense
2-class softmax

3.2 Datasets

The first dataset used in this experiment is the UoM dataset [2], a hyperspectral image dataset on plant stresses produced at the University of Manchester. This dataset consists of a spectral range between 400 nm and 720 nm with 10 nm of resolution. The dataset contains samples of Arabidopsis leaves in cold and heat stress conditions. There are 216 controlled (normal condition) samples and 465 abnormal samples (stressed) in total. 60% from the dataset used for training data while the 40% remaining used as the test set. The summaries of the UoM dataset and the spectral data of two classes are shown in Table 3 and Fig. 2, respectively.

Table 3. Sample distribution of UoM dataset

Train		Test	
Normal	Stress	Normal	Stress
259	130	173	86

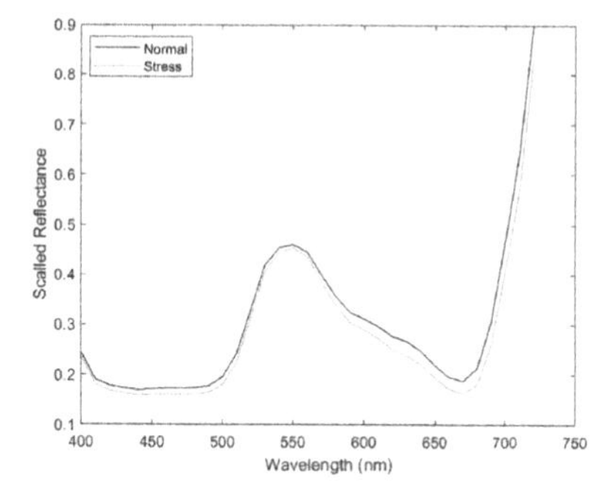

Fig. 2. Spectral average of two classes of UoM dataset.

The Indian pines dataset [9] is one of the most popular datasets used in hyperspectral image classification studies. Using this benchmark dataset, the classification performance of the proposed technique is compared with other classification algorithms or machine learning frameworks. This dataset has images of 145 × 145 pixel size with spatial resolution of 20 m per pixel covering 16 classes of different crops species. Originally, the Indian pines dataset covers 224 spectral bands between 360 nm and 2500 nm. This number of band is reduced to 200 after removing bands that covering water absorption. In this experiment, the dataset is split into trainset with 10% and 50% proportion from the whole dataset. Detail and sample distribution of the Indian pines train set are described in Fig. 3.

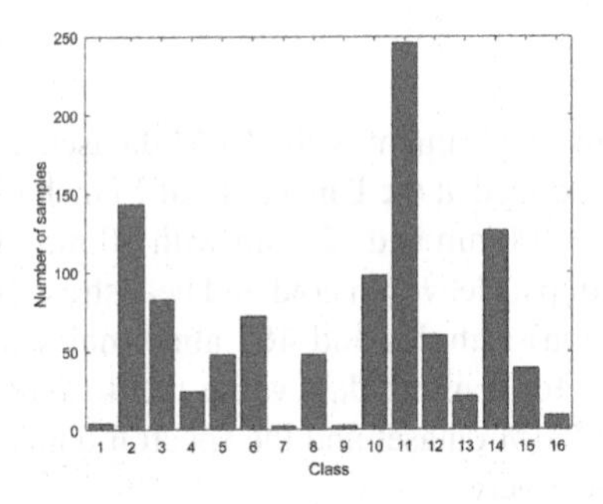

Fig. 3. Distribution of Indian Pines's data classes.

4 Results and Discussion

4.1 UoM Dataset

Classification results on the UoM dataset are shown in Table 4, along with that of some oversampling methods such as SMOTE, adaptive synthetic (ADASYN), original Safe-Level SMOTE, and ensemble learning based on novelty detection (Ensemble ND). The same CNN architecture in Table 2 used for the oversampling methods as a comparison. 10-fold cross-validation is used on this dataset and the experiment repeated for 3 times.

We used the Matthews correlation coefficient (MCC) to evaluate the effectiveness of the proposed model. The MCC was chosen because it is more robust and provides less misleading information in binary classification than the F1 Score [17].

Table 4. Classification results on UoM dataset.

	Sensitivity	Specificity	Accuracy	MCC
SMOTE	79.39	97.93	90.43	80.49
ADASYN	82.26	97.00	91.36	81.27
Safe-Level SMOTE	**86.44**	97.08	93.2	85.26
Ensemble ND [2]	–	–	91.58	–
Modified Safe-Level SMOTE	85.95	98.02	93.51	86.17
Further with ensemble	86.07	**98.51**	**93.82**	**86.93**

From Table 4, it can be seen that the proposed method outperformed other oversampling methods and the novelty detection methods. In terms of the ability to detect minority class (abnormal/stress), the proposed methods reach the similar level, 86.44% and 86.07% with the state-of-the-art methods. However, in other metrics the proposed methods had highest performances, in terms of detecting normal samples, i.e. specificity of 98.51%. Furthermore the proposed method (with ensemble) had the highest accuracy and MCC, 93.82% and 86.93%, respectively. In contrast, the SMOTE had the poorest classification performance as it only has an 80.49% MCC.

4.2 Indian Pines Dataset

Tables 5 and 6 show evaluation results on the Indian pines dataset with 50% and 10% training data respectively.

Table 5. Classification result on Indian Pines using 50% training data.

	Ensemble ND [2]	Hybrid resnet + inception [18]	**Modified Safe-Level SMOTE**	**Further with ensemble**
MCC	–	–	92.29	**95.13**
AA	91.70	–	93.21	**95.73**
OA	–	90.57	93.10	**94.75**

Table 6. Classification result on Indian Pines using 10% training data.

Class	CNN [19]	CNN + SMOTE [19]	OCSP [20]	SMOTERof [6]	**Modified Safe-Level SMOTE**	**Further with ensemble**
1	60.98	73.17	57.5	**96.09**	62.57	54.87
2	56.03	71.83	78.3	68.97	**80.75**	79.93
3	**76.44**	70.95	52.7	72.70	75.18	73.19
4	58.69	68.08	37.1	**86.55**	68.36	71.12
5	91.03	85.29	86.1	83.57	85.50	**87.81**
6	94.22	94.67	**97.6**	87.95	91.78	96.35
7	0	8	76	**100**	72.92	76
8	**99.3**	**99.3**	97.7	91.43	98.72	98.72
9	0	66.67	50	**100**	94.44	94.44
10	62.17	77.37	72.2	76.31	81.98	**82.55**
11	81.9	75.75	83.2	51.55	78.90	**82.79**
12	74.34	66.67	61.4	77.11	79.19	**79.31**
13	**99.46**	98.92	99.8	93.90	98.37	98.92
14	85.34	**97.81**	95.3	83.79	92.22	94.90
15	39.77	39.48	58.9	43.30	**65.85**	56.34
16	86.9	92.82	**89.2**	**97.87**	85.02	86.75
F1	–	–	–	72.82	81.51	**82.93**
AA	66.6	74.18	74.6	81.94	81.99	**82.13**
OA	75.88	78.94	78.8	–	82.57	**84.37**

From Tables 4, 5, and 6 it can be seen that the proposed Safe-Level SMOTE with modified distance and CNN performed better than other methods. On the UoM dataset, even though the proposed method has slightly lower specificity than the original Safe-Level SMOTE, the higher F1 score or MCC and accuracy show that the proposed approach improved the previous methods. It can be concluded that adaptation of Manhattan distance had the best performance, followed by the Safe-Level SMOTE.

On the Indian pines dataset with a multi-class problem, the single CNN had the poorest accuracy among all the methods. On other hand, Safe-Level SMOTE showed its superiority compared with another state of the art methods. Furthermore, the proposed method also compared with one of the state of the art CNN architecture, combination of ResNet and Inception [18]. Using the same dataset, it only had 82.4% and 90.57% accuracy while tested in 10% and 50% training data, slightly lower than the modified Safe-Level SMOTE with 0.2% and 2.5% difference, respectively and proposed method with 2% and 4.2%.

The experimental results show that the hyperspectral data of plant conditions and species have been successfully classified using a modified Safe-Level SMOTE algorithm, with markedly improved classification results over the original SMOTE algorithm. In addition to this, further extension with ensemble learning with various sampling rates provides even higher performance than previous studies. From these, we can conclude that the proposed method is more effective in handling imbalanced data, a common problem in agriculture applications.

5 Conclusions and Future Work

In this study, we provide an enhanced technique to improve classification accuracy on imbalanced multi/hyperspectral data situation by balancing its class distribution of training samples. Using oversampling, based on the SMOTE method, we have modified its distance function used to measure its nearest neighbours with the Manhattan distance. The Manhattan distance function is applied to the Safe-Level SMOTE algorithm to synthesize minority samples on multi/hyperspectral data. Using CNN as feature extraction and classifier, the experimental results show that the proposed method yields marked improvements in terms of classifying minority class compared to the other sampling methods, especially on binary classifications such as plant disease detection. Furthermore, after being tested on benchmark hyperspectral datasets, the proposed technique shows better performance than other state-of-the-art classification methods. For future work, we will explore how to perform oversampling to synthesis the hypercube datasets such that the spectral and spatial data is balanced simultaneously. In addition, combining oversampling and undersampling methods, to improve data synthesis and quality, will be studied further.

Acknowledgement. Tajul Miftahushudur would like to acknowledge the Scholarship provided by the Indonesian Endowment Fund for Education (LPDP).

References

1. Alsuwaidi, A., Veys, C., Hussey, M., Grieve, B., Yin, H.: Hyperspectral feature selection ensemble for plant classification. Hyperspectral Imaging Appl. (HSI 2016) (2016)

2. Alsuwaidi, A., Grieve, B., Yin, H.: Feature-ensemble-based novelty detection for analyzing plant hyperspectral datasets. IEEE J. Sel. Top. Appl. Earth Obs. Remote Sens. **11**(4), 1041–1055 (2018)
3. Sambasivam, G., Opiyo, G.D.: A predictive machine learning application in agriculture: Cassava disease detection and classification with imbalanced dataset using convolutional neural networks. Egypt. Inf. J. **22**(1), 27–34 (2020)
4. Hussein, B.R., Malik, O.A., Ong, W.-H., Slik, J.W.F.: Automated classification of tropical plant species data based on machine learning techniques and leaf trait measurements. In: Alfred, R., Lim, Y., Haviluddin, H., On, C.K. (eds.) Computational Science and Technology. LNEE, vol. 603, pp. 85–94. Springer, Singapore (2020). https://doi.org/10.1007/978-981-15-0058-9_9
5. Divakar, S., Bhattacharjee, A., Priyadarshini, R.: Smote-DL: a deep learning based plant disease detection method. In: 6th International Conference for Convergence in Technology (I2CT) (2021)
6. Feng, W., Huang, W., Ye, H., Zhao, L.: Synthetic minority over-sampling technique based rotation forest for the classification of unbalanced hyperspectral data. In: International Geoscience and Remote Sensing Symposium (IGARSS), vol. 12(7), pp. 2159–2169 (2018)
7. Zhang, X., Song, Q., Zheng, Y., Hou, B., Gou, S.: Classification of imbalanced hyperspectral imagery data using support vector sampling. In: International Geoscience and Remote Sensing Symposium (IGARSS) (2014)
8. Li, C., Qu, X., Yang, Y., Yao, D., Gao, H., Hua, Z.: Composite clustering sampling strategy for multiscale spectral-spatial classification of hyperspectral images. J. Sens. **2020** (2020). Article ID 9637839, 17 pages. https://doi.org/10.1155/2020/9637839
9. Baumgardner, M.F., Biehl, L.L., Landgrebe, D.A.: 220 Band AVIRIS hyperspectral image data set: June 12, 1992 Indian pine test site 3. Purdue Univ. Res. Repos. (2015)
10. Chawla, N.V., Bowyer, K.W., Hall, L.O., Kegelmeyer, W.P.: SMOTE: synthetic minority over-sampling technique. J. Artif. Intell. Res. **16**(1), 321–357 (2002)
11. Blagus, R., Lusa, L.: SMOTE for high-dimensional class-imbalanced data. BMC Bioinformatics **14**(106), 1471–2105 (2013)
12. Batista, G.E.A.P.A., Prati, R.C., Monard, M.C.: A study of the behavior of several methods for balancing machine learning training data. ACM SIGKDD Explor. Newsl. 6(**1**), 20–29 (2004)
13. Han, H., Wang, Wen-Yuan., Mao, Bing-Huan.: Borderline-SMOTE: a new over-sampling method in imbalanced data sets learning. In: Huang, De-Shuang., Zhang, Xiao-Ping., Huang, Guang-Bin. (eds.) ICIC 2005. LNCS, vol. 3644, pp. 878–887. Springer, Heidelberg (2005). https://doi.org/10.1007/11538059_91
14. Bunkhumpornpat, C., Sinapiromsaran, K., Lursinsap, C.: Safe-level-SMOTE: safe-level-synthetic minority over-sampling technique for handling the class imbalanced problem. In: Theeramunkong, Thanaruk, Kijsirikul, Boonserm, Cercone, Nick, Ho, Tu-Bao. (eds.) PAKDD 2009. LNCS (LNAI), vol. 5476, pp. 475–482. Springer, Heidelberg (2009). https://doi.org/10.1007/978-3-642-01307-2_43
15. Aggarwal, C.C., Hinneburg, A., Keim, D.A.: On the surprising behavior of distance metrics in high dimensional space. In: Van den Bussche, J., Vianu, V. (eds.) Database Theory — ICDT 2001. ICDT 2001. Lecture Notes in Computer Science, vol. 1973, pp. 420–434. Springer, Heidelberg (2001). https://doi.org/10.1007/3-540-44503-X_27
16. Feng, W., Huang, W., Bao, W.: Imbalanced hyperspectral image classification with an adaptive ensemble method based on SMOTE and rotation forest with differentiated sampling rates. IEEE Geosci. Remote Sens. Lett. **16**(12), 1879–1883 (2019)
17. Chicco, D., Jurman, G.: The advantages of the Matthews correlation coefficient (MCC) over F1 score and accuracy in binary classification evaluation. BMC Genomics. **21**(1) (2020). Article ID 6. https://doi.org/10.1186/s12864-019-6413-7

18. Alotaibi, B., Alotaibi, M.: A hybrid deep ResNet and inception model for hyperspectral image classification. PFG – J. Photogram. Remote Sens. Geoinformation Sci. **88**(6), 463–476 (2020). https://doi.org/10.1007/s41064-020-00124-x
19. Cai, L., Zhang, G.: Hyperspectral image classification with imbalanced data based on oversampling and convolutional neural network. In: AOPC: AI in Optics and Photonics (2019)
20. Li, J., Du, Q., Li, Y., Li, W.: Hyperspectral image classification with imbalanced data based on orthogonal complement subspace projection. IEEE Trans. Geosci. Remote Sens. **56**(7), 3838–3851 (2018)

AutoML Technologies for the Identification of Sparse Models

Aleksei Liuliakov(✉) and Barbara Hammer

Machine Learning Group, Bielefeld University, Bielefeld, Germany
{aliuliakov,bhammer}@techfak.uni-bielefeld.de

Abstract. Automated machine learning (AutoML) technologies constitute promising tools to automatically infer model architecture, meta-parameters or processing pipelines for specific machine learning tasks given suitable training data. At present, the main objective of such technologies typically relies on the accuracy of the resulting model. Additional objectives such as sparsity can be integrated by pre-processing steps or according penalty terms in the objective function. Yet, sparsity and model accuracy are often contradictory goals, and optimum solutions form a Pareto front. Thereby, it is not guaranteed that solutions at different positions of the Pareto front share the same architectural choices, hence current AutoML technologies might yield sub-optimal results. In this contribution, we propose a novel method, based on the AutoML method TPOT, which enables an automated optimization of ML pipelines with sparse input features along the whole Pareto front. We demonstrate that, indeed, different architectures are found at different points of the Pareto front for benchmark examples from the domain of systems security.

Keywords: AutoML · Feature selection · TPOT

1 Introduction

Machine learning (ML) methods play an increasingly important role in various domains including intrusion detection and cyber-security [4]. Yet the precise setup of ML pipelines often constitutes an exhaustive process, since the combination of optimum pre-processing, machine learning method, meta-parameter choice and strength of regularization severely depends on the specific setting and data set. In recent years, quite a number of approaches have been proposed to automate the design of ML pipelines or architecture search – commonly referred to as AutoML [10]. The technologies vary in the richness of their search space, ranging from model meta-parameters up to whole pipelines, the way in which possible choices are represented, ranging from vectorial representations up to complex relational representations, and in the optimization technology which

We gratefully acknowledge funding by the BMBF within the project HAIP, grant number 16KIS1212.

H. Yin et al. (Eds.): IDEAL 2021, LNCS 13113, pp. 65–75, 2021.
https://doi.org/10.1007/978-3-030-91608-4_7

they use to efficiently navigate through the search space, ranging from Bayesian optimization up to genetic algorithms [3,20,26,28]. Several AutoML methodologies directly offer APIs which link to popular machine learning libraries such as WEKA or scikit-learn [3,19,26].

So far, the vast majority of AutoML approaches relies on an optimization with respect to a single criterion, usually the generalization error of the proposed model. In practice, multiple different criteria are in the focus of the design of ML pipelines, including computational efficiency, sparsity, interpretability, or fairness, to name just a few [22]. In this work, we are interested in (global) interpretability and sparsity of the model in terms of input features, i.e. besides a high accuracy, we aim for a model which uses only few of the input signals. In classical ML, a large variety of feature selection technologies is readily available, commonly categorized as filter, wrapper or embedded methods [5]. Since feature selection constitutes an NP-hard problem by itself, these technologies are often realized via efficient but possibly sub-optimal approximations, which aim for one sparse and accurate solution. As an alternative, computationally complex meta-heuristics can aim for the full Pareto front of solutions with different degree of sparsity, whereby the ML model itself is usually not varied but included via wrapper technologies [2,7,11]. Current AutoML technologies, on the other side, implicitly integrate sparsity by incorporating feature selection techniques into the search space to explicitly improve overall performance objective. Hence current AutoML methods generate one single solution on the Pareto-front.

In this contribution, we are interested in possibilities to enrich AutoML methods such that they deliver representative solutions for the full Pareto-front of accurate and sparse models in the specific domain of intrusion detection. Thereby, we explicitly allow potentially different choices of the ML architectures for different points of the Pareto front. We put a particular focus on the compatibility of the technology with the popular ML toolbox scikit-learn, because of which we extend a Tree-based Pipeline Optimization Tool (TPOT) [20] by a wrapper which combines it with suitable feature selection methods. We demonstrate in theoretical data as well as two benchmarks from the domain of intrusion detection that a variation of the architecture increases the model performance along relevant parts of the Pareto front.

2 AutoML for Feature Selection

AutoML: Assume a training data set $\mathcal{D}$ is given and a loss function $\mathcal{L}$ is fixed such as the model accuracy. AutoML aims for an optimization of the Combined Algorithm Selection and Hyperparameter Optimization (CASH) problem [26].

$$A^*_{\lambda^*} \in \underset{A^{(j)}_\lambda \in \mathcal{A},\ \lambda \in \Lambda^{(j)}}{\operatorname{argmin}} \frac{1}{k} \sum_{i=1}^{k} \mathcal{L}(A^{(j)}_\lambda, \mathcal{D}^{(i)}_{train}, \mathcal{D}^{(i)}_{valid}). \tag{1}$$

where $\mathcal{A} = \{A^{(1)}, ..., A^{(f)}\}$ is the set of algorithms with associated hyperparameter spaces $\Lambda^{(1)}, ..., \Lambda^{(f)}$ which should be explored and $\mathcal{L}(A^{(j)}_\lambda, \mathcal{D}^{(i)}_{train}, \mathcal{D}^{(i)}_{valid})$

refers to the result of the loss function obtained by decision algorithm $A_\lambda^{(j)}$ with hyperparameters λ when trained on $\mathcal{D}_{train}^{(i)}$ and evaluated on $\mathcal{D}_{valid}^{(i)}$, where the sets result from a split of $\mathcal{D}$ in a k-fold cross validation scheme.

Several recent AutoML techniques [16,19,26] address the CASH problem 1, where the search space incorporates preprocessing operators, machine learning algorithms and hyperparameters. In the following, we will use the popular AutoML framework TPOT [20]. TPOT constitutes a wrapper for the Python machine learning packages, scikit-learn and XGBoost. Machine learning operators from scikit-learn are used as primitives for genetic programming, and the tree structures formed by these primitives are optimized using standard GP technology.

In this work the search space is restricted number of popular classifiers (e.g. *DecisionTreeClassifier, ExtraTreesClassifier, RandomForestClassifier, KNeighborsClassifier, LinearSVC, LogisticRegression, MLPClassifier, XGBClassifier*) and preprocessors (e.g. *PCA, StandardScaler, MinMaxScaler, PolynomialFeatures* etc.). Every operator itself depends on the range of its hyperparameters. For instance LogisticRegression parameter “penalty” is either {“*l1*”, “*l2*”}, parameter “C” is in a set $\{10^{-4}, 10^{-3}, 10^{-2}, 10^{-1}, 0.5, 1, 5, 10, 15, 20, 25\}$, parameter “dual” could be either $\{True, False\}$). TPOT performs search through this space of operators and its hyperparameters to return optimal architecture in the form of a machine learning pipeline, which corresponds to a solution of the problem 1.

Feature Selection: Assume input data $\mathcal{D}$ are referred to as tabular data X with features $\{1, \dots, n\}$. Feature selection refers to the challenge to identify a subset $I = \{i_1, \dots, i_m\}$ of features, where $m \ll n$, such that a model performs well on X^I instead of X, where X^I refers to the data set which is obtained by deleting all features in X which are not contained in the set I. There exists a large number of popular feature selection methods [6]. In this work, we will refer to two popular wrapper approaches, forward feature selection (FFS) and backward feature elimination (BFE), respectively. Essentially, these methods circumvent an exhaustive search through all possible sets of features by greedily singling out a single feature which is either added to the set of already selected features (forward selection) or deleted from the set of remaining features (backward elimination). In each case, the performance of the model on the resulting feature set serves as evaluation criterion for the greedy steps.

Pareto Optimization: In our work, we are interested in a generalization of the CASH problem, which aims for the optimization of a pair of objectives: the accuracy of the model as evaluated by the cross-validation error on X^I, and the sparsity of the model as evaluated by the size of the selected set of features $|I|$. As such, it constitutes and instance of a multi-objective or vector optimization [13]. More specifically, we address the following objectives

$$\begin{aligned} \underset{X^I \subset X,\ A_\lambda^{(j)} \in \mathcal{A},\ \lambda \in \Lambda^{(j)}}{\text{Optimize}} \quad & F(X) := (f_1(X^I), f_2(X^I)) \\ \text{where} \quad & f_1(X^I) = |I|, \\ & f_2(X^I) = \frac{1}{k} \sum_{i=1}^{k} \mathcal{L}(A_\lambda^{(j)}, {X^I}_{train}^{(i)}, {X^I}_{valid}^{(i)}) \end{aligned} \tag{2}$$

and f_1 needs to minimized, f_2, as before, refers to the accuracy of the classifier evaluated in a cross-validation, and needs to be maximized. We measure the accuracy as the ratio between correctly classified instances and the whole number of instances.

Within an AutoML framework, we can efficiently represent feature selection within the search space by a binary vector i of length n: I is then given by all features l with an entry 1 at position i_l, features l with an entry 0 at position i_l are deleted. Since the two objectives f_1 and f_2 might be contradictory, we aim for the Pareto front. This consists of all solutions $(X^I, A_\lambda^{(j)}, \lambda)$ which are not dominated by another choice of values, i.e. for all other instances of the search space, either the value f_1 is larger or the value f_2 is smaller.

Baseline: Current AutoML frameworks do not easily enable an optimization which yields the complete Pareto front, since they aim for a single criterion only. Starting from single criterion optimization, however, one obtains a natural (strong) baseline against which we will compare in our experimental section: we can first optimize the pipeline based on all features using TPOT; this way a strong model is obtained for the specific training problem. For this model, sparsity can be realized by subsequent BFE, where the overall model architecture stays fixed. As a result, models with different degree of sparsity are obtained. A limitation of this approach consists in the restriction to one fixed architecture optimized by TPOT once, hence a change of the ML architecture for different degrees of sparsity is not possible in this approach.

Sparsity Inducing Iterative TPOT: We aim for a solution of the full Pareto front for the problem 2, where different architectural choices can be made for different degrees of sparsity. Since a simultaneous search over features and architectural choices is exhaustive, we propose an iterative greedy scheme, which combines the very efficient approximation schemes BFE or FFS, respectively, with an AutoML scheme such as TPOT, to sample the Pareto front. Essentially, BFE and FFS, respectively, can be used as wrapper around the results obtained by TPOT. Since TPOT does not need to start from scratch in each iteration but can include the already found solution in its GP scheme, fast convergence of the TPOT optimization scheme reduces the computational load of the method.

The final pseudocode relies on the following functional modules:

pipe(X, y): Returns the accuracy for the current architecture **pipe(., .)** for data X, y.

FFS$(X_{current}, y, X_{candidates},$ **pipe(., .)**, $cv = 5)$ - one step of the forward feature selection heuristic: this loops through every feature $x_i \in X_{candidates}$, evaluates the accuracy with **pipe**$(X_{current} \cup x_i, y)$ with 5 fold cross validation and selects the feature $\mathbf{x}^*$ with largest accuracy. It returns $X_{current} \cup \mathbf{x}^*$.

BFE$(X_{current}, y,$ **pipe(., .)**, $cv = 5)$ is analogous with backward feature elimination heuristics and returns according $X_{current} \setminus X_i$.

TPOT$(X, y, cv = 5, time, config\_space)$ searches the best pipeline configuration using GP in $config\_space$, using 5 fold cross validation as a search objective, with "*time*" as stopping criterion, and returns the best pipeline.

MI(X, y) uses the mutual information to rank the features and returns the best feature X_i from X

These modules are combined towards a sparsity inducing iterative TPOT using FFS or BFE as follows:

FFS-TPOT(Input: Data X_{train}, X_{test}, y_{train}, y_{test}, with n features)
Initialize $X_1 :=$ MI(X_{train}, y_{train})
for i in range(1, n)
$sparsity_i :=$ i
$pipe_i$:=TPOT(X_i, y_{train})
$accuracy_i :=$ pipe(X_{test}, y_{test})
$X_{i+1} :=$ FFS$(X_i, y_{train}, X_{train} \setminus X_i, pipe_i)$
Return: Pareto solutions $(sparsity_i, accuracy_i)$ for all i

BFE-TPOT(Input: Data X_{train}, X_{test}, y_{train}, y_{test}, with n features)
Initialize $X_n := X_{train}$
for i in range(n, 1)
$sparsity_i :=$ i
$pipe_i$:=TPOT(X_i, y_{train})
$accuracy_i :=$ pipe(X_{test}, y_{test})
$X_{i-1} :=$ BFE$(X_i, y_{train}, pipe_i)$
Return: Pareto solutions $(sparsity_i, accuracy_i)$ for all i
Schematic description of FFS-TPOT is shown on the Fig. 1.

3 Experiments

Data Sets: We evaluate the method using four different synthetic data sets and two data sets from the domain of intrusion detection.

Synthetic data: Two data sets (**synthetic 1**, **synthetic 2**) for binary classification are designed in such a way that different classifiers are suited for different parts of the data, hence we expect that different requirements of sparsity are better treated by different architectural choices. More specifically, we randomly generate data sets and label those according to a fixed

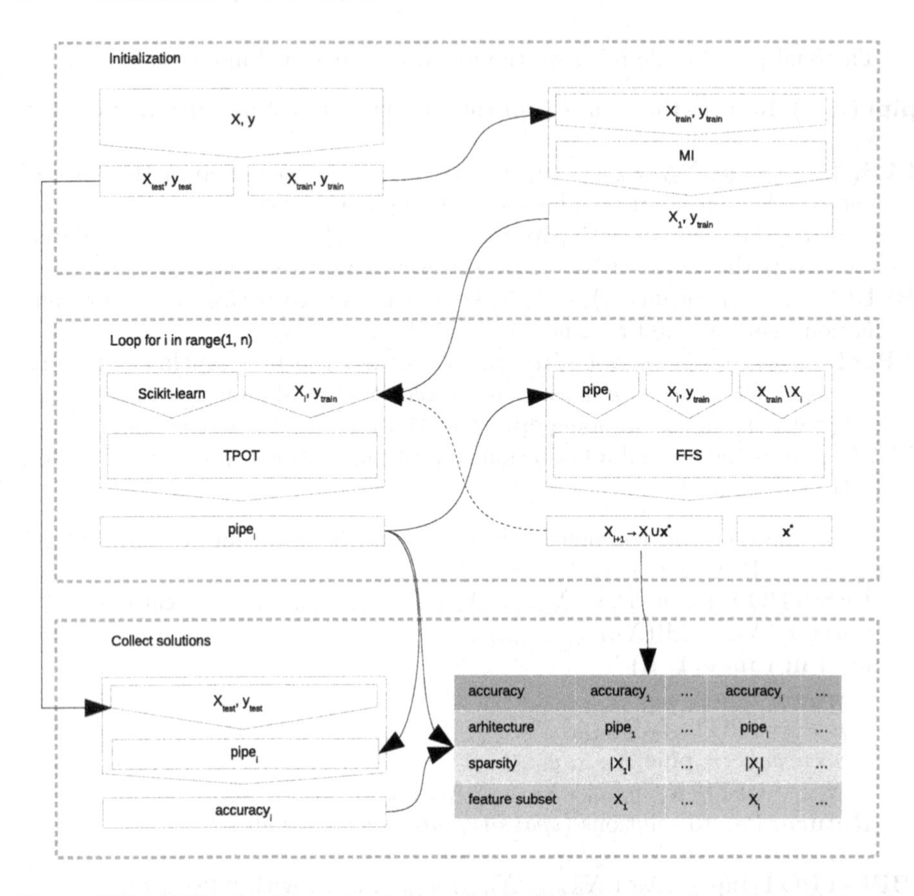

Fig. 1. FFS-TPOT algorithm description. In the Initialization phase data X, y splits to X_{test}, y_{test} and X_{train}, y_{train}. The first most relevant feature is selected by means of Mutual Information (MI). The Loop phase established by two functional blocks. On the left side TPOT evaluates and returns the most accurate ML pipeline $pipe_i$ with respect to input data from the initialization step and predefined configuration space of Scikit-learn models. The right block performs a greedy search of the most relevant feature from $X_{train} \setminus X_i$. The feature $\mathbf{x}^*$ which yields the best improvement in accuracy for $pipe_i$ joins to X_i and the updated feature subset $X_i \cup \mathbf{x}^*$ is used in the next loop to evaluate a new architecture. The algorithm loops through all features sequentially and for every step the resulting solutions (architecture, feature subset) and objectives values (accuracy, sparsity) are collected.

given classifier with different characteristics. These data sets are joined in a block diagonal form where, for each subset, additional features are filled randomly. All the features in the resulting data sets are relevant and non redundant. The specific classifier, number of features per classifier and data set size is depicted in Table 1. Another pair of data sets (**synthetic 3**, **synthetic 4**) were generated at random using make_classification module from scikit-learn library. The data sets have the size of 1500 instances and 20 features each. Synthetic 3 has six relevant, three redundant and 11 irrelevant features. Synthetic 4 has three relevant, three redundant and 14 irrelevant features.

UNSW-NB15 data set: The data have been created by the University of New South Wales as a benchmark in network intrusion detection system [23]. The data generation process is described in [17]: Raw traffic data is collected as packets in PCAP files, and flow based tabular data extracted thereof. An IP flow-based representation characterized by the five elements source IP, destination IP, source port, destination port and protocol is popular various approaches [9,12,18,24,27]. The single instances are then characterized by features which describe statistical properties and content, resulting in 47 features (30 integer, 10 float, 3 categorical, 2 binary, 2 timestamp). Labels distinguish attacks and normal class. The group of attacks is further differentiated into attack types (Fuzzers, Analysis, Backdoors, DoS, Exploits, Generic, Reconnaissance, Shellcode, Worms), yielding 82332 instances. While normal class instances versus attacks in the same order of magnitude, specific types of attacks are severely imbalanced (with extremes worms (44 cases) to generic (18871 cases)). Data are preprocessed by numerically encoding categorical features, feature scaling, and under-sampling of classes while training for imbalanced settings.

NSL-KDD data set: NSL-KDD is a publicly available data set for network intrusion detection produced by the University of New Brunswick [25] and based on the KDD CUP 99 data set [1]. It characterizes traffic connection records with 41 features each. The original data from KDD99 is cleaned as regards redundant records and artefacts as highlighted in [14,25]. The resulting data consists of 125973 train entries and 22544 test entries and attacks. Categorical data are numerically encoded and features are scaled. Training is done using 20% of the data by random sampling with replacement, as provided by authors.

Table 1. Characteristics of Synthetic_1 and Synthetic_2 data sets

Classifier	Features	Size	Features	Size
XGBoost	2	400	2	1000
SVC	2	600	4	500
MLP	2	800	4	250
Logistic regression	4	1000	4	250
Final data set	**10**	**2800**	**14**	**2000**

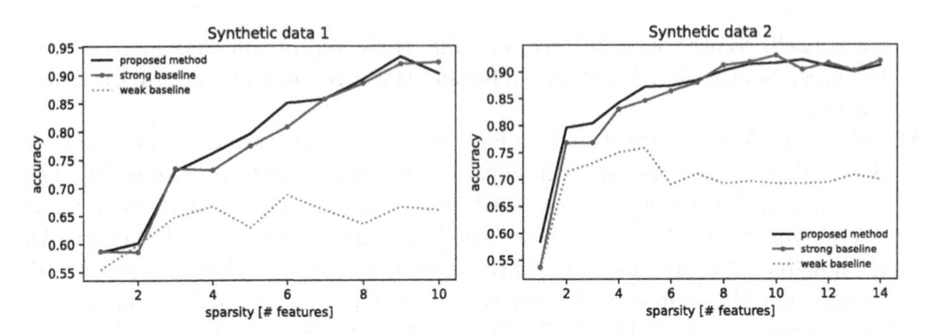

Fig. 2. Accuracy results obtained on the synthetic data 1 and synthetic data 2 for the two baselines and the proposed method for different levels of sparsity.

Experimental Setup: The proposed algorithms were encoded in the Python 3.8.8 environment using scikit-learn [21], TPOT [19], numpy [8], pandas [15], matplotlib. Computations were performed on a GPU cluster using 15 CPU cores. As a comparison, we use TPOT with subsequent BFE as descibed as a strong baseline above, and we use the performance of kNN ($k = 8$) with subsequent BFE as a simple but often suprisingly powerful approach as a weak baseline. We use FFS with MI for sparsity inducing iterative TPOT for the real data set, and BFE for all others, since these methods gave the respective best results for the individual settings.

Results: We report the results for the different methods as curves of the accuracy achieved by focusing on different degrees of sparsity. The plain black line shows the result of our approach. The dotted blue line shows the result of the weak kNN baseline, the red line with dots the strong baseline of BFE based on an optimized TPOT architecture for the synthetic data sets (Fig. 2, 3) and the intrusion detection data sets (Fig. 4).

As can be seen from the experiments, AutoML technologies can drastically improve the result as compared to a fixed benchmark protocol – here a kNN classifier. The same results holds for comparison with the strong baseline in some cases (synthetic data 3,4). But the results are similar for other default choices for all classifier from the considered set of classifiers with other established data sets. Using TPOT optimization once and subsequent BFE yields comparably good results over the whole range of sparsity. In particular for higher sparsity, the capability to change the architecture using iterative TPOT enables an improvement of the result, indicating that different architecture become relevant as soon as limited information is yet available. This is particularly pronounced for the real data sets from intrusion detection, as can be seen in Fig. 4.

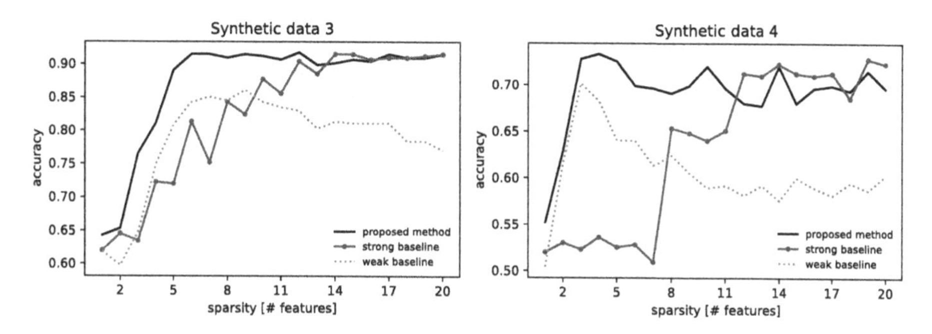

Fig. 3. Accuracy results obtained on the synthetic data 3 and synthetic data 4 for the two baselines and the proposed method for different levels of sparsity.

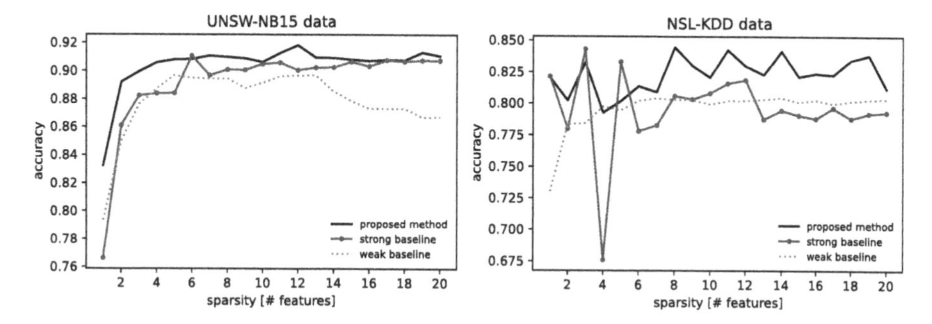

Fig. 4. Accuracy results obtained on the intrusion detection UNSW-NB15 and NSL-KDD data for the two baselines and the proposed method for different levels of sparsity.

4 Conclusion

We have presented a novel technology which combines AutoML methods with feature selection schemes to optimize the full Pareto front of sparse and accurate solutions. It turned out in a number of samples that iterative feature elimination together with a single architectural optimization yields acceptable results in a number of benchmarks, but an extension which enables repeated architecture search for different degrees of sparsity is able to improve the results. This finding demonstrates that, also for realistic data, different methods are best suited for different degrees of sparsity.

References

1. Kdd cup 1999 (1999). http://kdd.ics.uci.edu/databases/kddcup99/kddcup99.html
2. Al-Tashi, Q., Abdulkadir, S.J., Rais, H.M., Mirjalili, S., Alhussian, H.: Approaches to multi-objective feature selection: a systematic literature review. IEEE Access **8**, 125076–125096 (2020). https://doi.org/10.1109/ACCESS.2020.3007291

3. Feurer, M., Klein, A., Eggensperger, K., Springenberg, J.T., Blum, M., Hutter, F.: Auto-sklearn: efficient and robust automated machine learning. In: Hutter, F., Kotthoff, L., Vanschoren, J. (eds.) Automated Machine Learning. TSSCML, pp. 113–134. Springer, Cham (2019). https://doi.org/10.1007/978-3-030-05318-5_6
4. Guan, Z., Bian, L., Shang, T., Liu, J.: When machine learning meets security issues: a survey. In: 2018 IEEE International Conference on Intelligence and Safety for Robotics (ISR), pp. 158–165 (2018). https://doi.org/10.1109/IISR.2018.8535799
5. Guyon, I., Elisseeff, A.: An introduction to variable and feature selection. J. Mach. Learn. Res. **3**(Mar), 1157–1182 (2003)
6. Guyon, I., Gunn, S., Nikravesh, M., Zadeh, L.A.: Feature Extraction: Foundations and Applications, vol. 207. Springer, Heidelberg (2008). https://doi.org/10.1007/978-3-540-35488-8
7. Hamdani, T.M., Won, J.-M., Alimi, A.M., Karray, F.: Multi-objective feature selection with NSGA II. In: Beliczynski, B., Dzielinski, A., Iwanowski, M., Ribeiro, B. (eds.) ICANNGA 2007. LNCS, vol. 4431, pp. 240–247. Springer, Heidelberg (2007). https://doi.org/10.1007/978-3-540-71618-1_27
8. Harris, C.R., et al.: Array programming with NumPy. Nature **585**(7825), 357–362 (2020). https://doi.org/10.1038/s41586-020-2649-2
9. Hofstede, R., et al.: Flow monitoring explained: from packet capture to data analysis with netflow and ipfix. IEEE Commun. Surv. Tutor. **16**(4), 2037–2064 (2014)
10. Hutter, F., Kotthoff, L., Vanschoren, J. (eds.): Automated Machine Learning: Methods, Systems, Challenges. Springer, Heidelberg (2018). https://doi.org/10.1007/978-3-030-05318-5, http://automl.org/book
11. Kozodoi, N., Lessmann, S., Papakonstantinou, K., Gatsoulis, Y., Baesens, B.: A multi-objective approach for profit-driven feature selection in credit scoring. Decis. Supp. Syst. **120**, 106–117 (2019)
12. Lashkari., A.H., Gil., G.D., Mamun., M.S.I., Ghorbani., A.A.: Characterization of tor traffic using time based features. In: Proceedings of the 3rd International Conference on Information Systems Security and Privacy, vol. 1: ICISSP, pp. 253–262. INSTICC, SciTePress (2017). https://doi.org/10.5220/0006105602530262
13. Marler, R.T., Arora, J.S.: Survey of multi-objective optimization methods for engineering. Struct. Multidisc. Optim. **26**(6), 369–395 (2004)
14. McHugh, J.: Testing intrusion detection systems: a critique of the 1998 and 1999 darpa intrusion detection system evaluations as performed by lincoln laboratory. ACM Trans. Inf. Syst. Secur. (TISSEC) **3**(4), 262–294 (2000)
15. McKinney, W.: Data structures for statistical computing in python. In: van der Walt, S., Millman, J. (eds.) Proceedings of the 9th Python in Science Conference, pp. 56–61 (2010). https://doi.org/10.25080/Majora-92bf1922-00a
16. Mohr, F., Wever, M., Hüllermeier, E.: Ml-plan: automated machine learning via hierarchical planning. Mach. Learn. **107**(8), 1495–1515 (2018)
17. Moustafa, N., Slay, J.: Unsw-nb15: a comprehensive data set for network intrusion detection systems (unsw-nb15 network data set). In: 2015 Military Communications and Information Systems Conference (MilCIS), pp. 1–6 (2015). https://doi.org/10.1109/MilCIS.2015.7348942
18. Moustafa, N., Turnbull, B., Choo, K.R.: An ensemble intrusion detection technique based on proposed statistical flow features for protecting network traffic of internet of things. IEEE Internet Things J. **6**(3), 4815–4830 (2019). https://doi.org/10.1109/JIOT.2018.2871719
19. Olson, R.S., Bartley, N., Urbanowicz, R.J., Moore, J.H.: Evaluation of a tree-based pipeline optimization tool for automating data science. In: Proceedings of the Genetic and Evolutionary Computation Conference 2016, pp. 485–492 (2016)

20. Olson, R.S., Moore, J.H.: Tpot: A tree-based pipeline optimization tool for automating machine learning. In: Hutter, F., Kotthoff, L., Vanschoren, J. (eds.) Proceedings of the Workshop on Automatic Machine Learning. Proceedings of Machine Learning Research, vol. 64, pp. 66–74. PMLR, New York (2016)
21. Pedregosa, F., et al.: Scikit-learn: machine learning in python. J. Mach. Learn. Res. **12**, 2825–2830 (2011)
22. Pfisterer, F., Coors, S., Thomas, J., Bischl, B.: Multi-objective automatic machine learning with autoxgboostmc (2019). arXiv preprint arXiv:1908.10796
23. Ring, M., Wunderlich, S., Scheuring, D., Landes, D., Hotho, A.: A survey of network-based intrusion detection data sets (2019). CoRR abs/1903.02460, http://arxiv.org/abs/1903.02460
24. Sharafaldin, I., Lashkari, A.H., Ghorbani, A.: Toward generating a new intrusion detection dataset and intrusion traffic characterization. In: ICISSP (2018)
25. Tavallaee, M., Bagheri, E., Lu, W., Ghorbani, A.A.: A detailed analysis of the kdd cup 99 data set. In: 2009 IEEE Symposium on Computational Intelligence for Security and Defense Applications, pp. 1–6. IEEE (2009)
26. Thornton, C., Hutter, F., Hoos, H.H., Leyton-Brown, K.: Auto-weka: combined selection and hyperparameter optimization of classification algorithms. In: Proceedings of the 19th ACM SIGKDD International Conference on Knowledge Discovery and Data Mining, KDD '13, pp. 847–855. Association for Computing Machinery, New York (2013). https://doi.org/10.1145/2487575.2487629
27. Wang, W., Zhu, M., Zeng, X., Ye, X., Sheng, Y.: Malware traffic classification using convolutional neural network for representation learning. In: 2017 International Conference on Information Networking (ICOIN), pp. 712–717 (2017)
28. Wever, M.D., Mohr, F., Hüllermeier, E.: Ml-plan for unlimited-length machine learning pipelines. In: ICML 2018 AutoML Workshop (2018)

A Hierarchical Multi-label Classification of Multi-resident Activities

Hiba Mehri[1(✉)], Tayeb Lemlouma[1(✉)], and Nicolas Montavont[2]

[1] IRISA- University of Rennes 1, Lannion, France
{hiba.mehri,tayeb.lemlouma}@irisa.fr
[2] IRISA- IMT Atlantique, Rennes, France
nicolas.montavont@imt-atlantique.fr

Abstract. In this paper, we tackle the problem of daily activities recognition in a multi-resident e-health smart-home using a semi-supervised learning approach based on neural networks. We aim to optimize the recognition task in order to efficiently model the interaction between inhabitants who generally need assistance. Our hierarchical multi-label classification (HMC) approach provides reasoning based on real-world scenarios and a hierarchical representation of the smart space. The performance results prove the efficiency of our proposed model compared with a basic classification task of activities. Mainly, HMC highly improves the classification of interactive activities and increases the overall classification accuracy approximately from 0.627 to 0.831.

Keywords: Multi-resident · Recurrent neural networks · LSTM · CNN · Hierarchical multi-label classification · Activity recognition

1 Introduction

The rapid introduction of artificial intelligence (AI) and the Internet of Things (IoT) has been offering new opportunities to enhance the capacity of health monitoring systems and help professionals in decision-making. Some studies focus on recognizing the pattern of human activities in smart environments which is of high interest in various domains such as healthcare and elderly care. Human activities are extremely challenging when they involve many residents in the same environment. Comparing to single-resident activity recognition, multi-resident recognition is more complex and open to the interaction between activities.

In our studies, we aim to conceive a complete framework that recognizes and detects abnormal behaviour of individuals, eventually, the evolution of their health status. In this work, we focus as a first step on activity recognition in environmental data interpretation and the detection of abnormalities regarding the residents' behavior and occurrences of their activities of daily life (ADL). We propose a hybrid model based on neural networks (NN) to optimize activities' recognition in multi-resident smart homes while focusing on the interaction between residents. Single resident recognition is interesting to handle before stepping to more complex scenarios. Many approaches [4,13] study the recognition

H. Yin et al. (Eds.): IDEAL 2021, LNCS 13113, pp. 76–86, 2021.
https://doi.org/10.1007/978-3-030-91608-4_8

of ADL based on recurrent neural networks (RNN). The aim is to predict future activities based on a history of previous ones. RNN classifiers trained with time series data and specifically long short term memory (LSTM) improves the recognition accuracy [13]. In [9], smartphones accelerometer data have been used for various types of activities. The data are tested using Spiking NNs and applied on a real-world time series from the WISDM project [2] with a selection of the basic activities. The work shows the importance of the number of records fed to NNs which should be high enough to assure the stability of learning. Other recent studies in activities reconginition used hidden markov models (HMM) to identify the behavior's evolution [11] and detect abnormalities [3]. They show a lack in performance of HMM if compared to NN as they are not easily adaptable to data unless the number of hidden states is known. However, these approaches recall the coexistence of different constraints in smart homes as relationships between locations, activities and objects. Most of existing approaches still lack in prediction of some types of activities. As stated in [8], for multi-resident smart homes, it is more complex to recognize collaborative activities.

In [6], the potential of the concept of *multi-label classification* (MC) [1] is discussed in multi-resident spaces. Other specific approaches show the interest to use classifier chains along with the multi-label classification for the same purpose [7]. The study demonstrated the adaptability of multi-label classification for multi-resident problems that leads to identifying collaborative activities. However, as only basic features were included in the training and no correlation between labels is exploited, much still had to be done to optimize the activities recognition. In this work, we focus on multi-resident activity recognition to consider the users' interactions as well as the ambiguity of complex activities. We propose a new version of the multi-label classification by exploiting the conditional dependencies between the target labels. We aim to predict activities by considering the constraints imposed by the environment. The resulting approach involves a prediction task based on different types of NNs in order to identify the most performant hybrid model for activity recognition. The remainder of this paper is organized as follows. In Sect. 2, we define a hybrid model based on hierarchical multi-label classification and NNs used in the recognition of ADL. In Sect. 3, we present the results of the implementation of our model applied to real-world datasets. Section 4 concludes the work with some perspectives.

2 Methodology

In this section, we present our recognition framework and the theory behind the predictive model. We propose a new approach based on a deep neural networks (DNNs) to perform multi-label classification (MC). We propose to adapt the model to smart spaces by adding a hierarchy between location, activities, and objects.

2.1 Proposed Model

Multi-label classification (MC) is a predictive modeling task used in machine learning. In our context, we adopt MC in smart spaces by associating sensor events with a set of target labels. Each label represents an appropriate activity being achieved. It is worth noting that in AI, the difference in multi-label classification from basic classification problems lies in the fact that a sequence is generally associated with one specific class while with multi-labels, a given sequence can belong to different labels as different activities could be performed simultanously. Hence, MC are well suited to multi-resident environments. Formally, we define a MC problem with $\mathcal{MC} = (\mathcal{A}, \mathcal{X})$ where $\mathcal{A}$ is a finite set of classes : $\mathcal{A} = \{A_1, A_2, ..., A_n\}$ and $\mathcal{X}$ is a set of sensor readings (x, y), where y is the ground truth of x (the set of true labels associated with x). We associate a model m_A for each $\mathcal{MC}$ problem by using a function mapping every class $\mathcal{A}$ and every sensor event x to [0,1]. A sensor event x is predicted to belong to class $\mathcal{A}$ whenever $m_A(x)$ is greater than a pre-defined threshold θ. We define θ as the probability that the sensor event belongs to class $\mathcal{A}$. Besides, the MC problem is extended to be a hierarchical multi-label classification (HMC) problem by associating it with a finite set of constraints defined by $A_1 \rightarrow A$ in the case of two-levels hierarchy architecture. Logically, the model has to predict $\mathcal{A}$ whenever it predicts A_1. To design an HMC adapted to smart spaces, we define a selection of ADL performed in the context of smart homes that we aim to recognize. With this aim in mind, we associate each human activity and used object with a logical location to define hierarchical constraints as presented in Table 1. It should be underlined that some activities are strictly associated with specific obvious locations, while other activities may differ in other specific scenarios such as performing tasks remotely using the Internet. This type of activity can be performed anywhere in the house.

Table 1. Association of activities, objects to location

Location	Activities	Objects
Hall	Going out, other	House door, Hall
Kitchen	Preparing a meal, eating, having a snack, washing dishes	Fridge, kitchen drawer, chair
Bathroom	Toileting, having shower	Bathroom door, shower cabinet door, tap, water closet
Bedroom	Sleeping, studying, changing clothes	Bed, wardrobe
Living room	Watching TV, socializing, using internet	Couch, TV receiver, modem
Laundry	Washing clothes	Washing machine

Figure 1 presents a hierarchical structure for a selected set of ADL activities. Three levels of hierarchy are defined, every level groups the same classes. Level 1 groups the location classes such as kitchen, bathroom, bedroom, living room, and laundry. Level 2 groups the activities associated with each location. Finally, level 3 groups the objects associated with each activity and hierarchically with each location. To predict activities (level 2) from unlabeled data, objects related to each sensor sequence are considered the ground truth. The HMC model associated them with the appropriate location (level 1), and then, it determines the activity by applying a constraint resolution module. For instance, the HMC model handles the following constraint: *bed* $\rightarrow$ *sleeping* $\rightarrow$ *bedroom* as follows. If the sensor event is associated with the *bed*, the inhabitant is supposed to be in the bedroom. This constraint increases the probability of predicting *sleeping* as the recognized activity.

Algorithm 1 presents an overview of the implementation of the HMC model. Based on a dictionary of locations, activities, and objects extracted from real-world logical scenarios, the model determines the appropriate location of each sensor event, applies a first step of the hierarchical constraint module that resolves the constraints by connecting levels 3 and 1 of the structure. In the presented example of Algorithm 1, the model guarantees the satisfaction of the constraints with a post-processing step to enforce $m_A(x) > \theta$ whenever $m_{A_{11}}(x) > \theta$. In the final step of activities prediction, the model connects hierarchy levels 2 and 1 based on the input dictionaries and enforces the $m_{A_1}(x) > \theta$ condition to predict the appropriate activity. Note that the probability of predicting the correct activity is highly improved by the input hierarchical constraints.

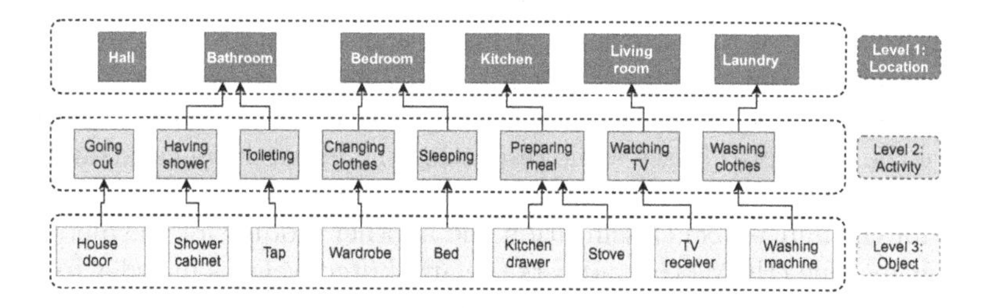

Fig. 1. Hierarchical structure of smart home

Algorithme 1 : HMC for recognition of activities

```
   Data : Sensor events, dictionary of activities associated with
          ⟨location, object⟩
   Result : Prediction of activities A1, A2...
 1 begin
 2   Location = {A, B, C, D...};
 3   Activities = {A1,A2,A3,B1,B2,C1...};
 4   Objects = {A11,A12,A21,B11,B22,C11...};
 5   for eachsensor_eventattimet do
 6     if Sensor_event.Sensor_value = 1 then
 7       Associate(Object)
 8       Impose(HierarchicalConstraintModule)
 9       Associate(Location)
10     end
11   end
12   Impose(HierarchicalConstraintModule)
13   if Sensor_event.Object = A11 and Sensor_event.Location = A then
14     Predict(Sensor_event.Activity = A1)
15   end
16 end
```

2.2 Hybrid HMC Model Using Recurrent and Convolutional NNs

In our prediction approach, we use DNNs that are considered more accurate if compared to other machine learning approaches where feature engineering is essential. To improve their capacity of prediction, DNNs are able to learn the feature encoding from data within each training iteration. To benefit from this advantage, we propose to associate the HMC task to a NN module during the prediction of the associate classes while respecting the hierarchical constraints. The NN module is split into two sub-modules: one input handling the pre-processed features for the multi-label classification and another module for the hierarchical constraint module (HCM) that adjusts the attributed weights in order to satisfy the constraints. In our reasoning on the type of NN used in the prediction module (HMC-NN), we take into consideration the dependencies that may exist between sensor events and the dependencies related to the occurrence time. In [10], the work confirmed that RNNs are suitable for time series analysis handling single and multi-resident contexts. RNNs consider more efficiently the interaction between activities and residents. Another candidate is the convolutional neural networks (CNNs) which are much more used in handling multi-label classification tasks especially for text classification [5]. Besides, they are also compatible with time-series data. Our hybrid HMC model consists of the following associations *HMC-CNN* and *HMC-LSTM*. **HMC-CNN** is the association of our HMC algorithm with a CNN for the prediction of activities. Using CNNs for time series classification brings many advantages. Mainly, they are highly resistant to noise in data and able to extract very informative and deep features along with the time dimension. They are also, very useful in capturing the correlation between variables which highly affects the accuracy of prediction. **HMC-LSTM** is the association of HMC with an LSTM that is a special kind of RNN with a learned memory state [12]. This enables a conditioned prediction even on events that happened in the past while still handling one event at a time thanks to their inherited recurrence propriety. In sum, thanks to LSTM, the model gains the ablility to remember information for long periods as a default behavior. This concept is very adaptable for activity recognition as memorizing the history of past activities is important to predict future ones.

3 Implementation and Performance Evaluation

In this section, we experiment our approach using different metrics for the classification problem. The challenge is to test the hybrid model on real world datasets specifically adapted for multi-resident activity recognition. Our implementation of the recognition model is performed using Python with other required libraries specific to data science as Pandas and Scikit-Learn. We use the Keras framework in the implementation of NNs, for the training and testing process. Our experiments are based on the data of ARAS project[1]. Collected data concern two-residents smart homes with several sequences triggered along with the performed activities during 30 days. The time gap between sensor events is one second.Sensed data relate to different types of information such as force, contact, distance, temperature, IR, and photocell. The sensor sequences are preprocessed and analyzed to apply the hierarchical model and then be fed to the NN module to predict on unlabeled data. As a first step, we test the evolution of our model's accuracy using different lengths of samples: 6, 12, 20 days...etc. Proportionally to the number of samples, we observe that the accuracy does not get significantly affected by the hierarchical optimization. However, changing the type of NN notably affects the accuracy especially using the LSTM network which is known to react easily to the variety of the training data. This has a positive impact on the final prediction accuracy compared to other NNs.

3.1 Evaluation Metrics and Results

For each used model, we measure the average classification accuracy for all the labels. Then, an individual evaluation of each label (activity) is performed to estimate the model's robustness in predicting specific activities. The average classification accuracy is a common metric that estimates the proportion of true predictions among the complete data. Initial data is composed of 25% for testing and 75% for training. We focus on the testing phase to assess the classification accuracy when facing unlabeled samples. Table 2 presents the average classification accuracy of the HMC model combined with a basic neural network (BNN), CNN, and LSTM. Also, we consider the comparison with a basic task of MC to evaluate the benefits of the hierarchical representation.

[1] https://www.cmpe.boun.edu.tr/aras/.

Table 2. Average classification accuracy for different tested models

Label	MC			HMC		
	MC-BNN	MC-CNN	MC-LSTM	HMC-BNN	HMC-CNN	HMC-LSTM
Resident 1	0.661	0.646	0.666	0.88	0.882	**0.885**
Resident 2	0.78	0.775	0.785	0.9	**0.964**	0.95
Interaction	0.49	0.42	0.431	0.57	0.63	**0.66**

The initial results show that HMC outperforms the MC task independently of the used NN and the type of recognition. For LSTM, the classification accuracy evolves from an average accuracy of 0.666 to 0.885 for resident 1, from 0.785 to 0.95 for resident 2, and from 0.431 to 0.66 for the recognition of interactive activities. On another hand, we observe that the classification accuracy is highly impacted by the variety of activities between the two residents. Indeed, the model provides better accuracy for resident 2. Our experimentations show that HMC-CNN proves a high performance in detecting individual activities of each resident. For all the models, it seems more difficult to detect the interaction, especially for the basic classifier. The results of HMC-LSTM are the more accurate with an mean prediction accuracy of 0.831.

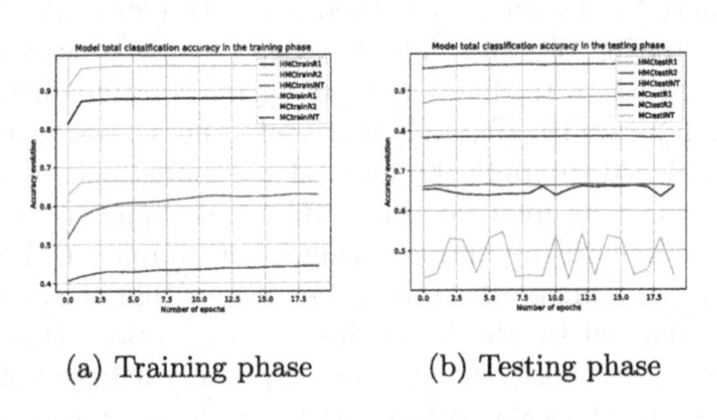

(a) Training phase (b) Testing phase

Fig. 2. Accuracy of HMC-LSTM and MC-LSTM

In Fig. 2, we compare the training and testing accuracy between HMC-LSTM and MC-LSTM. We evaluate the accuracy by considering the activities of residents 1 and 2 and their interactions. For all the situations, we observe that HMC outperforms MC. This is particularly true during the testing phase where it reaches its best accuracy for resident 2 with a value of 0.96 while reaching 0.66 for recognizing the interactive activities. Drawing on this observation of the HMC-LSTM model, we focus on the evaluation of its classification by common metrics precision, recall, and F1-score. Precision-Recall is the measure of prediction success while having very unbalanced classes. The precision is the measure of result relevancy, while recall measures the number of truly relevant results of the model. Finally, the F1-score is calculated by the following equation: $F_1 score = \frac{\text{Precision} \times \text{Recall}}{\text{Precision} + \text{Recall}}$. Table 3 provides the results for each activity.

Table 3. Per-label classification of activities

Label	Precision	Recall	F1-score	Label	Precision	Recall	F1-score
Going out	0.99	0.99	0.99	Having shower	0.85	0.8	0.82
Preparing meal	0.79	0.87	0.83	Toileting	0.85	0.78	0.82
Having a meal	0.84	0.92	0.88	Laundry	0.99	0.99	0.99
Washing dishes	0.82	0.39	0.52	Using internet	0.63	0.66	0.65
Having snack	0.36	0.83	0.5	Washing routine	0.47	0.79	0.59
Sleeping	0.99	0.99	0.99	Socializing	0.51	0.66	0.58
Watching TV	0.99	0.99	0.99	Changing clothes	0.39	0.68	0.54

According to the initial results of the F1-score, we observe that the model is highly affected by the variety of activities in the real-world dataset where it is impossible to control the scenarios of daily living in order to improve the accuracy of prediction. The frequency and duration of each activity affect highly the samples injected for the training. For some activities, such as going out, sleeping, watching TV, or having a meal, the model is more accurate because the capacity of learning is more important for such frequent and time-fixed daily activities. For other activities, that are less associated with a specific interval of time, the model lacks accuracy while interpreting activities due to the absence of routines. In order to evaluate the consistency of our model, we consider the evolution of accuracy in terms of iterations and the cross-entropy loss estimation.

Evolution of Accuracy in Terms of Number of Training Iterations: we propose to study the evolution of training and compare it with the testing accuracy to evaluate an over-fitting possibility affecting the model and determine the optimal number of training iterations. Figure 3a, 3b, 3c present the evolution of accuracy of HMC-LSTM for the activities of residents 1 and 2 and their interaction. We observe that the stabilization of the accuracy is similar during the training and testing for residents 1 and 2. However, for the residents' interaction, the training process takes longer to stabilize in approximately the 10th iteration. The training process lacks stability as there are not many interactive activities in the dataset. Moreover, we observe that the model is not committing over-fitting as the training and testing accuracy are synchronized. Such situations are detected if the training accuracy is significantly better than the testing. This means integrally that the model is perfectly memorizing the training dataset without being able to recognize new injected samples in the testing phase. Concerning the optimal number of iterations, the model is reaching its peak of performance with the selected parameters at the 20th training iteration for all the cases.

Cross-Entropy Loss Function Estimation: we use this function to measure the performance of the classification when the output is a probability. The loss increases when the predicted probability deviates from the true label (i.e. the right activity that should be predicted). Cross-entropy loss is interesting for possible optimizations. It is calculated by the following equation: $Cross\_entropy\_loss(\boldsymbol{y}, \hat{\boldsymbol{y}}) = -\sum_{i=1}^{N} y_i \log(\hat{y}_i)$ where $\boldsymbol{y}$ is a one-hot label vector and N is the number of classes (14 activities in our case). As a perfect model

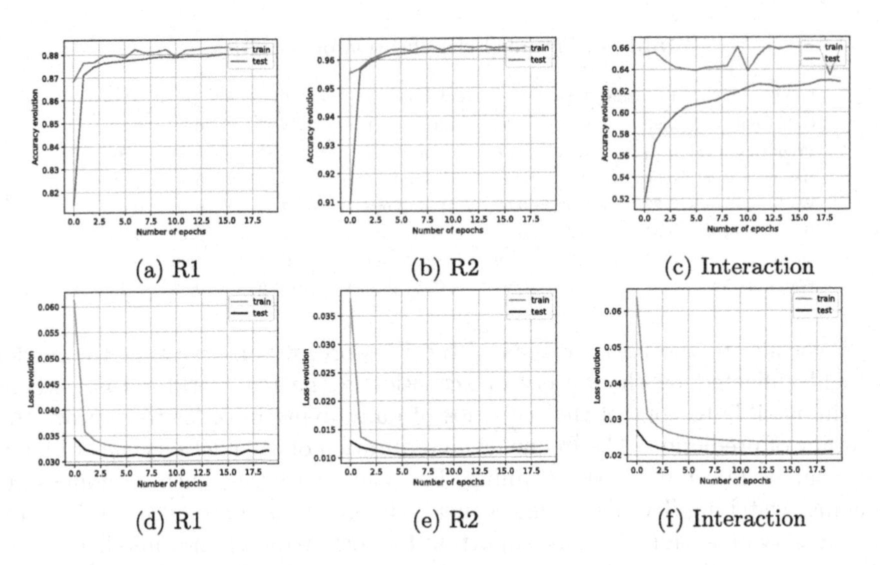

(a) R1 (b) R2 (c) Interaction

(d) R1 (e) R2 (f) Interaction

Fig. 3. Accuracy and cross-entropy loss of HMC-LSTM

has a cross-entropy loss of 0, we aim to minimize this loss. In Fig. 3d, 3e, 3f, for all the three classes, resident 1, resident 2, and interactive activities, the loss function decreases when the number of iterations increases until attending the final values that are in an interval of [0.01, 0.035]. We observe that the model provides a high prediction performance which minimizes the loss function. However, it still an object of optimization as it is possible to adjust the model weights during the training. the aim is to resolve the problem of unbalanced data and represent better the classes that are less present in the tested dataset. Figure 4 presents the evolution of loss function for the HMC and MC models for the testing and training phase. As we can observe, the HMC loss decreases to reach values under 0.04 for the two phases while the MC loss varies between 0.08 and 0.12. This result confirms the good prediction performance of the hierarchical proposed model.

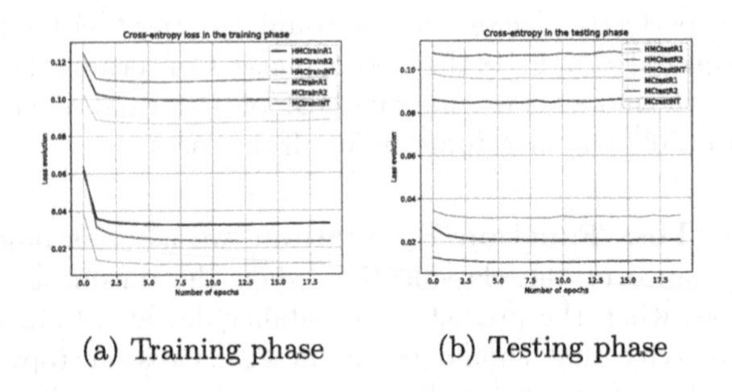

(a) Training phase (b) Testing phase

Fig. 4. Cross-entropy loss of HMC-LSTM and MC-LSTM

4 Conclusion

In this work, we focused in developing an activities' recognition model for multi-resident in smart environments. We proposed HMC: a hierarchical multi-label classification model that satisfies existing constraints between different elements of the environment such as location, activities, and objects. We evaluated our model with different types of NNs and along with a simple task of multi-label classification and a hierarchical one on a real-world dataset. The results reveal that the proposed HMC-LSTM model provides the best prediction accuracy while minimizing the loss function. For instance, HMC-LSTM model outperforms MC-LSTM by increasing the overall accuracy from 0.627 to 0.831 while optimizing the recognition of interactive activities. However, the model still lacking accuracy in determining interactive activities if compared to individual ones. In the future, we will explore the performance of our model on other real-world data in order to assure its generalizability. Other techniques of optimization may be applied such as a manual weighing methodology of features that could enable the consideration of activities that are not much represented in datasets.

References

1. Giunchiglia, E., Lukasiewicz, T.: Multi-label classification neural networks with hard logical constraints. Clin. Orthopaed. Related Res. **479** (2021)
2. Kwapisz, J.R., Weiss, G.M., Moore, S.A.: Activity recognition using cell phone accelerometers. ACM SigKDD Explor. Newsl. **12**(2), 74–82 (2011)
3. Liouane, Z., Lemlouma, T., Roose, P., Weis, F., Hassani, M.: A markovian-based approach for daily living activities recognition. In: 2016 5th International Conference on Sensor Networks (2016)
4. Natani, A., Sharma, A., Peruma, T., Sukhavasi, S.: Deep learning for multi-resident activity recognition in ambient sensing smart homes. In: 2019 IEEE 8th Global Conference on Consumer Electronics, pp. 340–341. IEEE (2019)
5. Qiu, M., Zhang, Y., Ma, T., Wu, Q., Jin, F.: Convolutional-neural-network-based multilabel text classification for automatic discrimination of legal documents. Sens. Mater. **32**(8), 2659–2672 (2020)
6. Raihani, M., Thinagaran, P., Md Nasir, S., Norwati, M.: Multi resident complex activity recognition in smart home: a literature review. Int. J. Smart Home **11**(6), 21–32 (2017)
7. Raihani, M., Thinagaran, P., Nasir, S.M., Norwati, M., Shah, Z.M.N.: Multi label classification on multi resident in smart home using classifier chains. Adv. Sci. Lett. **24**(2), 1316–1319 (2018)
8. Riboni, D., Murru, F.: Unsupervised recognition of multi-resident activities in smart-homes. IEEE Access **8**, 201985–201994 (2020)
9. Suriani, N.S., Rashid, F.N.: Smartphone sensor accelerometer data for human activity recognition using spiking neural network. Int. J. Mach. Learn. Comput **11**, 298–303 (2021)
10. Thapa, K., Al, A., Md, Z., Lamichhane, B., Yang, S.H.: A deep machine learning method for concurrent and interleaved human activity recognition. Sensors **20**(20), 5770 (2020)

11. Tran, S.N., Zhang, Q., Smallbon, V., Karunanithi, M.: Multi-resident activity monitoring in smart homes: a case study. In: 2018 IEEE International Conference on Pervasive Computing and Communications Workshops, pp. 698–703. IEEE (2018)
12. Xia, K., Huang, J., Wang, H.: LSTM-CNN architecture for human activity recognition. IEEE Access **8**, 56855–56866 (2020)
13. Zehtabian, S., Khodadadeh, S., Bölöni, L., Turgut, D.: Privacy-preserving learning of human activity predictors in smart environments. In: 2021 IEEE 40th International Conference on Computer Communications. IEEE (2021)

Fast and Optimal Planner for the Discrete Grid-Based Coverage Path-Planning Problem

Jaël Champagne Gareau(✉), Éric Beaudry, and Vladimir Makarenkov

Université du Québec à Montréal, Montreal, Canada
champagne_gareau.jael@courrier.uqam.ca,
{beaudry.eric,makarenkov.vladimir}@uqam.ca

Abstract. This paper introduces a new algorithm for solving the discrete grid-based coverage path-planning (CPP) problem. This problem consists in finding a path that covers a given region completely. Our algorithm is based on an iterative deepening depth-first search. We introduce two branch-and-bound improvements (Loop detection and Admissible heuristic) to this algorithm. We evaluate the performance of our planner using four types of generated grids. The obtained results show that the proposed branch-and-bound algorithm solves the problem optimally and orders of magnitude faster than traditional optimal CPP planners.

Keywords: Coverage path-planning · Iterative deepening depth-first search · Branch-and-bound · Heuristic search · Pruning · Clustering

1 Introduction

Path Planning (PP) is a research area aimed at finding a sequence of actions that allows an agent (e.g., a robot) to move from one state (e.g., position) to another [8] (e.g., finding an optimal path for an electric vehicle [1]). One of the problems studied in PP is the complete Coverage Path-Planning (CPP) problem. Basically, the objective of CPP is to find a complete coverage path of a given region, i.e., a path that covers every area in the region. This problem has many practical applications, including window washer robots, robotic vacuum cleaners, autonomous underwater vehicles (AUVs), mine sweeping robots, search and rescue planning, surveillance drones, and fused deposition modeling (FDM) 3D printers. All of them rely on efficient CPP algorithms to accomplish their task [3,6,7].

There exist many variants of the CPP problem. For example, the environment can be either discrete (e.g., grid-based, graph-based, etc.) or continuous, 2D or 3D, known *a priori* (off-line algorithms), or discovered while covering it (on-line algorithms), etc. Moreover, the coverage can be done by a single agent, or by the cooperation of multiple agents. Some variants also restrict the type of allowed movements or add different kinds of sensors to the agent (proximity sensor, GPS,

H. Yin et al. (Eds.): IDEAL 2021, LNCS 13113, pp. 87–96, 2021.
https://doi.org/10.1007/978-3-030-91608-4_9

gyroscopic sensor, etc.). Some variants even consider positional uncertainties and energetic constraints of the agent. In this paper, we focus on the classic variant consisting of a single agent in a 2D discrete grid-based environment with no specific constraints or uncertainties.

The objective of our study is to present an optimal CPP planner that runs orders of magnitude faster than a naive search through the state-space. Our research contributions are as follows:

1. A novel branch-and-bound optimal planner to the grid CPP problem;
2. An informative, admissible, efficient heuristic to the grid CPP problem;
3. Realistic environments for benchmark;

The rest of the paper is structured as follows. Section 2 presents a short overview of existing CPP solving approaches. Section 3 formally introduces the CPP variant we are focusing on. Section 4 and Sect. 5 present, respectively, our method and the obtained results. Finally, Sect. 6 describes the main findings of our study and presents some ideas of future research in the area.

2 Related Work

One of the most known algorithm which solves the CPP problem in a grid-based environment is the wavefront algorithm [14]. Given a starting position and a desired arrival (goal) position (which can be the same as the starting position), the wavefront algorithm propagates a *wave* from the goal to the neighboring grid cells (e.g., with a breadth-first search through the state-space). After the propagation, every grid cell is labeled with a number that corresponds to the minimum number of cells an agent must visit to reach the goal from the cell. The algorithm is then simple: the agent always chooses to visit the unvisited neighboring cell with the highest number first, breaking ties arbitrarily. One disadvantage of this strategy is the obligation to specify a goal state. In some applications, the ending location is not important, and not specifying it allows for finding shorter paths. There exists an *on-line* variant of the wavefront algorithm [12] that can be used if the environment is initially unknown. When a CPP algorithm should be applied to a road network (e.g., a street cleaning vehicle that needs to cover every street), a graph representation (instead of a grid) is advantageous. Many algorithms have been proposed to solve the discrete CPP in graph-based representation [13].

For the coverage of continuous regions, one major family of algorithms is the *cellular decomposition methods*. They consist in partitioning complex regions in many simpler, non-overlapping regions, called cells. These simpler regions don't contain obstacles, and are thus easy to cover. The most known algorithm that uses this strategy is the *boustrophedon decomposition* algorithm [4]. Another strategy that can be used for the complete coverage of continuous regions is to discretize the environment and use a discrete planner (e.g., the aforementioned wavefront algorithm).

The above algorithms are relatively fast, but provide no guarantee of optimality. This is expected, since a reduction exists between the CPP problem and the Travelling Salesman Problem, making the CPP problem part of the NP-Complete class of problems [10]. Thus, to the best of our knowledge, all CPP planners described in the literature either provide approximate solutions, or work only in a specific kind of environment.

For related works on more specific variants of the CPP problem, see the referred surveys [3,6,7].

3 Problem Modeling

This paper presents an optimal planner which, since the problem is NP-Complete, must have a worst-case exponential complexity. However, we propose two speed-up methods that, according to our evaluation, provide an orders of magnitude faster planner. We focus on the CPP variant consisting in a 2D grid that needs to be covered by a single agent with no particular goal position. Definition 1 to 3 describe more formally the environment to cover, the agent and the state-space model considered in our study.

Definition 1. *A 2D* environment *is an* $m \times n$ *grid represented by a matrix* $G = (g_{ij})_{m \times n}$, *where* $g_{ij} \in \{O, X\}$, *and:*

- *O indicates that a cell is accessible and needs to be covered;*
- *X means that the cell is inaccessible (blocked by an obstacle).*

Definition 2. *An* agent *is an entity with a position somewhere on the grid. It can move to neighboring grid cells by using an action a from the set of actions* $\mathcal{A} = \{(-1, 0), (+1, 0), (0, -1), (0, +1)\}$. *The effect of each action is as follows.*

If $p = (i, j)$ *denotes the agent's current position (i.e., the agent is on the grid cell* g_{ij}*) and it executes action* $a = (a_1, a_2)$, *then its new position* $\tilde{p}$ *is:*

$$\tilde{p} = \begin{cases} (i + a_1, j + a_2) & \text{if } g_{i+a_1, j+a_2} = O \\ p & \text{if } g_{i+a_1, j+a_2} = X. \end{cases}$$

Definition 3. *A* state *is a tuple* $s = (i_s, j_s, R)$, *where:*

- (i_s, j_s) *is the position (row, column) of the agent;*
- $R = \{(i, j) \mid g_{ij} = O$ *and position* (i, j) *has not yet been explored*$\}$.

We now formally define what we mean by a CPP problem instance, a solution to such an instance, and our optimization criterion in Definition 4 to 6, and give an example of a problem instance in Fig. 1.

Definition 4. *An* instance *of our CPP problem variant is given by a tuple* (G, s_0), *where G is an environment, as defined in Definition 1, and* $s_0 = (i_0, j_0, R_0)$ *is the initial state, where* $R_0 = \{(i, j) \mid g_{ij} = O\}$.

Definition 5. *A* solution *to such an instance* (G, s_0) *is an ordered list of actions* $p = \langle a_1, a_2, \ldots, a_k \rangle$ *(also called a plan) that moves the agent through positions:*

$$L = \langle (i_0, j_0), (i_1, j_1), \ldots, (i_k, j_k) \rangle,$$

with $R_0 \subseteq L$ *(i.e., the final state is* $(i_k, j_k, \emptyset)$*).*

Definition 6. *Let* $\mathcal{P}$ *be the set of solutions (plans) of a CPP problem instance. The* objective *is to find an optimal solution* $p^\star = \mathrm{argmin}_{\mathrm{p}\in\mathcal{P}}|\mathrm{p}|$. *Namely,* $p^\star$ *is a minimal ordered list of actions that solves the problem.*

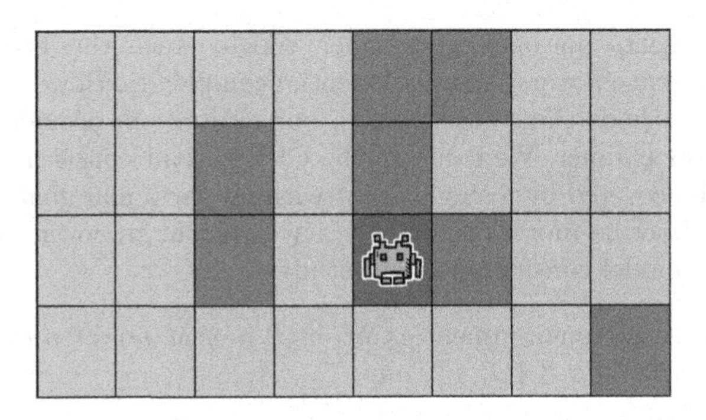

Fig. 1. A CPP instance. The dark green, light green and grey cells represent, respectively, the initial cell, the cells that remain to be covered and the inaccessible cells. (Color figure online)

4 Proposed Methods

In this section, we present an optimal CPP planner based on the ID-DFS algorithm, and describe two techniques (loop detection and an admissible heuristic) that preserve the optimality of the obtained solutions, while running orders of magnitude faster than an exhaustive search algorithm.

4.1 Iterative Deepening Depth-First Search (ID-DFS)

We begin by describing a naive planner first, since our branch-and-bound algorithm is based on it, and since no optimal CPP planners are explicitly mentioned in the literature. First, we observe that our problem can be viewed as a search in a graph where every node represents a state in the state-space (it is worth noting that such a graph is exponentially larger than the problem grid). Thus, every standard graph search algorithms can theoretically directly be used, including the well-known depth-first search (DFS) and breadth-first search (BFS) algorithms. However, because of the huge size of the search graph, BFS is impractical. Indeed, BFS needs in the worst-case scenario to store the complete state-space graph in the computer's memory, which is too large even for a small problem size (e.g., a 20×20 grid). Thus, every technique based on BFS, including the Dijkstra and A* algorithms, are non-applicable in practice.

On the other hand, the DFS algorithm can go arbitrarily deep in the search tree even though the solution is close to the root (e.g., it can get stuck by expanding nodes over and over indefinitely, never backtrack and thus never find a solution). A safeguard is to use a variant of DFS, called iterative deepening depth-first search (ID-DFS) [11], which uses DFS but has a depth limit. If a solution is not found within the depth limit k, DFS is carried out again with the depth limit $k + 1$ and continues until a solution is found. ID-DFS ensures that the algorithm never goes deeper than necessary and ensures that the algorithm terminates (if a solution exists). Algorithm 1 presents the details of a CPP planner based on ID-DFS. We recall from the previous section that R is the set of grid cells that remain to be covered (i.e., initially, it is equal to the total set of grid cells to cover). The rest of the pseudocode should be self-explanatory.

Algorithm 1. CPP planner based on ID-DFS

```
 1: global
 2:     p*: data-structure (eventually) containing the solution
 3:     s = (i_s, j_s, R): current agent position
 4: procedure ID-DFS-PLAN( )
 5:     for k ← |R_0| to ∞ do                              ▷ k is the depth limit
 6:         found ← ID-DFS-HELPER(k, 0)
 7:         if found then return
 8: procedure ID-DFS-HELPER(k: depth-limit, d: current depth) : boolean
 9:     if k = d then return |R| = 0       ▷ returns true iff the grid is fully covered
10:     for all a ∈ A do
11:         if a is a valid action in the current position then
12:             move agent by executing action a
13:             found ← ID-DFS-HELPER(k, d + 1)
14:             if found then
15:                 add a at the start of solution p*      ▷ p* is found in reverse order
16:                 return true
17:             else
18:                 backtrack one step in the search tree            ▷ undo last move
19:     return false
```

4.2 Pruning Using Loop Detection and Admissible CPP Heuristic

While Algorithm 1 yields an optimal solution, and guarantees it if such a solution exists, it has to analyze many branches of the search tree that are not promising (i.e., have little chance of leading to an optimal solution). Our branch-and-bound planner aims at alleviating this problem by cutting the unpromising parts of the search tree. One type of unpromising subtrees occur when the agent arrives in a grid cell (i, j), which has already been visited, without having covered any other grid cell since its last visit to position (i, j) (i.e., we detected a *loop* in the state-space). In order to take this into account, we introduce the matrix $M = (m_{ij})_{m \times n}$, where m_{ij} is the number of grid cells that remained to be covered the last time the agent was in position (i, j). We modify Algorithm 1 to consider and update this new matrix M. When a new recursive call begins, and the agent is in position (i, j), a condition is inserted to check if $m_{ij} \leq |R|$. If this condition is true, then the current path is clearly suboptimal, and the current subtree is thus pruned from the search space.

A second way to improve Algorithm 1 is to introduce an *admissible heuristic* cost function $h\colon \mathcal{S} \rightarrow \mathbb{N}$, i.e., a function that takes as input states $s = (i, j, R)$ from the set of states $\mathcal{S}$ and returns as output a lower bound $h(s)$ on the number of actions needed to cover the remaining uncovered grid cells (the cells in R). Such a heuristic can be used in two ways: (1) It allows pruning even more unpromising subtrees than with the previously mentioned method, and (2) it allows ordering the successors of a state by how much promising they are and thus finding a solution faster by exploring the most promising subtrees first.

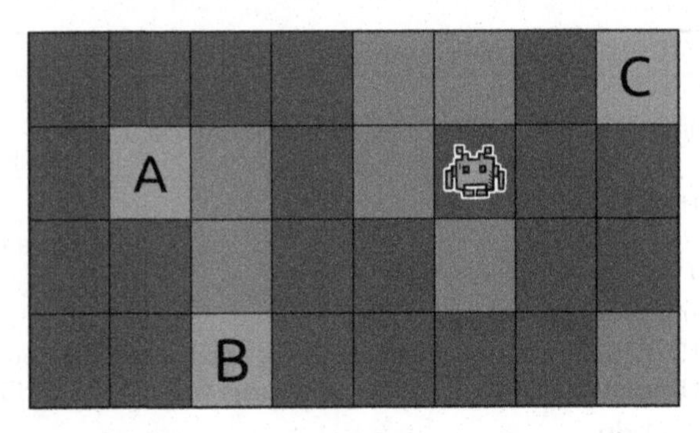

Fig. 2. Example showing how the proposed heuristic works in practice.

Algorithm 2. Heuristic cost computation

1: **procedure** MIN-REMAINING-MOVES((i, j, R): a state) : positive integer
2: $\quad left \leftarrow right \leftarrow up \leftarrow down \leftarrow 0$ ▷ Variables initialization
3: $\quad$ **for all** $(r_i, r_j) \in R$ **do** ▷ Loop on every remaining grid cell to cover
4: $\quad\quad$ **if** $r_i < i$ **then** ▷ The uncovered cell is above
5: $\quad\quad\quad up = \max(up, i - r_i)$
6: $\quad\quad$ **else** $down = \max(down, r_i - i)$ ▷ The uncovered cell is below
7: $\quad\quad$ **if** $r_j < j$ **then** ▷ The uncovered cell is to the left
8: $\quad\quad\quad left = \max(left, j - r_j)$
9: $\quad\quad$ **else** $right = \max(right, r_j - j)$ ▷ The uncovered cell is to the right
10: $\quad$ **return** $left + right + \min(left, right) + up + down + \min(up, down)$

We describe our novel heuristic by explaining how it computes the lower bound using an example presented in Fig. 2. In this figure, three grid cells (A, B and C) remain to be covered. Our heuristic computes the minimum number of every action in $\mathcal{A}$ that need to be done to cover the remaining cells. For example, the action corresponding to "go left" must be done at least $\max(4,3,0) = 4$ times and the action corresponding to "go right" must be done at least $\max(0,0,2) = 2$ times. Moreover, if the agent goes two cells to the right, the minimum number of moves to the left will now be two more than the previous minimum of 4, i.e., 6. Algorithm 2 shows more precisely how $h(s)$ is computed.

5 Results and Analysis

The algorithms described in Sect. 4 were implemented in C++. The tests were carried out on a PC computer equipped with an Intel Core i5 7600k processor and 32GB of RAM (our planner never used more than 10 MB even on the largest grids, thanks to ID-DFS, so the memory usage in not a problem). To measure the performance of our algorithms, we ran each of them 50 times on the same test grids and took the average of the results obtained. Every planner was tested with each of the four kinds of artificial grids shown in Fig. 3. Type (a) grids were generated with the Diamond-Square algorithm and have the shape of a coast [5]. Type (b) grids were generated by randomly placing simple shapes (triangles, discs and rectangles), whereas type (c) grids mimic a random walk on a grid, and type (d) grids include cells with randomly added "links" between neighboring positions on the grid. All grid types were generated with an inaccessible cells density of $(50 \pm 1)\%$.

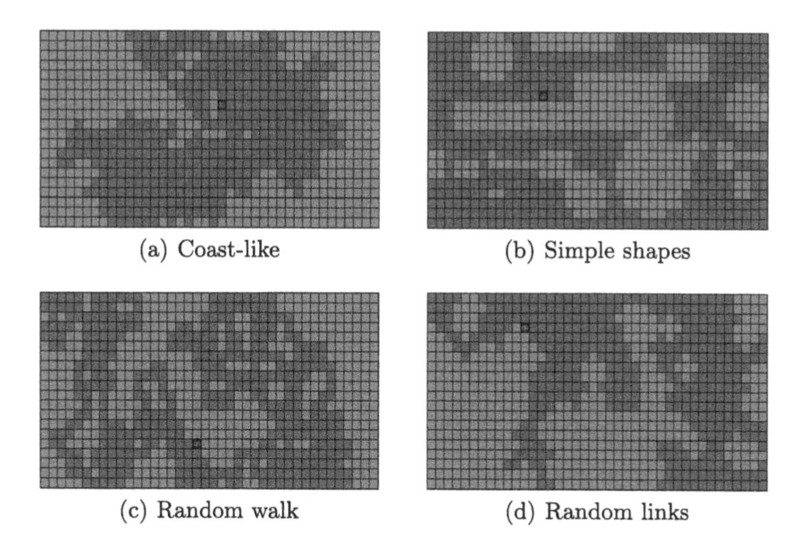

(a) Coast-like (b) Simple shapes

(c) Random walk (d) Random links

Fig. 3. The four types of generated grids in our benchmark

Table 1 reports the average running times measured for each planner on the considered types of test grids. Every generated grid had a square dimension, shown in column *Size*. The columns L and H, respectively, stand for our two improvements over the ID-DFS (Algorithm 1) planner, i.e., (L)oop detection and (H)euristic pruning. In the table, the character '-' means that the planner failed to solve the problem within 5 min. Note that we do not show solutions length in the table since all techniques yield optimal solutions. As we can see, both variants of our algorithm based on the branch-and-bound approach are orders of magnitude faster, on every type of grid, than the ID-DFS planner.

Table 1. Average running times (in ms) required by the ID-DFS planner and the proposed algorithms (Loop detection and Heuristic pruning)

Grid type	Size	ID-DFS	L	H	L+H
(a)	4×4	0.026	0.019	0.011	0.011
(a)	5×5	178.745	8.36	0.195	0.136
(a)	6×6	–	238154	333.692	97.341
(a)	7×7	–	–	767.201	233.994
(b)	4×4	0.004	0.003	0.002	0.002
(b)	5×5	0.34	0.052	0.016	0.014
(b)	6×6	–	6613.51	28.305	10.739
(b)	7×7	–	–	29249.8	527.177
(c)	4×4	0.01	0.006	0.006	0.006
(c)	5×5	13.498	2.126	0.142	0.1
(c)	6×6	74824	4589.35	22.353	10.841
(c)	7×7	–	–	45515.5	6485.34
(d)	4×4	0.158	0.073	0.017	0.016
(d)	5×5	3.541	0.389	0.058	0.045
(d)	6×6	26947.3	688.076	4.088	1.946
(d)	7×7	–	165167	383.875	70.261

6 Conclusion

This paper considers relevant but not very well studied problem of complete coverage path-planning (CPP). We showed how an exhaustive algorithm based on iterative deepening depth-first search (ID-DFS) can be effectively accelerated using a branch-and-bound approach. The proposed modifications allow the planner to find an optimal CPP solution orders of magnitude faster compared to ID-DFS, which makes it suitable for practical applications.

As future work, we plan to develop and test a method similar to particle swarm optimization (PSO) considering an initial particle that splits the problem into several sub-problems every time there is more than one eligible neighbor. The splitting process will take place until the number of particles reaches a certain threshold N. When this happens, a pruning process destroying the least promising particles can be carried out according to an evaluation heuristic to be determined. We also envisage to use clustering algorithms [9] to decompose a given grid into smaller, mostly independent sub-grids (i.e., similar to cellular decomposition, but for grid environments), which could be covered optimally one by one. Such an algorithm could be also easily parallelized. The use of clustering techniques has helped optimize the computation process in many different fields (see e.g., [2]).

Acknowledgments. We acknowledge the support of the Natural Sciences and Engineering Research Council of Canada (NSERC) and the Fonds de Recherche du Québec—Nature et Technologies (FRQNT). We would also like to thank Alexandre Blondin-Massé, Guillaume Gosset, and the anonymous authors for their useful advices.

References

1. Champagne Gareau, J., Beaudry, É., Makarenkov, V.: An efficient electric vehicle path-planner that considers the waiting time. In: Proceedings of the 27th ACM SIGSPATIAL International Conference on Advances in Geographic Information Systems, pp. 389–397. ACM, Chicago (2019). https://doi.org/10.1145/3347146.3359064
2. Champagne Gareau, J., Beaudry, É., Makarenkov, V.: A fast electric vehicle planner using clustering. In: Chadjipadelis, T., Lausen, B., Markos, A., Lee, T.R., Montanari, A., Nugent, R. (eds.) IFCS 2019. SCDAKO, pp. 17–25. Springer, Cham (2021). https://doi.org/10.1007/978-3-030-60104-1_3
3. Choset, H.: Coverage for robotics - a survey of recent results. Ann. Math. Artif. Intell **31**(1–4), 113–126 (2001). https://doi.org/10.1023/A:1016639210559
4. Choset, H., Pignon, P.: Coverage path planning: the boustrophedon cellular decomposition. In: Field and Service Robotics, pp. 203–209. Springer, Heidelberg (1998). https://doi.org/10.1007/978-1-4471-1273-0_32
5. Fournier, A., Fussell, D., Carpenter, L.: Computer rendering of stochastic models. Commun. ACM **25**(6), 371–384 (1982)
6. Galceran, E., Carreras, M.: A survey on coverage path planning for robotics. Rob. Auton. Syst. **61**(12), 1258–1276 (2013)
7. Khan, A., Noreen, I., Habib, Z.: On complete coverage path planning algorithms for non-holonomic mobile robots: survey and challenges. J. Inf. Sci. Eng. **33**(1), 101–121 (2017)
8. LaValle, S.M.: Planning Algorithms. Cambridge University Press, Cambridge (2006)
9. Mirkin, B.: Clustering for Data Mining. Chapman and Hall/CRC, Boca Raton (2005)
10. Mitchell, J.S.: Shortest paths and networks. In: Handbook of Discrete and Computational Geometry, 3rd edn., pp. 811–848. Chapman and Hall/CRC, Boca Raton (2017)
11. Russell, S., Norvig, P.: Artificial Intelligence: A Modern Approach, 3rd edn. Prentice Hall Press, Upper Saddle River (2009)
12. Shivashankar, V., Jain, R., Kuter, U., Nau, D.: Real-time planning for covering an initially-unknown spatial environment. In: Proceedings of the 24th International Florida Artificial Intelligence Research Society, FLAIRS - 24, pp. 63–68 (2011)

13. Xu, L.: Graph Planning for Environmental Coverage. Thesis, Carnegie Mellon University (2011). http://cs.cmu.edu/afs/cs/Web/People/lingx/thesis_xu.pdf
14. Zelinsky, A., Jarvis, R.A., Byrne, J., Yuta, S.: Planning paths of complete coverage of an unstructured environment by a mobile robot. In: Proceedings of the International Conference on Advanced Robotics & Mechatronics (ICARM), vol. 13, pp. 533–538 (1993)

Combining Encoplot and NLP Based Deep Learning for Plagiarism Detection

Ciprian Amzuloiu[1], Marian Cristian Mihăescu[1](✉), and Traian Rebedea[2]

[1] University of Craiova, Craiova, Romania
cristian.mihaescu@edu.ucv.ro
[2] University Politehnica of Bucharest, Bucharest, Romania
traian.rebedea@cs.pub.ro

Abstract. This paper tackles the classical problem of plagiarism detection by employing current state-of-the-art NLP methods based on Deep Learning. We investigate whether transformer models may be used along with existing solutions for plagiarism detection, such as *Encoplot* and clustering, to improve their results. Experimental results show that transformers represent a good solution for capturing the semantics of texts when dealing with plagiarism. Further efficiency improvements are needed as the proposed method is highly effective but also requires high computational resources. This prototype approach paves the way to further fine-tuning such that we may obtain a solution that scales well for a large number of source documents and large-sized documents.

Keywords: Transformers · Document embeddings · Plagiarism detection · Encoplot

1 Introduction

The PAN-PC-09[1] is probably the most used public datasets for assessing plagiarism detection solutions. It consists of 41,223 text documents in which were inserted 94,202 artificially created plagiarised sections. The entire corpus has been extracted from Project Guttenberg, and there were used 135 English books, 527 German books and 211 Spanish books. The artificially created plagiarisms are of four types: translations, random text operation, semantic word variation, and preserving word shuffling. The size of the documents range from small (i.e., one page) up to large documents (i.e., 100–1000 pages). The entire document corpus has been divided into two sets: 50% as source documents and 50% as suspicious documents. The plagiarized sections were created from passages in source documents and were inserted into the suspicious documents. Therefore, the proceedings of the workshop [9] and the results [6] represent the context of our research.

[1] PAN-PC-09, https://webis.de/data/pan-pc-09.html, last accessed 15 Sep 2021.

H. Yin et al. (Eds.): IDEAL 2021, LNCS 13113, pp. 97–106, 2021.
https://doi.org/10.1007/978-3-030-91608-4_10

The research problem regards determining the paragraphs from suspicious documents that plagiarize text passages from source documents. This problem is also known as external plagiarism and assumes the existence of a set of source documents (i.e., which were plagiarized) and a set of suspicious documents (i.e., which contain plagiarized text from source documents).

The key ingredients of the proposed approach regard usage of *Encoplot* [3] as one of the best-performing methods for plagiarism detection with more recent, state of the art transformer models such as BERT [1] and also with the classical DBSCAN [2] clustering algorithm. The proposed approach's novelty lies in integrating transformers into a custom-designed data analysis pipeline that tackles a classical problem for which we have bench-marked solutions.

2 Related Work

Our proposed approach combines methods and techniques previously used separately, either for text similarity or for plagiarism detection.

For example, Zubarev et al. [10] tackle the problem of text alignment as a subtask of plagiarism detection task for a dataset with cross-language Russian-English text. Few models were compared to detect translated plagiarism, and the BERT model showed excellent results and outperformed the custom model built for this task. However, it was noticed that it could be impractical to process a large number of sentences only for one suspicious document using BERT.

Further, Nguyen et al. [5] proposed the measurement of the similarity between short text pieces using word embeddings and an external knowledge source. The proposed method exploits the similarity between word contexts built by using word embeddings and semantic relatedness between concepts based on external sources of knowledge. It also performs both coreference resolution of named entities in two short texts to chain entities together and words segmentation to preserve the meaning of phrasal verbs and idioms. The experiments resulted in excellent results, except for texts without linguistic information. However, the proposed model is supervised and needs feature engineering. Unsupervised representation can automatically learn features for modelling pairs of sentences using neural networks with or without attention mechanisms.

Finally, Lo et al. [4] present a fully unsupervised cross-lingual semantic textual similarity (STS) metric based on contextual embeddings extracted from BERT. The results show that the unsupervised cross-lingual STS metric using BERT without fine-tuning achieves performance on par with supervised or weakly supervised approaches.

3 Proposed Approach

The key ingredients in the proposed data analysis pipeline are the selection of candidate plagiarized documents by the usage of *encoplot*, preprocessing, document embeddings computation and clustering. Figure 1 presents in detail the workflow of the data analysis pipeline.

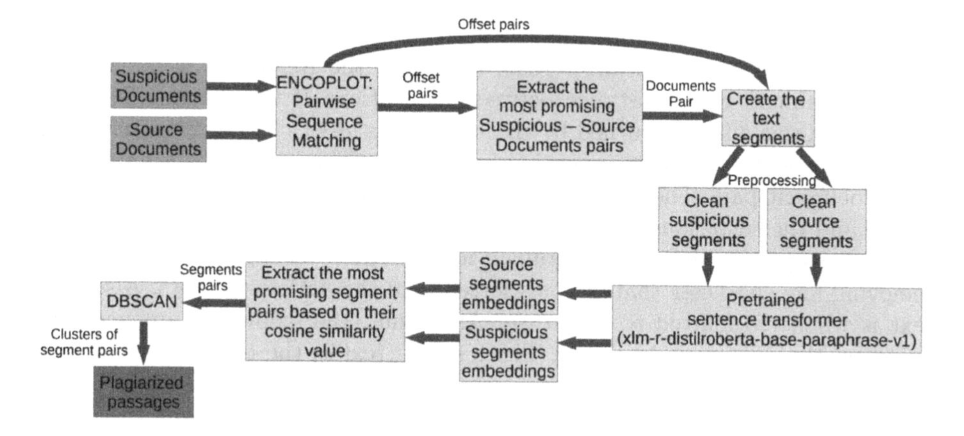

Fig. 1. Data analysis pipeline.

Pairwise Sequence Matching by *encoplot* is the first operation used to extract the first offset pairs list of segment candidates. It is less computationally expensive than the transformer model (used lately with the cosine similarity metric). The reason because *encoplot* is less computationally expensive is its implementation on a modified merge sort algorithm where it checks only the first occurrence of an N-gram in a document. That guarantees linear time, comparing to a dot-plot comparison of the N-grams, which would be quadratic. Because of that, it is an excellent way to use its results to select the input documents.

In the first instance, *encoplot* is receiving as input a suspicious document and a source document. The result consists of a list of positions (from the two input documents) with the same N-gram. If a document has more N-grams with the same content, the result consists only of the first occurrence of the N-gram. Having unique N-grams matching is good in most cases because a duplicate N-gram can be a part of a common phrase without any special informational content. An N-gram without duplicates more corresponds to a phrase with informational content, and matching it with an N-gram from the other document can be a sign of plagiarism.

We used *encoplot* with 16-grams of characters, and the output consisted of offset pairs is the input for the next step.

In order to extract **the most promising (Suspicious, Source) document pairs**, we verified the contiguity of the *encoplot* result for each pair. If there were at least 30 consecutive pairs with approximately the same jump between the suspicious offset and the source offset (the offset difference to be less than ten characters), we continue to the next steps with the current documents pair. Otherwise, there is no plagiarism between those two documents, and the next pair of suspicious and source documents can be compared.

After that, **the text segments are created** for the eligible document pairs, based on the offset hinted by *encoplot*. Each segment is build based on the offset position and the following *100* characters. After that, the segment is extended to completely capture the first and last word (necessary in order to give relevant words later to the transformer). Also, to reduce the redundant segments, we used a minimum range distance between two offsets. If the distance between two consecutive offsets returned by the *encoplot* is less than *25* characters, this offset is ignored from the segment's construction. This process is separate for each document pair. There will be different offsets for each comparison for the source and the suspicious document, so there will be different segments.

The preprocessing of the source and suspicious segments consists of removing any character that is not a letter from the list of segments. Also, the text is wholly formatted to lowercase. After that, the stopwords are removed based on the document language. The resulted text is the cleaned text of the initial segment.

The cleaned segments of the source and suspicious documents are the input used to **create the embeddings by the usage of the pre-trained sentence transformer**. After testing several pre-trained models, we chose to use *paraphrase-distilroberta-base-v1*, which is a multilingual model trained in more than 50 languages. The reason for choosing a multilingual model was translation plagiarism, which can occur in 10% of cases. For each cleaned text segment will result in a document embeddings vector computed using the transformer.

In plagiarism, there is a big chance to find at least one matching N-gram with the source text segment. Even for a text segment with a high level of obfuscation, there can still be no changed formulations, names, and even for translation plagiarism, there are words that keep similar forms from the original text that can result in matching N-grams. Although, one similar N-gram does not mean that the current segment is plagiarism. As it was mentioned, an N-gram can mean information with no informational content as well. The transformer, due to the attention mechanism, can provide semantic information about the text segment. Also, the cosine similarity between the embedding of two text segments provides a value of the semantic text similarity, which can prove if the initial matching N-gram was only a coincidence. The embeddings obtained in the previous input are used to **extract the most promising segment pairs based on their cosine similarity value**.

A similarity matrix can be obtained by applying the cosine similarity between the source and suspicious segment, filled with the resulting values. Each column of the matrix corresponds to a suspicious segment, and each row of the matrix corresponds to a source segment. Each position (i, j), where i represents the row number and j represents the column number, consists of the value of the cosine similarity between the source segment i and the suspicious segment j.

Extracting the most promising segment pairs requires sorting the values from each column of the similarity matrix in reverse order. From each sorted column, the corresponding segment pairs are extracted for the first *5* positions. In other words, for each suspicious segment, the segment pairs are obtained by making pairs with the most similar *5* source segments to the current suspicious segment.

By plotting the extracted segment pairs between a document with inserted plagiarism and the source document used for the artificial plagiarism insertion, we can notice peaks where the plagiarism is located. Those peaks are highlighted in Fig. 2, where the segments are plotted based on the similarity value and the offsets for the suspicious segment and the source segment.

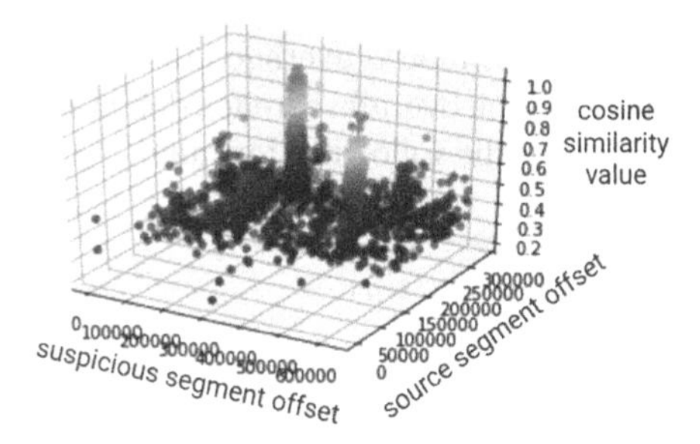

Fig. 2. Cosine similarity plot.

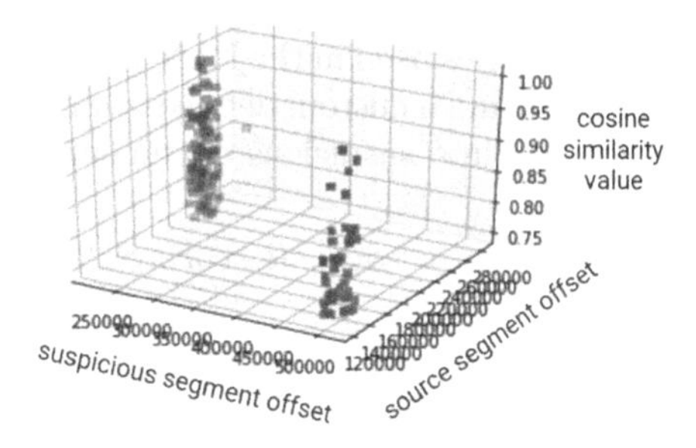

Fig. 3. Clustering result.

In order to prepare the list of pairs for the clustering process, we need to remove the unnecessary segments, and the first thought to do that would be by setting a threshold for the cosine similarity value of the segments. Although, this would not be enough due to the chance of obfuscation, where the cosine similarity value can be lower. The solution used for to set the threshold of *0.75* for the cosine similarity. We validate only the pairs with a similarity value larger than *0.75*. Also, we keep the segments that are at a distance of a maximum *100* characters to the pairs already validated (both for the source segment and the suspicious segment).

The clustering algorithm used for the obtained pairs is **DBSCAN**, due to the variation of clusters (a suspicious document can have from 0% to 100% plagiarism) and because of the detection based on density (if we have a pair separate from the others, it is considered noise and it is ignored to the clustering process). An example of the clustering result based on DBSCAN is in Fig. 3. The parameters used for DBSCAN were determinate experimentally. We used the *epsilon* as 100 * 100/1.5 and the minimum numbers of samples as 3.

Each cluster corresponds to a plagiarized text passage. The result is built by merging all the segment pairs from a cluster, with the offset to the smallest value of the suspicious/source segment from pairs and the length computed as a distance from the offset to the position of the last character from the cluster.

In terms of validation, the metrics are provided by the PAN competition. Each suspicious document is regarded as a sequence of characters, and all plagiarized sections are known in advance. Therefore, for each suspicious document, we have the sections that were automatically plagiarised, and the task of the system is to detect them accurately. The system will provide as a result the set of sections (i.e., sequence of characters) that the system detects as plagiarized in each suspicious document. Ideally, if all detected plagiarized sections correspond identically with actual plagiarized characters, the system is considered perfect. False-positive sections are those that have been marked as plagiarism, although they are not genuinely marked as plagiarism. False-negative sections are those that we not marked as plagiarism, although they are plagiarism.

Computing standard precision, recall, and granularity may have the advantage that they can be easily determined but has the disadvantage that they do not consider the length of the section marked as plagiarism. Therefore, the validation metrics used in the PAN-PC-09 competition use macro-averaged precision and recall and a custom defined granularity metric. The overall score for detected plagiarism in a suspicious document is presented in the following formula:

$$overall\_score(S, R) = \frac{F1}{\log_2(1 + granularity)}$$

where S is the suspicious document with actual marked plagiarized sections, R is the same suspicious document with our system determined plagiarized sections. The formulas for computing *F1*, *granularity* and micro-averaged precision and recall for plagiarism detection have been introduced and thoroughly motivated by Potthast et al. [6].

The fine-tuning of the presented plagiarism detection pipeline consists of choosing the parameters required for the processing of the results received from the *encoplot* algorithm and the document embeddings received from the transformer model. For the resulted document embeddings, the fine-tuning consists of choosing the parameters to build and filter the pairs of suspicious and source segments (cosine similarity threshold, the maximum number pairs, etc.) and the parameters for DBSCAN.

4 Experimental Results

The dataset that has been used is a subset of PAN-PC-09 [7], that is 14.429 source documents and 14.429 suspicious documents. Each suspicious document contains the characteristics of the plagiarized section: true or false for linguistic plagiarism, the level of obfuscation (i.e., none, low sau high), the *offset* (i.e., position where starts the plagiarized text), the *length* (i.e., the dimension of the plagiarized text)and the *reference offset* and *length* of the source document containing the original text which was plagiarized.

The suspicious documents contain two types of plagiarisms: linguistic and random. Linguistic plagiarism consists of a translated text and represents 10% of all plagiarized sections. Random plagiarism consists of random text operations (i.e., shuffle, deletion ore replacement of words or short phrases), semantic word variation (i.e., each word is replaced by a random chosen synonym or antonym) or by preserving word shuffling (i..e, the parts of speech are determined and remain in the same sequence while mixing the words).

In terms of dimensions of suspicions documents, 50% have small size (i.e., 1–10 pages), 35% have medium size (i.e., 10–100 pages) and 15% have large size (i.e., 100–1000 pages). Half of the suspicious documents are not plagiarised, while the remaining half has a plagiarized level between 0% and 100%. The dimension of a plagiarized section may be between 50 and 5000 words.

For the development of the pipeline we have used Google Colab[2] due to the large dataset, the computational power required and the variety.

The following libraries were used: *numpy*, *nltk*, *math*, *regex*, *pickle*, *matplotlib*, *xml*, *subprocess*, *glob*, *copy*, *sentece_transformers* and *sklearn*. The encoplot code was extracted from the *Pairwise sequence matching in linear time applied to plagiarism detection* [3] paper, with very less modifications on the output format. The code was compilled using GNU C++ compiller available inside the Colab shell.

To complete the preprocessing, we used the stopwords dataset provided by *nltk.corpus*. The stopwords corpus contains a collection of stopwords for 24 languages, including English, Spanish, and german. The pre-trained sentence transformer [8] was used from the sentence-transformers package provided by Hugging Face. The model was trained on parallel data for 50+ languages[3]. For the DBSCAN algorithm, we used the implementation provided by scikit-learn with the default Euclidean distance metric for calculating the distance between samples.

***Encoplot* with Greedy Approach.** The first experiment used *Encoplot* for detecting plagiarized sections and a greedy merge method by trying to form a segment where jumps are approximately equal (i.e., the difference is less than ten characters) by joining segments that have a minimum of 30 overlapping positions or have a jump greater than 200 characters. This method was ineffective for building the final plagiarized segments. This experiment was performed on the first 1025 suspicious documents and 14429 source documents. The small recall value is explained by the greedy way of building a plagiarized section. Considering the reasonable high precision, we concluded that the limitation lies in merging the segments. The considerable value of granularity means that for a single plagiarized section there were performed 29 distinct detections. The most significant limitation of this approach is that it does not employ any semantics similarity, and that is why we hypothesize that using transformers may detect similarities even if segments have a distinct structure.

Baseline Experiment. This experiment performs 272 comparisons (16 suspicious documents and 17 source documents). As a first experiment - as well as all the next ones - which uses *Transformer* and *cosine* similarity as a proof of concept with reasonable encouraging results.

Small Experiment. This experiment performs 9,030 comparisons (70 suspicious documents and 129 source documents). It has good encouraging results such as the first one, with minor score differences even though the number of comparisons was significantly higher.

Medium Experiment. This experiment performs 17,400 comparisons (100 suspicious documents and 174 source documents). We observe a decrease in the overall score as we increase the number of documents. Still, the granularity

[2] Google Colab, https://colab.research.google.com/.
[3] sentence-transformers, https://github.com/UKPLab/sentence-transformers.

remains around the 1.00 value - which is very good - and precision is still good. The lower value of recall is explained by the existence of small plagiarized sections and the fact that large detected sections do not fully cover the plagiarized section.

Large Experiment. This experiment performs 9,364,421 comparisons (649 suspicious documents and 14,429 source documents). Because it would be very intensive computational to check all source documents filtered by *Encoplot*, we decided to verify all suspicious documents against the first ten documents in terms of the number of offsets. This approach may be the reason for the lower overall score, while the precision is almost 70% indicates that it may represent a reasonable good solution.

Very Large Experiment. This experiment performs 14,789,725 comparisons (1,025 suspicious documents and 14,429 source documents) and the validation metrics are very similar with the results from the *Large experiment* (Table 1).

Table 1. Experimental results.

Experiment	Precision	Recall	Granularity	F1 score	Overall score
Encoplot+greedy	0.76	0.41	29.33	0.54	0.11
Baseline	0.99	0.68	1.025	0.81	0.79
Small	0.96	0.59	1.01	0.73	0.73
Medium	0.96	0.55	1.00	0.70	0.70
Large	0.69	0.42	1.00	0.52	0.52
Very large	0.67	0.44	1.00	0.53	0.53

In terms of running times, we mention that the overhead introduced by transformers is reduced by using encoplot as the first step of the pipeline. Detecting plagiarism in a suspicious document from a corresponding source document (i.e., suspicious-document00001 and source-document01409) takes 14 s without using encoplot and 8 s with encoplot. When the source document does not cotain plagiarized elements, it takes 10 s without the encoplot and 0.48 s with encoplot. For four suspicious documents and four source documents (16 comparisons) it takes 50 s with encoplot and 27 min without encoplot. Even though the time results depend on the size of the documents, usage of encoplot is a good solution for reducing the running time.

5 Conclusions

The first conclusion is that the integration of transformers represents a good option due to their attention mechanism. Specifically for the plagiarism detection problem, transformers proved to correctly detect similarity when the modifications inserted in the plagiarised segment are large.

However, the pipeline shows excellent results when *Encoplot* hints to the correct document pair. More work on fine-tuning the *Encoplot* selection process can reduce the work of the transformer such that it deals only with necessary text segments.

For many source documents that belong to specific topics, the topic detection mechanism is compulsory because transformers are computationally expensive and narrowing down the search space becomes mandatory.

Finally, we argue that a hybrid data analysis pipeline with *Encoplot* (for initial selection but without semantic analysis) and transformers (for semantic analysis but computationally expensive) represents a good solution for solving the problem of plagiarism detection.

The developed system is available as an open-source software freely available at https://github.com/CiprianAmz/Plagiarism_Detection.

Acknowledgements. This work was partially supported by the grant 135C/ 2021 "Development of software applications that integrate machine learning algorithms", financed by the University of Craiova.

References

1. Devlin, J., Chang, M.W., Lee, K., Toutanova, K.: Bert: pre-training of deep bidirectional transformers for language understanding (2018). arXiv preprint arXiv:1810.04805
2. Ester, M., Kriegel, H.P., Sander, J., Xu, X., et al.: A density-based algorithm for discovering clusters in large spatial databases with noise. In: KDD, vol. 96, pp. 226–231 (1996)
3. Grozea, C., Gehl, C., Popescu, M.: Encoplot: pairwise sequence matching in linear time applied to plagiarism detection. In: 3rd PAN Workshop. Uncovering Plagiarism, Authorship and Social Software Misuse, p. 10 (2009)
4. Lo, C.k., Simard, M.: Fully unsupervised crosslingual semantic textual similarity metric based on bert for identifying parallel data. In: Proceedings of the 23rd Conference on Computational Natural Language Learning (CoNLL), pp. 206–215 (2019)
5. Nguyen, H.T., Duong, P.H., Cambria, E.: Learning short-text semantic similarity with word embeddings and external knowledge sources. Knowl.-Based Syst. **182**, 104842 (2019)
6. Potthast, M., Stein, B., Eiselt, A., Barrón-Cedeño, A., Rosso, P.: Overview of the 1st international competition on plagiarism detection. In: 3rd PAN Workshop. Uncovering Plagiarism, Authorship and Social Software Misuse, p. 1 (2009)
7. Potthast, M., Stein, B., Eiselt, A., Barrón-Cedeño, A., Rosso, P.: Pan plagiarism corpus 2009 (pan-pc-09) (2009). https://doi.org/10.5281/zenodo.3250083
8. Reimers, N., Gurevych, I.: Making monolingual sentence embeddings multilingual using knowledge distillation. In: Proceedings of the 2020 Conference on Empirical Methods in Natural Language Processing. Association for Computational Linguistics (2020). https://arxiv.org/abs/2004.09813

9. Stein, B., Rosso, P., Stamatatos, E., Koppel, M., Agirre, E.: 3rd pan workshop on uncovering plagiarism, authorship and social software misuse. In: 25th Annual Conference of the Spanish Society for Natural Language Processing (SEPLN), pp. 1–77 (2009)
10. Zubarev, D., Sochenkov, I.: Cross-language text alignment for plagiarism detection based on contextual and context-free models. In: Computational Linguistics and Intellectual Technologies: Proceedings of the International Conference Dialogue, vol. 2019 (2019)

Drift Detection in Text Data with Document Embeddings

Robert Feldhans[1(✉)], Adrian Wilke[2], Stefan Heindorf[2], Mohammad Hossein Shaker[3], Barbara Hammer[1], Axel-Cyrille Ngonga Ngomo[2], and Eyke Hüllermeier[3]

[1] Bielefeld University, Bielefeld, Germany
{rfeldhans,bhammer}@techfak.uni-bielefeld.de
[2] DICE Group, Department of Computer Science, Paderborn University, Paderborn, Germany
{adrian.wilke,heindorf,axel.ngonga}@uni-paderborn.de
[3] University of Munich (LMU), Munich, Germany
mhshaker@mail.uni-paderborn.de, eyke@ifi.lmu.de

Abstract. Collections of text documents such as product reviews and microblogs often evolve over time. In practice, however, classifiers trained on them are updated infrequently, leading to performance degradation over time. While approaches for automatic drift detection have been proposed, they were often designed for low-dimensional sensor data, and it is unclear how well they perform for state-of-the-art text classifiers based on high-dimensional document embeddings. In this paper, we empirically compare drift detectors on document embeddings on two benchmarking datasets with varying amounts of drift. Our results show that multivariate drift detectors based on the Kernel Two-Sample Test and Least-Squares Density Difference outperform univariate drift detectors based on the Kolmogorov-Smirnov Test. Moreover, our experiments show that current drift detectors perform better on smaller embedding dimensions.

Keywords: Drift detection · Document embeddings · BERT · Word2Vec

1 Introduction

One of the key challenges when deploying machine learning models in practice is their degradation of performance after having been deployed [2]. In addition to technical issues, e.g., changes in the data format, performance degradation can be caused by (1) drift in the class distribution (virtual drift), and (2) drift in the labels (real drift). In the former case, the model might have insufficient training data for all classes and assumes a wrong prior probability. In the latter case, similar data points are labeled differently over time. For example, in the domain of natural language processing, this boils down to using language differently

H. Yin et al. (Eds.): IDEAL 2021, LNCS 13113, pp. 107–118, 2021.
https://doi.org/10.1007/978-3-030-91608-4_11

over time (virtual drift) due to novel words, grammatical constructs and writing styles. It also affects changing class labels over time (real drift) due to new human annotators joining the teams or updated annotation guidelines, e.g., changing the definition of disinformation [15], harassment [27], sexism [8], clickbait [7], etc.

To detect drift automatically and notify machine learning engineers to potentially update their models, automatic drift detectors have been proposed [12,17,22]. However, as we show in this paper, state-of-the-art drift detectors are hardly applicable to modern NLP applications: they assume the input data to be low-dimensional, hand-engineered vector spaces, while modern NLP applications employ complex language models such as transformers with high-dimensional, latent document embeddings. Towards this end, we empirically compare state-of-the-art drift detectors which have mainly been designed for low-dimensional vector spaces and we apply them to drift detection in natural language texts:

RQ1: Which drift detector works best for document embeddings?
RQ2: How does the performance of a drift detector depend on the embedding dimensions?

We perform our experiments using two benchmarking datasets into which we inject different amounts of drift. Moreover, we investigate how the predictive performance of a drift detector depends on the embedding dimensions. Our experiments show that multivariate approaches based on the Kernel Two-Sample test (KTS) and Least-Squares Density Difference (LSDD) outperform univariate drift detectors based on the Kolmogorov-Smirnov test (KS) and that most approaches perform better on low-dimensional data. The code underlying our research is publicly available.[1]

The remainder of this article is structured as follows: In Sect. 2, we briefly outline related work. Section 3 describes our methodology, i.e., datasets and drift detectors as well as our evaluation setup. Finally, Sect. 4 presents our results and Sect. 5 our final conclusions.

2 Related Work

Concept drift has become a highly researched field. There are several recommendable introductions from Lu et al. [18], Gama et al. [11], Nishida and Yamauchi [21], Gama and Castillo [10] and Basseville and Nikiforov [4]. Gama et al. [11] conducted a survey that provides an introduction to concept drift adaptation, which includes patterns of changes over time (e.g., incremental changes) and evaluation metric criteria (e.g., probability of true change detection and delay of detection), which are part of our injection experiments. Baier et al. [3] provided an analysis of 34 articles related to concept drift and a framework of 11 categories to characterize predictive services like drift detectors. With respect to

[1] https://github.com/EML4U/Drift-detector-comparison.

their overview of existing approaches, we fill the gap in the category *data input* (here: unstructured text); our paper can be classified as being *gradual* as well as *sudden* in the category *type of change*. The overview of concept drift by Tsymbal [24] distinguished between sudden and gradual concept drift. Additionally, batch systems (here: LSDD, MMD) and online systems (here: Confidence Distribution Batch Detection, CDBD) are distinguished. The classical FLORA systems [25] as well as the more recent MDEF [13], ADWIN [5], and EDDM Baena-Garcıa et al. [1] are also worth mentioning. Moreover, drift detection has successfully been used on image data [17,22] and various other applications [28].

3 Comparison of Drift Detectors

Given a classification problem, let X be a feature vector, y be the target variable and $P(y, X)$ their joint distribution. Following Gama et al. [11], *real concept* drift refers to changes in $P(y|X)$ and *virtual drift* refers to changes in $P(X)$, i.e., if the distribution of the incoming data changes. In our case, we focus on virtual drift. One of the particular challenges for natural language processing is that texts are typically transformed to high-dimensional feature vectors X which might only appear once in the dataset. Further, state-of-the-art NLP models, such as neural networks and transformers, follow a discriminative paradigm [20] instead of a generative one, making it difficult to estimate probability distributions.

For the experiments on drift detectors (Sect. 3.1), we use two real-world datasets that are transformed using three embedding models—one BERT model [9] with 768 dimensions, one Word2Vec model [19] with 768 dimensions, and another Word2Vec model with 50 dimensions—to explore the effect of different numbers of dimensions (Sect. 3.2). Finally, the data are arranged in four subsets in preparation for the following experiments (Sect. 3.3).

3.1 Drift Detectors in the Experiments

We selected the popular drift detectors KS, KTS and additionally, a more recent approach, LSDD (2018), as well as the semi-supervised CDBD.

Kolmogorov-Smirnov (KS) is a statistical test for agreement between two probability distributions using the maximum absolute difference between the distributions. We use the feature-wise two-sample implementation of the Alibi Detect[2] library. For multivariate data, Bonferroni correction is used to aggregate the p-values per dimension.

Kernel Two-Sample (KTS) [12] is a statistical independence test based on Maximum Mean Discrepancy (MMD). MMD is the squared distance between the embeddings of two distributions p and q in a *reproducing kernel Hilbert space*, $MMD(p, q) = ||\mu_p - \mu_q||^2_{\mathcal{H}}$, where μ denotes the mean embeddings. We use the implementation of Emanuele Olivetti[3] for our experiments.

[2] https://github.com/SeldonIO/alibi-detect.

[3] https://github.com/emanuele/kernel_two_sample_test.

Least-Squares Density Difference (LSDD) [6] is based on the least-squares density difference estimation method. For two distributions p and q, it is defined as $LSDD(p, q) = \int (p(x) - q(x))^2 \, dx$. To apply the test, we utilize the Alibi Detect(See Footnote 2) implementation.

Confidence Distribution Batch Detection (CDBD) [16] is an uncertainty-based drift detector using a two-window paradigm coupled with Kullback-Leibler divergence applied to a confidence score. It can be used with any classifier that produces a confidence (uncertainty) score about the classifier predictions. Given that CDBD requires labels in the beginning for training the classifier, it is the only semi-supervised detector in our comparison. CDBD compares the divergence between the distribution of confidence scores of a batch of reference instances to test instances. The higher the divergence, the more drift there is between the reference batch and the test batch. In this paper, we used Random Forest as the classifier and entropy of the output probability distributions as the confidence score.

3.2 Source Datasets and Embedding Models

Amazon Movie Reviews. The Amazon Movie Review dataset[4] consists of nearly 8 million user reviews, from 1997 to 2012, of movies purchasable on the Amazon website. In particular, this dataset contains the joined user reviews and summaries thereof in text form, which we will use to detect drift, and a score from one to five that the user gave the movie. The average length of each text is 172 words. It can be noted that this score should be directly correlated to the texts the user wrote and be indicative of the sentiment of said texts. As we suspect some (uncontrolled) drift over the twelve-year time frame of the dataset, we opted to use the data of one year (i.e., 2011) to reduce possible changes over time. We generally use the scores of the reviews as classes, especially for retraining the models.

The BERT model used with this dataset is a retrained version of the pre-trained model 'bert-base-uncased' provided by the *Hugging Face* [26] library. The pretrained BERT model was retrained for nine epochs, using all entries from 2011, and provides embeddings with 768 dimensions.

Both BoW models were computed using all tokenized texts contained in the dataset. For the training, we used 40 epochs and a minimum count of 2 for each word to be included. As the training algorithm, a distributed bag of words (PV-DBOW) of the *Gensim*[5] 3.8.3 software was applied. The computation took 32 h (50 dim) and 44 h (768 dim) for Amazon and 1.5 h for the two Twitter models on a 4x Intel(R) Xeon(R) CPU E5-2695 v3 @ 2.30 GHz machine with no GPU and 64 GB RAM.

---

[4] https://snap.stanford.edu/data/web-Movies.html.

[5] https://radimrehurek.com/gensim/.

Twitter Election. The Twitter Election dataset[6] is composed of tweets that refer to the two candidates of the 2020 US presidential election (i.e., these tweets use the hashtag #Biden and/or #Trump). All given data points were created during the last three weeks before the election, i.e., t_{min} = 2020-10-14 to t_{max} = 2020-11-07. The original dataset contains around 1.7 million singular data points, of which ca. 777k use hashtag #Biden and ca. 971k use #Trump.

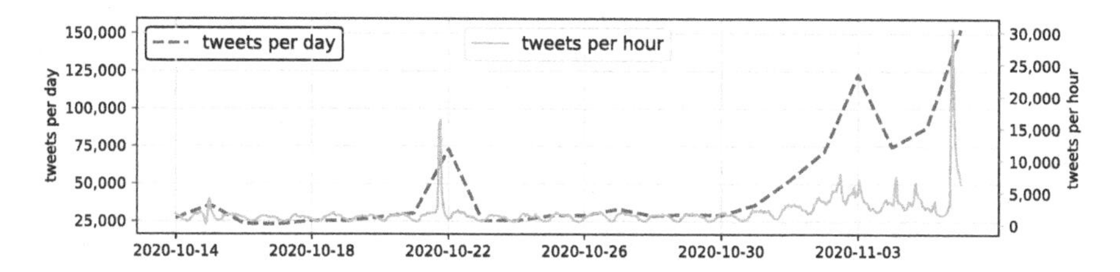

Fig. 1. Amount of tweets in the Twitter election dataset by day (red) and hour (blue). Note the sharp spike during the last TV debate (t_{debate} = 2020-10-22) and the overall increase during the election day ($t_{election}$ = 2020-11-03). Slight inaccuracies in the timeline can be attributed to time zone-related shifts. (Color figure online)

There are two distinct points in time which coincide with real-world events and are thus suspected to differ in distribution. These events are the last TV debate before the election (t_{debate} = 2020-10-22) and the election day itself ($t_{election}$ = 2020-11-03). The increased number of data points in both of these time frames supports this claim, as can be seen in Fig. 1. We generally distinguish #Biden and #Trump as the two classes of this dataset. As such, we removed all ambiguous tweets that contain both the #Biden and #Trump hashtags. Some data points in this set contain non-English language, most notably Spanish. To reduce the effect of non-English data points, we removed tweets from the dataset which were detected as non-English with the Python implementation[7] of *langdetect*.[8] Around 500k data points were removed this way, resulting in a dataset of ca. 521k #Biden tweets and 680k #Trump tweets.

The BERT model used with this dataset is a pretrained model provided by the *Hugging Face* library called 'bert-base-multilingual-cased'. It was chosen because of its ability to handle multilanguage input data, as the dataset contains traces of multi-language data. This model was not retrained.

The BoW models were computed analogously to the ones used in the Amazon movie dataset and took 1.5 h to compute on the same machine as used for the Amazon models.

[6] https://www.kaggle.com/manchunhui/us-election-2020-tweets.
[7] https://pypi.org/project/langdetect/.
[8] http://code.google.com/p/language-detection/.

3.3 Evaluation Setup: Sampling

To evaluate the drift detector's sensitivity and specificity, we created several subsets of the datasets with precisely controlled drift.

Drift Induction. This subset consists of two balanced sets of 2,000 samples of randomly chosen but class-balanced data points. The first set is then gradually injected with specifically chosen negative adjectives. In each step, a certain percentage $\gamma_i \in \{0.05 \cdot i | i = 0, 1, 2, ..., 20\}$ of texts is injected with one of these negative adjectives. With this subset, we want to evaluate the speed and confidence with which drift detectors detect gradually induced drift. Injection is done between two randomly chosen words of the text, and each text is injected with a maximum of one word over all steps. For example, in step γ_2, 0.1=10% of texts are injected with at most one negative adjective. This resembles the method of Shoemark et al. [23], with the difference that we are not replacing but adding real words instead of made-up ones.

The specific list of these adjectives was obtained by choosing the 22 adjectives that occur at least 500 times in all Amazon movie reviews of both one or five stars and appear at least twice as often in one star reviews. To ensure these adjectives are negatively connotated, we used a list of 4,783 negative opinion words,[9] mined from negative customer reviews [14]. The experiments using this subset are repeated ten times using unique data for each run.

Twitter Election Specifics. In processing this dataset, we also use the negative adjectives based on the Amazon Movie Review dataset for injection, to better compare the Amazon and Twitter variants. Additionally, there is a lack of a distinct gradient (akin to the score) between both classes of the dataset, so an analogous but distinct approach to the Amazon variant was discarded.

Same Distribution. Using this subset, we test the drift detectors against data drawn from the same distribution to explore whether the drift detectors abstain from identifying drift where there is none. This is of importance in general applications of drift handling, as retraining a model to compensate for drift is often expensive, so it should only be done if necessary. This subset consists of 500 random samples per class (i.e., 2,500 samples for the Amazon and 1,000 samples for the Twitter dataset). It is tested against twenty more subsets created with the same criteria, but distinct samples. We then present the mean of the results.

Different Classes. This subset is used to evaluate the drift between data of two different classes. As such, its main purpose is to establish a tangible maximum of (virtual) drift possible in each dataset, test whether or not the drift detectors are able to detect it, and to give context for the drift induction subset. With this in mind, any drift detector should be able to detect the virtual drift in this

[9] https://www.cs.uic.edu/~liub/FBS/sentiment-analysis.html#lexicon.

experiment. For this, 1,000 samples were taken from each class. The test was repeated ten times with unique data.

Amazon Movie Reviews Specifics. The classes used for this dataset were those defined by reviews of scores one and five.

Different Distribution. In this subset, we apply the prior knowledge of the Twitter dataset, thus no Amazon variant of this subset exists. In contrast to the other subsets, where drift is set up artificially, this one presents an application of a real-world example of drift. We check for drift between data of a typical point in time ($t_{reference} = t_{min} + 100h$) and three other points of interest, i.e., another typical point $t_{base} = t_{reference} + 24h$ to establish a baseline, t_{debate} and $t_{election}$. For each t, 1,000 class-balanced samples were taken between t and $t + 24h$. For this subset, we generate 8 permutations of this setup and present the mean of the results.

Semi-supervised Algorithm. The semi-supervised algorithm CDBD requires a deviant structuring of input data, as a portion of the data given to the detector is used to train the model. To avoid a small reference batch, the permutations of CDBD are different in that the entire data given to the detector is randomly shuffled to train different models and create different reference batches for each permutation.

4 Results

Results of the Drift Injection Experiments

Based on the drift induction setup (Sect. 3.3), we evaluate the speed and confidence with which drift detectors detect slowly induced drift in this experiment. An ideal drift detector would produce a high p-value with little to no injected words but quickly drop when more words are injected. The resulting p-values of the experiments are displayed in Fig. 2.

Regarding **RQ1**, the KTS and LSDD detectors generally produce similar results while outperforming CDBD and KS. KTS produced better results than LSDD, especially for the high-dimensional BoW experiment. In particular, KS behaves conservatively in its estimations and struggles to detect drift, even with considerable injection of words. CDBD shows erratic behavior with all BoW embeddings; it outperforms KS only with the BoW-768 embeddings and on the Twitter dataset with BoW-50 embeddings.

Regarding **RQ2**, the lower-dimensional BoW-50 data generally performs better than or equal to both higher-dimensional sets, and the best detectors (i.e., KTS and LSDD) reach a p-value of 0.05 with less induced drift than with the higher-dimensional data. However, it is important to note that almost all detectors start with a much higher p-value on the higher-dimensional BERT data, as described in Sect. 3.3.

Results of Same Distribution Experiments

In this experiment, we evaluate the drift detectors' ability to not falsely detect drift where none is present. A higher p-value suggests a better performance (see Table 1).

With respect to **RQ1**, the CDBD detector generally outperforms the other detectors across all embeddings ($p \approx 0.7$). The KS detector performs very consistently ($p \approx 0.5$). Depending on the model and embedding method used, the KTS yields the single best results in this test (on Amazon BERT) but has a high standard deviation. LSDD's performance is generally about as good as KTS's on each dataset, although slightly better on average. Its standard deviation is

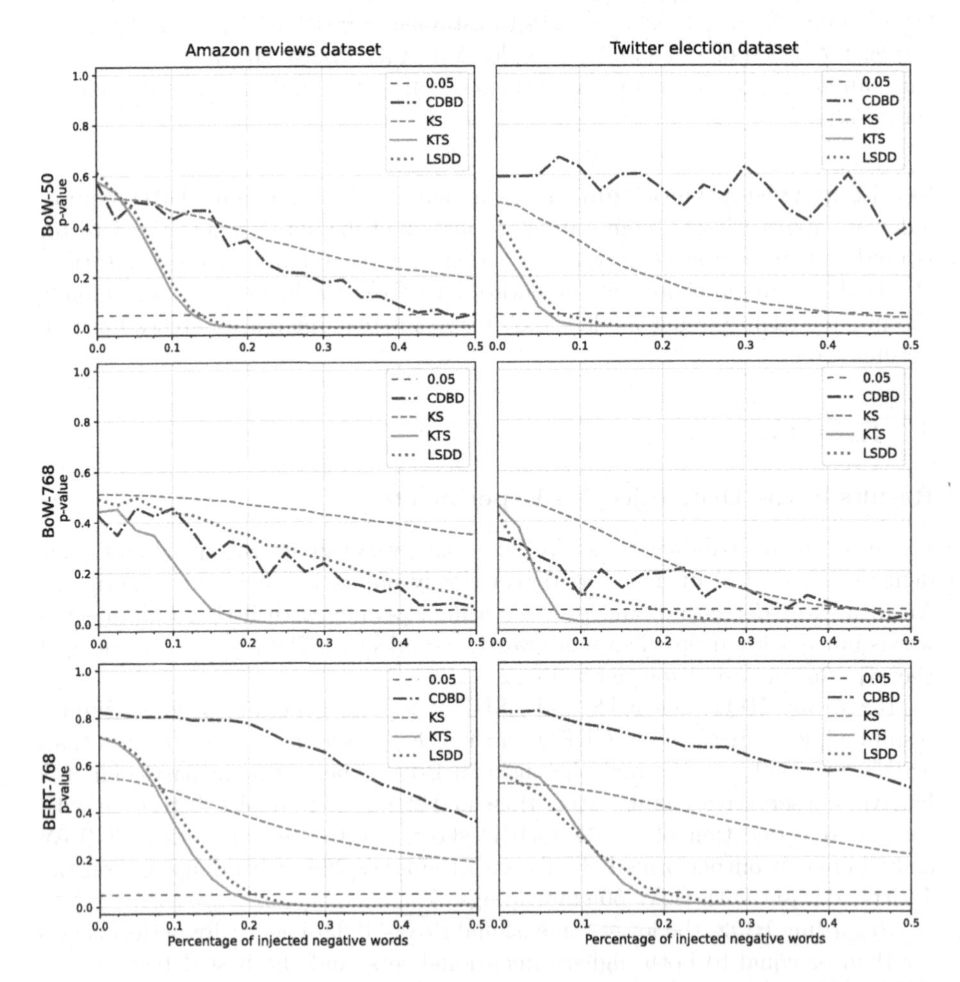

Fig. 2. Drift detection results on injection experiments: Amazon reviews dataset (left) and Twitter election dataset (right) as well as the models BoW-50 (top), BoW-768 (middle) and BERT-768 (bottom). Presented is the mean p-value of all runs.

comparable to KTS's. Regarding **RQ2**, we do not see a consistent effect of the embedding dimension.

Results of the Different Class Experiments

In this experiment, we evaluate the drift detectors based on their ability to detect the maximum possible (virtual) drift in each dataset. Here, a lower p-value suggests a better performance (see Table 2). The CDBD detector is omitted from this experiment since it is impossible to train a classifier on single-classed data given only the data used in this experiment.

With regard to **RQ1**, both LSDD and KTS perform as expected and consistently provide nearly perfect results across both datasets and all embeddings. The KS detector struggles to detect drift with statistical significance at the 0.05 significance level. It is able to do so only on the Amazon BERT embeddings.

Regarding **RQ2**, the results show a tendency of drift detectors to perform better for lower embedding dimensions, as exemplified by the KS detector.

Table 1. Drift detector scores of same distribution experiments (mean p-value of all runs, higher ∅ is better)

Dataset	Model	CDBD		KS		KTS		LSDD	
		∅	stdev	∅	stdev	∅	stdev	∅	stdev
Amazon	BoW-50	**0.6494**	0.2304	0.4678	0.0397	0.4294	0.2920	0.5060	0.2990
Amazon	BoW-768	**0.6338**	0.0727	0.5009	0.0224	0.5860	0.3010	0.4140	0.3492
Amazon	BERT-768	0.7148	0.1544	0.5676	0.0195	**0.8666**	0.1372	0.8160	0.2051
Twitter	BoW-50	**0.7863**	0.0785	0.5330	0.0419	0.5930	0.2398	0.5980	0.2392
Twitter	BoW-768	0.4257	0.1190	0.5204	0.0208	0.4828	0.3100	**0.6380**	0.2825
Twitter	BERT-768	**0.7720**	0.1448	0.5114	0.0190	0.3878	0.1940	0.4910	0.2048

Table 2. Results of the different class experiments (mean p-value of all runs, lower ∅ is better)

Dataset	Model	KS		KTS		LSDD	
		∅	stdev	∅	stdev	∅	stdev
Amazon	BoW-50	0.0709	0.0145	0.0020	0.0000	**0.0000**	0.0000
Amazon	BoW-768	0.0870	0.0160	0.0020	0.0000	**0.0000**	0.0000
Amazon	BERT-768	0.0126	0.0012	0.0020	0.0000	**0.0000**	0.0000
Twitter	BoW-50	0.1009	0.0207	0.0020	0.0000	**0.0000**	0.0000
Twitter	BoW-768	0.2522	0.0266	**0.0020**	0.0000	0.0180	0.0218
Twitter	BERT-768	0.1205	0.0047	0.0020	0.0000	**0.0000**	0.0000

Results of the Twitter Different Distribution Experiments

In this experiment, we evaluate the drift detectors in a controlled scenario with prior information about the dataset. A higher p-value in t_{base} and lower p-values in t_{debate} and $t_{election}$ suggest a better performance (see Fig. 3 and Table 3). For this experiment, we do not report CDBD results since large fluctuations in p-values render CDBD unusable in our experimental setup, emphasizing the necessity of large datasets for supervised drift detectors.

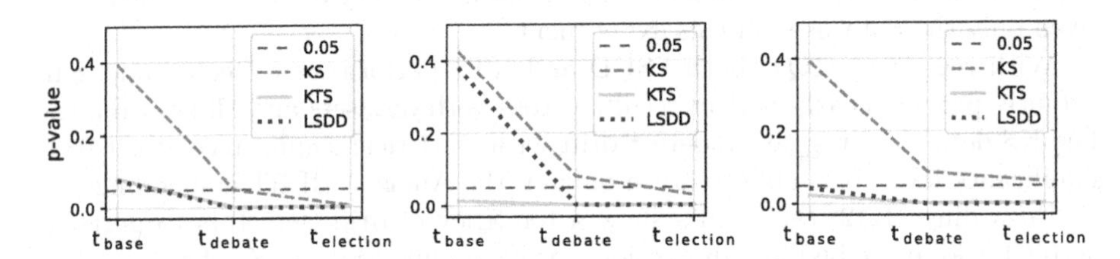

Fig. 3. Results of Twitter different distribution experiment (mean p-value of all runs): BoW-50 (left), BoW-768 (center) and BERT-768 (right)

Regarding **RQ1**, nine drift detection results are available per detector: three kinds of embeddings, each with three points in time. LSDD correctly predicted drift in eight of the nine cases, KTS in seven cases and KS in six cases. From a qualitative perspective, the KS detector produces the most pronounced curve (see Fig. 3), i.e., the largest difference in p-values with respect to t_{base}, and correctly predicts no drift in t_{base} across all embedding models. However, it struggles to detect drift in t_{debate} and $t_{election}$ at the 0.05 significance level. Both the KTS and LSDD detectors are capable of detecting drift that occurred in those points of time. However, their mean p-values are close to the 0.05 significance level in t_{base}, leading to fluctuating decisions considering their standard deviation.

For **RQ2**, all detectors produced correct results for the low-dimensional BoW-50 embeddings, whereas this is not guaranteed for higher dimensions.

5 Conclusion

Regarding our research questions, our conclusions are as follows:

RQ1. Our experimental results suggest LSDD and KTS as the best drift detectors with LSDD slightly outperforming KTS in the real-world Twitter election experiment. KS produced rather average results in all experiments due to its conservative estimation of p-values. CDBD, as a supervised drift detector, requires a large reference batch to produce robust results, questioning its usefulness in many practical applications.

RQ2. Our results indicate that lower embedding dimensions tend to produce better drift detection results.

Table 3. Results of the different distribution experiments (mean p-value of all runs)

t	Model	KS		KTS		LSDD	
		∅	stdev	∅	stdev	∅	stdev
t_{base}	BoW-50	**0.3961**	0.0697	0.0813	0.1444	0.0775	0.0935
t_{base}	BoW-768	**0.4259**	0.0225	0.0135	0.0304	0.3812	0.2382
t_{base}	BERT-768	**0.3933**	0.0274	0.0235	0.0187	0.0437	0.0394
t_{debate}	BoW-50	0.0493	0.0446	0.0020	0.0000	**0.0000**	0.0000
t_{debate}	BoW-768	0.0810	0.0853	**0.0020**	0.0000	0.0025	0.0066
t_{debate}	BERT-768	0.0873	0.0674	0.0020	0.0000	**0.0000**	0.0000
$t_{election}$	BoW-50	0.0052	0.0029	0.0020	0.0000	**0.0000**	0.0000
$t_{election}$	BoW-768	0.0281	0.0137	0.0020	0.0000	**0.0000**	0.0000
$t_{election}$	BERT-768	0.0618	0.0104	0.0020	0.0000	**0.0000**	0.0000

In future work, we would like to further explore the effect of different dimensionality reduction techniques on drift detectors and to devise novel drift detectors specifically tailored to text data with high-dimension document embeddings, e.g., based on the similarity metrics employed by the embedding approaches.

Acknowledgments. This work has been supported by the German Federal Ministry of Education and Research (BMBF) within the project EML4U under the grant no 01IS19080 A and B.

References

1. Baena-Garcıa, M., del Campo-Ávila, J., Fidalgo, R., Bifet, A., Gavalda, R., Morales-Bueno, R.: Early drift detection method. In: Fourth International Workshop on Knowledge Discovery from Data Streams, vol. 6 (2006)
2. Baier, L., Jöhren, F., Seebacher, S.: Challenges in the deployment and operation of machine learning in practice. In: ECIS (2019)
3. Baier, L., Kühl, N., Satzger, G.: How to cope with change? - preserving validity of predictive services over time. In: HICSS, ScholarSpace (2019)
4. Basseville, M., Nikiforov, I.V.: Detection of Abrupt Changes: Theory and Application. Prentice Hall, Hoboken (1993)
5. Bifet, A., Gavaldà, R.: Learning from time-changing data with adaptive windowing. In: SDM, pp. 443–448, SIAM (2007)
6. Bu, L., Alippi, C., Zhao, D.: A pdf-free change detection test based on density difference estimation. IEEE Trans. Neural Networks Learn. Syst. **29**(2), 324–334 (2018)
7. Chen, Y., Conroy, N.J., Rubin, V.L.: Misleading online content: recognizing clickbait as "false news". In: WMDD@ICMI, pp. 15–19. ACM (2015)
8. Chowdhury, A.G., Sawhney, R., Shah, R.R., Mahata, D.: #youtoo? Detection of personal recollections of sexual harassment on social media. In: ACL, pp. 2527–2537 (2019)

9. Devlin, J., Chang, M., Lee, K., Toutanova, K.: BERT: pre-training of deep bidirectional transformers for language understanding. CoRR (2018)
10. Gama, J., Castillo, G.: Learning with local drift detection. In: Li, X., Zaïane, O.R., Li, Z. (eds.) ADMA 2006. LNCS (LNAI), vol. 4093, pp. 42–55. Springer, Heidelberg (2006). https://doi.org/10.1007/11811305_4
11. Gama, J., Zliobaite, I., Bifet, A., Pechenizkiy, M., Bouchachia, A.: A survey on concept drift adaptation. ACM Comput. Surv. **46**(4), 44:1–44:37 (2014)
12. Gretton, A., Borgwardt, K.M., Rasch, M.J., Schölkopf, B., Smola, A.J.: A Kernel two-sample test. J. Mach. Learn. Res. **13**, 723–773 (2012)
13. Heit, J., Liu, J., Shah, M.: An architecture for the deployment of statistical models for the big data era. In: IEEE BigData (2016)
14. Hu, M., Liu, B.: Mining and summarizing customer reviews. In: KDD, pp. 168–177. ACM (2004)
15. Kumar, S., West, R., Leskovec, J.: Disinformation on the web: Impact, characteristics, and detection of Wikipedia hoaxes. In: WWW. ACM (2016)
16. Lindstrom, P., Namee, B.M., Delany, S.J.: Drift detection using uncertainty distribution divergence. Evol. Syst. **4**(1), 13–25 (2013)
17. Lopez-Paz, D., Oquab, M.: Revisiting classifier two-sample tests. In: ICLR (Poster), OpenReview.net (2017)
18. Lu, J., Liu, A., Dong, F., Gu, F., Gama, J., Zhang, G.: Learning under concept drift: a review. IEEE Trans. Knowl. Data Eng. **31**(12), 2346–2363 (2019)
19. Mikolov, T., Chen, K., Corrado, G., Dean, J.: Efficient estimation of word representations in vector space. In: ICLR (Workshop Poster) (2013)
20. Ng, A.Y., Jordan, M.I.: On discriminative vs. generative classifiers: a comparison of logistic regression and Naive Bayes. In: NIPS, pp. 841–848, MIT Press (2001)
21. Nishida, K., Yamauchi, K.: Detecting concept drift using statistical testing. In: Corruble, V., Takeda, M., Suzuki, E. (eds.) DS 2007. LNCS (LNAI), vol. 4755, pp. 264–269. Springer, Heidelberg (2007). https://doi.org/10.1007/978-3-540-75488-6_27
22. Rabanser, S., Günnemann, S., Lipton, Z.C.: Failing loudly: an empirical study of methods for detecting dataset shift. In: NeurIPS (2019)
23. Shoemark, P., Liza, F.F., Nguyen, D., Hale, S.A., McGillivray, B.: Room to Glo: a systematic comparison of semantic change detection approaches with word embeddings. In: EMNLP/IJCNLP, pp. 66–76. Association for Computational Linguistics (2019)
24. Tsymbal, A.: The problem of concept drift: definitions and related work. Comput. Sci. Dept. Trinity College Dublin **106**(2), 58 (2004)
25. Widmer, G., Kubat, M.: Learning in the presence of concept drift and hidden contexts. Mach. Learn. **23**(1), 69–101 (1996)
26. Wolf, T., et al.: Transformers: state-of-the-art natural language processing. In: EMNLP (Demos), pp. 38–45. ACL (2020)
27. Yin, D., Xue, Z., Hong, L., Davison, B.D., Kontostathis, A., Edwards, L.: Detection of harassment on web 2.0. In: Proceedings of the Content Analysis in the WEB 2, pp. 1–7 (2009)
28. Žliobaitė, I., Pechenizkiy, M., Gama, J.: An overview of concept drift applications. Big data analysis: new algorithms for a new society (2016)

Validation of Video Retrieval by Kappa Measure for Inter-Judge Agreement

Diana Bleoancă[1], Stella Heras[2], Javier Palanca[2], Vicente Julian[2], and Marian Cristian Mihăescu[1(✉)]

[1] Faculty of Automatics, Computers and Electronics, University of Craiova, Craiova, Romania
cristian.mihaescu@edu.ucv.ro
[2] Valencian Research Institute for Artificial Intelligence, Universitat Politècnica de València, Valencia, Spain
{sheras,jpalanca,vinglada}@dsic.upv.es

Abstract. Validation of information retrieval(IR) systems represents an inherently difficult task. We present a study that uses the *Kappa* measure for inter-judge agreement for establishing a reference quality benchmark for responses provided by a custom developed IR system in a comparative analysis with already existing search mechanism. Experiments show that it is difficult to assess the relevance of responses as human judges do not always easily agree on what is relevant and what is not. The results prove that when judges agree the responses from our system are mostly better than those returned by existing mechanism. This bench-marking mechanism opens the way for further detailed investigation of responses that were not relevant and possible improvement of the IR system design.

Keywords: Kappa measure · Relevance of response · Judge agreement

1 Introduction

Whenever designing and implementing an indexing and retrieval system it always arises the problem of assessing the relevance of obtained results. The most critical aspect in this aspect is the fact that relevance is a subjective aspect and therefore human evaluation is necessary. Still, evaluation itself is a critical aspect as it is the only way for benchmarking the implemented system, finding situations when it performs well and when it does not work well, and therefore give the possibility of further redesign and reevaluation in an attempt of improving the relevance of retrieved items.

Our study aims to employ *Kappa* measure for inter-judge agreement [13] as a method for evaluation the quality or retrieved videos by a previously implemented system [6] in a comparative analysis with already existing simple query mechanism.

H. Yin et al. (Eds.): IDEAL 2021, LNCS 13113, pp. 119–127, 2021.
https://doi.org/10.1007/978-3-030-91608-4_12

We have collected a number of real world queries and logged the responses provided by our retrieval mechanism and by already existing simple mechanism. The responses were given to judges for relevance evaluation such that no judge had any indication which mechanism produced the result. After marking each response as *relevant* or *non-relevant* we performed an analysis of *Kappa* measure for inter judge agreement such that we may end up with a comparative analysis and clear indications on the situation in which our system performed good or bad.

2 Related Work

This work is a step further of the work presented in [6]. In that work we proposed a mechanism that allows video retrieval based on the processing of the transcripts of such videos. The work is contextualized in a service of educational videos, mainly in Spanish, provided by the Universitat Politècnica de València(UPV).

For this purpose, several natural language processing (NLP) techniques have been used, such as lemmatization, for which a specific lemmatizer had to be built for the transcriptions of the Spanish videos in the dataset [1,3]. In addition, there are several techniques that allow us to transform the identified lemmas into real numbers that can be processed by the subsequent algorithms [8,9,11].

To analyze the relevance of each word in the transcript, we have employed LSI (Latent Semantic Indexing) [7], combined with TF-IDF [4,12], to better contextualize each word of the transcript. Other authors generate topic vectors by using an LDA algorithm instead [5,10].

Another important feature that our work presented in [6] was to argue the user's queries with relevant facts gathered from Wikipedia in order to better understand their context [14].

Finally, in this work we will focus on the validation of different retrieval engines (our proposal and the pre-defined search engine just based on titles) by using a human judgement process where several volunteers have classified a sample of videos by its relevance and hence an inter-judge agreement process to validate their classifications is needed [2,13].

3 Proposed Approach

Starting from our previous work, a search machine learning (ML) algorithm that returns most relevant videos based on a query by NLP analysis of their transcripts [6], our problem consisted into finding a method of validating and interpreting the results. Based on the *Kappa* measure we can conclude if content results can be marked as right and wrong (Fig. 1).

3.1 Procedure Pipeline

We further summarize the main steps from the procedure pipeline.

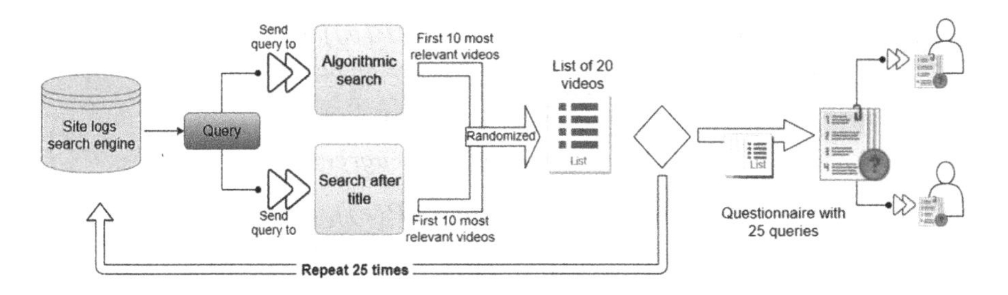

Fig. 1. Overview of the procedure pipeline.

Preparing the data required for this procedure could not be accomplished by selecting queries with random words, as it would not reflect a real life scenario. Thus, our first challenge was finding valid queries. For that, we have researched the logs from the in-place engine of searching, from the UPV website, for the last few days. After preprocessing them (deleting the duplicates and irrelevant queries, as teachers name or dummy words), we have retrieved the first 25 queries as input for our procedure.

For comparison purposes, each query from our database was sent to both engines in order to retrieve the 10 most relevant videos from each one. Because there were scenarios when the site has not returned enough responses (for few queries, the response consisted into only 2 or 3 videos), we chose to not consider them, being replaced by the next available inputs from the above explained list.

Now that we have all responses, we can go further with our experiment by computing the questionnaire. It will contain 25 tables, one for each query, along with the all videos name, directly hyperlinked to the video web page . All results' order was randomized in order to not interfere with the judges' opinion and perspective. For each video, we have two possible answers: relevant or non-relevant.

This procedure consists in involving two judges within the project that will analyze and provide a relevance score based on their expectation. In our case, two native Spanish persons, that are relatively close to the university and educational environment, were additionally involved in validating this algorithm. It is compulsory that the judges should be Spanish due to the fact that all videos are in Spanish. Furthermore, we have provided one sample to each of them. After the both feedback were done and all responses were judged from two different perspectives, we can compose the overall inter rater reliability, measuring agreement among the data collectors and what means for them "relevant" or not.

Overall *Kappa* was computed based on each table, where each data has the same weight, by computing:

- **The proportion of time judges agree (*P(a)*)** by dividing the actual number of agreements by the maximum number of agreements for each query

$$P_{(a)} = \frac{number\ of\ agreements}{total\ number\ possible\ agreements} \quad (1)$$

- **What agreement would be by chance (*P(e)*)**, representing the sum of the products between number of decisions per annotators divided by the total weight of all annotations, divided again by the total.

$$P_{(e)} = \frac{\frac{P^{judge1}_{(relevant)} * P^{judge2}_{(relevant)}}{Total\ annotations} + \frac{P^{judge1}_{(nonrelevant)} * P^{judge2}_{(nonrelevant)}}{Total\ annotations}}{Total\ annotations} \quad (2)$$

- **Cohen's kappa**, K calculation performed according to the following formula

$$\kappa = \frac{P_{(a)} - P_{(e)}}{1 - P_{(e)}} \quad (3)$$

3.2 Values Interpretation

Kappa value can range from −1 to +1, where 0 represents the amount of agreement that can be expected from random chance, and 1 represents perfect agreement between the raters.

Cohen suggested the *Kappa* result be interpreted as follows: values <0 as indicating no agreement and 0.01−0.20 as none to slight, 0.21−0.40 as fair, 0.41−0.60 as moderate, 0.61−0.80 as substantial, and 0.81−1.00 as almost perfect agreement. We have choose to analyze them in 3 categories: good agreement (> 0.80), tentative conclusion (> 0.60) and small value (> 0).

Any agreement less than perfect (1.0) is a measure not only of agreement, but also of the reverse, disagreement among the raters and any *Kappa* below 0.60 indicates inadequate agreement among the raters.

4 Experimental Results

We got both completed questionnaires and put them together in order to analyse the results. For each table, we have prepared 2 separate columns, one for each judge's response. In Table 1, we can visualize an example of annotations for query "calculo de probabilidades" (en. "calculation of probabilities"), where we have 18 responses with same agreement.

Next step is represented by the *P(a)*, *P(e)* and *Kappa* computation for each query.

Explained on the sample table, *P(a)* represents the division between the number of agreements, the number of rows where the first judge's response is in accordance with the second judge's response (18 such of answers) and the total number, which represents the number of rows (20 for sample).

Agreement by chance for p(relevant) is represented by the product between both numbers of relevant annotations by the total number. In the sample case,

Table 1. Sample table with annotations for query "calculo de probabilidades" (en. "calculations of probabilities") - R(Relevant) and NR(Nonrelevant).

No.	query 1 = "calculo de probabilidades"	J1 resp	J2 resp
1	Aprendizaje de HMMs: Algortimo Fordward-Backward *(en. "Learning from HMMs: Fordward-Backward algorithm")*	R	R
2	Cálculo de anclajes a posteriori: Factores que influyen en el cálculo *(en. "Post-hoc anchor calculation: Factors influencing the calculation")*	NR	NR
3	Cálculo de anclajes a posteriori: Normativa - cálculo tracción *(en. "Post-hoc anchor calculation: Regulations - traction calculation")*	NR	NR
4	Cálculo de coordenadas *(en. "Coordinates computing")*	NR	NR
5	Cálculo de Cubiertas *(en. "Covers computation")*	NR	NR
6	Cálculo de penalizaciones *(en. "Penalties computation")*	NR	NR
7	Cálculo de superficies. Método de coordenadas cartesianas *(en. "Calculation of surfaces. Cartesian coordinate method")*	NR	NR
8	CÁLCULO DEL SUPLEMENTO DE FATIGA PARA LA DEFINICIÓN DE ESTÁNDARES DE TRABAJO *(en. "CALCULATION OF THE FATIGUE SUPPLEMENT FOR THE DEFINITION OF WORK STANDARDS")*	NR	NR
9	Cálculo del VAN y la TIR de una inversión con la hoja de cálculo Excel *(en. " Calculation of the NPV and IRR of an investment with the Excel spreadsheet")*	R	NR
10	Convergencia de series: Convergencia absoluta y reordenación *(en. " Convergence of series: Absolute convergence and rearrangement")*	NR	NR
11	Eficacia en procesos de separación por etapas de equilibrio *(en. " Efficiency in equilibrium stage separation processes")*	NR	NR
12	Esquemas Algoritmicos *(en. "Algorithmic Schemes")*	NR	NR
13	Expresiones matemticas escritas en cdigo LaTeX en los exmenes de poliformaT con MathJax *(en. "Mathematical expressions written in LaTeX code in polyform tests with MathJax")*	NR	NR
14	Funcin Find *(en. "Find function")*	NR	NR
15	IMPUESTO SOBRE EL VALOR AADIDO: CLCULO DEL PORCENTAJE DE PRORRATA *(en. "VALUE ADDED TAX: CALCULATION OF THE PRORRATE PERCENTAGE")*	R	NR
16	Modelos de Markov ocultos: Estimacin de Viterbi. *(en. "Hidden Markov Models: Viterbi Estimation")*	R	R
17	Modelos ocultos de Markov: Traza del Algoritmo de Viterbi *(en. "Hidden Markov Models: Trace of the Viterbi Algorithm")*	R	R
18	Optmizacin heurstica mediante aceptacin por umbrales *(en. "Heuristic optimization through acceptance by thresholds")*	NR	NR
19	Resolucin de Balances de Materia con Hoja de Clculo *(en. "Resolution of Material Balances with Spreadsheet")*	NR	NR
20	Transformada de Fourier de Tiempo discreto: EJEMPLOS DE CLCULO DE TRANSFORMADAS *(en. "Discrete Time Fourier Transform: TRANSFORM CALCULATION EXAMPLES")*	NR	NR

there will be 5 points for judge 1 and 3 for second judge and the total will be divided by 40. So, we got a total of 0.375. With same approach for p(nonrelevant), we will have 6.375 as value.

After multiplying them and dividing them by the total number of annotations (in our case 40), that will result what agreement would be by chance (*P(e)*).

Now, applying formula (3), we have a *Kappa* equals to 0.8796. All these computations can be visualized on Table 2. As mentioned in the Chap. 3.2, the "Good agreement" rating provided in the Table 2 on the *Kappa* value is categorized as one of the most efficient values that can be obtained when applying this formula (> 0.80).

Table 2. Confusion matrix and computation details for sample query.

Confusion matrix

No responses	Judge 1	Judge 2
3	Relevant	Relevant
15	Nonrelevant	Nonrelevant
2	Relevant	Nonrelevant
0	Nonrelevant	Relevant

	Current weight	Total	Value
P(A)	18	20	0.9
P(R)	8	40	0.375
P(NR)	32	40	6.375
P(E)			0.1687
Kappa			**0.8796** Good agreement

After computing *Kappa* value for all queries, we can assemble a statistic. Overall, we have from a total of 25 kappa values 11 good agreements (44%), from which 2 of them with perfect score, 8 tentative conclusion (32%) and 6 small values (24%). Now that we have an overall understanding about the inter-rater reliability, we can go backwards, top to bottom, in order to analyze the ML based search algorithm's responses. Based on the fact that we have knowledge about each response belonging and the judges annotation, a relevance weight can be calculated.

In the Table 3, we can observe an overview of the weights provided by both judges on both search engines, sorted descendingly by their *Kappa* score. For observation purposes, only queries with the Kappa index higher than 0.6 were provided. The more annotations both judges agree on, the higher the Kappa score is. Thus, the first query "drenaje de carretera" has a *Kappa* index of 1 due to the fact that there were no relevant results returned on both algorithms, therefore both judges agreed on all annotations. Also, even though some queries return a higher number of relevant annotations, they are assigned a lower *Kappa* score because of their non-agreements (e.g. "recubrimiento textil" query, where there are 14 agreements out of 17).

Table 3. Backwards weights for each algorithm for queries with good agreements and tentative conclusions.

Query	*Kappa*	No of relevant videos by our algorithm	No of relevant videos by simple search
drenaje de carretera	1	0	0
placas alveolares	1	0	2
Control de calidad en ejecucion	0.9428	1	11
estructuras de acero	0.9423	3	12
teorema de ampere	0.9344	1	0
muros pantallas	0.8843	0	14
calculo de probabilidades	0.8796	6	2
transporte de masa	0.8773	0	6
integracion multiple	0.8744	0	4
forjados	0.8649	4	10
campo magnetico	0.8181	1	6
recubrimiento textil	0.7947	14	15
estatica de fluidos	0.75	0	2
electricidad	0.7139	16	12
paginas web	0.7093	12	15
teorema de minimos cuadrados	0.6923	0	5
tratamientos termicos	0.6923	2	3
ciencia de datos	0.6556	10	14
Circuitos	0.6546	13	13

5 Conclusions

In this work, we present a validation result for the output obtained from our natural language processing (NLP) search engine by following the *Kappa* inter-judge agreement methodology. By involving human judging, we were able to interpret the agreement about the relevance of the answers provided by two different search engines with an educational transcripts dataset. We have found that there are many situations in which judges do not agree and decided leave the analysis of this situation as future works. For situations when judges strongly agree (i.e., $kappa \geq 0.8$) and have tentative conclusions (i.e., $0.6 \leq kappa \leq 0.8$) we have observed better results in general for our search system, but not as good as expected. Best results occur when the query contains prepositions (i.e., *de*) which misleads the simple search, but is very well contextualized by our NLP based search mechanism.

Based on the current bench-mark results for good agreements and tentative conclusions values alongside the provided queries, we are ready to improve the search mechanism and provide better output overall. Moreover, the possible improvements can be reported not only on a better integration than the current mechanism of searching, but also understanding and diminishing as much as possible the existing disagreement relative to the algorithm's response. Future works may regard redesign of the data analysis pipeline by integrating alternative algorithms and performing hyper-parameter tuning, as the end goal of this would be to perform a comparison between this approach and the current mechanism.

Also, an area that we plan on developing in the future would be enlarging and diversifying the feedback provided using the *Kappa* inter-judge agreement methodology by increasing the number of judges willing to participate in this activity. This category of judges should be familiar with the project's goals and objectives and can be consisted of both students and teachers, but also persons that are frequently addressing this subject.

Finally, integration of state-of-the-art transformers as embedding method has the potential to better contextualize the queries and therefore produce more relevant responses.

Acknowledgements. This work was partially supported by the grant 135C/ 2021 "Development of software applications that integrate machine learning algorithms", financed by the University of Craiova.

References

1. State-of-the-art multilingual lemmatization. https://towardsdatascience.com/state-of-the-art-multilingual-lemmatization-f303e8ff1a8, Accessed 29 June 2021
2. Blackman, N.J.M., Koval, J.J.: Interval estimation for Cohen's kappa as a measure of agreement. Stat. Med. **19**(5), 723–741 (2000)
3. Aker, A., Petrak, J., Sabbah, F.: An extensible multilingual open source lemmatizer. In: Proceedings of the International Conference Recent Advances in Natural Language Processing, RANLP 2017, pp. 40–45. ACL (2017)
4. Bafna, P., Pramod, D., Vaidya, A.: Document clustering: TF-IDF approach. In: 2016 International Conference on Electrical, Electronics, and Optimization Techniques (ICEEOT), pp. 61–66. IEEE (2016)
5. Basu, S., Yu, Y., Singh, V.K., Zimmermann, R.: Videopedia: lecture video recommendation for educational blogs using topic modeling. In: Tian, Q., Sebe, N., Qi, G.-J., Huet, B., Hong, R., Liu, X. (eds.) MMM 2016. LNCS, vol. 9516, pp. 238–250. Springer, Cham (2016). https://doi.org/10.1007/978-3-319-27671-7_20
6. Bleoancă, D.I., Heras, S., Palanca, J., Julian, V., Mihăescu, M.C.: LSI based mechanism for educational videos retrieval by transcripts processing. In: Analide, C., Novais, P., Camacho, D., Yin, H. (eds.) IDEAL 2020. LNCS, vol. 12489, pp. 88–100. Springer, Cham (2020). https://doi.org/10.1007/978-3-030-62362-3_9
7. Deerwester, S., Dumais, S.T., Landauer, T.K., Furnas, G., Beck, F.D.L., Leighton-Beck, L.: Improvinginformation-retrieval with latent semantic indexing (1988)
8. Galanopoulos, D., Mezaris, V.: Temporal lecture video fragmentation using word embeddings. In: Kompatsiaris, I., Huet, B., Mezaris, V., Gurrin, C., Cheng, W.-H., Vrochidis, S. (eds.) MMM 2019. LNCS, vol. 11296, pp. 254–265. Springer, Cham (2019). https://doi.org/10.1007/978-3-030-05716-9_21

9. Gutiérrez, L., Keith, B.: A systematic literature review on word embeddings. In: Mejia, J., Muñoz, M., Rocha, Á., Peña, A., Pérez-Cisneros, M. (eds.) CIMPS 2018. AISC, vol. 865, pp. 132–141. Springer, Cham (2019). https://doi.org/10.1007/978-3-030-01171-0_12
10. Kastrati, Z., Kurti, A., Imran, A.S.: Wet: word embedding-topic distribution vectors for mooc video lectures dataset. Data Brief **28**, 105090 (2020)
11. Mikolov, T., Sutskever, I., Chen, K., Corrado, G.S., Dean, J.: Distributed representations of words and phrases and their compositionality. In: Advances in Neural Information Processing Systems, pp. 3111–3119 (2013)
12. Ramos, J., et al.: Using TF-IDF to determine word relevance in document queries. In: Proceedings of the First Instructional Conference on Machine Learning, vol. 242, pp. 29–48. Citeseer (2003)
13. Umesh, U.N., Peterson, R.A., Sauber, M.H.: Interjudge agreement and the maximum value of kappa. Educ. Psychol. Meas. **49**(4), 835–850 (1989)
14. Zhu, H., Dong, L., Wei, F., Qin, B., Liu, T.: Transforming wikipedia into augmented data for query-focused summarization (2019). arXiv:1911.03324

Multi Language Application of Previously Developed Transcripts Classifier

Theodora Ioana Dănciulescu[1], Stella Heras[2], Javier Palanca[2], Vicente Julian[2], and Marian Cristian Mihăescu[1](✉)

[1] Faculty of Automatics, Computers and Electronics, University of Craiova, Craiova, Romania
cristian.mihaescu@edu.ucv.ro

[2] Valencian Research Institute for Artificial Intelligence, Universitat Politècnica de València, València, Spain
{sheras,jpalanca,vinglada}@dsic.upv.es

Abstract. Developing classification models and using them on another similar data-set represents a challenging task. We have adapted an existing data analysis pipeline that classified Spanish educational video transcripts from Universitat Politècnica de València (UPV) to process English video transcripts from TedTalks. Performed adaptation and experimental results were performed also in educational context as in the initial study. We found that the process needs minor adaptations of the data analysis pipeline, but the overall results is highly dependent of the size of the data-set, and especially on the class balance. Parametrisation of the pipeline may open the way of deploying the model in other similar contexts by transfer learning.

Keywords: Classification of transcripts · TedTalks

1 Introduction

Over the last few years, the issue of proper keyword extraction has been tackled by academics from different areas (information retrieval [13], natural language processing [3], semantic web [8], etc.), each proposal attempting to tackle the challenge to elicit a strong collection of words that succinctly reflect a text's substance. This work is framed into this context and tries to provide a method to automatically detect the main topic of an e-learning object (e.g. video lectures, talks, articles, etc.).

The problem of transcript classification represents a particular case of text classification. We have previously developed a transcript classification model on a data-set from UPV media website [2,14] that performed very well.

We asked whether the same data analysis pipeline may be successfully used on another data-set which consists also from transcripts, but is in another language (i.e., English) and comes from another source (i.e., TedTalks). The solution was

H. Yin et al. (Eds.): IDEAL 2021, LNCS 13113, pp. 128–136, 2021.
https://doi.org/10.1007/978-3-030-91608-4_13

to add new steps in the data analysis pipeline and perform various adjustments that deal with the structure and specificity of the new data-set. The experimental approach followed almost the same workflow as previous one.

As stated in the previous paper [2], choosing the clusters represents a key point in a classification problem. Choosing a wide level of generalisation for the clusters: Biology & Sciences, Humanities and Engineering, we offered our experiment a big range of applicability regardless of the previous academic context. When choosing TedTalks as data-set source for adapting the existing pipeline, the focus is on the topics that are generally covered in TedTalks which can be easily classified in the same three clusters chosen for the context of UPV academic courses.

The current research is aiming to prove the applicability of an existing semi-supervised classification pipeline developed in a previous paper [2] by extending the applicability domain in terms of data-set context, language and data-set size. Sticking to the original pipeline idea, a new training data-set and validation data-set were formed by means of web-scrapping from Wikipedia articles (training data-set) and validating using Kaggle data-set from TedTalks (validation data-set).

Due to the fact that the data-sets emerge from open-source platforms or manual work versus the in-production environment source of the data from the previous work, the size of the current data-sets is much smaller. Besides the different contexts of the data-sets and their size, the major change brought to the pipeline developed in this work is that it classifies English transcripts in contrast with the Spanish transcripts used in the previous work.

The main contributions for this paper are:

1. Forming the training data-set by means of scrapping articles using open-source Wikipedia API. All the data was collected with respect to the three clusters of interest: Biology&Sciences, Humanities and Engineering.
2. Data analysis on the TedTalks data set in order to establish its suitability for classification using the three clusters mentioned above
3. Adapt specific libraries for pre-processing text from using Spanish to using English
4. Train the pipeline using the data-set gathered from Wikipedia and label the TedTalks transcripts with one of the three clusters.

2 Related Work

Among the different approaches proposed in the literature for automatic keyword extraction, a general classification [13] groups them into those that are based on rules, statistical methods, domain-specific expert knowledge (e.g. ontologies), and machine learning algorithms. For instance, [12] used a technique based on a regular expression grammar rule approach to identify the noun chunks in the text of the transcript of educational videos. The seminal work of [9] proposed a technique based on the co-occurrence statistical information of frequent terms.

[5] seeks to enhance the retrieval and accessibility of learning objects by using domain-specific ontologies to integrate semantic information.

From the machine learning approaches for keyword extraction, Support Vector Machines has been one of the most used and successful techniques [16]. [6] provided a solution that combines supervised and unsupervised learning and is built on a graph-based syntactic representation of text and online pages. Similarly, [4] presented an unsupervised keyword extraction strategy that incorporates many distinct approaches to the traditional TF-IDF model as well as acceptable heuristics. The Rapid Automatic Keyword Extraction (RAKE) work reported in [11], used unsupervised algorithms for extracting keywords that are domain- and language-independent.

Another of the most often used machine learning approaches for classifying documents according to a collection of themes is the Latent Dirichlet Allocation (LDA) model. One example is the work proposed in [1], which uses a non-Markov on-line LDA sampler topic model to automatically capture theme patterns and identify emergent topics. The LDA model has been utilized in the online educational area in works such as [15], where the authors use topic identification for the analysis of student comments supplied in online courses. Also, [7] tackled the difficulty of topic detection by finding terms that appear often in one topic but not in others.

Unsupervised learning approaches are widely employed for keyword extraction [10]. Unlike most related work, our technique is completely semi-supervised, requiring neither a previously labeled database nor an ontology as a source of ground truth for model training. Furthermore, to our knowledge, our model has been taught to classify a multilingual (English and Spanish) collection of e-learning objects automatically.

3 Data Analysis Pipeline

Taking as a starting point the pipeline described in [14] and continuing with the improvements brought by increasing the training set and more importantly the quality of the factors that influence the classification score, also described in [2], the general approach is suitable to be adapted to the English language. The libraries and the pre-processing methods used are open to extension by supporting the English language.

In Fig. 1 we contoured the extensions added to the original pipeline in the following way. The squares outlined in **black** represent the core ideas introduced in the initial pipeline from [14]. The squares outlined in **orange** represent the new steps added to the pipeline in order to improve previous results in [2]. The squares outlined in **green** represent the new approach regarding the adjustment of the pipeline for the English language.

The English data-set is represented by a free data-set extracted from Kaggle containing video data extracted from TedTalks. In contrast with the Spanish data-set that contained production data provided by inexperienced users, the English data-set contains much more relevant data, refined by experienced users.

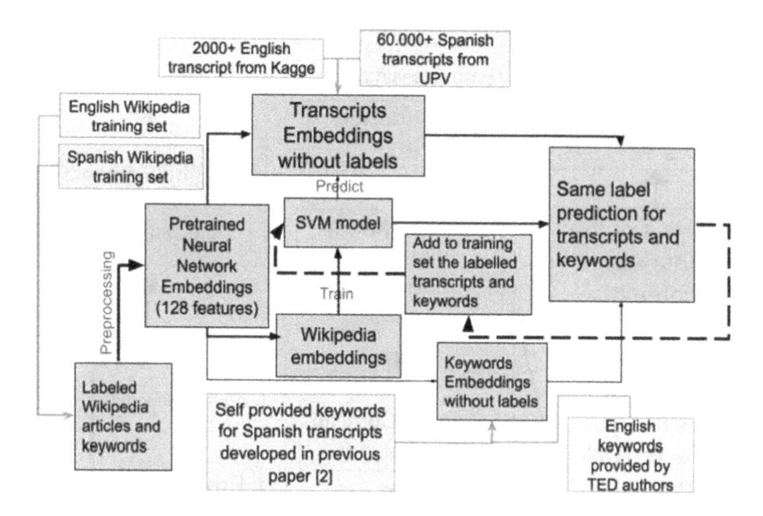

Fig. 1. Overview of the extensions added to the original pipeline (Color figure online)

First important observation is the much smaller size of the English data-set. This fact is directly reflected in the training phase of the model. The distribution on categories in the English data-set is also not balanced and this fact has to be taken into consideration when downloading and forming the training data-set with articles gathered from Wikipedia by using their research oriented API.

Offering a perspective of the training set of Wikipedia articles formed for the English data-set, we have a total of 3727 Wikipedia articles extracted from 97 Wikipedia categories. The distribution of articles in terms of balancing the training data-set is 612 articles from Biology, Natural Sciences, Anatomy topics, 907 Wikipedia articles from Engineering, Mathematics, Statistics topics and 2208 Wikipedia articles from Arts, Philosophy, Law, Psychology topics.

In order to have a relevant data-set for training, we used the information about the general purpose of TedTalks - they are adopting themes of general interest that can be approached by experienced listeners but also by the ordinary listener that does not have experience in the approached theme. This context information gives us a good sense about the terms used in this data-set which are not so technical oriented, nor very specific to a domain as they were in the academic context. Furthermore, from the previous research we noticed that the transcripts that were oriented to the Arts and Humanities cluster are harder to detect even in an academic context where the terms are very specific for the other two clusters: Biology and Engineering. By exploiting these findings from the TedTalks context and comparing it to the previous academic context, we concluded that there is a clear need of downloading a much bigger quantity of articles oriented to the Arts and Humanities cluster in order to diminish the collapse with the other two clusters which are lacking the previous advantage from the academic Spanish data-set: very domain specific terms in technical and scientific fields. The method used for classifying the transcripts, as described in

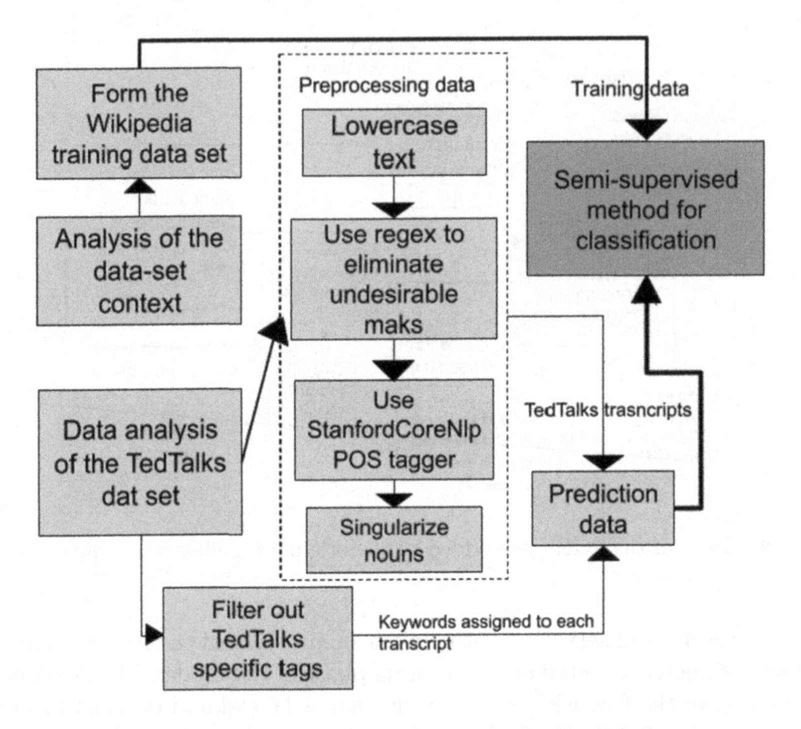

Fig. 2. Classification pipeline adjustment for English data-set

the previous works [14], is focused on two main factors: the transcript and the keywords of the specific transcript. In other words, for a certain learning object the classification cluster determination has equal influence from the transcript's text and from the transcript's keywords. Firstly a cluster is assigned with respect to the transcript's text and then the same process is repeated with respect to the transcript's keywords. If the two clusters assigned are identical in both cases, then the classification is validated and the transcript is added to the training set of the method.

Taking a look at the English data-set and analysing the tags associated to the videos, we deduced that a method for extracting better keywords from transcripts is not needed in this case as the keywords assigned to each video are very suggestive. In addition, methods for cleaning and pre-processing the transcripts and keywords were developed in order to obtain a set of tags general enough to satisfy the classification within the three clusters without losing the level of generalisation. Words associated to the TedTalks specific terms were removed, part of speech tagging techniques were applied on the transcripts, stop-words and unwanted symbols were eliminated.

4 Experimental Results

After running the experiments with the Spanish data-set from UPV and obtaining satisfactory results presented in a previous paper, the next step was to generalise and confirm the used pipeline with a different data-set. The main challenge is to prove the same pipeline's effectiveness is maintained when using English transcripts from a non-academic source.

The chosen data-set is acquired from Kaggle.com and consists of educational video transcripts from TedTalks conferences presented in English. The data-set contains 2550 entries which is a visible limitation from the very beginning, taking into consideration the big difference between the Spanish data-set size and English data-set size (15.000 vs 2.000).

The data-set is formed by two files: 'ted_main' - this file contains 17 columns, but the columns of interest in our algorithm are tags and URL. The tags column is one of the most important factors serving a major part in the classification algorithm. The URL column is also of great interest because it is used to link the data provided in this file to the transcripts that are provided in the second file, URL is used as a key in this case to correlate the data from the two files.

As a first step, it was checked if all the data of interest is present in every entry, such that we know if the data should be filtered in order to provide the pipeline with valid data. Analyzing the data-set we observed that all the entries contain values regarding URL and tags, which means that all the data can be safely used in the semi-supervised learning method. The second step was to analyse the topic distribution in the TedTalks data-set in order to have a brief validation that this data-set can be classified in the same three topics used for the academic data-set: Biology, Engineering and Humanities, not losing the level of generalisation.

For this we made use of the theme column from this file in order to have an overview of the most frequent themes approached in the current transcripts. As a first observation, we saw that there are 416 different themes defined in the data-set and their distribution. Performing data analysis methods we can observe that topics of technology, science, global issues and culture are the most frequent topics. Having this perspective, the themes distribution is feasible to the three clusters classification: Biology, Engineering and Humanities.

As stated, in our pipeline, at every iteration it is tried to classify the learning objects based on the transcript and on the keywords. If classification based on transcript and classification based on transcripts' keywords match, then the transcript is added with its now found label to the training data-set. In other words the training data-set is increased by each iteration with respect to the number of transcripts validated as labeled.

Given the much smaller size of the English data-set in comparison with the Spanish data-set - the English data-set represent 1/7 of the Spanish data-set in terms of processable number of learning objects, and also given the training set formed with the 3727 English Wikipedia which is very close to the 3747 Spanish Wikipedia articles set formed for training the pipeline for the Spanish data-set, it was expected that the pipeline for the English data-set will need less iterations to finish the classification. In Table 1 we can notice that the number of iterations was reduced to 3 instead of the average 10 iterations needed for the Spanish data-set. In the very first iteration most of the transcripts are already labelled. This fact is due to the very small size of the prediction data-set in contrast with the training set.

Table 1. Validation scores at each iteration for the English data-set

Iteration (valid /available)	Accuracy	Class	Precision	Recall	F1-score
#1 2386/2467	0.82 (+/− 0.03)	Biology & Sciences	0.77	0.56	0.65
		Engineering	0.84	0.70	0.76
		Humanities & Arts	0.82	0.94	0.87
#2 39/81	0.90 (+/− 0.03)	Biology & Sciences	0.78	0.55	0.65
		Engineering	0.85	0.68	0.76
		Humanities & Arts	0.81	0.94	0.87
#3 13/42	0.90 (+/− 0.03)	Biology & Sciences	0.75	0.53	0.62
		Engineering	0.85	0.68	0.76
		Humanities & Arts	0.80	0.94	0.87

The size of the training data is comparable to the size of the training data used for a prediction set 10 times bigger than the current data-set. The main advantage of the semi-supervised method is that it is learning from its own results: once some transcripts are validated as labeled, then the labeled transcript is added to the training set. In this case, when the prediction set is very small, the semi-supervised method is not benefiting from the learning in each iteration.

Having these stated, and considering the small size of the data-set, the mean accuracy obtained after the three iterations - 0.99% - is impressive. Despite the very small size of the data-set which is a serious impediment for the semi-supervised classification method, the results are satisfactory.

In this case, increasing the training data-set is not representing a valid solution as the training data-set is already appropriate to the prediction data. The transcripts are rapidly classified since the first iteration. Another factor here that directly influences the classification are the tags associated with each transcript. The fact that so many transcripts are labelled since the first iteration represents a strong confirmation that the tags are a very good reflection of the transcript's content.

Considering the limitation of the data-set size, which is a general limitation in the domain of trainable algorithms, the semi-supervised method results with the English data-set are comparable with the results with the Spanish data-set.

Therefore, the pipeline generalisation experiment in a context which touches the constant limitation of lacking sufficient data, proves to be a success. Not only the easiness to adapt to the English language but also the good results on small data validate the quality and potential of the presented pipeline.

5 Conclusions

Finally, as the experiments in the previous papers were presenting satisfactory results and outlined important findings and observations in the classification process for educational transcripts in Spanish, we continued with an extension for the English language. One of the main purposes for the extension was the need of validation for the developed classification pipeline. The strategy of extension consisted in running the same pipeline with a completely different data-set in terms of size, context, structure and language by simply adapting the used libraries for language processing to the English language. Considering that the set used in the semi-supervised learning used in the previous paper [2] is 20 times bigger than the set used in the current approach and that the current scores are close to the previous ones, this states the general applicability of the developed pipeline. Despite the changed parameters and the big differences found between the two data-sets, the experimental results confirmed that the classification method is easily adaptive and keeps its effectiveness.

Considering the major growth of online learning methodologies and the widely spread English language used for online teaching and online knowledge sharing, the need for generalisation in the classification method was natural. Moreover, as observed in the data analysis of the data-set extracted from UPV, we could notice a good quantity of videos presented in English. Since the last analysis of the UPV data-set, more videos were added, many of them being presented in English. Also, in the meantime, a recommending API was put in place for the UPV platform. This API uses a recommending system based on the classification method presented in the previous paper in order to generate more relevant results to the UPV students. This fact can represent a clear direction for future extensions as there can be developed a method for detecting the presentation language of a video and then using the proper classification method for the detected language.

Also, as a starting point for future works we may focus on different machine learning methods, other state-of-the-art methods in order to predict and assign labels to transcripts. Including also the analysis of how different methods would affect the results, what features of the both data-sets used are common and can be adapted to a different model will represent strong parts in the future accomplishments to improve the actual results keeping the multi language feature.

Open source software is freely available at https://github.com/Rec-Sys-for-Spanish-Educational-Videos/A-Semi-supervised-Method-to-Classify-Educational-Videos/tree/EnglishDataSet. The data-set can be found here https://www.kaggle.com/rounakbanik/ted-talks.

Acknowledgements. This work was partially supported by the grant 135C/ 2021 "Development of software applications that integrate machine learning algorithms", financed by the University of Craiova.

References

1. AlSumait, L., Barbará, D., Domeniconi, C.: On-line LDA: adaptive topic models for mining text streams with applications to topic detection and tracking. In: 2008 Eighth IEEE International Conference on Data Mining, pp. 3–12. IEEE (2008)

2. Danciulescu, T.I., Mihaescu, M.C., Heras, S., Palanca, J., Julian, V.: More data and better keywords imply better educational transcript classification? In: Proceedings of The 13th International Conference on Educational Data Mining (EDM 2020), pp. 381–387. ERIC (2020)
3. Hulth, A.: Combining machine learning and natural language processing for automatic keyword extraction. Ph.D. thesis, Institutionen för data-och systemvetenskap (tills m KTH) (2004)
4. Lee, S., Kim, H.J.: News keyword extraction for topic tracking. In: 2008 Fourth International Conference on Networked Computing and Advanced Information Management, vol. 2, pp. 554–559. IEEE (2008)
5. Lemnitzer, L., et al.: Improving the search for learning objects with keywords and ontologies. In: Duval, E., Klamma, R., Wolpers, M. (eds.) EC-TEL 2007. LNCS, vol. 4753, pp. 202–216. Springer, Heidelberg (2007). https://doi.org/10.1007/978-3-540-75195-3_15
6. Litvak, M., Last, M.: Graph-based keyword extraction for single-document summarization. In: Proceedings of the workshop on Multi-source Multilingual Information Extraction and Summarization, pp. 17–24. Association for Computational Linguistics (2008)
7. Liu, T., Zhang, N.L., Chen, P.: Hierarchical latent tree analysis for topic detection. In: Calders, T., Esposito, F., Hüllermeier, E., Meo, R. (eds.) ECML PKDD 2014. LNCS (LNAI), vol. 8725, pp. 256–272. Springer, Heidelberg (2014). https://doi.org/10.1007/978-3-662-44851-9_17
8. Martinez-Rodriguez, J.L., Hogan, A., Lopez-Arevalo, I.: Information extraction meets the semantic web: a survey. Semant. Web **11**(2), 255–335 (2020)
9. Matsuo, Y., Ishizuka, M.: Keyword extraction from a single document using word co-occurrence statistical information. Int. J. Artif. Intell. Tools **13**(01), 157–169 (2004)
10. Nasar, Z., Jaffry, S.W., Malik, M.K.: Textual keyword extraction and summarization: State-of-the-art. Inf. Process. Manage. **56**(6), 102088 (2019)
11. Rose, S., Engel, D., Cramer, N., Cowley, W.: Automatic keyword extraction from individual documents. Text Mining Appl. Theory **1**, 1–20 (2010)
12. Shukla, H., Kakkar, M.: Keyword extraction from educational video transcripts using NLP techniques. In: 2016 6th International Conference-Cloud System and Big Data Engineering (Confluence), pp. 105–108. IEEE (2016)
13. Siddiqi, S., Sharan, A.: Keyword and keyphrase extraction techniques: a literature review. Int. J. Comput. Appl. **109**(2) (2015)
14. Stoica, A.S., Heras, S., Palanca, J., Julian, V., Mihaescu, M.C.: A semi-supervised method to classify educational videos. In: Pérez García, H., Sánchez González, L., Castejón Limas, M., Quintián Pardo, H., Corchado Rodríguez, E. (eds.) HAIS 2019. LNCS (LNAI), vol. 11734, pp. 218–228. Springer, Cham (2019). https://doi.org/10.1007/978-3-030-29859-3_19
15. Unankard, S., Nadee, W.: Topic detection for online course feedback using LDA. In: International Symposium on Emerging Technologies for Education, pp. 133–142. Springer (2019). https://doi.org/10.1007/978-3-030-38778-5_16
16. Zhang, K., Xu, H., Tang, J., Li, J.: Keyword extraction using support vector machine. In: Yu, J.X., Kitsuregawa, M., Leong, H.V. (eds.) WAIM 2006. LNCS, vol. 4016, pp. 85–96. Springer, Heidelberg (2006). https://doi.org/10.1007/11775300_8

A Complexity Measure for Binary Classification Problems Based on Lost Points

Carmen Lancho[1(✉)], Isaac Martín de Diego[1], Marina Cuesta[1], Víctor Aceña[1,2], and Javier M. Moguerza[1]

[1] Data Science Laboratory, Rey Juan Carlos University, C/Tulipán, s/n, 28933 Móstoles, Spain
{carmen.lancho,isaac.martin,marina.cuesta,victor.acena, javier.moguerza}@urjc.es
[2] Madox Viajes, C/de Cantabria, 10, 28939 Arroyomolinos, Spain
https://www.datasciencelab.es
https://www.madoxviajes.com

Abstract. Complexity measures are focused on exploring and capturing the complexity of a data set. In this paper, the *Lost points* (*LP*) complexity measure is proposed. It is obtained by applying k-means in a recursive and hierarchical way and it provides both the data set and the instance perspective. On the instance level, the *LP* measure gives a probability value for each point informing about the dominance of its class in its neighborhood. On the data set level, it estimates the proportion of lost points, referring to those points that are expected to be misclassified since they lie in areas where its class is not dominant. The proposed measure shows easily interpretable results competitive with measures from state-of-art. In addition, it provides probabilistic information useful to highlight the boundary decision on classification problems.

Keywords: Complexity measures · Neighborhood measures · Binary classification · Supervised machine learning

1 Introduction

Supervised classification entails an important part of Machine Learning algorithms. When addressing a supervised classification problem, the more information is known from data, the better for the classification task and for the achievement of a performance as good as possible. There is a wide range of factors that may be detrimental to the performance of a classifier such as the overlap among classes, the quantity of noise, the density and distribution of classes, etc. These factors do not uniformly affect to all classifiers. A previous analysis of data complexity is really useful for the selection of the classifier to apply and to have suitable performance expectations. To deal with this previous

H. Yin et al. (Eds.): IDEAL 2021, LNCS 13113, pp. 137–146, 2021.
https://doi.org/10.1007/978-3-030-91608-4_14

exploratory analysis of data, several complexity measures have been proposed mainly from the original work in [5]. The set of existing complexity measures approaches the distinct aspects affecting the classification by focusing on the different data characteristics. However, a consolidated interpretation of a bunch of complexity measures is not straightforward [6].

In the present paper, the *Lost points* (LP) complexity measure is introduced. It measures the proportion of points in areas where its class is not dominant. That is, critical points that could potentially have an impact on the performance of classifiers. It is easily interpretable, classifier-independent and valuable as a guide for the classification task. In addition, LP is used, on an instance level, to provide a *dominance probability* value used to detect the most uncertain areas.

The rest of the paper is structured as follows. Section 2 introduces the neighborhood complexity measures where the LP measure fits in. The proposed method is described in Sect. 3. Experiments are detailed in Sect. 4. Finally, Sect. 5 concludes and states further research lines.

2 State of the Art

Following the classification presented in [6], data complexity measures can be grouped in 6 categories: feature-based measures, linearity measures, neighborhood measures, network measures, dimensionality measures and class imbalance measures. The present work is framed in the neighborhood ones that study the distribution of classes in local neighborhoods trying to capture information about class overlap and borderline or noisy points. The most common measures are:

- $N1$. Fraction of borderline points based on a Minimum Spanning Tree. It calculates the size and complexity of the decision boundary by identifying points in overlapping areas or on the border.
- $N2$. Ratio of intra/extra class nearest neighbor distance. First, it is computed the ratio r of the sum of the distances between each point and its closest neighbour (intra-class) and the sum of the distances between every point and its closest neighbor from other class (extra-class). $N2$ is defined as $r/(1+r)$.
- $N3$. A leave-one-out estimation of the 1 nearest neighbor classifier error rate.
- $N4$. Non-linearity of the nearest neighbor classifier. Test points are generated by linear interpolation between random pairs of points of the same class. Then, the error of nearest neighbor classifier is evaluated on that test set.

These measures take values between 0 and 1. Higher values indicate more complex problems. Notice that they were originally designed for the data set level but, for some of them, its instance level can be defined [2,6].

Regarding measures contemplating more than one level of information, the *instance hardness* defined as the likelihood, for a point, to be misclassified based on a set of learning algorithms using a leave-one-out procedure is presented in [9]. Furthermore, *hardness measures* are defined to identify why points are hard to classify. These measures can be averaged to get a global complexity data set value. The k-Disagreeing Neighbors (kDN) of a point is a *hardness measure* defined as the percentage of its k nearest neighbors that do not share its class.

The *R-value*, a measure focused on the existing overlap among classes is presented in [7]. A point is in an overlapping area if more than θ of its k nearest neighbors are from a different class. The *R-value* of a data set is the ratio of points lying in overlapping zones.

3 Proposed Method

The *LP* is a complexity measure able to provide information in two different levels: instance and data set. In the first one, a probability of dominance of a class from the perspective of every point is given. This will be called the *dominance probability.* In the second level, it is estimated the total proportion of lost points. Lost points are points expected to be misclassified because they are in areas where its class is not dominant: a point in an overlapping area where its class is not dominant or a borderline point more similar to the opposite class. Both overlapped and borderline points have a negative impact on classifier accuracy. Thus, the *LP* measure quantifies the relative importance of critical areas in a data set. Next, its calculation is detailed.

First, for every point, the *dominance probability* has to be obtained. For that purpose, a previous analysis of the data structure and the distribution of classes is needed. To tackle this task, the k-means algorithm is used. From an exploratory point of view, if the k-means algorithm is applied to a supervised data set and, for each cluster, the probability of every class is extracted, an informative class data structure map can be achieved [1]. Not only this map will reveal where a class is dominant, but it will also point out the most uncertain areas where classifiers tend to fail. To ensure a good selection of the number of clusters k and to guarantee robust partitions capturing the structure of data and interactions among classes, the k-means algorithm is hierarchically and recursively performed following [10] (see Fig. 1). In this recursive process, the different k-means applications will be denoted as *layers.* In the first layer, k-means are implemented using the whole data set and in following iterations, data input is the set of centroids from previous step. Every time the k-means is performed, a partition of clusters is achieved. The idea is to capture the behavior of classes through recursive partitions and to get successive clusters revealing how points from different classes are grouped.

In every layer, for every original data point, the probability of its class in its cluster is stored. This probability is the proportion of the class in a specific cluster based on the original points that belong to it. Since the procedure is hierarchical, it is straightforward to get to which cluster of any layer an original point belongs. In every iteration, the probability vector is averaged with the probability vector from previous layers. This vector is the *dominance probability.* Hence, it summarizes, from the perspective of every point, the dominance of its class through the variety of clusters where the point has been grouped. After this process, all points have an assigned probability value between 0 and 1. A high probability means the corresponding point has been, across all partitions, surrounded by points of its class. If its probability is low, the point lies in areas where its class is not dominant.

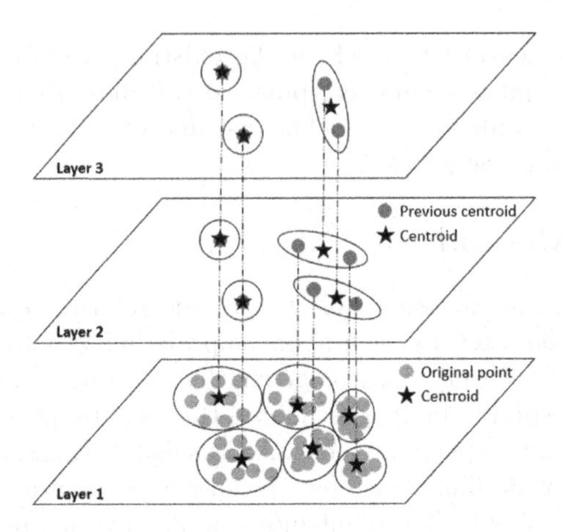

Fig. 1. Hierarchical and recursive application of k-means based on [10].

Algorithm 1. *LP* measure algorithm

Require: X, Y, σ, δ, k_{min}
Ensure: $p_{dominance}$, LP
1: $min_k = \max(2, k_{min})$
2: $p = (0, ..., 0)$ ▷ Initialization vector p to accumulate probabilities from all layers
3: $num\_layers = 1$
4: $k = \lfloor n/\sigma \rfloor$ ▷ k is the number of clusters
5: $X' = X$ ▷ X is always original data
6: **while** $k \geq min_k$ **do**
7: $kmeans(X', k) \rightarrow X_K, B_1, ..., B_k$ ▷ X_k: centroids, $B_1, ..., B_k$: clusters
8: $\forall x_i \in X : p_i += proportion\ of\ class\ y_i\ in\ cluster\ B(x_i)$
9: $X' = X_k$ ▷ New data X are centroids
10: $k = \lfloor k/\sigma \rfloor$ ▷ Update k
11: $num\_layers += 1$
12: **end while**
13: $p_{dominance} = p/num\_layers$ ▷ Dominance probability vector
14: $lost\_vector = I_{p_{dominance} \leq \delta}$ ▷ Binarization
15: $LP = mean(lost\_vector)$ ▷ LP calculation

In a second and final step, the *dominance probability* is binarized using a probability threshold δ: if the probability is equal or lower than δ, the point is expected to be lost and its binary value will be 1. Otherwise, if the probability is greater than δ, it will be binarized to 0 since the point lies in areas where its class is better represented. This binary information is averaged to achieve the *LP* measure for the whole data set. Thus, the *LP* measure expresses the proportion of points that are expected not to be correctly identified from its class since they are in regions where its class is not strong enough. It is defined between 0 and 1, which eases the interpretation. The maximum value is reached when the *dominance probability* of all points is lower or equal to δ.

Algorithm 1 presents the pseudocode to obtain the *LP* measure. Let be $X = \{x_i\}_{i=1}^n$ a data set, and $Y = \{y_i\}_{i=1}^n$ the corresponding binary labels. $\{B_1, .., B_k\}$ is a partition of k clusters, where $B(x_i)$ is the cluster to which point x_i belongs. Let be $p = (p_1, ..., p_n)$ the probability vector, where p_i is the probability of y_i in the cluster $B(x_i)$. Note that in all layers expect for the first one, the hierarchical structure has to be used to extract the belonging cluster of every original point. The method holds 3 parameters: (1) the probability threshold δ, (2) the proportion of grouped points per cluster σ which determines the number of clusters k in every layer and (3) k_{min}, which is the minimum number of clusters allowed by the user. The purpose of σ is to set the pace of grouping in the recursive k-means process. The iterative process stops when the following k is going to be lower than 2 (number of classes) or k_{min}. The final results come from this last layer.

4 Experiments

In this Section the *LP* measure is evaluated on a variety of simulated and real data sets for binary classification problems. In the first experiment, the performance of the *LP* measure has been compared with state-of-the-art alternatives. In the second experiment, the *dominance probability* is used to detect the most uncertainty areas on two simulated data sets. For all experiments, parameters' values are fixed: $k_{min} = 2$ since only binary classification problems are addressed, $\delta = 0.5$ as the default threshold for probabilities, and $\sigma = 5$ as an intermediate value. Smaller σ values do not correctly capture the data structure in the first layer and higher values minimize the number of layers. Higher values can also lose the data structure in the last layers due to the high number of clusters that they assemble. Moreover, all data sets are previously standardized and the Euclidean distance is used.

4.1 Performance of *LP* measure

To analyze the relative performance of the *LP* measure, 10 artificial and 10 real data sets have been used. The proposed measure have been compared with $N1$, $N2$, $N3$, $N4$ (obtained from [4]), *R-value* (using $k = 7$ and $\theta = 2$ following [7]) and kDN (with $k = 5$ following [9]). To evaluate the capacity of the measure to estimate complexity, a total of 8 machine learning algorithms have been used (*Support Vector Machine* (*SVM*) with linear kernel, *SVM* with *RBF* kernel, *Random Forest* (*RF*), *Multiple Layer Perceptron* (*MLP*), *XGBoost*, *k-Nearest Neighbour* (*kNN*), *Decision Tree* (*DT*) and *Logistic Regression* (*LR*)). The corresponding parameters have been selected using 5-fold cross validation and a grid search maximizing the balanced accuracy.

The 10 artificial data sets are composed of 2 bivariate normal distributions of 4500 points each with different degrees of overlap and density (see Fig. 2). In this case, the overlap of the distributions can be calculated. Thus, given two normal probability density functions $f(x)$ and $g(x)$, their overlap [11] is defined as:

$$overlap(f,g) = \int_{-\infty}^{\infty} \min\{f(x), g(x)\}dx.$$

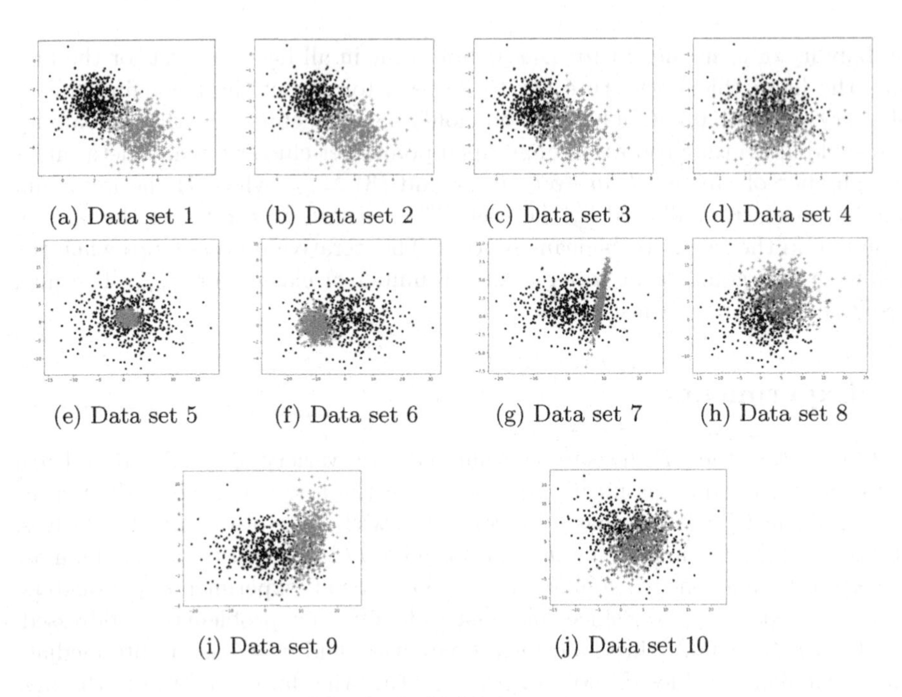

Fig. 2. Artificial data sets.

Table 1. Complexity measures, overlap and best classifier error for artificial data.

Data	*LP*	*N1*	*N2*	*N3*	*N4*	*R-value*	*kDN*	Overlap	Error	Best classifier
1	0.035	0.076	0.117	0.052	0.027	0.039	0.053	0.071	0.042	*Linear SVM*
2	0.058	0.121	0.162	0.084	0.049	0.063	0.083	0.114	0.063	*LR*
3	0.100	0.223	0.253	0.156	0.110	0.118	0.154	0.210	0.110	*MLP*
4	0.252	0.542	0.438	0.378	0.338	0.318	0.373	0.575	0.288	*MLP*
5	0.086	0.192	0.235	0.131	0.128	0.089	0.130	0.160	0.084	*MLP*
6	0.099	0.213	0.244	0.146	0.114	0.105	0.145	0.180	0.101	*SVM RBF/XGBoost*
7	0.048	0.096	0.161	0.061	0.049	0.040	0.065	0.065	0.045	*SVM RBF*
8	0.186	0.407	0.375	0.284	0.245	0.225	0.282	0.390	0.194	*MLP*
9	0.125	0.260	0.274	0.176	0.132	0.139	0.180	0.246	0.119	*Linear SVM*
10	0.253	0.539	0.421	0.371	0.351	0.320	0.367	0.565	0.285	*MLP*

The 10 real data sets have been obtained from [3] with a range of sizes from 306 up to 19020. In all cases, complexity measures are only applied to the training set [5]. Thus, for artificial data, 6000 points are destined to training and 3000 to testing. Real data sets are split in training (70%) and testing (30%). All results, except the classification error, are computed on the training set.

Tables 1 and 2 show the results for the complexity measures for artificial and real data sets, respectively. The best classifier and the corresponding error (1-accuracy) of classification are presented. A graphical comparison of all values is presented in Fig. 3.

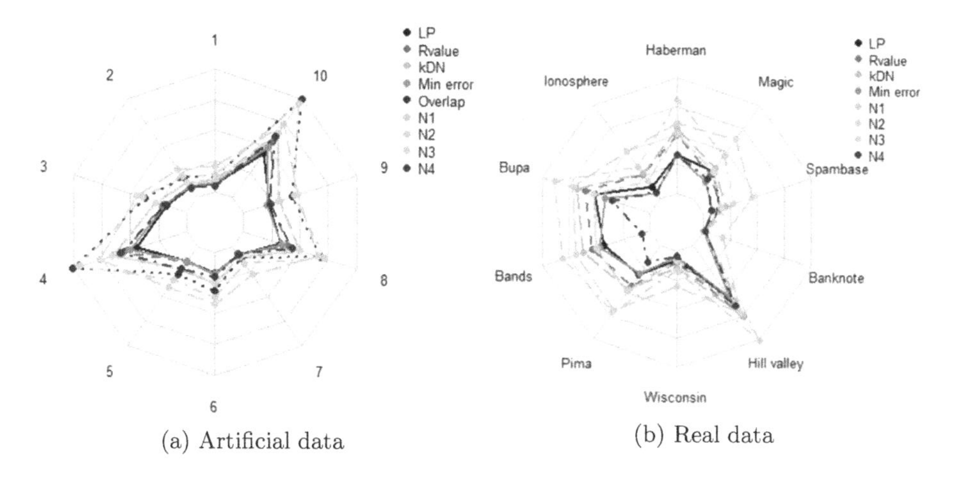

(a) Artificial data

(b) Real data

Fig. 3. Comparison of complexity measures, overlap and error.

Table 2. Complexity measures and best classifier error for real data.

Data	*LP*	*N1*	*N2*	*N3*	*N4*	*R-value*	*k*DN	Error	Best classifier
Haberman	0.215	0.519	0.394	0.332	0.215	0.327	0.362	0.217	*SVM RBF*
Ionosphere	0.078	0.220	0.329	0.122	0.041	0.167	0.171	0.038	*SVM RBF*
Bupa	0.332	0.573	0.464	0.340	0.228	0.390	0.441	0.269	*RF*
Bands	0.275	0.522	0.442	0.318	0.051	0.345	0.399	0.300	*RF*
Pima	0.207	0.456	0.454	0.324	0.116	0.283	0.309	0.195	*SVM RBF*
Wisconsin	0.052	0.111	0.200	0.052	0.031	0.065	0.092	0.034	*SVM RBF*
Hill valley	0.363	0.656	0.465	0.429	0.413	0.491	0.487	0.357	*LR*
Banknote	0.009	0.012	0.115	0.001	0.007	0.001	0.002	0.000	*SVM RBF/kNN*
Spambase	0.096	0.185	0.295	0.104	0.047	0.102	0.134	0.054	*MLP*
Magic	0.192	0.301	0.412	0.192	0.135	0.168	0.211	0.116	*XGBoost*

Table 1 reveals that *LP* measure offers, for all artificial data sets, a value that tightly adjusts to the error. It achieves to quantify the proportion of critical points that are affecting classifiers. Notice that the maximum difference between *LP* and the error is 0.036 (for data set 4). Thus, *LP* is able to anticipate what to expect from classifiers according to the distribution of classes. Regarding the other measures, $N1$ and $N2$ are the two measures that differ the most from the error and are, in fact, closer to the overlap. Even though *R-value* is defined as the ratio of elements in overlapping areas, it is more similar to the error than to the overlap. *R-value*, $N3$, $N4$ and kDN are quite close to the error (see Fig. 3a) but they do not reach the performance of *LP*.

Table 2 and Fig. 3b reflect that, for real data sets, similar patterns can be found but differences become more noticeable. The *LP* measure again stands out for its proximity to the error. In this case, the maximum difference between them is of 0.076 (for Magic data set) and, in general, *LP* subtly overestimates the proportion of critical points but remains an useful and interpretable reference information for the classification task. While *R-value*, $N3$ and kDN were quite

Table 3. Classification error and train size (%) for different uncertainty values.

	Data set 2		Data set 5	
Uncertainty	Linear SVM error	Train size	Random forest error	Train size
≥0 (all data)	0.065	100	0.094	100
>0.2	0.064	30.20	0.092	65.13
>0.4	0.064	22.35	0.097	56.18
>0.6	0.064	16.00	0.102	29.57
>0.8	0.063	9.43	0.165	15.58
>0.9	0.500	6.07	0.176	9.17

similar to the error for artificial data sets, they move away in real data sets. $N4$ is quite accurate but sometimes behaves oppositely to the error and the rest of measures. Also, by definition, $N4$ is more focused on the non-linearity based on interpolated points than in capturing those critical points.

Since complexity measures are expected to be well-correlated with the error, the Spearman correlation is studied among all of them and the error. LP measure presents the maximum correlation with the error (0.988) and $N4$ the minimum (0.867). All measures are highly correlated and are able to rank data sets matching the error. In addition, the high correlation (>0.9) of LP measure with the rest of neighborhood complexity measures reveals that it is in line with measures from state-of-art. Nevertheless, LP is the easiest to interpret and gives an accurate estimation of the proportion of critical points for the classification.

4.2 Uncertainty

This experiment is focused on the instance level information. The *dominance probability* achieved for every point can be used to discover and locate the most uncertainty areas on a data set. Consequently, it can also be used to identify the decision boundary. The most uncertainty areas are those where both classes are equally dominant. To transform the *dominance probability* into an uncertainty indicator, the Shannon entropy is used following [8]. By doing this, the *dominance probability* vector becomes an uncertainty vector taking values between 0 and 1, where 0 corresponds to no uncertainty (only one class is present) and 1 is the maximum level of uncertainty (the proportion of both classes is 0.5). Figure 4 shows the resulting uncertainty values for artificial data sets 2 and 5. In both cases, the decision boundary is recovered by the most uncertainty areas. Table 3 reflects the potential of this information: if data is filtered with different uncertainty thresholds to train with more problematic points as the uncertainty threshold increases, the error remains fairly stable with an important reduction of train size. For data set 2, an uncertainty threshold of 0.8 still maintains the error constant and only a 9.43% of the train data is used. For data set 5, a threshold of 0.6 shows a higher error (from 0.094 to 0.102) but just keeping 29.57% of the train data.

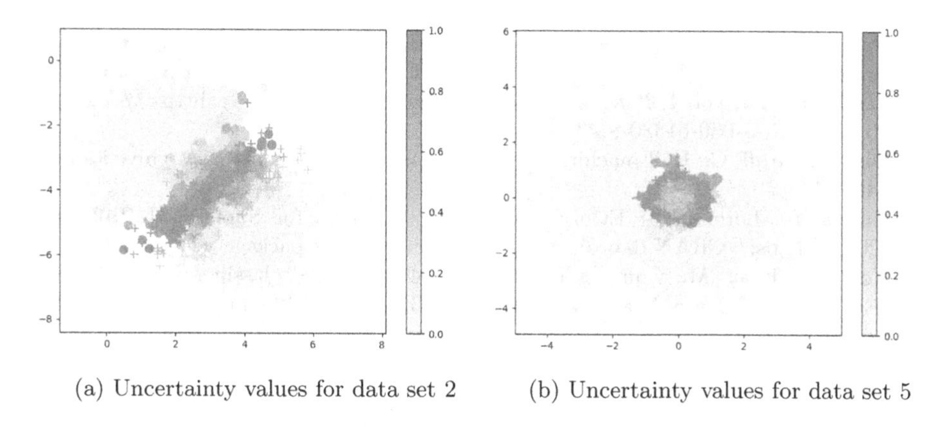

(a) Uncertainty values for data set 2 (b) Uncertainty values for data set 5

Fig. 4. Uncertainty based on *dominance probability.*

5 Conclusions

In the present work, the *LP* complexity measure for binary classification problems has been presented. It is a classifier-independent measure calculated by a hierarchical and recursive k-means procedure and, to the best of authors' knowledge, it is the only complexity measure offering a two-level analysis and that aggregates different layers on information, which increases the robustness of the method. As a neighborhood measure, it addresses the data complexity capturing overlapped, borderline and noisy points. It offers information about the proportion of points that are lost due to the distribution of the classes. The *LP* measure is closer to the best classification error than alternatives in the state-of-the-art. Thus, its values are easily interpretable and can be a reference in the classification task. It also gives a probabilistic value per points showing favorable results when used to detect the decision boundary and to carry out a data reduction.

Future work will mainly focus on improving the current method and its generalisation for multi-class problems. The first steps will be the analysis of different values for δ parameter and its application on the imbalanced case. Furthermore, it will be interesting to go deeper in the instance level and analyse its behaviour in instance selection: if a classifier is trained without lost points it could learn from cleaner data and increase its performance.

Acknowledgements. This research has been supported by grants from Rey Juan Carlos University (Ref: C1PREDOC2020), Madrid Autonomous Community (Ref: IND2019/TIC-17194) and the Spanish Ministry of Economy and Competitiveness, under the Retos-Investigación program: MODAS-IN (Ref: RTI-2018-094269-B-I00).

References

1. Algar, M.J., et al.: A quality of experience management framework for mobile users. Wirel. Commun. Mob. Comput. **2019**, 11 (2019). https://doi.org/10.1155/2019/2352941. Article ID 2352941

2. Arruda, J.L.M., Prudêncio, R.B.C., Lorena, A.C.: Measuring instance hardness using data complexity measures. In: Cerri, R., Prati, R.C. (eds.) BRACIS 2020. LNCS (LNAI), vol. 12320, pp. 483–497. Springer, Cham (2020). https://doi.org/10.1007/978-3-030-61380-8_33
3. Dua, D., Graff, C.: UCI machine learning repository (2017). http://archive.ics.uci.edu/ml
4. Garcia, L., Lorena, A.: ECoL: Complexity Measures for Supervised Problems (2019). https://CRAN.R-project.org/package=ECoL, r package version 0.3.0
5. Ho, T.K., Basu, M.: Complexity measures of supervised classification problems. IEEE Trans. Pattern Anal. Mach. Intell. **24**(3), 289–300 (2002)
6. Lorena, A.C., Garcia, L.P., Lehmann, J., Souto, M.C., Ho, T.K.: How complex is your classification problem? A survey on measuring classification complexity. ACM Comput. Surveys (CSUR) **52**(5), 1–34 (2019)
7. Oh, S.: A new dataset evaluation method based on category overlap. Comput. Biol. Med. **41**(2), 115–122 (2011)
8. Singh, S.: Prism-a novel framework for pattern recognition. Patt. Anal. Appl. **6**(2), 134–149 (2003)
9. Smith, M.R., Martinez, T., Giraud-Carrier, C.: An instance level analysis of data complexity. Mach. Learn. **95**(2), 225–256 (2013). https://doi.org/10.1007/s10994-013-5422-z
10. Wan, S., Zhao, Y., Wang, T., Gu, Z., Abbasi, Q.H., Choo, K.K.R.: Multi-dimensional data indexing and range query processing via Voronoi diagram for internet of things. Futur. Gener. Comput. Syst. **91**, 382–391 (2019)
11. Weitzman, M.S.: Measures of overlap of income distributions of white and Negro families in the United States, vol. 22. US Bureau of the Census (1970)

Linear Concept Approximation for Multilingual Document Recommendation

Vilmos Tibor Salamon[1], Tsegaye Misikir Tashu[1,2](✉), and Tomáš Horváth[1,3]

[1] ELTE - Eötvös Loránd University, Faculty of Informatics, Department of Data Science and Engineering, Telekom Innovation Laboratories, Pázmány Péter sétány 1/C, Budapest 1117, Hungary
{misikir,tomas.horvath}@inf.elte.hu
[2] Kombolcha Institute of Technology, College of Informatics, Wollo University, 208 Kombolcha, Ethiopia
[3] Faculty of Science, Institute of Computer Science, Pavol Jozef Šafárik University, Jesenná 5, 040 01 Košice, Slovakia

Abstract. In this paper, we proposed Linear Concept Approximation, a novel multilingual document representation approach for the task of multilingual document representation and recommendation. The main idea is in creating representations by using mappings to align monolingual representation spaces using linear concept approximation, that in turn will enhance the quality of content-based Multilingual Document Recommendation Systems. The experimental results on JRC-Acquis have shown that our proposed approach outperformed traditional methods on the task of multilingual document recommendation.

Keywords: Multilingual representation learning · Latent semantic indexing · Document recommendation · Multilingual NLP

1 Introduction

The amount of online information coming from various sources and communities has increased drastically. This growth of the Web is the most influential factor contributing to the increasing importance of information retrieval and recommender systems. The Webs in this era offer content in a language other than English and are becoming increasingly multilingual. Not only is the Web becoming multilingual, but users themselves are also increasingly proficient in more than one language [8]. The diversity of the Web and the explosive growth of the Internet are compelling reasons for the need for recommender systems that cross language boundaries. Therefore, intelligent information access platforms such as recommender systems need to evolve to effectively deal with this growing amount of multilingual information and communities [5,6].

H. Yin et al. (Eds.): IDEAL 2021, LNCS 13113, pp. 147–156, 2021.
https://doi.org/10.1007/978-3-030-91608-4_15

Indeed, recommender systems may suggest items of interest in a different language than the user explicitly used to express his interests. This problem is known in the literature as cross-language information access [5,6]. This motivates the need for efficient and effective techniques for recommender systems that cross the boundaries of language. In cross-language information access, relevant documents may be classified as irrelevant due to a small textual overlap between the query and the document, or interesting documents may be classified as not interesting due to the small overlap between the user interest and the item descriptions. Such an extreme case occurs when interesting documents are written in a different language than the query [7].

One way to overcome the language barrier is to use machine translation. However, the performance of translation-based approaches is limited by the quality of machine translation and translation ambiguity needs to be dealt with [12]. One possible solution is to consider the translation alternatives of individual words of queries or documents as in [10,11], which provides more opportunities for matching query words in relevant documents compared to using individual translations. But the alignment information is necessarily needed in the training phase of the cross-lingual system to extract target-source word pairs from parallel data, and this is not a trivial task. The second way to overcome the language barrier is to focus on the concepts associated with words. The meaning of words is inherently multilingual, in that the concepts remain the same in different languages, while the words used to describe those concepts change in each specific language. A concept-based representation of elements and user profiles could be an effective way to have a language-independent representation that could serve as a bridge between different languages [5,6,12].

The main idea presented in this paper is to exploit latent semantic indexing (LSI) and linear concept approximation as a bridge between different languages; construct language-independent representations that enable an effective cross-language content-based recommendation process.

2 Base Models

2.1 Latent Semantic Indexing

Given an $X \in \mathbb{R}^{m \times n}$ term-document matrix where m is the number of terms and n is the number of documents, the LSI tries to approximate the original matrix by a lower-dimensional concept-document matrix [2]. A computationally efficient way to compute the rank-k approximation for this application is a method called truncated singular value decomposition (SVD), where the largest k singular values and the corresponding columns of U and V will be computed. In this case, the factorization of the term-document matrix is $X_k = U_k \Sigma_k V_k^T$ where $U_k \in \mathbb{R}^{m \times k}$, $\Sigma_k \in \mathbb{R}^{k \times k}$ and $V \in \mathbb{R}^{k \times n}$.

The U_k and V_k matrices contain the eigenvectors of the XX^T and X^TX products of the term-document matrix. Since X is the term-document matrix it is easy to see that XX^T is the term-term matrix. Its elements are the correlation of each term pair or their co-occurrence in the documents. Similarly, X^TX is the document-document matrix whose elements are the correlations of each document. It has been proven that LSI encapsulates high-order term co-occurrence information [3]. Intuitively, it is assumed that the remaining k dimensions that have not been discarded now refer to k latent concepts. U_k is the term-concept matrix and V_k is the document-concept matrix with Σ_k giving weight to these concepts. It is possible to get the representation of the documents in this latent semantic space by multiplying the documents' concept-vector with their weights. Hence, the latent semantic representation of our documents are

$$\hat{X} = U_k^T X \in \mathbb{R}^{k \times n} \tag{1}$$

And to fold in a new document, we can do essentially the same with its $\boldsymbol{x} \in \mathbb{R}^m$ term-vector:

$$\hat{\boldsymbol{x}} = U_k^T \boldsymbol{x} \in \mathbb{R}^k \tag{2}$$

In this form, a new query will be the weighted sum of its constituent terms. The main advantage of LSI is that by reducing the term-document matrix to a rank-k approximation, the terms with more co-occurrence will be closer in the latent concept space. This will solve a few problems that conventional bag-of-words models face. LSI has become more popular as it addresses problems related to synonymy and polysemous words.

2.2 Cross-Lingual Latent Semantic Indexing

Cross-lingual Latent Semantic Indexing (CL-LSI) is the extension of the monolingual LSI to the cross-lingual domain [1,4]. Let our two languages be L_x and L_y and the term-document matrices of the parallel set of training documents $X \in \mathbb{R}^{m_x \times n}$ and $Y \in \mathbb{R}^{m_y \times n}$. Training the CL-LSI model begins by stacking the two matrices into a large one by keeping the documents aligned. Denote this matrix by Z.

$$Z = \begin{bmatrix} \boldsymbol{x}_1 \; \boldsymbol{x}_2 \; \dots \; \boldsymbol{x}_n \\ \boldsymbol{y}_1 \; \boldsymbol{y}_2 \; \dots \; \boldsymbol{y}_n \end{bmatrix} \in \mathbb{R}^{(m_x + m_y) \times n}$$

The idea is to train the LSI on this cross-lingual matrix. By using the k-truncated singular value decomposition, we will get

$$Z_k = U_k \Sigma_k V_k^T \tag{3}$$

where $U_k \in \mathbb{R}^{(m_x + m_y) \times k}$ is the multilingual term-concept matrix, $\Sigma_k \in \mathbb{R}^{k \times k}$ is the diagonal concept-weight matrix and $V_k \in \mathbb{R}^{n \times k}$ is the document-concept matrix. From this, the new latent semantic representation of the training data will be

$$\Sigma_k V_k^T \in \mathbb{R}^{k \times n} \tag{4}$$

To fold-in a test document of language L_x, $\boldsymbol{x} \in \mathbb{R}^{m_x}$, we have to extend it to zeros in the dimensions of the other language.

$$\tilde{\boldsymbol{x}} = \begin{bmatrix} \boldsymbol{x} \\ \mathbf{0} \end{bmatrix} \in \mathbb{R}^{m_x + m_y}$$

Then we can fold-in the extended vector to the CL-LSI as it was a monolingual LSI. The test document representation in the multilingual latent semantic space is then:

$$\hat{\boldsymbol{x}} = U_k^T \tilde{\boldsymbol{x}} \in \mathbb{R}^k \quad (5)$$

This way, all documents from L_x and L_y can be folded into a common vector space in which we can compare them using the cosine similarity.

3 The Linear Concept Approximation Model

The problem with LSI in multilingual environments is that the output dimensions of the decomposition are not guaranteed to be the same when we reduce the term-document matrices independently. Although there is likely to be some similarity due to language differences, the extracted concepts will not be equivalent, and cosine similarity is very sensitive to this difference.

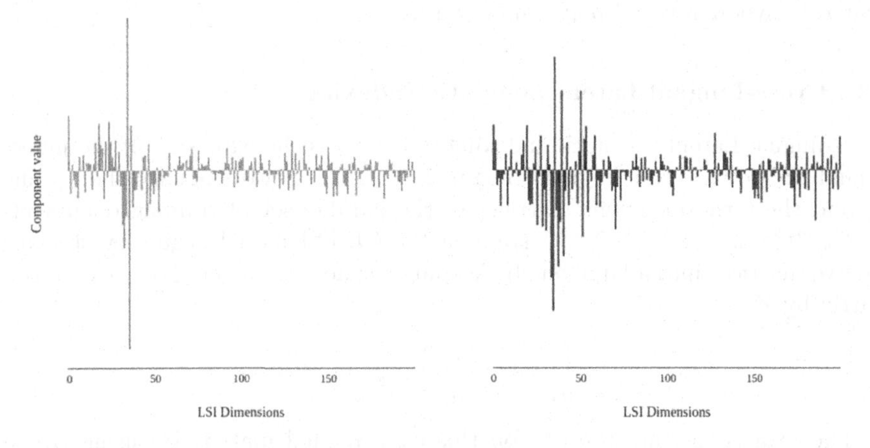

Fig. 1. Latent semantic difference between the same document in English (left) and Hungarian (right) language.

In Fig. 1, we can observe the difference between the first 200 latent concepts of a document written in (translated to) Hungarian and English. Although the documents are the same, it does not guarantee us that the first concept in Hungarian is also the first concept in English, even if there is a pair for each concept. However, the intuition is that there must be a connection between the two documents since they deal with (exactly) the same topics. Our proposed model is based on the assumption that this connection is linear and that each latent semantic dimension in one language can be approximated by the linear combinations of the latent semantic dimensions of the other language.

Linear Approximation of Concept Spaces: Let $X \in \mathbb{R}^{k_x \times n}$ and $Y \in \mathbb{R}^{k_y \times n}$ be our LSI-reduced concept-document matrices for languages L_x and L_y. In this representation, each column of these matrices represents a document, which is not beneficial for reasoning, so we denote their transposed document-concept matrices as $C_x = X^T \in \mathbb{R}^{n \times k_x}$ and $C_y = Y^T \in \mathbb{R}^{n \times k_y}$. In these matrices, each column is a concept in the respective language, i.e. $C_x = [\boldsymbol{c}_{x,1}, \ldots, \boldsymbol{c}_{x,k_x}]$ and $C_y = [\boldsymbol{c}_{y,1}, \ldots, \boldsymbol{c}_{y,k_y}]$. Our assumption is that a concept in L_x can be approximated fairly well by the linear combination of the concepts in L_y. This means that for every column $\boldsymbol{c}_{x,i}$ of C_x there are $\beta_{1,i}, \ldots, \beta_{k_y,i} \in \mathbb{R}$ for which

$$\boldsymbol{c}_{x,i} \approx \sum_{j=1}^{k_y} \beta_{j,i} \boldsymbol{c}_{y,j} \tag{6}$$

or, by defining the coefficient vector of $\boldsymbol{\beta}_i = [\beta_{1,i}, \ldots, \beta_{k_y,i}]^T \in \mathbb{R}^{k_y}$,

$$\boldsymbol{c}_{x,i} \approx C_y \boldsymbol{\beta}_i \tag{7}$$

For this is assumed to be true for every $\boldsymbol{c}_{x,i}$ $(i = 1, \ldots, k_x)$ concept of L_x, we can create a coefficient matrix $B \in \mathbb{R}^{k_y \times k_x}$ with columns $\boldsymbol{\beta}_i$. Using this matrix, we can linearly approximate the concept matrix C_x as

$$C_x \approx C_y B \tag{8}$$

Returning to the problem of only one concept at a time, we have a linear system with the matrix C_y having n rows and k_y columns

$$\boldsymbol{c}_{x,i} = C_y \boldsymbol{\beta}_i \tag{9}$$

Since we know that LSI is limited to the rank of the matrix, we can safely assume that $k_y < n$. This is an "overdetermined" system and, as such, does not have an exact solution. However, we can determine the coefficients $\beta_{j,i}$ such that the resulting vector best approximates $\boldsymbol{c}_{x,i}$ in the least squared sense. In other words, we are looking for the $\boldsymbol{\beta}_i$ vector that minimizes the squared error.

$$||C_y \boldsymbol{\beta}_i - \boldsymbol{c}_{x,i}||_2^2 \tag{10}$$

It has been shown that such an optimization problem can be solved by solving the normal equation using the pseudo-inverse:

$$\boldsymbol{\beta}_i^* = (C_y^T C_y)^{-1} C_y^T \boldsymbol{c}_{\boldsymbol{x},\boldsymbol{i}} \tag{11}$$

Denoting the pseudo-inverse as

$$P_{C_y} = ({C_y}^T C_y)^{-1} C_y^T \in \mathbb{R}^{k_y \times n} \tag{12}$$

we get a simpler expression

$$\boldsymbol{\beta}_i^* = P_{C_y} \boldsymbol{c}_{\boldsymbol{x},\boldsymbol{i}} \tag{13}$$

The whole B matrix can be rewritten as

$$B = P_{C_y} C_x \in \mathbb{R}^{k_y \times k_x} \tag{14}$$

and we get the best linear approximation of the concepts C_x by the concepts of C_y as

$$\hat{C}_y = C_y B \tag{15}$$

This means we can transform the k_y concept in the concept space of L_y into a linear approximation of the concepts of L_x. Therefore, the transformed documents of L_y will also be more close to their counterparts. By transposing the previous equation, we get the representation of the documents of L_y, what we are looking for, as

$$\hat{Y} = B^T Y \in \mathbb{R}^{k_x \times n} \tag{16}$$

and for a test document $\boldsymbol{y} \in \mathbb{R}^{k_y}$ as

$$\hat{\boldsymbol{y}} = B^T \boldsymbol{y} \in \mathbb{R}^{k_x} \tag{17}$$

The similarities of documents $\boldsymbol{x}$ and $\boldsymbol{y}$ are defined as

$$\text{sim}_1(\boldsymbol{x}, \boldsymbol{y}) = \text{cos-sim}(\boldsymbol{x}, \hat{\boldsymbol{y}}) \tag{18}$$

Since we have two languages, we can do two separate concept approximations: one from L_x to L_y and another the roles switched. The first approximate the C_x concepts in the span of C_y and the second the other way around. This way, every test document will have two representations and two similarities from each other. The final similarity will be the sum or the average of the two. The same way as we defined B and $\hat{Y}$ for L_y we can define an analogous to them for L_x as

$$A = P_{C_x} C_y \in \mathbb{R}^{k_x \times k_y} \tag{19}$$

This A matrix then transforms the k_y concept-vectors of L_y to the span of C_x. The new representation of X is then

$$\hat{X} = A^T X \in \mathbb{R}^{k_y \times n} \tag{20}$$

$$\hat{\boldsymbol{x}} = A^T \boldsymbol{x} \in \mathbb{R}^{k_y} \tag{21}$$

$$\text{sim}_2(\boldsymbol{x}, \boldsymbol{y}) = \text{cos-sim}(\hat{\boldsymbol{x}}, \boldsymbol{y}) \tag{22}$$

The final similarity between two documents is the average of the two

$$\text{sim}(\boldsymbol{x}, \boldsymbol{y}) = \frac{\text{sim}_1(\boldsymbol{x}, \boldsymbol{y}) + \text{sim}_2(\boldsymbol{x}, \boldsymbol{y})}{2} \tag{23}$$

4 Experimental Settings

4.1 Datasets

To evaluate the proposed approach, we used a publicly available parallel corpus, namely the JRC-Acquis corpus. The Corpus consists of a corpus of legal and legislative papers from European Unions. The corpus has over 4 million documents in total, with an average of over 18,000 parallel documents per language pair. We are interested in the documents available in both English, French, Hungarian, and Spanish. The intersection of documents available in the target language is 22,524 documents.

4.2 Evaluation Metrics

Mate Retrieval Rate: The only document that we are absolutely certain is relevant to a given document is its translation in another language. Therefore, the mainly used metric in cross-lingual information retrieval tasks is the mate retrieval rate, which is the proportion of test documents that are linked to their cross-language pair. This can be formulated as the number of rows/columns in the similarity matrix, where the diagonal element is the largest.

Mean Reciprocal Rank. This performance measure is commonly used in tasks where there is only one relevant document to retrieve. It defines the rank of a document i, denoted by r_i, as the order in which its parallel pair is retrieved

$$MRR = \frac{1}{n_{test}} \sum_{i=1}^{n_{test}} \frac{1}{r_i} \tag{24}$$

4.3 Results and Discussion

The proposed model was compared with other four baselines models such as coefficient approximation, K-Means centroid model, improved cross lingual latent semantic indexing and reference similarity. More about the baseline methods can be found at [9].

The best results obtained by the model with the optimal parameters are shown in Tables 1 and 2. Each model achieves better and higher performance on the other language pair than on Hungarian-English, Hungarian-French, or Hungarian-Spanish. The least sensitive model seems to be Linear Concept Approximation (LCA) and the most sensitive model is Reference Similarity, as can be seen from the Tables 1 and 2. The reference similarity has an increasing trend with respect to the dimensions, which could be due to the problem with LSI.

Tables 1 and 2 show the best achieved scores on JRC-Acquis for each model in terms of language pairs. The best model in each scenario and measurement is the LCA. The other proposed method, the coefficient approximation, also shows promising results. The others are behind, with the worst model being the reference similarity model. K-means is remarkably sensitive to the languages in mate retrieval, which could be due to the randomness of the method.

Table 1. Results on JRC-Acquis based on Mate retrieval rate

Models	Hu-En	Hu-Fr	Hu-Es	En-Fr	En-Es	Fr-Es
Kmeans	84.40	82.00	83.87	91.47	92.00	92.70
Improved CL-LSI	87.37	86.57	87.43	92.67	92.77	92.53
Reference sim.	70.33	66.43	70.10	85.10	86.83	87.33
Coefficient approx.	89.60	87.47	88.37	94.67	94.30	94.33
Linear concept approx.	91.47	91.00	91.93	95.57	95.60	95.70

Table 2. Results on JRC-Acquis based on Mean reciprocal rank

Models	Hu-En	Hu-Fr	Hu-Es	En-Fr	En-Es	Fr-Es
Kmeans	88.68	87.16	88.51	93.56	93.86	94.27
Improved CL-LSI	90.30	89.79	90.56	94.38	94.19	94.17
Reference sim.	77.39	74.45	77.04	88.95	89.54	90.01
Coefficient approx.	92.21	91.12	91.66	95.55	95.42	95.41
Linear concept approx.	93.31	93.06	93.70	96.07	96.09	96.15

The Effect of the Transformation: What this transformation does can be seen in Fig. 2 which shows the latent semantic vectors of a document in English and Hungarian as well as in transformed Hungarian. It can be clearly seen that by transforming the concepts of one language into the other, the most relevant features are very well-connected, and thus their cosine similarity is high.

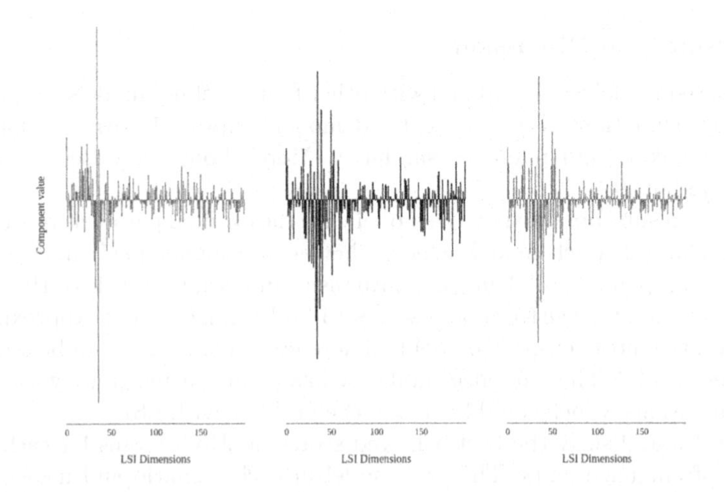

Fig. 2. Impact of the transformation. Left: original Hungarian document, middle: original English, right: transformed Hungarian.

Another way to show the effectiveness of these transformations is to show the first two components of some documents in the original and transformed semantic space, as shown in Fig. 3. Looking at the figure, we see that the original representation of the documents had little in common. The documents are scattered across the space spanned by the first two semantic dimensions. The biggest problem is that the nearest parallel documents are never their pairs. After the transformation, this problem is solved. The documents and their parallel pairs are clustered nicely together and are closer to each other than to any other document. This means that they are similar in both their semantic dimensions.

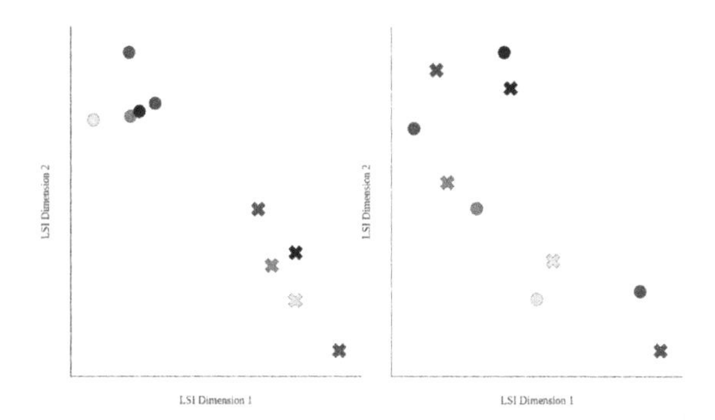

Fig. 3. Effect of the transformation on the documents in the first two latent semantic dimensions. The colors represent different documents, the "X"-s are the English and the circles are the Hungarian representations. Left - the original documents. Right - the transformed documents.

5 Conclusions

The main objective of this paper was to address the problems in representation learning for cross-lingual information retrieval and document recommendation.

The Linear Concept Approximation (LCA) model was proposed to address the issues. The experimental results showed the superiority of LCA over the basic methods. With an average increase of 4% in mate retrieval and 3% in reciprocal rank. The coefficient approximation has shown promising results with medium corpus size and length. The reference similarity model was by far the weakest of the methods studied. K-means has proven to be very resilient to the absence of data and can be a valuable alternative in situations with only a few training documents.

As the experiment was carried out on a limited dataset, further investigations might be required, but the results are promising. The future direction of the research will involve experimenting with more data and comparison with neural network-based models as well as investigating the impact of LCA on neural-based representation learning approaches.

Acknowledgment. "Application Domain Specific Highly Reliable IT Solutions" project has been implemented with the support provided from the National Research, Development and Innovation Fund of Hungary, financed under the Thematic Excellence Programme TKP2020-NKA-06 (National Challenges Subprogramme) funding scheme.

References

1. Cox, W., Pincombe, B.: Cross-lingual latent semantic analysis. In: Read, W., Larson, J.W., Roberts, A.J. (eds.) Proceedings of the 13th Biennial Computational Techniques and Applications Conference, CTAC-2006. ANZIAM J., vol. 48, pp. C1054–C1074 (2008)
2. Deerwester, S., Dumais, S.T., Furnas, G.W., Landauer, T.K., Harshman, R.: Indexing by latent semantic analysis. J. Am. Soc. Inf. Sci. **41**(6), 391–407 (1990)
3. Kontostathis, A., Pottenger, W.M.: A framework for understanding Latent Semantic Indexing (LSI) performance. Inf. Process. Manage. **42**(1), 56–73 (2006)
4. Littman, M.L., Dumais, S.T., Landauer, T.K.: Automatic cross-language information retrieval using latent semantic indexing. In: Cross-Language Information Retrieval, pp. 51–62. Springer, US, Boston, MA (1998). https://doi.org/10.1007/978-1-4615-5661-9_5
5. Lops, P., Musto, C., Narducci, F., De Gemmis, M., Basile, P., Semeraro, G.: MARS: a MultilAnguage recommender system. In: Proceedings of the 1st International Workshop on Information Heterogeneity and Fusion in Recommender Systems, pp. 24–31. HetRec 2010. ACM, New York, NY, USA (2010)
6. Narducci, F., et al.: Concept-based item representations for a cross-lingual content-based recommendation process. Inf. Sci. **374**, 15–31 (2016)
7. Sorg, P., Cimiano, P.: Exploiting Wikipedia for cross-lingual and multilingual information retrieval. Data Knowl. Eng. **74**, 26–45 (2012). applications of Natural Language to Information Systems
8. Steichen, B., Ghorab, M.R., O'Connor, A., Lawless, S., Wade, V.: Towards personalized multilingual information access - exploring the browsing and search behavior of multilingual users. In: Dimitrova, V., Kuflik, T., Chin, D., Ricci, F., Dolog, P., Houben, G.-J. (eds.) UMAP 2014. LNCS, vol. 8538, pp. 435–446. Springer, Cham (2014). https://doi.org/10.1007/978-3-319-08786-3_39
9. Tomáš, H., Tibor, S.V.: Content based recommendation in catalogues of multilingual documents (2018). MSc Thesis
10. Xu, J., Weischedel, R.: Cross-lingual information retrieval using hidden Markov models. In: Proceedings of the 2000 Joint SIGDAT Conference on Empirical Methods in Natural Language Processing and Very Large Corpora: Held in Conjunction with the 38th Annual Meeting of the ACL - Volume 13, pp. 95–103. EMNLP 2000, ACL, USA (2000)
11. Zbib, R., et al.: Neural-network lexical translation for cross-lingual IR from text and speech. In: Proceedings of the 42nd International ACM SIGIR Conference on Research and Development in Information Retrieval, pp. 645–654. SIGIR 2019. ACM, New York, NY, USA (2019)
12. Zhou, D., Truran, M., Brailsford, T., Wade, V., Ashman, H.: Translation techniques in cross-language information retrieval. ACM Comput. Surv. **45**(1), 1–44 (2012)

Unsupervised Detection of Solving Strategies for Competitive Programming

Alexandru Ştefan Stoica[1], Daniel Băbiceanu[1], Marian Cristian Mihăescu[1](✉), and Traian Rebedea[2]

[1] University of Craiova, Craiova, Romania
cristian.mihaescu@edu.ucv.ro
[2] University Politehnica of Bucharest, Bucharest, Romania
traian.rebedea@cs.pub.ro

Abstract. Transformers are becoming more and more used for solving various Natural Language Processing tasks. Recently, they have also been employed to process source code to analyze very large code-bases automatically. This paper presents a custom-designed data analysis pipeline that can classify source code from competitive programming solutions. Our experiments show that the proposed models accurately determine the number of distinct solutions for a programming challenge task, even in an unsupervised setting. Together with our model, we also introduce a new dataset called AlgoSol-10 for this task that consists of ten programming problems together with all the source code submissions manually clustered by experts based on the algorithmic solution used to solve each problem. Taking into account the success of the approach on small source codes, we discuss the potential of further using transformers for the analysis of large code bases.

Keywords: Transformers · Competitive programming · Source code analysis · Unsupervised learning

1 Introduction

This paper introduces a new dataset called AlgoSol-10 and an approach to determine distinct solutions in the context of competitive programming in terms of algorithmic approach and implementation. For consistency in description of works we will use the term *algorithmic solution* as the approach for solving a competitive programming problem. The term *source code solution* (or just *solution*) represents a particular implementation in C++ programming language of the algorithmic solution. The solution of a problem is represented by a source code file on the hard disk that has been coded by a competitor during a contest.

In competitive programming, it is expected that we do not know the number of distinct algorithmic solutions for a problem. We consider that a problem has different algorithmic solutions if they use different computer science methodology and concepts to correctly solve the same problem e.g. The problem of

H. Yin et al. (Eds.): IDEAL 2021, LNCS 13113, pp. 157–165, 2021.
https://doi.org/10.1007/978-3-030-91608-4_16

finding if a number is in a vector can be solved by different algorithmic solutions : 1) sorting the vector and binary search, 2) using a hash table, 3) using a balanced tree, 4)using a linear search, etc. The trivial approach is to go through the source code solutions of every participant and finally determine how many correct distinct algorithmic solutions were submitted. Considering that a problem may have hundreds of source code solutions, we want to provide a method that automatically determines the correct label for each solution for a fixed number of distinct algorithmic solutions.

We did not find any available dataset with labelled problems in terms of distinct algorithmic solutions to validate our method, so we created our own dataset. It has 10 problems from *Infoarena*[1] for which we manually labeled all source code solutions. To the best of our knowledge, we are the first to propose an approach to this problem in the context of competitive programming. We will provide a supervised method as well as a new unsupervised method that performs well in comparison with classical unsupervised methods.

One of the most challenging aspects of the tackled task is building the labelled dataset because it requires an expert in competitive programming who can correctly label various source code solutions that implement the same algorithmic solution. For example, regarding the problem of finding a value in a vector, the algorithmic solution that uses sort and binary search may have source code implementations in terms of sorting (i.e., bubble sort, quick sort, STL sort, etc.) and binary search (i.e., with while loop, with STL, etc.). The difficulty comes also from the fact that the code written in competitive programming does not respect the industry's coding standards, which implies that one cannot make any assumptions about specific rules. For example, Dijkstra's algorithm could be implemented in a function named *f*, variables could be named as letters etc. Finally, not every piece of code is compilable, even though at a particular time it was compiled, and that is because compilers usually change in time, and the code may have been written many years ago without knowing the version of the compiler used at that moment in time for each solution.

The proposed solution is based on unsupervised learning methodology, while there are used ten manually labelled problems for evaluation purposes. The embedding methods that are used within the workflow are classical Term Frequency-Inverse Document Frequency (Tf-IDF) [10], and Word2Vec (W2V) [12], as well as the more recent deep learning (DL) based Self-Attentive Function Embeddings (SAFE) [11].

2 Related Work

Natural Language Processing (NLP) generally deals with processing text information using various Machine Learning and rule-based techniques, with the former receiving wider interest lately. Recent developments target using state-of-the-art NLP techniques for processing source code as well, as it can also be regarded as *text*. Thus, Chen and Monperrus [6] present a literature study

[1] Infoarena, www.infoarena.ro, last accessed on 30th June 2021.

on diverse techniques for source code embeddings as a novel research direction within NLP. Further, Rabin et al. [13] aim to demystify the dimensions of the source embeddings as a new attempt to shed light on this new research domain.

The usage of DL for source code analysis has become more popular due to the attention mechanism from NLP that has been used in Code2Vec [1] and SAFE [11]. Code2Vec uses the attention mechanism on various paths in the abstract syntax tree (AST) of a function for generating corresponding embeddings. On the other hand, SAFE only uses the attention mechanism on assembly code obtained after compiling the source code and generates an embedding for each function. A classification problem (i.e., prediction of labels) that uses W2V and DL for the analysis of source code has been reported by Iacob et al. [8]. A usage of Code2Vec for recommending function names has been reported by Jiang et al. [9], although the proposed method has poor results on real-world code repositories as the domain is still in its infancy.

Other approaches build the profile of individual students based on their programming design [3], determine source code authorship [5], predict the bug probability of software [15], build source code documentation by summarization [2], classify source code [4] or detect the design pattern [7].

3 Proposed Approach

A particular aspect of the task is that in competitive programming, a problem usually has a small number of distinct algorithmic solutions which allows the possibility of manual labelling. This context has two implications: 1) we face a classification problem or an unsupervised learning situation with a known and small number of clusters, and 2) we may build a labelled dataset for validation purposes. Therefore, we can consider that for a particular problem, we know K (i.e., the number of distinct algorithmic solutions), and the task becomes to determine the label of each source code solution. Given that we do not have any labels for our instances (i.e., source code solutions to a given problem), we are clearly in the area of unsupervised learning.

As a proposed approach for this problem, we build a custom data analysis pipeline from the following components.

Preprocessing. The first step in the pipeline is preprocessing of source code solutions. This step is compulsory for all source code solutions that are given to *W2V* and *Tf-idf* for building the embeddings. The following steps are performed: 1) delete the *#include* directives; 2) delete comments; 3) delete functions that are not called; 4) replace all macro directives; 5) delete apostrophe characters; 6) delete all characters that are not ASCII (for example strings which may contain unicode characters); 7) tokenize the source code. We mention that *SAFE* embeddings do not require preprocessing because it takes as input the object code, not the source code. Thus, the only requirement for the source code to be provided as input to *SAFE* is that it needs to be compiled. As it can be seen here, some embeddings methods use source code as input (i.e., W2V, Tf-Idf) while SAFE uses object code. The main idea of this step is to do any preprocessing necessary in order to obtain more quality code embeddings, since they contain the main information.

Embedding Computation. For generating the source code embeddings with *W2V* and *SAFE* we use *AlgoLabel* [8]. The embeddings computation with *Tf-idf* is implemented separately and follows the classical bag-of-words approach.

W2V uses the tokens for each source code solution after preprocessing. Based on the obtained tokens, we build a neural network whose goal is to predict the current token given the nearby tokens in a C-BOW architecture [12]. The algorithm uses a window of five tokens, and the embedding is a vector of dimension 128. The resulted embedding of a source code solution is the average of embeddings for the tokens that make up the source code solution itself.

Tf-idf algorithm uses the tokens obtained for each source code solution after preprocessing. Those tokens are filtered by removing the stop words

```
"(", ")", ".", "#", ";", ",",
">>", "<<", "{", "}", "[", "]", ".", ""..""/
```

and by selecting the words which follow the following regular expression

```
\w+|\+|-|=|!=|\\*|\/|%|!|
```

The output of *Tf-idf* algorithm is a matrix whose number of columns is the cardinal of the vocabulary (i.e., the number of distinct filtered tokens).

SAFE uses only binary code that has been obtained after compilation of the source code solution. The *pretrained SAFE* model is used to compute one embedding for each function in the source code solution. The dimension of the obtained vector is 100, and the embedding of a source code solution is the average of the embeddings of all functions from the source code solution itself.

Building Clusters for Known Value of K. For determining the clusters (or patterns) within a set of source code solutions, we have used three flavours of unsupervised learning.

Clustering with a Single View. This approach is the simplest in the way that for each embedding technique, we run a specific clustering algorithm that partitions the available source code solutions in K groups.

Clustering with Multi-View. We consider the number of available embeddings methods used, V, as distinct views of the same dataset. This approach enables the usage of Multi View Spectral Clustering (i.e., MVSC) [14] algorithm, which also determines K groups.

Clustering by Voting. We assume that we have V views and for each view we have an associated embedding. For each view we use a clustering algorithm which will partition the embeddings for a particular view in K groups. Thus, for each source code solution we associate a vector of dimension V, where at each position i the value of x_i is in the range from 0 to $K-1$ and represents the cluster to which it belongs in the i-th view. More exactly, if we define *S(p)* as the set of all source code solutions associated to a problem p where $I_k = \{0, 1, ..., K-1\}$, $K \geq 2$ then $S(p) = \{s_i | s_i \in p\} \Rightarrow v^p(s_i) : S(p) \rightarrow I_k^v, v^p(s_i) = (x_0, x_1, ..., x_{V-1})$. We observe that the same vector may represent several source code solutions. Thus, we define the frequency of a vector as the number of source code solutions represented by the same vector.

$s^p(v) = \{s_i | s_i \in S(p), v = v(s_i)\}$, $f^p(v) = |s^p(v)|$ Having this function defined, we find vectors whose sum of frequencies is maximal and have no common coordinate. Furthermore, if we cannot find many vectors equal to K (i.e., known number of clusters), we must choose fewer clusters. We define $A = \{v_i | v_i \in I_k^v, v_i = \{x_0^{(i)}, x_1^{(i)}, ..., x_{V-1}^{(i)})\}, i \in \{1, .., K\}$

where $x_d^i \neq x_d^j$, $i \in \{1, .., K\}, j \in \{1, .., K\}, d \in \{0, .., V-1\}$ The task is to determine A such that $\sum f^p(v_i), v_i \in A$ has maximum value. The algorithmic solution itself is represented by the $s^p(v_i)$ vectors.

The vectors represent the K distinct algorithmic solutions that we extract from each view. Thus, each algorithmic solution contains similar source code solutions.

For solving this problem, we use a dictionary in which we compute the frequency of each vector. We extract the keys and choose all subsets of dimension K that meet the condition with no identical coordinate. For each subset, we compute the sum of key values and obtain the maximum sum subset. The key of the dictionary is a string that appends the coordinates separated by a separator. The complexity is exponential, but the algorithm is tractable because in our dataset $V \leq 3$ and $K \leq 3$.

Algorithm 1. Semi-supervised data analysis pipeline

Require: Solutions-Dataset = solutions for a problem
1: # Setup voters with their parameters: embedding, clustering algorithm and classification algorithm
2: # Build *Ground-Truth-Dataset* which maximizes $\sum f^{(p)}(v_i)$
3: # Sols-Train = Ground-Truth-Dataset
4: # Sols-Unlabeled = Solutions not in Sols-Train
5: **while** (# of valid solutions greater than *threshold* and # Sols-Unlabeled greater than 0) **do**
6: $Voter - X_i$ = Train classifier on Sols-Train based on the i-th view
7: **for all** (Sols-Unlabeled) **do**
8: #Predict the label of *solution* by all voters
9: **if** (*solution* has same label in all voters) **then**
10: #Append *solution* to *Sols-Train*
11: #Remove *solution* from *Sols-Unlabeled*
12: **end if**
13: **end for**
14: **end while**
15: Sols-Test = Sols-Train
16: #Validate Sols-Test

The algorithm presented above can be interpreted in the following way: each view represents a voter that partitions the items (i.e., the source code solutions) in k clusters. Each voter has its parameters in terms of used embedding, employed clustering algorithm and employed classification algorithm. The ideal situation occurs when voters perfectly agree on the distribution of items into clusters. The task is to correctly associate the clusters predicted by voter A with the cluster predicted by voter B. The problem becomes more complicated if we have more voters such that the matching becomes more difficult to be determined. The task is to determine the matching with the maximum number of identical items in coupled clusters. As initialization, we build a ground-truth matching that is given on the *1*-st coordinate (i.e., the cluster *id* to which the item belongs). After that, in the main loop, we predict the label (i.e., the cluster *id*) of the remaining items and consider that a label is correct if all classifiers predict it. These items are appended to the training dataset, and we re-train the voters only if the number of appended items is above a threshold and we still have unlabeled items. Finally, the items that could not be labelled are discarded.

Evaluation Methodology. We evaluate each method taking into account the optimal number of clusters for each problem. We denote the set of embeddings by $E = \{Word2Vec, Tf\text{-}idf, SAFE\}$, the set of clustering algorithms by $C = \{Kmeans, Spectral\ Clustering, Agglomerativ\ Clustering\}$, and the set of classification algorithms by $Clf = \{Random\ Forest, Svm, Xgboost\}$.

We define $[X]^n$ as the set of all subsets of dimension n with $|X|^n$ and $X \in \{E, C, Clf\}$. We mention that baseline results are obtained without hyperparameter tuning in either clustering and classification algorithms. As a general approach, the evaluation of a clustering algorithm will take into consideration all label permutations, and the one with the greatest *F1* score will be selected as the winner. Since problems may have algorithmic solutions with imbalanced number of source code solutions, the chosen quality metric is *F1-macro* because we want to treat the classes in the same way.

The evaluation of the method of building clusters with a single view takes into consideration all the combinations obtained by Cartesian product $[E]^1 \times [C]^1$ and evaluates each combination by the approach previously presented.

Similarly, the method that uses multi view spectral clustering will use as embeddings $[E]^2$ and $[E]^3$.

The method that uses clustering by voting is validated by determining the best results after co-training by employing classifiers and obtaining the best results after classification. These approaches will have as setup the Cartesian products $[E]^2 \times [C]^2 \times [Clf]^2$ and $[E]^3 \times [C]^3 \times [Clf]^3$, respectively.

After that, the supervised validation method is used to evaluate obtained models' capability to predict distinct algorithmic solutions. We have used a train-test split frequently used for smaller datasets: 80% of the data is used for training, and 20% is used for testing on the Cartesian product $[E]^1 \times [Clf]^1$. Finally, we determine for each problem the best method of clustering solutions.

4 Experimental Results

For evaluating the performance of the proposed algorithms with the employed code embeddings, it is compulsory to have the ground truth. Considering that there is no labelled dataset for distinct solutions for competitive programming, we decided to label several problems manually. The AlgoSol-10 dataset consists of ten problems from *Infoarena* and the criteria for selection are: 1) the problem must have at least two distinct algorithmic solutions; 2) the number of source code solutions for each class (i.e., algorithmic solution) should be large enough; 3) the classes should be as balanced as possible; 4) the algorithmic solutions to be as distinct as possible such that we may better evaluate how embeddings describe the algorithmic solutions for various implementations.

Table 1. Method with the highest F1-Macro score per problem.(UV - Unsupervised Voting, MVSC - Multi-View Spectral Clustering)

Method	Pb.	Embed.	Clustering	Estimator	F1	Size
UV	mst	w2v safe tf-idf	KMeans SpectralClustering KMeans	Random forest XGBClassifier XGBClassifier	Micro:0.9783 Macro:0.9731 Weight:0.9785	323
UV	cppp	w2v tf-idf	KMeans SpectralClustering	XGBClassifier SVC	Micro:0.5476 Macro:0.5176 Weight:0.4956	409
UV	gcd	w2v tf-idf	KMeans SpectralClustering	SVC SVC	Micro:0.7412 Macro:0.7163 Weight: 0.7350	429
UV	eval	w2v safe	SpectralClustering SpectralClustering	XGBClassifier Random forest	Micro:0.9871 Macro:.9796 Weight:0.9870	388
UV	inv	w2v safe tf-idf	KMeans KMeans KMeans	XGBClassifier XGBClassifier Random forest	Micro:0.9725 Macro:0.8447 Weight:0.8447	400
UV	invmod	w2v safe	SpectralClustering KMeans	SVC Random forest	Micro:0.9852 Macro:0.9849 Weight:0.9852	271
UV	schi	safe tf-idf	SpectralClustering SpectralClustering	Random forest Random forest	Micro:0.9953 Macro:0.9953 Weight:0.9953	429
UV	ancestors	w2v safe tf-idf	KMeans SpectralClustering KMeans	Random forest Random forest XGBClassifier	Micro:0.9316 Macro:0.9107 Weight:0.9287	161
UV	strmatch	w2v tf-idf	SpectralClustering KMeans	XGBClassifier Random forest	Micro:0.9792 Macro:0.9792 Weight:0.9791	434
UV	strmatch	w2v tf-idf	SpectralClustering KMeans	XGBClassifier Random forest	Micro:0.9792 Macro:0.9792 Weight:0.9791	434
MVSC	scc	safe tf-idf	None	None	Micro:0.7012 Macro:0.6651 Weight:0.6651	472

In terms of the size of the dataset, each problem has about 500 solutions written in C/C++ programming languages. The folders containing the source code solutions are hierarchically organized, such as the root folder has the name of the problem. Each problem's folder has one sub-folder for each distinct algorithmic solution, thus containing all the source code solutions belonging to a particular algorithmic approach. The name of the source code solution is the *id* from *infoarena* towards that solution.

Building the dataset has been performed by running a custom developed *scrapper*[2] which downloaded only correct source code solutions. The selected problems are: **strmatch** (ro., potrivirea sirurilor), **ancestors** (ro., stramosi), **schi**, **scc** (ro., componente tari conexe, en. strongly connected components), **gcd** (ro., cel mai mare divizor comun, en., greatest common divisor), **cppp** (ro., cele mai apropriate două puncte in plan, en., closest pair of points in plane), **eval**(ro., evaluare, en., evaluation), **invmod**(ro., invers modular), **mst**(ro., arbore partial de cost minim, en., minimum cost spanning tree), and **inv**. Detailed problem statements and source code solutions are open access on *infoarena* site.

From Table 1, it can be seen that the UV (i.e., unsupervised voting) has the highest score for all the problems, except for one problem. From here, we can conclude that the initialization of a dataset with the maximum frequency method improves, but it depends on the quality of the clustering methods and embeddings methods. In other words, if all the above methods yield a bad result, then the unsupervised voting method will not give perfect results. The implementation of proposed solution may be found in *AlgoDistinctSolutions*[3] git repository and contains the code for downloading source code solutions, the labeled dataset and the implementation itself.

5 Conclusions

The paper presents a custom-designed semi-supervised data analysis pipeline that correctly assigns source code solutions of problems from competitive programming to their correct algorithmic approach. For validation purposes, we have created a manually labelled dataset and have used several embedding methods in the context of different clustering algorithms integrated into an unsupervised learning pipeline. The fully unsupervised learning has also produced excellent results, such that we conclude the proposed method solves the tackled problem correctly. In the future, we plan to extend the number of problems in the dataset and make all the source code solutions compile. Further improvements may include designing a metric that provides information about the probability of the number of clusters from the dataset. Finally, we believe that this work opens the way towards other practical applications of Deep Learning based NLP on source code, such as generating alternative implementations in large codebases or plagiarism detection by comparing source code embeddings.

Acknowledgements. This work was partially supported by the grant 135C /2021 "Development of software applications that integrate machine learning algorithms", financed by the University of Craiova.

[2] InfoarenaScrappingTool, https://github.com/Arkin1/InfoarenaScrappingTool.
[3] AlgoDistinctSolutions, https://github.com/Arkin1/AlgoDistinctSolutions.

References

1. Alon, U., Zilberstein, M., Levy, O., Yahav, E.: code2vec: Learning distributed representations of code. In: Proceedings of the ACM on Programming Languages, vol. 3(POPL), pp. 1–29 (2019)
2. Arthur, M.P.: Automatic source code documentation using code summarization technique of NLP. Procedia Comput. Sci. **171**, 2522–2531 (2020)
3. Azcona, D., Arora, P., Hsiao, I.H., Smeaton, A.: user2code2vec: embeddings for profiling students based on distributional representations of source code. In: Proceedings of the 9th International Conference on Learning Analytics & Knowledge, pp. 86–95 (2019)
4. Barchi, F., Parisi, E., Urgese, G., Ficarra, E., Acquaviva, A.: Exploration of convolutional neural network models for source code classification. Eng. Appl. Artif. Intell. **97**, 104075 (2021)
5. Burrows, S., Uitdenbogerd, A.L., Turpin, A.: Application of information retrieval techniques for source code authorship attribution. In: Zhou, X., Yokota, H., Deng, K., Liu, Q. (eds.) DASFAA 2009. LNCS, vol. 5463, pp. 699–713. Springer, Heidelberg (2009). https://doi.org/10.1007/978-3-642-00887-0_61
6. Chen, Z., Monperrus, M.: A literature study of embeddings on source code. arXiv preprint arXiv:1904.03061 (2019)
7. Chihada, A., Jalili, S., Hasheminejad, S.M.H., Zangooei, M.H.: Source code and design conformance, design pattern detection from source code by classification approach. Appl. Soft Comput. **26**, 357–367 (2015)
8. Iacob, R.C.A., et al.: A large dataset for multi-label classification of algorithmic challenges. Mathematics **8**(11), 1995 (2020)
9. Jiang, L., Liu, H., Jiang, H.: Machine learning based recommendation of method names: how far are we. In: 2019 34th IEEE/ACM International Conference on Automated Software Engineering (ASE), pp. 602–614. IEEE (2019)
10. Jones, K.S.: A statistical interpretation of term specificity and its application in retrieval. J. Documentation (1972)
11. Massarelli, L., Di Luna, G.A., Petroni, F., Baldoni, R., Querzoni, L.: SAFE: self-attentive function embeddings for binary similarity. In: Perdisci, R., Maurice, C., Giacinto, G., Almgren, M. (eds.) DIMVA 2019. LNCS, vol. 11543, pp. 309–329. Springer, Cham (2019). https://doi.org/10.1007/978-3-030-22038-9_15
12. Mikolov, T., Chen, K., Corrado, G., Dean, J.: Efficient estimation of word representations in vector space. arXiv preprint arXiv:1301.3781 (2013)
13. Rabin, M.R.I., Mukherjee, A., Gnawali, O., Alipour, M.A.: Towards demystifying dimensions of source code embeddings. In: Proceedings of the 1st ACM SIGSOFT International Workshop on Representation Learning for Software Engineering and Program Languages, pp. 29–38 (2020)
14. Kanaan-Izquierdo, S., Andrey Ziyatdinov, A.P.L.: Multiview and multifeature spectral clustering using common eigenvectors. Pattern Recogn. Lett. 102, 30–36 (2018)
15. Shi, K., Lu, Y., Chang, J., Wei, Z.: Pathpair2vec: an ast path pair-based code representation method for defect prediction. J. Comput. Lang. **59**, 100979 (2020)

Application of Long Short-Term Memory Neural Networks for Electric Arc Furnace Modelling

Maciej Klimas(✉) and Dariusz Grabowski

Silesian University of Technology, Gliwice, Poland
{maciej.klimas,dariusz.grabowski}@polsl.pl

Abstract. Worldwide steelmaking industry strongly relies on the use of electric arc furnaces (EAFs). EAFs make use of electric arc phenomenon for melting scrap steel and consequently they can be sources of power quality issues, such as harmonics or voltage flickering. In order to design and implement effective systems for power quality improvement, it is necessary to dispose of an adequate model. Due to the complicated nature of the electric arc phenomenon, it is difficult to develop such an accurate model. Researchers around the world use different approaches, mostly relying on deterministic modelling with the addition of a stochastic ingredient. In this paper, we propose an approach which similarly is based on a deterministic equation enhanced with stochastic ingredients describing its coefficients. The identification of the time series of the equation is carried out by means of genetic algorithms. Next, we developed two models using long short-term memory artificial neural network (LSTM) for recreating the time series of the coefficients while remaining their stochastic properties. The second model also applies another LSTM for the reduction of stochastic-like residuals emerging from comparison of the first model with measurement data.

Keywords: Electric arc furnace · EAF · Long short-term memory · LSTM · Artificial neural network

1 Introduction

Operation of many branches of the industry rely strongly on the use of steel because this material is commonly used in many different applications. Environmental conditions and limited resources make recycling of the steel especially important. There are different methods for that, but one of the most widely used is via electric arc furnaces (EAFs). Although EAFs support ecology by their contribution to steel recycling, they can also cause some environmental problems. Electric arc which is used for scrap steel melting is a stochastic phenomenon and due to high power consumption it can negatively influence the power system. Disturbances caused by EAFs are harmonics, flickering or voltage

H. Yin et al. (Eds.): IDEAL 2021, LNCS 13113, pp. 166–175, 2021.
https://doi.org/10.1007/978-3-030-91608-4_17

sags and swells, among others. This consequently leads to excessive wear of the electrical equipment and additional power losses. In electrical engineering many different approaches are used to mitigate such problems. Optimal design and implementation of such systems improving power quality require an accurate model of the considered load. The goal of the development of new EAF models is to increase their accuracy in order to ensure that simulations of circuits including EAF models would provide better feedback to the engineers. Consequently, exploitation of the power system would be more optimal.

Nonlinear, dynamic and unpredictable nature of the electric arc makes it difficult to provide such accurate models. Literature shows that there are many different approaches to this problems. Some of them are based only on deterministic solutions connected with piecewise linear approximations of measurement data as in [4], or approaches based on other functions such as exponential or hyperbolic models [2]. Other researchers apply popular Cassie-Mayr models as in [6]. Many try to recreate the stochastic inregients for example with hidden Markov model [16] or Ornstein-Uhlenbeck process [15]. The nature of electric arc phenomenon is so complex that the modelling process can be simplified using machine learning methods. For instance artificial neural networks (ANNs) were used in modelling based on electric arc length [5] or directly on voltage and current measurement data [3]. In this paper we propose an analysis based on the widely used deterministic equation obtained from power balance, but with a stochastic ingredients recreated by long short-term memory (LSTM) networks. Mentioned equation is widely used in modelling of the electric arc phenomenon, and was also used by many researchers [8,11] or [10]. Such hybrid solutions were also proposed by other authors, as in [7], but our approach is not yet to be met in the literature, due to the application of deep learning methods in hybrid model based on deterministic equation obtained from the power balance. Deep learning modelling in terms of EAFs has been already used only for temperature prediction [12] or material engineering approach [14]. It is worth mentioning that presented models are especially oriented towards deep learning methods because, simpler neural networks have been already investigated in our earlier works, such as [9].

2 Deterministic Model of Electric Arc Furnace

2.1 Measurement Data

LSTM model is an example of an approach that needs relatively large amount of data for proper training and validation. In order to develop it, we have gathered measurement data of the EAF. It consists of voltage and current waveforms recorded during different stages of industrial size EAF work cycle. The area of interest of our analysis is the part of waveforms connected with the melting stage. The waveforms represent one-phase electric arc phenomenon as our goal is to develop accurate model of single arc column. As presented in Fig. 1 the waveforms are characterized by various level of deformation, especially voltage waveform strongly deviates from the sinusoidal shape. Additionally, the data depicts two kinds of stochastic behavior which are the main causes of inaccuracy of the deterministic models. One is connected with the change of the amplitude (more visible in long-term plot) while the second can be described as a high frequency ripples mostly visible on peaks of the voltage waveform (apparent in short-term plot). Those features of the data are the base for developing LSTM models which should reflect them accordingly.

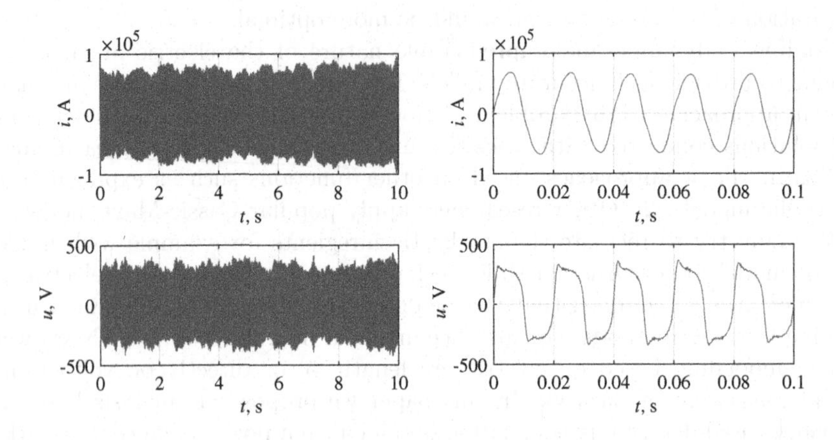

Fig. 1. Long and short-term samples of EAF measured current and voltage waveforms.

2.2 Power Balance Equation

Proposed concept bases on an equation describing the electric arc phenomenon, introduced in [1]. It originates from power balance approach and results in the following form:

$$k_1 r^n(t) + k_2 r(t) \frac{dr(t)}{dt} = \frac{k_3}{r^{m+2}(t)} i^2(t), \tag{1}$$

where: k_1, k_2, k_3 - model coefficients, m, n - parameters related to EAF work cycle, $r(t)$ - arc radius, $i(t)$ - arc current.

Additional equation allows calculating the arc voltage:

$$u(t) = \frac{k_3}{r^{m+2}(t)} i(t). \tag{2}$$

Equation (1) is widely used in deterministic approaches of EAF modelling and also as a starting point for development of stochastic models. Similarly, we apply this equation to the analysis of the measurement data.

2.3 Estimation of Equation Coefficients

Parameters of Eq. (1) can take different values depending of current stage of EAF work cycle. It is assumed that in the melting stage (corresponding with our measurement data) m and n parameters are constant and take values of $m = 0$ and $n = 2$. In that way only k coefficients remain unknown. Our approach assumes that those coefficients are stochastically changed in time in a way that would reflect stochastic changes of the EAF characteristic and that they remain constant for a single period of the measured waveforms, that is 20 ms. In order to estimate their values, the waveforms were divided into period-long frames. Current was then applied as an input to the Eq. (1) with assumed constant values of k coefficients. The output in form of the voltage waveform was then compared with measured voltage. We have used a genetic algorithm in order to minimize the RMSE error between those waveforms for each period-long window. The population size was equal to 50. Mutation was adaptive in terms of direction and step size with respect to the last generation. Crossover was based on a random binary vector, which selected genes from each of the parents. In that way we obtained three separate time series representing values of each k coefficients of Eq. (1) fitted to measurement data. Figure 2 presents those time series, which later are used for LSTM networks training and validation.

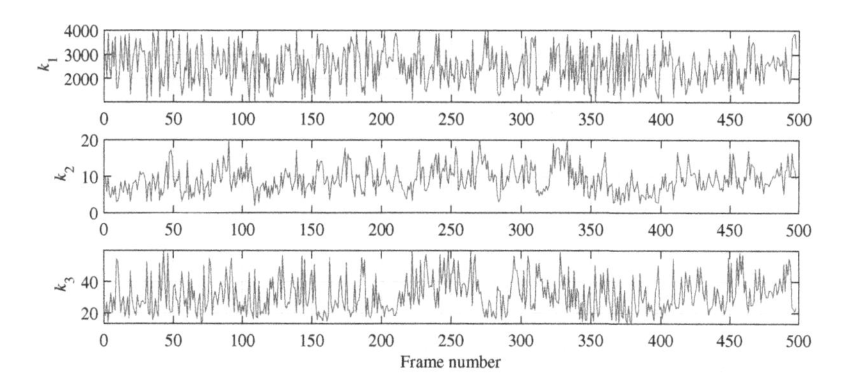

Fig. 2. k coefficients time series estimated from the measurement data.

3 Stochastic Model of the Electric Arc Furnace

3.1 Representation of Coefficients Time Series

Main idea behind LSTM choice for EAF modelling is to provide effective way to represent stochastic behavior of the k coefficients time series. The ability to learn and reproduce such behavior is the main advantage in favor of deep learning methods. Additionally it simplifies overall approach of the EAF modelling because it does not rely of detailed analysis of physical phenomena and reasons behind their stochasticity.

In order to model each time series, we propose using separate LSTM networks, one for each k coefficient. The structure of each network is the same - simple and suited for the regression problem. Figure 3 presents their topology.

Figure 4 additionally presents a more detailed view of the single LSTM cell structure, according to software documentation [13]. Various numbers of hidden units were tested and eventually the networks were fitted with 300 hidden units in the LSTM layer. Their training was conducted using Adam optimizer with a variable learning rate starting from 0.005. Number of training epochs was also investigated in order to obtain the best results and generally the training stopped after 500 epochs, with the batch size equal to 128. The training process was also visualized with the exemplary loss curve presented in Fig. 5. The other loss curves were not very different, and because of that we have limited ourselves to presenting an exemplary plot for LSTM network representing k_1 time sequence. The models were designed in Matlab software and the computing infrastructure included a portable computer with Intel Core i7 processor (4 cores, 1.8 GHz), with 16 GB RAM and Windows 10 operating system.

The developed networks allowed simulating each of the k coefficients time series, independently. Applying such simulated time series to the Eq. (1) and providing the current data as an input allows obtaining the final simulated output of the model – the voltage waveform.

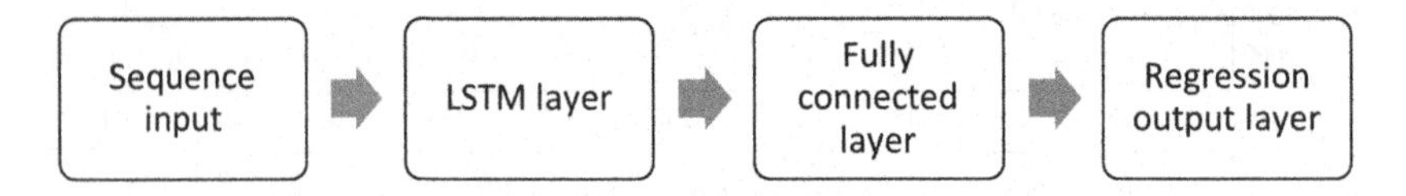

Fig. 3. Structure of used LSTM network.

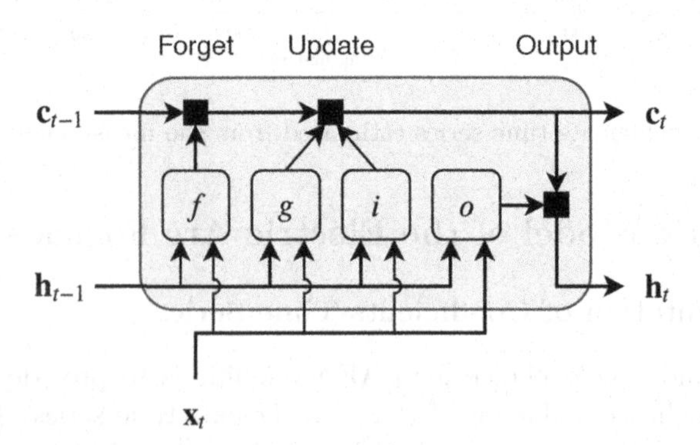

Fig. 4. Structure of single LSTM cell, where c - cell state, h - hidden (output) state and x - input data [13].

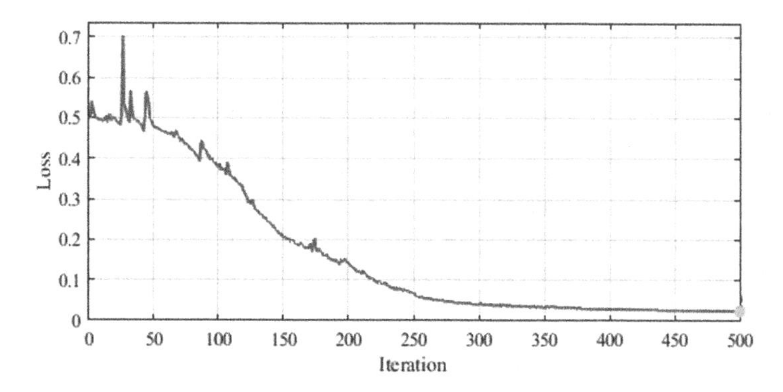

Fig. 5. Exemplary loss curve for training of the LSTM network representing k_1 time sequence.

3.2 Representation of High Frequency Ripples

Approach described in previous section can be considered as a complete model of the electric arc phenomenon. However, it only simulates the stochasticity of k coefficients, and assumes their constancy throughout each period of the signal. Changes in their values only cause variations in the shape of period-long frames of the EAF characteristic – the high frequency ripples around peak values are not modelled.

In order to enhance the previous model, we propose application of the second parallel path of signal analysis. The idea is to apply a high-pass filter to the voltage waveform and then develop separate deep learning model which will be trained and validated with such data. The waveform obtained after application of the high pass filter 600 Hz cut off frequency is presented in Fig. 6. In this way overall output voltage would be a sum of the output of the first LSTM model and the second LSTM model.

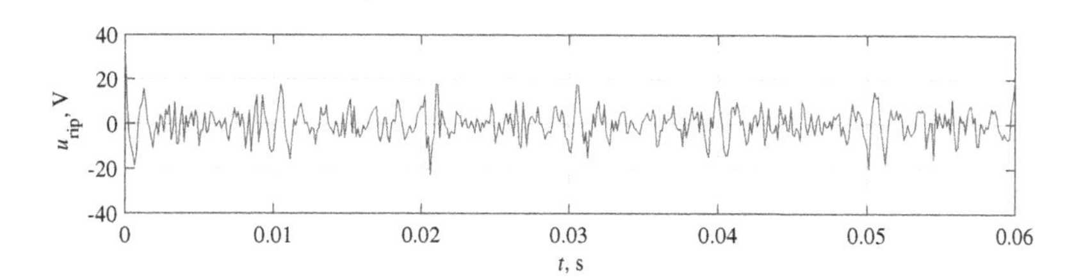

Fig. 6. Waveform of the high frequency ripples filtered from the measurement data.

4 Simulation Results

This section presents simulation results for each of the proposed models. Figure 7 shows exemplary realizations of k coefficients time series, as the first step of the modelling procedure. This data was then used to obtain output voltage waveform of the first LSTM model, which is presented in the same figure. The plot also presents whole $u-i$ characteristic of the EAF compared with the measurement data. In terms of qualitative assessment of the results, as presented, first model correctly reflects changes in the shape of the characteristic.

Previous section also justifies application of the next model which is specifically designed to recreate high frequency ripples of the voltage waveform. This waveform was added to the output of the first LSTM model resulting in overall output which reflects both kinds of stochastic behavior of the EAF characteristic – changes in the amplitude and high frequency ripples. Figure 8 shows output of the second LSTM network and the $u-i$ characteristic of EAF obtained as overall output of the second LSTM model compared with measurement data.

Problems related to stochasticity of the EAF modelling also extend to the evaluation aspects. Direct comparison of exemplary realizations with measurement data cannot provide any objective error measure. Although the realizations of the voltage waveform seem similar, as presented in Fig. 9, it is not correct to compare them directly with classic error measures used in signal processing. In this paper we limit our research to qualitative comparision of the exemplary realizations of the stochastic time series of k coefficients, the voltage ripples and the output of the models.

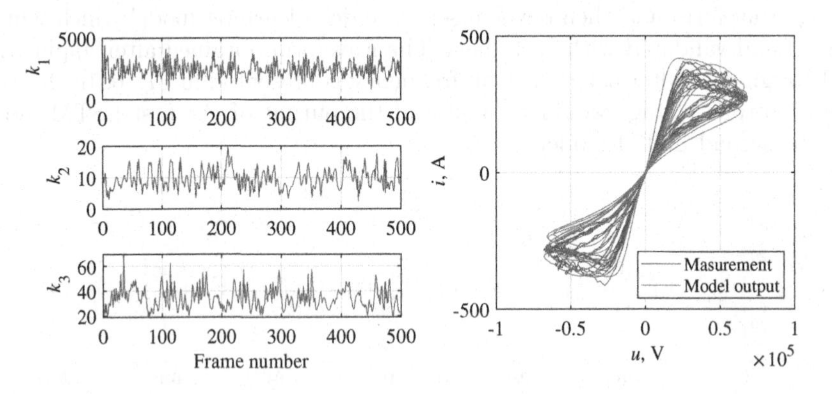

Fig. 7. Exemplary realization of k coefficients time series and comparison of the $u-i$ characteristic obtained from the LSTM model and measurements.

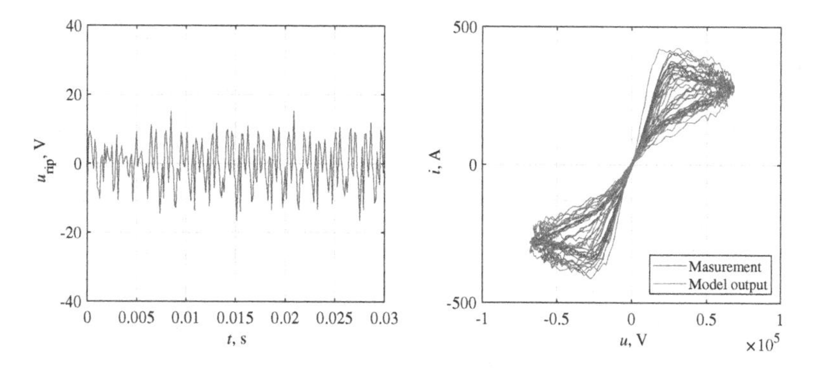

Fig. 8. Exemplary realization of voltage ripples and comparison of the $u-i$ characteristic obtained from the second LSTM model and measurements.

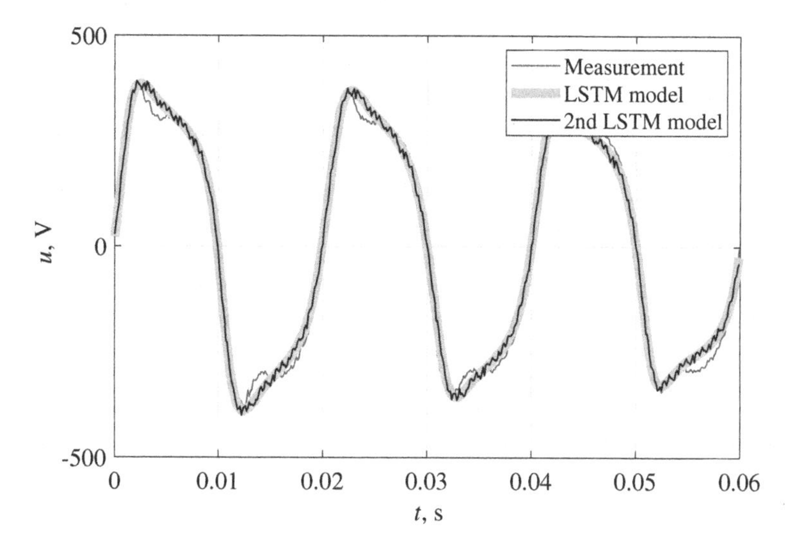

Fig. 9. Exemplary realizations of voltage obtained from both LSTM models compared with measurement data.

5 Conclusions

This paper proposes two models of the EAF, both based on LSTM networks. The first is based on the estimation of coefficients of the equation describing electric arc phenomenon. Variations of each coefficient are described with a time series representing their value in following periods of 20 ms. Those time series are recreated with three separate LSTM networks with the same topology. The second model is based on the first one. It uses additional LSTM network trained to recreate high frequency ripples which can be observed after applying high pass filter to measured voltage weaveform. Overall output of the second model is the sum of first model output and the ripples waveform simulated with the second model.

As presented in Sect. 4 the exemplary realizations of the stochastic signals are similar to the measurement data. Both estimated k coefficients time series, voltage ripples and output $u-i$ characteristic are characterized by the same features as the estimations from measured waveforms. Qualitatively, proposed approach gives promising results which meet the expectations of possible implementations based on deep learning models in the EAF modelling. Presented approach however, does not yet propose detailed optimization of LSTM models parameters. This idea requires an introduction of the objective measure of the discrepancy between two realizations of stochsatic processes. The stochasticity prevents from using classic measures such as root mean square error (RMSE). In order to correctly measure the accuracy of the new models and to provide a goal function for optimization of LSTM parameters, introduced formula should be based on the probability distributions of the signals and their autocorrelations. The performance of the presented models is good and comparable with the results obtained through other approaches. Due to the aforementioned issues, a direct comparison with other solutions is not trivial. However, the presented work reflects the first trials related to the application of deep learning in this context. In future, further improvements in terms of network topology or applied algorithms will be considered. Planned work includes the application of stochastic differential equations, fractional calculus, and chaos theory. Final conclusions based also on the performance comparision will consider not only new models, including the ones presented in this paper, but also existing approaches that can be found in the literature.

Acknowledgments. Research co-financed from government funds for science for years 2019–2023 as part of "Diamond Grant" programme and co-financed by the European Union through the European Social Fund (grant POWR.03.05.00-00-Z305).

References

1. Acha, E., Semlyen, A., Rajakovic, N.: A harmonic domain computational package for nonlinear problems and its application to electric arcs. IEEE Trans. Power Delivery **5**(3), 1390–1397 (1990). https://doi.org/10.1109/61.57981
2. Bhonsle, D.C., Kelkar, R.B.: New time domain electric arc furnace model for power quality study. In: 2014 IEEE 6th India International Conference on Power Electronics (IICPE), pp. 1–6 (2014). https://doi.org/10.1109/IICPE.2014.7115761
3. Chang, G.W., Shih, M.F., Chen, Y.Y., Liang, Y.J.: A hybrid wavelet transform and neural-network-based approach for modelling dynamic voltage-current characteristics of electric arc furnace. IEEE Trans. Power Delivery **29**(2), 815–824 (2014). https://doi.org/10.1109/TPWRD.2013.2280397
4. Chittora, P., Singh, A., Singh, M.: Modeling and analysis of power quality problems in electric arc furnace. In: 2015 Annual IEEE India Conference (INDICON), pp. 1–6 (2015). https://doi.org/10.1109/INDICON.2015.7443638
5. Garcia-Segura, R., Vázquez Castillo, J., Martell-Chavez, F., Longoria-Gandara, O., Ortegón Aguilar, J.: Electric arc furnace modeling with artificial neural networks and arc length with variable voltage gradient. Energies **10**(9) (2017). https://www.mdpi.com/1996-1073/10/9/1424

6. Golestani, S., Samet, H.: Generalised Cassie-Mayr electric arc furnace models. IET Gener. Transm. Distrib. **10**(13), 3364–3373 (2016). https://ietresearch.onlinelibrary.wiley.com/doi/abs/10.1049/iet-gtd.2016.0405
7. Golestani, S., Samet, H.: Polynomial-dynamic electric arc furnace model combined with ann. Int. Trans. Electr. Energy Syst. **28**(7), e2561 (2018). https://onlinelibrary.wiley.com/doi/abs/10.1002/etep.2561, e2561 ITEES-17-0706.R1
8. Gomez, A.l., Durango, J.J.M., Mejia, A.E.: Electric arc furnace modeling for power quality analysis. In: 2010 IEEE ANDESCON, pp. 1–6 (2010). https://doi.org/10.1109/ANDESCON.2010.5629655
9. Klimas, M., Grabowski, D.: Application of shallow neural networks in electric arc furnace modelling. In: 2021 IEEE International Conference on Environment and Electrical Engineering and 2021 IEEE Industrial and Commercial Power Systems Europe (EEEIC / I&CPS Europe), pp. 1–6 (2021). https://doi.org/10.1109/EEEIC/ICPSEurope51590.2021.9584512
10. Grabowski, D., Walczak, J.: Analysis of deterministic model of electric arc furnace. In: 2011 10th International Conference on Environment and Electrical Engineering, pp. 1–4 (2011). https://doi.org/10.1109/EEEIC.2011.5874805
11. Grabowski, D., Walczak, J.: Deterministic model of electric arc furnace - a closed form solution. COMPEL - Int. J. Comput. Math. Electr. Electron. Eng. **32**(4), 1428–1436 (2013). https://doi.org/10.1108/03321641311317220
12. Leon-Medina, J.X., et al.: Deep learning for the prediction of temperature time series in the lining of an electric arc furnace for structural health monitoring at Cerro Matoso (CMSA). In: Engineering Proceedings, vol. 2(1) (2020). https://www.mdpi.com/2673-4591/2/1/23
13. MATLAB: version 9.7.0.1261785 (R2019b). The MathWorks Inc., Natick, Massachusetts (2019)
14. Son, K., et al.: Slag foaming estimation in the electric arc furnace using machine learning based long short-term memory networks. J. Mater. Res. Technol. **12**, 555–568 (2021). https://www.sciencedirect.com/science/article/pii/S2238785421002118
15. Starkloff, H.J., Dietz, M., Chekhanova, G.: On a stochastic arc furnace model. Studia Universitatis Babeş-Bolyai, Mathematica **64**(2), 151–160 (2019). https://doi.org/10.24193/subbmath.2019.2.02
16. Torabian Esfahani, M., Vahidi, B.: A new stochastic model of electric arc furnace based on hidden Markov model: a study of its effects on the power system. IEEE Trans. Power Delivery **27**(4), 1893–1901 (2012). https://doi.org/10.1109/TPWRD.2012.2206408

An Empirical Study of the Impact of Field Features in Learning-to-rank Method

Hua Yang[1,2](✉) and Teresa Gonçalves[1]

[1] Department of Informatics, Universidade de Évora, Évora, Portugal
{huayang,tcg}@uevora.pt
[2] School of Computer Science, Zhongyuan University of Technology, Zhengzhou, China

Abstract. In learning-to-rank for information retrieval, a ranking model is usually learned using features that are extracted from the different fields of the documents and naively combined. However, such a conventional way to learn a ranking model does not accurately reflect the utility and contribution of the fields and may also risk joining highly correlated features from different fields. It lacks an empirical analysis of how field-grouped features determine or influence the performance of the ranking models learned. In this paper, we classify features by the fields they are extracted from and investigate the role of using field-grouped features in the learning-to-rank method, particularly to see whether using field-grouped features leads to a different and better performance than using the naively combined feature list. Our experiments, on two large scale publicly available learning-to-rank benchmark datasets, show that ranking models learned using field-grouped features have competitive advantages over the models learned using a naively combined feature list, and that aggregation results of different fields present a better performance. These results suggest that learning ranking models using field-grouped features can be useful to obtain more effective performances in learning-to-rank methods.

Keywords: Learning-to-rank · Field features · Aggregation

1 Introduction

Learning-to-rank has been extensively researched in web search in recent years. In the state-of-the-art approaches, most features are extracted from multiple fields of the documents [1,2]. For example, web documents are usually organized into multiple fields, such as *title*, *anchor*, *url*, *body*, etc. These features represent different perspectives and domain knowledge of the documents and may contain complementary information to be considered for improving the ranking performance. Conventionally, these features are naively combined into one single list and a ranking model is learned using the whole list.

H. Yin et al. (Eds.): IDEAL 2021, LNCS 13113, pp. 176–187, 2021.
https://doi.org/10.1007/978-3-030-91608-4_18

However, such a conventional way to learn a ranking model does not accurately reflect the utility and contribution of the different fields and can not provide insightful information about the domain. As such, there is a lack on empirical analysis about how ranking models using field-grouped features determine or influence the ranking performance. Therefore, in this paper, we investigate the effectiveness of using field-grouped features in the learning-to-rank method. Specifically, we try to address the following research questions: (i) How differently do these fields contribute in learning a ranking model, and how do the models learned using field-grouped features perform compared to the ones using the naively combined feature list? (ii) Will the aggregation of results obtained from the models learnt from different fields lead to a better performance when compared to using the naively combined feature list?

To address these research questions, we classify features by the fields they are extracted from and learn a set of ranking models using the field-grouped features. We also study the effectiveness of the aggregation results from different fields. We experimented on two large scale publicly available learning-to-rank benchmark datasets with eight popular learning-to-rank methods.

The results show that learning ranking models using field-grouped features can be useful to obtain better performances in the learning-to-rank experiments. To summarize, the main contributions are as follows: (i) A thorough comparison of the performance on different field-grouped features. The results show that using field-grouped features are competitive with the use of a naively combined feature list; (ii) A thorough analysis to show that using a naively combined feature list can have the potential risk of joining highly correlated features from different fields and may reduce the performance; (iii) An evidence presentation that aggregation results of different fields leads to better performance when compared to the use of a naively combined feature list.

The rest of this paper is outlined as follows. In Sect. 2 we review related work; in Sect. 3 we describe the method used; in Sect. 4 we describe the experiments performed and analyse the results; and finally in Sect. 5 we draw the conclusions.

2 Related Work

2.1 Learning-to-rank

In the information retrieval area, machine learning techniques can be applied to build ranking models for information retrieval systems, and this is known as learning-to-rank (LTR) [3]. Typically, the training data consists of three elements: training queries Q, the associated documents D, and the corresponding relevance judgments or the gold standard *qrel* file for the query and document pairs [3]. The learning algorithms are then used to generate a Learning to Rank (LETOR) model. The creation of a testing data for evaluation is very similar to the creation of the training data which includes the testing queries and the associated documents. To these testing queries, the Learning to Rank model is jointly used with a retrieval model to sort the documents according to their relevance to the query, and return a corresponding ranked list of the documents as the response to the query.

Learning-to-rank methods have been proposed making use of different machine learning algorithms. Typically, they can be categorized into three main groups: pointwise, pairwise, and listwise approaches [3]. These categories define different input and output spaces, use different hypotheses, and employ different loss functions. The pointwise approach regards the relevance degrees as numerical or ordinal scores, and the learning to rank problem is formulated as a regression or a classification problem. The pairwise approach deals with the ranking problem by treating documents pairs as training instances, and trains models via the minimization of related risks. The listwise approach regards an entire set of documents associated with a query as instances in the training, and trains a ranking function through the minimization of a listwise loss function.

2.2 Field Role in Information Retrieval

The impact and role of the document structure and fields on retrieval effectiveness have been studied for years [1,2,4–13]. Fielded extensions to different retrieval models have been proposed to satisfy the structured document retrieval. In this work, a number of related works are reviewed, and we categorize them depending on the models they used to explore the role of fields, including learning-to-rank models [1,4,5], information retrieval models [7–9], latent semantic models [10,11], and neural networks related models [2,12,13].

Fernando Diaz [4] trained a learning-to-rank model using a set of labelled features from domain experts. The labelled features were extracted from different fields, and the results showed that using feature labels outperformed the baselines. Chen *et al.* [1] investigated the effectiveness of the learning-to-rank approach for entity search. Their approach represented entities in multi-field representations. Field features presented different performance on different types of searches. Azarbonyad *et al.* [5] studied the learning-to-rank method for multi-label text classification and found that the titles were more informative than the body and other sources of information.

Wu [6] used the multiple linear regression technique to obtain weights to the component retrieval systems. The experiments showed that the weighting strategy outperformed the best component system and other data fusion methods. Kim and Croft [7] introduced a field relevance model to estimate term field weights for structured document retrieval. They found that field weighting strategy based on the proposed framework improved retrieval effectiveness. Zhiltsov *et al.* [8] proposed a fielded sequential dependence model. They found that different field weighting schemes were effective for different types of queries in ad-hoc entity retrieval. Jimmy *et al.* [9] investigated the effectiveness of boosting the title field of structured documents. They found that boosting of titles improved retrieval effectiveness for some queries.

Works by Shen *et al.* [10,11] explored title and body features using latent semantic models, and they observed that the title field was more effective than the body field for document retrieval.

Ogilvie and Callan [12] proposed a fielded extension of a standard language model. Their approach scored a structured document by linearly combining the probabilities of query terms calculated from multiple document fields. The best document representations were the full text, in-link text, and title. Mitra *et al.* [13] used a ranking model combining local and distributed representations. They studied body text from raw HTML content, and found that the combined model was better than either neural network individually and outperformed the baselines. Zamani *et al.* [2] used the neural network to learn multiple-field document representation for ad-hoc retrieval. Results showed that scoring the whole document jointly was better than generating a per-field score and aggregating for non-neural field weighting.

3 Methods

This section describes our methodology in studying the effectiveness of the field-grouped features in learning-to-rank. Conventionally, potentially informative features are extracted from different fields, and naively combined to create one single feature list which is then used to train a learning-to-rank model. We regard this way as the standard learning-to-rank method. To analyse the role of field features, we train models using field-grouped features. Thus, we propose an extension to the standard learning-to-rank method and note it as fLTR (field learning-to-rank) method.

Different from the standard learning-to-rank method, where one single feature list is built, the fLTR method tags the extracted features into multiple lists based on the document fields that they are originated from; then, a set of field-based learning-to-rank models is trained, with each model using one field-grouped features list. In such a way, we are able to investigate the contribution of each field in improving the ranking performance.

Supposing the training data is consisted of a number of k queries $q_{\mathrm{j}}(j = 1, ..., k)$, their associated documents $d^{(j)}$, the field F_{i} that features are extracted from, and the corresponding relevance judgments $y^{(j)}$. First, we use the field-grouped features from field F_{i} and employ the learning algorithm to learn a field learning-to-rank model H_{i}. Then, given a new query q, the model H_i is used to rank the documents according to their relevance to q. The corresponding result h_{i} is obtained. Assuming that n $(n \in m)$ field learning-to-rank models $(H_1, H_2, \cdots, H_n)$ are learned, n ranking lists will be obtained.

Since these results are obtained from different fields, we can compare them to the ones learnt using a standard learning-to-rank method and investigate how field-grouped features perform. Then, we use an aggregation method to combine the results. For each document pair, we take the score of the document d $(d \in D)$ with regard to query q for each ranking model H_i as $s_{F_i}(d)$ and combine the scores using an aggregation function *G[x]*. The aggregated score S can be defined as, $S(d) = G[\ s_{F_1}(d), \ldots, s_{F_i}(d), \ldots, s_{F_n}(d)\ ]$. Let the weight obtained by each field be denoted as w_i, the fused score S of the document d is, $S(d) = G[\ w_1 s_{F_1}(d), \ldots, w_i s_{F_1}(d), \ldots, w_n s_{F_n}(d)\ ]$. Being $G[x]$ a linear function, the fused score is denoted as, $S(d) = \sum_{i=1}^{n} w_i s_{F_1}(d)$

Following this idea, we use the aggregation method to aggregate the scores obtained from the set of ranking lists $(h_1, h_2, \cdots, h_n)$ which are correspondingly obtained from the field learning-to-rank models $(H_1, H_2, \cdots, H_n)$. And finally, the final ranking list H is obtained. We summarize the notations and their definitions in Table 1.

Table 1. Notations followed in our methods.

Notation	Meaning
D	Data collection in the context of search corpus
d	A document in D
q_j	Query contained in the training dataset, j=1...k
$d^{(j)}$	Associated document to q_j in the training dataset
$y^{(j)}$	Corresponding relevance judgment to document and query pair $(m^{(j)}, q_j)$
q	Query used for testing
m	Number of fields of d
n	Number of fields used for learning the ranking models, $n < m$
F_i	The ith field of a document d
H_i	Ranking model learnt using information from the field F_i
h_i	Ranking list obtained using model H_i for query q
$s_{F_i}(d)$	Score of the document d regard to query q and obtained using the model H_i
$G[x]$	Aggregation algorithm
S	Aggregated score
H	Final ranking list

4 Experiments and Results

In this section, we describe the experimental setup and results obtained on two publicly available large scale benchmarks.

4.1 Experimental Setup

Datasets. Microsoft LETOR 4.0 contains a package of benchmark datasets for research on learning-to-rank [14]. Our experiments are conducted on two of them: MQ2007 and MQ2008. Both datasets contain queries and documents sampled from real search engines. Each query-document pair is labelled by human annotators with relevance judgments: the larger the value, the more relevant the query-document pair. Both datasets contain 46 features for learning to rank. In MQ2007, there are about 1700 queries with tagged documents, and in MQ2008, there are about 800 queries with tagged documents.

Baselines. We compare the proposed fLTR method with eight state-of-the-art learning to rank baselines, including two point-wise algorithms (MART [15] and Random Forests [16]), three pair-wise algorithms (RankNet [17], RankBoost [18] and LambdaMART [19]), and three list-wise algorithms (AdaRank [20], Coordinate Ascent [21] and ListNeT [22]).

Field Learning-to-rank. We use the field learning-to-rank method to train ranking models on MQ2007 and MQ2008 datasets on five different fields: *anchor*, *body*, *title*, *url*, and *whole document* (abbr. *wholedoc*). In the original feature list, features *PageRank*, *Inlink number*, *Outlink number* and *Number of child page* [14] are not categorized into one of those fields, so these fields are removed for our experiments. Also, the features *Number of slash in URL* and *Length of URL* are excluded to maintain consistency in numbers and characteristics over the features for each field. Thus, each field is characterized by seven features extracted using the same information retrieval models. These features are listed in Table 2. We use the eight learning-to-rank algorithms introduced to train ranking models on each group of the field features.

Table 2. Selected features for each field.

Features	Fields				
	Body	Anchor	Title	Url	Wholedoc
TF of the field	✓	✓	✓	✓	✓
IDF of the field	✓	✓	✓	✓	✓
TF*IDF of the field	✓	✓	✓	✓	✓
DL of the field	✓	✓	✓	✓	✓
BM25 of the field	✓	✓	✓	✓	✓
LMIR.ABS of the field	✓	✓	✓	✓	✓
LMIR.DIR of the field	✓	✓	✓	✓	✓

Aggregation of Field Learning-to-rank Models. We use the aggregation method to combine the results from each field and obtain an overall result on all fields. We test with two widely used score aggregation methods, CombSUM, and CombMNZ [23]. All five fields have the same weights in our experiments. To make every field score have the same scale, the scores obtained with fLTR models are first normalized before aggregation. We test two normalization methods, Min-Max and Z-score normalization; they are shown to achieve better performance when compared to the other normalization methods [24].

Evaluation Measures. The popular information retrieval measure Normalized Discounted Cumulative Gain (NDCG) [25] is used to evaluate the performances in our experiments. The higher the NDCG score, the better the ranking quality. To observe performances of the ranking models at different positions, we report the NDCG values at the ranks of 5, 10, 15, 20, and 30. We use the original dataset splitting (including separate train, test, and validation) and conduct 5-fold cross-validation on the dataset. All models are tuned based on their performances on the validation set in terms of NDCG@10. The performance results are obtained over the test set and reported.

4.2 Experimental Results

All baselines and developed field learning-to-rank models are implemented in RankLib[1]. Figure 1 presents the experimental results on the two datasets: MQ2007 and MQ2008. The results are averaged scores over five runs on 5-fold cross validation. We categorize the results into eight groups, with each group representing one of the eight learning-to-rank algorithms: MART (MR), RankNet (RN), RankBoost (RB), AdaRank (AR), Coordinate Ascent (CA), LambdaMART (LM), ListNet (LN) and Random Forests (RF). For each algorithm and at different ranking positions of NDCG, we present the scores of the five fields using field learning-to-rank methods, the aggregation scores, and the baseline score. We have similar observations on both the MQ2007 and MQ2008 datasets, so we report them together.

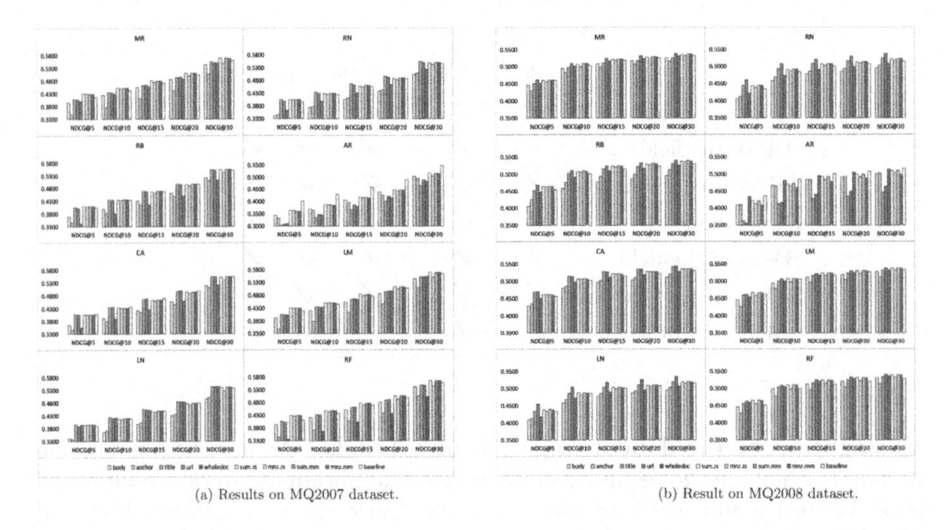

(a) Results on MQ2007 dataset.

(b) Result on MQ2008 dataset.

Fig. 1. Comparison of performances on fLTR models using field features and their aggregation.

For the field features, on the MQ2007 dataset, *title* field performs the best, is able to surpass all baselines with RN, RB and LN, achieves competitive results as the baselines with CA. *Url* field also shows better performance than the baselines with RN, LN (except NDCG@5) and RB measured on NDCG@15, achieves competitive results with CA. *Whole document* field surpasses the baselines with LN. *Body* and *anchor* field fail in all cases. On the MQ2008 dataset, *url* field performs the best, is able to surpass all baselines except the AR ones and LN measured on NDCG@5. *Title* field also shows better performance than the baselines on five learning-to-rank approaches: RN, RB, CA, LN, and RF. *Whole*

[1] https://sourceforge.net/p/lemur/wiki/RankLib/.

document field surpasses the baselines with LN and RF. *Body* field presents better performance with RF measured on NDCG@20 and NDCG@30. *Anchor* field fails in all cases.

Aggregated results further improve the performance. In detail, on the MQ2007 dataset, the aggregated results using the CombSUM method are able to surpass all baselines with MR, RN, LN and RF, and most baselines with RB and LN; fail in AR and CA; show better performances for MinMax than Z-score normalization in most cases. The results using the CombMNZ method and Min-Max also perform better than the baselines except for the AR and CA ones. Comparatively, using CombMNZ and Z-score shows slightly worse performance; surpasses the baselines in most cases with MR, LM, RF, and some cases with RN, RB, and LN; fails in AR and CA. And on the MQ2008 dataset, the aggregated results using the CombSUM method are able to surpass all baselines except the ones with AR, and they show similar performances for Min-Max and Z-score normalization. The results using the CombMNZ method and Min-Max also perform better than the baselines except for AR. Comparatively, using CombMNZ and Z-score show slightly worse performance, failing in AR as well as some in MR, LM, and LN.

4.3 Empirical Analyses

Analysis on Feature Correlations. We now investigate the reasons why the fLTR method is better than the baseline method by using fewer features obtained from a single field. An important study to do is to analyse the correlation between features. Using the features extracted on the MQ2008 dataset, we perform statistical significance tests between field features using pairwise comparison with the Pearson correlation. The comparison results are shown in Fig. 2, with (a), (b), (c) and (d) being, respectively, average, median, max, and sum on each field. By the analysis of the heatmap in all four measurements, the most remarkable point is the strong correlation between the *body* and *whole document* features. Another point is the high correlation between these two pairs: *title* and *anchor*, *title* and *url* features.

However, when applying machine learning approaches, more discriminant and diverse features with minimum correlations are expected [26]. These highly correlated features have commonalities, and naively combining these correlated features in learning-to-rank can lead to decreased ranking performance. For example, if we include one feature that is completely correlated with another feature, we obtain twice as much influence as the other features. This emphasis given to the redundant features reduces the influence of the remaining features. This can produce a biased result.

We conduct further experiments to investigate how highly correlated features affect the ranking performance. For example, we train a model using the combined features from *body* and *whole document* fields, and compare to the ones trained using either *body* or *whole document* fields. We use LambdaMART as the training method and evaluate the results at different positions of NDCG. The results are presented in Table 3 for the MQ2008 dataset. By the analysis,

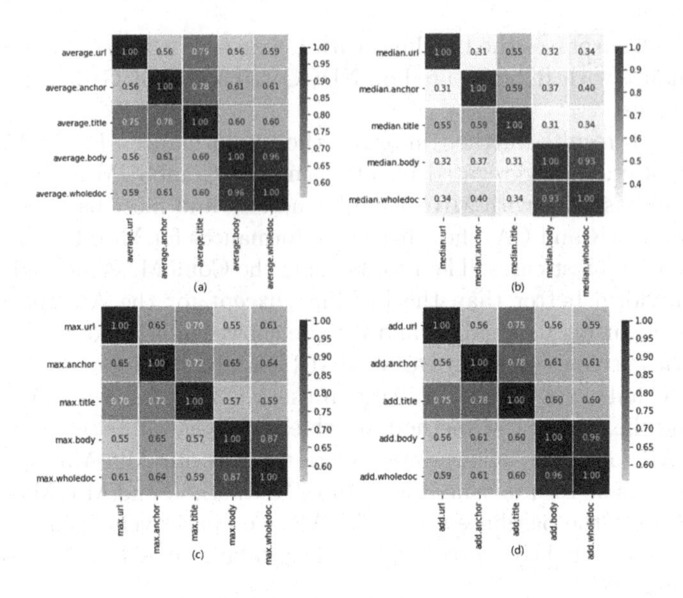

Fig. 2. Correlations between field features.

we can see that joining *body* and *whole document* features together decreases the performance compared to either using *body* or *whole document* features individually. Similarly, joining *anchor* or *url* features also shows worse performance than using *title* features alone.

Table 3. Highly correlated field-grouped features on MQ2008 dataset.

Features	NDCG@5	NDCG@10	NDCG@15	NDCG@20	NDCG@30
Comparison between body and wholedoc					
Body	0.4457	0.4944	0.5108	0.5184	0.5279
Wholedoc	0.4532	0.5002	0.5164	0.5248	0.5318
Body+wholedoc	0.4420	0.4893	0.5064	0.5131	0.5216
Comparison between title and anchor/url					
Title	0.4587	0.5014	0.5163	0.5240	0.5331
Title+anchor	0.4471	0.4929	0.5070	0.5143	0.5229
title+url	0.4465	0.4985	0.5109	0.5184	0.5273

These correlation analysis, along with the developed rankers' effectiveness presented in Sect. 4.2, provide useful insights for the interpretability of our field learning-to-rank method. Theoretically, the baseline method naively joins together all features from different fields; this risks joining highly correlated features, and can impact the performance of the model. In contrast, our approach trained several learning-to-rank models using features from one field only (each) and then aggregated these results; this can avoid the direct feature correlation interference in the learning to rank process. From the experimental results, the fLTR models show better performance than the baselines, proving that naively joining field features does interfere and negatively impact the ranking performance.

Overall Performance. Another significant point is considering the performance of fLTR and their aggregations in three types of learning-to-rank algorithms. To evaluate the overall ranking performance for all the algorithms in the experiments, we use the winning number measure [27] to count the winning numbers over both datasets. We calculate the winning numbers for eight standard learning-to-rank algorithms, and the results are shown in Fig. 3. For the two pointwise ranking algorithms, Random Forests (RF) performs much better than the MART (MR) algorithm. Considering the three pairwise algorithms, RankNet (RN) and RankBoost (RB) present better and similar performance when compared to the third algorithm LambdaMART (LM). For the listwise algorithms, ListNet (LN) performs the best with Coordinate Ascent (CA) sightly worse; AdaRank (AR) fails in all cases.

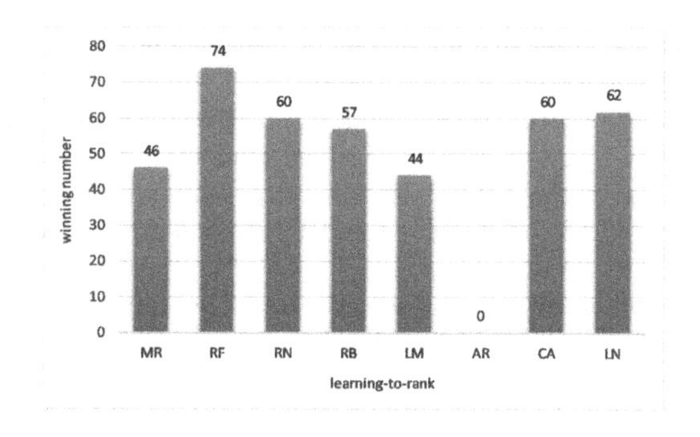

Fig. 3. Comparison cross the two datasets by different learning-to-rank methods.

5 Conclusion

In this paper, we extend the standard learning-to-rank approach and propose the field learning-to-rank method for information retrieval. We conduct experimental results on two large scale datasets with eight widely used learning-to-rank algorithms. Our experimental results show that field learning-to-rank is a competitive method compared to the standard learning-to-rank approach, and its aggregation results present much better performance than using the standard learning-to-rank approach in the overwhelming majority of cases. Our empirical analysis also shows that there are strong correlations between some features from different fields, like *body* and *whole document*, *title* and *anchor*, *title* and *url* fields. Moreover, the advantage of a field learning-to-rank approach is to avoid the potential risks of joining highly correlated features from different fields. Finally, our experiments also leads to the conclusion that *title* or *url* field contributes more than the remaining fields in the field level learning-to-rank.

References

1. Chen, J., Xiong, C., Callan, J.: An empirical study of learning to rank for entity search. In: Proceedings of the 39th International ACM SIGIR Conference on Research and Development in Information Retrieval, pp. 737–740 (2016)
2. Zamani, H., Mitra, B., Song, X., Craswell, N., Tiwary, S.: Neural ranking models with multiple document fields. In: Proceedings of the eleventh ACM International Conference on Web Search and Data Mining, pp. 700–708 (2018)
3. Liu, T.Y., et al.: Learning to rank for information retrieval. Found. Trends® Inf. Retrieval **3**(3), 225–331 (2009)
4. Diaz, F.: Learning to rank with labeled features. In: Proceedings of the 2016 ACM International Conference on the Theory of Information Retrieval, pp. 41–44 (2016)
5. Azarbonyad, H., Dehghani, M., Marx, M., Kamps, J.: Learning to rank for multi-label text classification: combining different sources of information. Nat. Lang. Eng. **27**(1), 89–111 (2021)
6. Wu, S.: Linear combination of component results in information retrieval. Data Knowl. Eng. **71**(1), 114–126 (2012)
7. Kim, J.Y., Croft, W.B., et al.: A Field Relevance Model for Structured Document Retrieval. In: Baeza-Yates, R. (ed.) ECIR 2012. LNCS, vol. 7224, pp. 97–108. Springer, Heidelberg (2012). https://doi.org/10.1007/978-3-642-28997-2_9
8. Zhiltsov, N., Kotov, A., Nikolaev, F.: Fielded sequential dependence model for ad-hoc entity retrieval in the web of data. In: Proceedings of the 38th International ACM SIGIR Conference on Research and Development in Information Retrieval, pp. 253–262 (2015)
9. Zuccon, G., Koopman, B.: Boosting titles does not generally improve retrieval effectiveness. In: Proceedings of the 21st Australasian Document Computing Symposium, pp. 25–32 (2016)
10. Shen, Y., He, X., Gao, J., Deng, L., Mesnil, G.: Learning semantic representations using convolutional neural networks for web search. In: Proceedings of the 23rd International Conference on World Wide Web, pp. 373–374 (2014)
11. Shen, Y., He, X., Gao, J., Deng, L., Mesnil, G.: A latent semantic model with convolutional-pooling structure for information retrieval. In: Proceedings of the 23rd ACM International Conference on Conference on Information and Knowledge Management, pp. 101–110 (2014)
12. Ogilvie, P., Callan, J.: Combining document representations for known-item search. In: Proceedings of the 26th annual international ACM SIGIR conference on Research and development in Informaion retrieval, pp. 143–150 (2003)
13. Mitra, B., Diaz, F., Craswell, N.: Learning to match using local and distributed representations of text for web search. In: Proceedings of the 26th International Conference on World Wide Web, pp. 1291–1299 (2017)
14. Qin, T., Liu, T.Y.: Introducing letor 4.0 datasets. arXiv preprint arXiv:1306.2597 (2013)

15. Friedman, J.H.: Greedy function approximation: a gradient boosting machine. Ann. Stat. **29**, 1189–1232 (2001)
16. Breiman, L.: Random forests. Mach. Learn. **45**(1), 5–32 (2001)
17. Burges, C., et al.: Learning to rank using gradient descent. In: Proceedings of the 22nd International Conference on Machine Learning, pp. 89–96 (2005)
18. Freund, Y., Iyer, R., Schapire, R.E., Singer, Y.: An efficient boosting algorithm for combining preferences. J. Mach. Learn. Res. **4**, 933–969 (2003)
19. Wu, Q., Burges, C.J., Svore, K.M., Gao, J.: Adapting boosting for information retrieval measures. Inf. Retrieval **13**(3), 254–270 (2010)
20. Xu, J., Li, H.: Adarank: a boosting algorithm for information retrieval. In: Proceedings of the 30th Annual International ACM SIGIR Conference on Research and Development in Information Retrieval, pp. 391–398. ACM (2007)
21. Metzler, D., Croft, W.B.: Linear feature-based models for information retrieval. Inf. Retrieval **10**(3), 257–274 (2007)
22. Cao, Z., Qin, T., Liu, T.Y., Tsai, M.F., Li, H.: Learning to rank: from pairwise approach to listwise approach. In: Proceedings of the 24th International Conference on Machine Learning, pp. 129–136. ACM (2007)
23. Fox, E., Shaw, J.: Combination of multiple searches. In: The Second Text Retrieval Conference (TREC-2), pp. 243–252. NIST Special Publication (1994)
24. Singh, D., Singh, B.: Investigating the impact of data normalization on classification performance. Appl. Soft Comput. **97**, 105524 (2020)
25. Järvelin, K., Kekäläinen, J.: Cumulated gain-based evaluation of IR techniques. ACM Trans. Inf. Syst. (TOIS) **20**(4), 422–446 (2002)
26. Das, A., Dasgupta, A., Kumar, R.: Selecting diverse features via spectral regularization. Adv. Neural Inf. Process. Syst. **25**, 1583–1591 (2012)
27. Qin, T., Liu, T.Y., Xu, J., Li, H.: Letor: a benchmark collection for research on learning to rank for information retrieval. Inf. Retrieval **13**(4), 346–374 (2010)

An Optimized Evidential Artificial Immune Recognition System Based on Genetic Algorithm

Rihab Abdelkhalek(✉) and Zied Elouedi

LARODEC, Institut Supérieur de Gestion de Tunis, Université de Tunis, Tunis, Tunisia
zied.elouedi@gmx.fr

Abstract. Artificial Immune Recognition System (AIRS) is a supervised learning algorithm inspired by the human immune system mechanisms. Many versions of AIRS have been established and have shown a great success in classification and prediction areas. However, these versions suffer from several problems such as the use of many random mechanisms in several levels and the requirement of a large number of user-predefined parameters. This may decrease the classification performance and affect the decision making process. Furthermore, the majority of AIRS versions are unable to deal with uncertainty spreading into the different phases of the classification process. Therefore, we propose in this paper, a new optimized AIRS version using Genetic Algorithm (GA) and Gradient Descent method under the belief function framework. In our approach, we aim to remove the randomness in AIRS by including a Genetic Algorithm as a search heuristic in the training phase. Besides, we propose a parameter optimization method under uncertainty to determine the optimal or near-optimal parameters values. Experiments performed on real world data sets proved that our approach beats traditional AIRS versions in terms of classification accuracy.

Keywords: Artificial immune recognition systems · Uncertain reasoning · Belief function theory · Parameter optimization · Genetic algorithm

1 Introduction

Artificial Immune Recognition System (AIRS) [16] is a supervised learning algorithm based on the human immune system mechanisms where the training data are considered as antigens. Such immunity-based technique has achieved exceptional performance on a broad range of classification problems. New versions have been suggested based on the original version of AIRS [4] in order to improve the classification accuracy namely, AIRS2 [17] and AIRS3 [10]. AIRS2 has been first introduced as a revision of the initially existing AIRS method. The classification process of AIRS2 is based on the k-nearest neighbors (kNN) where

H. Yin et al. (Eds.): IDEAL 2021, LNCS 13113, pp. 188–195, 2021.
https://doi.org/10.1007/978-3-030-91608-4_19

the antigen is assigned according to the majority vote of the selected neighbors. Based on the same intuition, an extended version of AIRS2 has also been proposed [11] where random mutations are replaced with deterministic crossovers. On the other hand, when it comes to AIRS3, its main purpose is to integrate a new component, denoted by *numRepAg*, holding the number of represented antigens for each cell. In such version, the value of k turns into *numRepAg* rather than the whole neighbors. The classification is then performed by deriving the total of *numRepAg* of all the chosen antigens with the same label. In this paper, we are interested in AIRS3 where an extended optimized version under uncertainty would be proposed. Actually, few approaches of AIRS have been introduced relying on uncertainty theories like the fuzzy set theories [5] and possibility theory [9]. The belief function theory (BFT) [6,14], also known as evidence theory or Dempster-Shafer theory offers a flexible tool for modelling imperfect information. Recently, two evidential versions of AIRS2 and AIRS3 [1,12] have been proposed under the BFT. Nevertheless, these approaches do not take into account the weight of the training antigens. That is why, authors in [2] presented a new evidential weighted AIRS which has improved the classification results. From there, an optimization of such method has been presented in [3] where a parameter optimization process is carried out by minimizing the error rate of the classification. Such evidential approach has achieved the best results compared to [1] and [2]. Although the efficiency of all these methods, they highly depend on a large number of user-predefined parameters and they involve many random techniques in diverse stages of the process, which may reduce the classification accuracy. Besides, the classical versions of AIRS show a deficiency related to the random mutation mechanisms in order to find a good candidate memory cell. More specifically, the attributes to mutate are picked in a random manner and their new assigned values are also randomly obtained. To alleviate this problem, many search techniques such as simulated annealing, hill climbing, tabu search and evolution-inspired techniques have been proposed. Among the aforementioned search techniques, we opt for one of the most widely used evolution-inspired search techniques, which is the Genetic Algorithm (GA). In this paper, we propose a new version of optimized evidential AIRS3 which incorporates a GA as a search method for the development of memory cells. The proposed Optimized EAIRS-GA approach produces excellent classification results with a minimum number of required parameters.

The rest of the paper is organized as follows: Sect. 2 recalls the fundamental concepts of the belief function theory. Section 3 represents an overview of the AIRS. Our hybrid approach, Optimized EAIRS-GA, is introduced in Sect. 4. Then, the experimental results carried out on real-world data sets are detailed in Sect. 5. Finally, Sect. 6 concludes this paper.

2 Belief Function Theory

Using the belief function theory [6,14,15], a problem domain is represented by a finite set of events called the frame of discernment Θ. It integrates the hypotheses related to the problem [15] such that: $\Theta = \{\theta_1, \theta_2, \cdots, \theta_n\}$. A basic belief assignment (*bba*) is the belief related to the elements of Θ such that $m : 2^\Theta \rightarrow [0,1]$ and $\sum_{A \subseteq \Theta} m(A) = 1$. Each mass of $m(A)$ quantifies the amount of belief assigned to an event A of Θ. Let m_1 and m_2 be two *bba*'s corresponding to reliable and independent information sources. The fusion of evidence can be ensured using Dempster's combination rule such that: $(m_1 \oplus m_2)(A) = k. \sum_{B,C \subseteq \Theta : B \cap C = A} m_1(B) \cdot m_2(C)$.

$$where\ (m_1 \oplus m_2)(\varnothing) = 0 \quad and \quad k^{-1} = 1 - \sum_{B,C \subseteq \Theta : B \cap C = \varnothing} m_1(B) \cdot m_2(C).$$

To make decisions, the *pignistic probability* denoted $BetP$ can be used, where $BetP$ is defined as follows: $BetP(A) = \sum_{B \subseteq \Theta} \frac{|A \cap B|}{|B|} \frac{m(B)}{(1 - m(\varnothing))}$ $for\ all\ A \subseteq \Theta$.

3 Related Works on AIRS

Artificial Immune Recognition System (AIRS) [4] is an immune-inspired supervised learning algorithm. It is well-known for its effectiveness in resolving machine learning problems. In literature, two main versions of AIRS have been suggested which are AIRS2 [17] and AIRS3 [10]. The first phase of AIRS is the initialization of the memory cell (MC) pool and the ARB pool, where a data normalization is performed and distances between all instances in the system are calculated. A selection of the best match memory cell, denoted by *mc_match*, is then achieved. The next step is the selection of the candidate memory cell denoted by *mc_candidate* as the ARB cell owning the highest stimulation value. Finally, an update of the MC pool is carried out by the introduction of memory cells. At the end of the training phase, we obtain a final MC pool, which will be employed during the classification process. The classification phase is accomplished using the k-nearest neighbors algorithm (k-NN) where the unlabeled antigen is classified by the majority-vote of the k nearest memory cells. In AIRS3 [10], a new component called the number of represented antigens (*numRepAg*) is computed for each memory cell in the MC pool during the training phase. Thanks to this parameter, the size of the MC pool is reduced and the k value becomes the number of represented antigens rather than the number of neighbors. Next, a computation of the sum of *numRepAg* of all the chosen cells with the same class as the training is performed. At the end, the testing antigen will take the class having the highest sum of *numRepAg*.

4 Optimized EAIRS-GA

In our work, we aim to improve the classification performance of AIRS3 by integrating one of the most known search and optimization techniques, which is Genetic Algorithm (GA) [8], under the belief function theory. The proposed approach is denoted by "Optimized EAIRS-GA". Its main goal is to optimize the traditional AIRS3 version by handling the randomness and reducing the number of required parameters under the belief function framework.

4.1 Phase 1: Initialization

During this step, a data normalization is carried out based on min max normalization in order to obtain attribute values between 0 and 1. Then, since attributes in datasets do not have the same relevance, information gains [13] is derived of each attribute A in the training set T denoted by $Gain(T, A)$. Accordingly, the weighted affinity threshold is calculated.

4.2 Phase 2: Learning Reduction

This procedure of the Optimized EAIRS-GA approach is made up of four steps:

1. **Computation of the number of represented antigens of MC**
   As in AIRS3, a computation of the number of represented antigens by each memory cell ($numRepAg$) is performed at the beginning of the learning reduction phase. As a result, we obtain a reduced MC pool containing a limited number of weighted antigens. This weight ($numRepAg$) will be taken into account during the classification process.
2. **Selection of the best-stimulated memory cell and parental selection**
   This phase plays an important role in determining the best candidate memory cell. First, for a given test antigen ag from the training set T, we select the best stimulated memory cell having the same class as ag and getting the lowest affinity value. Once the mc_match is identified, the parental selection is performed. This latter is inspired by the main operators of GA where we choose the mc_match and the training antigen ag as the parental chromosomes. Based on GA metaphor, each cell presents a piece of chromosomes.
3. **Crossover and mutation**
   Crossover and mutation is considered as one of the most significant stages in the genetic algorithm. During this phase, we rely on the whole arithmetic crossover as an efficient technique where the next generation is obtained from the results of arithmetic operations applied on the parental cells. In our approach, a linear combination of parents' chromosomes is proceeded based on a predefined weight parameter ranging between 0 and 1. So that, produced children has the weighted sum of two parental alleles. Two offsprings are produced by crossover, which represent the training antigen feature values and the mc_match feature values. Once the crossover is performed, a mutation is carried out in order to obtain more competent offsprings.
4. **Selection of a candidate memory cell and memory cell introduction**
   After the crossover and mutation, we will proceed to the selection of a candidate memory cell by computing affinities and testing the stopping criterion. In fact, we compare the affinity between the training antigen and the resulting offspring with the affinity threshold (AT). In the case where this latter is higher than the calculated affinity, we select this cell as a candidate memory cell denoted by $mc\_candidate$. If it is not the case, another production of new offsprings will be achieved until we reach the stopping criterion. After that, if the obtained $mc\_candidate$ matches the training antigen better than the actual

mc_match does, it will be integrated to the set of memory cells, becoming a long-lived memory cell. Furthermore, if the affinity between the *mc_match* and the *mc_candidate* is inferior to the product of the affinity threshold and affinity threshold scalar, the *mc_candidate* will take the place of the *mc_match* in the MC pool. Otherwise, the two cells will be maintained. Once this step is accomplished, the training on the present antigen ag is completed and the procedure resumes as described above for all remaining cells.

4.3 Phase 3: Evidential Optimization

We opt for the gradient descent to optimize one of the classification parameters which is γ. In fact, the classification process of the Evidential AIRS3 is based on two parameters α and γ. The value of α is set to 0.95 as introduced in [7], while the value of γ is computed using the optimization technique. Relying on the MC pool gotten in the learning reduction phase, we select the most stimulated cells to the current antigen and we compute the *bba* based on the EKNN formalism [3,7]. At this level, we define a cost function as the Mean Squared Error (MSE) in the final MC pool.

Once the cost function is computed, we opt for the gradient descent in order to find the optimum value of the parameter γ. Note that the optimal value of parameter γ after the tuning is the one provided with the minimum cost function.

4.4 Phase 4: Evidential Classification

During the classification phase, we rely on the optimal value of γ calculated during the optimization step and we create new optimized *bba*'s for the s selected memory cells. Unlike the traditional versions of AIRS, we do not consider all the k nearest neighbors as sources of evidence but only the resulting antigens developed in the previous step. According to these pieces of evidence, the selection of the right class will be processed and the *bba*'s of the chosen antigens will be generated [7]. Finally, we employ the Dempster rule of combination to fusion the collected *bba*'s created for each memory cell in the MC pool.

To make decisions, we use the pignistic probability ($BetP$) and we assign to the test sample the class having the highest $BetP$ value.

5 Experimental Study

In order to assess the effectiveness of our optimized evidential AIRS-GA approach denoted by EAIRS-GA, we carried out experiments on real world data sets chosen from the U.C.I repository[1]. During these tests, we compare the performance of our approach against five other AIRS methods namely, AIRS2 [17], AIRS3 [10], Fuzzy AIRS2 [5], Evidential AIRS2 (EAIRS2) [12] and Optimized Weighted Evidential AIRS3 [3].

[1] https://archive.ics.uci.edu/ml/index.php.

5.1 Framework

We relied on five real data sets which are described in Table 1. We mention that *InstancesNb*, *AttributesNb* and *ClassesNb* represent respectively the number of antigens, the number of attributes and the number of classes. During our tests, we used the following parameter values: Clonal rate = 10, Mutation rate = 0.4, HyperClonal Rate = 2, Number of resources = 200, Stimulation threshold = 0.3, Affinity threshold scalar = 0.2. Moreover, we have carried out our experiments with different values of k = [2, 3, 5, 7, 9, 10].

Table 1. Description of the data sets

Databases	*InstancesNb*	*AttributesNb*	*ClassesNb*
Cryotherapy (C)	90	6	2
Wine (W)	178	13	3
Pima Indians Diabetes (PID)	768	8	2
Fertility (F)	100	9	2

5.2 Evaluation Criterion

To evaluate the effectiveness of our approach, we opt for one of the most popular evaluation criterion which is the Percent Correctly Classified (PCC). This latter calculates the proportion of cases that have been correctly classified. Furthermore, we relied on Cross-Validation (CV) as a powerful evaluation technique. More specifically, we employed 10-fold Cross-Validation (CV) where we compute the average value of the resulting accuracies in the 10 cases of cross validation.

5.3 Experimental Results

To show the achievement of our approach, we did a comparison between our proposed AIRS with the traditional ones, which are AIRS2, AIRS3, Fuzzy AIRS2, EAIRS2 and optimized WE-AIRS3. This is done by comparing the Percent Correctly Classified (PCC) obtained from the different values of k. Figure 1 illustrates the results of our experiences through the varied selected k values. From these charts, we notice that our approach optimized EAIRS-GA surpasses the other versions of AIRS in terms of accuracy almost for all the data sets and for all the employed k values.

Table 2 depicts the mean PCC obtained using the different values of k. From the results given in Table 2, we can observe that our optimized EAIRS-GA offers interesting prediction accuracy. Indeed, for all the databases, the average PCC of our method reaches the highest value. For instance, for W database, the mean PCC reaches 96.99% with our approach while it represents only 62.13% for AIRS2, 85.43% for AIRS3, 79.45% for fuzzy AIRS2, 65.88% for EAIRS2 and 89.21% for optimized WE-AIRS3. These results confirm the effectiveness of our method against all the other classical AIRS algorithms under certain and uncertain environments.

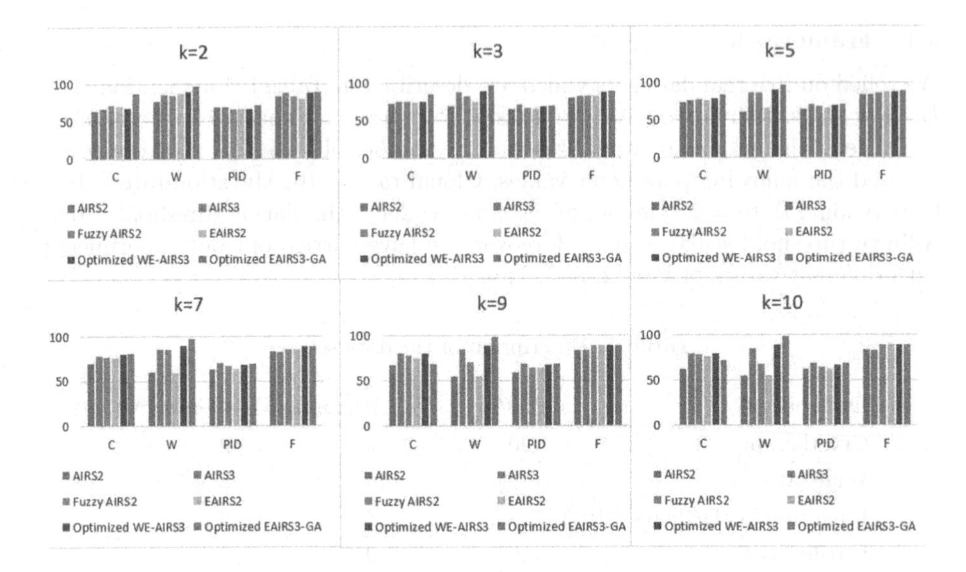

Fig. 1. *PCC* of used databases for various values of k

Table 2. The mean *PCC* (%)

Data sets	AIRS2	AIRS3	Fuzzy AIRS2	EAIRS2	Optimized WE-AIRS3	Optimized EAIRS3-GA
C	67.88	75.51	75.51	74.11	76.22	**79.55**
W	62.13	85.43	79.45	65.88	89.21	**96.99**
PID	63.51	69.43	65.88	64.25	67.48	**69.68**
F	82.50	83.43	85.32	84.84	87.77	**88.33**

6 Conclusion

In this paper, we proposed a new evidential AIRS based on one of the most efficient search and optimization algorithms, the Genetic Algorithm. Thanks to this method, we improved the existing AIRS3 by reducing the number of the predefined user parameters as well as the propagated randomness. We also employed the gradient descent for the selection of the optimal parameter value. Furthermore, our Optimized EAIRS-GA handles uncertainty relying on the belief function theory. A comparison between our approach and the other AIRS methods for different real data sets has confirmed the effectiveness of our work.

References

1. Abdelkhalek, R., Elouedi, Z., et al.: A belief classification approach based on artificial immune recognition system. In: Lesot, M.-J. (ed.) IPMU 2020. CCIS, vol. 1238, pp. 327–340. Springer, Cham (2020). https://doi.org/10.1007/978-3-030-50143-3_25
2. Abdelkhalek, R., Elouedi, Z.:WE-AIRS: a new weighted evidential artificial immune recognition system. In International Conference on Robotics and Artificial Intelligence, pp. 874–881 (2020)
3. Abdelkhalek, R., Elouedi, Z.: Parameter optimization and weights assessment for evidential artificial immune recognition system. In: Li, G., Shen, H.T., Yuan, Y., Wang, X., Liu, H., Zhao, X. (eds.) KSEM 2020. LNCS (LNAI), vol. 12275, pp. 27–38. Springer, Cham (2020). https://doi.org/10.1007/978-3-030-55393-7_3
4. Castro, L.N., De Castro, L.N., Timmis, J.: Artificial immune systems: a new computational intelligence approach. Springer Science and Business Media (2002)
5. Chikh, M.A., Saidi, M., Settouti, N.: Diagnosis of diabetes diseases using an artificial immune recognition system2 (AIRS2) with fuzzy k-nearest neighbor. J. Med. Syst. **36**(5), 2721–2729 (2012)
6. Dempster, A.P.: A generalization of bayesian inference. J. Roy. Stat. Soc. Series B (Methodological) **30**, 205–247 (1968)
7. Denoeux, T.: A k-nearest neighbor classification rule based on dempster-shafer theory. IEEE Trans. Syst. Man Cybern. **25**, 804–813 (1995)
8. Forrest, S.: Genetic algorithms. ACM Comput. Surv. **28**, 77–80 (1996)
9. Hentech, R., Jenhani, I., Elouedi, Z.: Possibilistic AIRS induction from uncertain data. Soft Comput. **20**(1), 3–17 (2015). https://doi.org/10.1007/s00500-015-1627-3
10. Jenhani, I., Elouedi, Z.: Re-visiting the artificial immune recognition system: a survey and an improved version. Artif. Intell. Rev. **42**(4), 821–833 (2012). https://doi.org/10.1007/s10462-012-9360-0
11. Jenhani, I., Elouedi, Z.: AIRS-GA: a hybrid deterministic classifier based on artificial immune recognition system and genetic algorithm. In: 2017 IEEE Symposium Series on Computational Intelligence, pp. 1–7, IEEE (2017)
12. Lahsoumi, A., Elouedi, Z.: Evidential artificial immune recognition system. In: Douligeris, C., Karagiannis, D., Apostolou, D. (eds.) KSEM 2019. LNCS (LNAI), vol. 11775, pp. 643–654. Springer, Cham (2019). https://doi.org/10.1007/978-3-030-29551-6_57
13. Quinlan, J.R.: Improved use of continuous attributes in C4. 5. J. Artif. Intell. Res. **4**, 77–90 (1996)
14. Shafer, G.: A Mathematical Theory of Evidence. Princeton University Press, New Jersey (1976)
15. Smets, P.: The transferable belief model for quantified belief representation. In: Quantified Representation of Uncertainty and Imprecision, pp. 267–301 (1998)
16. Watkins, A.: A resource limited artificial immune classifier. In: The 2002 Congress on Evolutionary Computation, pp. 926–931 (2002)
17. Watkins, A., Timmis, J.: Artificial immune recognition system (AIRS): revisions and refinements. In: AISB 2004 Convention, pp. 18 (2004)
18. Zouhal, L.M., Denoeux, T.: An evidence theoretic kNN rule with parameter optimization. IEEE Trans. Syst. Man Cybern. **28**(2), 263–271 (1998)

New Arabic Medical Dataset for Diseases Classification

Jaafar Hammoud(✉), Aleksandra Vatian, Natalia Dobrenko, Nikolai Vedernikov, Anatoly Shalyto, and Natalia Gusarova

ITMO University, Kronverksky pr. 49, Saint Petersburg, Russia

Abstract. The Arabic language suffers from a great shortage of datasets suitable for training deep learning models, and the existing ones include general non-specialized classifications.

In this work, we introduce a new Arab medical dataset, which includes two thousand medical documents collected from several Arabic medical websites, in addition to the Arab Medical Encyclopedia. The dataset was built for the task of classifying texts and includes 10 classes (Blood, Bone, Cardiovascular, Ear, Endocrine, Eye, Gastrointestinal, Immune, Liver and Nephrological) diseases.

Experiments on the dataset were performed by fine-tuning three pre-trained models: BERT from Google, Arabert that based on BERT with large Arabic corpus, and AraBioNER that based on Arabert with Arabic medical corpus.

Keywords: Arabic · Text classification · Medical

1 Introduction

These days, unstructured text is everywhere, from our conversations and comments on social media to emails, websites, etc. Their processing by means of artificial intelligence (AI) techniques and Natural Language Processing (NLP) is at a high level [1], including such an important type of text as medical records. Extracting useful information from medical texts and reports automatically plays a pivotal and important role in supporting medical decision-making [2–4].

Unfortunately, AI and NLP tools are used to a much lesser extent for processing texts in Arabic, especially in the medical field. This can be attributed to several reasons, the most important of which is that the use of electronic medical reports and records was not common in most Arabic countries. The handwritten and not digitally structured reports are still in common use in the majority of Arab countries. In addition, the Arab countries, which are somehow more advanced than their counterparts, use English as the main language in their medical records. Therefore, the Arabic language faces a clear shortage in the availability of appropriate data to train models capable of contributing to improving the quality of health care and helping doctors diagnose and treat [5].

In this work, we present a new Arabic dataset specialized in classifying medical texts and show its use on three NLP models.

H. Yin et al. (Eds.): IDEAL 2021, LNCS 13113, pp. 196–203, 2021.
https://doi.org/10.1007/978-3-030-91608-4_20

2 Related Work

In the period extending from the sixties of the last century until the end of the first decade of the current one, shallow classification algorithms were dominant, which depend on the process of extracting properties and features manually, which requires hiring experts in the field to deal with the data that the algorithms learn on.

Shallow learning algorithms are affected by the preprocessing process a lot, from the segmentation of words, data cleaning, to conducting an accurate statistical study about it, and then the representation of data in a way that is easier for computers, such as the Bag-of-words (BOW) [6], GloVe [7], word2vec [8], N-gram [9] and TF-IDF [10]. These methods generate a vector representing properties extracted from the text that will then feed the classifier.

The Bayesian network [11], and the hidden Markov network [12], is a combination of probability theory and graph theory. It is considered a probabilistic graphical model that expresses conditional dependencies between features in graphs.

Naïve-Bayes (NB) [13] is one of the simplest and commonly used models for classification. Usually, the NB algorithm is mainly based on the use of the former probability to calculate the later probability.

The hidden Markov model is suitable for dealing with data in the form of series such as texts and is considered one of the effective models in reducing computational complexity by redesigning the structure of the model, where researchers [14] used this model to implement the task of classifying medical text.

K-Nearest Neighbors algorithm [15] is used to classify unlabeled data and that is done without building the model, tune several parameters, or make additional assumptions. In [16] simple classification algorithms appropriate for annotating textual materials with partial information have been presented. As a result, they are suitable for large multi-label classification, especially in the biomedical domain.

Where the SVM method [17] turns text classification tasks into multiple binary classification tasks, the researchers [18] used Machine-generated regular expressions effectively for clinical text classification. They combined a regular expression-based classifier with SVM, to improve classification performance.

The researchers [19] proposed a HDTTCA approach to identify those who are eligible for telehealth services, which is a systematic approach that involves data set preprocessing, decision tree model building, and predicting and explaining the most important attributes in the data set for patients.

After 2010 the usage of deep learning models started to grow up, some research papers discussed models based on RNN and Long Short-Term Memory (LSTM) [20], Tree-LSTM [21], Multi-Timescale [22], Bidirectional-LSTM with two-dimensional max-pooling [23], some papers used the impact of CNN architectures with word embeddings, researchers [24] presented a Very Deep CNN model for text processing inspired by VGG [25] and ResNets [26]. One of the other interesting CNN-based models [27] presented a tree-based CNN to capture sentence level semantics. There is also a growing interest in applying CNNs to biomedical text classification [28–31].

Both RNN and CNN models suffer from dealing with long sequential, and the computational cost increases with the increase of the length of the sentence. Transformers [32] came to address this shortcoming, by applying self-attention by calculating the "attention score" to the impact of each word on the other in a parallel way. since 2018 a set of large-scale Transformer-based Pre-trained Language Model was launched [20] and they achieve today state-of-the-art results in NLP tasks, like BERT [33], BERT with nonlinear gradient method [34], openGPT [35, 36], RoBERTa [37], and ALBERT [38].

3 Dataset

The Arabic language suffers from a scarcity in the availability of appropriate datasets for training. There are datasets available for the task of classification, but it is not specialized in a particular field, but rather it is data collected from popular news websites in the Arab world, for example, the KACST [39] corpus was collected from the Saudi Press Agency, Arabic poems, and discussion forums, the BBC corpus[1] was collected from BBC Arabic website. This corpus has 4763 documents, with 7 categories, the CNN corpus[2] was collected from CNN Arabic website. This corpus has 5070 documents, with 6 categories and The OSAC [40] was collected with a crawler from dozens of Arabic websites. All these datasets provide records for classifying topics within general classes (such as politics, sports, economics, etc.)

In this work, we present a new Arabic dataset specialized in classifying medical texts. This dataset was collected semi-automatically by several libraries available in the Python language (Request, Beautiful Soup, and Selenium) from several Arab medical websites and from the Arabic Medical Encyclopedia. In its first version, the dataset includes 2,000 medical documents divided into 10 categories, the following Table 1 is a detailed description of this dataset.

Table 1. Summary of the dataset

Sources	Class	N	W	S
altibbi.com	Blood disease	215	1251.7	25.3
	Bone diseases	211	1325.3	26.8
webteb.com	Cardiovascular diseases	195	1749.5	27.1
	Ear diseases	180	1307.5	23.9
mayoclinic.org	Endocrine diseases	204	1184.6	22.4
	Eye diseases	190	1456.1	26.8
dailymedicalinfo.com	Gastrointestinal diseases	218	1381.6	25.9
	Immune diseases	203	1253.2	24.1
arab-ency.com.sy/medical/	Liver diseases	198	1386.7	27.3
	Nephrological diseases	186	1078.1	22.9

Where N denotes the number of samples, and W and S the average number of words and sentences per document, respectively.

[1] https://sourceforge.net/projects/newarabiccorpus/.

[2] https://osdn.net/projects/sfnet_ar-text-mining/downloads/Arabic-Corpora/cnn-arabic-utf8.7z/.

The process of verifying the correct classification of each document was carried out by three doctors working in Syrian university hospitals. This sample of the dataset is part of a project that aims to provide a medical Arabic dataset that supports the following tasks (Text Classification, Named Entity Recognition, Question-answering system), by the end of this project and when providing this dataset to the public, it will be available without any preprocessing operations on the text, that allows researchers to implement operations that serve their purposes, in addition, the pre-trained models having their own preprocessing operations (for example, removing stop words may have a negative impact on training accuracy), as these models are able to take advantage of the full context of a sentence.

4 Methodology and Experiments

To classify our medical documents presented in Table 1 into ten categories (Blood, Bone, Cardiovascular, Ear, Endocrine, Eye, Gastrointestinal, Immune, Liver, and Nephrological) diseases, three pre-trained models (BERT, AraBERT, BioAraBert) that based on Transformers and two shallow algorithms (SVM, Naive Bayes) were used.

4.1 BERT

BERT [33] employs a transformer that consists of two distinct mechanisms (encoder and decoder), with the encoder reading the text input and the decoder producing a task prediction. Only the encoder is required because the BERT's purpose is to construct a language model. The Transformer encoder is called bidirectional since it reads the full sequence of words at once, rather than (left-to-right or right-to-left).

BERT has two training techniques. The first is known as "Masked LM," in which the model substitutes 15% of the words in each sequence with a [MASK] token and attempts to predict the masked words' original value. The second technique is called "Next Sentence Prediction (NSP)" and it involves the model receiving pairs of phrases and attempting to predict whether the second sentence in the pair is the same as the second sentence in the original document.

4.2 Arabert

While BERT has been trained on 3.3 billion words extracted from the English Wikipedia and the Book Corpus [41], the Arabic Wikipedia is small compared to its English counterpart. The researchers in [42] scraped manually articles from Arabic news websites, and they used two large Arabic corpus (1.5 billion words Arabic [43], and OSIAN corpus [44]).

With using all these corpuses, the final size for pre-training dataset is 70 million sentences (~24 GB) of text.

4.3 ABioNER

The pretraining ABioNER [45] used the AraBERTv0.1-base in addition to Arabic biomedical literature corpus which is collected from (PubMed, MedlinePlus Health Information in Arabic[3], Journal of Medical and Pharmaceutical Sciences[4], Arab Journal for Scientific Publishing[5], Eastern Mediterranean Health Journal [46]). Figure 2 Shows AraBioNER structure.

Fine-Tuning

The dataset was divided into 80% for training, 10% for validation and 10% for testing. For fine-tuning we used (BERT-Base, Multilingual Cased model, Arabert v2, ABioNER that is based on Arabert v1).

For all three models we used Adam optimizer with learning rate $lr = 1e - 4$ and with two fully connected dense layers of size 1024 and 10 respectively, first one with "Relu" activation function, and second one with "Softmax" activation function, all that done for 4 epochs and 16 batch size.

We measure the effectiveness of the model using the traditional metrics:

$$F1 = 2 * \frac{Precision * Recall}{Precision + Recall}, Precision = \frac{TP}{TP + FP}, Recall = \frac{TP}{TP + FN}$$

Here TP denotes to true positive, FP to false positive, and FN to false negative.

While fine-tuning the three models, we optimized the batch size, maximum sequence length (MSL), number of epochs, number of tokens that documents are truncated to, and learning rate.

For our dataset, we find that using a batch size of 16, a learning rate of (2e–5), and an MSL of 512 tokens provides the best results. As is the case with [33, 47].

To compare with shallow algorithms, we took the most popular and efficient ones in literature SVM and Naïve Bayes. We mentioned in the introduction that shallow algorithms are greatly affected by the preprocessing operations, Therefore, we applied to our medical text a set of preprocessing operations, first tokenization by NLTK[6] library, then removing stop words, and then stemming by Snowball[7] stemmer, and finally, we selected TF-IDF that we mentioned in the Introduction too for word vectorization. The

Table 2. F1 score for the three models

	BERT	Arabert v2	ABioNER	SVM	NB
F1. Validation	94.1343	96.4327	**97.4331**	89.1308	87.6118
F1. Testing	92.2934	94.5415	**95.9124**	87.3473	85.6949

[3] https://medlineplus.gov/languages/arabic.html.
[4] https://www.ajsrp.com/journal/index.php/jmps.
[5] https://www.ajsp.net/.
[6] https://www.nltk.org/.
[7] https://snowballstem.org/

training for shallow algorithms was carried out using a scikit-learn library, the results came as shown in the Table 2.

The results show that the three models BERT, Arabert and ABioNER are more accurate and efficient than the shallow algorithms, and this is not inconsistent with the literature. The use of pre-trained models on a huge amount of data and fine-tuning them on a specific dataset is state-of-the-art these days.

From Table 2, the ABioNER model achieved the best accuracy, because this model is pre-trained on the same multilingual BERT's dataset and Arabert's Arabic dataset in addition to a large corpus of a medical dataset in Arabic.

For shallow algorithms, the good results shown in Table 2 show that the dataset that we have presented is balanced and suitable as a benchmark for future work aimed at dealing with Arabic medical texts.

Conclusion

In this work, we presented a new Arabic medical dataset, for the text classification task, the dataset includes 2000 articles, with 10 classes (Blood, Bone, Cardiovascular, Ear, Endocrine, Eye, Gastrointestinal, Immune, Liver, and Nephrological) diseases.

Through experiments, we have found that pre-trained models that have trained on large related corpus, and fine-tuned with specific datasets yield state-of-the-art results. This is evident by comparing the ABioNER model that has pre-trained on Arabic medical corpus before fine-tuned on our dataset with the original BERT model. This work was financially supported by Ministry of Science and Higher Education of the Russian Federation, Grant MK-5723.2021.1.6.

References

1. Li, Q., et al.: A text classification survey: from shallow to deep learning. arXiv preprint arXiv: 2008.00364 (2020)
2. Yao, L., Mao, C., Luo, Y.: Clinical text classification with rule-based features and knowledge-guided convolutional neural networks. BMC Med. Inform. Decis. Mak. **19**(3), 71 (2019)
3. Suzdaltseva, M., et al.: De-identification of medical information for forming multimodal datasets to train neural networks. In: Proceedings of the 7th International Conference on Information and Communication Technologies for Ageing Well and E-Health, pp. 163–170 (2021). https://doi.org/10.5220/0010406001630170
4. Hammoud, J., Dobrenko, N., Gusarova, N.: Named entity recognition and information extraction for Arabic medical text. In: Multi Conference on Computer Science and Information Systems, MCCSIS 2020-Proceedings of the International Conference on e-Health, pp. 121–127 (2020)
5. Alalyani, N., Marie-Sainte, S.L.: NADA: new Arabic dataset for text classification. Int. J. Adv. Comput. Sci. Appl. **9**(9) (2018)
6. Zhang, Y., Jin, R., Zhou, Z.H.: Understanding bag-of-words model: a statistical framework. Int. J. Mach. Learn. Cybern. **1**(1–4), 43–52 (2010)
7. Pennington, J., Socher, R., Manning, C.D.: Glove: global vectors for word representation. In: Proceedings of the 2014 conference on empirical methods in natural language processing (EMNLP), pp. 1532–1543, Oct 2014
8. Mikolov, T., Chen, K., Corrado, G., Dean, J.: Efficient estimation of word representations in vector space. arXiv preprint arXiv:1301.3781 (2013)

9. Cavnar, W.B., Trenkle, J.M.: N-gram-based text categorization. In: Proceedings of SDAIR-94, 3rd Annual Symposium on Document Analysis and Information Retrieval, vol. 161175, Apr 1994
10. Term frequency by inverse document frequency. In: Encyclopedia of Database Systems, p. 3035 (2009)
11. Zhang, M.L., Zhang, K.: Multi-label learning by exploiting label dependency. In: Proceedings of the 16th ACM SIGKDD International Conference on Knowledge Discovery and Data Mining, pp. 999–1008, July 2010
12. van den Bosch, A.: Hidden Markov models. In: Encyclopedia of Machine Learning and Data Mining, pp. 609–611 (2017)
13. Maron, M.E.: Automatic indexing: an experimental inquiry. J. ACM (JACM) **8**(3), 404–417 (1961)
14. O'Donnell, M.: Cataloging and classification: an introduction. Lois Mai Chan. Lanham, MD: Scarecrow Press, p. 580, 2007. ISBN 0-8108-6000-7. Tech. Serv. Q. **26**(1), 86–87 (2008)
15. Cover, T., Hart, P.: Nearest neighbor pattern classification. IEEE Trans. Inf. Theory **13**(1), 21–27 (1967)
16. Dramé, K., Mougin, F., Diallo, G.: Large scale biomedical texts classification: a kNN and an ESA-based approaches. J. Biomed. Semant. **7**(1), 1–12 (2016)
17. Cortes, C., Vapnik, V.: Support-vector Networks Machine learning, vol. 20, pp. 237–297. Kluwer Academic Publisher, Boston, MA (1995)
18. Bui, D.D.A., Zeng-Treitler, Q.: Learning regular expressions for clinical text classification. J. Am. Med. Inform. Assoc. **21**(5), 850–857 (2014)
19. Chern, C.C., Chen, Y.J., Hsiao, B.: Decision tree–based classifier in providing telehealth service. BMC Med. Inform. Decis. Mak. **19**(1), 1–15 (2019)
20. Minaee, S., Kalchbrenner, N., Cambria, E., Nikzad, N., Chenaghlu, M., Gao, J.: Deep learning–based text classification: a comprehensive review. ACM Comput. Surv. (CSUR) **54**(3), 1–40 (2021)
21. Tai, K.S., Socher, R., Manning, C.D.: Improved semantic representations from tree-structured long short-term memory networks. arXiv preprint arXiv:1503.00075 (2015)
22. Liu, P., Qiu, X., Chen, X., Wu, S., Huang, X.J.: Multi-timescale long short-term memory neural network for modelling sentences and documents. In: Proceedings of the 2015 Conference on Empirical Methods in Natural Language Processing, pp. 2326–2335, Sept 2015
23. Zhou, P., Qi, Z., Zheng, S., Xu, J., Bao, H., Xu, B.: Text classification improved by integrating bidirectional LSTM with two-dimensional max pooling. arXiv preprint arXiv:1611.06639 (2016)
24. Conneau, A., Schwenk, H., Barrault, L., Lecun, Y.: Very deep convolutional networks for text classification. arXiv preprint arXiv:1606.01781 (2016)
25. Simonyan, K., Zisserman, A.: Very deep convolutional networks for large-scale image recognition. arXiv preprint arXiv:1409.1556 (2014)
26. He, K., Zhang, X., Ren, S., Sun, J.: Deep residual learning for image recognition. In: Proceedings of the IEEE Conference on Computer Vision and Pattern Recognition, pp. 770–778 (2016)
27. Mou, L., et al.: Natural language inference by tree-based convolution and heuristic matching. arXiv preprint arXiv:1512.08422 (2015)
28. Karimi, S., Dai, X., Hassanzadeh, H., Nguyen, A.: Automatic diagnosis coding of radiology reports: a comparison of deep learning and conventional classification methods. In: BioNLP 2017, pp. 328–332, Aug 2017
29. Peng, S., You, R., Wang, H., Zhai, C., Mamitsuka, H., Zhu, S.: DeepMeSH: deep semantic representation for improving large-scale MeSH indexing. Bioinformatics **32**(12), i70–i79 (2016)

30. Rios, A., Kavuluru, R.: Convolutional neural networks for biomedical text classification: application in indexing biomedical articles. In: Proceedings of the 6th ACM Conference on Bioinformatics, Computational Biology and Health Informatics, pp. 258–267, Sept 2015
31. Hughes, M., Li, I., Kotoulas, S., Suzumura, T.: Medical text classification using convolutional neural networks. Stud. Health Technol. Inform. **235**, 246–250 (2017)
32. Vaswani, A., et al.: Attention is all you need. arXiv preprint arXiv:1706.03762 (2017)
33. Devlin, J., Chang, M.W., Lee, K., Toutanova, K.: Bert: Pre-training of deep bidirectional transformers for language understanding. arXiv preprint arXiv:1810.04805 (2018)
34. Hammoud, J., Eisab, A., Dobrenkoa, N., Gusarovaa, N.: Using a new nonlinear gradient method for solving large scale convex optimization problems with an application on Arabic medical text. arXiv preprint arXiv:2106.04383 (2021)
35. Radford, A., Narasimhan, K., Salimans, T., Sutskever, I.: Improving language understanding by generative pre-training (2018)
36. Radford, A., Wu, J., Child, R., Luan, D., Amodei, D., Sutskever, I.: Language models are unsupervised multitask learners. OpenAI Blog **1**(8), 9 (2019)
37. Liu, Y., et al.: Roberta: A robustly optimized BERT pretraining approach. arXiv preprint arXiv:1907.11692 (2019)
38. Lan, Z., Chen, M., Goodman, S., Gimpel, K., Sharma, P., Soricut, R.: Albert: a lite bert for self-supervised learning of language representations. arXiv preprint arXiv:1909.11942 (2019)
39. Marie-Sainte, S.L., Alalyani, N.: Firefly algorithm based feature selection for Arabic text classification. J. King Saud Univ. Comput. Inf. Sci. **32**(3), 320–328 (2020)
40. Saad, M.K., Ashour, W.M.: OSAC: open source Arabic corpora. In: 6th ArchEng International Symposiums, EEECS, vol. 10 (2010)
41. Zhu, Y., et al.: Aligning books and movies: towards story-like visual explanations by watching movies and reading books. In: Proceedings of the IEEE International Conference on Computer Vision, pp. 19–27 (2015)
42. Antoun, W., Baly, F., Hajj, H.: Arabert: transformer-based model for Arabic language understanding. arXiv preprint arXiv:2003.00104 (2020)
43. El-Khair, I.A.: 1.5 billion words Arabic corpus. arXiv preprint arXiv:1611.04033 (2016)
44. Zeroual, I., Goldhahn, D., Eckart, T., Lakhouaja, A.: OSIAN: Open source international Arabic news corpus-preparation and integration into the CLARIN-infrastructure. In: Proceedings of the Fourth Arabic Natural Language Processing Workshop, pp. 175–182, Aug 2019
45. Boudjellal, N., et al.: ABioNER: a BERT-based model for Arabic biomedical named-entity recognition. Complexity (2021)
46. WHO EMRO: EMHJ home. East. Mediterr. Health J. **27** (2021). http://www.emro.who.int/emhjournal/eastern-mediterranean-health-journal/home.html
47. Adhikari, A., Ram, A., Tang, R., Lin, J.: Docbert: Bert for document classification. arXiv preprint arXiv:1904.08398 (2019)

Improving Maximum Likelihood Estimation Using Marginalization and Black-Box Variational Inference

Soroosh Shalileh(✉)

Laboratory of Methods for Big Data Analysis (LAMBDA), HSE University, Pokrovsky Boulevard, 11, Moscow, Russian Federation

Abstract. Based upon Black Box Variational Inference, a new set of classification algorithms has recently emerged. The goals of this set of algorithms are twofold: 1) increasing generalization power; 2) decreasing computational and implementation complexity. To this end, we assume a set of latent variables during the generation of data points. We subsequently marginalize the conventional classification likelihood objective function w.r.t this set of latent variables and then apply black-box variational inference to estimate the marginalized likelihood. We evaluate the performance of the proposed method by comparing the results obtained from the application of our method to real-world datasets with those obtained using several classification algorithms. The experimental results prove that our proposed method is competitive.

Keywords: Black box variational inference classification · Variational inference classification · Variational inference · Classification

1 Introduction: Background, Previous Works, and Motivation

Classification is a popular field of machine learning, with various applications ranging from sociology to biology and computer science. To this date, several classification algorithms have been developed. Some approaches are designed to find a set of discriminating rules/parameters during the training process, while others are designed to learn a set of generation rules/parameters.

The in-common part of classification approaches is the objective function (or a set of objective functions in the case of, for instance, ensemble learning) to be optimized. Consequently, we can categorize them concerning either the methods used to define their objective functions or the core idea behind them in four categories.

The first category of classifiers aims to learn a scoring function so that the difference between the predicted values and the corresponding target values measured by a distance metric, e.g., Euclidean distance, is optimized [11,23].

H. Yin et al. (Eds.): IDEAL 2021, LNCS 13113, pp. 204–212, 2021.
https://doi.org/10.1007/978-3-030-91608-4_21

This category of approaches discriminates between various classes. The objective functions in the second category of classifiers are designed to maximize the likelihood/posteriori estimation of data generation process [1,3,7,18].

The third category of approaches is based on determining a set of if-then rules, which are defined in the form of decision trees during the learning process. The reader is encouraged to refer to [6,19,22] for more details. The fourth category of approaches is based on ensemble methods. Arcing classifier [5], Random Forest [4], Isolation Forest [17] are some the well-known examples of this category.

In this work, we proposed a marginalized objective function to consider the uncertainty in the data generation process and improve our classifier's generalization power.

More precisely, based on Variational Inference (VI), a new category of classification algorithms has emerged: the in-common assumption is latent/hidden variables during the generation of data points: forming the core concept of this category of classification algorithms. With this assumption, over the existence of latent variables, the goals are: 1) to provide more generalization power(and less over-fitting) and 2) to decrease the computational and implementation complexity.

Recently, the concept of black-box variational inference (BBVI) [21] has been proposed. The goal is to reduce the mathematical and implementation complexities of the classification algorithms while advancing their accuracy. The idea in BBVI is: instead of obtaining a closed-form solution of an objective function, a sampling technique is applied to computing the stochastic gradient of an objective function to estimate the parameters of the underlying distribution of a set of data points.

Our proposed method belongs to this group of methods. To be more precise, by assuming the existence of a set of latent variables and marginalizing the conventional likelihood estimation, we attempt to improve the classification power. However, Our proposed method differs from previous works in one or more aspects, as we will discuss shortly.

In [18] a Mean-Field variational inference is used to estimate the marginalized likelihood. Furthermore, to reduce the variance, the "control variates" method is applied. The authors obtained a closed-form solution for a specif model in their objective function in the above-mentioned work. Similarly, [7] is devoted to studying a set of parametric multivariate classification algorithms, with a specific medical application, by using the closed-form solutions of the defined objective function. On the contrary, we avoid obtaining any closed-form solution for optimizing our objective function. Instead, we apply a Monte Carlo estimator associated with a stochastic optimizer to optimize our objective function.

In [12] a variational Bayesian multinomial probit regression with Gaussian process priors is proposed. And, despite using the reparametrization technique, it completely differs from our proposed method because it obtains the closed-form solution of the objective function and uses the Gaussian process as a set of latent variables.

The most similar work from the literature to our proposed method is [1]. To be more precise, in both of these works, the objective function is marginalized over latent variables to obtain a marginalized likelihood estimation. Nevertheless, in [1] after applying the optimality condition, a discrete indicator function is used to reduce the variance of the gradient estimator. And based on this indicator function, they modified the objective function. On the contrary, we use the definition of the derivative of logarithm during the optimization of our objective function. More importantly, the algorithms used in [1] is a random sampling of the data points for computing the gradient of the objective function. Nevertheless, we applied the so-called BBVI (namely, the reparametrization technique with a stochastic optimizer).

The proposed methods in [3] and ours both utilize the reparametrization technique and a stochastic optimizer to estimate the marginalized likelihood. However, the objective function in [3] is a function of two arguments: 1) the underlying parameters of data points and 2) the underlying parameters of latent variables. On the contrary, our objective function has one argument (the underlying parameters for generating data points), and we marginalize it over the latent variables.

The rest of this paper is organized as follows. In Sect. 2, we explain our proposed method. Then, in Sect. 3, we provide the experimental results. Finally, we conclude the paper and explain our future works in Sect. 4.

2 A Marginalized Likelihood Estimation Using Black Box Variational Inference

Let $X = \{x_i\}_{i=1}^N$ be a set of N data points such that $x_i \in \mathbb{R}^V$ is a V-dimensional data point. And let $Y = \{y_i\}_{i=1}^N$ be the set of corresponding labels. Where y_i is in the form K dimensional one-hot vector, that is, $y_i \in \mathbb{R}^K$ where K is the number of classes. And let θ represents the model parameters to be estimated. We assume that during the generation of data points $X = \{x\}_{i=1}^N$, there exist a set of hidden/latent variables $Z = \{z_i\}_{i=1}^N$. Therefore we can define an objective function marginalized over the latent variables as follows:

$$\begin{aligned} \mathcal{L}_m(\theta) &= -\log \int p(Z|X,\theta)p(Y|X,Z,\theta)\, dZ \\ &= -\sum_{i=1}^{N} \log \sum_{j=1}^{K} p(z_j|x_i,\theta)p(y_i|x_i,z_j,\theta) \end{aligned} \tag{1}$$

In this case, each data point is marginalized with respect to all of the latent variables (for all classes). In order to optimize the proposed objective function (1), one can find an analytical solution for a specific distribution(s): yet there is no guarantee for the model to represent the data appropriately. More so, for some distributions $p(Z|X,\theta)$ might become intractable.

Therefore, in this work instead of finding an analytical solution, we adopt BBVI [21] to find an approximation for the proposed marginalized likelihood objective function.

However, applying optimality condition directly onto the Eq. (1) does not lead to a straightforward solution, and instead, we optimize its Variational Free Energy Lower Bound or Evidence Lower Bound (ELBO).

Let us recall the definition of ELBO first. For any two given distributions $r(x)$ and $q(x)$ over the same support, the evidence lower bound of $r(x)$ using $q(x)$ is: $ELBO := E_{q(x)}[\log r(x)] + H(x)$, where $H(x)$ is Shannon Entropy of distribution $r(x)$.

Regarding the ELBO definition and specifying $P(Y, Z|X, \theta)$ and $P(Z|X, \theta)$ as $r(x)$ and $q(x)$ respectively, and applying Jensen's inequality implies:

$$
\begin{aligned}
-\log \int p(Z|X,\theta)p(Y|X,Z,\theta)\ dZ \leq -\int p(Z|X,\theta)\log p(Y,Z|X,\theta)\ dZ \\
+ H(Z) \qquad \text{(2a)} \\
= -\int p(Z|X,\theta)\log p(Y|X,Z,\theta)\ dZ \qquad \text{(2b)}
\end{aligned}
$$

Where the RHS of Eq. (2a) is derived from the definition of ELBO. And clearly, $H(Z) = \int p(Z|X,\theta)\log p(Z|X,\theta)dZ$ is, indeed, the Shannon entropy. By substituting the definition of $H(Z)$ and opening the log of joint probability, i.e., $\log p(Y, Z|X, \theta)$, Eq. (2b) will be obtained. Let us denote Eq. (2b) with $\tilde{\mathcal{L}_m}(\theta)$.

Applying the optimality condition on it yields:

$$
\begin{aligned}
\nabla_\theta \tilde{\mathcal{L}}_m(\theta) &= \nabla_\theta[-\int p(Z|X,\theta)\log p(Y|X,Z,\theta)\ dZ] && \text{(3a)} \\
&= -\int \nabla_\theta[p(Z|X,\theta)\log p(Y|X,Z,\theta)]\ \ dZ && \text{(3b)} \\
&= -\int \nabla_\theta[p(Z|X,\theta)]\log p(Y|X,Z,\theta) \\
&+ p(Z|X,\theta)\nabla_\theta[\log p(Y|X,Z,\theta)]\ \ dZ && \text{(3c)} \\
&= -\int p(Z|X,\theta)\nabla\theta[\log p(Z|X,\theta)]\log p(Y|X,Z,\theta) \\
&+ p(Z|X,\theta)\nabla\theta[\log p(Y|X,Z,\theta)]\ \ dZ && \text{(3d)} \\
&= -\mathbb{E}_{p(Z|X,\theta)}[\nabla\theta[\log p(Z|X,\theta)]\log p(Y|X,Z,\theta) \\
&+ \nabla\theta[\log p(Y|X,Z,\theta)]] && \text{(3e)}
\end{aligned}
$$

Due to dominated convergence theorem [8], we can push the derivative inside the integral which justifies Eq. (3b). Obviously, the Eq. (3c) derived by taking derivatives of multiplication of two functions. The Eq. (3d) is obtained by replacing $\nabla\theta[p(Z|X,\theta)]$ with the definition of derivative of logarithm, that is, $\nabla \log f(.) = \frac{\nabla f(.)}{f(.)}$. And clearly, Eq. (3e) is derived from the definition of expectation.

The Eq. (3e) can be optimized using any Monte Carlo gradient estimators. In this work, we adopt path-wise Monte Carlo gradient estimator [15].

To adopt this estimator, we need to assume that there exists a transformation rule: such that we can draw samples ϵ_i $(i = 1, ..., N)$ from a simpler distribution $S(\epsilon_i)$ which is independent of θ and then we can transform this variate through a deterministic path $t(\epsilon_i; \theta)$. Concretely, $z_i = t(\epsilon_i, \theta)$ for $\epsilon_i \sim S(\epsilon)$ and $i = 1, ..., N$ where $S(.)$ is a parameter-free distribution, and this implies $Z \sim p(Z|X, \theta)$. This is equivalent to say that for a given ϵ that comes from a distribution S with no free parameters, it is possible to transform that noise source with a function that depends on the parameters to get a random variable that has the same distribution as the original one. As an example, assume $\epsilon \sim \mathcal{N}(0, I)$, and then do the location and scale transformation of ϵ, that is, $Z = \epsilon\sigma + \mu$ and this implies $Z \sim \mathcal{N}(\mu, \sigma^2)$. This technique is also called the reparametrization technique [15].

Applying the reparametrization technique, i.e., rewriting the Eq. (3e) using $Z = t(\epsilon, \theta)$ implies:

$$\begin{aligned}
\nabla_\theta \tilde{\mathcal{L}}_m(\theta) =& - \mathbb{E}_{p(Z|X,\theta)}[\nabla\theta[\log p(Z|X,\theta)] \log p(Y|X,Z,\theta) \\
&+ \nabla\theta[\log p(Y|X,Z,\theta)]] \qquad (4a) \\
=& - \mathbb{E}_{S(\epsilon)}[\nabla\theta[\log p(t(\epsilon,\theta)|X,\theta)] \log p(Y|X,t(\epsilon,\theta),\theta) \\
&+ \nabla\theta[\log p(Y|X,t(\epsilon,\theta),\theta)]] \qquad (4b)
\end{aligned}$$

For the sake of convenient we rewrite the Eq. (3e) as Eq. (4a). And Eq. (4b) is the result of applying reparametrization trick on it. In the next step, we can apply Monte Carlo gradient estimation on the Eq. (4b). Concretely, 1) we draw samples $\{\epsilon^l\}_{l=1}^{L}$; 2) we evaluate the argument of expectation using the above set; 3) and finally, we compute the empirical mean of evaluated quantities. The Eq. (5) explains the Monte Carlo estimate of gradient of objective function:

$$\begin{aligned}
\nabla_\theta \tilde{\mathcal{L}}_m(\theta) \approx& \frac{1}{L} \sum_{l=1}^{L} \nabla\theta[\log p(t(\epsilon^l,\theta)|X,\theta)] \log p(Y|X,t(\epsilon^l,\theta),\theta) \\
&+ \nabla\theta[\log p(Y|X,t(\epsilon^l,\theta),\theta)] \\
&where \;\; \epsilon^l \sim S(\epsilon); \; Z = t(\epsilon,\theta)
\end{aligned} \qquad (5)$$

Since, indeed, we Maximize a Marginalized Likelihood Estimation we name the base of our proposed algorithm as M-MLE. And for the ease of notation let $g(.)$ represents $\nabla\theta[\log p(Y|X, t(\epsilon^l, \theta), \theta)] + \nabla\theta[\log p(t(\epsilon^l, \theta)$ $|X, \theta)] \log p(Y|X, t(\epsilon^l, \theta), \theta)$. Where $t(\epsilon^l, \theta)$ is the Gaussian location and scale transform. With this new notation, we summarized our proposed algorithm for optimizing Eq. (5) in Algorithm (1).

Algorithm 1: Marginalized Maximum Likelihood Estimation (M-MLE)

Input: $X = \{(x_i\}_{i=1}^N$ and $Y = \{y_i)\}_{i=1}^N$: training set
Hyper-parameters: α: learning rate
Result: θ: learned parameters
θ, t initialize the model parameters and step counter respectively
while *not converged* **do**
  $\mathcal{M} = \{x_i, y_i\}_{i=1}^M \sim X, Y$; % draw mini-batch of samples $\mathcal{M}$
  $\varepsilon = \{\epsilon^l\}_{l=1}^M$ $(\epsilon^l \sim S(\epsilon)$; % draw samples, ϵ^l, from $S(.)$
  $\theta = \theta + \alpha \frac{1}{M} \sum_{l=1}^M g(.)$; % update rule from Eqn. (5)
  $t = t + 1$; % step counter
end

We have implemented our proposed algorithm using TensorFlow Probability. And using this library enables us to propose at least two versions of our algorithms: they differ from each other with respect to the activation function we used. Since we have applied the Path-Wise gradient estimator, we added the suffix PW to the name of our proposed algorithm. Finally, suppose the linear activation function is applied. In that case, we use M-MLE PW-Li to represent this version, and if we use the ReLu activation function, we use M-MLE PW-Re to represent this version of our proposed algorithm.

The M-MLE source code and other supplementary materials, can be found in our GitHub repository: https://github.com/Sorooshi/M-MLE-by-BBVI.

3 Experimental Results

3.1 Experimental Settings

Four real-world data sets, namely, IRIS [9], MNIST [16], Forest Cover Type (CovType) [2] and Wine data set [10] form this work's benchmarck.

Conventional Maximum Likelihood Estimation (C-MLE), AdaBoost [13], Classification with Path-Wise Gradient Estimator (CLS PW-Re) [3], are the chosen competitors. Noteworthy to add that for C-MLE, CLS PW-RE, and M-MLE methods, the learning rate is equal to 0.01, and the number of epochs is set to 1500. Adam optimizer [14] is used.

We use Area Under the Receiver Operating Characteristic Curve (ROC AUC) from prediction scores as our metric for comparison. Finally, it ought to mention that we run each algorithm five times, and we report the average and standard deviation.

3.2 Experimental Results at Multi-class Real-World Data Sets

It ought to emphasize that, in this work, we intend to evaluate the performance of the newly proposed objective function. Thus, we postpone our proposed method's extension and compare it with more complex methods to our future work. Table 1 compares the performance of the algorithm under consideration.

Table 1. Comparison of methods over Real-World Multi-Class data sets. The best results are highlighted in bold-face font.

Algorithms	IRIS	MNIST	CovType	Wine
	ave (std)	ave (std)	ave (std)	ave (std)
AdaBoost	0.979(0.010)	0.647(0.000)	0.802(0.002)	0.862(0.153)
C-MLE	0.995(0.003)	**0.761(0.000)**	0.922(0.001)	0.999(0.001)
CLS PW-Re	0.995(0.003)	0.708(0.010)	**0.972(0.001)**	0.997(0.003)
M-MLE PW-Li	**0.998(0.001)**	0.671(0.009)	0.928(0.002)	**1.000(0.000)**
M-MLE PW-Re	0.996(0.004)	0.496(0.002)	0.952(0.006)	**1.000(0.000)**

The winner of the IRIS and wine competitions is M-MLE. In the MNIST data set, the winner is the C-MLE algorithm. Furthermore, in the CovType data set, CLS PW-RE wins the competition. Nevertheless, the M-MLE PW-RE result is higher than the remaining algorithms and indeed acceptable at this data set.

3.3 Experimental Results at Moon-Shape Synthetic Data Sets

In order to study the ability of algorithms in classifying non-linear data sets, we consider a moon-shape data generator [20]. We generated two data set sizes, medium-size with 1000 data points and big-size with 100000. And for each of them, two different data representations are considered: balanced and imbalanced data representations. The latter represents the case in which the population of samples of a class is significantly larger than the other.

The results of applying the algorithms under consideration on moon-shape data set are recorded in Table 2.

Table 2. Comparison of methods over moon shape data set. The best results are highlighted in bold-face font.

Algorithms	Balanced		Imbalanced	
	Med	Big	Med	Big
	ave(std)	ave(std)	ave(std)	ave(std)
AdaBoost	0.881(0.011)	0.899(0.001)	0.873(0.018)	**0.899(0.001)**
C-MLE	**0.898(0.004)**	**0.903(0.001)**	**0.904(0.012)**	0.888(0.001)
CLS PW-Re	0.301(0.096)	0.256(0.122)	0.220(0.075)	0.892(0.001)
M-MLE PW-Li	0.886(0.008)	0.796(0.005)	0.891(0.013)	0.888(0.001)
M-MLE PW-Re	0.880(0.008)	0.795(0.006)	0.887(0.015)	0.882(0.002)

One can easily observe that C-MLE dominates this table. Although the proposed methods of this work do not win any competition in these experiments, the results still appear to be competitive and promising.

4 Conclusion and Future Work

This paper proposes a marginalized likelihood objective function by assuming latent variables during the data generation process. We optimally estimate the data distribution's parameters by applying black-box variational inference. The determination of the parameters of marginalized likelihood estimation allows for the classification of the test data points. We evaluated the performance of our proposed methods using the fundamental algorithms at four real-world data sets and four synthetic moon-shape data sets. The obtained results led us to conclude that the proposed method is indeed effective and competitive. More so, there are some cases in which marginalization leads to higher generalization power and accuracy. However, the proposed methods, despite obtaining relatively good accuracy at moon-shape datasets, do not improve the classification results obtained by C-MLE.

The list of our future works is as follows:

1. extending the current implementation of the proposed objective function with a more complicated network structure.
2. investigating the impact of applying different transformation rules –instead of path-wise gradient estimator;
3. investigating the impact of applying different Monte Carlo gradient estimators on our proposed objective function;
4. modifying our proposed method, i.e., the objective function and corresponding algorithm for the task of regression;
5. extending the list of competitors.

Acknowledgment. The research leading to these results has received funding from Russian Science Foundation under grant agreement n° 19-71-30020.

References

1. Ba, J., Mnih, V., Kavukcuoglu, K.: Multiple object recognition with visual attention. arXiv preprint, page arXiv:1412.7755 (2014)
2. Blackard, J.A., Dean, D.J.: Comparative accuracies of artificial neural networks and discriminant analysis in predicting forest cover types from cartographic variables. Comput. Electron. Agric. **24**(3), 131–151 (1999). https://archive.ics.uci.edu/ml/datasets/covertype
3. Blundell, C., Cornebise, J., Kavukcuoglu, K., Wierstra, D.: Weight uncertainty in neural networks. arXiv preprint, page arXiv:1505.05424 (2015)
4. Breiman, L.: Random forests. Mach. Learn. **45**(1), 5–32 (2001)
5. Breiman, L.: Arcing classifier (with discussion and a rejoinder by the author). Ann. Stat. **26**(3), 801–849 (1998)
6. Breiman, L., Friedman, J., Stone, C.J., Olshen, R.A.: Classification and Regression Trees. CRC Press, Boca Raton (1984)
7. Brodersen, K.H., Daunizeau, J., Mathys, C., Chumbley, J.R., Buhmann, J.M., Stephan, K.E.: Variational bayesian mixed-effects inference for classification studies. Neuroimage **76**, 345–361 (2013)

8. Cinlar, E.: Probability and stochastics. 261. Springer Science & Business Media (2011)
9. Fisher, R.A.: The use of multiple measurements in taxonomic problems. Ann. Eugen. **7**(2), 179–188 (1936)
10. Forina, M.: An extendible package for data exploration, classification and correlation (1998). Institute of Pharmaceutical and Food Analysis and Technologies
11. Friedman, J., Hastie, T., Tibshirani, R.: Regularization paths for generalized linear models via coordinate descent. J. Stat. Softw. **33**(1), 1 (2010)
12. Girolami, M., Rogers, S.: Variational bayesian multinomial probit regression with gaussian process priors. Neural Comput. **18**(8), 1790–1817 (2006)
13. Hastie, T., Rosset, S., Zhu, J., Zou, H.: Multi-class adaboost. Stat. Interface **2**(3), 349–360 (2009)
14. Kingma, D.P., Ba, J.: A method for stochastic optimization. arXiv preprint, cs.LG,:arXiv 1412.6980 (2014)
15. Kingma, D.P., Welling, M.: Auto-encoding variational bayes. arXiv preprint, page arXiv:1312.6114 (2014)
16. LeCun, Y., Cortes, C., Burges, C.J.: Mnist handwritten digit database. ATT Labs, February 2010. http://yann.lecun.com/exdb/mnist
17. Liu, F.T., Ting, K.M., Zhou, Z.-H.: Isolation-based anomaly detection. ACM Trans. Knowl. Disc. Data (TKDD) **6**(1), 3 (2012)
18. Paisley, J., Blei, D., Jordan, M.: Variational bayesian inference with stochastic search. arXiv preprint, page arXiv:1206.6430 (2012)
19. Pandya, R., Pandya, J.: C5.0 algorithm to improved decision tree with feature selection and reduced error pruning. Int. J. Comput. Appl. **117**(16), 18–21 (2015)
20. Pedregosa, F., et al.: Scikit-learn: machine learning in python (2011). Moon data set: https://scikit-learn.org/stable/modules/generated/sklearn.datasets.make_moons.html
21. Ranganath, R., Gerrish, S., Blei, D.M.: Black box variational inference. In: Proceedings of the Seventeenth International Conference on Artificial Intelligence and Statistics (2014)
22. Rutkowski, L., Jaworski, M., Pietruczuk, L., Duda, P.: The cart decision tree for mining data streams. Inf. Sci. **266**, 1–15 (2014)
23. Schölkopf, B., Smola, A.J., Williamson, R.C., Bartlett, P.L.: New support vector algorithms. Neural Comput. **12**(5), 1207–1245 (2000)

A Deep Learning-Based Approach for Train Arrival Time Prediction

Bas Jacob Buijse[1], Vahideh Reshadat[2(✉)], and Oscar Willem Enzing[3]

[1] Department of Cognitive Science and Artificial Intelligence or Data, Jheronimus Academy of Data Science, 5211 DA's-Hertogenbosch, The Netherlands
bas.buijse@walnutdata.nl

[2] Department of Industrial Engineering and Innovation Sciences, Eindhoven University of Technology, P.O. Box 513, 5600 Eindhoven, MB, The Netherlands
v.reshadat@tue.nl

[3] ProRail, 3511 Utrecht, EP, The Netherlands
oscar.enzing@prorail.nl

Abstract. Level crossings have a function to let the traffic cross the railroad from one side to the other. In the Netherlands, 2300 level crossings are spread out over the country, playing a significant role in daily traffic. Currently, there isn't an accurate estimation of the arrival time of trains at level crossings while it plays an important role in traffic flow management in intelligent transport systems. This paper presents a state-of-the-art deep learning model for predicting the arrival time of trains at level crossings using spatial and temporal aspects, external attributes, and multi-task learning. The spatial and temporal aspects incorporate geographical and historical travel data and the attributes provide specific information about a train route. Using multi-task learning all the information is combined and an arrival time prediction is made both for the entire route as for sub-parts of that route. Experimental results show that on average, the error is only 281 s with an average trip time of one hour. The model is able to accurately predict the arrival time at level crossings for various time steps in advance. The source code is available at https://github.com/basbuijse/train-arrival-time-estimator.

Keywords: Train arrival time prediction · Deep learning · Spatial-temporal neural networks · Multi-task learning

1 Introduction

Every metropolis with an extensive transportation network commonly experiences problems with traffic flow management. These problems may lead to the delay of public transport, poor emergency services, increased fuel consumption, environmental pollution, etc. [2]. Providing accurate and timely traffic information such as arriving time of vehicles plays an important role in intelligent transport systems. In this term, level crossings also play a big role as they can hinder traffic for a specific time period when a train passes.

H. Yin et al. (Eds.): IDEAL 2021, LNCS 13113, pp. 213–222, 2021.
https://doi.org/10.1007/978-3-030-91608-4_22

A railroad is divided into sections separated by electrical separation welds used to track whether a particular train is on the railroad. Whenever a train crosses an electrical separation weld, an electrical circuit is closed, and the electrical signal is converted to a digital signal. The signal is received, and the train position is updated by a train traffic controller in a control room. This helps the train traffic controller to guide the train traffic across the railroad safely. Being informed of the train cross passing time at level crossings in advance can have several advantages for multiple stakeholders. Those benefits are direct or indirect and include but are not limited to enhanced traffic flow, improved decision making for emergency services, and additional safety enhancements for (non-secured) level crossings.

Forecasting methods have large influence on the development of different artificial intelligent branches consists of Fuzzy Systems [10], Natural Language Processing [12–15], Expert Systems [19] etc. Numerous methods for predicting the arrival times of vehicles have been proposed in recent years. These methods support a variety of different approaches from traditional machine learning and statistical based models (e.g. support vector machines [7] and Kalman filter [1,6]) to neural network based architectures (e.g. long short term memory (LSTM) [3,11,17,20]. Predicting the arrival times with the help of deep learning models has been done in some recent works [9,22]. However, most of these works aim to predict the arrival time for road vehicles [5,16], and the railroad industry lagged behind that in terms of development. This paper presents a deep neural network-based architecture for predicting the train arrival time at level crossings. It is inspired by the Deep Travel Time Estimator model of Wang et al. [18]. The main contributions of this paper are as follows:

- In this paper, a state-of-the-art deep arrival time estimator is presented that predicts the arrival time of trains at level crossings. The model uses the spatial-temporal features, external attributes, and a special Convolutional Neural Network (CNN) layer called Geo-Conv layer.
- To the best of our knowledge, there is no previous study in the domain of arrival time predictions for trains that have utilized a deep learning model as an arrival time estimator to predict the arrival time of trains at level crossings.
- The experiments show that the proposed model can accurately predict the arrival time for various timesteps in advance.

The rest of this paper is organized as follows: In Sect. 2, an overview is given of the datasets that are used. Section 3 presents our proposed model in detail; Sect. 4 is dedicated to the details of the experiments and results; and finally, Sect. 5 concludes the paper.

2 Datasets

The dataset for training and testing of the proposed model is described in more detail in this section. In Listing 1.1 the required input format of the model is shown.

Listing 1.1. The data in JSON-format


```
{
    ....
}
{
    "Sections":["AH$2001T", ...., "HDR$1119BT"],
    "time_gap":[0.0, ...., 8507.0],
    "dist_gap":[0.0, ...., 185.56992499610186],
    "lats":[51.98704004887645, ...., 52.94988504678495],
    "lngs":[5.873883222270905, ...., 4.762389563964237],
    "TrainID":3066,
    "timeID":1142,
    "weekID":5,
    "dateID":1,
    "time":8507.0,
    "dist":185.56992499610186,
    "trainTypeID":3
}
{
    ....
}
```


In the listing, the *Sections* key is a list with all the sections that have been traversed for a particular train route. The *time_gap* and *dist_gap* keys respectively hold a list with the cumulative time and distance since the start of the route. Every new value in these lists correspond to the entry of a new section. The $timeID$ key/value pair represents the start time of a trip in minutes from 12:00 PM and thus is a number between 0 and 1440 (1440/60 = 24 h). The $weekID$ is a number between 0 (Monday) and 6 (Sunday) that represents the day in the week. The $dateID$ is the day of the month and last of all the *time* and *dist* keys hold the values for the total time and total distance traveled. Last of all, the $trainTypeID$ key holds the type of the train and the $TrainID$ key holds the unique identifier for that train.

In order to predict the arrival time for trains at level crossings, an additional dataset is used that contains the coordinates of all the level crossings in the Netherlands. The coordinates of a level crossings often interfere with the coordinates of the end of a section. Thus, whenever the coordinates of a level crossing are known the travel time prediction can simply be made up until that point.

3 Proposed Model

We propose a new model for predicting the arrival time of trains at level crossings using a deep neural network architect. The model uses the spatial-temporal features, external attributes, and a special CNN layer namely, the Geo-Conv layer. These three main components are described in more detail in this section. An overview of the model is shown in Fig. 1.

The attribute component of the model processes the external factors and includes the start time of the trip (TimeID), day of the week (DateID), the train type (TrainTypeID), and a specific train identifier (TrainID). An embedding method is used to transform the categorical attributes into a lower dimensional vector. This means that each value of a categorical attribute $v \in [V]$ is mapped to an embedding vector $\mathbb{R}^{Ex1}$. The influence of the above four attributes on the arrival time have been investigated while training the model.

Besides, the travel distance (Dist) is also incorporated as an attribute. The output of this component is a concatenation of all the attributes.

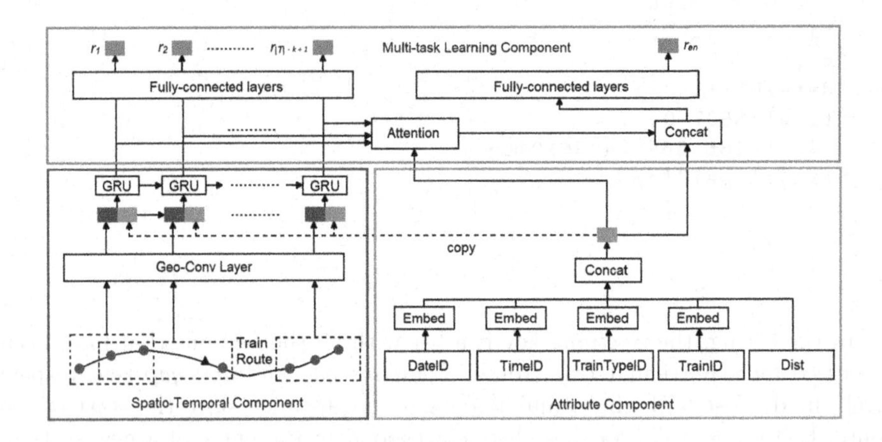

Fig. 1. The architecture of the proposed model

The spatio-temporal component in the model consists of two parts. The first part consists of a so-called GEO-convolutional network (GEO-Conv) that converts the coordinates of the sections into feature maps. The second part consists of gated recurrent units (GRUs) that learn the temporal aspects from the feature maps created by the GEO-convolutional neural network. GRUs use less training parameters and therefore use less memory. Moreover, in comparison with LSTMs they are executed faster and trained faster; so they are computationally efficient [21].

The GEO-Conv is similar to a standard CNN. Since a CNN usually needs a grid of equal partitioned cells in order to convolve over this grid, a modified GEO-Conv is applied. For instance, a CNN is applied to an image of $N \times M$ pixels. This is in contrast to GPS points that could take every position on a map and thus are not structured at all. In order to grasp this fine granularity, the GEO-Conv is introduced. This layer uses a convolutional filter with kernel size k and a 1D-window that is applied over the sequence to generate a convoluted location sequence. The kernel size is a hyperparameter that can be adjusted by the user. A bigger kernel size means that more subsequent segments are pooled together into a single embedding. For example in Fig. 1 the kernel size is illustrated to be equal to three. This means three subsequent segments are pooled together. Because it is difficult for the 1D-window to extract the distance directly from the raw coordinates, the final output of the GEO-convolution layer is concatenated with the distance of each segment.

The multi-task learning component is the final part of the model. This component combines all the previous output data and predicts a travel time for both the entire route and every segment or section.

To estimate the travel time for the segments, the spatio-temporal features of the $i'th$ subsequence of the route are used. Then, two fully connected layers are used to map every feature vector to a variable called r_i which represents the travel time for the $i'th$ segment. Figure 1 shows this fact. The prediction of the entire path is done based on the same feature sequence. For this, an attention pooling method is used that combines the spatial information of a subsequence with the external factors such as the start time of the route. Then, the attention vector is fed to a fully connected layer connected to the residual connections that enable the network to skip layers. This robust technique allows the training of very deep neural networks with multiple layers [8]. Finally, the network outputs predictions for the entire path as well as the segments. For the entire path, a single neuron is used to output the final prediction denoted as r_n.

4 Experiments

In this section, the performance of the proposed model is evaluated using the datasets described in Sect. 2. The dataset used for training consists of approximately 350.000 train routes that 70% is used as training data and 15% as validation- and test data. Applying grid search, the combination of hyperparameters that achieves the best performance is gained. The setting of the model hyperparameters used in the experiments is shown in Table 1. Cross-validation is not used in this research as the data is collected over time and thus it contradicts the fundamental assumptions of cross-validation that the data is independent and identically distributed [4].

Table 1. The best model hyperparameters after grid search

Hyperparameter	Description	Value
Batch size	The size of the batch to train	8
Epochs	The amount of training iterations	10
Kernel size	The kernel size of the Geo-Conv layer	2
Alpha	The weight of combination in multi-task learning	0.8

Table 2 provides the results of the best performing model. Both the error for the collective estimation (prediction for the entire routes) as well as the error for the individual estimation (prediction per section) are reported in Table 2. The individual estimations are used to provide the arrival time prediction for the various timesteps before the level crossing. Considering the fact that the average trip time is approximately one hour, an MAE (Mean Absolute Error) of 281 s for all the train trips in the test set reflects very accurate predictions.

Table 2. The results of the best model after grid search

Results best model after grid search		
	MAE (sec)	RMSE (sec)
Collective estimation	281.67	583.88
Individual estimation	7.18	34.48

Figure 2 and Fig. 3 give a visual insight on how the model performs on the test data. The MAE for the travel time estimation per sections is shown in Fig. 2. The smaller the MAE is, the greener a section.

Figure 3 shows the number of times that a section appears in the test data. The more a section appears in the dataset, the more its color turns to the blue spectrum. The color turns grey if a section only appears a few times in the dataset. Comparing Fig. 2 and Fig. 3, it can be concluded that fewer section data is available in the parts where the model has a high MAE. Due to the intrinsic of the model, this is expected. Therefore, the model most likely doesn't perform well for some parts in the Netherlands with less data available.

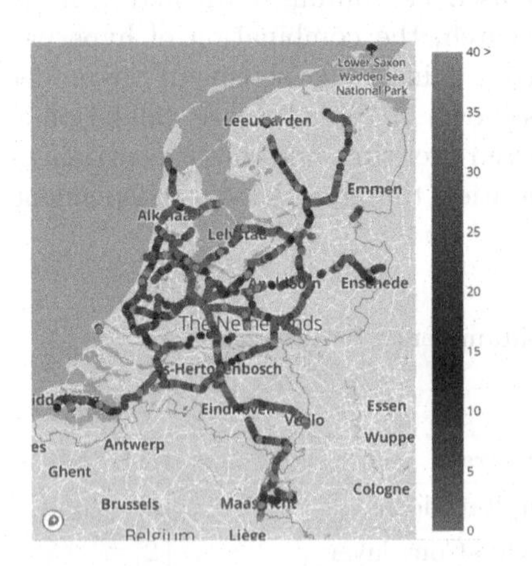

Fig. 2. The MAE per section (Color figure online)

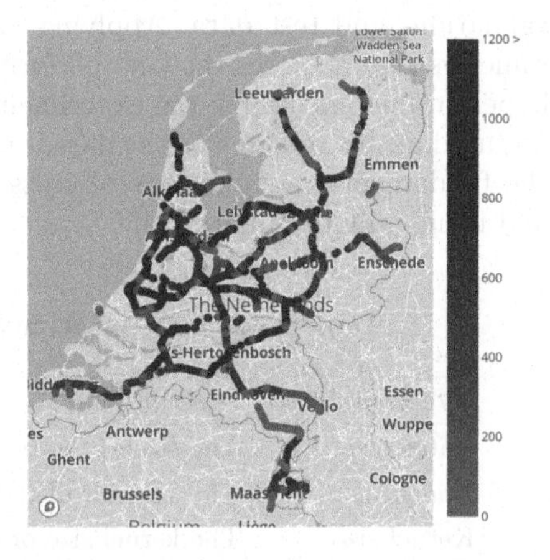

Fig. 3. Number of routes traversing a section (Color figure online)

As an instance of how the model performs in the prediction of train arrival times at level crossing, two sample level crossings are investigated.

The first sample belongs to a level crossing at the Jachtlaan in the city of Apeldoorn of the Netherlands. Since it is situated close to the station of

Apeldoorn, trains are approaching this level crossing at different speeds (e.g., freight trains do not stop at the Apeldoorn station and cross the level crossing with a much higher speed than intercities and sprinters that stop at the station).

The second level crossing is a crossing at Jonkheer Ramweg near the Schalkwijk village. This level crossing is not positioned close to a station. However, because it is positioned in a more rural area, fewer train routes in the dataset crossed this level crossing. The arrival time prediction (Prediction Period) for both level crossings is made up to 20 min before the trains arrive at the level crossings. In this case, a farmer crossing with his herd or emergency services looking for the fastest route will have enough time to make a plan.

Table 3. Accuracy of the predictions for a level crossing in Apeldoorn

Predictions for the LC at the Jachtlaan, Apeldoorn			
Prediction Period	Accuracy of the prediction (in seconds)		
	98%	90%	Average
19–21 min	[−50 ∼ −20]	[−45 ∼ −25]	−35
9–11 min	[−7.6 ∼ 7.8]	[−5.3 ∼ 5.6]	0.1
0–1 min	[−2.9 ∼ −1.9]	[−2.8 ∼ −2.1]	−2.4

Table 3 shows the arrival time predictions for the level crossing at Jachtlaan. The table reports three prediction periods and the confidence intervals of 90% and 98%. The average prediction per period is displayed in the last column of the table. The results show that the prediction accuracy increases whenever the train is closer to the level crossing. This is mainly because a shorter route consists of less sections and therefore the accumulation of the error is less. Considering the 19–21 min prediction period, it can be seen that the 98% prediction interval is still only 50 s off. This means that even 20 min before the train arrives at the level crossing, the model is able to make an accurate prediction which is at most one minute off in 98% of the cases.

Table 4. Accuracy of the predictions for a LC in Schalkwijk

Predictions for the LC at the Jonkheer Ramweg, Schalkwijk			
Prediction Period	Accuracy of the prediction (in seconds)		
	98%	90%	Average
19–21 min	[−108 ∼ −91]	[−106 ∼ −94]	−100
9–11 min	[−27 ∼ −20]	[−27 ∼ −21]	−24
0–1 min	[0 ∼ 0.9]	[0.2 ∼ 0.8]	0.5

The predictions for the level crossing at the Jonkheer Ramweg in Schalkwijk can be seen in Table 4. It can be noted that the predictions are less accurate as compared to the predictions made for the level crossing at the Jachtlaan. This can be explained due to the fact that less data is available for the railroad traversing this level crossing. However, the error is still manageable.

5 Conclusion

The paper presents a deep learning-based architecture for predicting train arrival times at level crossings that consists of three main components, namely Spatio-Temporal, Multi-task learning, and Attribute. The Spatio-Temporal component consists of a Geo-Conv layer which extracts the spatial information from a train route and a GRU which extracts the temporal information. The Attribute component extracts information from the external attributes. Several useful attributes are embedded in the Attribute component, such as a unique train identifier (trainID) and the type of the train (trainTyepID). Using a multi-task learning component, prediction for the entire route and for the sections of that route is possible. This enables us to provide the most actual arrival time predictions possible at various points in time. Finally, the Multi-task learning component applies attention and predicts both the travel time per segment and the travel time for the entire route. The results of the experiments show that the model is capable of predicting the arrival time with great precision for various timesteps. This model can function as a basis for other applications built upon the predictions of this model in order to predict the arrival time for the trains accurately.

For future work, it would be interesting to investigate if more attributes could be included in the attribute component. Think for example of including the weather as a categorical variable in the attribute component. Moreover, it could be interesting to test whether using an LSTM instead of the GRU would obtain better results. Besides, it is interesting to test the model performance for different level crossings. Finally, the running time of a train arriving at a level crossing and leaving it again is not considered in this research and thus could be embedded into future work.

References

1. Achar, A., Bharathi, D., Kumar, B.A., Vanajakshi, L.: Bus arrival time prediction: a spatial Kalman filter approach. IEEE Trans. Intell. Transp. Syst. **21**(3), 1298–1307 (2019)
2. Agafonov, A., Yumaganov, A.: Spatial-temporal K nearest neighbors model on mapreduce for traffic flow prediction. In: Yin, H., Camacho, D., Novais, P., Tallón-Ballesteros, A.J. (eds.) IDEAL 2018. LNCS, vol. 11314, pp. 253–260. Springer, Cham (2018). https://doi.org/10.1007/978-3-030-03493-1_27

3. Agafonov, A., Yumaganov, A.: Bus arrival time prediction with LSTM neural network. In: Lu, H., Tang, H., Wang, Z. (eds.) ISNN 2019. LNCS, vol. 11554, pp. 11–18. Springer, Cham (2019). https://doi.org/10.1007/978-3-030-22796-8_2
4. Bergmeir, C., Benítez, J.M.: On the use of cross-validation for time series predictor evaluation. Inf. Sci. **191**, 192–213 (2012)
5. Chen, C.H.: An arrival time prediction method for bus system. IEEE Internet Things J. **5**(5), 4231–4232 (2018)
6. Emami, A., Sarvi, M., Bagloee, S.A.: Using Kalman filter algorithm for short-term traffic flow prediction in a connected vehicle environment. J. Modern Transp. **27**(3), 222–232 (2019)
7. Hashi, A.O., Hashim, S.Z.M., Anwar, T., Ahmed, A.: A robust hybrid model based on Kalman-SVM for bus arrival time prediction. In: Saeed, F., Mohammed, F., Gazem, N. (eds.) IRICT 2019. AISC, vol. 1073, pp. 511–519. Springer, Cham (2020). https://doi.org/10.1007/978-3-030-33582-3_48
8. He, K., Zhang, X., Ren, S., Sun, J.: Deep residual learning for image recognition. In: Proceedings of the IEEE Conference on Computer Vision and Pattern Recognition, pp. 770–778 (2016)
9. Jabamony, J., Shanmugavel, G.R., et al.: IoT based bus arrival time prediction using artificial neural network (ANN) for smart public transport system (SPTS). Int. J. Intell. Eng. Syst. **13**(1), 312–323 (2020)
10. Khetarpaul, S., Gupta, S.K., Malhotra, S., Subramaniam, L.V.: Bus arrival time prediction using a modified amalgamation of fuzzy clustering and neural network on spatio-temporal data. In: Sharaf, M.A., Cheema, M.A., Qi, J. (eds.) ADC 2015. LNCS, vol. 9093, pp. 142–154. Springer, Cham (2015). https://doi.org/10.1007/978-3-319-19548-3_12
11. Petersen, N.C., Rodrigues, F., Pereira, F.C.: Multi-output bus travel time prediction with convolutional LSTM neural network. Expert Syst. Appl. **120**, 426–435 (2019)
12. Reshadat, V., Faili, H.: A new open information extraction system using sentence difficulty estimation. Comput. Inf. **38**(4), 986–1008 (2019)
13. Reshadat, V., Feizi-Derakhshi, M.R.: Studying of semantic similarity methods in ontology. Res. J. Appl. Sci. Eng. Technol. **4**(12), 1815–1821 (2012)
14. Reshadat, V., Hoorali, M., Faili, H.: A hybrid method for open information extraction based on shallow and deep linguistic analysis. Interdiscip. Inf. Sci. **22**(1), 87–100 (2016)
15. Reshadat, V., Hourali, M., Faili, H.: Confidence measure estimation for open information extraction. Inf. Syst. Telecommun. 1 (2018)
16. van der Spoel, S., Amrit, C., van Hillegersberg, J.: Predictive analytics for truck arrival time estimation: a field study at a European distribution centre. Int. J. Prod. Res. **55**(17), 5062–5078 (2017)
17. Treethidtaphat, W., Pattara-Atikom, W., Khaimook, S.: Bus arrival time prediction at any distance of bus route using deep neural network model. In: 2017 IEEE 20th International Conference on Intelligent Transportation Systems (ITSC), pp. 988–992. IEEE (2017)
18. Wang, D., Zhang, J., Cao, W., Li, J., Zheng, Y.: When will you arrive? estimating travel time based on deep neural networks. In: Proceedings of the AAAI Conference on Artificial Intelligence, vol. 32 (2018)
19. Wentworth, J.: Expert systems in transportation. Technical report, AAAI Technical Report WS-93-04 (1993)
20. Xu, H., Ying, J.: Bus arrival time prediction with real-time and historic data. Cluster Comput., 1–8 (2017). https://doi.org/10.1007/s10586-017-1006-1

21. Yang, S., Yu, X., Zhou, Y.: Lstm and gru neural network performance comparison study: taking yelp review dataset as an example. In: 2020 International Workshop on Electronic Communication and Artificial Intelligence (IWECAI), pp. 98–101. IEEE (2020)
22. Zhang, X., Yan, M., Xie, B., Yang, H., Ma, H.: An automatic real-time bus schedule redesign method based on bus arrival time prediction. Adv. Eng. Inform. **48**, 101295 (2021)

A Hybrid Approach for Predicting Bitcoin Price Using Bi-LSTM and Bi-RNN Based Neural Network

Sunanda Das(✉), Md. Masum Billah, and Suraiya Akter Mumu

Department of Computer Science and Engineering, Khulna University of Engineering & Technology, Khulna 9203, Bangladesh
sunanda@cse.kuet.ac.bd

Abstract. Bitcoin is an electronic or digital currency. However, unlike government-issued currencies, there is no single entity that issues bitcoin or is in charge of processing transactions. That's why bitcoin has become popular in the recent era. As bitcoin's price fluctuates a lot in a short period, it is very challenging to predict the bitcoin price accurately. In this paper, we proposed a method by merging different highest-level building blocks in deep learning such as Convolutional Neural Networks (CNN), Long Short Term Memory (LSTM), Bi-LSTM, Recurrent Neural Network (RNN), and Bi-RNN to predict bitcoin price as accurately as possible. Though CNN, LSTM, Bi-LSTM, RNN, Bi-RNN or ARIMA independently produce an acceptable result, our proposed hybrid method is fairly reliable compared to individual building block of the network as the proposed method outperforms other individual models and achieves RMSE, MAE, MAPE, MedAE of 2.69%, 1.78%, 2.20%, and 1.23% respectively.

Keywords: Bitcoin · Prediction · Deep learning · CNN · LSTM · Bi-LSTM · RNN · Bi-RNN · Time series forecasting

1 Introduction

Bitcoin is the world's first and most valuable cryptocurrency, which is used worldwide for digital payment or investment purposes. Satoshi Nakamoto, who is a mysterious person or group of individuals, founded bitcoin in 2008 [1]. Bitcoin is a decentralized digital currency that used a blockchain-based network. Bitcoin transaction data is stored on its blockchain, and bitcoin miners generate new coins by adding new transaction data to the blockchain, which tends to keep the network running.

Bitcoin is open-source, its design is public, and no one can control it. Bitcoin is available to anyone who wants to transaction bitcoin and every confirmed transaction is incorporated in the blockchain [2]. Blockchain is a set of blocks where each block is a set of transactions. Bitcoin is a digital file deposited in a

H. Yin et al. (Eds.): IDEAL 2021, LNCS 13113, pp. 223–233, 2021.
https://doi.org/10.1007/978-3-030-91608-4_23

digital wallet similar to a virtual bank deposit. There have many bitcoin wallets for smartphones or computers. People can send or receive bitcoin by using their digital wallets. Bitcoin is not controlled by the government or central banks, or single administrators. Hence, bitcoin can be sent from user to user on the peer-to-peer technology allowing immediate payments without any central authority [3]. A bitcoin transaction is recorded publicly in the blockchain list. All computers running the blockchain have the same list of blocks and transactions, allowing them to see new blocks filled with new bitcoin transactions in real-time. So it is difficult to make any fake transaction of bitcoin, and it is very secure.

In this paper, a Bi-LSTM and Bi-RNN based hybrid model has been proposed, which predicts the closing price of bitcoin in a 1-day interval. For evaluating the performance of the proposed model, a comparison has been made with CNN, LSTM, Bi-LSTM, Bi-RNN, and several other models using different errors such as RMSE, MAE, MAPE, and MedAE.

2 Related Work

As bitcoins are becoming more popular day by day, many researchers have developed several methods to predict the bitcoin price. McNally et al. [4] proposed deep learning-based models such as recurrent neural network (RNN), LSTM for predicting the closing price of bitcoin where the simple moving average (SMA) is used for feature engineering. They applied a temporal window size of 20 days and concluded LSTM network produces better results compared to RNN and ARIMA model.

Ji et al. [5] demonstrated a comparative analysis employing deep learning-based networks such as deep neural network (DNN), LSTM, CNN, deep residual network, and their combinations for forecasting bitcoin price. They shared that for the regression task, i.e., for the prediction of bitcoin price, LSTM networks perform better and DNN-based networks are better suited for predicting the direction of price. Aggarwal et al. [6] investigated socio-economic factors like gold price, Twitter sentiment, and different cryptocurrencies for bitcoin price prediction and proposed root LSTM and gated recurrent unit (GRU) based network. They conferred gold price has less impact on the bitcoin price prediction while Twitter sentiment may give false information about the up-down of price. Also, the LSTM model performs more reliably compared to CNN or GRU-based models.

In [7], Chen et al. analyzed different statistical methods, i.e., logistic regression, linear discriminant analysis, and various machine learning models such as random forest, support vector machine (SVM), LSTM, etc., for bitcoin price prediction. For daily price prediction, they utilized the statistical methods while machine learning-based models are adopted for forecasting the price for 5-minutes intervals. It is observed from their results that for high-frequency data, machine learning models perform better than the statistical methods.

Velankar et al. [8] attempted to predict the daily price of bitcoin using bayesian regression and generalized linear model (GLM). Their study concentrates on finding the optimal features for predicting bitcoin price. However, they didn't provide any evaluations to compare the effectiveness of their models. Datta et al. [9] used the gated recurring unit (GRU) with recurrent dropout for estimating bitcoin price. They consider features that have financial linkage, such as Transaction Fees, Money Supply, US$ Index, etc., in their study. Although recurrent dropout enhances the performance capability of the GRU architecture, further investigation is required to verify its reliability.

3 Methodology

The proposed method is divided into several stages, including dataset collection, normalization, dataset splitting, and so on, as shown in Fig. 1.

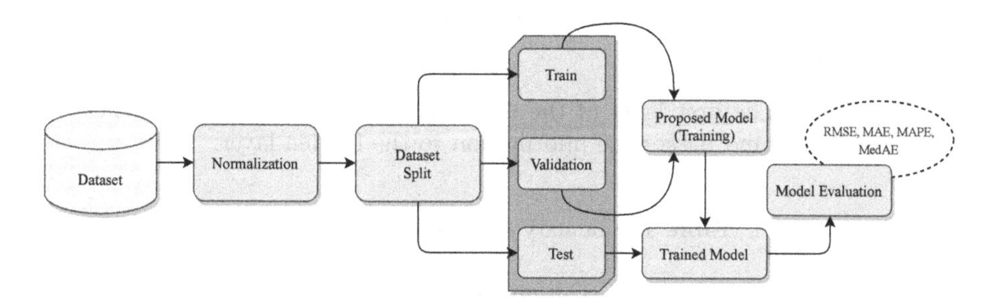

Fig. 1. Work flow of the proposed method.

3.1 Dataset Collection and Description

We collected the Bitcoin dataset from yahoo finance [10]. The dataset consists of seven attributes, namely Date, Open, High, Low, Close, Adj Close, and Volume. We have taken the 'Close' column, which includes bitcoin closing prices in USD from 1st January 2017 to 31st December 2019.

3.2 Normalization

The min-max normalization technique has been applied to the dataset to keep the close price between the range of 0 to 1. To normalize a range between an arbitrary set of values $[l, r]$, the min-max normalization formula is

$$p' = l + \frac{p - min(p)(r - l)}{max(p) - min(p)} \tag{1}$$

Where p' is the normalized price, p is the actual price.

3.3 The Proposed Model

The proposed hybrid model consists of an input layer, 1D-CNN layer, Bi-LSTM, LSTM, Bi-RNN, RNN, and Dense layer that work in a sequential manner. The proposed model is illustrated in Fig. 2 and the summary of the model is given in Table 1.

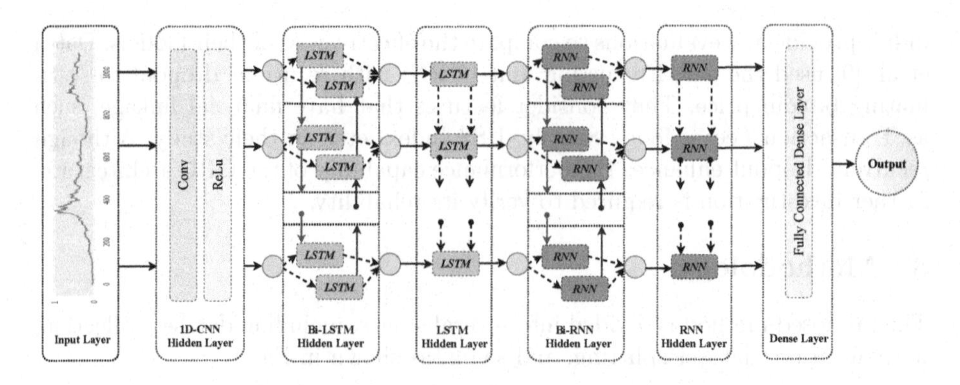

Fig. 2. The proposed model architecture.

Input Layer: The first layer of the model is the input layer which takes the normalized data and passes the information to the hidden layer.

Table 1. Summary of the proposed model.

Layers	Output shape	Parameters
Conv1D	1×32	1,632
Bi-LSTM	1×64	16,640
LSTM	1×32	12,416
Bi-RNN	1×64	4,160
RNN	32	3,104
Dense	1	33
Total parameters		**37,985**

Convolutional Neural Network (CNN): 1D-CNN [11] has shown higher performance in applications with small labeled data and high signal variations gained from various sources. We applied 1D-CNN in the proposed model with Rectified Linear Unit (ReLU) as an activation function that decides when the neurons of the network have to be activated. The mathematical formula for 1D-CNN can be presented as

$$H(j) = \sum_{n=1}^{N_k} I(j+n) \times K(n) \tag{2}$$

where $1 \leq j \leq N_i$, I and H denote the input and output feature map with N_i dimension and K is the convolutional kernel with a dimension size of N_k.

Recurrent Neural Network (RNN): RNN [12] is a type of neural network that makes decisions based on the current input as well as what it has learned from prior inputs. With the help of internal memory, it can loop back the produced output back into the network. RNN follows a precise temporal order when processing inputs that mean the current input is only contextualized by previous inputs, but not by future inputs. The RNN processing chain is duplicated in bidirectional RNN (Bi-RNN) [13], which processes inputs in both forward and reverse time order thus enabling the network to consider the future context as well.

Long Short Term Memory (LSTM): The LSTM layer [14], which is typically a gated Recurrent Neural Network, is quite effective for solving the vanishing and exploding gradient problem. By keeping a more stable error rate, LSTM allows the network to understand several time stages. Again the LSTM contains forget and remember gates, which allow the neuron to determine which information to forget or distribute depending on its significance. The necessary equations for the LSTM network are defined in Eq. (3)–Eq. (8). A bidirectional LSTM (Bi-LSTM) is made up of two LSTMs, one of which takes input in forward direction while the other takes it in a backward direction. The standard architecture of RNN and LSTM are shown in Fig. 3.

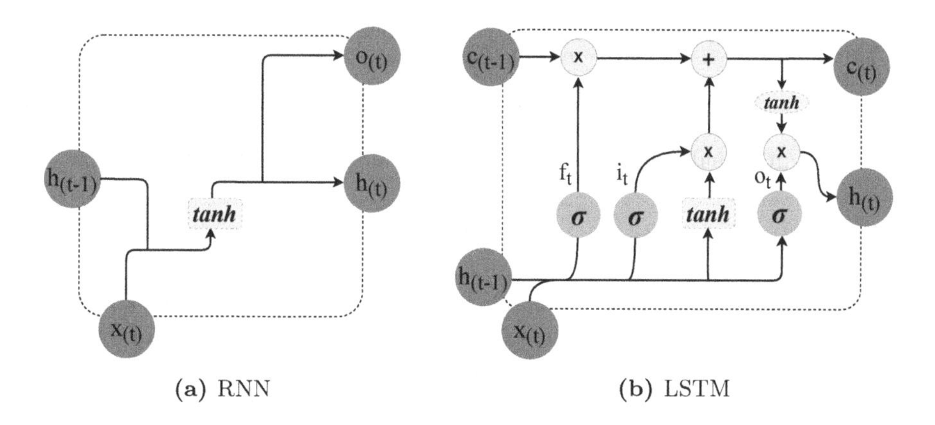

(a) RNN **(b)** LSTM

Fig. 3. Standard architecture of RNN, and LSTM.

$$f_t = \sigma(W_f x_t + W_f h_{t-1} + b_f) \tag{3}$$

$$i_t = \sigma(W_i + W_i h_{t-1} + b_i) \tag{4}$$

$$\tilde{c}_t = tanh(W_c x_t + W_c h_{t-1} + b_c) \tag{5}$$

$$o_t = \sigma(W_\circ x_t + W_\circ h_{t-1} + b_\circ) \tag{6}$$

$$c_t = f_t * c_{t-1} + i_t * \tilde{c}_t \tag{7}$$

$$h_t = o_t * tanh(c_t) \tag{8}$$

here, f_t symbolises the activation vector for forget gate, x_t is the LSTM unit's input vector, i_t signifies the activation vector of the input/update gate, o_t indicates the activation vector of the output gate, h_t indicates the vector of hidden states, $\tilde{c}_t$ denotes the activation vector for cell input, c_t denotes a vector that represents the current state of a cell. Here W_x denotes weight matrices of respective gate(x), b_x denotes bias vectors for gate(x) and subscript t denotes timestamps.

3.4 Performance Evaluation

To compare the proposed model with other models, we have used four types of errors, i.e., Root Mean Square Error (RMSE), Mean Absolute Error (MAE), Mean Absolute Percentage Error (MAPE), and Median Absolute Error (MedAE). The formulas for calculating the errors are given below:

$$RMSE = \sqrt{\frac{1}{n}\sum_{i=1}^{n}(p_i - \hat{p}_i)^2} \tag{9}$$

$$MAE = \frac{1}{n}\sum_{i=1}^{n}|p_i - \hat{p}_i| \tag{10}$$

$$MAPE = \frac{1}{n}\sum_{i=1}^{n}\frac{|p_i - \hat{p}_i|}{|p_i|} \tag{11}$$

$$MedAE = median(|p_1 - \hat{p}_1|, ..., |p_n - \hat{p}_n|) \tag{12}$$

where p is an actual value, $\hat{p}$ is the predicted value and n is the sample size.

4 Result Analysis

After developing the model, we applied the Adam as the optimizer and Mean Absolute Error as the loss function for compiling the model. The model is trained for 512 epochs with a batch size of 64. The loss vs epoch curve, as shown in Fig. 4a, indicates that the model is free from overfitting. For estimating the best

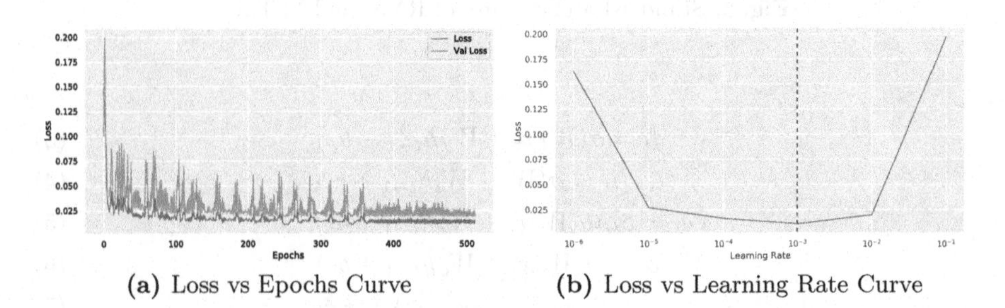

(a) Loss vs Epochs Curve (b) Loss vs Learning Rate Curve

Fig. 4. Loss vs epochs and loss vs learning rate curve of the proposed model.

learning rate for the Adam optimizer, we vary the learning rate and calculate the respective minimum loss of that rate as indicated by Fig. 4b. From the plot, it is apparent that for learning rate 10^{-3}, the model produced the lowest loss.

In our proposed model, we vary different hyper-parameters such as window size, CNN filters, CNN kernel size, LSTM units, and RNN units to analyze how MAE dependent on them. The details of this experiment can be found in Fig. 5. If we take a closer look at the plot, according to Fig. 5a and Fig. 5b, the model produces adequate results when the window size varies between 5–10, and both LSTM units and RNN units varies between 16–64. Similarly, we can determine the hyper-parameter ranges from the other subplots. From Fig. 5, it is evident that the value of MAE is lowest when the window size is 10, CNN filters is 32, CNN kernel size is 5, and both the LSTM units and RNN units are 32.

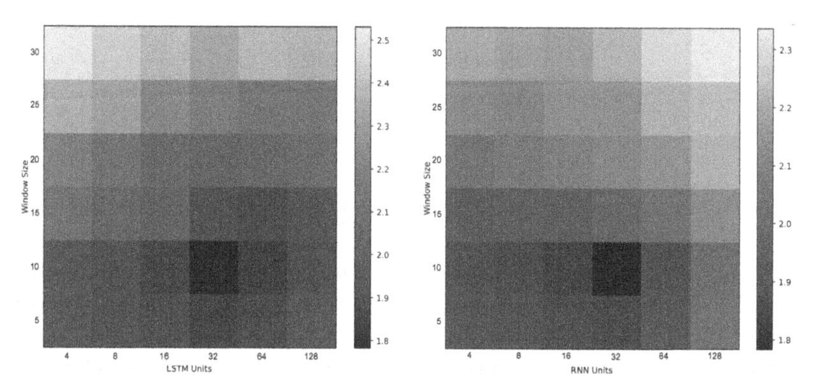

(a) MAE dependent on Window Size and LSTM Units.

(b) MAE dependent on Window Size and RNN Units.

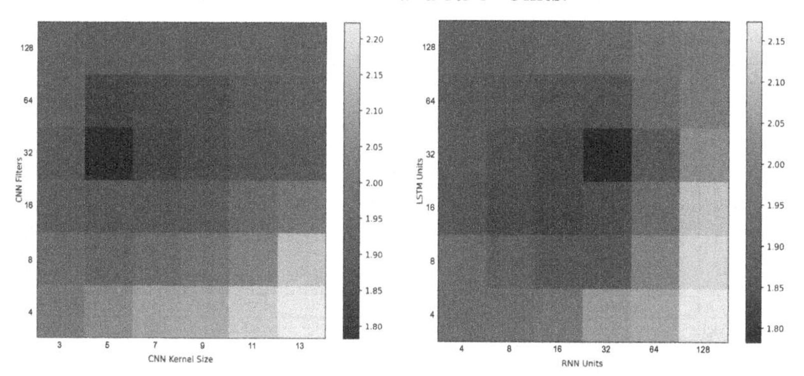

(c) MAE dependent on CNN Filters and CNN Kernel Size.

(d) MAE dependent on LSTM Units and RNN Units.

Fig. 5. MAE dependent on hyper-parameters.

The summary of the fine-tuned hyper-parameters used in our proposed hybrid model can be found in Table 2.

Table 2. Hyper-parameters of the proposed model.

Parameters	Value
Window size	10
CNN filters	32
CNN kernel size	5
LSTM units	32
RNN units	32
Learning rate	10^{-3}
Epochs	512
Batch size	64

In Fig. 6, a comparative analysis has been illustrated between different models with respect to several errors, i.e., RMSE, MAE, MAPE, and MedAE. From the illustration, it is comprehensible that the proposed hybrid model offers the lowest error in every category compared to other models as the proposed model achieves RMSE, MAE, MAPE, and MedAE of 2.69%, 1.78%, 2.20%, and 1.23% respectively and successfully surpasses the other individual models in performances especially the popular ARIMA model as the ARIMA model delivers the RMSE, MAE, MAPE, and MedAE of 2.99%, 2.10%, 2.59%, and 1.54% respectively.

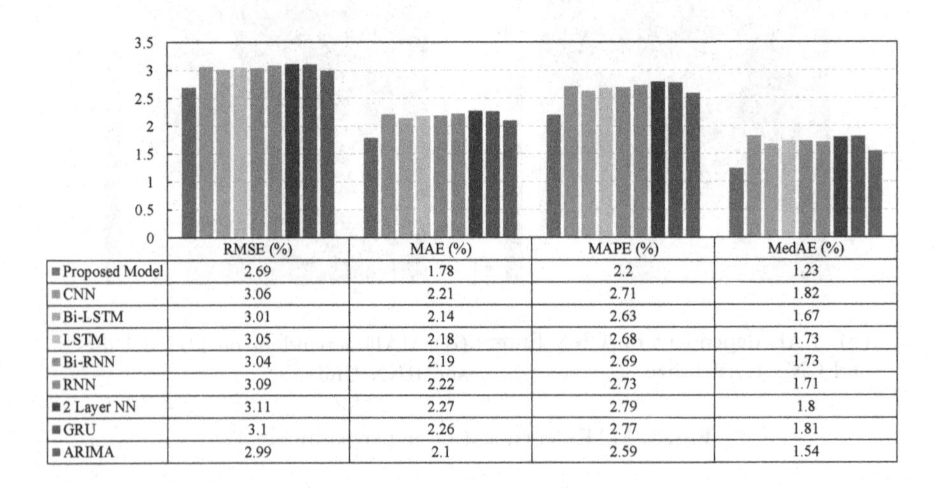

	RMSE (%)	MAE (%)	MAPE (%)	MedAE (%)
Proposed Model	2.69	1.78	2.2	1.23
CNN	3.06	2.21	2.71	1.82
Bi-LSTM	3.01	2.14	2.63	1.67
LSTM	3.05	2.18	2.68	1.73
Bi-RNN	3.04	2.19	2.69	1.73
RNN	3.09	2.22	2.73	1.71
2 Layer NN	3.11	2.27	2.79	1.8
GRU	3.1	2.26	2.77	1.81
ARIMA	2.99	2.1	2.59	1.54

Fig. 6. Comparative analysis of different models.

Cumulative distribution function (CDF) plots with respect to the different errors for all models can be inspected in Fig. 7. In comparison to other individual models, the proposed model has a larger area under the curve for each error, as

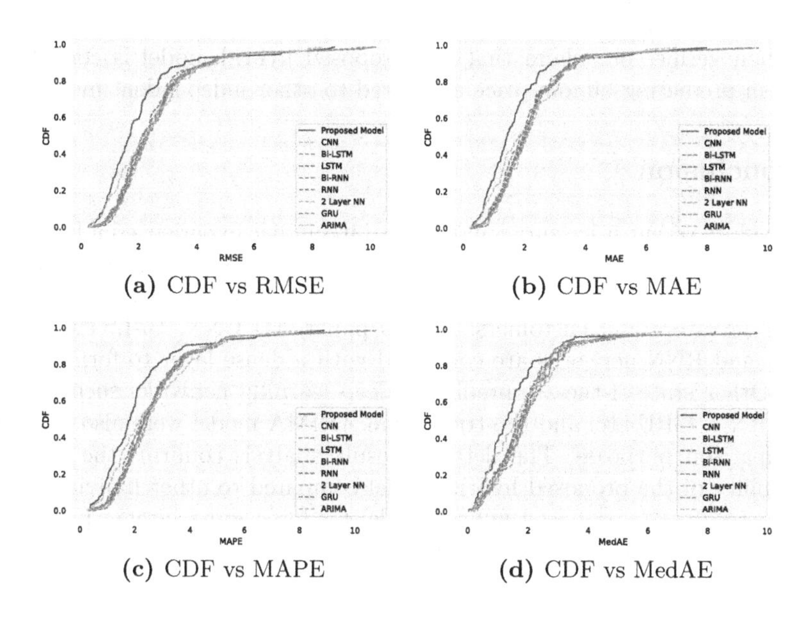

(a) CDF vs RMSE (b) CDF vs MAE

(c) CDF vs MAPE (d) CDF vs MedAE

Fig. 7. Analysis of CDF plots for different errors.

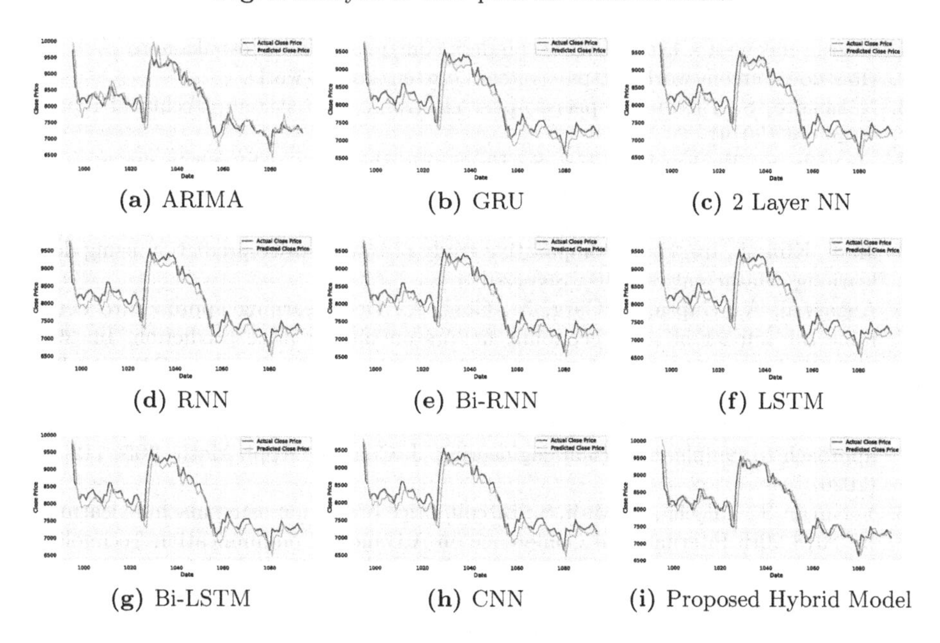

(a) ARIMA (b) GRU (c) 2 Layer NN

(d) RNN (e) Bi-RNN (f) LSTM

(g) Bi-LSTM (h) CNN (i) Proposed Hybrid Model

Fig. 8. Prediction curve of various models.

seen in Fig. 7. Since a greater area under the curve indicates that the model has a lesser error, it is understandable that our proposed hybrid model outperforms the other individual models.

The prediction curve for various models is exhibited in Fig. 8 that represents the actual close price and the predicted close price for specific models. The visual illustration verifies our claim that the proposed hybrid model is considerably reliable in predicting bitcoin price compared to other independent models.

5 Conclusion

Because of its recent price surge and crash, bitcoin has experienced a lot of press and public interest. This paper aimed to develop an adequate method for forecasting the price of bitcoin using the hybrid deep learning model while reducing risks for investors and customers. In this paper, 1D CNN, Bi-LSTM, LSTM, Bi-RNN and RNN networks are combined with a dense layer to form a hybrid model. Other state-of-the-art predictive deep learning networks such as CNN, LSTM, RNN, GRU etc. and the traditional ARIMA model were also developed for comparison purposes. The detailed result analysis confirms the reliability and stability of the proposed hybrid model compared to other individual models. As a result, the proposed hybrid model for forecasting bitcoin price would aid consumer growth while also significantly reducing risk for potential investors.

References

1. Satoshi nakamoto. https://www.coindesk.com/people/satoshi-nakamoto
2. How does bitcoin work? https://bitcoin.org/en/how-it-works
3. Nakamoto, S.: Bitcoin: A peer-to-peer electronic cash system. Technical report, Manubot (2019)
4. McNally, S., Roche, J., Caton, S.: Predicting the price of bitcoin using machine learning. In: 2018 26th Euromicro International Conference on Parallel, Distributed and Network-Based Processing (PDP), pp. 339–343. IEEE (2018)
5. Ji, S., Kim, J., Im, H.: A comparative study of bitcoin price prediction using deep learning. Mathematics **7**(10), 898 (2019)
6. Aggarwal, A., Gupta, I., Garg, N., Goel, A.: Deep learning approach to determine the impact of socio economic factors on bitcoin price prediction. In: 2019 Twelfth International Conference on Contemporary Computing (IC3), pp. 1–5. IEEE (2019)
7. Chen, Z., Li, C., Sun, W.: Bitcoin price prediction using machine learning: an approach to sample dimension engineering. J. Comput. Appl. Math. **365**, 112,395 (2020)
8. Velankar, S., Valecha, S., Maji, S.: Bitcoin price prediction using machine learning. In: 2018 20th International Conference on Advanced Communication Technology (ICACT), pp. 144–147. IEEE (2018)
9. Dutta, A., Kumar, S., Basu, M.: A gated recurrent unit approach to bitcoin price prediction. J. Risk Financ. Manage. **13**(2), 23 (2020)
10. Bitcoin usd (btc-usd). https://finance.yahoo.com/quote/BTC-USD/history?p=BTC-USD

11. Kiranyaz, S., Avci, O., Abdeljaber, O., Ince, T., Gabbouj, M., Inman, D.J.: 1d convolutional neural networks and applications: a survey. Mech. Syst. Signal Process. **151**, 107,398 (2021)
12. Mikolov, T., Karafiát, M., Burget, L., Černocký, J., Khudanpur, S.: Recurrent neural network based language model. In: Eleventh Annual Conference of the International Speech Communication Association (2010)
13. Schuster, M., Paliwal, K.K.: Bidirectional recurrent neural networks. IEEE Trans. Signal Process. **45**(11), 2673–2681 (1997)
14. Hochreiter, S., Schmidhuber, J.: Long short-term memory. Neural Comput. **9**(8), 1735–1780 (1997)

DC-Deblur: A Dilated Convolutional Network for Single Image Deblurring

Boyan Xu and Hujun Yin(✉)

The University of Manchester, Manchester M13 9PL, UK
boyan.xu@postgrad.manchester.ac.uk, hujun.yin@manchester.ac.uk

Abstract. Single image deblurring is a significant and challenging task in image processing vision and machine learning. Convolutional Neural Network (CNN) based models for deblurring often have a complex structure and a considerable number of parameters compared with those for other image restoration tasks such as image denoising, dehazing and super-resolution. The main reason is the requirement of large reception fields, which are important to image deblurring due to possible large blur kernels. Dilated convolution is a useful way to increase reception field without adding extra parameters. In this paper, we propose a novel network by adopting a dilated convolution structure, and we further improve the training process by combining L1 loss, MS-SSIM loss and MSE loss. The proposed network is light and fast. Quantitative and qualitative experiments indicate that our method outperforms state-of-the-art models, in terms of performance and speed.

Keywords: Image deblurring · Convolutional neural network · Dilated convolution · Image restoration · Loss function

1 Introduction

Motion blur is a common type of degradation resulted from long exposure in image acquisition of moving objects or with moving cameras. It occurs frequently while using a handheld device, such as mobile phones and compact cameras. However, many image processing algorithms, including classification, tracking and detection, rely on working with sharp images. Image deblurring aims to recover latent sharp image from a degraded input caused by motion-induced smearing. It has attracted a great deal of attention in computer vision and image processing communities over the past decades and remains a challenging problem due to its ill-posed nature.

With rapid development of Deep Convolutional Neural Networks (DCNNs) [12], learning based methods have become the mainstream for image restoration, e.g., image denoising [24], image dehazing [25], super-resolution [22] and image deblurring [15]. In some cases, a network that performs well on one of the aforementioned tasks can also work on other image inverse problems. For instance, the

H. Yin et al. (Eds.): IDEAL 2021, LNCS 13113, pp. 234–245, 2021.
https://doi.org/10.1007/978-3-030-91608-4_24

method in [17] performs well on super-resolution and edge-preserving filtering, and has been shown in [1] to work well on image dehazing and deraining. However, in many cases such transferable capability is hard to find. Image deblurring is a common restoration problem, yet networks that perform well on denoising or dehazing cannot be adopted for deblurring directly. For instance, adopting [1,17] for deblurring has led to poor performance. To investigate the reason, we have observed that deblurring networks often require larger model sizes and more parameters than other image restoration networks. For example, a state-of-the-art dehazing and deraining network Gated Context Aggregation Network (GCANet) [1] has a size of 2.7 MB, yet typical deblurring networks like PSSNSC [3] have a size of 46.5 MB. Note that PSSNSC adopts parameter sharing, thus, when we consider the full computational complexity, PSSNSC would be more expensive. Earlier deblurring networks such as DeepDeblur [15] have an even larger size of 303.6 MB.

The requirement for large reception field is the core reason for this condition. Due to the mechanism of image blurring, blur kernels could be very large, and usually dozens of pixels involving in blur kernels. Thus, deblur networks need large reception fields both in feature extraction and restoration in order to model and handle the blur kernels. Generally, previous work increases the reception field by using more convolution layers [15]. It is undeniable that deeper networks could increase the performance, but it would also increase the complexity, making the forward processing slow. Many researchers have tried to develop faster and better networks, but architectures of these networks are still very complex [10].

To make a network more efficient, the network structure should be able to cover a large reception field without adding too many parameters. To this end, we adopt a dilated convolution network approach. Dilated convolution (or atrous convolution) was originally developed for wavelet decomposition [5]. The main idea is to insert "holes" (zeros) between pixels in convolutional kernels to increase the convolution range, thus enabling dense feature extraction in DCNNs. In this paper, we propose a dilated convolution based network for single image deblurring, termed as DC-Deblur. For the framework and backbones of the network, we adopt an encoder-decoder structure and use densely connected structure to extract features and help feature restoration. The remaining paper is organised as follows. A review of most recent work in learning based image deblurring and dilated convolution is given in Sect. 2. We introduce the proposed network in Sect. 3, and compare with other state-of-the-art methods in Sect. 4. Conclusions are given in Sect. 5.

2 Related Work

2.1 Deep Image Deblurring

Neural networks have become the driving force in many image processing tasks including image deblurring due to their outstanding performances. A landmark progress was made by the DeepDeblur [15], as an end-to-end kernel-free approach that does not need to consider blur kernels. DeepDeblur adopts residual blocks and multi-scale structure, achieving good results but also resulting in a large model size. Many end-to-end methods have been proposed to pursue even better performances, including SRN-Deblur [19], PSSNSC [3], DSHMN [11], Deblur-GAN [9] and DeblurGAN-v2 [10]. DeblurGAN-v2 has a smaller size of 13.5 MB, but has to sacrifice much in quantitative performance.

2.2 Dilated Convolution

Dilated convolution, also known as atrous convolution, was first proposed in [21], and has been widely used in many computer vision tasks such as object detection [2], audio generation [16], and video modelling [7]. It can significantly increase the reception field and extract global information by inserting zeros between weights, without requiring additional parameters. Consider an one-dimension input x, the output y for a dilated convolution at location i is defined as

$$y[i] = \sum_{j=1}^{J} f[i + r * j]x[j] \tag{1}$$

where f is the filter implemented by convolutional layer with kernel size J, and dilation rate r. For image deblurring, if we treat standard convolutions as dilated convolutions with a dilation rate of $r = 1$, we can remove a downsampling layer with a subsampling rate of 2 by letting the dilation rate of all subsequent layers be 2. This results in dilated convolutional layers with dilation rates of $r = 2, 4, 8,$ etc.

However, when using dilated convolutions to restore an image, the gridding artifacts affect the models significantly. In [20], smoothed dilated convolution was proposed by using the original dilated convolution parameters (with the hole 0 removed) to perform convolutions and then periodically sampling the original feature map into 4 feature maps with reduced resolution, and then the convolution results were combined by upsampling.

3 Proposed Method

In this section, we introduce the proposed DC-Deblur network. The overall structure of the network is shown in Fig. 1. Given a blurry image I_{in}, we encode it by an encoder, and process the information by a dilated convolution structure and gated fusion. Then the gated feature map will be decoded to a blur residue by a decoder structure, and finally add to the restored sharp image by using residual link. We use densely connected structure blocks [6] for the encoder and the decoder, and in the rest of the paper, we term them as DenseBlocks.

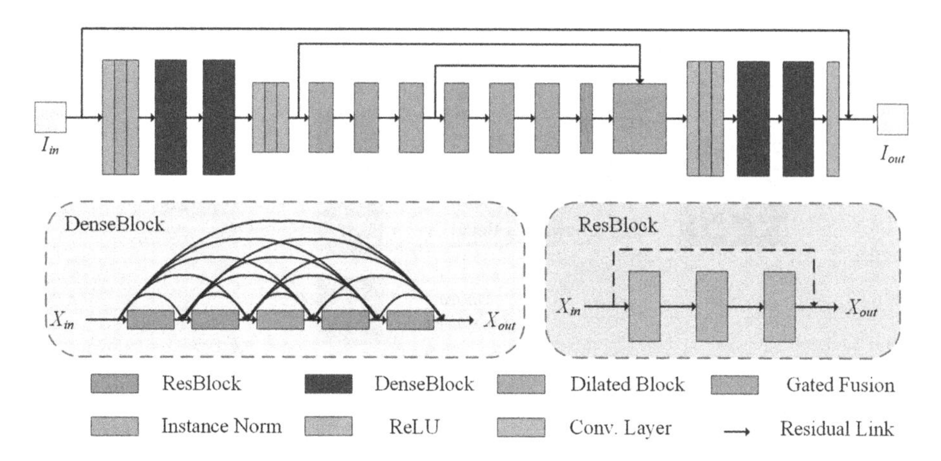

Fig. 1. Overall network structure of the proposed network, following an auto-encoder structure. It consists of two convolution blocks and two DenseBlocks as the encoder part, and one deconvolution block and DenseBlocks as the decoder part. Several smoothed dilated resblocks are inserted between them to aggregate context information without gridding artifacts. To fuse the features from different levels, an extra gate fusion sub-network is leveraged. The proposed network predicts the residues between target sharp image and blurry input image in an end-to-end way.

3.1 Efficient Dilated Convolution

As discussed in Sect. 2, dilated convolutions often suffer from the gridding artifacts. To handle this problem, smoothed dilated convolution proposed by Wang *et al.* [20], was successfully adapted for image dehazing and deraining in GCANet [1]. However, some details can be further improved for deblurring. Specifically, separable and shared (SS) convolutions are used for technical implementation. Consider a convolution layer with input and output channels C and kernel $n \times n$, separable convolution handle each channel separately. In contrast with a standard convolution that connects all C channels in input to all C channels in outputs, leading to $n^2 \times C^2$ parameters, a separable convolution only connects the ith output channel to the ith input channel, resulting in only $n^2 \times C$ parameters. Based on separable convolutions, shared convolution means the same $n^2 \times C$ parameters are shared by all pairs of input and output channels. SS convolutions only have one filter scanning all spatial locations for input and output of C channels and share this filter across all channels. We use 3×3 in SS convolution, and thus can help fusing the information of neighbour pixels to help avoid gridding artifacts, as is shown in Fig. 2

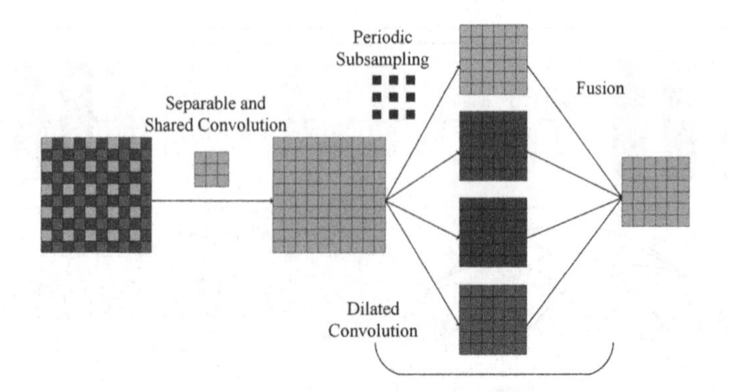

Fig. 2. Illustration of the smoothed dilated convolution method. Separable and Shared convolution is used to produce a smoothed feature map, and then periodic subsampling is used to implement dilated feature extraction. The four outputs are then fused into a new smoothed dilated layer. The grey map represents smoothed feature maps.

We further analyse the difference between image dehazing/deraining and image deblurring, and modify the dilated convolution blocks by removing normalization layers, as is shown in Fig. 3. In DeepDeblur [15], it was found that removing the batch normalization (BN) unit in the original residual building block benefited the convergence and training time. Since normalization layers normalize the features, they reduce the range flexibility by normalizing the features [13]. Hence it is better to remove them in image restoration. Based on the previous analysis, we believe that removing these normalization layers would benefit the performance. Experiments verified this idea, and further details are given in Sect. 4.

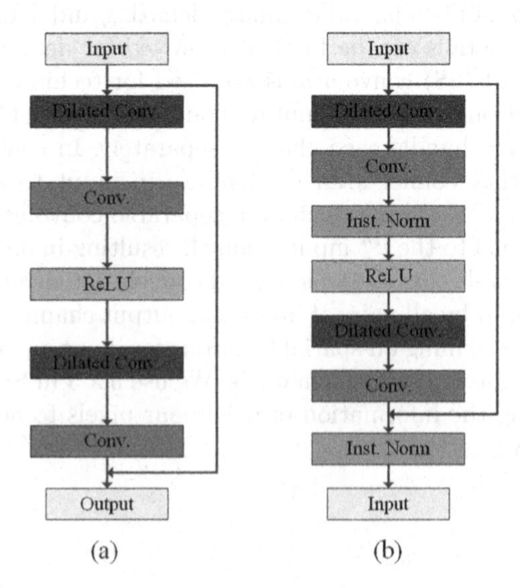

Fig. 3. Comparison of our dilated convolution blocks and GCANet [1]: (a) dilated convolution blocks of GCANet, and (b) modified dilated convolution blocks of proposed DS-Deblur network.

3.2 Network Structure

Following the similar network design principles in [3,15,19], our overall network is designed as a simple encoder-decoder structure, as shown in Fig. 1, where I_{in} denotes the input blurry image, and I_{out} is the output sharp image. We adopted DenseBlocks [6] in our network, and also used the modified ResBlocks [15]. These ResBlocks consist of two convolution layers and a ReLU layer. Note that the residual links are implemented by the densely connected links when using ResBlocks in DenseBlock, as shown in Fig. 1. Each DenseBlock consists of five ResBlocks. We use DenseBlocks with 64 channels, two DenseBlocks in the encoder, two others in the decoder. We use a limited number of IN layers out of dilated blocks to make the training stable. Dilated blocks are used after the first downsample. By the using of dilated convolution structure, the second downsample is avoided, which means that we use only one downsample and one upsample in our deblurring network. In contrast, previous work such as DeepDeblur [15], SRN-Deblur [19], PSSNSC [3] all used two or more subsamples.

In this paper, we further adopt a gated fusion structure inspired by [1] via incorporation of an extra gated fusion sub-network G, as shown in Fig. 1. We first extract the feature maps from different levels, denoting them as F_1, F_2, and F_3, and feed them into the gated fusion sub-network. The output of the gated fusion sub-network are three different importance weights M_1, M_2, and M_3, corresponding to each feature level. Finally, these three features maps F_1, F_2, and F_3 from different levels are linearly combined with the regressed importance weights,

$$(M_1, M_2, M_3) = G(F_1, F_2, F_3) \tag{2}$$

$$F_{out} = M_1 \times F_1 + M_2 \times F_2 + M_3 \times F_3 \tag{3}$$

3.3 Loss Function

Loss function is an other key part in image restoration and deblurring. In [19], l_2 loss was used, which also known as the mean squared error (MSE) loss. MSE loss has close relationship with peak signal to noise ratio (PSNR), an important image quality assessment (IQA) index. The relationship can be denoted as

$$PSNR = 10 \log_{10}(\frac{255^2}{MSE}). \tag{4}$$

Another important loss function is the l_1 loss. In [26], the authors explored loss functions in image restoration, especially in denoising, super-resolution and JPEG artifacts removal by proposing a mixed loss by combining l_1 loss and Multi-Scale Structural SIMilarity (MS-SSIM) loss. MS-SSIM is based on Structural SIMilarity (SSIM), which is a well-known criterion and brings IQA from pixel-based to structure-based.

In our experiments, MSE loss gave similar or even better performances compared with the l_1 loss in image deblurring. Details are given in Sect. 4. This might be because that blur kernels are larger and complex. Thus, the euclidean

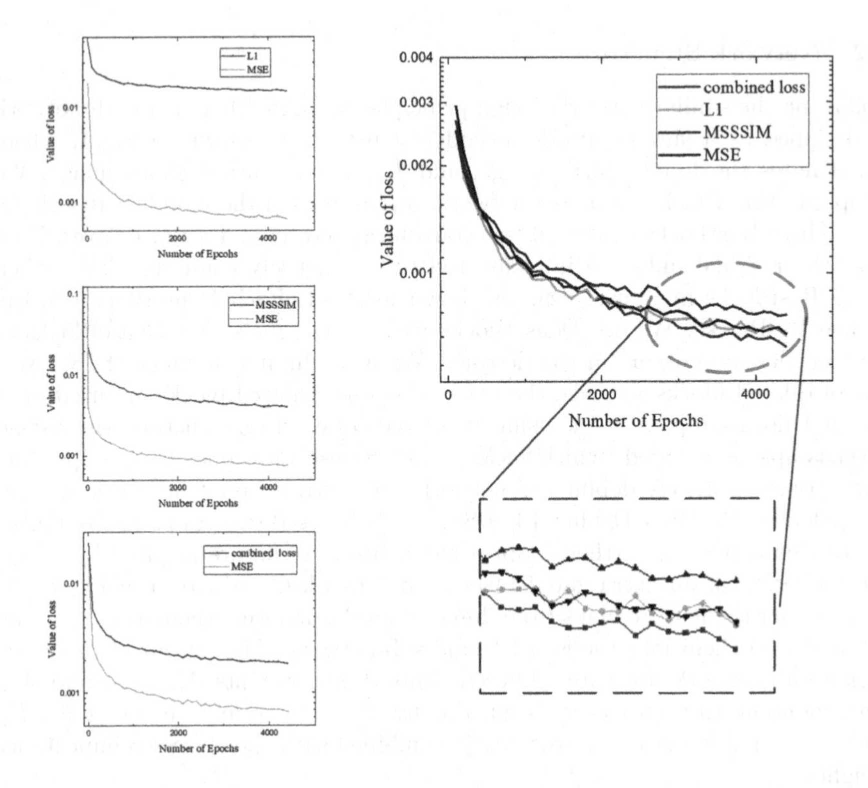

Fig. 4. Visual comparison on GoPro dataset, with Dual Residual Network [14], SRN [19], DeblurGAN-v2 [10], PSSNSC [3].

distance between a blurry image and sharp image is greater than that of a low-resolution and high-resolution image. In this work, we further combine l_1 loss, MSE loss and MS-SSIM loss. The loss function of our network is denoted by

$$\mathcal{L}^{Mix} = \omega \times \mathcal{L}^{MS-SSIM} + \gamma \mathcal{L}^{l_1} + \lambda \times \mathcal{L}^{MSE}. \tag{5}$$

l_1 loss, MSE loss and MS-SSIM loss have different numerical factors. For instance, the value of l_1 loss is often 10 times bigger than that of MSE loss. This is negative for the convergence of the network. To balance these parts in the loss, we set $\omega = 0.1$, $\gamma = 0.05$ and $\lambda = 1$.

4 Implementation and Experiments

4.1 Implementation

We implemented the proposed DC-Deblur network by Pytorch on a NVIDIA Tesla P100 GPU. During training, a 256 × 256 region from a blurred and its ground truth image at the same location were randomly cropped out as the training input. The batch size was set to 18 during training. All weights were initialized using the Xavier method [4], and biases were initialized to zero. The network was optimized using the Adam method [8] with default setting $\beta_1 = 0.9$, $\beta_2 = 0.999$ and $\epsilon = 10^{-8}$. The learning rate was initially set to 0.0001 and linearly decayed to 0. The number of training epochs was set empirically by the experiments to ensure convergence of the network.

4.2 Experiments

Datasets: GoPro dataset [15] and HIDE dataset [18] were used.

GoPro Dataset: GoPro dataset [15] consists of large-scale training and testing blurry image pairs synthesised from real videos, and is the most used kernel-free deblurring dataset. GoPro dataset contains 2,103 pairs of images for training (which is the training set) and 1,111 pairs for evaluation (i.e. the testing set). In this paper, we trained our model on the training set, and tested the trained model on the testing set.

HIDE Dataset: The images of HIDE dataset are subsequently organised into two categories, including long-shot (HIDE I) and regular pedestrians (close-ups, HIDE II). The training set combines HIDE I and HIDE II, leading to 6397 for training, 1063 of HIDE I and 962 of HIDE II for testing. HIDE dataset has a large number of human faces and complex objects, including high-speed moving vehicles and strong lighting interference. Thus, it is useful in evaluating real-world applications.

Quantitative and Qualitative Evaluations: We compared our method with the state-of-the-art image deblurring methods both quantitatively and qualitatively. Five state-of-the-art methods were implemented for quantitative evaluation: DeepDeblur [15], SRN-Deblur [19], PSSNSC [3], DeblurGANv2 [10] and SVRNN [23]. The testing results on the GoPro dataset are shown in Table 1. Our method outperformed others in all the benchmarks, including PSNR, SSIM, and FSIM. In addition, our method was also better in speed of forward processing due to its smaller model size. By using dilated convolutions, we reduced the network size by about 60% compared with the state-of-the-art methods. Note that SRN-Deblur and PSSNSC used parameter sharing when training. For instance,

Table 1. Testing results on GoPro dataset.

Algorithm	DeepDeblur [15]	SRN-Deblur [19]	PSSNSC [3]	DeblurGANv2 [10]	SVRNN [23]	Ours
PSNR	29.08	30.26	30.92	29.55	29.19	**31.21**
SSIM	0.9135	0.9432	0.9421	0.9340	0.9306	**0.9448**
FSIM	0.9633	0.9653	0.9637	0.9527	0.9446	**0.9743**
RunTime	3.09 s	0.86 s	0.68 s	0.35 s	1.40 s	**0.29 s**
Size	303.6 MB	27.5 MB	46.5 MB	244.5 MB	37.1 MB	**10.1 MB**

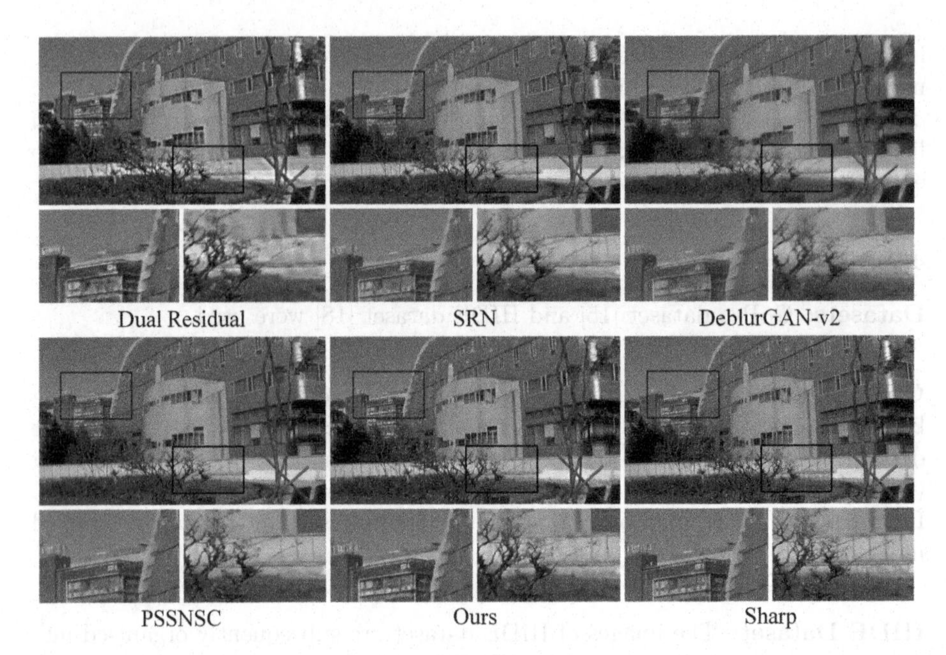

Fig. 5. Comparisons of the influence of various loss functions in network convergence. When l_1 loss, MS-SSIM loss, and combined loss are used to optimise the network, calculate the value of MSE, in order to make a fair comparison.

PSSNSC [3] shared the parameters of each DenseBlocks six times, thus the number parameters involved in processing was much larger. Visual results are shown in Fig. 5. Our approach yields sharper restoration.

We also tested our network on the HIDE dataset. The results are shown in Table 2. Our method also achieved better performance in PSNR, SSIM, and FSIM. Since we used the same model as that on GoPro dataset, the processing time and model size are not listed.

Table 2. Testing results on HIDE datasets.

Algorithm	HIDE I (long-short)			HIDE II (close-ups)		
	PSNR	SSIM	FSIM	PSNR	SSIM	FSIM
DeepDeblur [15]	27.43	0.9020	0.9622	26.18	0.8780	0.9339
SRN-Deblur [19]	29.41	0.9137	0.9681	27.54	0.9070	0.9393
PSSNSC [3]	29.98	0.9234	0.9728	28.14	0.9021	0.9470
DeblurGANv2 [10]	28.29	0.8960	0.9591	26.64	0.8722	0.9248
SVRNN [23]	28.69	0.9038	0.9614	26.68	0.8702	0.9231
Ours	**30.26**	**0.9297**	**0.9801**	**28.86**	**0.9133**	**0.9545**

Table 3. Testing results of ablation studies on GoPro dataset. Our adopted network with dilated convolution and combined loss performs the best.

Method	PSNR	SSIM	Size
DenseNet Deblur	30.26	0.9345	27.0 MB
Ours with Norm	31.05	0.9403	10.1 MB
Ours with MSE loss	31.14	0.9433	10.1 MB
Ours with L1 loss	31.09	0.9426	10.1 MB
Ours with MS-SSIM loss	30.65	0.9388	10.1 MB
Ours (adopted)	**31.21**	**0.9448**	**10.1 MB**

4.3 Ablation Study

We further experimented on different settings of our network to study its effectiveness and efficiency. Generally, our ablation study focused on the network framework and loss functions.

Network Framework: To measure the ability of our model, we did two more experiments on the network structure. Firstly, we tested the network performance without dilated convolution. However, directly removing the dilated blocks would lead to performance deterioration obviously. To solve this problem, we designed a new network called DenseNet Deblur. We used DenseBlocks to replace the dilated convolutions. Specially, we used 6 DenseBlocks in the encoder, and 6 DenseBlocks for the decoder. This network was similar to PSSNSC deblurring network to some degrees, while we did not use parameter sharing and multi-scale structure, and all of our DenseBlocks were 64-channel. The results are shown in Table 3. Our method have better performance than DenseNet Deblur, even though DenseNet Deblur had a much larger size. This shows that our method can achieve better performance and reduce the complexity of the model. Secondly, we tested the network with normalisation layers. To this end, we used the dilated convolution structure in Fig. 3(b). The results in Table 3 show that our network performed better when normalisation layers were removed.

Loss Function: Loss function was also further examined. In the previous section, we point out that multi-loss function can help training the network better rather than using single loss. In this section, we also give further comparisons of various loss functions. Our experiments on loss functions were based on the network structure in Fig. 1. First, we used MSE loss only, and the results are shown in Table 3, denoted as *Ours with MSE loss*. We also tested our model based on L1 loss only and MS-SSIM loss only. Their results are also shown in Table 3, denoted as *Ours with L1 loss* and *Ours with MS-SSIM loss*, respectively. The training process is shown in Fig. 4. Based on these experiments, the MS-SSIM only training had the worst performance, conforming with the result in

[26]. The network trained by L1 loss only was slightly worse than that of MSE loss only; this is different with the result of [26], mainly because of the difference between image deblurring and super-resolution. It is obvious that our method with combined loss had the best performance. In Fig. 4, the speed of convergence of L1 loss was slightly faster, but the final result was nearly the same as that of MSE loss. The curve of MS-SSIM loss was obviously higher than others, especially after 500 epochs. However, based on our experiments, the loss function combined with MSE, L1 and MS-SSIM gave better overall ability.

5 Conclusion

In this paper, a novel image deblurring network termed as DC-Deblur is proposed. The network is based on a dilated convolution structure, which is able to increase the reception field without incurring additional parameters. The requirement of large reception field is unavoidable for image deblurring due to the ill-posed nature of deblurring and possible large blur kernels. DC-Deblur can reduce the number of parameters, making the network efficient. We further improve the network performance by combining L1 loss, MS-SSIM loss and MSE loss. Experiments and quantitative and qualitative evaluations indicate that the proposed method outperforms the state-of-the-art models, in both performance and speed.

References

1. Chen, D., et al.: Gated context aggregation network for image dehazing and deraining. In: IEEE Winter Conference on Applications of Computer Vision (WACV), pp. 1375–1383. IEEE (2019)
2. Dai, J., Li, Y., He, K., Sun, J.: R-FCN: object detection via region-based fully convolutional networks. In: Advances in Neural Information Processing Systems (NIPS), pp. 379–387 (2016)
3. Gao, H., Tao, X., Shen, X., Jia, J.: Dynamic scene deblurring with parameter selective sharing and nested skip connections. In: Proceedings of the IEEE Conference on Computer Vision and Pattern Recognition (CVPR), pp. 3848–3856 (2019)
4. Glorot, X., Bengio, Y.: Understanding the difficulty of training deep feedforward neural networks. In: Proceedings of the International Conference on Artificial Intelligence and Statistics, pp. 249–256 (2010)
5. Holschneider, M., Kronland-Martinet, R., Morlet, J., Tchamitchian, P.: A real-time algorithm for signal analysis with the help of the wavelet transform. In: Combes, J.M., Grossmann, A., Tchamitchian, P. (eds.) Wavelets, pp. 286–297. Springer, Heidelberg (1990). https://doi.org/10.1007/978-3-642-75988-8_28
6. Huang, G., Liu, Z., Van Der Maaten, L., Weinberger, K.Q.: Densely connected convolutional networks. In: Proceedings of the IEEE Conference on Computer Vision and Pattern Recognition (CVPR), pp. 4700–4708 (2017)
7. Kalchbrenner, N., et al.: Video pixel networks. In: International Conference on Machine Learning, pp. 1771–1779 (2017)
8. Kingma, D.P., Ba, J.: Adam: a method for stochastic optimization. arXiv preprint arXiv:1412.6980 (2014)

9. Kupyn, O., Budzan, V., Mykhailych, M., Mishkin, D., Matas, J.: Deblurgan: blind motion deblurring using conditional adversarial networks. In: Proceedings of the IEEE Conference on Computer Vision and Pattern Recognition (CVPR), pp. 8183–8192 (2018)
10. Kupyn, O., Martyniuk, T., Wu, J., Wang, Z.: Deblurgan-v2: Deblurring (orders-of-magnitude) faster and better. In: Proceedings of the IEEE International Conference on Computer Vision (ICCV), pp. 8878–8887 (2019)
11. Lazebnik, S., Schmid, C., Ponce, J.: Beyond bags of features: Spatial pyramid matching for recognizing natural scene categories. In: Proceedings of the IEEE Conference on Computer Vision and Pattern Recognition (CVPR), vol. 2, pp. 2169–2178 (2006)
12. LeCun, Y., Bengio, Y., Hinton, G.: Deep learning. Nature **521**(7553), 436 (2015)
13. Lim, B., Son, S., Kim, H., Nah, S., Mu Lee, K.: Enhanced deep residual networks for single image super-resolution. In: Proceeding of the IEEE Conference on Computer Vision and Pattern Recognition (CVPR), pp. 136–144 (2017)
14. Liu, X., Suganuma, M., Sun, Z., Okatani, T.: Dual residual networks leveraging the potential of paired operations for image restoration. In: Proceedings of IEEE Conference on Computer Vision and Pattern Recognition (CVPR), pp. 7007–7016 (2019)
15. Nah, S., Hyun Kim, T., Mu Lee, K.: Deep multi-scale convolutional neural network for dynamic scene deblurring. In: Proceedings of the IEEE Conference on Computer Vision and Pattern Recognition (CVPR), pp. 3883–3891 (2017)
16. Oord, A.v.d., et al.: Wavenet: a generative model for raw audio. arXiv preprint arXiv:1609.03499 (2016)
17. Pan, J., et al.: Learning dual convolutional neural networks for low-level vision. In: Proceedings of IEEE Conference on Computer Vision and Pattern Recognition (CVPR), pp. 3070–3079 (2018)
18. Shen, Z., et al.: Human-aware motion deblurring. In: Proceedings of the IEEE Conference on Computer Vision (ICCV), pp. 5572–5581 (2019)
19. Tao, X., Gao, H., Shen, X., Wang, J., Jia, J.: Scale-recurrent network for deep image deblurring. In: Proceedings of the IEEE Conference on Computer Vision and Pattern Recognition (CVPR), pp. 8174–8182 (2018)
20. Wang, Z., Ji, S.: Smoothed dilated convolutions for improved dense prediction. In: Proceedings of the ACM SIGKDD International Conference on Knowledge Discovery and Data Mining, pp. 2486–2495 (2018)
21. Yu, F., Koltun, V.: Multi-scale context aggregation by dilated convolutions. arXiv preprint arXiv:1511.07122 (2015)
22. Zeng, Y., van der Lubbe, J.C., Loog, M.: Multi-scale convolutional neural network for pixel-wise reconstruction of van gogh's drawings. Mach. Vis. Appl. **30**(7–8), 1229–1241 (2019)
23. Zhang, J., et al.: Dynamic scene deblurring using spatially variant recurrent neural networks. In: Proceedings of the IEEE Conference on Computer Vision and Pattern Recognition (CVPR), pp. 2521–2529 (2018)
24. Zhang, K., Zuo, W., Chen, Y., Meng, D., Zhang, L.: Beyond a gaussian denoiser: Residual learning of deep CNN for image denoising. IEEE Trans. Image Process. **26**(7), 3142–3155 (2017)
25. Zhang, S., He, F., Ren, W.: Photo-realistic dehazing via contextual generative adversarial networks. Mach. Vis. Appl. **31**(5), 33 (2020)
26. Zhao, H., Gallo, O., Frosio, I., Kautz, J.: Loss functions for image restoration with neural networks. IEEE Trans. Comput. Imaging **3**(1), 47–57 (2016)

Time-Series in Hyper-parameter Initialization of Machine Learning Techniques

Tomáš Horváth[1,4](✉), Rafael G. Mantovani[2], and André C. P. L. F. de Carvalho[3]

[1] Faculty of Informatics, ELTE - Eötvös Loránd University, Pázmány Péter sétány 1/C, Budapest 1117, Hungary
tomas.horvath@inf.elte.hu

[2] Federal Technology University - Paraná, Campus of Apucarana, Rua Marcílio Dias, 635 - Jardim Paraíso, Apucarana, PR 86812-460, Brazil
rafaelmantovani@utfpr.edu.br

[3] Institute of Mathematical and Computer Sciences, University of São Paulo, Avenida Trabalhador São Carlense, 400 - Centro, São Carlos, SP 13566-590, Brazil
andre@icmc.usp.br

[4] Institute of Computer Science, Faculty of Science, Pavol Jozef Šafárik University, Jesenná 5, 040 01 Košice, Slovakia
tomas.horvath@upjs.sk

Abstract. Initializing the hyper-parameters (HPs) of machine learning (ML) techniques became an important step in the area of automated ML (AutoML). The main premise in HP initialization is that a HP setting that performs well for a certain dataset(s) will also be suitable for a similar dataset. Thus, evaluation of similarities of datasets based on their characteristics, named meta-features (MFs), is one of the basic tasks in meta-learning (MtL), a subfield of AutoML. Several types of MFs were developed from which those based on principal component analysis (PCA) are, despite their good descriptive characteristics and relatively easy computation, utilized only marginally. A novel approach to HP initialization combining dynamic time warping (DTW), a well-known similarity measure for time series, with PCA MFs is proposed in this paper which does not need any further settings. Exhaustive experiments, conducted for the use-cases of HP initialization of decision trees and support vector machines show the potential of the proposed approach and encourage further investigation in this direction.

Keywords: Automated ML · Metalearning · PCA · DTW

1 Introduction

The growing popularity of machine learning (ML) in various application domains and the shortage of data scientists has raised the demand for automated ML (AutoML) [7]. A special focus of AutoML is on configuring the hyper-parameters

H. Yin et al. (Eds.): IDEAL 2021, LNCS 13113, pp. 246–258, 2021.
https://doi.org/10.1007/978-3-030-91608-4_25

(HPs) of ML techniques, a task belonging to the family of black-box optimization problems [11], often approached by heuristics ranging from very simple methods, such that random search, to more advanced ones, such that sequential model-based optimization (SMBO) or biologically inspired techniques, e.g. particle swarm optimization (PSO). It is important to note that these heuristics need to perform a certain number of iterations to arrive at recommendation in case of a new dataset. Moreover, the performance of these heuristics depends on the initial selection of the HP settings from which these models start their computations [15]. Another important aspect is that these heuristics possess their own HPs (e.g. the population size or the surrogate model) which would also need some fine-tuning to perform in the most optimal way.

Another approach to recommend optimal HP setting for a ML technique on a given dataset is to use meta-learning (MtL) [5]. The main premise of MtL is that knowledge and experience gained from previous applications of various ML techniques on different datasets can be employed to recommend the optimal HP setting in case of a new dataset. In this context, "knowledge" and "experience" are represented by a so-called meta-model that captures the relation between the characteristics of a dataset and the predictive performance of ML techniques w.r.t. various HP settings.

The first step in the use of MtL is the representation of each dataset by a vector of features, often named meta-features (MFs) which describe important aspects of a dataset and are used as input attributes in MtL. The next step is to record the predictive performance of ML techniques w.r.t. their HP settings on these datasets. This performance measure will be the target attribute for MtL. Finally, a meta-model induced by employing classification or regression techniques on the created meta-data can be used to predict the most adequate HP setting for a ML technique on a new dataset. However, not only the ML techniques deriving the meta-model but, also, many MF extraction techniques have their own HP settings which should be fine-tuned as well.

This study proposes a simple, yet efficient MtL approach based on Principal Component Analysis (PCA) MF vectors, denoted PCA-MF here, for real-time recommendation and initialization of HP settings of ML techniques. The proposed approach utilizes dynamic time warping (DTW), a similarity measure well-known in time-series analysis (see [4]), to compute the similarity of two PCA-MFs. The use of DTW is, according to our knowledge, new in MtL. The proposed approach has no own HPs, thus, there is no need to an additional fine-tuning. To empirically evaluate the proposed approach, we simulated the use-case of recommending HP settings for Support Vector Machine (SVM) and Decision Tree Induction (DT) algorithms applied to a new dataset. Experiments using 50 real-world datasets are performed comparing the proposed approach (employing a simple k-Nearest Neighbor (kNN) meta-model) with various baseline approaches.

2 PCA Meta-Features

Eight main MF types (categories) are documented in the literature as introduced in Table 2. Another type of MFs explores statistical correlations in the dataset's predictive attribute values. For such, PCA can be used and the resulting MFs are called PCA-MF in this paper. Since PCA-MF is in the main focus of this paper, the underlying model will be introduced in a more formal way in this section.

2.1 Principal Component Analysis

According to [13], PCA finds a linear transformation of attributes in a dataset such that the information present in the data is maximally preserved in the sense of minimum squared error. Thus, PCA can be seen as a feature extraction method assigning simple weights to the extracted (latent) features.

To formally show how PCA works, let $\mathbf{D} \in \mathbf{X} \times \mathbf{Y}$ be a pre-processed (i.e. standardized and binarized) dataset such that $\mathbf{X} \in \mathbb{R}^{r \times c}$ represents the predictive attributes and $\mathbf{Y} \in \mathbb{R}^{r \times t}$, the target attribute. For simple regression, binary classification and multi-class classification tasks $t = 1$ while in case of multi-label classification $t > 1$. PCA takes $\mathbf{X}$ as input and returns a matrix $\mathbf{W} \in \mathbb{R}^{c \times c}$, the i-th column $\mathbf{w}_i = (w_{i_1}, w_{i_2}, \dots, w_{i_c}) \in \mathbb{R}^c$ of which is the i-th transformation vector of weights that maps $\mathbf{X} \in \mathbb{R}^{r \times c}$ to the i-th principal component $\mathbf{p}_i = \mathbf{X}\mathbf{w}_i^T \in \mathbb{R}^r$ $(1 \leq i \leq c)$. Together with $\mathbf{W}$, PCA returns a vector $\mathbf{e} = (e_1, e_2, \dots, e_c)$ which elements, the eigenvalues, express the importance of the respective principal components. Eigenvalues are related to the standard deviation in $\mathbf{X}$ projected to the corresponding principal components, and represent the proportion of variance in $\mathbf{X}$ explained by the corresponding principal components. The values in $\mathbf{e}$ are in decreasing order, i.e. for each $a, b \in \{1, \dots, c\}$ such that $a < b$, $e_a \geq e_b$. The proportion of variance π_i in the data, explained by the corresponding ith principal component, is computed as $\pi_i = \frac{e_i}{\sum_{i=1}^{c} e_i}$ and the approach proposed in this paper is based on a MF vector

$$\mathbf{m}^{pca} = (\pi_1, \pi_2, \dots, \pi_c) \tag{1}$$

2.2 Related Work on PCA-MF

To explain how related approaches work, let us consider that two MF vectors $\mathbf{m}_i^{pca} = (\pi_1, \pi_2, \dots, \pi_{c_i})$ and $\mathbf{m}_j^{pca} = (\pi_1, \pi_2, \dots, \pi_{c_j})$ extracted from datasets $\mathbf{D}_i$ and $\mathbf{D}_j$, respectively, may have different lengths, i.e. $c_i \neq c_j$. However, the use of MtL requires[1] a fixed number of predictive attributes, therefore, the characterization of a dataset must result in a fixed number of MFs. The usual solution when the number of MFs vary for different datasets is to aggregate the MFs of the same type into one or more values resulting in one or more MF(s).

[1] Since a meta-model is learned by traditional ML techniques.

The ratio d/c was used in [2] as a MF, where d is the number of principal components that explains 95% of the variance in the dataset[2]. The skewness and the kurtosis of the first principal component $\mathbf{p}_1$ are added to the previous PCA MF d/c in [8]. The use of the minimum and the maximum eigenvalue (e_1 and e_c) and the proportion of variance π_1 explained by the first principal component were proposed as MFs in [9].

The price of aggregation to only one, two or three values, as described above, is a possible loss of useful meta-information about the dataset. The use of 22 PCA-based MFs was proposed in [1], such that the 10 histogram bin values of the proportion of variance explained by each principal component[3], together with their normalized values, the proportion of variance explained by the first eigenvalue (π_1 from the Eq. 1) and the ratio d/c from [2].

A more general definition of the *histogram MF vector* introduced in [1], by allowing an arbitrary number b of bins for the histograms, can be defined as

$$\mathbf{m}^{his} = (\theta_1, \theta_2, \dots, \theta_b) \tag{2}$$

such that $\theta_i = \sum_{j=1}^{c} \delta\left(\frac{i-1}{b} < \pi_j \leq \frac{i}{b}\right)$ for $1 \leq i \leq b$ and $\delta(x) = 1$ if the expression x is true, otherwise $\delta(x) = 0$.

We propose another method, slightly different variation of the histogram MF vector, called the *cut-point MF vector* in this paper which, first, computes the vector $\mathbf{m}^c = (\vartheta_1, \vartheta_2, \dots, \vartheta_c)$ of cumulative proportion of variances from the vector $\mathbf{m}^{pca}$ where $\vartheta_i = \sum_{j=1}^{i} \pi_j$ for $1 \leq i \leq c$. Next, $\mathbf{m}^c$ is aggregated into the vector

$$\mathbf{m}^{cup} = (\kappa_1, \kappa_2, \dots, \kappa_b) \tag{3}$$

where $\kappa_i = \arg\min_j \vartheta_j \geq \frac{i}{b}$, for $1 \leq i \leq b$. In other words, the indices of those eigenvalues are returned which correspond to the cumulative proportion of variance that first reach the given cut points $\frac{1}{b}, \frac{2}{b}, \dots, \frac{b}{b}$ of the interval $[0, 1]$. Since κ_b is, in general, equal to the number of eigenvalues (principal components), the introduced cut-point aggregation preserves the meta-information about the number of predictive attributes in the dataset. As a result, a MF vector $\mathbf{m}^{pca}$ of length c is aggregated into a MF vector $\mathbf{m}^{cup}$ of length b such that b is the same for all datasets.

As far as we know, the histogram MF vector method has not become widely used in MtL. Also, we could not find any approach similar to the cut-point MF method in the literature.

3 The Proposed Approach

Let $\mathbf{h} = (h_1, \dots, h_k) \in \mathcal{H}$ be a particular HP setting for a ML algorithm (often called as base-learner) where $\mathcal{H} = \mathcal{H}_1 \times \dots \times \mathcal{H}_k$ is the admissible domain of

[2] According to the vector $\mathbf{m}^c = (\vartheta_1, \vartheta_2, \dots, \vartheta_c)$, $d = min\{i \,|1 \leq i \leq c,\ \vartheta_i \geq 0.95\})$, and c is the number of attributes in the dataset.

[3] Basically, a 10-bin histogram of the values of $\mathbf{m}^{pca}$ from the Eq. 1.

Algorithm 1. kNN based recommendation of initial hyper-parameters

1: **procedure** RecommendHP($mfe, \mathbf{D}, \mathcal{H}^\star, \mathcal{M}, k, sim$)
2: $\mathbf{m} \leftarrow$ getMetaFeatures($mfe, \mathbf{D}$)
3: $\{i_1, i_2, \ldots, i_k\} \leftarrow$ getNearestNeighbors($\mathbf{m}, \mathcal{M}, k, sim$)
4: $\mathbf{h} \leftarrow @(\mathbf{h}^\star_{i_1}, \mathbf{h}^\star_{i_2}, \ldots, \mathbf{h}^\star_{i_k}),\ \{\mathbf{h}^\star_{i_1}, \mathbf{h}^\star_{i_2}, \ldots, \mathbf{h}^\star_{i_k}\} \subseteq \mathcal{H}^\star$ ▷ aggregation @
5: **return h**

possible HP settings. Let the function $f : \mathcal{H} \times \mathcal{D} \rightarrow \mathbb{R}$ represent the accuracy of the base learner with the HP setting $\mathbf{h} \in \mathcal{H}$ on the dataset $\mathbf{D} \in \mathcal{D}$, where $\mathcal{D}$ is the set of datasets. Let $\mathcal{D}^\star = \{\mathbf{D}_1, \mathbf{D}_2, \ldots, \mathbf{D}_n\} \subset \mathcal{D}$ be the set of those datasets, called train datasets for MtL, for which the "best" HP settings $\mathbf{h}^\star_i = \underset{\mathbf{h}\in\mathcal{H}}{argmax}\ f(\mathbf{h}, \mathbf{D}_i)$ with relation to the base-learner are already known, computed off-line by an arbitrary HP tuning technique and stored in $\mathcal{H}^\star = \{\mathbf{h}^\star_1, \mathbf{h}^\star_2, \ldots, \mathbf{h}^\star_n\}$. In this off-line phase, the MF vectors related to each $\mathbf{D}_i \in \mathcal{D}^\star$ of the form $\mathbf{m}^{pca}$, $\mathbf{m}^{his}$ or $\mathbf{m}^{cup}$ (Eqs. 1, 2 or 3, respectively), denoted here as $\mathbf{m}_i = (m_{i_1}, m_{i_2}, \ldots, m_{i_q})$, are extracted and stored in $\mathcal{M} = \{\mathbf{m}_1, \mathbf{m}_2, \ldots, \mathbf{m}_n\}$ where q is the number of MFs.

The k-Nearest Neighbor (kNN) approach is a commonly used algorithm in MtL [5], mainly because it is simple and works fast when the amount of data, number of datasets and MFs is not large, which is the usual case in MtL.

The generic kNN based recommendation process of initial HP values consists of three major steps, as illustrated in the Algorithm 1: First, the MF vector $\mathbf{m} = (m_1, m_2, \ldots, m_q)$ is extracted from a "new" dataset $\mathbf{D} \in \mathcal{D}$ according to a given MF extraction approach mfe[4] corresponding to a certain MF type. Second, the indexes $i_1, i_2, \ldots, i_k$ of the k-nearest neighbors $\mathbf{D}_{i_1}, \mathbf{D}_{i_2}, \ldots, \mathbf{D}_{i_k}$ of $\mathbf{D}$, using a MF similarity function sim[5], are found. Finally, the best HP settings $\mathbf{h}^\star_{i_1}, \mathbf{h}^\star_{i_2}, \ldots, \mathbf{h}^\star_{i_k}$ recorded for the k-nearest neighbors of $\mathbf{D}$ are aggregated (line 4) using some aggregation function[6] @ to get the returned results.

3.1 Utilizing Dynamic Time Warping

As pointed out before, aggregating MFs brings a risk of losing possibly useful information. However, the use of the standard MtL approaches, such as the kNN with vector similarity measures (see footnote 5) or any other ML meta-learner (e.g. Random Forest), would work only with MF vectors of equal lengths. Moreover, MF aggregation would require to specify (i.e. fine-tune) the length of the MF vectors. Finally, the choice of the optimal MF types to extract and use is data dependent.

[4] We are using all the eight MF types described above as well as their different combinations in our experiments as baselines.

[5] In case of MF vectors of the same size (e.g. $\mathbf{m}^{his}$ and $\mathbf{m}^{cup}$) we were experimenting with well-known vector similarity measures such as Euclidean distance, inner product, cosine similarity and Pearson correlation.

[6] This study uses a simple average as an aggregation function, however, any other aggregation function, like a weighted average, can be used, too.

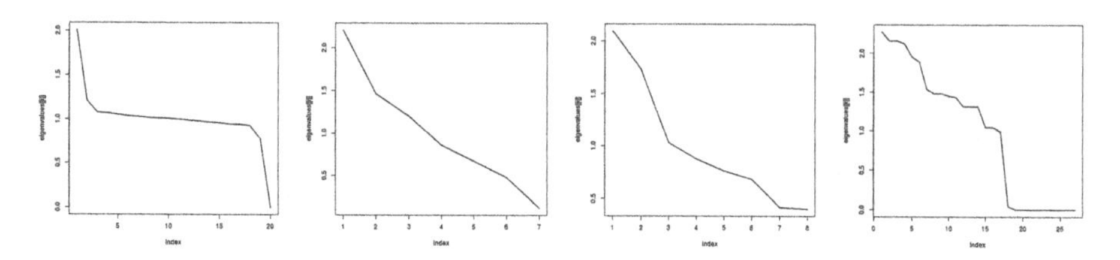

Fig. 1. An example of PCA-MFs $\mathbf{m}^{pca}$ for four datasets. The graphs show the eigenvalues (y-axis) for the principal components (x-axis) which numbers correspond to (different) numbers of attributes in these datasets.

Thus, we propose a novel approach, utilizing Dynamic Time Warping (DTW) [12], to measure the similarity between two MF vectors $\mathbf{m}_i^{pca}$ and $\mathbf{m}_j^{pca}$, defined in the Eq. 1. DTW is an edit-distance like algorithm for measuring the similarity of two time series by finding the best matching alignment between them even if the two time series on its input have different lengths. This is the main motivation of our proposal to employ DTW for measuring the similarity between two datasets represented by $\mathbf{m}_i^{pca}$ and $\mathbf{m}_j^{pca}$ of different lengths resulting from different number of attributes in the corresponding datasets $\mathbf{D}_i$ and $\mathbf{D}_j$.

The two MF vectors $\mathbf{m}_i^{pca}$ and $\mathbf{m}_j^{pca}$ are, basically, two decreasing sequences expressing the (speed of) down-grade of the proportion of variance along the latent directions (ordered w.r.t. their importance) computed via PCA in the corresponding two datasets $\mathbf{D}_i$ and $\mathbf{D}_j$. Since each sequence is composed by the principal components extracted from a dataset and the principal components usually preserve the main characteristics of a dataset, similar sequences are expected to be mapped to the same meta-target label.

An example situation with four datasets is illustrated in Fig. 1 where, according to the "shapes" of the PCA-MFs, the middle two datasets are the most similar to each other. Thus, it is assumed that the optimal HP settings (i.e. meta-target labels) would be also similar for these two datasets.

Summarizing, in our proposal, the kNN meta-learner (Algorithm 1) is applied to the PCA-MF vectors $\mathbf{m}^{pca}$ (Eq. 1) with the *sim* function being the DTW.

The similarity of two datasets is a subjective matter and can be seen, expressed and measured in various ways what can be seen also in the colorful palette of various MF types (see Table 2). PCA-MFs represent another view on dataset similarity looking at this matter in different context. It is important that nor DTW nor PCA, in its basic form, has no HPs to set up beforehand. The only HP to fine-tune would be the k in the kNN model, however, this would have to be set up using other types of MFs as well.

Table 1. (Multi-class) Classification datasets used in the experiments and their main characteristics: number of rows (r), columns (c) and classes (t).

No.	Name	r	c	t	No.	Name	r	c	t
1	acute-infl.-nephr.	99	6	2	2	analcatdata_lawsuit	263	4	2
3	appendicitis	106	7	2	4	autoUniv-au6-cd1-400	400	40	8
5	banknote-authentication	1348	4	2	6	breast-cancer-wisconsin	463	9	2
7	breast-tissue-4class	105	9	4	8	bupa	341	6	2
9	climate-simul.-craches	540	20	2	10	cloud	108	6	4
11	connect.-mines-vs-rocks	208	60	2	12	dermatology	366	34	6
13	ecoli	336	7	8	14	fertility-diagnosis	100	9	2
15	glass	213	9	6	16	habermans-survival	289	3	2
17	hayes-roth	93	4	3	18	heart-dis.-proc.-hun.	293	13	2
19	hepatitis	155	19	2	20	horse-colic-surgical	300	27	2
21	indian-liver-patient	570	10	2	22	ionosphere	350	33	2
23	iris	147	4	3	24	leaf	340	15	30
25	led7digit	146	7	10	26	leukemia-haslinger	100	50	2
27	mammographic-mass	689	5	2	28	monks3	438	6	2
29	movement-libras	330	90	15	30	parkinsons	195	22	2
31	pima-ind.-diab.	768	8	2	32	planning-relax	176	12	2
33	prnn_crabs	200	7	2	34	qualitative-bankr.	103	6	2
35	robot-failure-lp4	116	90	3	36	saheart	462	9	2
37	seeds	210	7	3	38	spect-heart	228	22	2
39	statlog-heart	270	13	2	40	teaching-assist.-eval.	110	5	3
41	thoracic-surgery	470	16	2	42	thyroid-newthyroid	215	5	3
43	tic-tac-toe	958	9	2	44	user-knowledge	403	5	5
45	volcanoes-e3	1276	3	5	46	voting	281	16	2
47	wdbc	569	30	2	48	wholesale-channel	440	7	2
49	wine	178	13	3	50	wpbc	198	33	23

4 Experiments

The experiments use two base-learners which differ in their learning paradigms as well as the number and types of HPs. These base-learners are i) Support Vector Machine (SVM) with the RBF kernel with two HPs (both real valued) to initialise, and, ii) Decision Tree (DT) with nine HPs (real-valued and Boolean) to initialise.

Since the motivation of this study is a MtL approach suitable for on-line initialization of HP values, and class imbalance can occur, the experiments measure the predictive performance regarding balanced accuracy [6] and runtime. A total of 50 multi-class classification datasets, from the UCI Machine Learning Repository[7] are used in the experiments as introduced in Table 1.

Various approaches were used as baselines for the experiments, differing in the simplicity and strategy adopted to initialize or tune HPs. The first batch of used baselines contains four HP tuning techniques:

- The default HP setting for SVM and DT provided by `R` packages `e1071` and `RWeka`, respectively, denoted as "DF". Although this is the simplest baseline, it can present a good performance in some situations [10].

[7] http://archive.ics.uci.edu/ml/.

Table 2. MF types, and their abbreviations (Abbr), identified in the literature. # denotes the number of different MFs belonging to a given MF type/category.

MF type	Abbr	#	Description
Simple	SL	17	Simple measures
Statistical	ST	7	Statistics measures
Inf. theoretic	IT	8	Information theory measures
Landmarking	LM	9	Performance of some ML algorithms
Model-based	MB	17	Features extracted from decision trees
Time	TI	5	Execution time of some ML algorithms
Complexity	CO	14	Measures analyzing complexity
Complex Network	CN	9	Complex network property measures

- The best HP setting found by an exhaustive Random Search in the HP space, denoted by "RS", suggested in [3] as a good alternative for HP tuning. The number of trials in RS was set to 2500 for SVM and 5000 for DT.
- The best HP setting found by Particle Swarm Optimization [16], denoted as "PSO". For SVM, the maximum number of evaluations was set to 2500. For DT, the maximum number of evaluations was 5000. The default HP settings recommended by the corresponding R libraries for SVM and DT, respectively, were added to the initial populations.
- The best HP setting found by Sequential Model-based Optimization [14], denoted here as "SMBO". For both SVM and DT, the maximum number of evaluations (budget) was set to 200 and Random Forest (RF) was used as surrogate learning model.

These baselines are for HP tuning and not HP initialisation since they are computationally too expensive for on-line scenarios. However, since they perform exhaustive (random or sophisticated) search, they can be good references to measure the performance of HP initialisation approaches, such that, we can see how close the initialised HPs are to the tuned ones. The other family of baselines involves the following, traditional MtL approaches for HP initialisation:

- HP settings recommended by a kNN based meta-learner applying traditional vector similarity measures (see footnote 5) to MF vectors $\mathbf{m}^{cmf} = (m_1, m_2, \ldots, m_l)$, representing various MF type combinations, having the same length l for all the 50 datasets. Here, all the $2^8 - 1 = 255$ different combinations of the 8 MF types presented in the Table 2 are considered corresponding to 255 different MF vectors[8]. The lengths l of these MF vectors vary from 5 (only the TI MFs are used) to 86 (all the MFs are used). These baselines are denoted as "NN-CMF", an acronym for kNN based meta-learner with a certain Combination of MF types.

[8] Where either all or none of the MFs belonging to a certain MF type were present in a MF vector, according to the given combination of MF types.

Algorithm 2. One complete cycle of experiments

1: HP tuning using SMBO, PSO, RS, DF ▷ Computing $\mathcal{H}^{\star}$
2: extraction of MF ($\mathbf{m}^{pca}$,$\mathbf{m}^{his}$, $\mathbf{m}^{cup}$, $\mathbf{m}^{cmf}$) ▷ Computing $\mathcal{M}$
3: recommendation of HP settings for the base-learners ▷ Computing $\mathbf{h}$
4: computation of the accuracy of the base-learners with the recommended HP settings averaged on the folds of 5-fold cross-validation of a given dataset.

- HP settings recommended by a kNN based meta-learner applying traditional vector similarity measures (see footnote 5) to MF vectors $\mathbf{m}^{his}$ and $\mathbf{m}^{cup}$, defined in Eqs. 2 and 3, respectively, having the same length b for all the 50 datasets. These baselines are denoted as "NN-HIS" and "NN-CUP", respectively.

The experiments compare the performance of the proposed approaches with the baselines. They also investigate the influence of different HP settings and extensions of the compared approaches on their predictive accuracy. For such, an experimental procedure performing a complete cycle of recommendation is adopted for the experiments. The steps of this cycle are listed in the Algorithm 2 (the comments in the lines refer to the given parts and notations in Algorithm 1).

This cycle is computed for all datasets in a leave-one-out manner, i.e. for each dataset the cycle is performed such that the other 49 datasets are used as "train" datasets for MtL and HP tuning. The balanced accuracy averaged over the 5 folds will be denoted as "cycle-accuracy" of a dataset w.r.t. some base-learner and HP setting recommender. For each dataset, nested cross-validation is utilized for tuning the HPs of a base-level learner (SVM and DT) on this dataset in case of SMBO, PSO and RS with 10 outer and 3 inner folds. The whole cycle presented in the Algorithm 2 is run 30 times for each "test" dataset, and the average of the 30 computed[9] "cycle-accuracy" values is returned what will be denoted as "overall-accuracy" values.

The proposed approach, i.e. using $\mathbf{m}^{pca}$ MF vectors and the DTW similarity measure, is denoted as "DTW" here.

The datasets and the source codes for the proposed approach as well as for the experiments (e.g. Algorithms 1 and 2) are publicly available[10] for further use.

4.1 Results

The relative performance of the tested approaches, when compared to each other, is similar for all the four choices for the HP tuning strategy. Figure 2 and 3 illustrate the averaged "overall-accuracies" across all the different settings of

[9] This was done because of the stochastic nature of the used HP tuning algorithms (SMBO, PSO and RS), thus, to get more accurate statistics about the performance of the used approaches.

[10] https://github.com/rgmantovani/TimeSeriesHPInitialization.

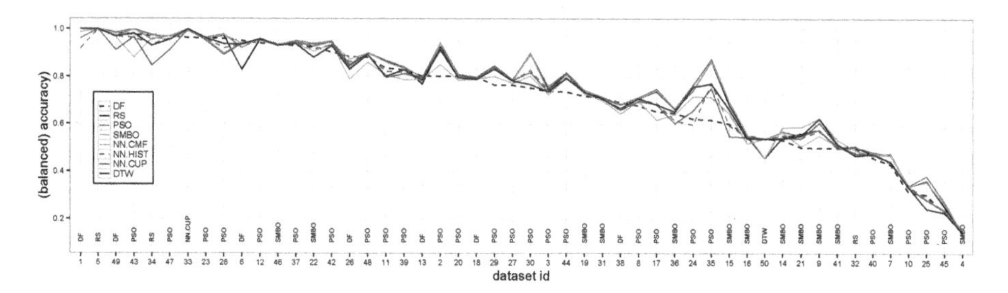

Fig. 2. Performance of different approaches for SVM base-learner: Averaged "overall-accuracies" across all the different settings of complete cycles (see Algorithm 2). Clear winner approaches are marked in the bottom line (abbreviations are placed in the vertical).

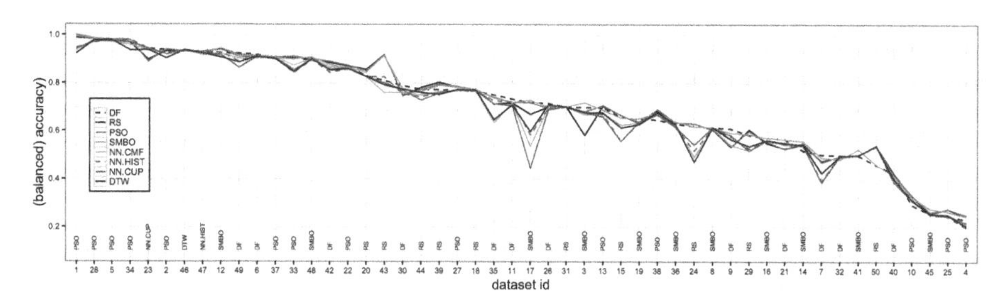

Fig. 3. Performance of different approaches for DT base-learner: Averaged "overall-accuracies" across all the different settings of complete cycles (see Algorithm 2). Clear winner approaches are marked in the bottom line (abbreviations are placed in the vertical).

complete cycles (see Algorithm 2). The proposed approach (DTW) as well as the aggregated PCA MF approaches (NN-HIS and NN-CUP) performed slightly better than the CMF approaches with regard to the number of clear wins.

In the Fig. 4, the extraction times for different MF extraction approaches related to the used datasets are illustrated. SL MFs are the most faster to extract, however, these are very simple MFs and usually not performing well if only these are used alone without any other MF types. The following most fastest to extract MFs are the proposed PCA-MFs.

4.2 Discussion

There are two important issues, mentioned in the introduction, regarding the traditional MtL, approaches: First, as showed in the experiments, the choice of a suitable combination of MF types for HP recommendation seems to be data dependent. Second, the extraction of some MF types usually has a high computational cost. It is important to remember that in the experiments, either all the MF of a certain MF type were considered or were excluded completely from the consideration. No MF selection procedure was performed on the set of all the

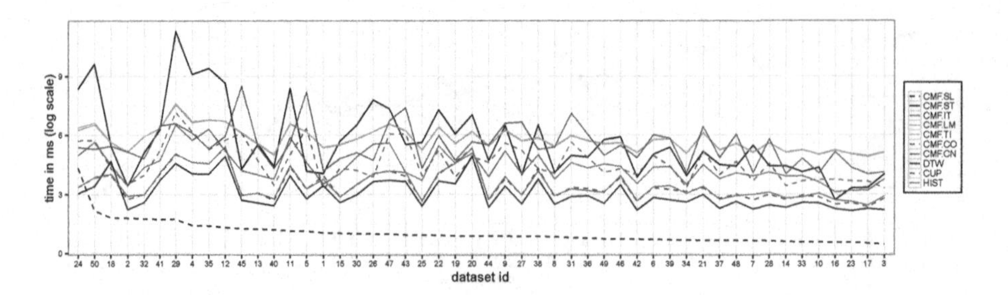

Fig. 4. MF Extraction times, on a logarithmic scale, for the used datasets and MF extraction approaches.

86 MFs listed in Table 2 what would probably result in better recommendation, but with additional computational cost. The most time-consuming process is to find the suitable combination of MFs which has the larger range of values, i.e. 2^n where n is the number of MFs to be considered (86 in this case). The use of MF selection methods can significantly increase the performance, however, is time-consuming and not appropriate for on-line scenarios where the (initial) HP settings need to be delivered in real time.

Regarding the number of bins in case of NN-CUP and NN-HIS approaches, the b parameter, experiments indicate that a tuning is required, since, in general, there is no clear better choice. However, since the reasonable values to use are the dividers of 100, the tuning has to deal with a small set of values (i.e. 1, 2, 4, 5, 10, 20, 25, 50 and 100) compared to the before mentioned range of 2^n for the possible combinations of MF types. It is also worth to mention that PCA needs to be performed only once for a dataset to obtain various MF vectors for different values of b.

Finally, DTW does not need to tune any own HPs what, considering it's competitive performance, makes it a good choice for scenarios where HP settings need to be initialized fast.

5 Conclusions

This paper proposed and investigated a fast approach to HP initialisation of ML algorithms utilizing PCA-MFs and DTW. To the best of authors' knowledge, the proposed DTW approach is novel.

The proposed approach was evaluated and compared to various baselines in the use-case scenario of initialisation of HPs for SVM and DT using kNN meta-learner. Experiments were performed on 50 real-world datasets.

The results showed that the proposed approach presents a good predictive accuracy when compared with various baseline approaches with a run time faster than the used baselines. Also, the performance of the base-learners using the initialised HPs are very close to the performance of the base-learners using optimized HPs via exhaustive HP tuning approaches.

The proposed DTW approach has no parameters to tune and is sufficiently accurate and fast to be used in on-line scenarios where HP values need to be recommended in real time.

Acknowledgment. "Application Domain Specific Highly Reliable IT Solutions" project has been implemented with the support provided from the National Research, Development and Innovation Fund of Hungary, financed under the Thematic Excellence Programme TKP2020-NKA-06 (National Challenges Subprogramme) funding scheme.

References

1. Amasyali, M.F., Ersoy, O.K.: A study of meta learning for regression. Technical report, ECE Technical Reports. Paper 386, Purdue e-Pubs, Purdue University (2009)
2. Bardenet, R., Brendel, M., Kégl, B., Sebag, M.: Collaborative hyperparameter tuning. In: Proceedings of the 30th International Conference on Machine Learning, vol. 28, pp. 199–207. JMLR Workshop and Conference Proceedings (2013)
3. Bergstra, J., Bengio, Y.: Random search for hyper-parameter optimization. J. Mach. Learn. Res. **13**, 281–305 (2012)
4. Berndt, D.J., Clifford, J.: Using dynamic time warping to find patterns in time series. In: AAAI Workshop on Knowledge Discovery in Databases, pp. 359–370 (1994)
5. Brazdil, P., Giraud-Carrier, C., Soares, C., Vilalta, R.: Metalearning: Applications to Data Mining, 1st edn. Springer, Heidelberg (2009)
6. Brodersen, K.H., Ong, C.S., Stephan, K.E., Buhmann, J.M.: The balanced accuracy and its posterior distribution. In: Proceedings of the 2010 20th International Conference on Pattern Recognition, pp. 3121–3124. IEEE Computer Society (2010)
7. Feurer, M., Klein, A., Eggensperger, K., Springenberg, J., Blum, M., Hutter, F.: Efficient and robust automated machine learning. In: Advances in Neural Information Processing Systems 28, pp. 2944–2952. Curran Associates, Inc. (2015)
8. Feurer, M., Springenberg, J.T., Hutter, F.: Using meta-learning to initialize bayesian optimization of hyperparameters. In: International Workshop on Meta-learning and Algorithm Selection co-located with 21st European Conference on Artificial Intelligence, pp. 3–10 (2014)
9. Janssens, J.H.: Outlier Selection and One-Class Classification. Ph.D. thesis, Tilburg University, Netherlands (2013), TiCC PhD Dissertation Series No.27
10. Mantovani, R.G., Rossi, A.L.D., Vanschoren, J., Bischl, B., de Carvalho, A.C.P.L.F.: To tune or not to tune: Recommending when to adjust svm hyper-parameters via meta-learning. In: 2015 International Joint Conference on Neural Networks, pp. 1–8 (2015)
11. Mantovani, R.G., Horváth, T., Cerri, R., Junior, S.B., de Vanschoren, J., de Carvalho, L.F.: An empirical study on hyperparameter tuning of decision trees, A.C.P. (2019)
12. Ratanamahatana, A., Keogh, E.: Everything you know about dynamic time warping is wrong. In: 3rd Workshop on Mining Temporal and Sequential Data, in conjunction with 10th ACM SIGKDD International Conference Knowledge Discovery and Data Mining (2004)
13. Sharma, A., Paliwal, K.K.: Fast principal component analysis using fixed-point algorithm. Pattern Recogn. Lett. **28**(10), 1151–1155 (2007)

14. Snoek, J., Larochelle, H., Adams, R.P.: Practical bayesian optimization of machine learning algorithms. In: Pereira, F., Burges, C., Bottou, L., Weinberger, K. (eds.) Advances in Neural Information Processing Systems 25, pp. 2951–2959. Curran Associates, Inc. (2012)
15. Wistuba, M., Schilling, N., Schmidt-Thieme, L.: Hyperparameter search space pruning – a new component for sequential model-based hyperparameter optimization. In: Appice, A., Rodrigues, P.P., Santos Costa, V., Gama, J., Jorge, A., Soares, C. (eds.) ECML PKDD 2015. LNCS (LNAI), vol. 9285, pp. 104–119. Springer, Cham (2015). https://doi.org/10.1007/978-3-319-23525-7_7
16. Yang, X.S., Cui, Z., Xiao, R., Gandomi, A.H., Karamanoglu, M.: Swarm Intelligence and Bio-Inspired Computation: Theory and Applications, 1st edn. Elsevier Science Publishers B. V. (2013)

Prediction of Maintenance Equipment Failures Using Automated Machine Learning

Luís Ferreira[1,2(✉)], André Pilastri[1], Vítor Sousa[3], Filipe Romano[3], and Paulo Cortez[2]

[1] EPMQ - IT Engineering Maturity and Quality Lab, CCG ZGDV Institute, Guimarães, Portugal
{luis.ferreira,andre.pilastri}@ccg.pt

[2] ALGORITMI Centre, Department Information Systems, University of Minho, Guimarães, Portugal
pcortez@dsi.uminho.pt

[3] Valuekeep, Braga, Portugal
{vitor.sousa,filipe.romano}@valuekeep.com

Abstract. Predictive maintenance is a key area that is benefiting from the Industry 4.0 advent. Recently, there have been several attempts to use Machine Learning (ML) in order to optimize the maintenance of equipments and their repairs, with most of these approaches assuming an expert-based ML modeling. In this paper, we explore an Automated Machine Learning (AutoML) approach to address a predictive maintenance task related to a Portuguese software company. Using recently collected data from one of the company clients, we firstly performed a benchmark comparison study that included four open-source modern AutoML technologies to predict the number of days until the next failure of an equipment and also determine if the equipments will fail in a fixed amount of days. Overall, the results were very close among all AutoML tools, with AutoGluon obtaining the best results for all ML tasks. Then, the best AutoML predictive results were compared with a manual ML modeling approach that used the same dataset. The results achieved by the AutoML approach outperformed the manual method, thus demonstrating the quality of the automated modeling for the predictive maintenance domain.

Keywords: Automated machine learning · Predictive maintenance · Supervised learning

1 Introduction

The Industry 4.0 phenomenon allowed companies to focus on the analysis of historical data to obtain useful insights. In particular, predictive maintenance is a crucial application area that emerged from this context, where the goal

H. Yin et al. (Eds.): IDEAL 2021, LNCS 13113, pp. 259–267, 2021.
https://doi.org/10.1007/978-3-030-91608-4_26

is to optimize the maintenance and repair process of equipments through the usage of Machine Learning (ML) algorithms [17]. Indeed, some ML studies try to anticipate the failure of equipments (typically, manufacturing machines), aiming to reduce the costs of repairs [4,6,11,16]. Other approaches [2,3,5,19] use ML algorithms to predict the behavior of the manufacturing process.

Despite all potential Industry 4.0 benefits, many organizations do not currently apply ML to enhance maintenance activities. And those who do rely mostly on data science experts, the ML models are tuned manually, often requiring a large number of trial-and-error experiments. In effect, we have only found one study that applied AutoML for the maintenance domain [18]. Yet, we note that such study only used synthetic data, which might not reflect the complexities of real industrial maintenance data. In contrast with the "traditional" ML expert design approach, in this paper we apply Automated Machine Learning (AutoML), aiming to automate the ML modeling phase and thus reduce the data to maintenance insights process cycle. Moreover, we apply AutoML using real-world data, collected from the client of a Portuguese software company in the area of maintenance management.

The AutoML was explored for two specific prediction tasks: the number of days until an equipment fails and if the equipments will fail in a fixed number of days. We designed a large set of computational experiments to assess the AutoML predictive performance of four open-source tools. To provide a baseline comparison, we also measure the best AutoML results with a manual ML modeling that was made previously by one of the company's professionals. The comparison clearly favors the AutoML results, thus attesting the potential of the AutoML approach for the predictive maintenance domain.

2 Materials and Methods

2.1 AutoML Tools

In this article, we apply and compare four modern open-source AutoML tools, based on a recent benchmark study performed in [9]. In order to achieve a more fair comparison, we did not tune the hyperparameters of the AutoML tools. Table 1 summarizes the main characteristics of the four explored AutoML tools:

Table 1. Description of the AutoML tools.

Tool	Framework	API	Version
AutoGluon	Gluon	Python	0.2.0
H2O AutoML	H2O	Java, Python, R	3.32.1.3
Rminer	Rminer	R	1.4.6
TPOT	Scikit-Learn	Python	0.11.7

- **AutoGluon** is an AutoML toolkit based on the Gluon framework [1]. In this work, we only considered the tabular data module, which runs several algorithms and returns a Stacked Ensemble with multiple layers [8].
- **H2O AutoML** is the AutoML module from the H2O framework. H2O AutoML runs several algorithms from H2O and two Stacked Ensembles, one with the best ML model of each family and another with all models [10,13].
- **rminer** is a library for the R programming language, focused on facilitating the usage of ML algorithms [7]. Rminer also provides AutoML functions that can be highly customized. In this paper, we used the `"automl3"` template[1], which runs several ML algorithms and one Stacked Ensemble.
- **TPOT** is a Python AutoML tool that uses Genetic Programming to automate several phases of the ML workflow [12,15]. It uses the Python Scikit-Learn framework to produce ML pipelines.

2.2 Data

The provided data has a large number of datasets related to predictive maintenance, which are detailed in Fig. 1.

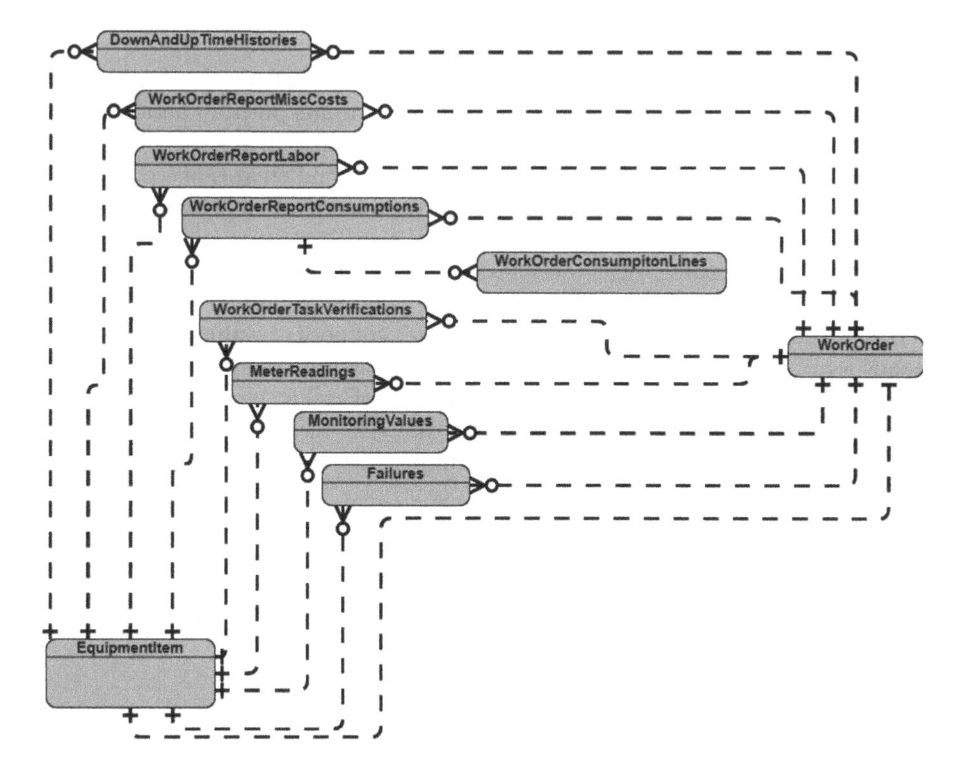

Fig. 1. Entities and relationships between the datasets.

[1] https://CRAN.R-project.org/package=rminer.

For the context of this work, we assume a tabular dataset composed of the aggregation of several attributes from each entity. Overall, the data includes 2,608 records and 21 input attributes. Each record represents an action (e.g., a work order) related to one of the company's equipments (e.g., industrial machine). Each record includes diverse inputs attributes, such as the tasks performed by the machine, consumption of material, and meter readings.

The data also includes five target variables for regression or binary classification tasks. The regression task target (attribute DaysToNextFailure) describes the number of days between that record and the failure of the respective equipment. As for the binary classification targets (attributes FailOnxDays), these describe if the equipment will fail or not in a certain amount of days (e.g., in three days).

Table 2 details the input and output variables (**Attribute**), their description (**Description**), data type (**Type**), number of levels (**Levels**), domain values (**Domain**), and example values from one of the records (**Example**). Half (12) of the 21 input attributes are categorical. Among these, most present a low cardinality (e.g., RecordType, Brand). However, some of the attributes present a very high cardinality (e.g., Part).

2.3 Data Preprocessing

Since several data attributes are of type String, which is not accepted by some AutoML tools, we opted to encode all String attributes into numerical types. For the String attributes that presented a low cardinality (five levels or less), we applied the known One-Hot encoding. Since this method creates one binary column for each level of the original attribute, we applied a different transformation for the columns with a higher cardinality.

Indeed, for the categorical variables with more than five levels, we used the Inverse Document Frequency (IDF) technique [14]. This method is used to convert a categorical column into a numerical column of positive values, based on the frequency of each level of the attribute. IDF uses the function $f(x) = log(n/fx)$, where n is the length of x and fx is the frequency of x. The benefit of IDF, when compared with One-Hot Encoding, is that the IDF technique does not generate new columns, which is useful for attributes with high cardinality (e.g., the attribute Part has 161 levels).

The remaining attributes (of Integer and Float types) were not altered because most AutoML tools already apply preprocessing techniques to the numerical columns (e.g., normalization, standardization). After applying the transformations, the final dataset had 42 inputs and 5 target columns.

2.4 Evaluation

In order to evaluate the results from the AutoML tools, we adopted a similar approach to the benchmark developed in [9]. For every predictive experiment, we divided the dataset into 10 folds for an external cross-validation and adopted an internal 5-fold cross-validation (i.e., over the training data) to select the best

Table 2. Description of the equipment maintenance dataset attributes.

Attribute	Description	Type	Levels	Domain	Example
RecordType	Type of record	String	5	–	Failure
Brand	Brand of the equipment	String	2	–	Rossi
WOType	Type of work order	String	4	–	Corrective
PriorityLevel	Priority Level of the work order	String	4	–	Urgent
Responsible	Responsible for the work order	String	3	–	R4
Employee	Employee that performed the action	String	12	–	E100
TotalTime	Duration of the action (in hours)	Float	17	[0, 8]	8
Quantity	Consumption quantity	Float	32	[0, 300]	90
Part	Part that was consumed	String	161	–	T-1073
Meter	Meter associated to meter reading	String	11	–	L-0002
MeterCumulative reading	Cumulative reading of meter	Float	1477	[0, 73636]	22767
IncrementValue	Increment compared to last reading	Float	475	[0, 54570]	168
MaintenancePlan	Maintenance Plan associated to task	String	5	–	P-000011
Task	Executed task	String	5	–	T-0001
AssetWithFailure	Identification of the equipment	String	15	–	A577
ParentAsset	Parent equipment of AssetWithFailure	String	11	–	LINHA2
Day	Day of the month of the record	Integer	31	[1, 31]	4
DayOfWeek	Day of the week of the record	Integer	7	[1, 7]	6
Month	Month of the record	Integer	12	[1, 12]	2
Year	Year of the record	Integer	6	[2015, 2019]	2019
DaysAfterPurchase	Age of the equipment (in days)	Integer	852	[0, 6309]	4479
DaysToNext failure	Number of Days until the next failure of the equipment	Integer	1015	[0, 1550]	3
FailOn3Days	Indication whether the equipment Will fail in the next 3 days	Integer	2	{0,1}	1
FailOn5Days	Indication whether the equipment Will fail in the next 5 days	Integer	2	{0,1}	1
FailOn7Days	Indication whether the equipment Will fail in the next 7 days	Integer	2	{0,1}	1
FailOn10Days	Indication whether the equipment Will fail in the next 10 days	Integer	2	{0,1}	1

algorithm and hyperparameters (executed automatically by the AutoML tools). To evaluate the test set (from external 10-fold validation) predictions we used the Mean Absolute Error (MAE) ($\in [0.0,\infty[$, where 0.0 represents a perfect model) for the regression task and the Area Under the Curve (AUC) of the Receiver Operating Characteristic (ROC) analysis ($\in [0.0,1.0]$, where 1.0 indicates an ideal classifier) for the binary classification targets. We also used MAE and AUC for the internal validation, responsible for choosing the best ML model.

For all four AutoML tools we defined a maximum training time of one hour (3,600 s) and an early stopping of three rounds, when available. The maximum time of one hour was chosen since it is the default value for most of the AutoML tools. We computed the average of the evaluation measures, computed on the test sets of the 10 external folds, to provide an aggregated value. Additionally, we use confidence intervals based on the t-distribution with 95% confidence to verify the statistical significance of the experiments. In order to identify the best results for each target, we choose the AutoML tool that had the best average predictive performance (with maximum precision of 0.01). All experiments were executed using an Intel Xeon 1.70GHz server with 56 cores and 64GB of RAM.

3 Results

All the experiments were implemented in Python or R (when the tool did not have a Python API) using the AutoML libraries detailed in Sect. 2.1. For each AutoML tool, we executed five experiments, one for each target variable (DaysToNextFailure and FailOnxDays). Table 3 shows the average external test scores for all 10 folds and the respective confidence intervals (near the ± symbol).

Table 3. Average results obtained by the AutoML tools (best values in **bold**).

Target	Measure	AutoGluon	H2O AutoML	Rminer	TPOT
DaysToNextFailure	MAE	**4.95** ± 0.57	5.53 ± 0.62	8.89 ± 0.75	7.05 ± 0.57
FailOn3Days	AUC	**0.98** ± 0.02	**0.98** ± 0.01	0.95 ± 0.05	0.97 ± 0.03
FailOn5Days	AUC	**0.97** ± 0.02	0.96 ± 0.03	0.93 ±0.04	0.96 ± 0.02
FailOn7Days	AUC	**0.98** ± 0.01	**0.98** ± 0.01	0.97 ± 0.03	**0.98** ± 0.01
FailOn10Days	AUC	**0.99** ± 0.01	0.98 ± 0.01	0.98 ± 0.02	**0.99** ± 0.01

Overall the best AutoML tool was AutoGluon, which produced the highest AUC values for the binary classification tasks and the lowest MAE value for the regression task. For the regression task (DaysToNextFailure), besides AutoGluon, the best AutoML tools were H2O AutoML, TPOT, and rminer. The maximum predictive difference was 3.94 points (days). As for the binary classification, the predictive test set results are more similar: maximum difference of 3 *percentage points (pp)* for FailOn3Days, 4 *pp* for FailOn5Days, 1 *pp* for FailOn7Days, and 1 *pp* for FailOn10Days. AutoGluon was the best tool for all four binary classification targets, followed by H2O AutoML and TPOT (best in two targets each).

Finally, we compare the best AutoML results for each target with the best result achieved by a human ML modeling (held before this study). For each target, Table 4 shows the best predictive result and the respective AutoML tool in rounded brackets. For each AutoML tool, we also show the algorithm that was most often the leader, across the external folds. For the human modeling, we show the best obtained result and the used algorithm (also in rounded brackets).

It should be noted that the human modeling used a distinct preprocessing procedure since it applied the One-Hot encoding to all categorical attributes (and not IDF for the high cardinality ones, as we adopted for the AutoML tools). Nevertheless, the comparison clearly favors the AutoML results for all predicted target variables. In particular, for the binary classification task, the human modeling achieved only slightly better results than a random model,

Table 4. Comparison between the best AutoML result and human ML modeling result for each target (best values in **bold**).

Target	Measure	Best results	
		AutoML	Human modeling
DaysToNextFailure	MAE	**4.948** (AutoGluon: Ensemble)	68.361 (Random Forest)
FailOn3Days	AUC	**0.979** (H2O AutoML: GBM)	0.500 (Random Forest)
FailOn5Days	AUC	**0.971** (AutoGluon: Ensemble)	0.529 (Random Forest)
FailOn7Days	AUC	**0.982** (TPOT: Random Forest)	0.581 (Random Forest)
FailOn10Days	AUC	**0.988** (AutoGluon: Ensemble)	0.563 (Random Forest)

GBM - Gradient Boosting Machine.

while all AutoML tools achieved results that can be considered excellent (e.g., AUC higher than 0.90). For regression, the human modeling achieved an average error of 68.36 d, while the highest MAE obtained by the AutoML tools was 8.89 (achieved by rminer).

4 Conclusions

Predictive maintenance is a key industrial application that is being increasingly enhanced by the adoption of ML. Yet, most ML related works assume an expert ML model design that requires manual effort and time. In this paper, we explore the potential of AutoML to automate predictive maintenance ML modeling. We used real-world data provided by a Portuguese software company within the domain of maintenance management to predict equipment malfunctions.

Our goal was to anticipate failures from several types of equipments (e.g., industrial machines), using two ML tasks: regression - to predict the number of days until the next failure of the equipment; and binary classification - to predict if the equipment will fail in a fixed amount of days (e.g., in three days). For the ML modeling and training, we used four recent state-of-the-art Automated Machine Learning (AutoML) tools: AutoGluon, H2O AutoML, rminer, and TPOT.

Several computational experiments were held, assuming five predictive tasks (one regression and four binary classifications). For all ML tasks, AutoGluon presented the best average results among the AutoML tools. The AutoML results were further compared with a human ML design, performed previously by a professional of the Portuguese company. The comparison favored all AutoML tools, which provided better average results than the manual approach by a large margin. These results confirm the potential of the AutoML modeling, which can automatically provide high quality predictive models. This is particularly valuable for the predictive maintenance domain since industrial data can arise with a high velocity, thus the predictive models can be dynamically updated through time, reducing the effort of the data analysis.

In future work, we intend to perform experiments with more AutoML tools and from the domain of predictive maintenance datasets. In particular, we intend to experiment with AutoML technologies that can automatically perform feature engineering and selection tasks.

Acknowledgments. This work was executed under the project Cognitive CMMS - Cognitive Computerized Maintenance Management System, NUP: POCI-01-0247-FEDER-033574, co-funded by the Incentive System for Research and Technological Development, from the Thematic Operational Program Competitiveness of the national framework program - Portugal2020.

References

1. Auto-Gluon: AutoGluon: AutoML Toolkit for Deep Learning - AutoGluon Documentation 0.2.0 documentation (2021). https://auto.gluon.ai/
2. Ayvaz, S., Alpay, K.: Predictive aintenance system for production lines in manufacturing: a machine learning approach using IoT data in real-time. Expert Syst. Appl. **173**, 114598 (2021). https://doi.org/10.1016/j.eswa.2021.114598
3. Butte, S., Prashanth, A., Patil, S.: Machine learning based predictive maintenance strategy: a super learning approach with deep neural networks. In: 2018 IEEE Workshop on Microelectronics and Electron Devices (WMED), pp. 1–5. IEEE (2018)
4. Carvalho, T.P., Soares, F.A.A.M.N., Vita, R., da Piedade Francisco, R., Basto, J.P.T.V., Alcalá, S.G.S.: A systematic literature review of machine learning methods applied to predictive maintenance. Comput. Ind. Eng. 137 (2019). https://doi.org/10.1016/j.cie.2019.106024
5. Çınar, Z.M., Abdussalam Nuhu, A., Zeeshan, Q., Korhan, O., Asmael, M., Safaei, B.: Machine learning in predictive maintenance towards sustainable smart manufacturing in industry 4.0. Sustainability **12**(19), 8211 (2020)
6. Cline, B., Niculescu, R.S., Huffman, D., Deckel, B.: Predictive maintenance applications for machine learning. In: 2017 Annual Reliability and Maintainability Symposium (RAMS), pp. 1–7. IEEE (2017)
7. Cortez, P.: Data mining with neural networks and support vector machines using the R/rminer tool. In: Perner, P. (ed.) ICDM 2010. LNCS (LNAI), vol. 6171, pp. 572–583. Springer, Heidelberg (2010). https://doi.org/10.1007/978-3-642-14400-4_44
8. Erickson, N., et al.: Autogluon-tabular: Robust and accurate automl for structured data. arXiv preprint arXiv:2003.06505 (2020). https://arxiv.org/abs/2003.06505
9. Ferreira, L., Pilastri, A., Martins, C.M., Pires, P.M., Cortez, P.: A comparison of AutoML tools for machine learning, deep learning and XGBoost. In: International Conference on Joint Conference on Neural Networks, IJCNN 2021, IEEE (July 2021)
10. H2O.ai: H2O AutoML (2021). http://docs.h2o.ai/h2o/latest-stable/h2o-docs/automl.html, h2O version 3.32.1.3

11. Kanawaday, A., Sane, A.: Machine learning for predictive maintenance of industrial machines using IoT sensor data. In: 2017 8th IEEE International Conference on Software Engineering and Service Science (ICSESS), pp. 87–90. IEEE (2017)
12. Le, T.T., Fu, W., Moore, J.H.: Scaling tree-based automated machine learning to biomedical big data with a feature set selector. Bioinformatics **36**(1), 250–256 (2020)
13. LeDell, E., Poirier, S.: H2O AutoML: scalable automatic machine learning. In: 7th ICML Workshop on Automated Machine Learning (AutoML) (July 2020). https://www.automl.org/wp-content/uploads/2020/07/AutoML_2020_paper_61.pdf
14. Matos, L.M., Cortez, P., Mendes, R., Moreau, A.: A comparison of data-driven approaches for mobile marketing user conversion prediction. In: Jardim-Gonçalves, R., Mendonça, J.P., Jotsov, V., Marques, M., Martins, J., Bierwolf, R.E. (eds.) 9th IEEE International Conference on Intelligent Systems, IS 2018, Funchal, Madeira, Portugal, September 25–27, 2018. pp. 140–146. IEEE (2018). https://doi.org/10.1109/IS.2018.8710472
15. Olson, R.S., Urbanowicz, R.J., Andrews, P.C., Lavender, N.A., Kidd, L.C., Moore, J.H.: Automating biomedical data science through tree-based pipeline optimization. In: Squillero, G., Burelli, P. (eds.) EvoApplications 2016. LNCS, vol. 9597, pp. 123–137. Springer, Cham (2016). https://doi.org/10.1007/978-3-319-31204-0_9
16. Paolanti, M., Romeo, L., Felicetti, A., Mancini, A., Frontoni, E., Loncarski, J.: Machine learning approach for predictive maintenance in industry 4.0. In: 14th IEEE/ASME International Conference on Mechatronic and Embedded Systems and Applications, MESA 2018, Oulu, Finland, July 2–4, 2018. pp. 1–6. IEEE (2018). https://doi.org/10.1109/MESA.2018.8449150
17. Silva, A.J., Cortez, P., Pereira, C., Pilastri, A.: Business analytics in industry 4.0: a systematic review. Expert Syst. p. e12741. https://doi.org/10.1111/exsy.12741
18. Tornede, T., Tornede, A., Wever, M., Mohr, F., Hüllermeier, E., et al.: AutoML for predictive maintenance: one tool to RUL them All. In: Gama, J. (ed.) ITEM/IoT Streams -2020. CCIS, vol. 1325, pp. 106–118. Springer, Cham (2020). https://doi.org/10.1007/978-3-030-66770-2_8
19. Vazan, P., Janikova, D., Tanuska, P., Kebisek, M., Cervenanska, Z.: Using data mining methods for manufacturing process control. IFAC-PapersOnLine **50**(1), 6178–6183 (2017)

Learning Inter-Lingual Document Representations via Concept Compression

Marc Lenz[1(✉)], Tsegaye Misikir Tashu[1,2(✉)], and Tomáš Horváth[1,3(✉)]

[1] Faculty of Informatics, Department of Data Science and Engineering, Telekom Innovation Laboratories, ELTE - Eötvös Loránd University, Pázmány Péter Sétány, Budapest 1117, Hungary
{cr2jd9,misikir,tomas.horvath}@inf.elte.hu

[2] College of Informatics, Wollo University, Kombolcha Institute of Technology, 208 Kombolcha, Ethiopia

[3] Faculty of Science, Institute of Computer Science, Pavol Jozef Šafárik University, Jesenná 5, 040 01 Košice, Slovakia

Abstract. In this work, we proposed a novel approach to derive inter-lingual document representations. The introduced methods aim to enhance the quality of content-based Multilingual Document Recommendation and information retrieval Systems. The main idea centers around creating inter-lingual representations by using mappings to align monolingual representation spaces. According to the experimental results carried out on JRC-Acquis and EU bookshop multilingual corpora, the proposed concept compression approach has outperformed the traditional cross-lingual retrieval and recommendations methods.

Keywords: Cross-lingual information retrieval · Cross-lingual document representation · Multilingual NLP

1 Introduction

While a lot of existing Natural Language Processing techniques are language-specific, there is a growing body of literature concerning multi-lingual representations and language models. There has been a large interest in the multilingual capabilities of pre-trained language models [18] such as mBERT [5] or XLM [7]. Also, a considerable amount of literature has been published on interlingual-lingual word-vectors [2,3,9].

In this work, we shift the focus to learning document representations in a language-invariant manner. The intention is to present a method that enables to extension well known (vector-based) monolingual document retrieval methods to the cross-lingual domain via inter-lingual representations. Using such inter-lingual representations enables a direct comparison of documents of any language without the need for machine translation. This is tremendously useful for cross-lingual retrieval systems [8]. Traditional search engines and recommender systems are often monolingual [10], ignoring search results written

H. Yin et al. (Eds.): IDEAL 2021, LNCS 13113, pp. 268–276, 2021.
https://doi.org/10.1007/978-3-030-91608-4_27

in other languages and therefore discarding many documents with potentially relevant content. Using inter-lingual representations allows cross-lingual recommender and retrieval solely based on content rather than language. Beyond information retrieval and recommender systems, it provides further useful applications, such as cross-lingual transfer learning [1] or plagiarism detection [11]. Furthermore, avoiding the usage of Machine Translation also circumvents related problems such as the costliness of Machine Translation Systems and difficulties with resource-poor languages [20].

The approach presented in this paper focuses on the alignment of monolingual representations to obtain inter-lingual representations. It means that the novelty lies in the alignment of the representations than the representations themselves.

2 Related Work

Generally, there has been much interest in extending vector-based language representations to multiple languages. While many of the approaches focus on inter-lingual word-embeddings, fewer publications are concerned with inter-lingual representations on a document level. However, on both levels, word and document representations, there is an existing research that is related to the work presented in this paper.

2.1 Inter-Lingual Document Representations

Various approaches have been proposed for obtaining inter-lingual document representations. Some base their document representations on word representations by combining (inter-lingual) the vectors into a single vector by summing or averaging them [8]. Others try to reach language invariance by representing a document via its relation to concepts in the cross-lingual knowledge base, such as Wikipedia [12] or BabelNet [6]. Others again aim to represent longer passages of text directly in a language-invariant way, such as [14] that utilize deep neural network architectures to obtain cross-lingual embeddings.

This work is concerned with the latter. One of the best-known methods which embed documents directly into an inter-lingual representation space is cross-lingual semantic indexing (CL-LSI), an extension of LSI which was proposed by [4]. Similar to LSI, it computes embeddings by SVD of the Document-Term matrix, with the sole modification that the Document-term matrix consists of the concatenation of the monolingual document-term matrices [13].

2.2 Further Related Research

There has been much interest in extending vector-based word representations to multiple languages, such that their representations are inter-lingual [3]. Mikolov et al. [9] first noticed that continuous word embedding spaces exhibit similar structures across languages. They proposed to exploit this similarity by learning a linear mapping from a source to a target embedding space. They employed a parallel vocabulary of five thousand words as anchor points to learn this mapping and evaluated their approach on a word translation task. Since then, several studies aimed at improving these cross-lingual word embeddings [3,16,19]. Generally, the approaches to improve the results [9] can be roughly categorized as either modifying the operator between spaces or modifying the method used for retrieval. Approaches that have been proven to be successful are adding an orthogonal constraint to the linear mapping or introduce hub-scaling to the retrieval task. Furthermore, there have been several approaches attempting to align word representations without using parallel data. One approach which yielded comparable results to the supervised models using parallel data was proposed by Conneau et al. [2]. It uses a domain-adversarial approach to align monolingual representation spaces (Fig. 1).

3 Proposed Approach

Given two monolingual document collections, $D_x = \{d_{x,1}, \ldots, d_{x,n}\}$ and $D_y = \{d_{y,1}, \ldots, d_{y,n}\}$, first a representation learning algorithm is applied to the sets. We used Latent Semantic Indexing (LSI), which is based on the k-truncated Singular-Value-Decomposition of the Document term matrix. However, any representation learning model which maps the document sets D_x and D_y to vectors within the $\mathbb{R}^k$ is suitable. We obtain sets of vectors, $C_x = \{\hat{d}_{x,1}, \ldots, \hat{d}_{x,n}\} \subset \mathbb{R}^k$, $C_Y = \{\hat{d}_{y,1}, \ldots, \hat{d}_{y,n}\} \subset \mathbb{R}^k$. One can think of C_x, C_y as "Concept Spaces", which encode more general concepts of language and meaning. While the vectors in C_x, C_y might capture concepts and information, which are similar across languages, it is likely that they encode it in different ways. Therefore, a direct comparison of $\hat{d}_{x,k}, \hat{d}_{y,k}$ is yet unlikely to reveal similarities on a content level.

3.1 Linear Concept Compression (LCC)

We proposed to use a novel approach called a Linear Concept Compression in this study. The motivation behind linear concept compression is to find mappings into an inter-lingual space, E_x, E_y, such that the comparison of $E_x(\hat{d}_{x,k}), E_y(\hat{d}_{y,k})$, provides a measure of content similarity.

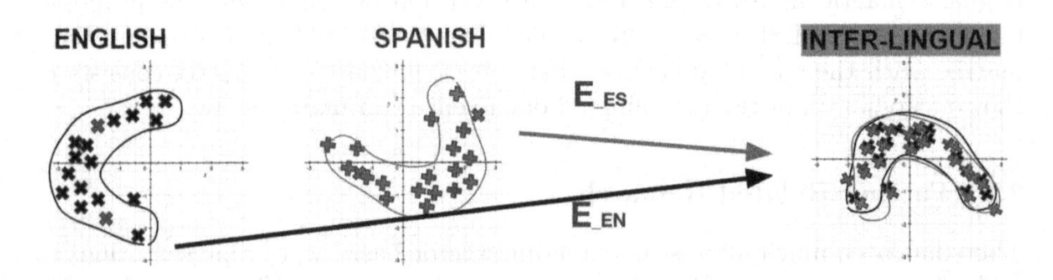

Fig. 1. An illustration of concept compression, the separate document representations in each language space (here English and Spanish) are mapped into a common inter-lingual representation space by E_{EN} and E_{Es}.

For two monolingual representations, we want to find their inter-lingual representations which encode the same information as the different monolingual spaces do. More precisely, for a given document d and its representations in each respective language, $\hat{d}_{x,k}$ and $\hat{d}_{y,k}$, we want to find mappings E_x and E_y, respectively, such that $E_x(\hat{d}_{x,k}) = E_y(\hat{d}_{y,k})$ and the information of $\hat{d}_{x,k}$ and $\hat{d}_{y,k}$ is preserved. While the problem statement certainly seems similar to the idea of Canonical Correlation Analysis, another solution should be proposed.

The intuition is to find an Encoder-Decoder approach. The purpose of the Encoder is to encode monolingual representations in a language-independent way. The purpose of the decoder is to reconstruct the monolingual representations of multiple languages from that encoding.

Linear Solution via Reduced Rank Regression: Introducing the constraint that the Encoder and Decoder are linear, this can be done via minimizing the following system of linear equations

$$|| \begin{bmatrix} C_x & 0 \\ 0 & C_y \end{bmatrix} \begin{bmatrix} E_x \\ E_y \end{bmatrix} \begin{bmatrix} D_x & D_y \end{bmatrix} - \begin{bmatrix} C_x & C_x \\ C_y & C_y \end{bmatrix} ||_2 \tag{1}$$

which is equivalent to minimizing

$$min_{rg(A)=d} || \begin{bmatrix} C_x & 0 \\ 0 & C_y \end{bmatrix} A - \begin{bmatrix} C_x & C_x \\ C_y & C_y \end{bmatrix} ||_2$$

and then finding a decomposition of A such that $A = ED$. Generally, the problem of reduced rank regression, $min_{B \in \mathbb{R}^{N \times N}, rg(A) \leq k} ||XB - Y||_F$, can be solved by the k-truncated SVD of $X(X^\top X)^{-1} X^\top Y$. Furthermore, for $B_k = U \Sigma V^T$, we can define $E = U\Sigma$ and $D = V^T$ which provides a solution for the equation above.

Extension to n Concept Spaces: If we have n different Concept spaces, C_1, $\ldots, C_n$, the minimization problem introduced above can be extended to solving

$$min_{rg(A)=d} || \begin{bmatrix} C_1 & \ldots & 0 \\ \vdots & \vdots & \vdots \\ 0 & \ldots & C_n \end{bmatrix} A - \begin{bmatrix} C_1 & \ldots & C_1 \\ \vdots & \vdots & \vdots \\ C_n & \ldots & C_n \end{bmatrix} ||_2$$

3.2 Neural Network Concept Compression (NNCC)

Instead of using linear functions for the encoder and the decoder, also neural networks can be used. Similar to Eq. 1 we define an "encoder" function E, $E : \mathbb{R}^{m_k + l_k} \to \mathbb{R}^h$ and a "decoder" function D, $E : \mathbb{R}^h \to \mathbb{R}^{m_k + l_k}$, both being neural networks. If we chain E and D, defining $f := D \circ E$, f is a neural network and $f : \mathbb{R}^{m_k + l_k} \to \mathbb{R}^{m_k + l_k}$. We then minimize

$$\mathbb{E}(||f(\begin{bmatrix} c_x \\ 0 \end{bmatrix}) - \begin{bmatrix} c_x \\ c_y \end{bmatrix} ||_2 + ||f(\begin{bmatrix} 0 \\ c_y \end{bmatrix}) - \begin{bmatrix} c_x \\ c_y \end{bmatrix} ||_2)$$

by computing

$$\nabla_\theta \frac{1}{N} \sum_{k=1}^{N} (\|f(\begin{bmatrix} c_x \\ 0 \end{bmatrix}) - \begin{bmatrix} c_x \\ c_y \end{bmatrix} \|_2 + \|f(\begin{bmatrix} c_x \\ 0 \end{bmatrix}) - \begin{bmatrix} c_x \\ c_y \end{bmatrix} \|_2) \tag{2}$$

And updating $\boldsymbol{\theta}_{t+t}$ stepwise with $\nabla_{\boldsymbol{\theta}_t}$. Then, in order to obtain the embeddings, we can compute (Fig. 2)

$$Emb(d) = E(\begin{bmatrix} LSI(d) \\ 0 \end{bmatrix}) \in \mathbb{R}^h$$

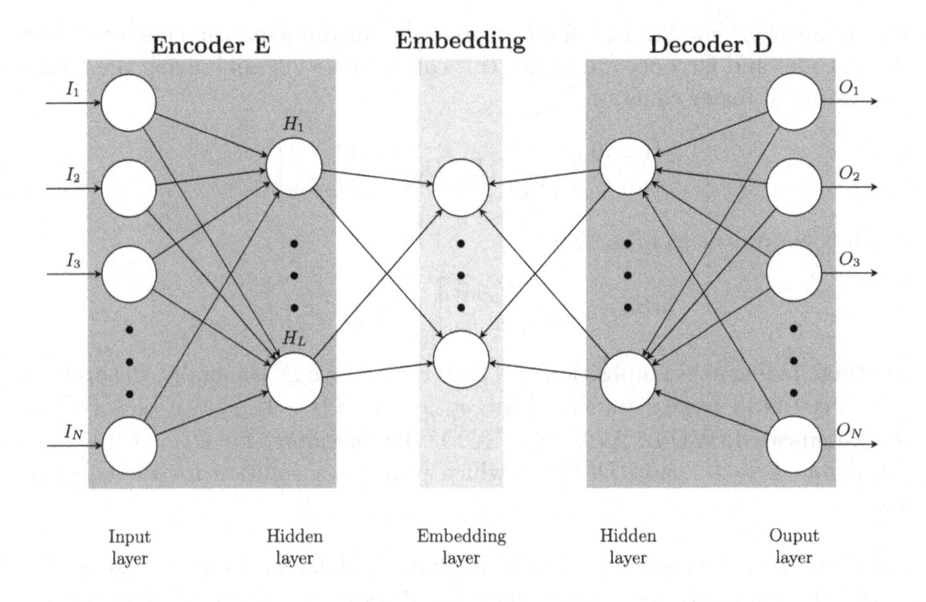

Fig. 2. Structure of the neural network used for the concept compression approach.

4 Experiments

The quality of the inter-lingual document representations obtained by the introduced methods of Linear Concept Compression (LCC) and Neural Network Concept Compression (NNCC) should be evaluated on a cross-lingual retrieval task. While both methods could be applied to any given document representation algorithm, here three methods are compared to put the quality of the obtained embeddings into perspective.

4.1 Methods Used

- **CL-LSI:** Representations obtained via Cross-Lingual LSI, the natural extension of the LSI method

- **LSI+LCC:** Representations obtained via the alignment of monolingual representations (derived by using LSI) by using LCC
- **LSI+NNCC:** Representations obtained via the alignment of monolingual representations (derived by using LSI) by using LCC

A specific requirement to measure retrieval success in this context is the availability of data sets containing parallel documents in multiple languages. Potential sources for large amounts of parallel documents are international organizations and governments. For this work, document collections of the European Union were used.

4.2 Datasets

- **JRC-Acquis:** The Acquis Communautaire (AC) is the total body of European Union (EU) law applicable in the EU Member States. This collection of legislative text changes continuously and currently comprises selected texts written between the1950 s and now. As of the beginning of the year 2007, the EU had 27 Member States and 23 official languages.
- **EU-Bookshop:** The EU Bookshop is an online service and archive of publications from various European institutions. The service contains a large body of publications in 24 official languages of the European Union. Skadins et al. [15] used the online service to build a large parallel corpus. The corpus was made available within the OPUS-Project [17], a collection of translated texts from the web.

4.3 Evaluation Metrics

- **Mate Retrieval Rate:** The mate (the same document in an other language) retrieval rate is defined as
$$\frac{\#MatesRetrieved}{\#Documents} \tag{3}$$
or, more generally, we want to know how high each mate has been ranked. For that, we will compute the rank of each mate within the search results.
- **Mean reciprocal rank:** The mean reciprocal rank is defined as
$$\frac{1}{\#Documents} \sum_{d \in Documents} \frac{1}{r(d)} \tag{4}$$

Both scores lie between 0 and 1, 1 being the best possible value for both definitions.

4.4 Experimental Setup

For the experimental setup of the JRC-Arquis, 8 languages of a total of 23 languages were selected. These were: English, French, German, Hungarian, Portuguese, Polish, Czech, and Dutch. The intersection of documents that are available in all these 8 languages contains approximately 20000 documents. Out of those, 5000 documents were randomly sampled and split into a training, validation, and test dataset. The training dataset contains 3000 documents, the validation, and test dataset contains every 1000 samples. The training setup consists of 2 stages:

- **Hyper-parameter selection:** To select a good set of hyper-parameters for each model, the mate-retrieval scores for different sets of parameters will be computed on the validation dataset for a subset of the language pairs. This subset consists of all combinations of the languages Hungarian, French, and English. Taking both language directions into consideration, 6 different scores are computed. For each set of hyper-parameters, the average of those 6 scores are computed. After that, the hyper-parameters with the best average score are selected for the second stage of the training.
- **Comparison of Models on Test Data:** During the second stage of the training, the models will be trained using the hyper-parameters which were selected during the first training stage. The models will be trained on all language combinations of the 8 selected languages. The experimental set-up for the EU-Bookshop was exactly the same, with the difference that only 3 out of 24 languages were chosen. Those are English, Portuguese, and French.

5 Results

Table 1 shows averaged mate-retrieval and reciprocal rank scores for all language pairs on the JRC-Acquis for each respective model. The Cross-Lingual LSI provides a strong baseline, with a 90.9% mate retrieval rate and a reciprocal rank of 93.8. However, the representations derived by LSI+LCC provide even better retrieval performance, with 92.2% Mate Retrieval Rate and 94.7 reciprocal rank. On the contrary, LSI+NCC performs slightly worse than the baseline of CL-LSI with 89.5% Mate Retrieval Rate and 92.3%.

Table 1. Scores for the different models evaluated on a cross-lingual retrieval task on the JRC-Acquis.

Inter-lingual representations	Mate retrieval rate (%)	Reciprocal rank(%)
CL-LSI (Baseline)	90.9	93.8
LSI + LCC	92.2	94.7
LSI + NNCC	89.5	92.3

Table 2 shows averaged mate-retrieval and reciprocal rank scores for all language pairs on the EU-Bookshop for each respective model. Generally, the retrieval success of all models is worse than for the JRC-Acquis dataset. The baseline of Cross-Lingual LSI achieves a 56.7% mate retrieval rate and a reciprocal rank of 64.0. As for the JRC-Arquis, the representation derived by LSI+LCC provides the best retrieval performance, with a 67.9% Mate Retrieval Rate and 70.3 reciprocal rank. LSI+NCC performs better than the baseline of CL-LSI and slightly worse than LSI+LCC with 64.0% Mate Retrieval Rate and 67.9%.

Table 2. Scores for the different Models evaluated on a cross-lingual retrieval task on the EU-Bookshop on 3 language pairs.

Inter-lingual representations	Mate retrieval rate (%)	Reciprocal rank(%)
CL-LSI (Baseline)	56.7	64.0
LSI + LCC	67.9	70.3
LSI + NNCC	64.0	67.9

6 Conclusion

In conclusion, it has been demonstrated that aligning monolingual document representations can be a very promising approach to learn inter-lingual representations. The alignments of the monolingual document representations derived by LSI outperform the multilingual counterpart of LSI, CL-LSI. This suggests that an alignment of monolingual document representations spaces may result in better inter-lingual representations than learning language-invariant representations directly.

While the results are promising, the evaluation should be extended. This concerns multiple aspects. First and foremost, the proposed methods of LCC and NNCC should be also applied to other types of monolingual document representation methods such as Doc2Vec, to gain a better understanding how the representation method influences the final results in realistic use-cases, only very few or no parallel data is available.

Therefore, an interesting topic for follow-up research would be on the performance of the approaches given limited amounts of parallel data. Besides that, there are still various approaches that worked well for word vectors but were not tested in this context. This includes mainly unsupervised methods [2] or using different methods for using hub-scaling within the retrieval task [16]. If those methods also provide a use for document representations has yet to be determined.

Acknowledgment. "Application Domain Specific Highly Reliable IT Solutions" project has been implemented with the support provided from the National Research, Development and Innovation Fund of Hungary, financed under the Thematic Excellence Programme TKP2020-NKA-06 (National Challenges Subprogramme) funding scheme.

References

1. Buys, J., Botha, J.A.: Cross-lingual morphological tagging for low-resource languages. arXiv:1606.04279 (2016)

2. Conneau, A., Kiela, D.: Senteval:an evaluation toolkit for universal sentence representations. arXiv:1803.05449 (2018)
3. Conneau, A., Lample, G., Ranzato, M., Denoyer, L., Jégou, H.: Word translation without parallel data. arXiv vol. 1710, p. 04087 (2017)
4. Deerwester, S., Dumais, S.T., Furnas, G.W., Landauer, T.K., Harshman, R.: Indexing by latent semantic analysis. J. Am. Soc. Inform. Sci. **41**(6), 391–407 (1990)
5. Devlin, J., Chang, M.W., Lee, K., Toutanova, K.: Bert: Pre-training of deep bidirectional transformers for language understanding. arXiv:1810.04805 (2018)
6. Franco-Salvador, M., Rosso, P., Navigli, R.: A knowledge-based representation for cross-language document retrieval and categorization. In: Proceedings of the 14th Conference of the European Chapter of the Association for Computational Linguistics (2014)
7. Lample, G., Ott, M., Conneau, A., Denoyer, L., Ranzato, M.: Phrase-based & neural unsupervised machine translation. arXiv:1804.07755 (2018)
8. Litschko, R., Glavaš, G., Ponzetto, S.P., Vulić, I.: Unsupervised cross-lingual information retrieval using monolingual data only. In: The 41st International ACM SIGIR Conference on Research & Development in Information Retrieval (2018)
9. Mikolov, T., Le, Q.V., Sutskever, I.: Exploiting similarities among languages for machine translation. arXiv:1309.4168 (2013)
10. Nie, J.Y.: Cross-language information retrieval. Synth. Lect. Human Lang. Technol. **3**(1), 1–125 (2010)
11. Potthast, M., Barrón-Cedeno, A., Stein, B., Rosso, P.: Cross-language plagiarism detection. Lang. Resour. Eval. **45**(1), 45–62 (2011)
12. Potthast, M., Stein, B., Anderka, M.: A Wikipedia-based multilingual retrieval model. In: Macdonald, C., Ounis, I., Plachouras, V., Ruthven, I., White, R.W. (eds.) ECIR 2008. LNCS, vol. 4956, pp. 522–530. Springer, Heidelberg (2008). https://doi.org/10.1007/978-3-540-78646-7_51
13. Saad, M., Langlois, D., Smaïli, K.: Cross-lingual semantic similarity measure for comparable articles. In: Przepiórkowski, A., Ogrodniczuk, M. (eds.) NLP 2014. LNCS (LNAI), vol. 8686, pp. 105–115. Springer, Cham (2014). https://doi.org/10.1007/978-3-319-10888-9_11
14. Schwenk, H., Douze, M.: Learning joint multilingual sentence representations with neural machine translation. arXiv:1704.04154 (2017)
15. Skadiņš, R., Tiedemann, J., Rozis, R., Deksne, D.: Billions of parallel words for free: Building and using the EU bookshop corpus. In: Proceedings of LREC (2014)
16. Smith, S.L., Turban, D.H., Hamblin, S., Hammerla, N.Y.: Offline bilingual word vectors, orthogonal transformations and the inverted softmax. arXiv:1702.03859 (2017)
17. Tiedemann, J.: Parallel data, tools and interfaces in opus. In: Chair, N.C.C., et al. (eds.) Proceedings of the Eight International Conference on Language Resources and Evaluation (LREC 2012), European Language Resources Association (ELRA), Istanbul, Turkey (2012)
18. Wu, S., Conneau, A., Li, H., Zettlemoyer, L., Stoyanov, V.: Emerging cross-lingual structure in pretrained language models. arXiv:1911.01464 (2019)
19. Xing, C., Wang, D., Liu, C., Lin, Y.: Normalized word embedding and orthogonal transform for bilingual word translation. In: Proceedings of the North American Chapter of the ACL: Human Language Technologies, pp. 1006–1011 (2015)
20. Yarmohammadi, M., et al.: Robust document representations for cross-lingual information retrieval in low-resource settings. In: Proceedings of Machine Translation Summit XVII Volume 1: Research Track (2019)

Mixture-Based Probabilistic Graphical Models for the Partial Label Ranking Problem

Juan C. Alfaro[1,3](✉), Juan A. Aledo[2,3], and José A. Gámez[1,3]

[1] Departamento de Sistemas Informáticos, Universidad de Castilla-La Mancha, 02071 Albacete, Spain
{JuanCarlos.Alfaro,Jose.Gamez}@uclm.es

[2] Departamento de Matemáticas, Universidad de Castilla-La Mancha, 02071 Albacete, Spain
JuanAngel.Aledo@uclm.es

[3] Laboratorio de Sistemas Inteligentes y Minería de Datos, Instituto de Investigación en Informática de Albacete, 02071 Albacete, Spain

Abstract. The *Label Ranking* problem consists in learning *preference models* from training datasets labeled with a *ranking* of class labels, and the goal is to predict a ranking for a given unlabeled instance. In this work, we focus on the particular case where both, the training dataset and the prediction given as output allow *tied* labels (i.e., there is no particular preference among them), known as the *Partial Label Ranking* problem. In particular, we propose *probabilistic graphical models* to solve this problem. As far as we know, there is no probability distribution to model rankings with ties, so we transform the rankings into discrete variables to represent the precedence relations (*precedes*, *ties* and *succeeds*) among pair of class labels (*multinomial distribution*). In this proposal, we use a *Bayesian network* with *Naive Bayes* structure and a *hidden variable* as *root* to collect the interactions among the different variables (predictive and target). The inference works as follows. First, we obtain the *posterior-probability* for each pair of class labels, and then we input these probabilities to the *pair order matrix* used to solve the corresponding *rank aggregation problem.* The experimental evaluation shows that our proposals are competitive (in accuracy) with the state-of-the-art *Instance Based Partial Label Ranking* (*nearest neighbors paradigm*) and *Partial Label Ranking Trees* (*decision tree induction*) algorithms.

Keywords: Mixture-based models · Bayesian networks · Naive bayes · (Partial) Label ranking

1 Introduction

In recent years, the *non-standard supervised classification* problems have grown significantly. In particular, the *Label Ranking* (*LR*) problem [9] consists in learning *preference models* able to predict *rankings* (a.k.a. *total orders* or *permutations*) defined over a finite set of class labels. An important difference between

H. Yin et al. (Eds.): IDEAL 2021, LNCS 13113, pp. 277–288, 2021.
https://doi.org/10.1007/978-3-030-91608-4_28

this problem and other non-standard supervised classification problems (e.g., *ordinal classification*) is that the instances of the training dataset are labeled with rankings, and these rankings are used during model learning.

In this paper, we focus on the *Partial Label Ranking* problem [6]. In this problem, the rankings associated with the instances of the training dataset and the predictions given as output are *partial rankings* (a.k.a. *total orders with ties* or *bucket orders*), that is, *rankings* with (possibly) *tied* class labels.

Based on [22], we rely on the use of a *hybrid Bayesian network* [14] where the root is a *hidden discrete variable* to jointly model the probability distributions for the discrete (*multinomial*) and *continuous* (*Gaussian*) attributes, and for the rankings. Although permutations (complete rankings without ties) may be modeled with the *Mallows distribution* [20], as far as we know, there is no probability distribution for bucket orders (complete rankings with ties). Therefore, we transform the ranking global preferences into a set of pairwise local preferences (*precedes*, *ties* and *succeeds*). By doing that, these variables can be modeled by a multinomial distribution (discrete variables), and so they can be easily integrated in the hybrid Bayesian network. The prediction for these variables are used to solve the associated *rank aggregation problem* [12], so outputting a partial ranking for a given unlabeled instance.

The paper is structured as follows. In Sect. 2, we review some basic notions concerning rankings. In Sect. 3, we formally describe the proposed model. In Sect. 4, we extend this model to allow interactions between the (continuous) attributes by means of a *multivariate Gaussian distribution*. In Sect. 5, we set out the experimental evaluation conducted to assess the proposed models. Finally, in Sect. 6, we provide the conclusions and future research lines.

2 Rank Aggregation Problem

Given a set of *items* $[[n]] = \{1, \ldots, n\}$, a *ranking* π represents a precedence relation among them. In particular, rankings may be *without ties* (a.k.a. *total orders* or *permutations*) or *with ties* (a.k.a. *partial rankings*, *total orders with ties* or *bucket orders*) if there is no preference among some of the ranked items.

The *rank aggregation problem* (*RAP*) [1] consists in obtaining a *consensus order* from a set of rankings. In particular, the *Kemeny Ranking Problem* (*KRP*) [19] is probably the most well-known, whose goal is to obtain the *consensus permutation* (a.k.a. *central permutation*) that minimizes a particular distance measure (e.g., the *Kendall distance*) with respect to a set of permutations. The KRP is usually solved with the *Borda count* algorithm because of its good trade-off between accuracy and efficiency.

In addition to the KRP, another well-known RAP is the *Optimal Bucket Order Problem* (*OBOP*) [12]. The goal of the OBOP is to obtain the *bucket matrix* B (associated with a bucket order π) that minimizes the distance D

$$D(B, P) = \sum_{u,v \in [[n]]} |B(u, v) - P(u, v)|$$

where P is the *pair order matrix* associated with a set of bucket orders (see [12] for the details).

Although there are several heuristic methods to solve the OBOP [2–4,17], we use a particular instance of the *Bucket Pivot Algorithm with least indecision assumption* [2] named LIA_G^{MP2}, because of the good trade-off it achieves between accuracy and efficiency.

3 Hidden Naive Bayes

Given that the goal is to output a partial ranking for an unlabeled instance, we have to obtain the pair order matrix C for solving the corresponding RAP. In our proposal, we use a Bayesian network to get the entries for this matrix, which codifies the preferences of a class label c_u over c_v, with $u < v$. Since $P(u, v) = 1 - P(v, u)$, we model both entries with a variable $Z_{u,v}$, and also the probability that c_u is tied with c_v. Thus, for each pair of class labels, we create the discrete variable

$$Z_{u,v} = \begin{cases} z_1, \text{ if } c_u \succ c_v \\ z_2, \text{ if } c_u \sim c_v \\ z_3, \text{ if } c_u \prec c_v \end{cases}$$

The advantage of this approach is that we only manage discrete variables. However, the complexity of the model grows quadratically with the number of labels according with $n_L = (n \cdot (n-1))/2$.

We propose a Bayesian network with Naive Bayes structure and a hidden variable to capture the interactions between the predictive variables and the target variable, and so obtain the *a-posteriori* probabilities for an unlabeled instance.

3.1 Model Definition

Figure 1 shows the *Plateau* representation of the proposed model. Note that the (hybrid) Bayesian network contains the discrete and continuous predictive variables, the discrete target variables and the discrete hidden variable H. In particular:

- *Discrete predictive variables*, denoted by X_j, $j = 1, \ldots, n_J$ with $dom(X_j) = \{x_{j_1}, \ldots, x_{j_{r_j}}\}$. They are observed both in the learning and inference stages.
- *Continuous predictive variables*, denoted by Y_k, $k = 1, \ldots, n_K$. They are observed both in the learning and inference stages.
- *Target variables*, denoted by $Z_{u,v}$, $u = 1, \ldots, n-1$ and $v = u+1, \ldots, n$, with domain $dom(Z_{u,v}) = \{z_1, z_2, z_3\}$, being $z_1 = c_u \succ c_v$, $z_2 = c_u \sim c_v$ and $z_3 = c_u \prec c_v$. They are observed in the learning stage.
- *Hidden variable*, denoted by H with $dom(H) = \{h_1, \ldots, h_{r_H}\}$, where r_H is the number of mixtures. This variable is never observed.

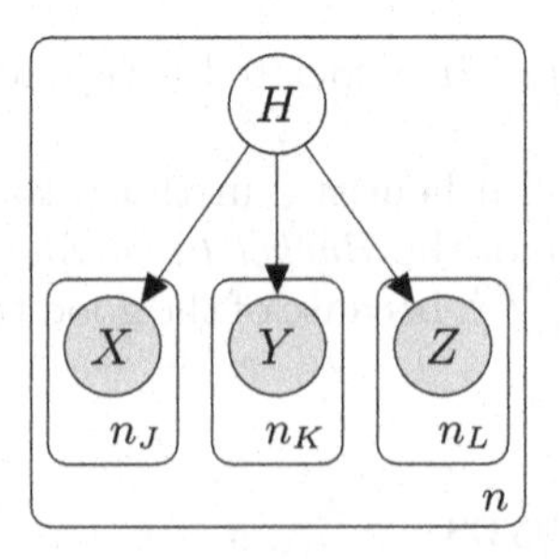

Fig. 1. Proposed HNB model

The discrete attributes, the target variables and the hidden variable follow a (conditional) multinomial distribution, and the continuous attributes follow a (conditional) Gaussian distribution. The joint probability distribution is given by

$$P(H, X_1, \ldots, X_{n_J}, Y_1, \ldots, Y_{n_K}, Z_{1,2}, \ldots, Z_{n-1,n}) = p(h_w) \cdot \prod_{j=1}^{n_J} P(X_j|H) \cdot \prod_{k=1}^{n_K} P(Y_k|H) \cdot \prod_{u=1,v=u+1}^{u=n-1,v=n} P(Z_{u,v}|H)$$

3.2 Parameter Estimation

We assume *complete* and *i.i.d.* data in both, the attributes and in the ranking variable. Therefore, we only deal with the hidden variable H, and we use the *Expectation-Maximization* (*EM*) to estimate jointly the parameters of both, the observed and hidden variables:

- **E step**: Under the assumption that the parameters of the discrete attributes $p(x_{ji}^{h_w})$, continuous attributes $\mu_k^{h_w}, \sigma_k^{h_w}$, ranking variable $p(z_{u,v}^{h_w})$ and hidden variable $p(h_w)$, $j = 1, \ldots, n_J$, $i = 1, \ldots, r_j$, $k = 1, \ldots, n_K$, $l = 1, \ldots, n_L$, $u = 1, \ldots, n-1$, $v = u+1, \ldots, n$, $w = 1, \ldots, r_H$ are known, the probability of an instance $e_t = (x_{1,t}, \ldots, x_{n_J,t}, y_{1,t}, \ldots, y_{n_K,t}, \pi_t)$ being in a mixture is

$$\begin{aligned} P(h_w|x_{1,t}, \ldots, x_{n_J,t}, y_{1,t}, \ldots, y_{n_K,t}, \pi_t) &= \frac{1}{C} \cdot p(h_w) \cdot \prod_{k=1}^{n_K} \frac{1}{\sigma_{k,t}^{h_w}\sqrt{2\pi}} \\ \cdot e^{-\frac{1}{2}\left(\frac{y_{k,t}-\mu_{k,t}^{h_w}}{\sigma}\right)^2} &\cdot \prod_{j=1}^{n_J} p(x_{j,t}^{h_w}) \cdot \prod_{u=1,v=u+1}^{u=n-1,v=n} p(z_{u,v,t}^{h_w}) \end{aligned} \quad (1)$$

  where C is a normalization constant.
- **M step**: Under the assumption that the probabilities of belonging to each mixture for all examples are known, the parameters of the model can be estimated as follows:
  - Multinomial parameters for the discrete attributes and the target variables. Each multinomial parameter is estimated by means of *maximum likelihood estimation* (*MLE*), where the count for each instance is weighted by the probability of $H = h_w$ given the instance.

- Gaussian parameters for the continuous attributes. The Gaussian parameters $\mu_k^{h_w}$ y $\sigma_k^{h_w}$ are estimated by means of MLE for each $H = h_w$, weighting each instance by the probability of it being in the mixture.

Stopping Condition: We use the *log-likelihood* of the model given the data with $\alpha = 0.001$ as convergence value. Moreover, we fix a maximum of $\beta = 100$ iterations.

3.3 Model Learning and Selection

We use the following procedure to compute the number of mixtures of the hidden variable H:

1. The dataset is divided in training Tr (0.8) and validation Tv (0.2), and the τ_X *rank correlation coefficient* is used to evaluate the models goodness in the search procedure.
2. The search for the number of mixtures is carried out greedily. First, we evaluate the model with $r_H = \{2^i\}_{i=1}^{10}$, and we select the best number of mixtures r'_H according with τ_X^{Tv}. Second, we apply a binary search in $[\frac{r'_H}{2}, r'_H]$, and we use the best number of mixtures r^*_H to train the model with the whole dataset.

Each time a new value for r_H is tried, the process starts from scratch, that is, all the parameters of the components are initialized (probabilities and weights) using the *k-means clustering algorithm* [23] with $k = r_H$ and $\gamma = 10$ different centroid seeds. Then, the EM algorithm is executed.

3.4 Inference

In the inference process, the method needs to compute the partial ranking π_t of the values in $dom(C)$ associated with an unlabeled instance e_t. Although the standard approach would select the (partial) ranking that maximizes the *a-posteriori probability* given the instance e_t, the cardinality of the search space $n!/2 \cdot \ln 2^{n+1}$, is too high, so we need to use an approximate method:

1. We compute the a-posteriori probability $P(Z_{u,v}|e_t)$ for each target variable using the Bayesian network.
2. We use these probabilities to populate the pair order matrix P_t associated with the instance e_t

$$P_t(u,v) = P(Z_{u,v} = z_1|e_t) + \tfrac{1}{2} \cdot P(Z_{u,v} = z_2|e_t)$$

$$P_t(v,u) = P(Z_{u,v} = z_3|e_t) + \tfrac{1}{2} \cdot P(Z_{u,v} = z_2|e_t) = 1 - P_t(u,v)$$

with $u < v$ and $P_t(u,v) = 0.5$ if $u = v$, and we solve the OBOP using P_t to obtain the (partial) ranking π_t for the instance e_t.

4 Gaussian Mixture Semi Naive Bayes

In this section, we assume that all the attributes are continuous, and we allow interactions among them.

4.1 Definition

We propose a *Semi-Naive Bayes* (*SNB*) [8,15] structure to model the continuous attributes using a multivariate normal distribution, while the rest of interactions are still managed by the hidden variable H.

We assume two variants of the *Gaussian mixture model* (*GMM*) [21]: *full*, where each component has its own general covariance matrix, and *tied*, where all components share the same general covariance matrix.

4.2 Estimation

The differences of the GMSNB model with respect to the HNB model are:

- **E step**: Similarly to the HNB model, we use Eq. 1, but, instead of the product of the conditional Gaussian distributions, we use the probability density function of the multivariate normal distribution $\mathcal{MN}(\boldsymbol{y}_t|\boldsymbol{\mu}^{h_w}, \Sigma^{h_w})$, where $\boldsymbol{y}_t$ is the configuration of values for the continuous attributes in e_t, $\boldsymbol{\mu}^{h_w}$ is the vector of means and Σ^{h_w} is the covariance matrix.
- **M step**: In the same way that the parameters of the HNB model, the means and empirical covariances of the continuous attributes are weighted by $w_t^{h_w} = P(h_w|\boldsymbol{y}_t, \pi_t)$.

4.3 Learning and Inference

The learning and inference stages of the HNB model and the GMSNB model are the same, but we model the continuous attributes with the multivariate normal distribution.

5 Experimental Evaluation

In this section, we detail the datasets used, the algorithms tested, the methodology adopted and the results obtained in the evaluation of our proposal.

5.1 Datasets

Table 1 shows the main characteristics of the 15 (semi-synthetic) datasets used as *benchmark* for the PLR problem [6]. The columns #rankings y #buckets stand for the mean number of different (partial) rankings in the dataset and the mean number of buckets per ranking, respectively. The datasets (and their description) are provided at: https://www.openml.org/u/25829.

Table 1. Description of the datasets

Datasets	#Instances	#Attributes	#Labels	#Rankings	#Buckets
Authorship	841	70	4	47	3.063
Blocks	5472	10	5	116	2.337
Breast	109	9	6	62	3.925
Ecoli	336	7	8	179	4.140
Glass	214	9	6	105	4.089
Iris	150	4	3	7	2.380
Letter	20000	16	26	15014	7.033
Libras	360	90	15	356	6.889
Pendigits	10992	16	10	3327	3.397
Satimage	6435	36	6	504	3.356
Segment	2310	18	7	271	3.031
Vehicle	846	18	4	47	3.117
Vowel	528	10	11	504	5.739
Wine	178	13	3	11	2.680
Yeast	1484	8	10	1006	5.929

5.2 Algorithms

We tested the following algorithms:

- *Instance Based Partial Label Ranking* (*IBPLR*) [6]. The Euclidean distance was used to identify the k nearest neighbors, and the (partial) rankings associated with these neighbors were weighted according to the (inverse) distance. The number of nearest neighbors was adjusted with a five-fold cross validation (5-cv) over the training dataset (see [6] for details).
- *Partial Label Ranking Trees* (*PLRT*) using the four criteria described in [6].
- *HNB-PLR* (Sect. 3). We considered four alternatives: Gaussian distribution (HNB-PLR-G) for the continuous attributes and multinomial distribution for the *equal-width* (HNB-PLR-W), *equal-frequency* (HNB-PLR-F) and *entropy-based* [13] (HNB-PLR-E) discretized versions. The number of bins for the equal-width and equal-frequency binning was fixed to 5.
- *GMSNB-PLR* (Sect. 4). We used a different covariance matrix for each mixture (full, GMSNB-PLR-F) and the same covariance matrix for all the mixtures (tied, GMSNB-PLR-T).

5.3 Methodology

We decided to apply the following design decisions:

- A 5×10 cross validation method was used (standard in the PLR problem).
- The accuracy was measured with the τ_X rank correlation coefficient [11].

- We used the standard statistical analysis procedure [10,16] by using the tool `exreport` [7] to analyze the results:
  1. First, a *Friedman test* is carried out with a significance level of $\alpha = 0.05$. If the obtained p-value is less than or equal to $\alpha = 0.05$, we reject the null hypothesis H_0, and so at least one algorithm is not equal to the rest.
  2. Second, a post-hoc test using the *Holm procedure* [18] is applied to discover the outstanding methods. This method compares all the algorithms with respect to the one ranked first by the Friedman test (control algorithm).

5.4 Reproducibility

The source code is provided at: https://github.com/alfaro96/scikit-lr. The experiments were executed in computers running the CentOS Linux 7 operating system, with CPU Intel(R) Xeon(R) E5–2630 a 2.40 GHz, and 16 GB of RAM memory.

5.5 Results

In this section, we provide and analyze the accuracy and CPU time results.

Accuracy. Table 2 shows the accuracy of the mixture-based models. Each cell contains the average and standard deviation over the test datasets of the 5×10 cv for the τ_X rank correlation coefficient between the real and predicted (partial) rankings. The boldfaced values correspond to the algorithms leading to the best accuracy for each dataset.

Table 2. Accuracy for the HNB-PLR and GMSNB-PLR algorithms

Dataset	HNB-PLR-G	HNB-PLR-F	HNB-PLR-W	HNB-PLR-E	GMSNB-PLR-F	GMSNB-PLR-T
Authorship	**0.814 ± 0.020**	0.797 ± 0.023	0.793 ± 0.023	0.797 ± 0.027	0.724 ± 0.022	0.806 ± 0.023
Blocks	0.931 ± 0.005	0.926 ± 0.005	0.899 ± 0.008	**0.942 ± 0.005**	0.922 ± 0.007	0.926 ± 0.006
Breast	0.736 ± 0.057	**0.760 ± 0.055**	0.648 ± 0.088	0.729 ± 0.058	0.641 ± 0.109	0.717 ± 0.076
Ecoli	**0.758 ± 0.035**	0.728 ± 0.033	0.727 ± 0.032	0.740 ± 0.034	0.714 ± 0.035	0.757 ± 0.028
Glass	0.692 ± 0.061	0.757 ± 0.045	0.707 ± 0.055	0.759 ± 0.049	0.662 ± 0.062	**0.761 ± 0.043**
Iris	0.874 ± 0.058	0.860 ± 0.076	0.887 ± 0.043	0.802 ± 0.105	0.871 ± 0.046	**0.897 ± 0.041**
Letter						
Libras	0.579 ± 0.031	0.545 ± 0.027	0.558 ± 0.034	**0.613 ± 0.034**	0.289 ± 0.030	0.578 ± 0.039
Pendigits	0.807 ± 0.005	0.804 ± 0.006	0.806 ± 0.005	0.805 ± 0.005	0.793 ± 0.007	**0.809 ± 0.006**
Satimage	0.870 ± 0.006	0.857 ± 0.007	0.843 ± 0.007	0.857 ± 0.007	0.813 ± 0.009	**0.875 ± 0.006**
Segment	0.866 ± 0.013	0.867 ± 0.009	0.870 ± 0.011	**0.883 ± 0.009**	0.846 ± 0.013	0.871 ± 0.012
Vehicle	0.731 ± 0.030	0.727 ± 0.028	0.709 ± 0.029	0.697 ± 0.028	0.606 ± 0.043	**0.781 ± 0.021**
Vowel	0.707 ± 0.032	0.728 ± 0.021	0.725 ± 0.024	0.562 ± 0.063	0.596 ± 0.027	**0.756 ± 0.014**
Wine	0.835 ± 0.042	0.822 ± 0.051	0.821 ± 0.055	0.826 ± 0.047	0.824 ± 0.055	**0.850 ± 0.047**
Yeast	0.747 ± 0.017	0.740 ± 0.014	0.715 ± 0.014	0.707 ± 0.027	0.731 ± 0.017	**0.775 ± 0.011**

We used the standard statistical analysis procedure described in Sect. 5.3:

1. The p-value obtained in the Friedman test was 4.124e^{-6}, so we rejected the null hypothesis (H_0), and, at least, one algorithm was different.
2. Table 3 shows the results of the post-hoc test, taking as control the GMSNB-PLR-T algorithm. The columns *ranking* and p-valor represent the ranking obtained by the Friedman test and the p-value adjusted by the Holm procedure, respectively. The columns *win*, *tie*, *loss* contain the number of times that the control algorithm win, tie and losses with respect to the row-wise one. The boldfaced p-values are non-rejected null hypotheses (H_0).

Table 3. Results of the post-hoc test for the mean accuracy of the HNB-PLR and GMSNB-PLR algorithms

Method	Ranking	p-value	Win	Tie	Loss
GMSNB-PLR-T	1.90	–	–	–	–
HNB-PLR-G	2.47	**4.068e^{-1}**	9	0	6
HNB-PLR-E	3.27	**9.087e^{-2}**	10	0	5
HNB-PLR-F	3.70	2.525e^{-2}	13	1	1
HNB-PLR-W	4.40	1.010e^{-3}	15	0	0
GMSNB-PLR-F	5.27	4.148e^{-6}	14	0	1

In the statistical analysis, we can observe that the GMSNB-PLR-T algorithm is ranked first by the Friedman test, and it is statistically different from HNB-PLR-F, HNB-PLR-W and GMSNB-PLR-F algorithms. However, there is no statistical difference with respect to the HNB-PLR-G and HNB-PLR-E algorithms.

Let us compare the best algorithms with respect to the IBPLR and PLRT algorithms. Table 4 shows the results of this comparison.

Table 4. Mean accuracy for the IBPLR and PLRT algorithms

Conjunto de datos	IBPLR	PLRT-A	PLRT-D	PLRT-E	PLRT-G
Authorship	**0.829 ± 0.018**	0.757 ± 0.025	0.763 ± 0.019	0.780 ± 0.023	0.776 ± 0.023
Blocks	0.937 ± 0.005	0.940 ± 0.004	0.941 ± 0.005	0.944 ± 0.004	**0.946 ± 0.004**
Breast	0.751 ± 0.058	0.770 ± 0.075	**0.777 ± 0.067**	0.766 ± 0.058	0.763 ± 0.064
Ecoli	0.759 ± 0.027	0.758 ± 0.027	**0.765 ± 0.026**	0.763 ± 0.033	0.755 ± 0.031
Glass	0.756 ± 0.045	**0.764 ± 0.048**	0.761 ± 0.048	0.761 ± 0.037	0.758 ± 0.034
Iris	0.900 ± 0.051	0.912 ± 0.046	0.909 ± 0.044	**0.916 ± 0.045**	0.905 ± 0.040
Letter	**0.689 ± 0.005**	0.667 ± 0.004	0.667 ± 0.004	0.669 ± 0.005	0.670 ± 0.005
Libras	**0.648 ± 0.029**	0.584 ± 0.029	0.588 ± 0.030	0.575 ± 0.026	0.583 ± 0.029
Pendigits	**0.819 ± 0.006**	0.799 ± 0.006	0.801 ± 0.006	0.813 ± 0.006	0.811 ± 0.005
Satimage	**0.881 ± 0.005**	0.834 ± 0.006	0.839 ± 0.006	0.846 ± 0.006	0.848 ± 0.005
Segment	0.890 ± 0.008	0.886 ± 0.010	0.889 ± 0.009	0.894 ± 0.009	**0.896 ± 0.009**
Vehicle	0.739 ± 0.020	0.747 ± 0.031	0.757 ± 0.027	**0.793 ± 0.021**	0.778 ± 0.021
Vowel	**0.745 ± 0.017**	0.673 ± 0.031	0.680 ± 0.028	0.679 ± 0.023	0.682 ± 0.023
Wine	**0.845 ± 0.036**	0.841 ± 0.050	0.837 ± 0.046	0.824 ± 0.046	0.825 ± 0.060
Yeast	**0.790 ± 0.010**	0.769 ± 0.010	0.774 ± 0.009	0.775 ± 0.010	0.774 ± 0.009

For these set of algorithms, the p-value obtained by the Friedman test was $1.5e^{-2}$, so we rejected the null hypothesis H_0 and at least one algorithm is different to the rest. Table 5 shows the results of the post-hoc test adjusted with the Holm procedure, taking as control the IBPLR algorithm (ranked first by the Friedman test).

Table 5. Results of the post-hoc test for the mean accuracy of the compared algorithms

Method	Ranking	p-value	Win	Tie	Loss
IBPLR	3.07	–	–	–	–
PLRT-E	3.70	$\mathbf{5.763e^{-1}}$	8	0	7
PLRT-D	4.13	$\mathbf{5.763e^{-1}}$	9	0	6
PLRT-G	4.23	$\mathbf{5.763e^{-1}}$	9	0	6
PLRT-A	4.40	$\mathbf{5.442e^{-1}}$	9	0	6
GMSNB-PLR-T	4.63	$\mathbf{3.992e^{-1}}$	11	0	4
HNB-PLR-G	5.77	$1.523e^{-2}$	15	0	0
HNB-PLR-E	6.07	$5.574e^{-1}$	13	0	2

According with these results, we can conclude that:

- The IBPLR algorithm is ranked first by the Friedman test, without statistical difference with respect to the PLRT and GMSNB-PLR-T algorithms. These results show that the model-based methods are competitive with respect to the instance-based in the PLR problem, which is not the case for the LR problem [5,9].
- The main disadvantage of the HNB-PLR and GMSNB-PLR algorithms is the memory requirements, as they are not able to deal with datasets generating a high number of target variables. For instance, there are no results for the `letter` dataset (325 target variables and 20000 instances).

Time. Our proposals are slower than the instance-based (IBPLR) and tree-based methods (PLRT) because of the high number of mixtures (see Table 6) and so EM iterations required to properly model the joint probability distribution. Furthermore, since we have a high number of target variables (due to the ranking transformation), the EM algorithm takes too much time to converge. For instance, taking the `pendigits` dataset (10992 instances and 10 class labels, that is, 45 target variables), the GMSNB-PLR-T is 120 times slower than the IBPLR algorithm and 850 slower than the PLRT-G.

Table 6. Mean number of mixtures for each PGM

Dataset	HNB-PLR-G	HNB-PLR-F	HNB-PLR-W	HNB-PLR-E	GMSNB-PLR-F	GMSNB-PLR-T
Authorship	36.340 ± 36.988	24.58 ± 18.377	31.380 ± 21.813	40.840 ± 52.281	3.020 ± 0.141	35.420 ± 41.952
Blocks	167.440 ± 76.169	238.400 ± 168.220	81.300 ± 33.121	341.660 ± 147.979	68.520 ± 23.693	213.200 ± 97.692
Breast	15.960 ± 7.284	29.120 ± 19.157	19.800 ± 19.078	17.720 ± 12.795	4.520 ± 2.288	20.320 ± 8.110
Ecoli	31.000 ± 14.321	25.820 ± 21.930	31.140 ± 26.869	39.900 ± 20.928	12.920 ± 6.552	47.040 ± 27.871
Glass	17.220 ± 9.951	66.020 ± 32.922	27.920 ± 23.357	45.520 ± 23.603	5.180 ± 2.164	39.760 ± 16.577
Iris	17.020 ± 15.946	34.820 ± 20.457	24.180 ± 17.253	9.560 ± 4.739	7.960 ± 4.000	32.320 ± 22.709
Letter						
Libras	47.660 ± 12.967	40.940 ± 14.621	43.960 ± 13.425	121.760 ± 38.154	218.080 ± 39.608	56.200 ± 10.900
Pendigits	388.800 ± 108.298	204.680 ± 67.601	266.480 ± 95.088	261.520 ± 98.865	93.800 ± 27.355	405.840 ± 102.542
Satimage	326.660 ± 101.758	283.660 ± 96.820	198.720 ± 108.040	272.060 ± 108.377	29.620 ± 14.246	392.460 ± 110.909
Segment	140.400 ± 56.351	202.580 ± 158.267	196.300 ± 154.706	230.320 ± 141.072	42.680 ± 21.920	337.380 ± 121.548
Vehicle	73.320 ± 35.361	292.600 ± 151.414	346.300 ± 144.204	172.420 ± 126.264	12.260 ± 2.448	66.480 ± 60.716
Vowel	75.320 ± 26.250	169.260 ± 55.564	186.260 ± 49.278	95.640 ± 42.740	7.640 ± 2.884	174.580 ± 52.597
Wine	6.700 ± 9.033	11.980 ± 15.213	17.120 ± 19.256	24.960 ± 23.206	3.800 ± 1.030	14.480 ± 17.117
Yeast	103.500 ± 56.536	46.000 ± 18.553	127.660 ± 97.812	159.040 ± 113.482	30.300 ± 16.656	219.880 ± 93.535

6 Conclusions and Future Work

In this paper, we have proposed an algorithm based on Bayesian networks and rank aggregation to solve the PLR problem. In particular, we have transformed the ranking variable into several target variables to model the preferences among pair of class labels with a multinomial distribution. Our proposal is based on a SNB structure with a hidden variable as root to model the interaction between the predictive and target variables. Thus, we only need to estimate the parameters with the EM algorithm. Given an unlabeled instance, the a-posteriori probabilities for the target variables are computed and input to the pair order matrix used to solve the OBOP, and so obtain the output (partial) ranking.

From the experimental evaluation, we have concluded that the GMSNB-PLR-T algorithm is competitive with the IBPLR and PLRT algorithms. Note that, although the GMSNB-PLR-T requires more computational resources during the learning phase than the IBPLR algorithm, this is not the case during the inference phase.

As future research, we plan to reduce the problem using clustering techniques to solve the memory problems, which we expect that also reduces the CPU time required in the learning phase.

Acknowledgement. This work has been funded by the Government of Castilla-La Mancha and "ERDF A way of making Europe" through the project SBPLY/17/180501/000493. It is also part of the projects PID2019–106758GB–C33 and FPU18/00181 funded by MCIN/AEI/ 10.13039/501100011033 and "ESF Investing your future".

References

1. Aledo, J.A., Gámez, J.A., Molina, D.: Approaching the rank aggregation problem by local search-based metaheuristics. J. Comput. Appl. Math. **354**, 445–456 (2019)
2. Aledo, J.A., Gámez, J.A., Rosete, A.: Utopia in the solution of the bucket order problem. Decis. Support Syst. **97**, 69–80 (2017)

3. Aledo, J.A., Gámez, J.A., Rosete, A.: Approaching rank aggregation problems by using evolution strategies: the case of the optimal bucket order problem. Eur. J. Oper. Res. **270**, 982–998 (2018)
4. Aledo, J.A., Gámez, J.A., Rosete, A.: A highly scalable algorithm for weak rankings aggregation. Inf. Sci. **570**, 144–171 (2021)
5. Aledo, J.A., Gámez, J.A., Molina, D.: Tackling the supervised label ranking problem by bagging weak learners. Inf. Fusion **35**, 38–50 (2017)
6. Alfaro, J.C., Aledo, J.A., Gámez, J.A.: Learning decision trees for the partial label ranking problem. Int. J. Intell. Syst. **36**, 890–918 (2021)
7. Arias, J., Cózar, J.: ExReport: Fast, reliable and elegant reproducible research. http://jacintoarias.github.io/exreport 27 Oct 2021 (2016)
8. Bielza, C., Larrañaga, P.: Discrete bayesian network classifiers: a survey. ACM Comput. Surv. **47**, 1–43 (2014)
9. Cheng, W., Hühn, J., Hüllermeier, E.: Decision Tree and Instance-Based Learning for Label Ranking. In: Proceedings of the 26th Annual International Conference on Machine Learning, pp. 161–168 (2009)
10. Demšar, J.: Statistical comparisons of classifiers over multiple data sets. J. Mach. Learn. Res. **7**, 1–30 (2006)
11. Emond, E.J., Mason, D.W.: A new rank correlation coefficient with application to the consensus ranking problem. J. Multi-Criteria Decis. Anal. **11**, 17–28 (2002)
12. Fagin, R., Kumar, R., Mahdian, M., Sivakumar, D., Vee, E.: Comparing and aggregating rankings with ties. In: Proceedings of the Twenty-third ACM SIGMOD-SIGACT-SIGART Symposium on Principles of Database Systems, pp. 47–58 (2004)
13. Fayyad, U.M., Irani, K.B.: Multi-interval discretization of continuous-valued attributes for classification learning. Artif. Intell. **13**, 1022–1027 (1993)
14. Fernández, A., Gámez, J.A., Rumí, R., Salmerón, A.: Data clustering using hidden variables in hybrid Bayesian networks. Prog. Artif. Intell. **2**, 141–152 (2014)
15. Flores, M.J., Gámez, J. A., Martínez, A.: Supervised classification with bayesian networks: a review on models and applications. In: Intelligent Data Analysis for Real-Life Applications: Theory and Practice, pp. 72–102. IGI Global (2012)
16. García, S., Herrera, F.: An extension on "Statistical comparisons of classifiers over multiple data sets" for all pairwise comparisons. J. Mach. Learn. Res. **9**, 2677–2694 (2008)
17. Gionis, A., Mannila, H., Puolamäki, K., Ukkonen, A.: Algorithms for discovering bucket orders from data. In: Proceedings of the 12th ACM SIGKDD International Conference on Knowledge Discovery and Data Mining, pp. 561–566 (2006)
18. Holm, S.: A simple sequentially rejective multiple test procedure. Scand. J. Stat. **6**, 65–70 (1979)
19. Kemeny, J., Snell, J.: Mathematical Models in the Social Sciences. The MIT Press, Cambridge (1972)
20. Mallows, C.L.: Non-null ranking models. Biometrika **44**, 114–130 (1957)
21. Reynolds, D.: Gaussian Mixture Models. In: Li, S.Z., Jain, A. (eds.) Encyclopedia of Biometrics. Springer, Boston (2009) https://doi.org/10.1007/978-0-387-73003-5_196
22. Rodrigo, E.G., Alfaro, J.C., Aledo, J.A., Gámez, J.A.: Mixture-based probabilistic graphical models for the label ranking problem. Entropy **23**, 420 (2021)
23. Wu, X., Kumar, V.: The Top Ten Algorithms in Data Mining. Chapman and Hall, London (2009)

From Classification to Visualization: A Two Way Trip

Marina Cuesta[1(✉)], Isaac Martín de Diego[1], Carmen Lancho[1], Víctor Aceña[1,2], and Javier M. Moguerza[1]

[1] Data Science Laboratory, Rey Juan Carlos University, C/Tulipán, S/n, 28933 Móstoles, Spain
{marina.cuesta,isaac.martin,carmen.lancho,victor.acena, javier.moguerza}@urjc.es
[2] Madox Viajes, C/de Cantabria, 10, 28939 Arroyomolinos, Spain
https://www.datasciencelab.es
https://www.madoxviajes.com

Abstract. *High Dimensional Data* (*HDD*) is one of the biggest challenges in Data Science arising from Big Data. The application of dimensionality reduction techniques over *HDD* allows visualization and, thus, a better problem understanding. In addition, these techniques also can enhance the performance of *Machine Learning* (*ML*) algorithms while increasing the explanatory power. This paper presents an automatic method capable of obtaining an adequate representation of the data, given a previously trained *ML* model. Likewise, an automatic method is introduced to bring a *Support Vector Machine* (*SVM*) model based on an adequate representation of the data. Both methods provide an Explanaible Machine Learning procedure. The proposal is tested on several data sets providing promising results. It significantly eases the visualization and understanding task to the data scientist when a *ML* model has already been trained, as well as the *ML* selection parameters when a reduced representation of data has been achieved.

Keywords: High dimensional data · Visualization · Classification · Explanaible machine learning

1 Introduction

The rapid growth and progress of the Internet of Things (IoT) technology and its inclusion in everyday life are leading to the generation and constant collection of large volumes of data. Indeed, it is expected to continue to grow exponentially in the upcoming years. Big data is in charge of the massive analysis of these data and it is becoming one of the most recurrent topics in Data Science research. Among all the challenges associated with Data Science about Big Data, one of the most important ones are the *High Dimensional Data (HDD)* [1]. This is data with a large number of variables measured for each observation.

H. Yin et al. (Eds.): IDEAL 2021, LNCS 13113, pp. 289–299, 2021.
https://doi.org/10.1007/978-3-030-91608-4_29

The application of dimensionality reduction techniques is a common step when dealing with a high dimensionality problem. These techniques offer a reduced number of variables resulting from a transformation of the original ones, preserving as much information as possible. Furthermore, the transformed data with a reduced number of features eases the visualization of the original data. This task is a recurrent problem that emerges in several phases of a Data Science project and it is essential when understanding the underlying problems of the data. Moreover, when the dimensionality reduction technique is reproducible for test data, it is possible to train *Machine Learning (ML)* algorithms over the new transformed and reduced data. Reducing the number of features is intended to improve the performance of *ML* algorithms while increasing the explanatory power and understandability of the results.

In the current work, it is presented an automatic method to obtain a visualization of *HDD*, derived from a satisfactory *ML* model previously validated by the user. Following the same idea, it is introduced an automatic method to obtain a classification drawn from an adequate visualization that the user previously selects. Notice that both methods lie within the field of Explainable Machine Learning [2]. On the one hand, since a visualization is gathered from a trained *ML* model, it offers a post-hoc understandability of the model. On the other hand, the information from an adequate representation is used to train a *ML* model so that it can be more easily explained. To accomplish the task here exposed, a *Support Vector Machine (SVM)* with a *Radial Basis Function* (*RBF*) Kernel is used as the classification method. For visualization, it is applied the *T-Distributed Stochastic Neighbor Embedding* (*t-SNE*) algorithm [8]. The Euclidean metric resulting from the reduced *t-SNE* coordinates is approximated by the metric that arises from a *RBF* Kernel with a fine selection of parameters. This is performed through the *Symmetric Procrustes Problem* (*SPP*) [7].

The rest of the paper is structured as follows. Section 2 introduces the foundations mentioned above of the presented approach. The proposed method is described in Sect. 3. Section 4 addresses the experiments carried out to evaluate the performance of the proposal. Finally, Sect. 5 concludes and proposes future research lines.

2 State of the Art

HDD is in the spotlight of the field of Data Science. The application of *ML* algorithms over *HDD* can derive into a series of challenges such us over fitting or data sparsity. This is known as the curse of dimensionality [3]. Moreover, training in a high-dimensional space increases rapidly the computational cost. There are various techniques to mitigate the issues arising from the curse of dimensionality. Among them, bringing in more data, feature selection or dimensionality reduction techniques. These last techniques can be classified into linear and non-linear according to the transformation function. A global perspective of dimensionality reduction techniques and a comparative study of various of them can be found in [3]. *Classic Multidimensional Scaling* (*cMDS*) is one of the most popular linear dimensionality reduction techniques [11]. *cMDS* offers a visualization of *HDD* based on the information about pairwise distances of the data set. Non-linear dimensionality reduction techniques are capable of modeling more complex structures. *t-SNE* algorithm [8] and *Uniform Manifold Approximation and Projection* (*UMAP*) algorithm [9] are two commonly used non-linear techniques. Both are presented as alternative methods to classical techniques, demonstrating adequate visual representations. They are non-parametric methods with a high stochastic component. This is, an explicit transformation function from high-dimensional to low-dimensional space is not built. Hence, the provided transformations are not reproducible either for new data or another execution with the same data. In this work, *t-SNE* is used for visualization of *HDD*. It models high-dimensional data by projecting them in two dimensions so that data close in the high dimension remain close in the low dimension, while distant data are represented, with a high probability, by distant points.

Artificial Intelligence (*AI*) and *ML* have exponentially grown during the last years, developing very high accurate models. However, the improvement in the prediction performance often comes with a high model complexity and a lack of explainability and interpretability. Explainable Machine Learning is a branch of *ML* that aims to create techniques and models that can be understood by humans, maintaining a high level of learning performance [2]. This eases the understanding of the results offered by the *AI* technique so that humans can make informed and trusted decisions based on the achieved outputs.

2.1 Preliminary Concepts

SVM is a well-known technique of supervised *ML* that has proven very successful results when applied to real-world data [10]. In the context of classification, *SVM* finds the hyperplane that best distinctly classifies the data points. In general, it is necessary a previous step where the data is mapped into a higher-dimensional space where the classes are linearly separable with high probability by using the well-known kernel trick [12]. The *RBF* Kernel is one of the most popular kernel function used in non-linear *SVM* classification. Let $\mathbf{x}_i$ and $\mathbf{x}_j$ be two data points of the data set X, then the *RBF* kernel is defined as

$$K(\mathbf{x}_i, \mathbf{x}_j) = e^{-\gamma \cdot ||\mathbf{x}_i - \mathbf{x}_j||^2} \tag{1}$$

where $\gamma > 0$. The adequate selection of parameter γ is fundamental in the performance of the *SVM* model [6]. Among all the parameter selection techniques, in this work, a grid search method will be performed to compare with the proposed method.

As mentioned in Sect. 1, the approximation of two matrices is carried on via the *SPP*. Given a matrix $A \in \mathcal{M}_{nxp}$ and a reference matrix B of the same size, the *SPP* finds the symmetric transformation matrix $Q \in \mathcal{M}_{nxn}$ so that $Q \cdot A$ is the closest to B following the Frobenius norm.

3 Proposed Method

In this Section, the automatic method to obtain a visualization of *HDD* from a previously *ML* trained model, and a *ML* model from a previously selected visualization, is presented. The approach underlying both ways is based on two metrics estimated over the *HDD*. The first one is the Euclidean metric that the *t-SNE* algorithm induces over the *HDD*. Let $X \in \mathbb{R}^P$ be the original data in a high $P - dimensional$ space. X is reduced to a two-dimensional space embedding through the *t-SNE* algorithm. The Euclidean distance matrix is then calculated over the new reduced coordinates. This metric depends on the *perplexity* parameter selected in the *t-SNE*. The resulting matrix is denoted as $D_{t-SNE}(X, perplexity)$. The second metric is the dissimilarity that a *RBF* kernel induces over the *HDD*. Let $K_{RBF}(X, \gamma)$ be the *RBF* kernel over the X data. It is possible to calculate the *RBF* dissimilarity matrix, $D_{RBF}(X, \gamma)$, by computing its complementary. That is, $D_{RBF}(X, \gamma) = 1 - K_{RBF}(X, \gamma)$. Both metrics $D_{t-SNE}(X, perplexity)$ and $D_{RBF}(X, \gamma)$ are approximated by means of the *SPP* selecting the adequate parameters γ and *perplexity*.

3.1 From Classification to Visualization

In this way, a visualization is automatically offered using a satisfactory classification previously trained. For classification, a *SVM* with a *RBF* Kernel is applied. The best γ parameter for the classification ($\gamma_{selected}$) is chosen by the user through a grid search, maximizing the accuracy in the train data set and in the test data set simultaneously. Thus, the *RBF* metric $D_{RBF}(X, \gamma_{selected})$ is obtained. Next, for the previously chosen $\gamma_{selected}$, the optimum *perplexity* parameter ($perplexity_{optimum}$) in the *t-SNE* algorithm is obtained. This is achieved by bringing $D_{t-SNE}(X, perplexity)$ as close as possible to $D_{RBF}(X, \gamma_{selected})$ following the *SPP*. That is, $perplexity_{optimum}$ is the value that minimizes:

$$||D_{RBF}(X, \gamma_{selected}) - Q \cdot D_{t-SNE}(X, perplexity_{optimum})||_2^2 \quad (2)$$

where the distance matrices are previously standardized so that all their elements lie between 0 and 1, and Q is the resulting Procrustes symmetric transformation matrix. Then, a visualization of the data set is offered by applying the *t-SNE* algorithm over X with $perplexity_{optimum}$ to obtain a reduced two-dimensional embedding. Finally, these two new reduced coordinates are plotted into the plane, achieving a visualization of the *HDD* in a two-dimensional space that preserves as much information as possible.

For classification and visualization tasks, it is of great interest to add a new data point x_{new} to the existing plot obtained applying *t-SNE* to X. This task is not straightforward since the *t-SNE* algorithm is not reproducible for new data or even for a new execution. To fulfill the visualization, it is necessary a previous step where the *cMDS* algorithm is performed over $D_{t-SNE}(X, perplexity_{optimum})$ in order to estimate reproducible coordinates

Table 1. Data sets characteristics

Data set	# Variables	# Instances	# Classes
Wine	13	178	3
Breast cancer wisconsin	30	569	2
Waveform	20	5000	3
Simulated normal	300	2000	2
Musk version 1	166	476	2
Winnipeg	171	2000	7

of the *t-SNE* embedding. When new data points come, the Euclidean metric induced by the *t-SNE* algorithm is approximated by $D_{RBF}(X, \gamma_{selected})$ and the *t-SNE* coordinates are estimated by means of the trained *cMDS*.

3.2 From Visualization to Classification

Here, a classification is automatically drawn from an satisfactory visualization. A visualization of the HDD is performed through *t-SNE* algorithm, with an adequate *perplexity* parameter selected by the user ($perplexity_{selected}$). The resulting visualization induces the $D_{t-SNE}(X, perplexity_{selected})$ metric. As in the opposite way, a *SVM* with a *RBF* kernel function is employed for the classification. In this case, $\gamma_{optimum}$ is selected minimizing the Symmetric Procrustes error:

$$||D_{t-SNE}(X, perplexity_{selected}) - P \cdot D_{RBF}(X, \gamma_{optimum})||_2^2 \quad (3)$$

where the distance matrices are previously standardized so that all their elements lie between 0 and 1, and P is the resulting Procrustes symmetric transformation matrix. Using $\gamma_{optimum}$ as the parameter value for γ in a *SVM* with a *RBF* kernel, a classification is provided.

4 Experiments

The performance of the two proposed methods is illustrated in this section through six data sets: two artificial and six real data sets. Concerning the real data sets, the classical Wine, Breast Cancer Wisconsin, Musk Version 1 and Winnipeg data sets, gathered from the UCI Repository [5], are considered. A Simulated Normal data set and the well-known Waveform data set [4] are contemplated as artificial data sets. The Simulated Normal data set is assembled from two 300-dimensional Normal distribution of 1000 points and zero mean each. The pairwise co-variances of variables in both distributions are 0, while the variances of variables are 100 in one distribution and 1000 in the other one. Table 1 contains the main features of the data sets. Note that the Winnipeg data set originally has 325834 rows, but a subsample of 2000 rows is here used for computational cost reasons.

Table 2. $\gamma_{selected}$ for each data set in a *SVM* classification with *RBF* kernel and their corresponding $perplexity_{optimum}$ values (range in square brackets).

Data set	$\gamma_{selected}$	$\mathbf{perplexity_{optimum}}$
Wine	0.002	150 [140,160]
Breast cancer wisconsin	0.03	425 [410,440]
Waveform	0.003	3800 [3700,3900]
Simulated normal	0.0001	1775 [1700,1850]
Musk version 1	0.006	350 [320,380]
Winnipeg	0.002	1800 [1750,1850]

4.1 From Classification to Visualization Experiments

Following the method detailed in Sect. 3.1, a grid search method to select the *RBF* Kernel in a *SVM* model is performed over the six data sets. Each data set has been partitioned into a train data set with 95% of data and a test data set with 5% of the data. Applying the proposed method, the $\gamma_{selected}$ parameter values lead to a $perplexity_{optimum}$ values for visualization. These are the ones that best approximate the metric that the corresponding *RBF* kernel induces by the one that is derived from the *t-SNE* algorithm. Concerning the stability of the method, 20 replicates of the experiment for each data set are performed. For each data set, the chosen $perplexity_{optimum}$ is the average of the 20 obtained values. Table 2 displays the obtained $\gamma_{selected}$ and $perplexity_{optimum}$ for each data set. The $perplexity_{optimum}$ column comprises both the chosen value and the range of values obtained throughout the 20 replicates. Notice that the range of the obtained $perplexity_{optimum}$ values within each data set are quite narrow leading to similar visualizations. Hence, the stability of the method is proven.

The estimated $perplexity_{optimum}$ values prompt to a visualization of the data set. The automatic representation of the data has been obtained for the corresponding training data sets. Train data points in Fig. 1 contains the achieved graphs for the six data sets. Those are the ones represented by circles. All of them turn out to be a good representation of data. The automatic feature of the method is very useful because it eases significantly the visualization phase to the user, automating the selection of parameters in this task. Moreover, the representation enhance the understandability of the previously trained model, aiding in the decision-making through the results.

Given the trained *ML* models, it is possible to get predictions for new data. In addition, to increase the explainability of the predictions obtained, it is informative to plot new data into the previously reached representations. As explained before, a previous step applying *cMDS* is needed to get the representation of the new points. Figure 1 shows the visualization of test data in the six data sets. The test data points are represented by stars. The trained *ML* model returns a probability of belonging to the corresponding true class of each test data point. This probability is symbolized in the graph by a color gradient: the darker the

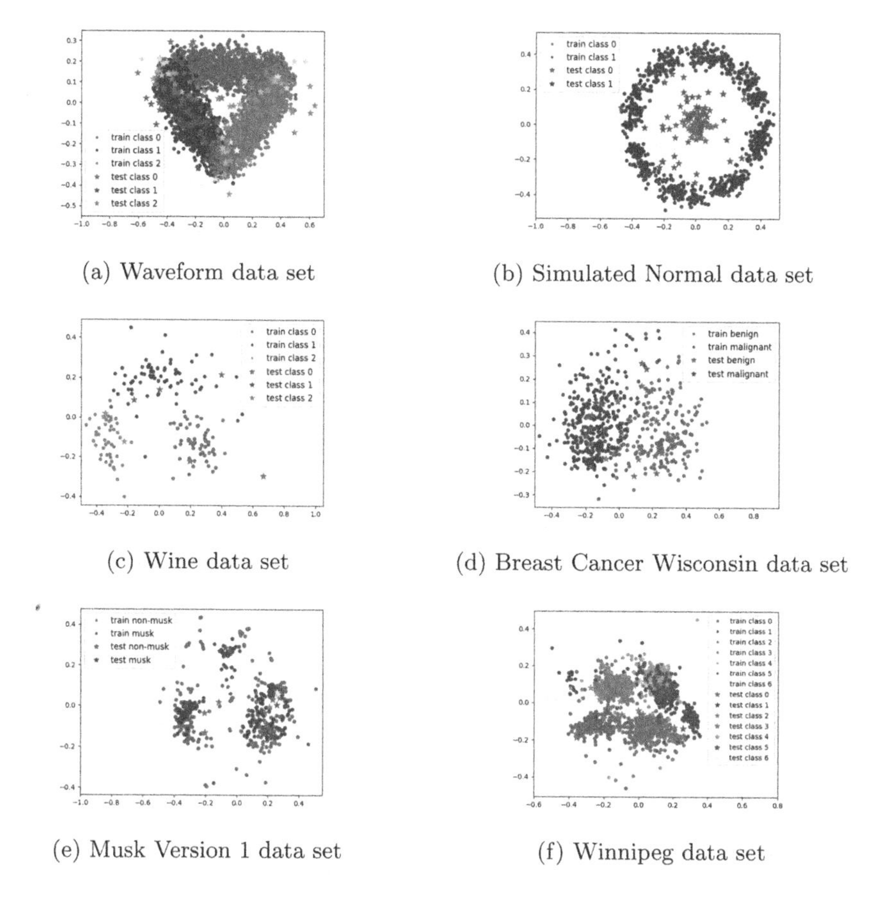

(a) Waveform data set

(b) Simulated Normal data set

(c) Wine data set

(d) Breast Cancer Wisconsin data set

(e) Musk Version 1 data set

(f) Winnipeg data set

Fig. 1. Data visualization from classification models. A color gradient is used to represent the probability of belonging to the corresponding true class of each test data point: from white (probability 0) to the original color (probability 1). (Color figure online)

color is, the higher the corresponding probability is. The resulting graphs are very satisfactory for the real data sets examples and the Waveform data set. In the Simulated Normal data set, all of the class 0 data test points are correctly plotted, while only some of the class 1 data test points are a little bit misplaced.

4.2 From Visualization to Classification Experiments

Next, experiments showing the performance of the proposed method to automatically obtain a classification derived from a satisfactory visualization are presented. The first step is to select an adequate $perplexity_{selected}$ parameter for the correct visualization of each data set through the *t-SNE* algorithm. The chosen $perplexity_{selected}$ values for the Wine, Breast Cancer Wisconsin, Waveform, Simulated Normal, Musk Version 1 and Winnipeg data sets are 110, 220, 1500,

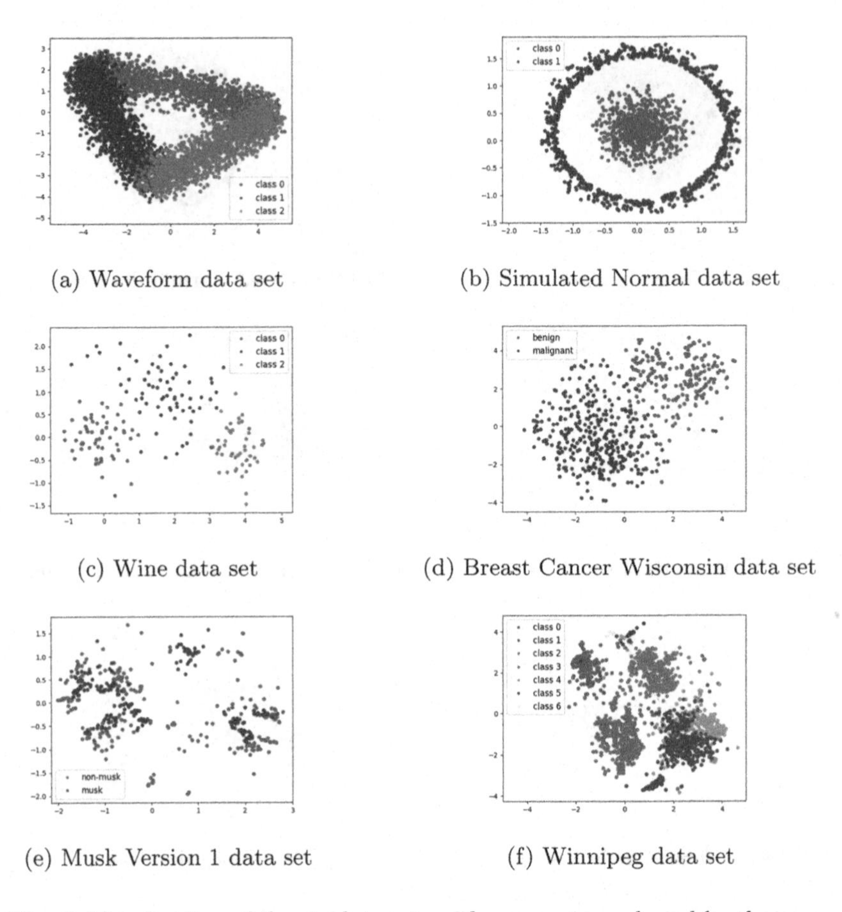

(a) Waveform data set
(b) Simulated Normal data set
(c) Wine data set
(d) Breast Cancer Wisconsin data set
(e) Musk Version 1 data set
(f) Winnipeg data set

Fig. 2. Visualization of the six data sets with parameters selected by the user.

500, 300, and 1000 respectively. The representations drawn with those values are shown in Fig. 2. Following the method exposed in Sect. 3.2, the $\gamma_{optimum}$ parameter value for the training of a *RBF SVM* model can be estimated. For each data set, 20 replicates of the following experiment have been performed. The data set is randomly partitioned into train and test data sets. The train data set is used to calculate $\gamma_{optimum}$ from the value of $perplexity_{selected}$ chosen by the user. Then, a *SVM* with a *RBF* kernel with parameter $\gamma_{optimum}$ is trained using the train data set. This leads to an accuracy of the model over the training data. The test data set is predicted with the trained *SVM* model, offering an accuracy value in the test data set. Table 3 contains the range of $\gamma_{optimum}$ values, and the range of achieved accuracy values in train and test, along the 20 replicates of the experiment for each data set. In all cases, the performance of the *SVM* is very high. For comparison, Table 4 contains the accuracy values achieved with $\gamma_{selected}$ of Table 2, which are found by a grid search method maximizing both accuracy in train and test simultaneously. It is then proven that the results

Table 3. Accuracy achieved in the train and test data set through a *SVM* with *RBF* Kernel with parameter $\gamma_{optimum}$.

Data set	$\gamma_{\mathbf{optimum}}$	Accuracy in train	Accuracy in test
Wine	[0.02,0.03]	[0.98,1]	[0.93,1]
Breast Cancer Wisconsin	[0.02,0.03]	[0.98,0.99]	[0.96,0.99]
Waveform	0.02	[0.88,0.89]	[0.86,0.88]
Simulated Normal	[0.002,0.003]	1	1
Musk Version 1	0.002	[0.87,0.91]	[0.78,0.89]
Winnipeg	0.002	[0.98,0.99]	[0.96,0.98]

Table 4. *SVM* model accuracy with $\gamma_{selected}$ using a grid search method.

Data set	Train	Test
Wine	0.97	1
Breast Cancer Wisconsin	0.99	0.99
Waveform	0.88	0.87
Simulated Normal	1	1
Musk Version 1	0.96	0.95
Winnipeg	0.98	0.99

obtained through the method here proposed are very close to those shown in Table 4. Notice that the worth of the proposed method is that the *ML* model results have been automatically obtained from an understandable representation of data in the two-dimensional space. In this sense, the proposed method is more explainable and comprehensible to the user, avoiding an uncontrolled search based exclusively on computational power.

5 Conclusions

The present paper presents an automatic and understandable method to obtain a visualization of *HDD*, derived from a satisfactory *ML* classification that the user previously selects. Furthermore, it is also introduced the opposite way of an automatic and understandable method to obtain a *ML* model, drawn from an adequate visualization that the user previously selects. The underlying idea of both ways consists in the approximation of two metrics through the *SPP*. The first metric is the dissimilarity that is induced from a *RBF* Kernel. The other one is the Euclidean distance calculated over the reduced *t-SNE* coordinates of a *HDD*. The method is understandable in the "from classification to visualization way" since the achieved visualization eases the understanding of the trained Machine Learning model. It enables to see graphically the numeric outputs of the model. Then, it is possible to detect problematic points, for example.

The other way "from visualization to classification" is also an understandable method because it facilitates the training of the model with a metric that the user considers as adequate. The results obtained in the experiments validate the performance of the proposed method in both two ways. All the visualizations reached from the classification are very satisfactory, as well as the inclusion of new data points in them. In addition, the classifications achieved from visualization offer high accuracy values, which are close to the ones obtained through a grid search method. The fact that both the visualization and the classification are obtained automatically and providing a high explanatory power is the worth of the proposed method here presented. In this way, the proposal supports data scientists and users in general in these two fundamental stages of a Data Science project.

In the future, other dimensionality reduction techniques such as *UMAP* will be considered in the automatic method to obtain a visualization from a classification. Additionally, other classification algorithms will be included in the automatic achievement of a classification derived from a visualization. Furthermore, alternative methods to approximate different dissimilarity matrices will be studied. Finally, the bias-variance trade-off will be studied from a theoretical point of view.

Acknowledgements. This research has been supported by grants from Rey Juan Carlos University (Ref: C1PREDOC2020), Madrid Autonomous Community (Ref: IND2019/TIC-17194) and the Spanish Ministry of Economy and Competitiveness, under the Retos-Investigación program: MODAS-IN (Ref: RTI-2018-094269-B-I00).

References

1. Amaratunga, D., Cabrera, J.: High-dimensional data. Journal of the National Science Foundation of Sri Lanka 44(1) (2016)
2. Arrieta, A.B., et al.: Explainable artificial intelligence (xai): Concepts, taxonomies, opportunities and challenges toward responsible ai. Information Fusion **58**, 82–115 (2020)
3. Ayesha, S., Hanif, M.K., Talib, R.: Overview and comparative study of dimensionality reduction techniques for high dimensional data. Information Fusion **59**, 44–58 (2020)
4. Breiman, L., Friedman, J.H., Olshen, R.A., Stone, C.J.: Classification and regression trees. Routledge (2017)
5. Dua, D., Graff, C.: UCI machine learning repository (2017), http://archive.ics.uci.edu/ml
6. Han, S., Qubo, C., Meng, H.: Parameter selection in svm with rbf kernel function. In: World Automation Congress 2012. pp. 1–4. IEEE (2012)
7. Higham, N.J.: The symmetric procrustes problem. BIT Numer. Math. **28**(1), 133–143 (1988)
8. Van der Maaten, L., Hinton, G.: Visualizing data using t-sne. J. Mach. Learn. Res. **9**(11), 2579–2605 (2008)
9. McInnes, L., Healy, J., Melville, J.: Umap: Uniform manifold approximation and projection for dimension reduction. arXiv preprint arXiv:1802.03426 (2018)

10. Moguerza, J.M., Muñoz, A.: Support vector machines with applications. Stat. Sci. **21**(3), 322–336 (2006)
11. Torgerson, W.S.: Multidimensional scaling: i. theory and method. Psychometrika **17**(4), 401–419 (1952) https://doi.org/10.1007/BF02288916
12. Vert, J.P., Tsuda, K., Schölkopf, B.: A primer on kernel methods. Kernel Methods Comput. Biol. **47**, 35–70 (2004)

Neural Complexity Assessment: A Deep Learning Approach to Readability Classification for European Portuguese Corpora

João Correia and Rui Mendes(✉)

Centro Algoritmi, Departamento de Informática,
Universidade do Minho, Braga, Portugal
azuki@di.uminho.pt

Abstract. The following paper describes a Deep Learning model capable of classifying the inherent readability complexity of a piece of text in European Portuguese. The model was developed using modern Natural Language Processing techniques, featuring a highly-fine-tuned Neural Network which takes as input both the text as well as multiple metrics relating to it. This classifier was trained on a dataset featuring texts divided in 5 CEFR categories, obtaining an accuracy of 73%, a top 2 accuracy of 90% and an adjacent accuracy of 94%.

Keywords: Deep learning · Natural language processing · Textual complexity

1 Introduction

A text's readability defines the amount of language mastery needed by a reader to correctly interpret it. There are many contexts where it is useful or even necessary to match readers with texts adequate to their proficiency level. A task which perfectly fits the above description is language teaching. For this task, choosing texts that fit the current teaching level intended for the students leads to a better and more fit learning experience [21].

This paper puts forward a new system whose purpose is to aid choosing the right teaching material by classifying a given text in its corresponding CEFR [10] level: A1, A2, B1, B2 or C1. These are the 5 proficiency levels that the Camões Institute, the official Portuguese language institute, offers exams and certifications for, as specified in the *Framework for Teaching Portuguese Abroad* [17]. The used dataset, which was kindly provided by Doctor Rui Vaz of the Camões Institute, features texts used in real exams labeled by the CEFR level of the exam they were present in. Since these exams are not a day-to-day occurrence,

This research was supported by FCT—Fundação para a Ciência e Tecnologia within the R&D Units Project Scope: UIDB/00319/2020.

H. Yin et al. (Eds.): IDEAL 2021, LNCS 13113, pp. 300–311, 2021.
https://doi.org/10.1007/978-3-030-91608-4_30

the dataset is rather small, as well as heavily class imbalanced, which, in itself, presents a substantial challenge when designing the classifying system.

The resulting model is a Neural Network which employs a myriad of features in order to best adapt to the aforementioned challenging dataset. The model takes in 2 inputs: the text, tokenized, and a list of computed metrics regarding it.

This paper is structured along the following sections: initially, related work is discussed in Sect. 2, then, the dataset is presented and analysed in Sect. 3. Section 4 details the developed model, explaining its architecture and implementation. Later, in Sect. 5, the obtained results are presented and discussed and, in Sect. 6, other areas where the model could be applied are discussed. Section 7 concludes this paper and proposes future work.

2 Related Work

Early work in the area of classifying texts by their readability mainly focused in the creation of indices based on features of the text. The Flesch Reading Ease formula [13], published in 1943 and directed towards the English language, was both one of the earliest as well as one of the more impactful formulas from this era. This formula is based on some basic text metrics, namely the total number of sentences, words and syllables of the text. In 1975, whilst using the same features, this formula was updated to present a score corresponding to a U.S. grade level, giving birth to the Flesch–Kincaid Grade Level [23]. Other relevant English formulas are the Gunning Fog Index [18] and the Coleman–Liau Index [6]. Some other formulas, based around the rise of cognitivism in the 70's, would rather focus on the syntactic aspect of a text [14], such as the Golub Syntactic Density Score [16], published in 1974, which takes into account clause dependency for classifying readability. Striving away from exclusivity to the English language, two indices were designed with Western European languages in mind: first, in 1968, the Lesbarhets Index (LIX) [3] and, in 1983, the Rate Index (RIX) [2].

Recent attempts at classifying readability came in the form of machine learning systems. In 2005, Schwarm and Ostendorf [29] used support vector machines to develop a readability classifying system based on textual features. In 2010, Feng et al. [11] used classification models and better features to improve on the state of art. Regarding specifically classifying readability for texts intended to be used as material for second language learning, in 2010 Kotani et al. present a regression model based on linguistic features [24] of teaching material and, in 2012, François and Miltsakaki [14] compare multiple classic readability formulas to newer machine learning models in the classification of French texts for second language learning. The advent of the Bidirectional Encoder Representation from Transformers (BERT) model [8], published in 2018, set a new state-of-the-art performance for multiple Natural Language Processing (NLP) tasks. For readability assessment, multiple articles have used this model successfully for multiple languages, such as Mandarin [32], English and Philipino [22].

For the European Portuguese, two publications must be highlighted: the first, LX-CEFR [4], published by Branco et al. in 2014, is a system developed for the

Camões Institute, which is capable of classifying texts in their corresponding CEFR level. This is a linear regression model which can use one of four different metrics, depending on user choice: the Flesch Reading Ease index, the text's noun density, the average sentence length or the average word length, in syllables. When trained on 125 texts, it obtained a max accuracy of 22%, corresponding to the use of the Flesch Reading Ease index as the chosen metric. The second paper, published by Curto et al. in 2015 [7] also tries to classify European Portuguese texts by their CEFR level, using an updated version of the Camões Institute dataset, now composed of 237 texts. Through the extraction of 52 features, the article compares multiple classic machine learning algorithm, achieving a max accuracy of 75% on 5 levels classification by using the LogitBoost boosting algorithm [15]. The present paper seeks to follow the footsteps of these two publications by using a newer and more balanced version of the Camões Institute dataset, now composed of 500 texts. This dataset shall be fully described in the following section. In what regards the model, it departs from the previous methodology, as it will use Deep Learning techniques, instead of classical Machine Learning algorithms.

3 Dataset

The dataset used for training and testing the model is a private corpus kindly provided by the Camões Institute. It is an European Portuguese corpus containing texts labeled by their CEFR level, being composed by 500 textual entries used in real exams created by the Camões Institute. Table 1 describes the dataset by its class distribution.

Table 1. Dataset distribuition by CEFR level

	A1	A2	B1	B2	C1
Number of texts	80	135	184	45	56
% of dataset	16%	27%	36.8%	9%	11.2%

Two evident problems quickly arise from observing the dataset distribution: The first one regards the dataset size. Neural networks usually require great amounts of data in order to generalize [9], with datasets commonly having thousands or even millions of data points. A dataset with 500 data points sits beneath what would be desirable for such a data hungry algorithm. Therefore, it will be necessary to implement mechanisms to offset the lack of data and achieve generalization. The second problem is class imbalance. As it can be seen, texts in the class A2 or B1 compose 63,8% of the dataset. Both training as well as evaluating the model will have to take into account this imbalance and apply adequate measures to reduce its impact.

Besides the two aforementioned challenges, a third problem is also present in the dataset, one that is inherent to all types of CEFR classification: class ambiguity. The CEFR guidelines [10] define the CEFR levels only in a qualitative sense, giving no quantitative information on what constitutes a text belonging to each level. The guidelines leave a lot up to interpretation, not providing hard boundaries into what vocabulary should be understood neither providing syntactical information about the texts that should be understood, such as average number of sentences, average sentence length or average syllable count per word. The *Framework for Teaching Portuguese Abroad* also does not clarify, in quantitative terms what is expected of each level.

As such, the selection of reading materials for teaching European Portuguese as a second language is done manually by a group of linguistic experts, making the selection process both cumbersome and prone to human bias. This is brilliantly showcased by Branco et al. in the LX-CEFR paper [4]. The team untagged the texts from their labels and recruited five European Portuguese language instructors, who were trained experts at this task, to classify all the 125 texts according to their CEFR level. From the 125 texts, only one text (0.90%) received an unanimous classification. When talking about consensus from at least 4 experts, the percentage was 17%, that is, only 21 texts. These facts must be taken into consideration when later evaluating the performance of the developed model.

Despite its limitations, this is, as far as our knowledge goes, the most complete and updated European Portuguese readability corpus, having been assembled by experts in the area of linguistics and teaching. This, in conjunction with the fact that earlier versions of this dataset have been used in both the LX-CEFR [4] and Automatic Classifier [7] articles, has led to the conclusion that it is the most fitting dataset for the task at hand.

4 Model Architecture

The proposed model was developed using the TensorFlow python library [1], mainly the Keras module [5] included in it.

The neural network takes two inputs: 1. The texts in their tokenized version (Textual branch); 2. 14 Metrics about each of the texts (Metrics branch).

Figure 1, obtained using keras' plot_model function, represents the model's architecture. The None keyword in the shape tuples indicates a variable size of input batches, which was necessary to allow both training and testing on the dataset, but can be assumed as the batch size for training, which is 32. Each of the inputs will be processed in their branch of the neural network, only being connected at the tail of the network. Therefore, each branch will be detailed separately in the following subsections. The final architecture was a result of hyperparameter tuning through Grid Search, where the dimension of the embeddings, the number of layers, the dropout % and the number of both LSTM cells and neurons were all tested.

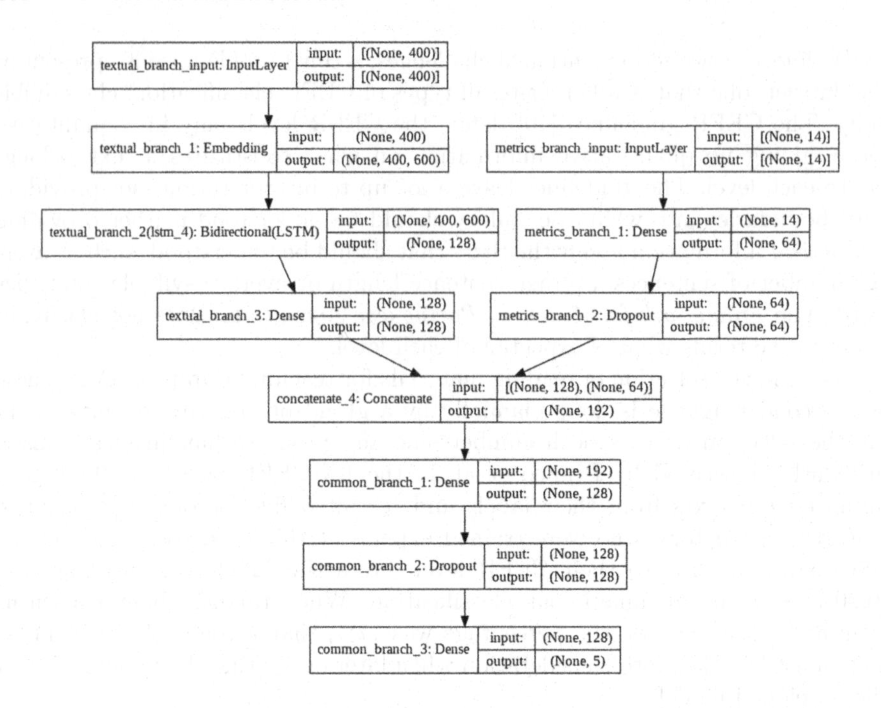

Fig. 1. Model architecture

4.1 Textual Branch

Data Preprocessing. This branch takes as input the tokenized version of each text from the dataset. As such, the texts are previously all tokenized using Keras' TextVectorization functionality.

In the tokenization process, each word in a text will be converted to token which is represented by an integer. Each tokenized text will contain a maximum length of 400 tokens. If the text does not have enough words for it to be converted to a 400 token long list, the token '0' is used as padding. If the text is longer, the remaining words will be disregarded.

Architecture. The first layer of the textual branch is an Embedding layer. This layer is responsible for creating **word embeddings**. Word embeddings are vectors that encode the meaning and context of a word in an n-dimensional space. Words that are close in meaning should be close to each other in embedding space [31]. This dataset is too small for a reliable customized embedding layer, since most words are only captured in one or two contexts. Thus, pre-trained embeddings were used. These are embeddings that are trained on other much larger datasets, where each word is captured in multiple contexts and the embedding algorithm is able to fully capture a word's meaning. The used pre-embeddings are provided by the *NILC - Interinstitutional Center for Computational Linguistics*, being trained in 17 corpus, both in European Portuguese as well as

Brazilian Portuguese, resulting in 1,395,926,282 tokens [19]. The chosen embedding algorithm was Skip-gram [27] and the chosen vector dimension was 600, which means each word is represented by an array of 600 values.

The second layer is composed of 64 bidirectional LSTM cells [20,28]. A unidirectional LSTM preserves only information from past inputs. Meanwhile, in a bidirectional LSTM the input is read both from the beginning to the end as well as from the end to the beginning. This allows the network to get context from both sides of the input, improving its understanding of the current input. A dropout of 10% is also applied.

The final layer of the textual branch is a densely connected layer composed of 128 neurons. The activation function is Rectified Linear Unit (ReLU).

4.2 Metrics Branch

Data Preprocessing. The inputs for this branch are a group of metrics calculated for each text. These metrics will be composed of both simple text features, such as the number of words, as well as a group of existing reading formulas.

The group of metrics will be composed of the following 14 different values: 1. Total number of words; 2. Total number of unique words; 3. Total number of sentences; 4. Total number of syllables; 5. Total number of characters; 6. Total number of long words (>6 characters); 7. (EN) Flesch Reading Ease [13]; 8. (EN) Simple Measure of Gobbledygook (SMOG) [26]; 9. (EN) Automated Readability Index (ARI) [30]; 10. (EN) Flesch–Kincaid Grade Level [23]; 11. (EN) Coleman–Liau index [6]; 12. (EN) Gunning Fog Index [18]; 13. (EUW) Rate Index (RIX) [2]; 14. (EUW) Lesbarhets Index (LIX) [3].

Metrics marked with EN are meant for the English language while ones marked with EUW are meant for Western European Latin languages. Even though most of the metrics are meant for English, the LX-CEFR paper showed that some of them, specifically the Flesch Reading Ease metric, outperform simple metrics such as the number of words. The large number of metrics is meant to help the network diversify, making it applicable to more areas other than readability classification. Metrics such as SMOG have shown to perform particularly well in classifying medical texts by their readability [12].

Architecture. The first layer that composes the metrics branch is a dense layer. This layer is composed by 64 densely connected neurons and the ReLU activation function. A dropout of 20% is applied afterwards, meaning that 20% of the inputs will be randomly set as 0. This layer is meant to prevent overfitting, something a neural network is very prone to doing when training on such a small dataset. The dropout value is higher on this branch when compared to the textual branch to prevent the model from overfitting to the metrics and ignoring the text input.

4.3 Common Branch

The final part of the network consists of a concatenation layer meant to concatenate the output of the textual branch with the outputs from the metrics branch. This layer is followed by a densely connected layer composed of 128 neurons and ReLU as the activation function. After that, another 20% dropout layer is applied. The final layer, responsible for the classification, is composed of 5 neurons, one for each class, and uses the Softmax activation function, as the task at hand is a multiclass classification one.

4.4 Model Compilation

As referred in the last section, this model was designed with multiclass classification in mind. Therefore, the used loss function was Keras' CategoricalCrossentropy. The AdamW algorithm [25] was used as optimizer. This is a custom version of the famous Adam optimizer which, on top of the normal Adam update step, applies weight decay regularization. As stated in its paper, this is different from normal L2 regularization, as it better regularizes larger gradients. This optimizer was made available by the Tensorflow-addons library. The learning rate hyperparameter was set to 10^{-3} and the weight decay to 10^{-8}. Furthermore, a learning rate scheduler was used, which, after 40 epochs of training, starts decaying the learning rate by e^{-1} in each subsequent epoch, where e represents Euler's number. This helps the model stabilize in the last few epochs of training. The model was trained for 50 epochs.

5 Evaluation

In this section, the performance of the developed model will be compared against several algorithms. This comparison will be based on the dataset presented in Sect. 3. Besides the proposed model, the other algorithms are:

– A linear regression based on the Flesch Reading Ease index, as it was the best performing method on the LX-CEFR paper [4]; – The LogitBoost boosting algorithm [15], as it was the best performing model in the paper by Curto et al. [7]; – The metrics branch, composed of a layer of 64 neurons and a dropout of 20%; – The textual branch, composed of the pre-trained embeddings, a recurrent layer of 128 bidirectional LSTM cells and a dense layer of 128 neurons.

Both the Logitboost algorithm as well as the metrics branch use as input the same group of 14 metrics described in Sect. 4. The BERT algorithm was also tested, however, the dataset proved too small for the model to be able to converge without overfitting immensely.

The chosen evaluation procedure was Stratified K-Fold, as provided by the python Scikit-learn library. Stratified K-Fold is a variation of the classical K-fold method, which preserves class distribution, contributing to a more accurate performance assessment. The number of splits was $K = 5$, as it provided a safe balance between a higher K, which would contribute to a smaller testing sample and, therefore, less reliable results, and a smaller K, which would heavily reduce the number of training samples, hampering generalization. A $K = 5$ will produce training sets of 400 texts and testing sets of 100. The same random seed was used in the evaluation of all algorithms, to ensure validity.

Table 2 displays three metrics for each algorithm: its accuracy, its top 2 accuracy and its adjacent accuracy. Top 2 accuracy measures if the target label was in the top 2 predictions of the algorithm. Adjacent accuracy, considers a prediction correct if it matches the correct label or one of its neighbours. E.g.: while predicting the CEFR level of a B2 text, a prediction would be considered correct if it was either B1, B2 or C1. A1 and C1 are both edge cases were only the upper or lower neighbour are considered accordingly. This metric was used in the article by Curto et al. [7] as a less strict version of the normal accuracy. The choice of also using top 2 accuracy comes down to using a metric which is not as permissive as the adjacent accuracy but does not take into account the neighbouring classes. Since linear regression only outputs one value, top 2 accuracy cannot be measured, nor can adjacent accuracy reliably.

Table 2. Algorithm evaluation on the dataset using a stratified K-fold cross validation with 5 folds. The measures displayed in the columns are accuracy, top 2 accuracy and adjacent accuracy.

Algorithm	Acc.	Top 2 acc.	Adj. acc.
Linear regression	43%	–	–
LogitBoost	67%	90%	91%
Metrics branch	60%	90%	91%
Textual branch	57%	70%	78%
Proposed model	**73%**	**90%**	**94%**

As can be observed, the proposed model achieves the best single target accuracy as well the best adjacent accuracy, while it ties with LogitBoost and the metrics branch in top 2 accuracy. The models who rely solely on textual metrics had a good performance, mainly since readability is heavily linked with the morphological characteristics of a text. The textual branch does not obtain an impressive score by itself. One possible explanation may have to deal with the dataset's reduced size, which can hamper the ability of the textual branch to capture contexts and map words to their inherent readability, even with the use of pre-trained embeddings, as well as prevent it from deducting some textual metrics which could prove useful.

The better performance achieved by the proposed model against both branches when performing separately highlights the importance of combining textual metrics with an analysis of the actual content of the teaching material, since word embeddings and recurrent layers can capture meaning and context, something very impactful for readability.

Dataset augmentation did not have a positive impact on the performance of any algorithm. Since most augmentation techniques for text rely on changing some words for other words close in meaning, this can actually deteriorate the quality of the dataset, since nothing guarantees these new words represent the same reading difficulty as the previous ones.

In order to better assess the proposed model's performance and behaviour, its performance on a specific fold was dissected. Table 3 shows the confusion matrix of the model's performance on said fold, where an accuracy of 73% was achieved. Each row corresponds to the actual class while each column to the prediction.

Table 3. Confusion matrix of a fold

	A1	A2	B1	B2	C1
A1	14	2	0	0	0
A2	8	16	3	0	0
B1	1	1	34	1	0
B2	0	0	2	5	2
C1	0	0	4	3	4

As can be observed, the highest number of predictions always corresponded to the correct class, except for the C1 level, where the B1 class got the same number of predictions.

Performance was worse in classes with fewer samples, but adjacent accuracy was acceptable, being always above 60% for these classes, meaning the model can be trusted to provide a basis prediction which can aid with the selection of adequate reading material.

6 Other Areas of Application

The presented model may be applicable to more readability classification tasks other than second language teaching. On one hand, as stated on Sect. 4, the diverse group of metrics allows it to adapt to different fields of application, since different metrics are tuned for different contexts. On the other hand, the textual branch allows the model to better perform in contexts where the readability of the text is not directly correlated with visible syntactical features but is rather dependent on the meaning of the words that compose the text, their arrangement and context. A clear example of such a task would be evaluating the readability of poetry, since many poets chose different types of rhyme schemes and syllable metres and, yet, are capable of creating both simple and very complex works within such restrictions. The use of pre-trained embeddings also proves to be a helpful element, since they were not trained only on teaching material but rather a diverse group of corpora, capturing meaning in very diverse contexts.

The combination of both branches is what gives the model its edge over classical models who rely solely on syntactical metrics and gives it the ability to capture the fine nuances that sometimes dictate in which category a text falls under.

7 Concluding Remarks and Future Work

The model proposed in this article presents better results than previous systems by achieving an accuracy of 73%, a top 2 accuracy of 90% and adjacent accuracy of 94% in the task of classifying texts by the CEFR level of reading comprehension. Even though these are promising results, it is advised to experiment with larger and more balanced datasets before labelling this algorithm as state-of-the-art.

By taking as input both the source text as well as metrics extracted from it, the model is capable of better capturing nuances and contexts, being more flexible than previous models that only relied on metrics. Thus, the model is more adaptable to other tasks where syntactical metrics do not pose much importance. As stated in Sect. 5, training the BERT algorithm when performing K-Fold validation occurs in a great amount of overfitting, due to the small nature of the dataset. However, if a more complete version of the dataset is ever gathered, BERT may show promising results, as is has done in many other natural language processing tasks.

Acknowledgments. This research was supported by FCT—Fundação para a Ciência e Tecnologia within the R&D Units Project Scope: UIDB/00319/2020.

References

1. Abadi, M., et al.: TensorFlow: Large-scale machine learning on heterogeneous systems (2015). https://www.tensorflow.org/
2. Anderson, J.: Lix and rix: variations on a little-known readability index. J. Reading **26**(6), 490–496 (1983)
3. Björnsson, C.H.: Läsbarhet. stockholm: Liber (1968)
4. Branco, A., Rodrigues, J., Costa, F., Silva, J., Vaz, R.: Rolling out text categorization for language learning assessment supported by language technology. In: Baptista, J., Mamede, N., Candeias, S., Paraboni, I., Pardo, T.A.S., Volpe Nunes, M.G. (eds.) PROPOR 2014. LNCS (LNAI), vol. 8775, pp. 256–261. Springer, Cham (2014). https://doi.org/10.1007/978-3-319-09761-9_29
5. Chollet, F., et al.: Keras. https://github.com/fchollet/keras (2015)
6. Coleman, M., Liau, T.L.: A computer readability formula designed for machine scoring. J. Appl. Psychol. **60**(2), 283 (1975)
7. Curto, P., Mamede, N., Baptista, J.: Automatic text difficulty classifier. Assisting the selection of adequate reading materials for European Portuguese teaching. In: Proceedings of CSEDU, pp. 36–44 (2015)
8. Devlin, J., Chang, M.W., Lee, K., Toutanova, K.: Bert: Pre-training of deep bidirectional transformers for language understanding. arXiv preprint arXiv:1810.04805 (2018)

9. Edwards, C.: Growing pains for deep learning. Commun. ACM **58**(7), 14–16 (2015)
10. Council of Europe. Common European framework of reference for languages: Learning, teaching, assessment - companion volume. https://rm.coe.int/common-european-framework-of-reference-for-languages-learning-teaching/16809ea0d4
11. Feng, L., Jansche, M., Huenerfauth, M., Elhadad, N.: A comparison of features for automatic readability assessment (2010)
12. Fitzsimmons, P.R., Michael, B., Hulley, J.L., Scott, G.O.: A readability assessment of online parkinson's disease information. J. R. Coll. Phys. Edinb. **40**(4), 292–296 (2010)
13. Flesch, R.: Marks of a readable style. contributions to education# 897 (1943)
14. François, T., Miltsakaki, E.: Do NLP and machine learning improve traditional readability formulas? In: Proceedings of the First Workshop on Predicting and Improving Text Readability for target reader populations, pp. 49–57 (2012)
15. Friedman, J., Hastie, T., Tibshirani, R.: Additive logistic regression: a statistical view of boosting (with discussion and a rejoinder by the authors). Ann. Stat. **28**(2), 337–407 (2000)
16. Golub, L.S., Kidder, C.: Syntactic density and the computer. Elementary English **51**(8), 1128–1131 (1974)
17. Grosso, M.J., Soares, A., Sousa, F.D., Pascoal, J.: Quadro de referência para o ensino português no estrangeiro. Documento orientador. DGE MEC Portugal (2011)
18. Gunning, R., et al.: Technique of clear writing (1952)
19. Hartmann, N., Fonseca, E., Shulby, C., Treviso, M., Rodrigues, J., Aluisio, S.: Portuguese word embeddings: Evaluating on word analogies and natural language tasks. arXiv preprint arXiv:1708.06025 (2017)
20. Hochreiter, S., Schmidhuber, J.: Long short-term memory. Neural Comput. **9**(8), 1735–1780 (1997)
21. Howard, J., Major, J.: Guidelines for designing effective English language teaching materials. TESOLANZ J. **12**(10), 50–58 (2004)
22. Imperial, J.M.: Knowledge-rich bert embeddings for readability assessment. arXiv preprint arXiv:2106.07935 (2021)
23. Kincaid, J.P., et al.: Derivation of new readability formulas (automated readability index, fog count and flesch reading ease formula) for navy enlisted personnel. Technical Report, Naval Technical Training Command Millington TN Research Branch (1975)
24. Kotani, K., et al.: A machine learning approach to measurement of text readability for efl learners using various linguistic features. Online Submission (2011)
25. Loshchilov, I., Hutter, F.: Decoupled weight decay regularization. arXiv preprint arXiv:1711.05101 (2017)
26. Mc Laughlin, G.H.: Smog grading-a new readability formula. J. Read. **12**(8), 639–646 (1969)
27. Mikolov, T., Chen, K., Corrado, G., Dean, J.: Efficient estimation of word representations in vector space. arXiv preprint arXiv:1301.3781 (2013)
28. Schuster, M., Paliwal, K.K.: Bidirectional recurrent neural networks. IEEE Trans. Sig. Process. **45**(11), 2673–2681 (1997)

29. Schwarm, S.E., Ostendorf, M.: Reading level assessment using support vector machines and statistical language models. In: Proceedings of the 43rd Annual Meeting of the Association for Computational Linguistics (ACL 2005), pp. 523–530 (2005)
30. Smith, E.A., Senter, R.: Automated readability index. AMRL-TR. Aerospace Medical Research Laboratories (US), pp. 1–14 (1967)
31. Teller, V.: Speech and language processing: an introduction to natural language processing, computational linguistics, and speech recognition (2000)
32. Tseng, H.-C., Chen, H.-C., Chang, K.-E., Sung, Y.-T., Chen, B.: An innovative BERT-based readability model. In: Rønningsbakk, L., Wu, T.-T., Sandnes, F.E., Huang, Y.-M. (eds.) ICITL 2019. LNCS, vol. 11937, pp. 301–308. Springer, Cham (2019). https://doi.org/10.1007/978-3-030-35343-8_32

Countering Misinformation Through Semantic-Aware Multilingual Models

Álvaro Huertas-García[1,2(✉)], Javier Huertas-Tato[1], Alejandro Martín[1], and David Camacho[1]

[1] Department of Computer System Engineering, Universidad Politécnica de Madrid, Calle de Alan Turing, 28031 Madrid, Spain
alvaro.huertas.garcia@alumnos.upm.es,
{javier.huertas.tato,alejandro.martin,david.camacho}@upm.es

[2] Department of Computer Sciences, Universidad Rey Juan Carlos, Calle Tulipán, 28933 Madrid, Spain

Abstract. The presence of misinformation and harmful content on social networks is an emerging problem that endangers public health. One of the most successful approaches for detecting, assessing, and providing prompt responses to this misinformation problem is Natural Language Processing (NLP) techniques based on semantic similarity. However, language constitutes one of the most significant barriers to address, denoting the need to develop multilingual tools for an effective fight against misinformation. This paper presents an approach for countering misinformation through a semantic-aware multilingual architecture. Due to the specificity of the task addressed, which involves assessing the level of similarity between a pair of texts in a multilingual scenario, we built an extension of the well-known Semantic Textual Similarity Benchmark (STSb) to 15 languages. This new dataset allows to fine-tune and evaluate multilingual models based on Transformers with a siamese network topology on monolingual and cross-lingual Semantic Textual Similarity (STS) tasks, achieving a maximum average Spearman correlation coefficient of 83.60%. We validate our proposal using the Covid-19 MLIA @ Eval Multilingual Semantic Search Task. The results reported demonstrate that semantic-aware multilingual architectures are successful at measuring the degree of similarity between pairs of texts, while broadening our understanding of the multilingual capabilities of this type of models. The results and the new multilingual STS Benchmark data presented and made publicly in this study constitute an initial step towards extending methods proposed in the literature that employ semantic similarity to combat misinformation at a multilingual level.

Keywords: Misinformation · Natural Language Processing · Transformers · Semantic textual similarity · Multilingualism

H. Yin et al. (Eds.): IDEAL 2021, LNCS 13113, pp. 312–323, 2021.
https://doi.org/10.1007/978-3-030-91608-4_31

1 Introduction

Technology and social media keep us informed, productive, and connected. However, these platforms enable and amplify the quick spread of harmful, manipulative and false information. Different types of false information are described in the literature, such as disinformation, misinformation, and fake news. In the context of this paper, only the terms misinformation and disinformation will be used, following the recommendations of media experts and institutions, such as the Poynter Institute[1] and the Council of Europe [6].

Disinformation and misinformation counteract, undermine, and affect information's quality, providing false information that only differs in the purpose or motivation behind the information spreading. Whereas misinformation is unintended, disinformation pursues to deceive people intentionally. Additionally, it is not easy to distinguish them in social media where the amount of information is vast. In particular, since COVID-19, the infectious disease caused by the SARS-CoV-2 virus, emerged in Wuhan, China, in December 2019, the public has been bombarded with vast quantities of information, much of which is not checked. Information disorders are undermining the public health response during the management of disease outbreaks, leading the World Health Organization (WHO) to coin this situation as the term *infodemic* [2,17].

The detection, assessment, response, and damage control on public health are a challenge. Due to the amount of information that must be evaluated, Natural Language Processing (NLP) can help address these limitations. NLP is an interdisciplinary field of Artificial Intelligence that uses computational techniques for the automatic analysis and representation of human language [28]. One of the "levels of language" that NLP systems deal with is *semantics*. Semantics determines the possible meanings of a sentence by focusing on the interactions between word-level meanings. Therefore, NLP models encode sentences' meaning as vectors in a high-dimensional space [19]. These vectors are referred to as *embeddings*.

As described in [6], one of the requirements to identify reliable and valuable information is that there must be a common code between the sender and recipient (i.e., a common language). As well as spoken languages, there is not a uniform distribution on the Internet. English is far more used since 56.8% of all the websites use it, while the second and third languages most used by websites are Russian and Spanish with 7.6% and 4.6%, respectively [14]. Consequently, it is a fact that the language spoken determines access to information online and is one of the targets for fighting misinformation.

In this work, we propose the use of a multilingual semantic-aware transformer-based architecture for semantic similarity evaluation with the goal of fighting against misinformation and disinformation. In order to train this architecture, we also propose a new multilingual extension of the renowned English Semantic Textual Similarity benchmark (STSb) [1]. We use this extended version to evaluate, fine-tune, and break down the multilingual capabilities of multilingual NLP models on multilingual semantic textual similarity (STS) tasks. Finally, the potential

[1] https://www.poynter.org/.

usefulness of these models in combating misinformation is demonstrated in the Covid-19 MLAI @ Eval Semantic Search task.

2 Related Work

The emergence of the transformer [23] architecture, implemented by models such as BERT [5] or RoBERTa [15], set new state-of-the-art performances on various NLP tasks like semantic textual similarity [19] and in a plethora of domains [16]. Traditionally, these attention-based models use cross-encoders, that is to say, two sentences are passed to the transformer architecture as input while a target value is predicted [12,19]. This topology requires to feed the model with both sentences, causing a massive computational overhead because the attention is quadratic to the sequence length [5]. An alternative solution for this problem is the use of siamese architectures for training [11,19]. Siamese architectures consist of two pre-trained Transformer-based models with tied weights that can be fine-tuned for a specific task like computing similarity scores. This approach is also called dual-encoder or bi-encoder [12].

Although solving NLP tasks with these powerful Transformer-based models is interesting, it does not overcome the language bottleneck. Most studies have focused on monolingual models, usually English, which is a high-resource language [20]. Extending existing embedding models to new languages with lower resource levels is a new challenge in NLP. One possibility to address this issue is to apply distillation techniques [10] that transfer knowledge from large-scale state-of-the-art models to other models. Reimers and Gurevych [20] used a knowledge distillation technique, named teacher-student strategy, to transfer knowledge from monolingual models to other languages through multilingual models that act as student models. This strategy uses the original model (monolingual teacher model) to generate sentence embeddings of the source language and then trains a new system (a multilingual student model) on translated sentences from several languages to mimic the original respective teacher models.

Regarding multilingual model availability, one of the main issues for developing this kind of model lines in the lack of multilingual data needed for fine-tuning models for a specific domain. Considering this limitation, research to extend monolingual datasets to other languages can be found in the literature. An illuminating example is an attempt made by Ham et al. [9] for translating English STSb, among other datasets, to Korean. Likewise, Reimers and Gurevych [20] used Google Translator to extend the 2017 task of the STSb to German, French, Italian, and Dutch languages. As far as we know, no one has fully extended STSb, a selection of the English datasets used in the STS tasks between 2012 and 2017, to a multilingual scenario.

Concerning previous work for countering misinformation, the study by Jwa et al. [13] introduced the use of a model based on BERT in the misinformation detection field, outperforming previous results. Nonetheless, there is still considerable uncertainty about the use of a fully automatic data-driven decision-making algorithm to establish false and true conclusions. On the other hand, Vijjali et al. [24] proposed a different approach using a two-stage automated pipeline for fighting COVID-19 misinformation, where the first step consists of retrieving the most relevant facts database using semantic similarity and analyzing the entailment between the claim and the facts as the second step. A similar strategy is described in [7], where the authors present an approach to automatically analyze the information credibility of messages propagated through WhatsApp, comparing these messages with news articles using semantic similarity. Guo et al. [8] introduced the CORD19STS dataset, which includes 13,710 annotated sentence pairs collected from the COVID-19 open research dataset (CORD-19) to achieve a COVID-19 domain-specific STS task. These applications rely on measuring semantic textual similarity (STS), making STS a crucial instrument in the fight against misinformation.

3 A Semantic-Aware Multilingual Architecture for Cross-Lingual Semantic Textual Similarity Evaluation

Our approach consists on building a semantic-aware multilingual architecture for cross-lingual semantic similarity assessment. A model trained on this task is highly useful in the fight against misinformation as it allows to search for highly relevant documents, thus providing the correct information from validated sources when necessary. The first step to build this architecture involves evaluating a series of candidate multilingual pre-trained models. In a second step, and due to the lack of specific multilingual datasets for this task, we generate a new extension to 15 languages of the well-known STSb dataset.

3.1 Pre-trained Multilingual Bi-Enconder Models

There exist a plethora of pre-trained multilingual bi-encoder models that can be used as base models to later apply a specific fine-tuning process. The selected models were extracted from Sentence Transformers [19,20]:

- **paraphrase-xlm-r-multilingual-v1**: Distilled version of RoBERTa [15] trained on large-scale paraphrase data using XLM-R [3] as the student model.
- **stsb-xlm-r-multilingual**: Distilled BERT [5] version trained in NLI [27] and STSb [1] using XLM-R as the student model.
- **quora-distilbert-multilingual**: Distilled version of monolingual distilled BERT using multilingual distilled BERT as the student. Monolingual distilled BERT was first tuned on NLI and STSb data, then fine-tuned for Quora Duplicate Question detection retrieval,
- **paraphrase-multilingual-MiniLM-L12-v2**: Multilingual version of the MiniLM model from Microsoft [26] trained on large-scale paraphrase data.
- **paraphrase-multilingual-mpnet-base-v2**: Distilled version of the MPNet model from Microsoft [22] fine-tuned with large-scale paraphrase data using XLM-R as the student model.

The pre-trained model with the best performance in the STSb test split is fine-tuned in two different ways. Firstly, using the English STSb data, and secondly using the full extended multilingual STSb data to assess the impact of the multilingual extension.

3.2 An Extension of the Multilingual STS Benchmark

The original STS Benchmark [1] consists of train-dev-test splits of 5749, 1500, and 1379 pairs of sentences, respectively, labelled with a similarity score between 0 and 5, from less to more similar. To extend the original English STSb to a multilingual dataset, we translate the STS Benchmark splits from English to 15 languages[2]. The Google Translator python package[3] is used for this purpose, following the same procedure that has been largely used in existing related literature [9,20]. In addition, to maintain data quality, translated sentence pairs with a confidence value below 0.7 were dropped. As a result, Dutch is the language with the lowest sentence pairs in development (1483 sentence pairs) and test (1358 sentence pairs) sets. Besides, Google Translator distinguishes two variants of Chinese: simplified and using Mainland Chinese terms (*zh-CN*), and traditional and using Taiwanese terms (*zh-TW*). The multilingual STS Benchmark is public available at GitHub[4].

3.3 Fine-Tuning Pre-trained Architectures Through the Extended STS Benchmark Extension

As described above, for enhancing the multilingual semantic textual similarity performance, the best pre-trained model is fine-tuned on the extended STSb version using Cosine Similarity Loss from Sentence Transformers [19,20]. To obtain the best results and avoid overfitting, we optimized the following hyperparameters using the grid search method: learning rate, epochs, batch size, scheduler, and weight decay. The selected hyperparameter values and the resulting model are openly available in HuggingFace[5].

It is worth noting that in addition to the machine-translated version of STS Benchmark splits on 15 languages, as described above, the languages have also been combined into monolingual and cross-lingual tasks, giving a total of 31 tasks. Monolingual tasks have both sentences from the same language source (e.g., Ar-Ar, Es-Es), while cross-lingual tasks have two sentences, each in a different language being one of them English (e.g., En-Ar, En-Es).

4 Experimentation

The evaluation of our approach requires first to assess the performance of each pre-trained architecture considered in the STSb benchmark. Then, the architecture showing the better similarity evaluation capabilities is fine-tuned using the extension of STSb proposed in this manuscript.

[2] ar, cs, de, en, es, fr, hi, it, ja, nl, pl, pt, ru, tr, zh-CN, zh-TW.
[3] Google Translator python package: https://pypi.org/project/google-trans-new/.
[4] Multilingual STSB available at https://github.com/Huertas97/Multilingual-STSB.
[5] Fine-tuned model available in Hugging Face hub.

4.1 Pre-trained Architectures Performance in STSb

To evaluate the performance in STSb, the sentence embeddings for each pair of sentences are computed and the semantic similarity is measured using the cosine similarity metric. Then, the Spearman correlation coefficient (ρ) is computed between the scores obtained and the gold standard scores. The motivation for the choice of Spearman correlation coefficient in this work are twofold. Firstly, as mentioned in [4], the Spearman's rank correlation coefficient is non-parametric since it replaces the observations by their rank and calculates the correlation. This metric is therefore insensitive to outliers, non-linear relationships, or non-normally distributed data [18]. Secondly, it is the official metric used for semantic textual similarity on the GLUE Benchmark [25].

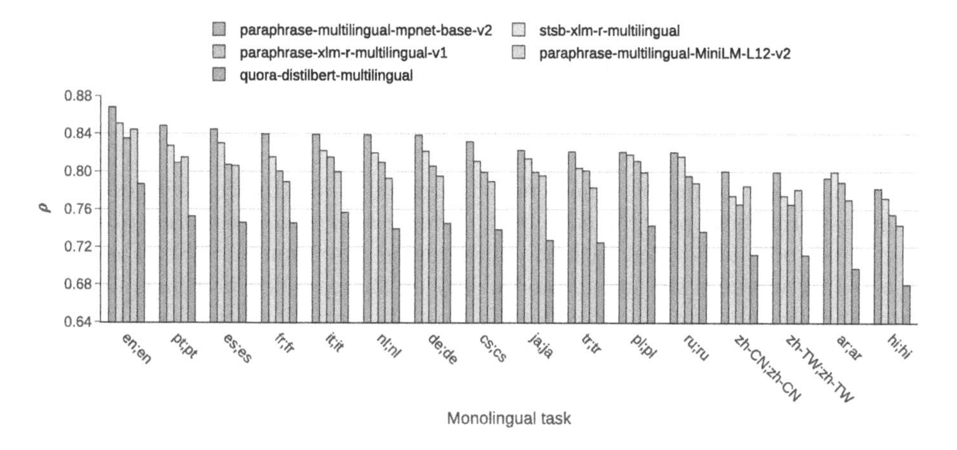

Fig. 1. Spearman ρ correlation coefficient in the monolingual task of the multilingual extended version STS Benchmark test between the sentence representation from multilingual pre-trained models and the gold labels.

Figures 1 and 2 depict the multilingual STSb performance according to the monolingual and cross-lingual tasks of the pre-trained models. As may be seen in Fig. 1, the *paraphrase-multilingual-mpnet-base-v2* model followed by *stsb-xlm-r-multilingual* model have the best monolingual results. As expected, the best performance is obtained in the original English monolingual task. Interestingly, the multilingual capability of the models is demonstrated since English-close languages, also called Germanic languages (i.e., Dutch and German), and non-English-close languages (e.g., Japanese, Russian, Polish) have a score over 0.8. Conversely, the performance of the pre-trained models was not ideal in other non-English-close languages such as Chinese, Arabic, and Hindi languages. This same pattern is visualized for the cross-lingual performance of the pre-trained models shown in Fig. 2.

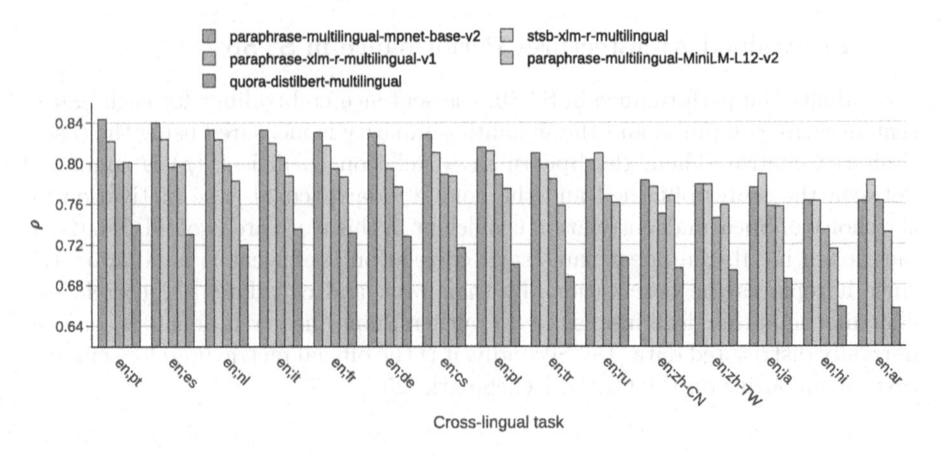

Fig. 2. Spearman ρ correlation coefficient in the cross-lingual task of the multilingual extended version STS Benchmark test between the sentence representation from multilingual pre-trained models and the gold labels.

In an effort to improve the semantic textual similarity capabilities across languages, the best pre-trained model, *paraphrase-multilingual-mpnet-base-v2*, is selected to be fine-tuned with the multilingual extension of STSb. Therefore, to correctly measure the impact and utility of the extended multilingual STSb data, this model is fine-tuned following two different approaches, as explained in the previous section. Firstly, model *stsb-EN-paraphrase-multilingual-mpnet-base-v2* represents the model fitted with English STSb train data only, and *mstsb-paraphrase-multilingual-mpnet-base-v2* the model fitted with multilingual train data from the STSb extended version. The comparison of these models and the base pre-trained model is illustrated in Figs. 3 and 4. The results clearly show that fine-tuning with the extended version of STSb presented in this study improves the semantic similarity capability across different languages. Not only improving in English-close languages, but also in non-English-close languages.

The numerical comparisons are presented in Table 1. Only English, Spanish, and average results across monolingual and cross-lingual tasks and all languages are reported for reasons of space. The average of correlation coefficients is computed by transforming each correlation coefficient to a Fisher's z value, averaging them, and then back-transforming to a correlation coefficient. From this table, one can see that all models decrease ρ in cross-lingual tasks and how multilingual extension fine-tuning outperforms only English fine-tuning. Surprisingly, *mstsb-paraphrase-multilingual-mpnet-base-v2* outperforms the model fine-tuned with the original English STSb data in the original monolingual English task. On the other hand, *stsb-EN-paraphrase-multilingual-mpnet-base-v2* shows better results for tasks where Spanish is involved. Overall, these results suggest that fine-tuning a model with the multilingual STSb data presented in this study improves the semantic similarity performance across languages.

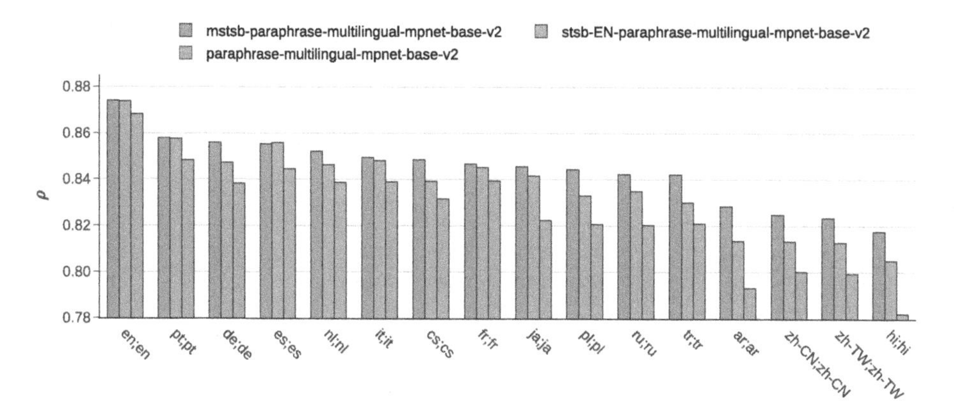

Fig. 3. Spearman ρ correlation coefficient in monolingual tasks from the multilingual extended version STS Benchmark test between the sentence representation from multilingual models and the gold labels.

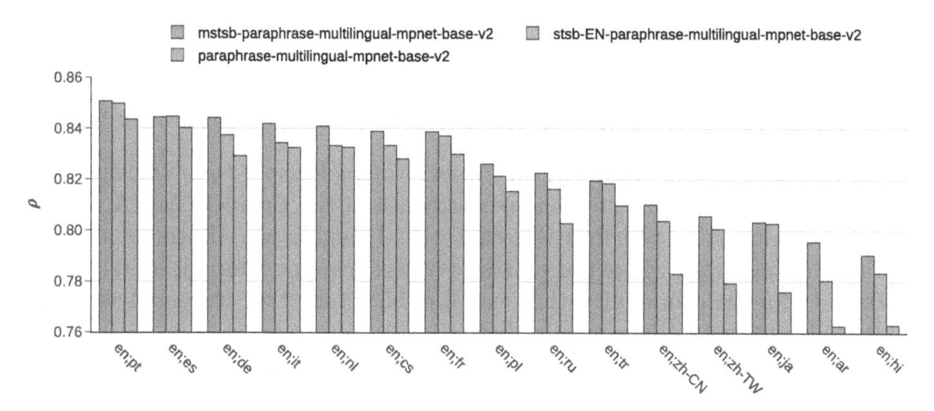

Fig. 4. Spearman ρ correlation coefficient in cross-lingual tasks from the multilingual extended version STS Benchmark test between the sentence representation from multilingual models and the gold labels.

Besides the multilingual bi-encoder models, monolingual bi-encoders and cross-encoders are evaluated in the original English STSb (see Table 2). As anticipated, cross-encoders obtain great results while sacrificing computation time. What is noticeable here is that *mstsb-paraphrase-multilingual-mpnet-base-v2* exceeds the monolingual bi-encoders pre-trained on NLI and STSb and can compete with the monolingual cross-encoders. Nevertheless, the multilingual models' results confirm that there is still room for improvement.

Table 1. Spearman ρ correlation coefficient between the sentence representation from multilingual bi-encoders, and the gold labels for the extended STS Benchmark test set. Performance is reported by convention as $\rho \times 100$.

Model	EN-EN	ES-ES	EN-ES	Avg. Mono lingual	Avg. Cross lingual	Avg.
paraphrase-xlm-r-multilingual-v1	83.50	80.72	79.65	79.84	77.88	78.91
Stsb-xlm-r-multilingual	85.04	82.99	82.36	81.14	80.37	80.77
Quora-distilbert-multilingual	78.66	74.57	73.00	73.46	70.67	72.14
paraphrase-multilingual-MiniLM-L12-v2	84.42	80.59	79.94	79.31	77.06	78.25
paraphrase-multilingual-mpnet-base-v2	86.82	84.45	84.03	82.67	81.02	81.89
stsb-EN-paraphrase-multilingual-mpnet-base-v2	87.38	**85.58**	**84.47**	83.83	82.09	83.00
mstsb-paraphrase-multilingual-mpnet-base-v2	**87.40**	85.55	84.45	**84.49**	**82.58**	**83.60**

4.2 Fighting Misinformation Through Multilingual Semantic Search

To determine whether *mstsb-paraphrase-multilingual-mpnet-base-v2* helps fight misinformation, we also tested the performance in the Covid-19 MLIA @ Eval initiative Multilingual Semantic Search Task[6]. This task is a Multilingual Information Retrieval task where the goal is to collect relevant information related to COVID-19 for the community, the general public, and other stakeholders across multiple languages to prevent misinformation.

The overall task consists of a collection of documents and a set of 30 topics, where the systems have to find relevant articles containing answers to the questions in the topics. The languages of the document collections are English, French, German, Greek, Italian, Spanish, and Swedish. As test data, we selected the most extensive subset of documents from round 1, the Medical Information System (MEDISYS) from March 2020. A ratio of 5 non-relevant documents for every relevant document is maintained for all languages and topics, making a total of 1 million documents with an average of $16{,}688 \pm 3{,}910$ documents per language. Performance is determined using the average of R-precision and P@5 of the topics.

The results of these tests are presented in Table 3 for each language. For a better comparison, the well-known BM25 Okapi algorithm [21] is included as the baseline reference. As can be seen, the fine-tuned model is on par with the BM25 algorithm, including languages that were not included in the STSb extended data (e.g., Greek, Swedish). Nevertheless, the combination of both approaches leads to the best results. One reasonable explanation is that this hybrid approach takes advantage of the semantic-awareness of *mstsb-paraphrase-*

[6] Covid-19 MLIA @ Eval initiative, http://eval.covid19-mlia.eu/task2/.

Table 2. Spearman ρ correlation coefficient between the sentence representation from monolingual bi-encoders and cross-encoders, and the gold labels for the original English STS Benchmark test set. Performance is reported by convention as $\rho \times 100$.

	Model	EN-EN
Monolingual Bi-encoders	roberta-large-nli-stsb-mean-tokens	86.39
	roberta-base-nli-stsb-mean-tokens	85.44
	bert-large-nli-stsb-mean-tokens	85.27
	bert-base-nli-stsb-mean-tokens	85.05
Monolingual Cross-encoders	ce-roberta-large-stsb	91.47
	ce-roberta-base-stsb	90.17

Table 3. R-precision (RPrec) and precision metric computed at cu-off rank 5 (P@5) from the selected subset of Covid-19 MLIA Semantic Search Task. The transformer model corresponds to *mstsb-paraphrase-multilingual-mpnet-base-v2*.

	BM25								Transformer								BM25 + Transformer							
Lang	de	el	en	es	fr	it	sv	Avg	de	el	en	es	fr	it	sv	Avg	de	el	en	es	fr	it	sv	Avg
RPrec	0.42	0.52	0.52	0.53	0.49	0.46	0.46	**0.49**	0.51	0.53	0.50	0.54	0.52	0.49	0.46	**0.51**	0.45	0.54	0.54	0.57	0.52	0.49	0.48	**0.51**
P@5	0.75	0.89	0.93	0.91	0.86	0.81	0.63	**0.82**	0.75	0.80	0.91	0.91	0.86	0.80	0.62	**0.81**	0.80	0.92	0.96	0.94	0.91	0.87	0.67	**0.87**

multilingual-mpnet-base-v2 to re-rank the extracted documents by the BM25 algorithm. In summary, these results demonstrate the utility of semantic similarity to improve the development of tools for fighting misinformation.

5 Conclusion

We have demonstrated how pre-trained multilingual models are able to approximate and improve the performance of monolingual bi-encoders in Semantic Textual Similarity (STS) tasks. Additionally, it has been proven that fine-tuning a pre-trained multilingual model, *paraphrase-multilingual-mpnet-base-v2*, with the extended version of STSb to 15 languages enhances the performance across languages increasing the average Spearman correlation coefficient from 81.89% to 83.60%. The results support that cross-encoders have better performance than bi-encoders at the expense of computational cost. Finally, applying this fine-tuned model in Covid-19 MLIA @ Eval Multilingual Semantic Search Task proved the benefit of including multilingual semantic-aware Transformer models to fight misinformation.

Multilingual models provide a powerful instrument in the fight against misinformation by introducing multilingualism, overcoming the language bottleneck. Based on the evaluation results, the multilingual models analyzed can be used in monolingual and cross-lingual semantic textual similarity tasks, performing better on monolingual tasks and English-closed languages such as Portuguese, Spanish, Dutch, or German. Nonetheless, these results also show that multilingual models can deal with non-English-closed languages such as Russian or

Japanese. This work has presented a novel multilingual extended version of the train-dev-test splits of Semantic Textual Similarity Benchmark that can be readily used in practice to develop multilingual tools for countering misinformation.

Acknowledgements. This research is funded by the project CIVIC: Intelligent characterisation of the veracity of the information related to COVID-19, granted by BBVA FOUNDATION GRANTS FOR SCIENTIFIC RESEARCH TEAMS SARS-CoV-2 and COVID-19, by the Ministry of Science and Education under PID2020-117263GB-100 (FightDIS) project, by Comunidad Autónoma de Madrid under S2018/ TCS-4566 (CYNAMON), S2017/BMD-3688 grant and by European Commission, 2020-EU-IA-0252 IBERIFIER - Iberian Digital Media Research and Fact-Checking Hub.

References

1. Cer, D., Diab, M., Agirre, E., Lopez-Gazpio, I., Specia, L.: SemEval-2017 task 1: Semantic textual similarity multilingual and crosslingual focused evaluation. In: Proceedings of the 11th International Workshop on Semantic Evaluation (SemEval-2017), pp. 1–14. Association for Computational Linguistics, Vancouver, Canada (August 2017)
2. Cinelli, M., et al.: The COVID-19 social media infodemic. Sci. Rep. **10**(1), 16598 (2020)
3. Conneau, A., et al.: Unsupervised cross-lingual representation learning at scale (2020)
4. Dalgaard, P.: Introductory Statistics with R. Statistics and Computing, Springer, New York (2008). https://doi.org/10.1007/978-0-387-79054-1
5. Devlin, J., Chang, M.W., Lee, K., Toutanova, K.: Bert: Pre-training of deep bidirectional transformers for language understanding (2019)
6. Estrada-Cuzcano, A., Alfaro-Mendives, K., Saavedra-Vásquez, V.: Disinformation y misinformation, posverdad y fake news: precisiones conceptuales, diferencias, similitudes y yuxtaposiciones. Información, cultura y sociedad **42**, 93–106 (2020)
7. Gaglani, J., Gandhi, Y., Gogate, S., Halbe, A.: Unsupervised Whatsapp fake news detection using semantic search. In: 2020 4th International Conference on Intelligent Computing and Control Systems (ICICCS), pp. 285–289 (2020)
8. Guo, X., Mirzaalian, H., Sabir, E., Jaiswal, A., Abd-Almageed, W.: Cord19sts: Covid-19 semantic textual similarity dataset (2020)
9. Ham, J., Choe, Y.J., Park, K., Choi, I., Soh, H.: Kornli and korsts: new benchmark datasets for Korean natural language understanding (2020)
10. Hinton, G., Vinyals, O., Dean, J.: Distilling the knowledge in a neural network (2015)
11. Huertas-Tato, J., Martín, A., Camacho, D.: Sml: a new semantic embedding alignment transformer for efficient cross-lingual natural language inference. arXiv preprint arXiv:2103.09635 (2021)
12. Humeau, S., Shuster, K., Lachaux, M.A., Weston, J.: Poly-encoders: Transformer architectures and pre-training strategies for fast and accurate multi-sentence scoring (2020)
13. Jwa, H., Oh, D., Park, K., Kang, J.M., Lim, H.: Exbake: automatic fake news detection model based on bidirectional encoder representations from transformers (bert). Appl. Sci. **9**(19), 4062 (2019)

14. Kemp, S.: Digital 2020: october global statshot (2020). https://datareportal.com/reports/digital-2020-october-global-statshot
15. Liu, Y., et al.: Roberta: a robustly optimized bert pretraining approach (2019)
16. Martín, A., González-Carrasco, I., Rodriguez-Fernandez, V., Souto-Rico, M., Camacho, D., Ruiz-Mezcua, B.: Deep-sync: a novel deep learning-based tool for semantic-aware subtitling synchronisation. Neural Comput. Appl. 1–15 (2021). https://doi.org/10.1007/s00521-021-05751-y
17. Naeem, S.B., Bhatti, R.: The Covid-19 'infodemic': a new front for information professionals. Health Inf. Libr. J. **37**(3), 233–239 (2020)
18. Reimers, N., Beyer, P., Gurevych, I.: Task-oriented intrinsic evaluation of semantic textual similarity. In: Proceedings of COLING 2016, the 26th International Conference on Computational Linguistics: Technical Papers, pp. 87–96. The COLING 2016 Organizing Committee, Osaka, Japan (2016)
19. Reimers, N., Gurevych, I.: Sentence-bert: sentence embeddings using siamese bert-networks (2019)
20. Reimers, N., Gurevych, I.: Making monolingual sentence embeddings multilingual using knowledge distillation (2020)
21. Robertson, S., Walker, S., Hancock-Beaulieu, M.M., Gatford, M., Payne, A.: Okapi at trec-4. In: The Fourth Text REtrieval Conference (TREC-4), pp. 73–96. Gaithersburg, MD: NIST (January 1996)
22. Song, K., Tan, X., Qin, T., Lu, J., Liu, T.Y.: Mpnet: masked and permuted pre-training for language understanding (2020)
23. Vaswani, A., et al.: Attention is all you need (2017)
24. Vijjali, R., Potluri, P., Kumar, S., Teki, S.: Two stage transformer model for Covid-19 fake news detection and fact checking (2020)
25. Wang, A., Singh, A., Michael, J., Hill, F., Levy, O., Bowman, S.: GLUE: a multi-task benchmark and analysis platform for natural language understanding. In: Proceedings of the 2018 EMNLP Workshop BlackboxNLP: Analyzing and Interpreting Neural Networks for NLP, pp. 353–355. Association for Computational Linguistics, Brussels, Belgium (November 2018). https://doi.org/10.18653/v1/W18-5446, https://aclanthology.org/W18-5446
26. Wang, W., Wei, F., Dong, L., Bao, H., Yang, N., Zhou, M.: Minilm: deep self-attention distillation for task-agnostic compression of pre-trained transformers (2020)
27. Williams, A., Nangia, N., Bowman, S.: A broad-coverage challenge corpus for sentence understanding through inference. In: Proceedings of the 2018 Conference of the North American Chapter of the Association for Computational Linguistics: Human Language Technologies, Volume 1 (Long Papers), pp. 1112–1122. Association for Computational Linguistics, New Orleans, Louisiana (June 2018)
28. Yang, Y., et al.: Multilingual universal sentence encoder for semantic retrieval (2019)

Genetic and Ant Colony Algorithms to Solve the Multi-TSP

Sílvia de Castro Pereira[1(✉)], E. J. Solteiro Pires[1,2(✉)], and Paulo Moura Oliveira[1,2(✉)]

[1] Universidade de Trás-os-Montes e Alto Douro, Vila Real, Portugal
silvia.pereira@ipb.pt,
{oliveira,epires}@utad.pt
[2] INESC TEC - INESC Technology and Science, Porto, Portugal

Abstract. Multiple traveling salesman problem (mTSP) is a variant of the famous and standard traveling salesman problem, an NP-hard problem in combinatorial optimization. This kind of problem can be solved using exact methods but usually results in high exponential computational complexities. Heuristics and metaheuristics are required to overcome this shortcoming. This study proposes a hybrid method based on the Genetic Algorithm, Ant Colony Optimization, and 2-opt to improve the solution. Computational results with some benchmark instances are provided and compared with other published studies. In three instances, the proposed technique provides better results than the best-known solutions reported in the literature.

Keywords: Multiple traveling salesman problem · Genetic algorithm · Ant colony optimization · 2-opt

1 Introduction

The Traveling Salesman Problem (TSP) is a well-known optimization problem that has been attracting the research efforts of various domains due to its wide application in real-world problems [3]. This problem consists of a set of vertices V connected by edges E that can be represented on a graph $G = (V, E)$. The objective is to find the shortest minimum distance circuit (Hamiltonian circuit) passing through each vertice only once. In this problem with a small number of vertices, exact methods can be used to find the optimal solution. However, as the number of vertices increases, computing time grows exponentially, making it very difficult to solve the problem in an acceptable timeframe. As the name implies, heuristic methods do not guarantee to find the optimal solution, although eventually, it does find it. When the problem considers more than one traveling salesman it is called multi-TSP (mTSP). The computational complexity of the mTSP increases significantly concerning single traveling salesmen problems. Whereas TSP is a standard optimization problem, mTSP has not received the same attention in the scientific literature. Therefore, some papers published in the literature are described in the sequel.

H. Yin et al. (Eds.): IDEAL 2021, LNCS 13113, pp. 324–332, 2021.
https://doi.org/10.1007/978-3-030-91608-4_32

In 2006, Bektas [4] reported a review of mTSP problem and its practical applications to highlight some formulations and to describe exact and heuristic solution procedures proposed to solve this type of problem. In the same year, Carter and Ragsdale [6] developed a genetic algorithm (GA) and evaluated its performance on mTSP problems. They proposed a particular GA chromosome representation and related operators and compared the proposed technique's theoretical properties and computational performance to previous work.

In 2007, Brown et al. [5] proposed a grouping GA (GGA) to indicate which salesperson is assigned to each tour and the ordering of the cities within each tour and compared the new method to a standard GA. In 2009, Liu et al. [15] used an ant colony (AC) algorithm for the mTSP considering two minimization objectives: the maximum tour length of all the salesmen and the maximum tour length of each salesman. Singh and Baghel [20] presented a new grouping GA also to solve last both minimization objectives.

In 2012, Sedighpour et al. [19] proposed a hybrid metaheuristic GA2OPT to solve mTSP. In a first approach, they apply the GA in each iteration and then the 2-opt to improve the solution. Also in 2011, Ghafurian and Javadian [10] implemented an AC algorithm to solve the multiple depot variant of the mTSP, in which more than one salesman depart from several starting cities and having returned to the starting city, so that each city is visited by exactly one salesman, and the tour lengths stay within certain limits.

In 2014, Aray et al. [2] proposed a modified GA, using a local search technique to improve the solution. Larki and Yousefikhoshbakht [13] implemented a new hybrid algorithm - the MICA (*Modified Imperialist Competitive Algorithm*) tested in 26 benchmark functions with 20 to 150 nodes. In six instances, this new algorithm found a better value than the *Best-Known Solution* (BKS) for minsum mTSP.

Rostami et al. [18] proposed a *Gravitational Emulation Local Search Algorithm*- GELS based on Newton's laws. The GELS algorithm is based on the local search concept and uses two main parameters in physics, velocity and gravity. Performance of the modified GELS has been compared with well-known optimization algorithms such as the GA and ant colony optimization (ACO). Soylu [21] proposed a general variable neighborhood search heuristic for mTSP and tested its performance on some benchmark problems. More recently, Alaidi and Mahmood [1] proposed a hybrid method that transforms the mTSP in multiple TSP standard problems and then applies a parallel ACO (Ant Colony Optimization) with different constraints in pheromone evaporation.

Harrath et al. [11] presented a hybrid approach called AC2OptGA, a combination of three algorithms: Modified Ant Colony, 2-Opt, and Genetic Algorithm, to solve the mTSP. ACO was used to generate population for TSP, then 2-Opt was applied to improve solutions and GA was applied to improve the mTSP solutions. This approach was evaluated using various data instances from standard benchmarks and presents better results than M-GELS, the current best-known approach, for large-sized instances.

A hybrid algorithm is proposed in this work, combining two metaheuristics, GA and ACO. Then, to improve the finals results, a 2-opt algorithm is applied. The technique is tested in multiple traveling instances. The remaining paper is divided into the following sections. Section 2 defines the mTSP problem, its variants, and formulations. Section 3 describes the generic operation of each of the three algorithms used in the construction of the new proposal. Next, Sect. 4 presents the proposed algorithm. Section 5 provides the computational test results. Finally, Sect. 6 draws the concluding remarks and highlights some research directions.

2 Multiple Traveling Salesman Problem

The mTSP consists of finding a set of tours for m salesmen who start in a city and return to the same city (depot). All cities are visited only once by a unique salesman. There are several versions of this problem reported in the literature:

- Single depot: Initially proposed by Miler, Tucker, and Zemlin [16] in which the m salesmen start and end their routes in the same city;
- Multiple depots: This variant considers several depots, and one or several traveling salesmen start from a depot. Each traveling salesman returns to the starting depot or any other depot at the end of the route.
- Minmax: The problem goal minimizes the most costly route for m paths [9];
- Minsum: The problem goal minimizes the total distance of m salesmen [16].

2.1 Problem Definition

The Traveling Salesman Problem can be represented by a graph $G = (V, A)$ where $V = \{v_1, v_2, ..., v_n\}$, in which V is the set of n nodes and an edge set $A = \{(i, j) : i, j \in V, i \neq j\}$. The nodes represent the cities and the edges the possible paths between cities. Each path has a distance or cost associated. For each edge (i, j), c_{ij} is the cost associated. The binary variable x_{ij} takes the value 1 if edge (i, j) is included in a tour and takes the value 0 if edge (i, j) is not included. Multiple TSP in the single depot version is usually formulated in a similar way to TSP, differing only in the initial depot's constraint (first edge) and in the number of salesmen m. The equations to define the single depot mTSP are the following:

$$\min \sum_{j=1}^{n} \sum_{i=1}^{n} c_{ij} x_{ij} \tag{1}$$

$$\sum_{j=1}^{n} x_{1j} = \sum_{j=1}^{n} x_{j1} = m \tag{2}$$

$$\sum_{j=2}^{n} x_{ij} = 1, \quad i = 1, 2, ..., n \tag{3}$$

$$\sum_{i=2}^{n} x_{ij} = 1, \quad j = 1, 2, ..., n \tag{4}$$

$$u_i - u_j + n x_{ij} \leq n - 1, \quad 2 \leq i \neq j \leq n \tag{5}$$

$$x_{ij} \in \{0, 1\}, \quad i, j = 1, 2, ..., n \tag{6}$$

The objective function is described in (1) with the aim of minimizing the total traveling cost. The restrictions are then formulated. The constraint (2) ensures that there are precisely m arcs corresponding to the m traveling salesmen of the problem, in and out of the initial depot. Restrictions (3) and (4) guarantee that from the i depot enters and exits just a single arc. The sub tour elimination constraint (5) defines an integer variable u_i for the position of node i in a tour and n representing the maximal number of nodes visited by any salesman. Finally, (6) defines that the variable x_{ij} is binary.

3 Description of Algorithms

The new approach proposed here combines a GA, ACO, and 2-opt algorithms. In the first step, these three algorithms are explained and then analyzed the new method in further detail in the next section.

3.1 The Genetic Algorithm

In the1970 s, a new category of evolutionary algorithms, known as Genetic Algorithms (GA), was proposed by John Holland [12]. The GA is a metaheuristic inspired by Charles Darwin's theory of natural evolution. It is an optimization and search algorithm based on the evolution of species to create the diversity found in nature. GA uses some terms like chromosome to define an array of values representing a candidate solution to optimize a function called fitness function. In the first step, the initial population (a set of chromosomes) is created randomly and evaluated by the fitness function. The evaluation measures the quality of the solution. Then the selection operator is applied. This operator allows choosing some chromosomes for reproduction based on a probability defined on GA's initialization parameters. After the selection operator, the crossover operator is used to swap a subsequence of two-parent chromosomes to create two offspring. Classical crossover operation occurs at one point where a random crossover point is selected, and the tails of its two parents are swapped to get new offspring. In the last step, the mutation operator randomly flips individual values of chromosomes with a usually very low probability.

3.2 The Ant Colony Optimization Algorithm

The Ant Colony Optimization (ACO) algorithm was proposed by Marco Dorigo in 1992 [8]. It is a multi-agent heuristic inspired by real ants' behavior in nature. In the ACO algorithm, each ant finds a possible solution. Initially, ants are all in a nest and only go out to find food. Ants search food leaving a certain amount of pheromone on the way. The edge's pheromone amount, τ_{ij}, between the points i and j of the graph, is evaluated according to the formula:

$$\tau_{ij} = (1-\rho)\tau_{ij} + \sum_{k=1}^{na} \Delta\tau_{ij}^{k} \tag{7}$$

where na is the number of ants that visited the edge (i, j), ρ is the pheromone evaporation rate, and $\Delta\tau_{ij}^{k}$ is the amount of pheromone deposited by ant k in the edge (i, j). This amount of pheromone is calculated by:

$$\Delta_{ij}^{k} = \begin{cases} Q/L_k, & \text{if the ant } k \text{ has passed through this edge} \\ 0, & \text{otherwise} \end{cases} \tag{8}$$

where L_k measures the path distance visited by ant k, and Q is a constant value. After the ants have passed a particular edge of the graph, they need to choose the next edge to follow in order to reach the food source. The amount of pheromone in the next edge to choose will be decisive, so the greater this amount, the greater the probability that the ant will select the corresponding edge. The probability of movement of ant k from node i to node j, which is not visited yet, is calculated by the equation:

$$p_{ij}^{k} = \frac{[\tau_{ij}]^{\alpha}[\eta_{ij}]^{\beta}}{\sum_{l \in N_i^k} [\tau_{ij}]^{\alpha}[\eta_{ij}]^{\beta}} \tag{9}$$

where α is the parameter that represents the importance of the pheromone edge in deciding the subsequent movement of the ant; β the parameter that represents the importance of distance in deciding the subsequent movement of the ant; N_i^k the set of nodes that have not been visited by ant k yet; and η_{ij} the inverse distance between node i and j. A cycle is completed when all ants have finished their movement. As time goes by, it turns out that ants will tend to follow the shortest path, thus converging to the optimal solution of the problem.

3.3 The 2-Opt Algorithm

To solve hard combinatorial optimization problems, local search methodologies like the 2-opt heuristic (see Fig. 1) can be used and produces good quality solutions for a large set of problems [17]. Initially proposed by Croes [7] for solving TSP standards, it is used to improve the founded solutions removing two nonadjacent arcs from a tour and adding two new arcs while the tour structure is maintained.

4 The Proposed Hybrid Algorithm

In this section, our approach denominated by GACO2 is described. In the first step, it's necessary to define the instance and the number of salesmen used.

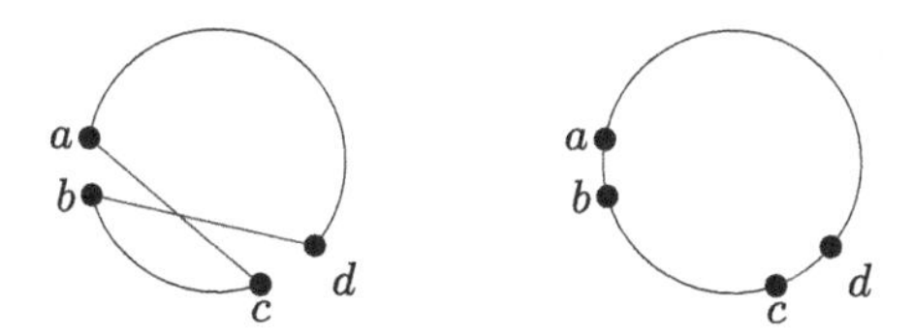

Fig. 1. Heuristic 2-opt

Next, the GA control parameters are defined. The GA is run to create the new initial population for the ACO construction method. The representation of a GA's chromosome is illustrated in Fig. 2 by an array.

TSP	1	2	2	1	0	0	1	2	1
City:	*0*	*1*	*2*	*3*	*4*	*5*	*6*	*7*	*8*

Fig. 2. Chromosome representation

The chromosome length is the problem instance number of cities. The values inside the array can be generated randomly or using multiple methods. In this case, the nearest neighbor algorithm is used and represent the number of salesmen used. In the nearest neighbor algorithm, the salesman starts at a random city and repeatedly visits the nearest city until all have been visited. Consider the example of Fig. 2 in which three salesmen $T = \{0, 1, 2\}$ are set to travel by 9 cities $C = \{0, 1, 2, 3, 4, 5, 6, 7, 8\}$. For this case, the distribution of each city by salesmen are TSP-0 ={4,5}, TSP-1 ={0,3,6,8}, and TSP-2 ={1,2,7}. The value 0 is found in the fourth and fifth index of the array. The value 1 in position 0, 3, 6, and 8. Finally, the value 2 is found in position 1, 2, and 7.

Then, the selection method is used to select two parents, each time, based on their fitness scores. In this stage, a tournament selection technique was used, which compares two individuals based on the value of each individual's fitness. Next, the one-point crossover technique is applied, followed by the mutation operator with a low random probability.

In every GA generation, each solution is evaluated by calling the objective function (10), where S_j is a set of $|S_j|$ arcs travelled by the salesman j. I.e., to evaluate the fitness of a chromosome, the GACO2 executes an ACO algorithm for each salesman generating a solution with path and distance. Next, a local search 2-opt method is applied to improve the solutions for each cities path. Finally, all salesman path distances are summed.

$$f = \sum_{k=1}^{m} \sum_{j=1}^{|S_j|} c_j^2, \qquad j \in S_j \tag{10}$$

5 Experiments and Computational Results

The proposed algorithm is coded in Python and implemented on an i3 2 GHz (16GB RAM) notebook PC on a Windows 10 operating system. It was tested on a group of instances proposed by Carter and Ragsdale [6] that includes six problems with n = 51 and 100 cities and m = 3, 5, and 10 salesmen.

The algorithm parameter setting is crucial for its performance. Appropriate parameters can enhance the algorithm global search ability and improve the convergence speed considerably [14]. In GA, the probability of crossover used was 90% and the probability of mutation 5%. The population size used was 83 and the number of iterations applied 1500. The ACO has more parameters to set: α and β were 1, evaporation rate 95%, population size 30 and iterations was 17.

The global test results obtained with the proposed algorithm are presented in Table 1. The first column indicates the number of cities n, in the second the number of salesmen m. The problem instance is given by mTSPn. The next columns are the best-known solutions founded actually in literature. According to this table, optimal results are found by the GACO2 algorithm for most of the problems.

Table 1. GACO2 results and comparison with other algorithms for instances $n = \{51, 100\}$.

n	m	GA	GGA	GA2PC	TCX	ACO	NMACO	GACO2
51	3	460	924	543	492	449	447	**436**
	5	499	882	586	519	481	473	**464**
	10	669	1001	723	670	592	583	**576**
100	3	22959	79347	26653	26130	22241	22051	**22028**
	5	24559	70871	30408	28612	23901	23678	**23424**
	10	33136	89778	31227	30988	27981	**27643**	31377

Table 2. TSPs effectively used in GACO2.

n	m	TSP used	%TSP Not used
51	3	2	33%
	5	4	20%
	10	6	40%
100	3	2	33%
	5	4	20%
	10	6	40%

Table 2 shows the number of TSP used. This technique shows that is possible to use, for example, only two of three TSP and assign one to another task. Next figures presents the paths of all TSP for mTSP51 instance (Figs. 3–5).

The Fig. 3 shows that using only two TSP the total distance traveled can be minimized. The hybrid approach GACO2 reduce the TSP used in category of 3, 5 and 10 salesmen. Figure 4 shows that four travelers are used and one can avail to execute another task. With 10 TSP this approach used only 60% of total TSP and obtained better results but only with small-scale TSP problems.

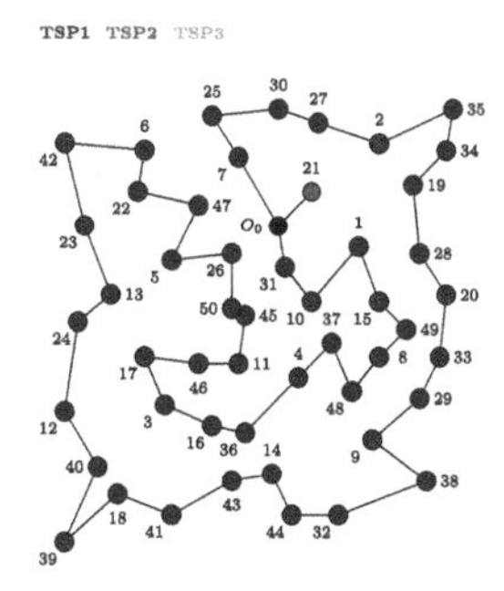

Fig. 3. Resulting tour for 3TSP51

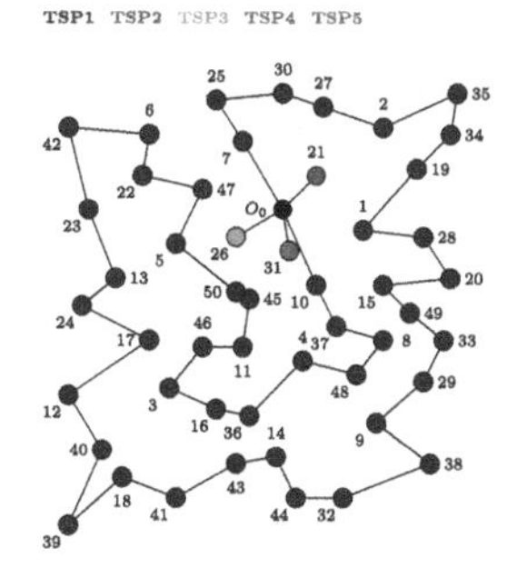

Fig. 4. Resulting tour for 5TSP51

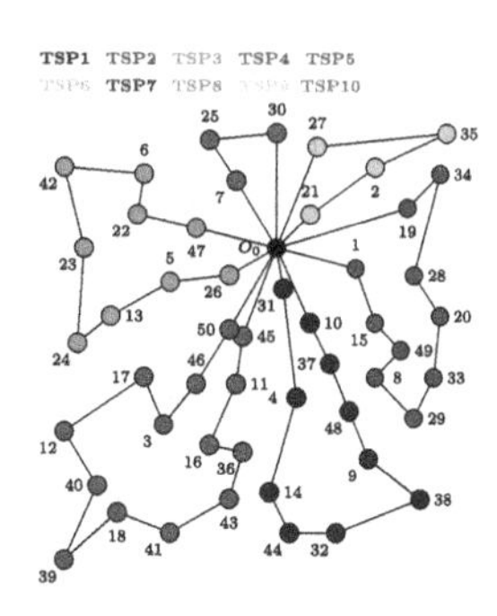

Fig. 5. Resulting tour for 10TSP51

6 Conclusion

This study focused on the multiple traveling salesman problem with minsum objective and single depot. A combination of three different algorithms was used for solving some mTSP instances. This approach used GA, ACO, and 2-Opt and tried to minimize the number of travelers used, keeping the categories of numbers of TSP (3, 5, and 10). The performance of the proposed algorithm is compared with others methods from the literature. For small-scale TSP problems, the proposed algorithm obtained better solutions than the ones known from the literature. However, for large-scale TSP problems, the proposed GACO2 does not get such good results when many TSP are used. In future work, the proposed algorithm will be used with a different set of parameters in order to found better solutions.

Acknowledgments. This work is financed by National Funds through the Portuguese funding agency, FCT – Fundação para a Ciência e a Tecnologia, within project UIDB/50014/2020.

References

1. Alaidi, A.H.M., Mahmood, A.: Distributed hybrid method to solve multiple traveling salesman problems. In: 2018 International Conference on Advance of Sustainable Engineering and its Application (ICASEA), pp. 74–78. IEEE (2018)
2. Aray, V., Goyal, A., Jaiswal, V.: An optimal solution to multiple travelling salesperson problem using modified genetic algorithm. J. Appl. Innov. Eng. Manage. **3**(1), 425–430 (2014)
3. Bektas, T.: The multiple traveling salesman problem: an overview of formulations and solution procedures. Omega **34**(3), 209–219 (2006). https://doi.org/10.1016/j.omega.2004.10.004

4. Bektas, T.: The multiple traveling salesman problem: an overview of formulations and solution procedures. Omega **34**(3), 209–219 (2006)
5. Brown, E.C., Ragsdale, C.T., Carter, A.E.: A grouping genetic algorithm for the multiple traveling salesperson problem. Int. J. Inf. Technol. Decis. Making **6**(02), 333–347 (2007)
6. Carter, A.E., Ragsdale, C.T.: A new approach to solving the multiple traveling salesperson problem using genetic algorithms. Eur. J. Oper. Res. **175**(1), 246–257 (2006)
7. Croes, G.A.: A method for solving traveling-salesman problems. Oper. Res. **6**(6), 791–812 (1958)
8. Dorigo, M.: Optimization, learning and natural algorithms [ph. d. thesis]. Politecnico di Milano, Italy (1992)
9. Frederickson, G.N., Hecht, M.S., Kim, C.E.: Approximation algorithms for some routing problems. SIAM J. Comput. **7**(2), 178–193 (1978). https://doi.org/10.1137/0207017
10. Ghafurian, S., Javadian, N.: An ant colony algorithm for solving fixed destination multi-depot multiple traveling salesmen problems. Appl. Soft Comput. **11**(1), 1256–1262 (2011)
11. Harrath, Y., Salman, A.F., Alqaddoumi, A., Hasan, H., Radhi, A.: A novel hybrid approach for solving the multiple traveling salesmen problem. Arab J. Basic Appl. Sci. **26**(1), 103–112 (2019). https://doi.org/10.1080/25765299.2019.1565193
12. Holland, J.: Adaption in Natural and Artificial Systems 1992 Editi. MIT Press, Cambridge (1975)
13. Larki, H., Yousefikhoshbakht, M.: Solving the multiple traveling salesman problem by a novel meta-heuristic algorithm. J. Optim. Indus. Eng. **7**(16), 55–63 (2014)
14. Liu, L.q., Dai, Y.t., Wang, L.H.: Ant colony algorithm parameters optimization. Comput. Eng. **34**(11), 208–210 (2008)
15. Liu, W., Li, S., Zhao, F., Zheng, A.: An ant colony optimization algorithm for the multiple traveling salesmen problem. In: 2009 4th IEEE Conference on Industrial Electronics and Applications, pp. 1533–1537. IEEE (2009)
16. Miller, C.E., Tucker, A.W., Zemlin, R.A.: Integer programming formulation of traveling salesman problems. J. ACM (JACM) **7**(4), 326–329 (1960)
17. Reeves, C.: Modern Heuristic Techniques for Combinatorial Problems, (ed.) Mcgraw-Hill, New York (1993)
18. Rostami, A.S., Mohanna, F., Keshavarz, H., Hosseinabadi, A.: Solving multiple traveling salesman problem using the gravitational emulation local search algorithm. Appl. Math. Inf. Sci. **9**(2), 1–11 (2015)
19. Sedighpour, M., Yousefikhoshbakht, M., Mahmoodi Darani, N.: An effective genetic algorithm for solving the multiple traveling salesman problem. J. Optim. Ind. Eng. **4**, 73–79 (2012)
20. Singh, A., Baghel, A.S.: A new grouping genetic algorithm approach to the multiple traveling salesperson problem. Soft Comput. **13**(1), 95–101 (2009)
21. Soylu, B.: A general variable neighborhood search heuristic for multiple traveling salesmen problem. Comput. Ind. Eng. **90**, 390–401 (2015)

Explainable Artificial Intelligence in Healthcare: Opportunities, Gaps and Challenges and a Novel Way to Look at the Problem Space

Petra Korica(✉), Neamat El Gayar, and Wei Pang

School of Mathematical and Computer Sciences, Heriot-Watt University, Edinburgh, UK
{pk2005,n.elgayar,w.pang}@hw.ac.uk

Abstract. Explainable Artificial Intelligence (XAI) is rapidly becoming an emerging and fast-growing research field; however, its adoption in healthcare is still at the early stage despite the potential that XAI can bring to the application of AI in this industry. Many challenges remain to be solved, including setting standards for explanations, the degree of interaction between different stakeholders and the models, the implementation of quality and performance metrics, the agreement on standards for safety and accountability, its integration into clinical workflows, and IT infrastructure. This paper has two objectives. The first one is to present summarized outcomes of a literature survey and highlight the state-of-the-art for explainability including gaps, challenges, and opportunities for XAI in healthcare industry. For easier comprehension and onboarding to this research field we suggest a synthesized taxonomy for categorizing explainability methods. The second objective is to ask the question if applying a novel way of looking at explainability problem space, through a specific problem/domain lens, and automating that approach in an AutoML similar fashion, would help mitigate the challenges mentioned above. In the literature there is a tendency to look at the explainability of AI from model-first lens, which puts concrete problems and domains aside. For example, the explainability of a patient's survival model is treated the same as explaining a hospital cost procedure calculation. With a well-identified problem/domain that XAI should be applied to, the scope is clear and well-defined, enabling us to (semi-) automatically find suitable models, optimize their parameters and their explanations, metrics, stakeholders, safety/accountability level, and suggest means of their integration into clinical workflow .

Keywords: Artificial intelligence · Machine learning · Interpretability · Explainability · Explainable AI · XAI · AI in Healthcare

1 Introduction

The lack of explainability and transparency of state-of-the-art artificial intelligence (AI) systems is one of the main reasons why AI is not yet (fully) trusted and hence widely deployed in many real-world problems, including healthcare problems. World Health Organization (WHO) recently published a set of guidelines for the future of AI in healthcare, which include "ensuring explainability" [1]. Healthcare can uniquely benefit from

H. Yin et al. (Eds.): IDEAL 2021, LNCS 13113, pp. 333–342, 2021.
https://doi.org/10.1007/978-3-030-91608-4_33

the applications of AI, e.g., saving a significant number of lives through early diagnosis or drug development; however, it also poses unique challenges and risks if AI is not carefully implemented and used. Healthcare benefits from AI models working together with humans, assisting humans in decision making, for example, in diagnosis and prognosis. Explainable AI models can be of high value in holding AI models and assisted decision-making systems accountable [2], which is vital for their applications in healthcare. Accountable Machine Learning is another emerging topic that is related to the explainability of AI, and the machine learning (ML) systems. ML systems encompass ML models, supporting infrastructure, hardware, and processes etc. Explainability in this context is used to enable the system to justify its outputs, meaning its decisions, predictions and even the reasoning process in some cases, which is an important part of accountability of a system [3].

The first objective and main contribution of this paper is to present the outcomes of the literature survey that we have done and to create a taxonomy that summarizes as many aspects of XAI as possible into a "simple" structure. The guiding idea is to categorize and make the vast knowledge quicker addressable, foster joint understanding and easier comprehensibility of the field as well as to highlight the state-of-the-art for explainability as well as the gaps, challenges, and opportunities explainable artificial intelligence faces in the healthcare industry.

In the literature there is a tendency to look at the explainability from model-first lens approach, putting the concrete problems and domains aside. Examples of such approach would be: 1) applying black-box XAI methods after training without considering how to reformulate the problem to be explainable, 2) without considering the users of the model, 3) without considering the recipients of the explanation and 4) applying the same XAI methods to different problems in different domains. Considering this, our second objective is to pose a question whether applying a novel way of looking at the explainability problem space by applying a so-called "problem/domain lens" approach would help the adoption of explainable AI in healthcare. The proposed approach is looking at the domain and problem that the ML models are trying to solve and based on that offers guidance on what XAI methods with what parameters, what form of explanations, what stakeholders, what regulations should be looked at. Furthermore, automating that approach in an AutoML similar way, would help mitigate the challenges mentioned above.

The rest of the paper is organised as follows: after looking at current research in the Sect. 2, we will identify the challenges in Sect. 3. This is followed by Sect. 4, in which we will show an example of the proposed approach for the healthcare industry.

2 Current Outlook on XAI

Explainable AI (XAI) and its application to healthcare is an emerging research field influenced by regulatory frameworks and research programs such as DARPA XAI [4], GDPR [5] and impending Artificial Intelligence Act [6]. Although this field is not completely new, it has been exponentially growing since the launch of DARPA XAI program [4] in 2017. For example, between February and July of 2021, there were around 6,500 new papers published mentioning "Explainable AI" and around 8,500 mentioning "Interpretable AI" [7, 8]. This is an incredible rate of increase of publications which shows the difficulties to keep track of for the researchers in the field.

XAI is a multi-disciplinary research field influenced by psychology, philosophy, sociology, human-computer interface, and education, where the recipient of the explanation is important. However, this field is still immature in its applications to real-life situations, and there is work to be done by the research community to establish mutual understanding and standards for definitions, safety, metrics, etc. Let's look at the proposed definitions for XAI. In the literature, e.g. [9], the "Interpretability" is defined as the quality or feature of a model providing enough expressive data to understand how the model works. In literature the term domain is noted as relevant for the interpretability [9, 10], with some authors such as [10] are arguing that interpretability is inherently domain specific.

On the other hand, "Explainability" can be defined as the ability of models to summarize the reasons for their behavior, gain the trust of users or produce insights into the causes of their decisions [11]. Explainability entails the definition of an explanation. "Explanation" is defined in the literature [12, 13] as a statement, fact or situation that tells one why something happened. There are several desired facts to explanations such as accuracy, fidelity, consistency, stability, comprehensibility, relevance, and certainty [14]. A good explanation needs to be produced responsibly and evaluated rigorously. In [15] the authors bring an important aspect in addition to the definitions above: "The explainability of an AI systems behavior needs to consider different dimensions: 1) who is the receiver of that explanation, 2) why that explanation is needed, and 3) in which context and other situated information the explanation is presented." This again brings us to the domain and problem relevancy.

Model interpretability can be intrinsic or post-hoc (external). Intrinsic means that the interpretability is achieved by restricting the complexity of the ML model at the time of building while post-hoc means that interpretability is achieved by applying methods that analyze the model after training and generate an explanation [14]. In case of intrinsically interpretable models, the explanation is the model itself.

In the literature, we found somewhat different but related categorizations of the methodology used for applying XAI. We consolidated the different points of view and combined them into following categorization of XAI that is adapted from [9, 14, 16–21, 29]:

- **Model Design:** Intrinsic, post-hoc.
- **Scope:** Local (explaining just one instance of prediction), global (explaining entire model behavior).
- **Relevancy to Model:** Model-specific, model-agnostic.
- **Methodology:** Example-based, simplification-based, feature relevance-based, perturbation-based, back propagation-, gradient-based, ontology-based.
- **Timing:** Pre-model, in-model, post-model.
- **Presentation:** Visualization, text explanation, mathematical explanation.
- **Data Type:** Text, tabular, image, graphs.

So far, many applications of XAI are on deep learning models due to their powerful but opaque nature with the goal to provide easy model-agnostic explanation of already designed models. In the literature, most used approaches are post-hoc model-agnostic, which means that they can be applied to "any" black-box models after they have been trained (e.g., in literature we see high usage of XAI methods focused on local instances such as LIME [22], feature relevance-based SHAP [23] or Partial Dependence Plot [24], backpropagation or gradient-based such as Attribution Maps or DeepLIFT [25]). The main benefit of post-hoc specific models is their flexibility to be applied to any existing or new machine learning models without the need to understand the structure and internal functioning of the model. The opposite of post-hoc methods are intrinsically interpretable models where the model itself can be used for explanation, see [10] for details. An example of those are the visual transformers which are gaining traction due to the explanations given by their built-in attention map, and we see their usage increasing in healthcare applications, for example in [38, 39]. We also see increased usage of model-specific methods (such as Grad-CAM [26]) for deep learning networks. Model-specific XAI methods are generally using some of the model-specific architecture to create explainability such as removing the last layer and creating a class activation heatmap, as described in [26].

Explainability of a machine learning model needs to be presented in different ways to different stakeholders as they will need a personalized view and level of depth. Following a workflow of a usual ML project, the stakeholders could be summarized into the following categories which are synthetized from [12, 20, 27]: Model Builder (data scientist/ML developer, ML ops or IT/Dev ops engineer), Model Breakers (domain experts, business decision makers, auditors) and Model Consumers (provider of the solution/service using the ML model, end user of the model).

3 Opportunities, Gaps and Challenges

With reference to above mentioned points, below is an attempt to outline the most important areas that the future research in XAI should address. We haven't classified the points in opportunities, gaps, and challenges as we believe that each of the points represents a gap and offers an opportunity or challenge based on how well the point is addressed. All below-mentioned points are relevant for applying XAI to the healthcare domain:

- **Definition of the terms**: Agree upon definitions of interpretability and explainability, agree upon vocabulary and taxonomy of XAI methods, see [9–11, 14, 16–21, 29].
- **Explanations:** Reach an agreement what are good, human-friendly explanations, perform more exploration of the human-computer-interaction aspect of explanations such as visualization, using concepts or ontologies, etc. Create benchmarking models for the quality of explanations, automatic generations of explanations from the ML model, and reducing human subjectivity. Investigate legal implications of explanations provided, etc. See [12–15, 17, 29] for details.
- **Quality and performance metrics:** Create frameworks for the evaluation of performance and definition of agreed upon metrics for measuring and benchmarking XAI

methods, see [9, 10, 13, 14]. An interesting interpretability challenge was found in the user study carried out with data scientists by [28] with the results of the study indicating that data scientists tend to over-trust and not correctly use the interpretability tools which can have dangerous implications.
- **ML Ops:** Integration and automation of XAI methods into ML model life cycle and deployment model, see [12, 20, 27, 28].
- **Safety:** We need to further research on security of explainability of AI, including methods to prohibit the fooling of XAI methods through perturbations and randomized input as well as methods to mitigate inferring private and sensitive information through explanations, see [9, 10, 13, 17, 19–21]. In addition, XAI methods could be relevant to expose risks entailed in the large parameter natural language models such as GPT-3 [34] whose usage is growing. We are just becoming aware of additional security risks that could be impactful in healthcare. In [35] the authors claim that such models can memorize parts of the training data within their parameters. This means that it is possible to carry out attacks to retrieve potentially privacy-sensitive information that were present in the training data.
- **Regulations:** Create regulatory and legislative framework for XAI involving fairness, protection of privacy, truthfulness of explanations and accountability, see [10, 14, 18].
- **Human - ML model collaboration:** Are there ways to work on creating explainability jointly and interactively using novel XAI methods and create a dialog and collaboration between human and the ML model to be interpreted? There is a lack of methods to generate feedback from human to ML model (the so-called machine teaching). Define a clear framework of integration into human workflows. See [9, 10, 14, 18] for details.

In addition, we foresee the need for specific adjustments in healthcare such as specialized interfaces for various medical situations, e.g., for emergency room or surgery preparations, more rigorous definitions of data privacy, more comprehensive processing guidelines to ensure patient safety as well as a more rigorous definition of the level of explanation to be provided. The topics of integrating XAI into healthcare workflows, accountability, and safety of the XAI methods used are very important for healthcare. Furthermore, creating and implementing a special auditing process or framework with safeguards checks for XAI for healthcare could be beneficial to facilitate fairness, safety, stability, and fidelity of XAI.

Another potential restriction is to audit the quality of the data used with the goal to ensure that the data that the machine learning systems have been trained on is not of low quality, not biased or perhaps even wrongly annotated data and to ensure that there is no privacy-sensitive information leak. We believe that data access is a bigger issue in XAI in healthcare, and healthcare research overall as it is very difficult to obtain data in the first place. In fact, an interesting area for the future research could be the explainability of the synthetic data in healthcare. As there are difficulties in obtaining the access and sharing health data for research, we believe that next to federated learning, there will be more realistic looking synthetic health data produced. We believe this could be an interesting area of research due to difficulties with obtaining and cleaning training data as the research community will need to prove the quality of training the model using synthetic data and that models trained in such a fashion generate well on real patients' data without the risk of exposing any privacy-sensitive information.

4 Recommendations for Applying a Problem/Domain Approach to Interpretability and Explainability

In Sect. 3, we highlighted the main gaps/issues for XAI which are slowing down the adoption of AI in healthcare. As stated previously, the state-of-the-art approach is looking at explainability through "model lens": model-agnostic interpretability, model-specific interpretability, and intrinsically interpretable models. The same "model lens" is then applied to any problem and any domain. See [16, 18, 36–39] for examples of such previous work. We believe that this way of applying XAI often creates unnecessary complex situations as different problems require different levels of explainability, and they have different stakeholders and regulations that often might require different explanations. For example, in healthcare domain itself, the scenarios can be very different: explainability of a model in an emergency room scenario, dealing with life-or-death situations, will have different requirements compared to explainability of a model in drug discovery. The importance of the domain for XAI has also been noted by authors in e.g. [10, 18]. To address this, we suggest applying a "problem/domain lens" approach to the explainability from the start and using this approach to (semi-) automatically find the most suitable XAI model for a problem in a domain and if none is found, develop a new model by combining or redesigning existing ones or developing completely new problem/domain explainable models considering relevant explanations, performance, and safety metrics, etc.

While this approach will not mitigate all issues stated, we expect improvements by capitalizing on the knowledge of the problem/domain in our approach. This makes it easier to design quality, performance, safety level, needed explanations level and facilitate human and model collaboration. Let us look at an example in medical imaging:

Problem definition: Using AI to detect and classify type and potential abnormality of white blood cells in patients' blood smear images with suspicion of haematological malignancy.

Potential machine learning algorithms: We know the common machine learning algorithms applied for such problems – e.g., CNN, Capsule Nets, Visual Transformers, etc. Note that we can also design intrinsically interpretable deep learning models for image recognition such as adapted versions of CNN like [30] or different architectures like [31, 38, 39].

Potential XAI methods: If we are not using an intrinsically interpretable model, we can use some of the common XAI methods like SHAP [23], LIME [22], gradient-based Grad-CAM [26] or attribution-propagation such as layer-wise relevance propagation (LRP) [32], attention map [38, 39] and others to verify that the model is looking at the right pixels/patches in the image for its decision. This is something that our approach could (semi-) automatically suggest given the problem/domain. Figure 1 below shows one of the experiments we conducted in exploring suitable XAI methods for the problem we defined above. The figure shows the results of using CNN and Grad-CAM for Acute Lymphoblastic Leukaemia (ALL) classifying lymphoblasts on a blood-smear image of ALL-IDB database [40].

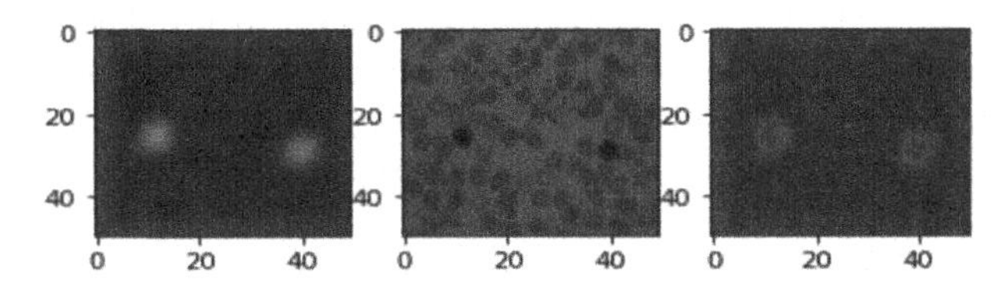

Fig. 1. An example of Grad-CAM implementation on ALL blood smears

Metrics: We can identify the KPIs that are relevant for this problem and domain, such as accuracy, number of patients that pathologist can perform this analysis per day, turnaround time from patient reception to diagnosis, response time for blood smear analysis. See [33] for more examples.

Stakeholders: For this problem the types of stakeholders such as doctors (pathologist, haematologist), nurses, patients, and insurance companies. Also, we need to consider that we might need to discuss with stakeholders what XAI methods would be acceptable/preferable to them.

Safety and regulatory level: We need to be aware of the regulations, laws, and standards such as hospital regulations, insurance regulations, GDPR or future Artificial Intelligence Act, etc.

Similar to automated machine learning (AutoML) [41], we are working on expanding this idea to conceptualize a framework/algorithm to automatize XAI for practitioners. The example above is just a short example, however having the knowledge of the problem and the domain lets us tailor a unique approach of the explainability and explanations that are relevant, using industry language and standards which makes them easy to use for the stakeholders. Using the AutoML or AutoXAI idea we could add a degree of automatism in selecting the right XAI method, parameter optimization, metrics, stakeholders, and regulatory suggestions, and selecting the right level and type of explanations.

5 Conclusion and Future Work

In conclusion, it is important to address the opportunities and the challenges of explainable AI thus enabling wider use and deployment of explainable machine learning assisted decision-making support systems in healthcare. As stated in the introductory chapter, the main objective and contribution of this paper is two-fold: we communicated the summary of our field survey and for easier comprehension and onboarding to this research field we suggest a synthesized taxonomy for categorizing explainability methods and a summary of opportunities, gaps, and challenges for applying explainability of AI to healthcare.

Secondly, we are investigating with our research question how looking at the problem, the context, and the domain where the ML model will be applied can simplify, streamline, and personalize the applications of XAI methods. Furthermore, we are exploring whether this approach be done in an AutoML similar fashion, as AutoXAI, to support the ML practitioners in applying the right "configuration settings" to a XAI application.

Our initial experience of the work in progress using this approach shows promise. In our future work, we will be assessing the results of applying this problem/domain lens on explainability and pointing out gaps as well as suggesting future areas of research.

References

1. World Health Organization (WHO). https://www.who.int/news/item/28-06-2021-who-issues-first-global-report-on-ai-in-health-and-six-guiding-principles-for-its-design-and-use. Accessed 14 July 2021
2. Aurangzeb, A.M., Eckert, C., Teredesai, A.: Interpretable machine learning in healthcare. In: Proceedings of the 2018 ACM International Conference on Bioinformatics, Computational Biology, and Health Informatics, pp. 559–560 (2018)
3. Pang, W., Markovic, M., Naja, I., Fung, C.P., Edwards, P.: On evidence capture for accountable AI systems. In: SICSA Workshop on eXplainable Artificial Intelligence (XAI) (2021)
4. Gunning, D., Aha, D.: Explainable artificial intelligence (XAI) program. AI Mag. **40**(2), 44–58 (2019)
5. European Law General Data Protection Regulation. https://eur-lex.europa.eu/legal-content/EN/TXT/?uri=CELEX%3A02016R0679-20160504&qid=1532348683434. Accessed 27 July 2021
6. European Commission Artificial Intelligence Act. https://eur-lex.europa.eu/legal-content/EN/TXT/?qid=1623335154975&uri=CELEX%3A52021PC0206. Accessed 18 July 2021
7. Dimensions query "Explainable AND Artificial Intelligence". https://app.dimensions.ai/analytics/publication/overview/timeline?search_mode=content&search_text=explainable%20AND%20%22artificial%20intelligence%22&search_type=kws&search_field=full_search. Accessed 14 July 2021
8. Dimensions query "Interpretable AND Artificial Intelligence". https://app.dimensions.ai/analyics/publication/overview/timeline?search_mode=content&search_text=interpretable%20AND%20%22artificial%20intelligence%22&search_type=kws&search_field=full_search. Accessed 14 July 2021
9. Das, A., Rad, P.: Opportunities and challenges in explainable artificial intelligence (XAI): a survey. ArXiv preprint arXiv:2006.11371 (2020)
10. Rudin, C.: Stop explaining black box machine learning models for high stakes decisions and use interpretable models instead. Nat. Mach. Intell. **1**(5), 206–215 (2019)
11. Gilpin, L.H., Bau, D., Yuan, B.Z., Bajwa, A., Specter, M., Kagal, L.: Explaining explanations: an overview of interpretability of machine learning. In: 2018 IEEE 5th International Conference on Data Science and Advanced Analytics (DSAA), pp. 80–89, IEEE (2018)
12. Derek, D., Schulz, S., Besold, T.R.: What does explainable AI really mean? A new conceptualization of perspectives. ArXiv preprint arXiv:1710.00794 (2017)
13. Doshi-Velez, F., Kim, B.: Towards a rigorous science of interpretable machine learning. ArXiv preprint arXiv:1702.08608 (2017)
14. Molnar, C.: Interpretable Machine Learning, A Guide for Making Black Box Models Explainable. Leanpub, Monee, IL, USA (2020)

15. Ferreira, J.J., Monteiro, M.S.: What are people doing about XAI user experience? A survey on AI explainability research and practice. In: Marcus, A., Rosenzweig, E. (eds.) HCII 2020. LNCS, vol. 12201, pp. 56–73. Springer, Cham (2020). https://doi.org/10.1007/978-3-030-49760-6_4
16. Tjoa, E., Guan, C.: A survey on explainable artificial intelligence (XAI): toward medical XAI. IEEE Trans. Neural Netw. Learn. Syst. 1–21 (2020)
17. Longo, L., Goebel, R., Lecue, F., Kieseberg, P., Holzinger, A.: Explainable artificial intelligence: concepts, applications, research challenges and visions. In: Holzinger, A., Kieseberg, P., Tjoa, A.M., Weippl, E. (eds.) CD-MAKE 2020. LNCS, vol. 12279, pp. 1–16. Springer, Cham (2020). https://doi.org/10.1007/978-3-030-57321-8_1
18. Adadi, A., Berrada, M.: Peeking inside the black box: a survey on explainable artificial intelligence (XAI). IEEE Access (6), 52138–52160 (2018)
19. Carvalho, D.V., Pereira, E.M.: Cardoso: machine learning interpretability: a survey on methods and metrics. Electronics **8**(8), 832 (2019)
20. Arrieta, A.B., et al.: Explainable Artificial Intelligence (XAI): concepts, taxonomies, opportunities and challenges toward responsible AI. Inf. Fus. **58**, 82–115 (2020)
21. Linardatos, P., Papastefanopoulos, V., Kotsiantis, S.: Explainable AI: a review of machine learning interpretability methods. Entropy **23**(1), 18 (2021)
22. Ribeiro, M.T., Singh, S., Guestrin, C.: Why should i trust you? Explaining the predictions of any classifier. In: Proceedings of the 22nd ACM SIGKDD International Conference on Knowledge Discovery and Data Mining, pp. 1135–1144 (2016)
23. Lundberg, S., Lee, S.I.: A unified approach to interpreting model predictions. In: Proceedings of the 31st Conference on Neural Information Processing Systems (NIPS), pp. 4765–4774 (2017)
24. Friedman, J.H.: Greedy function approximation: a gradient boosting machine. Ann. Stat. 1189–1232 (2001)
25. Avanti, S., Greenside, P., Kundaje A.: Learning important features through propagating activation differences. In: International Conference on Machine Learning. PMLR, pp. 3145– 3153 (2017)
26. Selvaraju, R.R., Cogswell, M., Das, A., Vedantam, R., Parikh, D., Batra, D.: Grad-cam: visual explanations from deep networks via gradient-based localization. In: Proceedings of the IEEE International Conference on Computer Vision, pp. 618–626. IEEE (2017)
27. Hong, S.R., Hullman, J., Bertini, E.: Human factors in model interpretability: Industry practices, challenges, and needs. In: Proceedings of the ACM on Human-Computer Interaction 4 CSCW1, pp. 1–26 (2020)
28. Kaur, H., Nori, H., Jenkins, S., Caruana, R., Wallach, H., Wortman Vaughan, J.: Interpreting interpretability: understanding data scientists' use of interpretability tools for machine learning. In: Proceedings of the 2020 CHI Conference on Human Factors in Computing Systems, pp. 1–14 (2020)
29. Carrilo, A., Cantu, L.F., Noriega, A.: Individual explanations in machine learning models: a survey for pratictioners. arXiv preprint arXiv:2104.04144 (2021)
30. Chen, C., Li, O., Tao, C., Barnett, A.J., Su, J., Rudin, C.: This looks like that: deep learning for interpretable image recognition. arXiv preprint arXiv:1806.10574 (2018)
31. Singh, G., Yow, K.C.: These do not look like those: an interpretable deep learning model for image recognition. IEEE Access **9**, 41482–41493 (2021)
32. Eitel, F., et al.: Uncovering convolutional neural network decisions for diagnosing multiple sclerosis on conventional MRI using layer-wise relevance propagation. NeuroImage Clin. (24), 102003 (2019)

33. Royal College of Pathologists, Key Performance Indicators in Pathology. https://www.rcpath.org/uploads/assets/e7b7b680-a957-4f48-aa78e601e42816de/Key-Performance-Indicators-in-Pathology-Recommendations-from-the-Royal-College-of-Pathologists.pdf. Accessed 25 July 2021
34. Floridi, L., Chiriatti, M.: GPT-3: Its nature, scope, limits, and consequences. Mind. Mach. **30**(4), 681–694 (2020)
35. Carlini, N., et al.: Extracting training data from large language models. arXiv preprint arXiv:2012.07805 (2020)
36. Shaban-Nejad, A., Michalowski, M., Buckeridge, D.L.: Explainability and interpretability: keys to deep medicine. In: Shaban-Nejad, A., Michalowski, M., Buckeridge, D.L. (eds.) Explainable AI in Healthcare and Medicine. SCI, vol. 914, pp. 1–10. Springer, Cham (2021). https://doi.org/10.1007/978-3-030-53352-6_1
37. Harsha, N., Jenkins, S., Koch, P., Caruana R: Interpretml: a unified framework for machine learning interpretability. arXiv preprint arXiv:1909.09223 (2019)
38. Matsoukas, Christos, M., Haslum, J.F., Söderberg, M., Smith, K.: Is it time to replace CNNs with transformers for medical images? arXiv preprint arXiv:2108.09038. Accepted at ICCV-2021: Workshop on Computer Vision for Automated Medical Diagno-sis (CVAMD) (2021)
39. Wenqi, S., Tong, L., Zhu, Y., Wang, M.D.: COVID-19 automatic diagnosis with ra-diographic imaging: explainable attention transfer deep neural networks. IEEE J. Biomed. Health Inf. (25), 2376–2386 (2021)
40. Labati, R.D., Piuri, V., Scotti, F.: All-IDB: the acute lymphoblastic leukemia image database for image processing. In: 2011 18th IEEE International Conference on Image Processing, pp. 2045–2048. IEEE (2011)
41. Hutter, F., Kotthoff, L., Vanschoren, J.: Automated Machine Learning: Methods, Systems, Challenges. Springer, Heidelberg (2019). https://doi.org/10.1007/978-3-030-05318-5

End-to-End Deep Learning for Detecting Metastatic Breast Cancer in Axillary Lymph Node from Digital Pathology Images

Turki Turki[1(✉)], Anmar Al-Sharif[1], and Y-h. Taguchi[2(✉)]

[1] Department of Computer Science, King Abdulaziz University, Jeddah 21589, Saudi Arabia
tturki@kau.edu.sa

[2] Department of Physics, Chuo University, Tokyo 112-8551, Japan
tag@granular.com

Abstract. Metastatic breast cancer is one of the attributed leading causes of women deaths worldwide. Accurate diagnosis to the spread of breast cancer to axillary lymph nodes (ALNs) is done by breast pathologist, utilizing the microscope to inspect and then providing the biopsy report. Because such a diagnosis process requires special expertise, there is a need for artificial intelligence-based tools to assist breast pathologists to automatically detect breast cancer metastases. This study aims to detect breast cancer metastasized to ALN with end-to-end deep learning (DL). Also, we utilize several DL architectures, including DenseNet121, ResNet50, VGG16, Xception as well as a customized lightweight convolutional neural network. We evaluate the DL models on NVIDIA GeForce RTX 2080Ti GPU using 114 processed microscopic images pertaining to ALN metastases in breast cancer patients. Of note, we included in the evaluation several machine learning (ML) algorithms coupled with histogram features, extracted from the same studied images. These ML include Random Forest, AdaBoost and SVM. Compared to all models employed in this study, experimental results show that DenseNet121 generates the highest performance results based on measures used in class-imbalance problems.

Keywords: Breast cancer · Axillary lymph node status · Computational pathology · Artificial intelligence · Deep learning

1 Introduction

Breast cancer metastasis is associated with increased mortality rates attributed to the possible spread of cancer cells outside the tumor through lymphatic vessel to other areas in the lymph system, including lymph nodes in axilla, near collarbone and breast (see Fig. 1) [1–3]. To accurately diagnose breast cancer metastasized to lymph nodes, a biopsy specimen is obtained from the patient and is provided to a breast pathologist, using a microscope associated with camera to inspect the specimen via a computer software through manual whole slide imaging (see Fig. 2). Such a diagnosis process requires experience and is time consuming when dealing with many patients [4–8].

H. Yin et al. (Eds.): IDEAL 2021, LNCS 13113, pp. 343–353, 2021.
https://doi.org/10.1007/978-3-030-91608-4_34

Accurate and early detection of breast cancer metastases can help pathologists to avoid examining breast cancer under the microscope and rapidly reporting the status of breast cancer patients [9]. Therefore, patients can immediately start the treatment plan with medical oncologists and surgeons [10]. Because the treatment plan provided by oncologists and surgeons depends on the diagnosis and report obtained from pathologists, researchers have proposed computational methods using artificial intelligence to assist pathologists [11–15]. For example, Lee et al. [16] utilized convolutional neural networks to predict breast cancer metastases to axillary lymph nodes (ALNs). They utilized DenseNet-121 to build a deep learning (DL) model, which was then compared against other machine learning algorithms (including logistic regression [17], SVM [18, 19], and XGBoost [20]) with hand-crafted features. Experimental results demonstrated that DenseNet-121 performed better than machine learning algorithms.

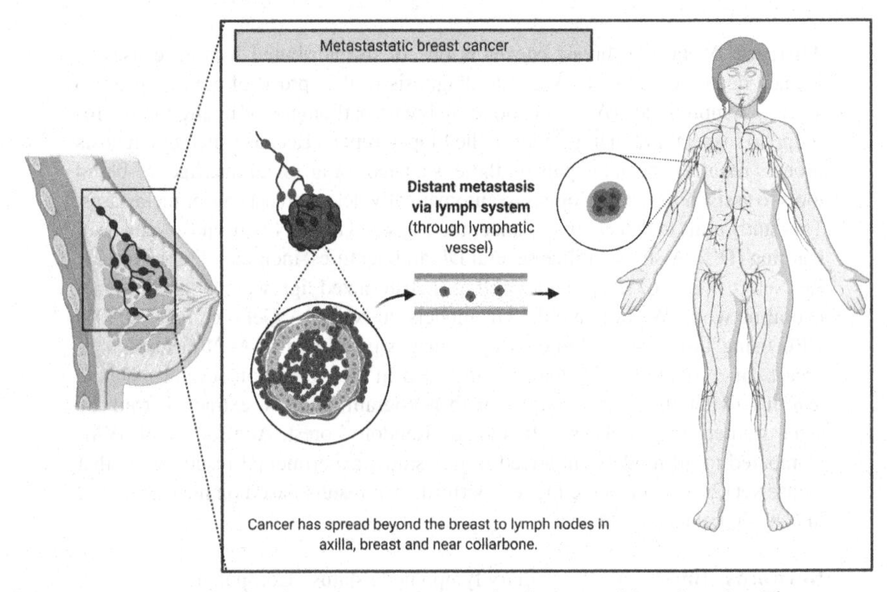

Fig. 1. An overview showing breast cancer metastasized to several lymph nodes. Figure created with Biorender.com.

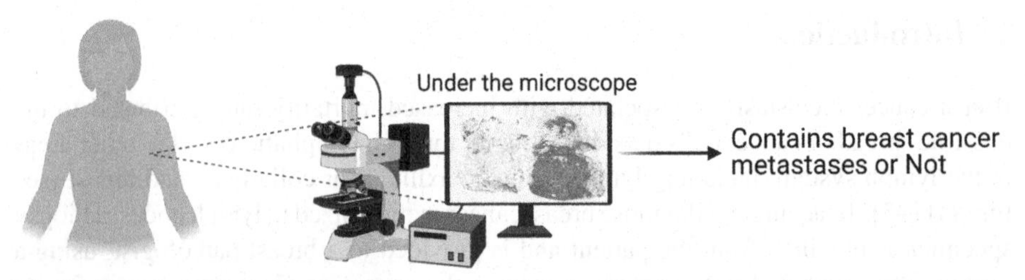

Fig. 2. Whole slide imaging is done by a pathologist in which a biopsy with H&E stain is placed under the microscope, including an attached camera. Figure created with Biorender.com.

Zhou et al. [21] proposed a DL approach for the automatic prediction of lymph node metastasis in breast cancer patients, working as follows. A training and validation sets consisting of 877 images, where 441 images belong to patients with lymph-node metastases while 436 images are for patients with no lymph-node metastases. Several DL architectures were adapted and utilized for this binary classification tasks, including Inception-ResNet V2, ResNet-101, and Inception V3. The resulted models were evaluated using internal and external test sets (labeled by 5 radiologists), consisting of 97 and 81 images, respectively. Reported results show that Inception V3 generated the highest performance results. Zheng et al. [22] proposed a DL approach incorporating breast cancer images and clinical parameters (obtained from clinical texts such as tumor type, age estrogen receptor (ER) status and others characteristics pertaining to tumor and patients) to assist clinical oncologists in deciding if there is a need or not to perform ALN dissection or sentinel lymph node biopsy of patients. Therefore, avoiding such an unnecessary surgery and associated complications. Experimental results demonstrated the good performance results of the presented DL approach when is used with ResNet50 DL. Other researchers have proposed other DL approaches for various learning tasks pertaining to breast cancer metastases [23–31].

Although previous studies aim to provide computational methods to detect breast cancer metastasized to lymph nodes, this study is unique in several respects. First, we aim to utilize and adapt additional DL architectures including a customized convolutional neural network. Examples of such DL architectures include DenseNet121, ResNet50, VGG16 and Xception. Second, we process 114 gigapixel images pertaining to ALN metastases on NVIDIA GeForce RTX 2080Ti GPU. Then, resizing each image to 256 * 256 pixels to be used in DL experiments. We include these images in supplementary

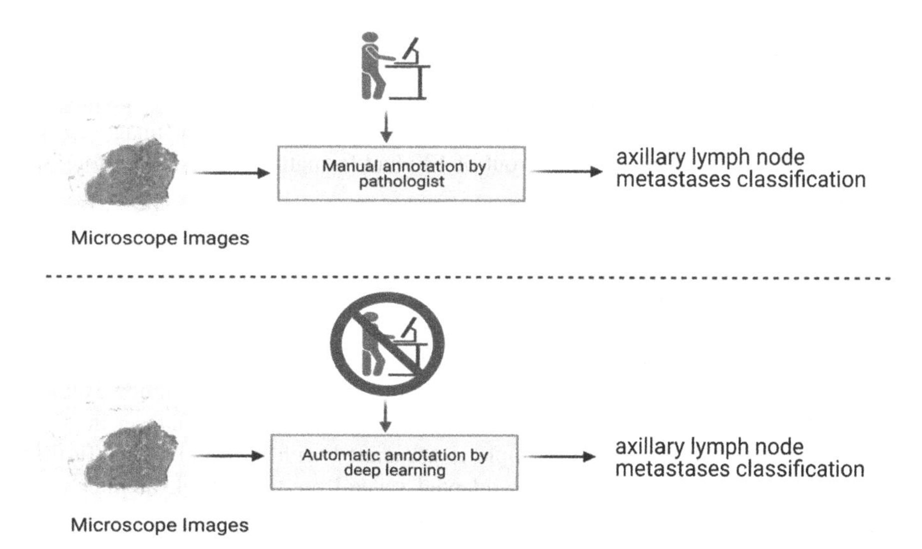

Fig. 3. The upper panel demonstrates a manual annotation by a pathologist to determine the outcome (i.e., axillary lymph node metastases (ALN) or no ALN metastases) of microscope images in breast cancer patients. The lower panel displays an automatic annotation by deep learning (DL) technology. Figure created with Biorender.com.

material. Third, we evaluate and report performance differences among DL models on the whole image dataset using 5-fold cross-validation. Moreover, we include machine learning (ML) algorithms coupled with hand-crafted features pertaining to histogram. These features are extracted from the same image dataset provided to DL models. Examples of such ML include Random Forest [32], AdaBoost [33] and SVM. Fourth, our reported results on NVIDIA GeForce RTX 2080Ti GPU show that DensNet121 outperformed other DL models including a convolutional neural network consisting of few layers on manual whole slide images pertaining to ALN metastases. Fifth, the presented study provides a pathological process automation using DL as shown in Fig. 3.

The remaining of this study is organized as follows. The second section describes the image dataset pertaining to ALN metastases, as well as presenting the DL approach to address the binary classification task. The third section reports performance results related to predicting ALN metastasis. Then, presenting a discussion of the results in this study. Finally, we conclude the study and point out future research directions.

2 Methods

2.1 Axillary Lymph Node Metastases Image Dataset

Microscopic images for ALN metastases in breast cancer patients were downloaded from the Cancer Imaging Archive at https://wiki.cancerimagingarchive.net/pages/viewpage.action?pageId=52763339 [27, 34, 35]. The dataset used in our study consisted of 114 whole slide images, where 84 had no ALN metastases and 30 images had ALN metastases. We used python script on NVIDIA GeForce RTX 2080Ti GPU to process the 114 gigapixel whole slide images, then resize them to (256 * 256 pixels). We include these processed images in supplementary material (see medRxiv at https://www.medrxiv.org/content/medrxiv/early/2021/04/13/2021.04.09.21255183/DC1/embed/media-1.zip?download=true) and include some pictures in Figures throughout the manuscript. It is worth noting that the original image dataset had 130 images. However, 16 images out of 130 had formatting issues. Therefore, we excluded them.

2.2 Deep Learning Approach

In Fig. 4, we show the DL approach used in this study to discriminate between positive and negative ALN in breast cancer patients. For a given training set $S = \{(x_i, y_i)\}_{i=1}^{m}$ composed of m labeled ALN images obtained from a pathologist. If a training example is associated with class label of 1, then that indicates the presence of cancer in ALN. On the other hand, if a training example is associated with a class label of 0, then cancer is not present in the ALN. We adapted several pre-trained DL architectures, including DensNet121. ResNet50, VGG16, Xception and a lightweight convolutional neural network (named LCNN). For the pre-trained DL models, we replaced the dense layers in the densely connected classifier of each pre-trained model by two dense layers. The first dense layer includes a ReLu activation, followed by a dropout. Because we have a binary classification task, the second dense layer had 1 unit and a sigmoid activation.

The LCNN had 4 layers as two pairs of interleaved Convolutional and MaxPooling layers for extracting useful features. Then, features are flattened, and the outcome is provided to a densely connected classifier consisting of three dense layers, where the last layer has 1 unit and a sigmoid activation as in the adapted pre-trained DL models. For all DL models, we trained each DL model on a balanced training set of ALN images by undersampling from the majority class examples. In this study, majority class examples correspond to images of negative ALN. Then, we applied the trained models to unseen ALN images to generate predictions correspond to probabilities. If predicted probability is greater than 0.5, then it is mapped to 1; otherwise, it is mapped to 0. Prediction of 1 (i.e., considered positive) means ALNs have cancer while 0 (i.e., considered negative) means ALNs are free of cancer.

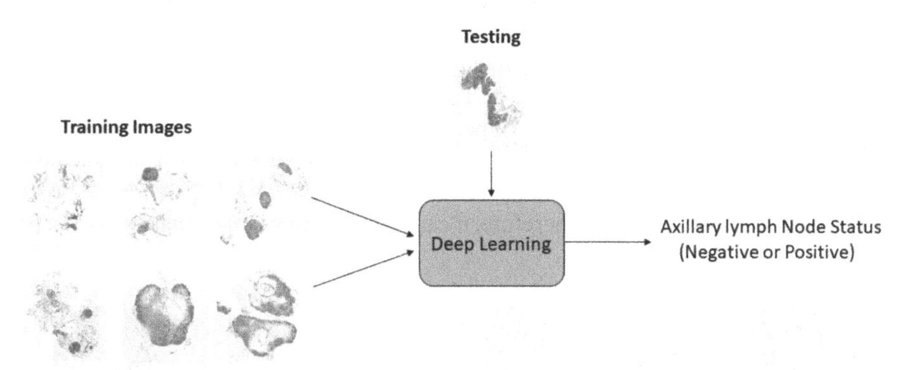

Fig. 4. A demonstration of the deep learning (DL) approach employed in this study for discrimination between positive axillary lymph node (ALN) and negative ALN in metastasized breast cancer.

3 Results

3.1 Classification Methodology

We adapted five DL architectures—LCNN, DensNet121. ResNet50, VGG16, Xception—for predicting ALN metastases in breast cancer patients. All architectures work under the supervised learning scenario. That is, a training set is provided to train each DL model. Then, a test set is provided to each trained DL model to generate predictions, mapped to 1 (i.e., positive ALN) or 0 (i.e., negative ALN). We employed four performance measures: Accuracy (ACC), F1-score (F1), area under curve (AUC) and Matthews correlation coefficient (MCC) [36]. Also, we utilized 5-fold cross-validation as in [37].

3.2 Implementation Details

We utilized Anaconda Python 3.7.4 (with Jupyter Notebook), Keras and Sklearn libraries to run and evaluate DL experiments. For the hardware, we used NVIDIA RTX 2080Ti

GPU consisting of 4352 CUDA cores with 11 GB memory. In terms of ML algorithms, we implemented the following: AdaBoost using adaboost function in JOUSBoost package [38], Random Forest using randomForest function in randomForest package [39], SVM using svm function coupled with radial kernel in e1071 package [40], histogram of oriented gradients to extract features using HOS function in OpenImageR package [41].

3.3 Classification Results

Figure 5 displays the combined confusion matrices of 5 test splits during the cross-validation process. If numbers in each confusion matrix is added, then we get 114, corresponding to the whole image dataset in this study. As we have 84 images from the negative class label and 30 images of the positive class label, DenseNet121 accurately predicts 18 out of 30 positive class examples (i.e., 60%). For the 84 negative class examples, DenseNet121 accurately predicts 60 out of these 84 images (i.e., ~ 71%). Although RandomForest performs the best on the negative class examples by accurately predicting 72 out of 84 (i.e., ~85.7%), it performs poorly on the positive class examples by accurately predicting 7 out of 30 (i.e., ~ 23.33%) as shown in Fig. 5. The same holds for SVM as it performs well on the negative class examples, while performing poorly on the positive class examples. Therefore, it can be seen from Table 1 that DenseNet121

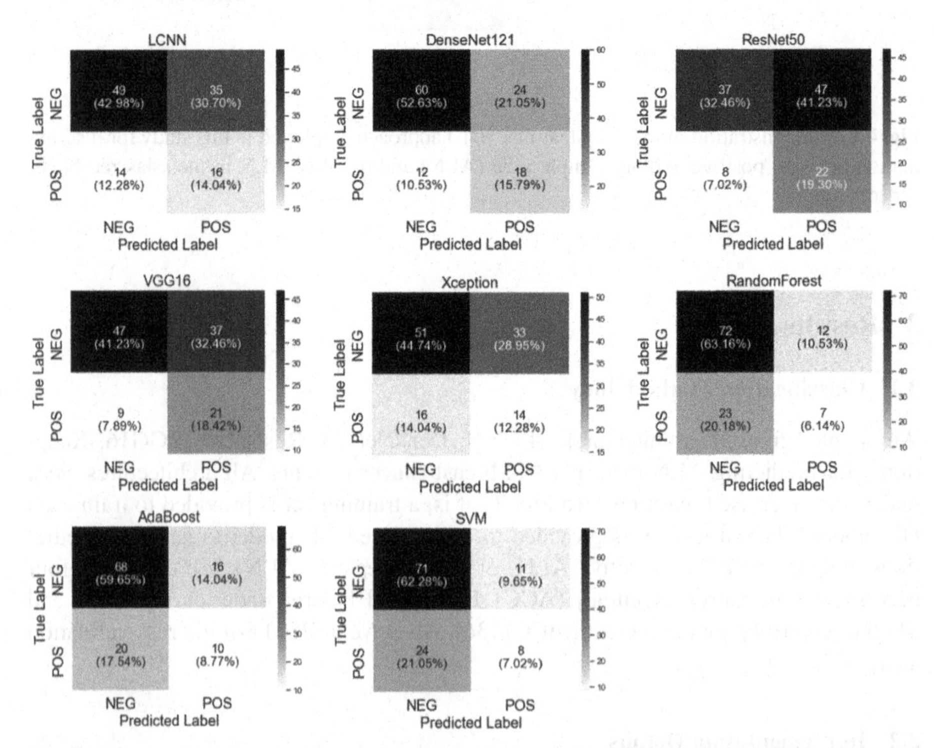

Fig. 5. The combined confusion matrices of predicting axillary lymph node (ALN) metastases using all machine learning (ML) and deep learning (DL) models during 5-fold cross-validation. LCNN is lightweight convolutional neural network.

(when compared to other models) generates the highest results (as shown in bold) in terms of several measures for imbalanced classification. These results demonstrate the good performance of DenseNet121.

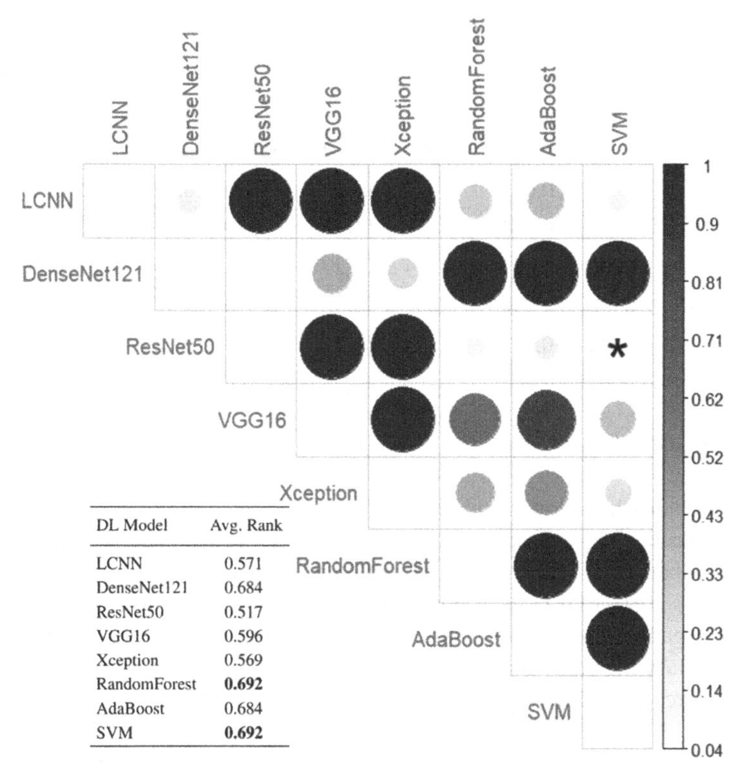

DL Model	Avg. Rank
LCNN	0.571
DenseNet121	0.684
ResNet50	0.517
VGG16	0.596
Xception	0.569
RandomForest	**0.692**
AdaBoost	0.684
SVM	**0.692**

Fig. 6. *P*-value and average ranking between machine learning (ML) and deep learning (DL) models using the Friedman test with Bergmann and Hommel's correction. The highest average ranking (shown in bold) of a model indicates its performance is better than remaining models. The asterisk between two models shows that the model with highest ranking generates significant results as $P < 0.05$.

Figure 6 reports *P*-values and average ranking between studied DL and ML models according to Friedman post-hoc test with Bergmann and Hommel's correction. Although SVM and RandomForest models generated the highest average ranking (0.692) when accuracy performance measure is considered, the asterisk between each pair of all models indicates that the performance difference is not significant when compared against DenseNet121.

Table 1. Performance comparison between different machine learning (ML) and deep learning (DL) models (during 5-fold cross-validation) for predicting positive axillary lymph nodes (ALNs) and negative ALNs. MACC is the mean accuracy. MF1 is the mean F1. MMCC is the mean MCC. SD is the standard deviation. The highest performance result is shown in bold.

Model	MACC ± SD	MF1 ± SD	MAUC ± SD	MMCC ± SD
LCNN	0.571 ± 0.058	0.377 ± 0.139	0.568 ± 0.091	0.112 ± 0.146
DenseNet121	0.684 ± 0.042	**0.466** ± 0.173	**0.646** ± 0.068	**0.274** ± 0.131
ResNet50	0.517 ± 0.055	0.406 ± 0.167	0.584 ± 0.122	0.133 ± 0.174
VGG16	0.596 ± 0.032	0.440 ± 0.162	0.610 ± 0.065	0.208 ± 0.134
Xception	0.569 ± 0.073	0.355 ± 0.136	0.573 ± 0.120	0.110 ± 0.171
RandomForest	**0.692** ± 0.043	0.274 ± 0.127	0.566 ± 0.069	0.140 ± 0.150
AdaBoost	0.684 ± 0.056	0.338 ± 0.143	0.586 ± 0.072	0.160 ± 0.146
SVM	**0.692** ± 0.053	0.270 ± 0.173	0.533 ± 0.070	0.088 ± 0.147

4 Discussion

Our study includes the following: (1) evaluating several ML algorithms and adapting several DL architectures including a customized convolutional neural network to address the binary classification task pertaining to detecting ALN metastases, (2) utilizing NVIDIA GeForce RTX 2080Ti GPU to evaluate DL models using 5-fold cross-validation and resize 114 gigapixel images to (256 * 256 pixels). We made these images available in supplementary material, and (3) reporting performance results using several performance measures as well as *p*-values using Friedman test with Bergmann and Hommel's correction.

When training ML models, we used SMOTE [42, 43] to oversample minority class examples. As a result, we created a balanced training set. Then, providing the balanced training set to a machine learning algorithm, to obtain a model. This model is subsequently used to perform prediction on the test set. It is worth noting that we reported results of ML models using 5-fold cross-validation as DL models in this study. The only difference is that the dataset provided to ML models consisted of 114 examples, each consisting of 54 features. Although SVM and RandomForest models generated the highest MACC of 69.2% (Table 1), DenseNet121 achieved the highest performance results based on performance measures for imbalances classification (Table 1). These results may demonstrate the superiority of DensNet121 when dealing with such an imbalanced and small image dataset.

For all pre-trained DL models in this study, we used the feature extraction technique where we use all layers (excluding dense layers) for the feature extraction. Then, we tuned parameters of new dense layers during the training, because we changed densely connected layers to deal with the binary classification problem as in [37]. For the LCNN model, we build it from scratch with data augmentation.

For the optimization process configuration during training of DL models, we used RMSprop optimizer. Also, we used the binary_crossentropy loss function and the accuracy metric as in [37].

It is worth noting that DenseNet121 generated the highest results in terms of imbalanced performance measures (Table 1) when compared to all other models. This coincides with recent work by Lee et al. [16], where DenseNet121 (compared against machine learning models coupled with hand-crafted features) also generated the highest performance results when assessed on predicting ALN metastasis of another dataset.

5 Conclusion and Future Work

This paper presents a DL approach for detecting ALN metastases from digital pathology images. The DL approach adapts several DL architectures, including DenseNet12, ResNet50, VGG16, Xception and a customized lightweight convolutional neural network. A training set is provided to train DL models. Then, a test set is provided to each trained model to generate predictions. We used NVIDIA GeForce RTX 2080Ti GPU to process 114 gigapixel images and evaluate models using 5-fold cross-validation; these images are available in supplementary material. Experimental results show that DenseNet121 outperforms all models in this study, including hand-crafted features coupled with ML algorithms such as Random Forest, AdaBoost and SVM.

Future work includes: (1) to further boost the performance results via incorporating clinical information combined with data augmentation techniques; and (2) to utilize other learning scenarios such as semi-supervised and unsupervised learning for exploiting unlabeled data [44, 45].

Acknowledgment. This study was supported by KAKENHI 19H05270, 20K12067, 20H04848.

References

1. Davis, R.T., et al.: Transcriptional diversity and bioenergetic shift in human breast cancer metastasis revealed by single-cell RNA sequencing. Nat. Cell Biol. **22**(3), 310–320 (2020)
2. Liu, L., Zhang, Y., Lu, J.: The roles of long noncoding RNAs in breast cancer metastasis. Cell Death Dis. **11**(9), 1–14 (2020)
3. Tan, B.-S., et al.: LncRNA NORAD is repressed by the YAP pathway and suppresses lung and breast cancer metastasis by sequestering S100P. Oncogene **38**(28), 5612–5626 (2019)
4. Balkenhol, M.C., et al.: Deep learning assisted mitotic counting for breast cancer. Lab. Invest. **99**(11), 1596–1606 (2019)
5. Magnusson, R., Gustafsson, M.: LiPLike: towards gene regulatory network predictions of high certainty. Bioinformatics **36**(8), 2522–2529 (2020)
6. Ibrahim, A., et al.: Artificial intelligence in digital breast pathology: techniques and applications. The Breast **49**, 267–273 (2020)
7. Zhou, S., et al.: Intense basolateral membrane staining indicates HER2 positivity in invasive micropapillary breast carcinoma. Mod. Pathol. **33**, 1–12 (2020)
8. Sethy, C., et al.: Nectin-4 promotes lymphangiogenesis and lymphatic metastasis in breast cancer by regulating CXCR4-LYVE-1 axis. Vasc. Pharmacol **140**, 106865 (2021)

9. Anglade, F., Milner, D.A., Jr., Brock, J.E.: Can pathology diagnostic services for cancer be stratified and serve global health? Cancer **126**, 2431–2438 (2020)
10. Biganzoli, L., et al.: The requirements of a specialist breast centre. The Breast **51**, 65–84 (2020)
11. Browning, L., et al.: Digital pathology and artificial intelligence will be key to supporting clinical and academic cellular pathology through COVID-19 and future crises: the PathLAKE Consortium perspective. J. Clin. Pathol. **74**, 443–447 (2020)
12. Sobhani, F., et al.: Artificial intelligence and digital pathology: Opportunities and implications for immuno-oncology. Biochim. Biophys. Acta (BBA)-Rev. Cancer **1875**, 188520 (2021)
13. Boor, P.: Artificial intelligence in nephropathology. Nat. Rev. Nephrol. **16**(1), 4–6 (2020)
14. Steiner, D.F., Chen, P.-H.C., Mermel, C.H.: Closing the translation gap: AI applications in digital pathology. Biochim. Biophys. Acta (BBA)-Rev. Cancer **1875**, 188452 (2020)
15. Simić, S., et al.: Deep convolutional neural networks on automatic classification for skin tumour images. Logic J. IGPL (2021)
16. Lee, Y.-W., et al.: Axillary lymph node metastasis status prediction of early-stage breast cancer using convolutional neural networks. Comput. Biol. Med. **130**, 104206 (2021)
17. Kleinbaum, D.G., et al.: Logistic Regression. Springer, Heidelberg (2002). https://doi.org/10.1007/978-1-4419-1742-
18. Turki, T., Wei, Z.: Boosting support vector machines for cancer discrimination tasks. Comput. Biol. Med. **101**, 236–249 (2018)
19. Schölkopf, B., Smola, A.J., Bach, F.: Learning with Kernels: Support Vector Machines, Regularization, Optimization, and Beyond. MIT Press (2002)
20. Chen, T., Guestrin, C.: Xgboost: a scalable tree boosting system. In: Proceedings of the 22nd ACM sigkdd International Conference on Knowledge Discovery and Data Mining (2016)
21. Zhou, L.-Q., et al.: Lymph node metastasis prediction from primary breast cancer US images using deep learning. Radiology **294**(1), 19–28 (2020)
22. Zheng, X., et al.: Deep learning radiomics can predict axillary lymph node status in early-stage breast cancer. Nat. Commun. **11**(1), 1–9 (2020)
23. Pan, C., et al.: Deep learning reveals cancer metastasis and therapeutic antibody targeting in the entire body. Cell **179**(7), 1661–1676. e19 (2019)
24. Steiner, D.F., et al.: Impact of deep learning assistance on the histopathologic review of lymph nodes for metastatic breast cancer. Am. J. Surg. Pathol. **42**(12), 1636 (2018)
25. Moreau, N., et al.: Deep learning approaches for bone and bone lesion segmentation on 18FDG PET/CT imaging in the context of metastatic breast cancer. In: 2020 42nd Annual International Conference of the IEEE Engineering in Medicine & Biology Society (EMBC). IEEE (2020)
26. Hu, Y., et al.: Deep learning system for lymph node quantification and metastatic cancer identification from whole-slide pathology images. Gastric Cancer **24**(4), 868–877 (2021). https://doi.org/10.1007/s10120-021-01158-9
27. Campanella, G., et al.: Clinical-grade computational pathology using weakly supervised deep learning on whole slide images. Nat. Med. **25**(8), 1301–1309 (2019)
28. Wang, H., et al.: Prediction of breast cancer distant recurrence using natural language processing and knowledge-guided convolutional neural network. Artif. Intell. Med. **110**, 101977 (2020)
29. Levine, A.B., et al.: Rise of the machines: advances in deep learning for cancer diagnosis. Trends Cancer **5**(3), 157–169 (2019)
30. Wang, J., et al.: Boosted EfficientNet: detection of lymph node metastases in breast cancer using convolutional neural networks. Cancers **13**(4), 661 (2021)
31. Jin, Y.W., et al.: Integrative data augmentation with U-Net segmentation masks improves detection of lymph node metastases in breast cancer patients. Cancers **12**(10), 2934 (2020)

32. Breiman, L.: Random forests. Mach. Learn. **45**(1), 5–32 (2001)
33. Schapire, R.E., Freund, Y.: Boosting: Foundations and Algorithms. Cambridge, MA: MIT Press (2012)
34. Clark, K., et al.: The cancer imaging archive (TCIA): maintaining and operating a public information repository. J. Digital Imaging **26**(6), 1045–1057 (2013). https://doi.org/10.1007/s10278-013-9622-7
35. Campanella, G., Hanna, M.G., Brogi, E., Fuchs, T.J.: Breast metastases to axillary lymph nodes. Cancer Imaging Arch. (2019)
36. Japkowicz, N., Shah, M.: Evaluating learning algorithms: a classification perspective. Cambridge University Press (2011)
37. Turki, T., Taguchi, Y.: Discriminating the single-cell gene regulatory networks of human pancreatic islets: a novel deep learning application. Comput. Biol. Med. **132**, 104257 (2021)
38. Olson, M.: JOUSBoost: An R package for improving machine learning classifier probability estimates. (2017)
39. RColorBrewer, S., Liaw, M.A.: Package 'randomForest'. University of California, Berkeley: Berkeley, CA, USA (2018)
40. Meyer, D., et al.: Package 'e1071'. R J. (2019)
41. Mouselimis, L.: OpenImageR: an image processing Toolkit. R package version. **1**(5) (2017)
42. Sáez, J., Luengo, J., Stefanowski, J., Herrera, F.: Managing borderline and noisy examples in imbalanced classification by combining SMOTE with ensemble filtering. In: Corchado, E., Lozano, J.A., Quintián, H., Yin, H. (eds.) IDEAL 2014. LNCS, vol. 8669, pp. 61–68. Springer, Cham (2014). https://doi.org/10.1007/978-3-319-10840-7_8
43. Sáez, J.A., et al.: SMOTE–IPF: Addressing the noisy and borderline examples problem in imbalanced classification by a re-sampling method with filtering. Inf. Sci. **291**, 184–203 (2015)
44. Chapelle, O., Schölkopf, B., Zien, A.: Semi-supervised learning. Adaptive Computation and Machine Learning Series. The MIT Press (2006)
45. Hastie, T., Tibshirani, R., Friedman, J.: Unsupervised learning. In: The elements of statistical learning. SSS, pp. 485–585. Springer, New York (2009). https://doi.org/10.1007/978-0-387-84858-7_14

SFU-CE: Skyline Frequent-Utility Itemset Discovery Using the Cross-Entropy Method

Wei Song[(✉)] and Chuanlong Zheng

School of Information Science and Technology, North China University of Technology, Beijing 100144, China
songwei@ncut.edu.cn

Abstract. Considering both frequency and utility, skyline frequent-utility itemsets (SFUIs) increase the amount of actionable information available for decision-making. In this paper, we propose an algorithm for mining SFUIs called skyline frequent-utility mining based on cross-entropy (SFU-CE). We first model the SFUI mining problem using CE with utility as the optimization object. We then propose critical utility pruning for unpromising itemsets with utility and its upper bound. Furthermore, we also design a random mutation strategy to make itemsets within each sample more diverse. We demonstrate the efficiency and accuracy of SFU-CE by comparing it with state-of-the-art algorithms.

Keywords: Skyline frequent-utility itemsets · Cross-entropy · Critical utility · Random mutation

1 Introduction

As an important sub-field of data mining, itemset mining is used to discover interesting and useful patterns in transaction databases. To evaluate the interestingness or usefulness of itemsets, various measures have been proposed, among which frequency and profit are the two most influential measures.

Considering both frequency and profit, the utility-frequent itemset (UFI) was designed and two algorithms were proposed in [14]. For the UFI model, both minimum utility and support are required. However, it is difficult for users to set appropriate thresholds and the model incurs a high computational cost.

To solve this problem, Goyal et al. introduced SFUIs and proposed the SKYMINE algorithm for mining them [4]. SFUIs are itemsets that are not dominated by any other itemsets in terms of both frequency and utility. Thus, neither the frequency threshold nor utility threshold is required. Using both frequency and utility is meaningful in practical decision-making. For example, parents may choose to rent houses for the students according to the price and distance of the houses from the school to save the time that it takes for students to travel to and from school. In this case, the distance from the house to the school and the price of the house are contrasted; that is, a house that is close to the school normally has a higher price than a house that is far from the school.

H. Yin et al. (Eds.): IDEAL 2021, LNCS 13113, pp. 354–366, 2021.
https://doi.org/10.1007/978-3-030-91608-4_35

In the first SKYMINE algorithm [4], the UP-tree structure was used to transform the original information for mining SFUIs. In the recent algorithms SFU-Miner [8], SKYFUP-D [5], and SKYFUP-B [5], the utility-list (UL) structure was used to improve efficiency. In addition to the UL structure, different arrays, such as the umax array and utilmax array, have also been designed to record the utility and frequency of intermediate results. Because mining SFUIs does not require either a support threshold or utility threshold, the algorithm's efficiency remains the most challenging issue for the SFUI mining (SFUIM) problem.

Inspired by biological and physical phenomena, heuristic methods have been used to discover itemsets within an acceptable time. Typical heuristic methods include the genetic algorithm [2], particle swarm optimization [6], and artificial bee colony algorithm [9]. Recently, the cross-entropy (CE) [11, 13] method has also been used to discover top-*k* high utility itemsets (HUIs) and has demonstrated high efficiency.

Different from TKU-CE and TKU-CE+ [11, 13] that use CE for mining HUIs, we propose a CE-based algorithm called skyline frequent-utility mining based on cross-entropy (SFU-CE) for mining SFUIs. First, we model the SFUIM problem using the typical CE method so that items within itemsets with higher utilities have more opportunities to appear in the next sample. Thus, the sample can filter itemsets with a lower utility, which tend to be dominated by other itemsets. Second, we focus on the maximal utility of items with the highest frequency among all single items, and propose critical utility pruning. Thus, the search space is narrowed using utility and transaction-weighted utilization. Third, we design a random mutation (RM) strategy to improve the diversity of each sample. Different from the typical CE method that generates a new sample strictly following the probability vector, the RM strategy generates new itemsets according to probabilities higher than 0.5. Experimental results demonstrated that SFU-CE discovered sufficient SFUIs in less time.

2 Preliminaries

2.1 SFUIM Problem

Let $\boldsymbol{I} = \{i_1, i_2, \ldots, i_m\}$ be a finite set of items. Set $X \subseteq \boldsymbol{I}$ is called an *itemset*, or a k-itemset if it contains k items. Let $\boldsymbol{D} = \{T_1, T_2, \ldots, T_n\}$ be a transaction database. Each transaction $T_d \in \boldsymbol{D}$ ($1 \leq d \leq n$), where d is a unique identifier, is a subset of $\boldsymbol{I}$. The *frequency* of itemset X, denoted by $f(X)$, is defined as the number of transactions in which X occurs as a subset.

The *internal utility* $q(i_p, T_d)$ represents the quantity of item i_p in transaction T_d. The *external utility* $p(i_p)$ is the unit profit value of item i_p. The *utility* of item i_p in transaction T_d is defined as $u(i_p, T_d) = p(i_p) \times q(i_p, T_d)$. The utility of itemset X in transaction T_d is defined as $u(X, T_d) = \sum_{i_p \in X} u(i_p, T_d)$. The utility of itemset X in $\boldsymbol{D}$ is defined as $u(X) = \sum_{X \subseteq T_d \wedge T_d \in \boldsymbol{D}} u(X, T_d)$. The *transaction utility* (TU) of transaction T_d is defined as $TU(T_d) = u(T_d, T_d)$. The *transaction-weighted utilization* (TWU) model is generally used as the upper bound of utility, and it has been proven that an itemset is not an HUI if its TWU is lower than the minimum utility threshold [7]. Formally, the TWU of itemset X is the sum of the TUs of all the transactions containing X, which is defined as $TWU(X) = \sum_{X \subseteq T_d \wedge T_d \in \boldsymbol{D}} TU(T_d)$.

An itemset X *dominates* another itemset Y in the database if $f(X) \geq f(Y)$ and $u(X) > u(Y)$, or $f(X) > f(Y)$ and $u(X) \geq u(Y)$. An itemset X is an SFUI if it is not dominated by any other itemsets in $\boldsymbol{D}$ considering both frequency and utility factors. The SFUIM problem is to discover all the non-dominated itemsets in the database.

Table 1. Example database

TID	Transactions	TU
1	(A, 1) (B, 3) (C, 1) (E,3) (G,1)	30
2	(B, 1) (C, 1) (F, 2)	10
3	(B, 2) (C, 1) (D, 1) (E, 2)	20
4	(A, 1) (B, 1) (C, 1) (D, 1) (E, 1) (F, 1) (G,1)	20
5	(D, 2) (E, 2)	10
6	(G,1)	4

Table 2. Profit table

Item	A	B	C	D	E	F	G
Profit	2	5	3	3	2	1	4

As a running example, consider the database in Table 1 and the profit table in Table 2. For convenience, an itemset {B, C} is denoted by BC. In the example database, the utility of item B in transaction T_1 is $u(\text{B}, T_1) = 5 \times 3 = 15$, the utility of itemset BC in transaction T_1 is $u(\text{BC}, T_1) = 15 + 3 = 18$, and the utility of itemset BC in the transaction database is $u(\text{BC}) = u(\text{BC}, T_1) + u(\text{BC}, T_2) + u(\text{BC}, T_3) + u(\text{BC}, T_4) = 47$. The TU of T_1 is $TU(T_1) = u(\text{ABCEG}, T_1) = 30$, and the utilities of the other transactions are shown in the third column of Table 1. Because transactions T_1, T_2, T_3, and T_4 contain BC, $f(\text{BC}) = 4$, and $TWU(\text{BC}) = TU(T_1) + TU(T_2) + TU(T_3) + TU(T_4) = 80$. We can further verify that BC is not dominated by any other itemsets, so it is an SFUI. The other SFUI in this example is BCE, where $f(\text{BCE}) = 3$ and $u(\text{BCE}) = 51$.

2.2 Cross-Entropy Method

The CE method is an effective heuristic method that can be used for solving difficult combinatorial optimization problems (COPs) [1]. In this paper, we determine the SFUIs following the COP methodology.

Let $Y = (y_1, y_2, \ldots, y_n)$ be an n-dimensional binary vector, that is, the value of y_i ($1 \leq i \leq n$) is either zero or one. The goal of the CE method is to reconstruct the unknown vector Y by maximizing the function $S(X)$ using a random search algorithm:

$$S(X) = n - \sum\nolimits_{j=1}^{n} \left| x_j - y_j \right|. \tag{1}$$

A naive approach to determine Y is to repeatedly generate binary vectors $X = (x_1, x_2, \ldots, x_n)$ until a solution is equal to Y, which leads to $S(X) = n$. Elements of the trial binary vector X, that is, $x_1, x_2, \ldots, x_n$, are independent Bernoulli random variables with success probabilities, and these probabilities comprise a probability vector (PV) $P = (p_1, p_2, \ldots, p_n)$. The CE method for COP consists of creating a sequence of PVs $P_0, P_1, \ldots$ and levels $\gamma_1, \gamma_2, \ldots$ such that the sequence $P_0, P_1, \ldots$ converges to the optimal PV and the sequence $\gamma_1, \gamma_2, \ldots$ converges to the optimal performance.

Initially, $P_0 = (1/2, 1/2, \ldots, 1/2)$. For a sample $X_1, X_2, \ldots, X_N$ of Bernoulli vectors, calculate $S(X_i)$ for all i and sort the elements in descending order of $S(X_i)$. Let γ_t be a ρ sample quantile of the performances, that is,

$$\gamma_t = S(X_{\lceil \rho \times N \rceil}), \tag{2}$$

where $\lceil \rho \times N \rceil$ is the smallest integer that is greater than or equal to $\rho \times N$. Then each element of the probability vector is updated by

$$P_{t,j} = \sum_{i=1}^{N} I_{\{S(X_i) \geq \gamma_t\}} \times I_{\{X_{ij}=1\}} \Big/ \sum_{i=1}^{N} I_{\{S(X_i) \geq \gamma_t\}}, \tag{3}$$

where $j = 1, 2, \ldots, n$, $X_i = (x_{i1}, x_{i2}, \ldots, x_{in})$, t is the iteration number, and

$$I_E = \begin{cases} 1, & \text{if } E \text{ is true} \\ 0, & \text{otherwise} \end{cases}, \tag{4}$$

where E is an event.

Equation 3 is used iteratively to update the PV until the stopping criterion is met. There are two possible stopping criteria: γ_t does not change for a number of subsequent iterations or the PV has converged to a binary vector.

3 SFU-CE Algorithm

3.1 Modeling SFUI Discovery Using the CE

We use bitmap [12] in SFU-CE to identify transactions that contain the target itemsets. We calculate the frequency and utility values of the target itemsets efficiently using bitwise operations.

Specifically, SFU-CE uses a *bitmap cover* representation for itemsets. In a bitmap cover, there is one bit for each transaction in the database. If item i appears in transaction T_j, then bit j of the bitmap cover for item i is set to one; otherwise, the bit is set to zero. This naturally extends to itemsets. Let X be an itemset, $Bit(X)$ corresponds to the bitmap cover that represents the transaction set for the itemset X. Let X and Y be two itemsets. $Bit(X \cup Y)$ can be computed as $Bit(X) \cap Bit(Y)$, that is, the bitwise-AND of $Bit(X)$ and $Bit(Y)$.

In this paper, a binary vector representing an itemset is called an *itemset vector* (IV), and its dimensions are the same as those of the PV. The IVs within the kth iteration are determined by the kth PV. Then, the $(k + 1)$th PV is determined by the IVs within the kth iteration according to Eq. 3. This procedure is repeated until the termination conditions are reached.

To discover the SFUIs from the transaction database, we use the utility of the itemset to replace Eq. 1 directly; that is, for an itemset X,

$$S(X) = u(X) \tag{5}$$

In each iteration t, we sort a sample $X_1, X_2, \ldots, X_N$ in descending order of $S(X_i)$ ($1 \leq i \leq N$), and update the sample quantile γ_t and the PV P_t accordingly.

3.2 Critical Utility Pruning

Because the frequency measure follows the downward-closure property, the frequencies of 1-itemsets are typically high, which can be used as a pruning strategy to narrow the search space.

Definition 1. The critical utility of SFUIs (CUS) in transaction database $\boldsymbol{D}$ is the maximal utility of single items that have the highest frequency, and is defined as.

$$\text{CUS} = max\{u(i) \,|f(i) = f_{max}\} \tag{6}$$

where i is an item in $\boldsymbol{D}$ and f_{max} is the maximal frequency of all 1-itemsets in $\boldsymbol{D}$.

In the running example, three items have the highest frequency: f(B) = 4, f(C) = 4, and f(E) = 4. Because u(B) = 35, u(C) = 12, and u(E) = 16, CUS = 35.

Theorem 1. Let X be an itemset. If $u(X) <$ CUS, X is not an SFUI.

Proof. Let i_c be the 1-itemset, where $f(i_c) = f_{max}$ and $u(i_c) =$ CUS, and i_x be the 1-itemset such that $i_x \in X$. Then, $f(X) \leq f(i_x)$. Because $f(i_x) \leq f_{max}$, $f(X) \leq f_{max}$ holds. Considering $u(X) <$ CUS, X is dominated by i_c and X is not an SFUI. □

Using Theorem 1, once a 1-itemset is found to have utility lower than the CUS, it can be identified immediately as not an SFUI. As the upper bound of utility, it is obvious that TWU can also be used for pruning with the CUS.

Corollary 1. Let X be an itemset. If $TWU(X) <$ CUS, X and all itemsets containing X are not SFUIs.

Proof. Let i_c be the 1-itemset with $f(i_c) = f_{max}$ and $u(i_c) =$ CUS. Similar to Theorem 1, we can easily prove that X is not an SFUI.

Consider an arbitrary itemset Y containing X: $u(Y) \leq TWU(Y) \leq TWU(X) <$ CUS $= u(i_c)$, and $f(Y) \leq f(X) \leq f_{max} = f(i_c)$. Thus, Y is dominated by i_c and Y is not an SFUI. □

Using Corollary 1, once a 1-itemset is found to have TWU lower than the CUS, this itemset and all its supersets can be pruned safely. In the running example, because $TWU(\mathrm{F}) = 30 <$ CUS, F is deleted from the database.

3.3 Random Mutation

For the typical CE-based itemset discovery algorithm [11], the IVs strictly rely on the PV. Because of the randomness of CE, the algorithm tends to fall into the local minimum, which leads to it missing the correct SFUIs. To increase the number of correct SFUIs, a mutation (a classical concept in the genetic algorithm) is used in the proposed algorithm. Within each iteration, in addition to the IVs following the PVs, SFU-CE also generates part of the new IVs randomly.

For each iteration t, let $P = (p_1, p_2,..., p_n)$ be the current PV with j $(1 \leq j \leq n)$ probabilities higher than 0.5. These j probabilities are denoted by $p_{r1}, p_{r2}, ..., p_{rj}$ $(1 \leq r_1 \leq r_2 \leq ... \leq r_j \leq n)$. To perform RM, a positive integer *num* $(1 \leq num \leq j)$ is generated first, then *num* bits (among the r_1-th, r_2-th, ..., r_j-th bits) of an IV performing RM are selected randomly, and their values are set to one and the other bits are set to zero.

Let α $(0 < \alpha < 1)$ be the *mutation factor*. In each iteration, $\lfloor N \times \alpha \rfloor$ IVs are generated by RM, and the remaining $N - \lfloor N \times \alpha \rfloor$ IVs are still generated according to the PV. It should be noted that only the bits corresponding to probabilities strictly higher than 0.5 perform RM because if the probability of 0.5 is also considered, all the IVs in the first iteration are all generated by RM.

Consider the running example, suppose the PV in one iteration is <0.3, 0.8, 0.7, 0.2, 0.6, 0.1>. There are only six probabilities because F is deleted using Corollary 1. In this PV, three probabilities are higher than 0.5: the second probability, third probability, and fifth probability. To generate an IV using this PV, a positive integer no higher than three is generated first. For example, let this number be two. Then, two bits within the second, third, and fifth bits are set to one, and the other bits are set to zero. We can see that <010010> is such an IV corresponding to itemset BE.

3.4 Algorithm Description

The proposed SFU-CE algorithm for mining SFUIs is described in Algorithm 1.

Algorithm 1	**SFU-CE**
Input	Transaction database ***D***, sample numbers N, quantile parameter ρ, maximum number of iterations *max_iter*, mutation factor α
Output	SFUIs
1	Initialization( );
2	**while** $t \leq max\_iter$ **and** P_t is not a binary vector **do**
3	Calculate P_t using Eq. 3;
4	Perform RM;
5	Generate the remaining IVs according to P_t;
6	Sort the N itemsets in utility-descending order and denote them by $X_1, X_2, \ldots, X_N$;
7	**for** $i = 1$ to N **do**
8	$CSFUI$ = SFUI-Filter(X_i, $CSFUI$);
9	**end for**
10	Calculate γ_t using Eq. 2;
11	t++;
12	**end while**
13	Output all SFUIs in $CSFUI$.

In Algorithm 1, the procedure Initialization, described in Algorithm 2, is called in Step 1. The loop from Step 2 to Step 12 discovers the SFUIs when the iteration number is no higher than the threshold and the PV is not a binary vector. For a binary PV, all the N itemsets are the same in one iteration because each item in each itemset is definitely one or zero. In Step 3, the PV is updated according to the current sample. Within the current sample, $\lfloor N \times \alpha \rfloor$ IVs are generated by RM (Step 4), whereas other IVs are produced with the probabilities in PV (Step 5). In Step 6, the itemsets are sorted in descending order of utility. In the loop from Step 7 to Step 9, each itemset is checked by the function SFUI-Filter (described in Algorithm 3). Then, in Step 10, the sample quantile is updated, and in Step 11, the iteration number is incremented by one. Finally, all itemsets in $CSFUI$ are output as SFUIs in Step 13.

Algorithm 2	**Initialization( )**
1	Scan ***D*** once to delete items with TWU values lower than the CUS;
2	Represent the database using a bitmap;
3	$P_0 = (1/2, 1/2, \ldots, 1/2)$;
4	Generate all itemsets of the first iteration with P_0;
5	Sort the N itemsets in utility-descending order and denote them by $X_1, X_2, \ldots, X_N$;
6	$CSFUI = \varnothing$;
7	**for** $i = 1$ to N **do**
8	$CSFUI$ = SFUI-Filter(X_i, $CSFUI$);
9	**end for**
10	Calculate γ_t using Eq. 2;
11	$t = 1$;

In Algorithm 2, in Step 1, the database is scanned and unpromising items are pruned using Corollary 1. Next, the database is represented by a bitmap in Step 2. In Step 3, all the probabilities in the PV are initialized to 1/2. Using this PV, the itemsets of the first iteration are generated in Step 4. In Step 5, the itemsets are sorted in utility-descending order and in Step 6, *CSFUI*, the set of candidate SFUIs is initialized as an empty set. In the loop from Steps 7 to 9, each itemset is checked by the function SFUI-Filter (described in Algorithm 3). In Step 10, the sample quantile is calculated. Finally, the iteration number is set to one in Step 11.

Algorithm 3	**SFUI-Filter(*X*, CSFUI)**
Input	Itemset *X*, set of candidate SFUIs *CSFUI*
Output	*CSFUI*
1	**if** $u(X) \geq$ CUS **then**
2	Remove all itemsets dominated by *X* in *CSFUI*;
3	**if** *X* is dominated by any itemset in *CSFUI* **then**
4	**return** *CSFUI*;
5	**end if**
6	$X \rightarrow$ *CSFUI*;
7	**end if**
8	**return** *CSFUI*;

According to Theorem 1, Algorithm 3 does not perform a meaningful operation unless the utility of the enumerating itemset is no lower than the CUS (Steps 1 to 7). In Step 2, the itemsets in *CSFUI* are deleted if it is dominated by the enumerating itemset. If the enumerating itemset is dominated by an itemset in *CSFUI*, in Steps 3 to 5, the algorithm is terminated and *CSFUI* is returned. If the enumerating itemset is not dominated by any itemset in *CSFUI*, it is inserted into *CSFUI* in Step 6. Finally, if the algorithm is not terminated before, *CSFUI* is returned in Step 8.

4 Performance Evaluation

We evaluate the performance of SFU-CE algorithm and compare it with SKYMINE [4], SFU-Miner [8], SKYFUP-D [5], and SKYFUP-B [5]. We downloaded the source code of SKYMINE and SFU-Miner from the SPMF data mining library [3], and obtained the source code of SKYFUP-D and SKYFUP-B from the author.

We performed the experiments on a computer with a quad-core 3.40 GHz CPU and 8 GB memory running 64-bit Microsoft Windows 10. We wrote our programs in Java. We used four datasets for performance evaluation, two synthetic datasets generated using the transaction utility database generator provided on the SPMF [3], and two real datasets downloaded from the SPMF [3]. The characteristics of the datasets are presented in Table 3.

For all experiments, we set the sample size to 2,000, maximum number of iterations to 2,000, quantile ρ to 0.2, and mutation factor α to 0.3.

Table 3. Characteristics of the datasets

Datasets	Avg. trans. Length	No. of items	No. of trans
T25I50D10k	25	50	10,000
T35I10050k	35	100	50,000
Chess	37	75	3,156
Connect	43	129	67,557

4.1 Runtime

First, we consider the efficiency performance of these algorithms, and show the comparison results in Table 4.

Table 4. Execution times of the five algorithms

Unit (Sec)	SKYMINE	SFU-Miner	SKYFUP-D	SKYFUP-B	SFU-CE
T25I50D10k	20.03	1915.78	77.75	–	7.82
T35I10050k	163.67	–	1006.42	–	69.28
Chess	–	167.53	37.92	36.22	16.20
Connect	–	–	6523.72	–	673.28

For the four datasets, the proposed SFU-CE algorithm and SKYFUP-D algorithm returned results in all cases, whereas the other three algorithms could not. Specifically, SKYMINE ran out of memory on two datasets, SKYFUP-B ran out of memory on three datasets, and SFU-Miner did not return any results after 20 h on two datasets. We use "-" for these seven entries.

Table 4 shows that the proposed SFU-CE algorithm always ran faster than the compared algorithms. On T35I100D50k, SFU-CE was one order of magnitude faster than SKYFUP-D. This shows that CE improved the efficiency of SFUIM. There are three reasons for the efficiency improvement. First, different from the four exact algorithms, SFU-CE only considered itemsets that had high utility values with gradual PV convergence and RM rather than all combinations of items. Second, the critical utility pruning strategies using Theorem 1 and Corollary 1 were also effective for omitting unpromising itemsets. Third, SFU-CE did not record the utilities of different frequencies with different array structures, such as umin in SKYMINE [4], umax in SFU-Miner [8], and utilmax in SKYFUP-D and SKYFUP-B [5]. Accordingly, the time cost of dominant relation checking within these arrays was avoided.

4.2 Accuracy

A heuristic SFUIM algorithm cannot ensure the discovery of all the correct itemsets within a certain number of iterations; that is, some itemsets discovered by SFU-CE may not correspond to the actual SFUIs of the entire dataset. So we compare the percentage of SFUIs discovered by SFU-CE to the actual SFUIs. We used the SKYFUP-D algorithm to discover the actual SFUIs from the four datasets, and used the following equation to calculate the accuracy of SFUIs discovered by SFU-CE:

$$Acc = Num\_CE/Num \times 100\%, \tag{7}$$

where *Num_CE* is the number of actual SFUIs discovered by SFU-CE and *Num* is the total number of actual SFUIs.

Table 5. Accuracy for the four datasets

Datasets	*Num_CE*	*Num*	*Acc* (%)
T25I50D10k	6	6	100
T35I10050k	4	4	100
Chess	35	35	100
Connect	42	46	91.30

Table 5 shows that SFU-CE discovered SFUIs with 100% accuracy, except on the Connect dataset, for which the average accuracy was 97.825% on the four datasets. These results demonstrate that the CE-based algorithm discovered most of actual SFUIs in less time.

4.3 Diversity

To verify the effect of RM proposed in Sect. 3.3, we evaluated the degree of diversity of the mining results using the bit edit distance (BED) [10], which is defined as.

$$BED(V, V_t) = NBits, \tag{8}$$

where V and V_t are two IVs and *NBits* is the number of bitwise-complement operations transformed from V to V_t. We can see from Eq. 8 that the greater the value of $BED(V, V_t)$, the more obvious the difference between V and V_t. For example, transforming $V = < 110100 >$ to $V_t = < 101110 >$ requires three bitwise-complement operations: transform the second bit from 1 to 0, transform the third bit from 0 to 1, and transform the fifth bit from 0 to 1. Thus, $BED(V, V_t) = 3$.

Following [10], we used two types of BED in our experiments: the greatest degree of diversity of all pairs of IVs, *maximal BED* (Max_BED), and the average degree of diversity of all pairs of IVs, *average BED* (Ave_BED). Let $V_1, V_2, \ldots, V_N$ be the IVs in one iteration. Max_BED is defined as.

$$Max\_BED = max\{BED(V_i, V_j) |\ 1 \le i \le N,\ 1 \le j \le N, i \ne j\}, \tag{9}$$

and Ave_BED is defined as

$$Ave\_BED = \sum_i \sum_{j \ne i} BED(V_i, V_j) \Big/ N \times (N-1)\cdot \tag{10}$$

We also implemented an algorithm that strictly generates all the IVs within each iteration according to the PV, without RM, and refer to this algorithm as SFU-Base. Because the number of iterations for which the two algorithms converged was inconsistent, we used the percentage of the total number of iterations when comparing their diversity.

Table 6. BED for the T25I50D10k dataset

Percentage of total number of iterations (%)	SFU-Base		SFU-CE	
	Ave_BED	*Max_BED*	*Ave_BED*	*Max_BED*
25	4.58	18.0	4.89	18.0
50	2.46	13.0	2.73	15.0
75	1.56	9.0	1.64	9.0
100	0.0	0.0	0.27	1.0

Table 7. BED for the T35I10050k dataset

Percentage of total number of iterations (%)	SFU-Base		SFU-CE	
	Ave_BED	*Max_BED*	*Ave_BED*	*Max_BED*
25	11.82	23.0	10.85	24.0
50	1.97	9.0	2.92	13.0
75	1.06	4.0	1.55	6.0
100	0.0	0.0	0.0	0.0

Tables 6 and 7 show the diversity comparison results on the two synthetic datasets. Although the diversity of SFU-CE was better, its advantage over SFU-Base was not obvious. The reason behind these results is that all SFUIs on these two synthetic datasets were 1-itemset. This means that there was only one probability in the PV that was higher than 0.5; hence, the effect of RM was not obvious.

The diversity comparison results on the two real datasets are shown in Tables 8 and 9. In the first few iterations, the effect was similar to that on the synthetic datasets: the difference between the two algorithms was not obvious. As the number of iterations increased, the diversity advantage of SFU-CE became increasingly significant. The results demonstrated that the proposed RM strategy improved the diversity of the samples. Consequently, the execution speed increased and accuracy improved.

Table 8. BED for the chess dataset

Percentage of total number of iterations (%)	SFU-Base		SFU-CE	
	Ave_BED	*Max_BED*	*Ave_BED*	*Max_BED*
25	13.40	28.0	11.46	28.0
50	5.83	16.0	6.36	16.0
75	1.97	7.0	3.95	13.0
100	0.0	0.0	2.91	12.0

Table 9. BED for the connect dataset

Percentage of total number of iterations (%)	SFU-Base		SFU-CE	
	Ave_BED	*Max_BED*	*Ave_BED*	*Max_BED*
25	12.73	26.0	8.22	24.0
50	7.62	19.0	7.39	19.0
75	2.43	9.0	5.10	18.0
100	0.0	0.0	3.92	16.0

5 Conclusion

In this paper, we studied the SFUIM problem from the perspective of CE and proposed an SFUIM algorithm called SFU-CE. Because SFUIs are itemsets that are not dominated by other itemsets when considering both frequency and utility constraints, we used utility as the optimization object of CE that checked intermediate results using utility, and proposed critical utility pruning that filtered intermediate results by frequency and utility. To improve the diversity of each sample, we designed RM to generate some new itemsets in addition to the PV. Experiments on publicly available datasets demonstrated that the SFU-CE algorithm was efficient, accurate, and produced diverse samples.

Acknowledgments. We thank Prof. Jerry Chun-Wei Lin for providing the source code of the SKYFUP-D and SKYFUP-B algorithms. This work was partially supported by the National Natural Science Foundation of China (61977001) and the Great Wall Scholar Program (CIT&TCD20190305).

References

1. de Boer, P.-T., Kroese, D.P., Mannor, S., Rubinstein, R.Y.: A tutorial on the cross-entropy method. Ann. OR **134**(1), 19–67 (2005)
2. Djenouri, Y., Comuzzi, M.: GA-apriori: combining apriori heuristic and genetic algorithms for solving the frequent itemsets mining problem. In: Kang, U., Lim, E.-P., Yu, J.X., Moon, Y.-S. (eds.) PAKDD 2017. LNCS (LNAI), vol. 10526, pp. 138–148. Springer, Cham (2017). https://doi.org/10.1007/978-3-319-67274-8_13

3. Fournier-Viger, P., et al.: The SPMF open-source data mining library version 2. In: Berendt, B., et al. (eds.) ECML PKDD 2016. LNCS (LNAI), vol. 9853, pp. 36–40. Springer, Cham (2016). https://doi.org/10.1007/978-3-319-46131-1_8
4. Goyal, V., Sureka, A., Patel, D.: Efficient skyline itemsets mining. In: Proceedings of the Eighth International C* Conference on Computer Science and Software Engineering, pp. 119–124 (2015)
5. Lin, J.C.-W., Yang, L., Fournier-Viger, P., Hong, T.-P.: Mining of skyline patterns by considering both frequent and utility constraints. Eng. Appl. Artif. Intell. **77**, 229–238 (2019)
6. Lin, J.-W., Yang, L., Fournier-Viger, P., Hong, T.-P., Voznak, M.: A binary PSO approach to mine high-utility itemsets. Soft. Comput. **21**(17), 5103–5121 (2016). https://doi.org/10.1007/s00500-016-2106-1
7. Liu, Y., Liao, W.-K., Choudhary, A.: A two-phase algorithm for fast discovery of high utility itemsets. In: Ho, T.B., Cheung, D., Liu, H. (eds.) PAKDD 2005. LNCS (LNAI), vol. 3518, pp. 689–695. Springer, Heidelberg (2005). https://doi.org/10.1007/11430919_79
8. Pan, J.-S., Lin, J.C.-W., Yang, L., Fournier-Viger, P., Hong, T.-P.: Efficiently mining of skyline frequent-utility patterns. Intell. Data Anal. **21**(6), 1407–1423 (2017)
9. Song, W., Huang, C.: Discovering high utility itemsets based on the artificial bee colony algorithm. In: Phung, D., Tseng, V.S., Webb, G.I., Ho, B., Ganji, M., Rashidi, L. (eds.) PAKDD 2018. LNCS (LNAI), vol. 10939, pp. 3–14. Springer, Cham (2018). https://doi.org/10.1007/978-3-319-93040-4_1
10. Song, W., Li, J.: Discovering high utility itemsets using set-based particle swarm optimization. In: Yang, X., Wang, C.-D., Islam, M.S., Zhang, Z. (eds.) ADMA 2020. LNCS (LNAI), vol. 12447, pp. 38–53. Springer, Cham (2020). https://doi.org/10.1007/978-3-030-65390-3_4
11. Song, W., Liu, L., Huang, C.: TKU-CE: cross-entropy method for mining top-k high utility itemsets. In: Fujita, H., Fournier-Viger, P., Ali, M., Sasaki, J. (eds.) IEA/AIE 2020. LNCS (LNAI), vol. 12144, pp. 846–857. Springer, Cham (2020). https://doi.org/10.1007/978-3-030-55789-8_72
12. Song, W., Liu, Y., Li, J.: Vertical mining for high utility itemsets. In: Proceedings of the 2012 IEEE International Conference on Granular Computing, pp. 429–434 (2012)
13. Song, W., Zheng, C., Huang, C., Liu, L.: Heuristically mining the top-k high-utility itemsets with cross-entropy optimization. Appl. Intell. (2021). https://doi.org/10.1007/s10489-021-02576-z
14. Yeh, J.-S., Li, Y.-C., Chang, C.-C.: Two-phase algorithms for a novel utility-frequent mining model. In: Proceedings of the International Workshops on Emerging Technologies in Knowledge Discovery and Data Mining, pp. 433–444 (2007)

Evaluating Football Player Actions During Counterattacks

Laurynas Raudonius(✉) and Richard Allmendinger

The University of Manchester, Manchester, UK
laurynas.raudonius@protonmail.com, richard.allmendinger@manchester.ac.uk

Abstract. Evaluation of actions made by footballers has a wide range of applications in the field of football analytics as it would help compare players and consequently aid scouting, team selection and many other aspects of professional football. Although there is a substantial amount of research aiming to solve this particular problem, most existing research in this domain focuses on just one aspect of the game, namely ball passes. Even though passes are a vital part of football, they do not provide a holistic view on evaluating player actions. This project presents a model that paints a more complete picture and values player contributions (dribbles and passes) to counterattacks by employing four separate performance indicators (which are then combined into one output): distance, danger, outplayed players and space control. The model was tested and optimised on datasets of 10 matches that were provided by the Stats Perform group. It was found that the four indicators are generally complementary in that they tell us different information about how player actions impacted the counterattack. Another finding of this project is that the way separate indicators and combined scores are distributed closely models the distribution of footballers based on their quality - most are average and the better they get, the less of them there are. Finally, we discuss a use case of the proposed model for player recruitment.

Keywords: Association football · Counterattacks · Positional data analysis · Player action valuation

1 Introduction

Association football is the most popular sport in the world by fan base, so it is fairly obvious that it generates the most revenue - money generated just by the five biggest football leagues is more than any other sport generates globally [1–3]. Keeping that in mind it does not come as a surprise that football analytics is a fast growing field [4] in the global sports market with many competing companies, as teams are prepared to dedicate substantial resources to gain an edge over their opponents. There is also a considerable amount of scientific research

H. Yin et al. (Eds.): IDEAL 2021, LNCS 13113, pp. 367–377, 2021.
https://doi.org/10.1007/978-3-030-91608-4_36

carried out on various aspects of the game by scientists and R&D departments across the world [10].

A challenge that comes up rather frequently is valuing footballers' actions. Since a pass is the most frequent event that appears in football and often increases danger, understanding and valuing this aspect of the game has been primarily the focus of existing football analytics research. For example, Bransen et al. [5] measure the danger of passes in football by looking at similar passes that have been done before. A distinction between this and other studies is that the study used event data only and then took start and finish positions of a pass into account. Chawla et al. [6] propose an automated pass classifier that can tell if a pass is good, bad or in the middle for any given scenario. Their model was trained using data labeled by human observers. The trained classifier is able to achieve a classification accuracy of over 90%. Håland et al. [11] propose a generalized additive mixed model that determines the accuracy, game overview and efficiency of a pass. Based on these predictors, ratings were assigned to players as a result of passes they performed. The work of Power et al. [17] focuses on measuring risk (likelihood of a pass reaching the target) and reward (likelihood of a pass resulting in a shot) of a pass in football; the researchers then used a logistic regression model to convert continuous likelihoods into binary outputs (e.g. risky vs non-risky pass). Rein et al. [18] took a different approach; instead of leaving the assessment of a pass to a machine learning model, they proposed two indicators without the need for explicit model training: how many defenders were outplayed by a pass and how did controlled space change during the pass. The researchers then investigated pass type (defense to midfield, midfield to attack etc.) effects on the two indicators and the relationship between the indicators on game success. Even though the aforementioned studies have been well-received and contributed a lot to football analytics, one shortcoming they share is that they valued just one subset of all the attacking actions (related to passing only) a player can take. Only accounting for passes can only provide a very one-sided player rating and while it is useful and applicable in football, it is very limiting. A study by Decroos et al. [8] overcame this limitation and looked at how any of the actions a player takes could impact the scoreline both positively and negatively and rated the players based on which actions they took.

The model described in this study, on the other hand, takes a more pragmatic approach and values players' attacking contributions based on actions that change the position of the ball - mostly passes and dribbles (ball carries).

2 Data

The data needed to carry this study out was provided by Stats Perform. It consists of 10 pairs of datasets, each corresponding to one full match and consisting of two parts: tracking data, which is positional information of all players on the pitch and the ball throughout the match sampled 10 Hz frequency, and event data, which is a list of all the events that were observed in the match (shots, passes, tackles etc.), each with a timestamp and additional information on it.

A filtering algorithm was applied to extract meaningful episodes of counterattacks from the full match data. A successful counterattack was defined as a chance that occurred in 20 s after an opposing team's action in their attacking half. In total, 11 episodes were found, 7 of which resulted in a goal.

3 Methodology

To enable a holistic valuation of football player contribution during counterattacks, we propose a model based on four performance indicators: distance, danger, outplayed defenders and space control. After presenting the contribution of this methodology, this section motivates and formally defines these indicators, and then suggests how they can be combined into a single player value.

3.1 Contributions

In this study we define a contribution as game states (positions of the ball and players of both teams on the pitch at a single moment in the match) at two timestamps: the moment a player got hold of the ball (after a successful tackle on the opposition, pass reception etc.), and the moment a different player got hold of it (most of the time after a pass). For the sake of simplicity, the model does not differentiate between different actions taken but is merely concerned with these two game states and the difference between them. Neglecting player actions removes bias from the model as it allows measuring their influence indirectly via game states (rather than measuring how difficult/flashy a particular player action was). Moreover, most existing research that quantified passes did not take into account actions that were taken before the pass. Since the use of game states accounts also for dribbles, the proposed model can be seen as more complete (at least in that regard).

3.2 Performance Indicators

Distance. This is likely the most intuitive indicator of the four. It is obvious that if a player carried the ball from, for example, their own third to the opposing team penalty box, they have contributed a lot to make the attack more dangerous and therefore their contribution should have a high valuation. To measure it exactly, the model calculates the Euclidean distance between the ball and the goal when a player received/won the ball and the same distance when they finished their actions (i.e. a pass). The difference between those two is how much closer to the goal the ball was brought by the player and therefore is the value of the indicator. The mathematical definition is given by

$$C_{distance}(x_{0b}, y_{0b}, x_b, y_b) = \sqrt{(x_{0b} - x_g)^2 + (y_{0b} - y_g)^2} - \sqrt{(x_b - x_g)^2 + (y_b - y_g)^2},$$

where x_{0b}, y_{0b} and x_b, y_b is the ball position at the start and end of the contribution, respectively, and x_g, y_g the center of the defending team's goal. We will be using these variables in the definition of the other indicators too.

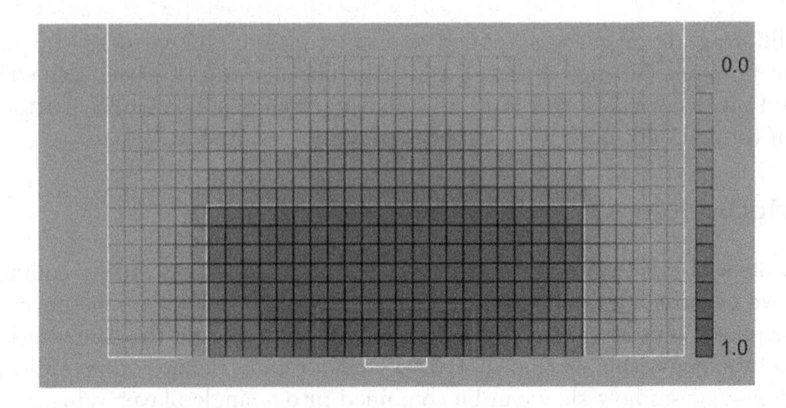

Fig. 1. Danger values of the pitch. Each square is 2 × 2 m. Green and transparent squares have smaller danger values, red and opaque have the larger ones. The exact danger values are taken from Link et al. [12]. (Color figure online)

Assuming players respect the constraints of the pitch and cannot control the ball outside of it, the indicator can take values in the range $[-110.36; 110.36]$ as 110.36 m is the distance from the corner of the pitch to the goal at the opposite side - the largest possible distance a player can cover in one contribution. Negative values are possible as players can take the ball further from goal.

Danger. A more advanced approach than the above is to also account for what position on the pitch the ball is in after the player's actions, not just how far it is to the goal. Assessing the danger of a point on the pitch is a topic that was investigated thoroughly by Link et al. [12] in their study on danger in football attacks. Link et al. proposed to divide an attacking third into $2 \times 2\,\text{m}^2$ and then assign a danger score in the range $[0; 1]$ to each square (see Fig. 1).

To calculate the exact values they followed five key rules:

1. As the distance from goal decreases and centrality increases, the danger rises.
2. Moving into the penalty area brings about a sudden increase in the danger because of the risk of a penalty kick.
3. There is a homogeneous area in front of goal in which the danger does not increase further.
4. An acute angle to the goal reduces the danger.
5. Areas to the side of the penalty area are dangerous because of the possibility of a cross with little risk of offside.

Since the derivation rules are sensible, we adopt the division proposed in [12] in our model. Consequently, we compute the danger indicator as follows:

$$C_{danger}(x_{0b}, y_{0b}, x_b, y_b) = danger(x_b, y_b) - danger(x_{0b}, y_{0b}),$$

where $danger(x, y)$ is the danger value (as taken from [12]) of point (x, y) on the pitch. This indicator measures the difference between the position of the ball when the player finished and started their contribution with respect to the danger of it. Since the danger values are in $[0;1]$ and it is possible for a player to make an attack less dangerous, the value range of this indicator is $[-1;1]$.

Outplayed Players. Although the previous two indicators are informative, they are fundamentally based on the ball position only. Consider a player that had a clear path to goal from halfway line; based on the previous two indicators the contribution of that player would be valued greatly, even though the player was not challenged much. An important point to consider when valuing player actions is how did the actions impact defense, and this is exactly what the third indicator measures. To be more specific, it calculates how many players were left behind the ball during the player's actions. This approach to valuing player passes was discussed in [18], and was chosen to be implemented in this model too. The value of this indicator for a contribution is the difference between the number of defenders behind the ball at the end and the start of the contribution:

$$C_{outplayed}(x_{0b}, t_0, x_b, t_b) = outplayed(x_b, t_b) - outplayed(x_{0b}, t_0),$$
$$outplayed(x, t) = |O|,$$

where t_0 and t_b is the time at the start and end of a contribution respectively, O the set of defending outfield players at time t that are between the goal and the ball, and $|O|$ the number of players in O. Function *outplayed*() then returns how many defenders are between the ball and the goal at a moment in time. Since goalkeepers stay behind the ball most of the time, they are not accounted for in this indicator. That leaves us with 10 outfield players, so the value of this indicator is in $[-10;10]$.

Increase in Space Control. This last indicator is inspired by the work of Memmert and Raabe [13]. The authors explored the relationship between a team's success and how much space they controlled in the opponents' 30 m area, and found that it has a somewhat direct correlation; many successful teams (e.g., FC Barcelona, Borussia Dortmund in the early 2010s) controlled large spaces in their opponents' half, so it only makes sense to give higher valuation to players whose contributions increased the controlled space. A point to note here is that controlled space is measured using Voronoi cells [16,19]. A player's Voronoi cell is the set of the points on the pitch they are closer to than any other player.

Calculating space in the whole pitch would not be as informative because a player that increases the space in their own half does not necessarily contribute to the counterattack. Hence, we have set a threshold (area) on the pitch determining when space control is accounted for. Experimentation with different areas in front of the goal have shown that if controlled space is measured within the final quarter of the pitch only, then variance in space control is greatest across the players hence allowing us to differentiate better the significance of a player contribution. Hence, we have decided to this as our threshold. We compute this indicator by taking the difference between the percentage of space controlled by the attacking team at the start and the end of a contribution:

$$C_{space}(t_0, t_b) = space(t_b) - space(t_0)$$
$$space(t) = \sum_{p \in T_A} vor(p)/S,$$

where T_A is the set of all players in the attacking team, $vor(p)$ player p's Voronoi cell's area, and S the area the controlled space is measured over (in our case, the final 25% of the pitch). This means that the *space*() function returns the amount of space controlled by the attacking team in the opponent quarter of the pitch at time t. As this indicator measures the percentage of space control in the final 25% of the pitch, which is in range [0; 100], and a player's contribution can also reduce the controlled space, the value range of the indicator is [−100; 100].

3.3 Counterattack Score

The four indicators can be combined into a single score for ease of computing an overall Player Counterattack Score. A similar transformation is done by the Passer Rating [9] in American Football, which takes 5 separate statistics into account. We use a normalized sum to combine our indicators ensuring that all four indicators are within the same range (of [−1; 1]). Formally, we compute the overall Player Counterattack Score as

$$C_{combined} = 2.5(C_{distance}/110.36 + C_{danger} + C_{outplayed}/10 + C_{space}/100),$$

where the scalar of 2.5 ensures that scores are in [−10; 10], which is potentially easier to interpret by the end users. It is important to note that $C_{combined}$ assumes that all indicators contribute equally to the Player Counterattack Score. If available, it is easy to adjust the importance (weightings) of the indicators based on expert preferences (e.g. provided by the coach or talent scout).

4 Results

This section investigates the complementary nature of the proposed indicators and then discusses some use cases.

4.1 Distribution of Performance Indicator Values

The purpose of our first analysis is to understand the distribution of observed performance indicator values and the overall counterattack score, and their alignment with the actual distribution of quality of players (see Fig. 2). We would expect the model to report some very small (even negative) indicator scores, a large number of low to medium scores, and a few large scores. This follows from the fact that we expect the presence of a few players with very low performance, most will be mediocre, and a small number of players will stand out. This is backed up by how other stats are distributed in football; the bottom right plot of Fig. 2 is a good example of such a distribution: this plot shows the distribution of the number of goals that players scored in Premier League's 2019/2020 season. It is evident that the intuition fits: most players (75%) have scored 0–5 goals whereas the better the players get, the less of them there are (indicated by the long tail of the distribution).

With a minor difference of also having negative scores, the combined counterattack score distribution (bottom left plot in Fig. 2) matches the intuition too. There are a few contributions with a very low score, most (57%) are in the interval [0; 1.5] and there is a number of contributions with a higher combined score. The way it is distributed proves that the model is balanced as it does not report very high/low scores for many contributions.

The distribution is also similar for the individual performance indicators, though the distribution is skewed for the danger indicator, which is expected because most contributions do not happen in the final third of the pitch and, consequently, do not improve the danger score. Additionally relatively many contributions fall within 0 and 10 in terms of space control. This can be attributed to contributions that are short in time because the controlled space does not change that much during these.

4.2 Similarity of the Measures

The purpose of having multiple indicators to value a player is to obtain a holistic view on the contribution. Consequently, we need indicators that minimize overlap in terms of information measured. To validate this, Fig. 3 shows the correlation levels between pairs of indicators using Pearson [15] correlation score.

Looking at all pairwise correlation scores, it is apparent that the pairs of performance indicators are either positively correlated or not correlated. The absence of indicators with negative correlation is expected as no two indicators are, so to say, opposite in that when one increases the other should decrease. The indicators, distance and space control, have the largest positive correlation. This can be attributed to contributions with large distance scores generally taking longer to complete; in turn, this allows players and their teammates to advance into more attacking positions and, consequently, control more space in their opponents' areas. On the other end of the spectrum, there is almost no correlation between the indicators, outplayed players and danger. This is because many contributions with high danger scores were crosses into the penalty box which increases the danger considerably but does not outplay any players as the pass is horizontal. The correlation analysis results are positive because the lack of very positive and negative correlations between pairs of indicators indicates that the indicators can measure complementary information to a certain degree.

4.3 Towards Multi-objective Optimization for Player Selection

The availability of multiple performance indicators opens up opportunities in areas such as holistic player recruitment and performance monitoring.

For instance, we can use the indicators to identify the set of (Pareto optimal [7]) players that provide the best trade-off in terms of the indicator scores.

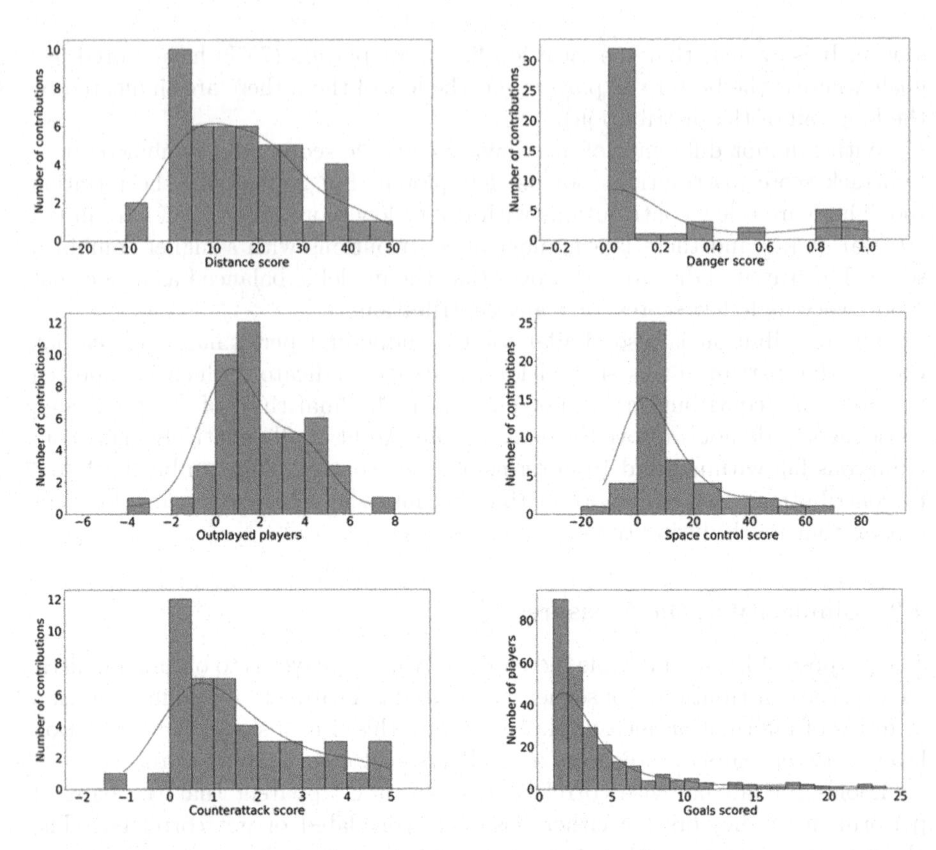

Fig. 2. Distributions of considered performance indicators (from left to right, and top to bottom): distance, danger, outplayed players, and space control. The bottom left plot shows the overall counterattack score, and the bottom right plot the distribution of the number of goals scored by players in Premier League's 2019/2020 season.

Figure 4 shows the set of all players (left plot) and the Pareto set of players (right plot).

Several observations can be made from the figure: (i) There is no single player that is best in terms of all four indicators, which may indicate different counterattack styles across players; (ii) the more indicators a player is contributing to, the larger the overall counterattack score; (iii) doing very well in several indicators can compensate poor indicator scores for other indicators. This is what we want our model to do because it means we are not punishing players too severely if they are doing badly in some aspects of a counterattack.

Thinking ahead, if we look at the four indicators as objective functions (which they are), then the model could helps us in conjunction with multi-objective optimization [14] to, for example, recruit players that are complementary to existing players in terms of counterattacking skills.

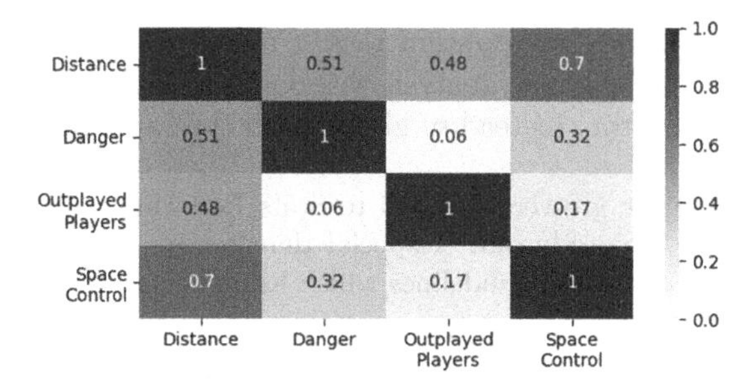

Fig. 3. Heatmap of Pearson correlation coefficient matrix between pairs of player performance indicators.

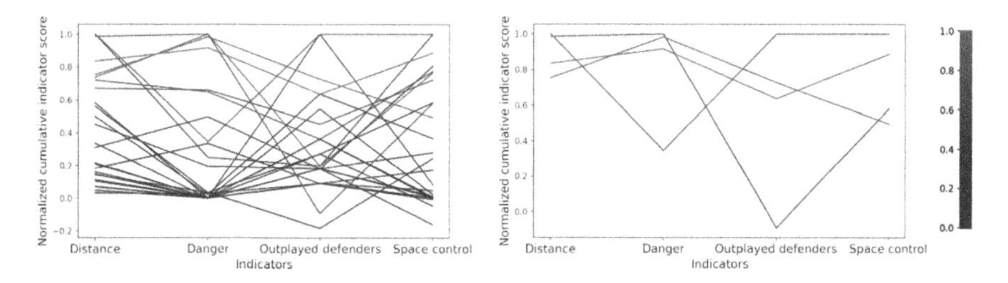

Fig. 4. Normalized cumulative indicator scores for all players with at least one contribution to a counterattack (left plot) and Pareto set of players obtained via non-dominated sorting (right plot). For players that are associated with more than one contribution, we have summed up the scores for the individual indicators. The normalization involved normalizing each player's (cumulative) indicator score by the largest (cumulative) score for that indicator across all players. The line color indicates the overall counterattack score. (Color figure online)

5 Discussion and Conclusions

We have presented a way to value footballer contributions in counterattacks. Although the issue of assigning numerical values to actions in football has been discussed in the literature, most studies have one characteristic shortcoming: they focus on just one subset of actions made by footballers, mostly passes. This paper, on the other hand, proposes a model that does not differentiate between different actions taken by players but is only concerned with how the game state changed while the player controlled the ball. The valuation model described here succeeds in producing an unprejudiced and complete score. It is based on four separate performance indicators – distance of keeping a ball, danger induced by a player, number of outplayed players, and space control – which measure scores derived from tracking and event data. We have shown that the indicators are complementary to a certain degree, thus allowing the proposed model to be more holistic than existing models. We also discussed the use of the proposed model for player recruitment. Regarding future work, with some adjustments the model could be tailored to value all attacking actions players can take, not only in counterattacks. With more data and opinions of experts in football performance analysis we could also assign weights to each of the performance indicators instead of giving each the same weighting.

Acknowledgments. We are very grateful to Stats Perform group for providing the data needed to carry the study out. A special thanks goes to Dr. Thomas Seidl for providing valuable feedback and guidelines which helped complete the study.

References

1. America only has 4 of the most profitable sports leagues in the world. https://www.sportscasting.com/america-only-has-4-of-the-most-profitable-sports-leagues-in-the-world/
2. European soccer posts record revenues with the English premier league leading the way. https://www.cnbc.com/2019/05/30/european-soccer-posts-record-revenues-as-epl-dominates-deloitte.html
3. The most popular sports in the world. https://www.worldatlas.com/articles/what-are-the-most-popular-sports-in-the-world.html
4. Sports analytics market - growth, trends, Covid-19 impact, and forecasts (2021–2026). https://www.mordorintelligence.com/industry-reports/sports-analytics-market
5. Bransen, L., Van Haaren, J., van de Velden, M.: Measuring soccer players' contributions to chance creation by valuing their passes. J. Quant. Anal. Sports **15**(2), 97–116 (2019). https://doi.org/10.1515/jqas-2018-0020
6. Chawla, S., Estephan, J., Gudmundsson, J., Horton, M.: Classification of passes in football matches using spatiotemporal data. ACM Trans. Spat. Algorithms Syst. **3**(2), 1–30 (2017). https://doi.org/10.1145/3105576
7. Debreu, G.: Valuation equilibrium and pareto optimum. Proc. Natl. Acad. Sci. **40**(7), 588–592 (1954). https://doi.org/10.1073/pnas.40.7.588
8. Decroos, T., Bransen, L., Van Haaren, J., Davis, J.: Actions speak louder than goals: valuing player actions in soccer. In: Proceedings of the 25th ACM SIGKDD International Conference on Knowledge Discovery & Data Mining, pp. 1851–1861. ACM (2019). https://doi.org/10.1145/3292500.3330758
9. von Dohlen, P.: Tweaking the NFL's quarterback passer rating for better results. J. Quant. Anal. Sports **7**(3) (2011). https://doi.org/10.2202/1559-0410.1359
10. Gudmundsson, J., Horton, M.: Spatio-temporal analysis of team sports. ACM Comput. Surv. **50**(2) (2017). https://doi.org/10.1145/3054132
11. Håland, E.M., Wiig, A.S., Stålhane, M., Hvattum, L.M.: Evaluating passing ability in association football. IMA J. Manag. Math. **31**(1), 91–116 (2019). https://doi.org/10.1093/imaman/dpz004
12. Link, D., Lang, S., Seidenschwarz, P.: Real time quantification of dangerousity in football using spatiotemporal tracking data. PLOS ONE **11**(12) (2016). https://doi.org/10.1371/journal.pone.0168768
13. Memmert, D., Raabe, D.: Data Analytics in Football: Positional Data Collection, Modelling and Analysis. Routledge (2018). https://doi.org/10.4324/9781351210164
14. Miettinen, K.: Nonlinear Multiobjective Optimization, vol. 12. Springer, Heidelberg (2012). https://doi.org/10.1007/978-1-4615-5563-6

15. Pearson, K.: Note on regression and inheritance in the case of two parents. Proc. R. Soc. Lond. **58**(347), 3 (1895). https://doi.org/10.1098/rspl.1895.0041
16. Perl, J., Memmert, D.: A pilot study on offensive success in soccer based on space and ball control - key performance indicators and key to understand game dynamics. Int. J. Comput. Sci. Sport **16**(1), 65–75 (2017). https://doi.org/10.1515/ijcss-2017-0005
17. Power, P., Ruiz, H., Wei, X., Lucey, P.: "Not all passes are created equal:" objectively measuring the risk and reward of passes in soccer from tracking data. In: KDD 2017 (2017). https://doi.org/10.1145/3097983.3098051
18. Rein, R., Raabe, D., Memmert, D.: "Which pass is better?" Novel approaches to assess passing effectiveness in elite soccer. Hum. Mov. Sci. **55**, 172–181 (2017). https://doi.org/10.1016/j.humov.2017.07.010
19. Voronoi, G.: Nouvelles applications des paramètres continus à la théorie des formes quadratiques. Deuxième mémoire. Recherches sur les parallélloèdres primitifs. https://doi.org/10.1515/crll.1908.134.198

An Intelligent Decision Support System for Production Planning in Garments Industry

Rui Ribeiro[1,2(✉)], André Pilastri[1], Hugo Carvalho[1], Arthur Matta[1], Pedro José Pereira[2], Pedro Rocha[3], Marcelo Alves[3], and Paulo Cortez[2]

[1] EPMQ - IT Engineering Maturity and Quality Lab, CCG ZGDV Institute, Guimarães, Portugal
{rui.ribeiro,andre.pilastri,hugo.carvalho,arthur.matta}@ccg.pt
[2] ALGORITMI Centre, Department of Information Systems, University of Minho, Guimarães, Portugal
id6927@alunos.uminho.pt, pcortez@dsi.uminho.pt
[3] INFOS - Informática e Serviços, Leça da Palmeira, Portugal
{pedro.rocha,marcelo.alves}@infos.pt

Abstract. In this paper, we propose an Intelligent Decision Support System (IDSS) that combines prediction and optimization for production planning. We worked with a company that provides software for the garments Industry and that had access to real-world data related with a client that works with subcontractors. Using an Automated Machine Learning (AutoML) approach, we firstly target four predictive tasks that are crucial to estimate production planning indicators. Then, we use historical data and one of the predicted indicators to search for the best subcontractor allocation plan, which minimize both the cost and production time via an Evolutionary Multiobjective Optimization (EMO) algorithm (NSGA-II), achieving interesting results.

Keywords: Textile production planning · AutoML · NSGA-II

1 Introduction

Currently, there is a pressure in industries to increase efficiency (e.g., reduce operating costs and time) in order to compete in their markets. One way is to adopt an Intelligent Decision Support Systems (IDSS), which incorporate Artificial Intelligence techniques to provide actionable knowledge from raw data [2]. In this paper, we assume an IDSS for the garments Industry and that is based in the concept of Adaptive Business Intelligence (ABI) [11], which combines Machine Learning (ML), to predict relevant decision context variables, with Modern Optimization (MO) [7], to search for the best decision choices (according to one or more objectives).

H. Yin et al. (Eds.): IDEAL 2021, LNCS 13113, pp. 378–386, 2021.
https://doi.org/10.1007/978-3-030-91608-4_37

There are some related works that employ MO methods to support production plans in the textile industry. For instance, in [1] Genetic Algorithms (GAs) were used to create production orders involving the spinning and weaving areas of the fabrication process. GAs were also adopted in [12] to optimize job orders of textile production lines. A combination of GA with Simulated Annealing was used by [13] to create energy efficient production orders. In another study, NSGA-II was used by [10] to solve a multi-objective multi-site order scheduling problem in the production planning stage with the consideration of multiple plants, multiple production departments and multiple production processes. More recently, [14] used a mathematical programming model to optimize textile production considering diverse "green" goals (e.g., waste reuse, energy recycling) and [3] optimized the master production scheduling using GA. Within our knowledge, none of these works adopted a data-driven ABI approach that combines predictive and prescriptive analytics, as provided by ML and MO algorithms. In this paper, we follow such innovative ABI combination by using an Automated Machine Learning (AutoML) [9] to first predict four important garment subcontractor decision variables. Then, we adopt historical data and one of the predicted variables (production time) to feed an Evolutionary Multiobjective Optimization (EMO) that searches for the best subcontractor allocation plan, simultaneously minimizing the total allocation cost and time.

2 Materials and Methods

2.1 Garment Data

The data was provided by INFOS, which is Portuguese software company that works with several textile industry clients. The company developed an Enterprise Resource Plan (ERP) that supports the production of garments. The goal of this research is to develop an IDSS based on the ABI concept and that will be integrated into the INFOS ERP system, allowing it to automatically design garment subcontractor plans regardless of size of the company and the complexity of the production order. The subcontractor selection is a non trivial task, since is a large range of textile operations, each involving costs and delivery dates. We collected all company garment related records, including purchase and manufacturing orders, from 2016 to 2020. The data was then divided in three major groups: purchase of raw material, manufacturing and subcontractor. Next, we implemented an Extraction, Transformation, Load (ETL) process to select and clean the data (e.g., removal of missing features and records with wrong dates). All data processing procedures (including the ABI system) were implemented in the Python language by the authors.

Table 1 describes the input features (**Attribute**), their description (**Description**), data **Type**, number of **Levels** and **Domain** values separated by objective (four predictive targets and one optimization task). The final set of input features was obtained after several iterations of predictive task executions. The datasets for the predictive (regression) tasks include: Lead Time – 3,315 records; Production Time – 25,449 examples; Production Waste – 24,425 instances; and

Table 1. Description of the input features by objective.

Objective	Attribute	Description	Type	Levels	Domain
Lead time	Supp_cod	Supplier identification	Integer	102	[14, 2265]
	Date_purch	Date of purchase order	Date	873	–
	Rmat_cod	Raw material code	String	188	–
	Qty	Quantity to buy	Integer	876	[1, 11650]
Production time	Subc_cod	Subcontractor identification	Integer	275	[0, 9999]
	Mat_cod	Final product code	String	846	–
	Oper_desc	Textile operation	String	93	–
	Qty	Quantity to produce	Float	9356	[1, 97512]
Production waste	Mat_cod	Final product code	String	864	–
	Qty	Quantity to produce	Float	4646	[1, 13448.3]
	Subc_cod	Subcontractor identification	Integer	41	[8, 9996]
	Rmat_cod	Raw material code	String	1386	–
Delivery delays	Plan_endate	Planned date to end production	Date	940	–
	Mat_cod	Code of the final product	String	205	–
	Qty	Quantity to produce	Integer	3156	[3, 75838]
Production plan	Oper_desc	Textile operation	String	94	–
	Avgp_cost	Average cost of textile operation	Float	47	[0.01, 1.75]
	Price	Cost of textile operation for given product per subcontractor	Float	247	[0.01, 11.60]
	Subc_cod	Subcontractor identification	Integer	293	[24, 2254]
	Mat_cod	Code of the final product	String	169	–
	Capacity	Subcontractor production capacity by textile operation	Integer	14	[100, 3000]

Delivery Delays – 6,016 records. Finally, the optimization objective (Production Plan) contains 5,500 records related with subcontractors.

Regarding the target output target variables for the predictive tasks, we detected that the company does not have records of them, being necessary to calculate them: *Ldtime* was obtained by subtracting the receiving date of a order from the placement order date and if resulting value was negative that row was discarded; for *Prod_days* we create a function that subtracts the production finish date from the planned production start date and outputs the number of working days between the two dates and the if the number of days was negative that row was discarded; in the case if *Waste_ratio* we first subtracted the produced quantity from the quantity to produce and if the resulting value was positive it was changed to zero, afterwards we divided the absolute result by the quantity to produce, multiplying the final result by 100; finally for *Delay_days* we create a function that subtracts the scheduled delivery date finish date from the delivery date and outputs the number of working days between the two dates and the if the number of days was negative it was changed to zero. Table 2 describes the four output target variables with their description (**Description**), data type (**Type**), number of levels (**Levels**) and domain values (**Domain**).

In terms of preprocessing, since the String variables had a high cardinally, we employed a Label Encoder, in order to transform each level into a distinct

Table 2. Description of the output target variables.

Target	Description	Type	Levels	Domain
Ldtime	Days between delivery date and purchase date	Integer	94	[0, 59]
Prod_days	Working days to produce a certain quantity	Integer	47	[1, 63]
Waste_ratio	Percentage of wasted material	Integer	247	[0, 100]
Delay_days	Working days between scheduled delivery date and delivery date	Integer	293	[0, 64]

numeric value. This option provided better results when compared with the known One-Hot encoding, which created a very high number of input features. As for the Date features, we adopted the proleptic Gregorian ordinal of a date, allowing to provide a simpler numeric value. Then, all numeric inputs were normalized by using a z-score standardization.

2.2 Intelligent Decision Support System

The proposed IDSS contains three main modules (Fig. 1): data extraction and processing, prediction and optimization. The first module is responsible for receiving the garment data, selecting the features for each objective and then creating the necessary input for prediction. The prediction module receives the data separated by predictive task splitting it into training and test sets (data separation, according to the adopted cross-validation method). Then, it trains the predictive models (model training), evaluating the models performance (model evaluation), selecting and storing the best prediction model (model selection). Then, the user inserts the data related to the lead time, using the respective model to predict the number of days that will take to receive the raw materials and can define a starting and end date for production. Finally the optimization module receives the subcontractors data (Table 1), filtered by the product to manufacture and the textile operations to execute, the quantity to produce and maximum allowed dates (all provided by the user). Then, the MO algorithm uses this data and also one of the predicted indicators (production time) to search for the best subcontractor quantity allocation, aiming to reduce the total costs and time.

To reduce the modeling effort during the development of the prediction module, we adopted the `H2O` AutoML tool that provided good results in recent AutoML benchmark study [9]. The AutoML was configured to automatically select the best regression model and its hyperparameters based on the best Mean Absolute Error (MAE), using a internal 10-fold cross-validation applied over the training data. Five different ML algorithms were searched by the tool: Random Forest, Extremely Randomized Trees, Generalized Linear Models, Gradient Boosting Machine and two Stacked Ensembles, one with best model of each family and other with all trained models. An external 10-fold cross-validation was executed to evaluate the ML models and the quality of the regression was accessed by using the MAE and Normalized MAE (NMAE) metrics. The lower the values, the better are the predictions. The NMAE measure normalizes the

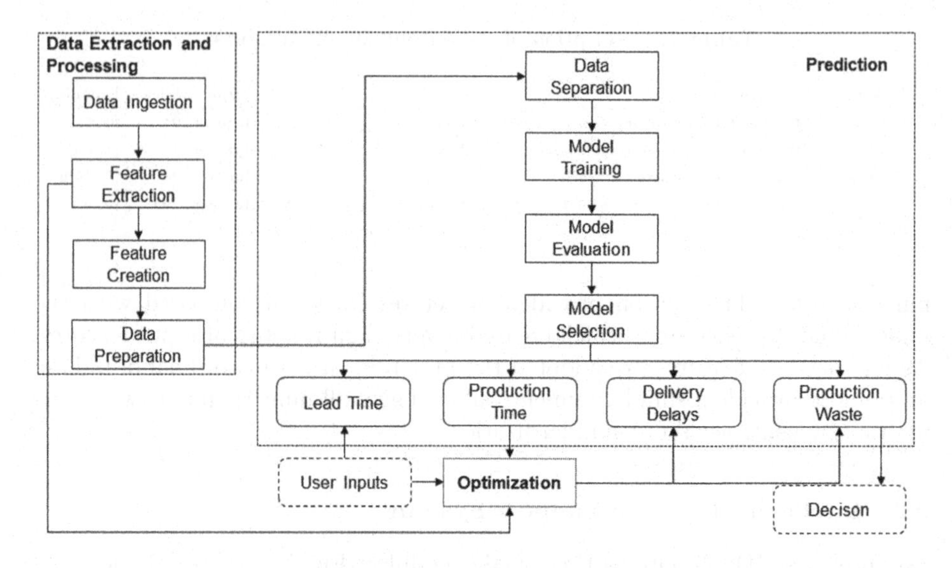

Fig. 1. Flow diagram describing the behaviour of the IDSS.

MAE by the range of the output target on the test set, thus it provides a percentage that is easy to interpret and that is scale independent.

A production order can be defined as a composition of tasks that are executed sequentially. Each task can be represented by a set of candidate subcontractors offering similar services, where each service can have a different value in price and quantity per subcontractor. The subcontractor allocation is defined as a multi-objective task (i.e., reduce both cost and time), thus we employ a Pareto approach via an EMO algorithm, namely NSGA-II [7], as implemented in the pymoo Python module [4]. NSGA-II is a multi-objective optimization algorithm with three distinctive features: fast non-dominated sorting approach, fast crowded distance estimation procedure and usage of a simple crowded comparison operator [8]. When compared with other hypervolume based algorithms (e.g., SMS-EMOA), the NSGA-II algorithm tends to obtain competitive results when only two or three objectives are optimized [6]. The algorithm returns a population of non dominated solutions, each representing a different subcontractor allocation and that is associated with a distinct cost-time trade-off. The full subcontractor optimization can be defined in terms of x textile sequential operations that need to be executed. For each operation, there are y candidates (subcontractors) with different price and capacity parameters. Each solution is naturally represented as a sequence of q_i integer values ($0 \leq q_i \leq q_{max}$), denoting the quantity assigned for each subcontractor i, where q_{max} denotes the total required quantity for operation x, and $i \in \{1, ..., M\}$ and M represent the number of available subcontractors for operation x. We repair solutions by ignoring any excess of subcontractor allocation (first allocated subcontractor is served first) or by randomly distributing the deficit allocation to any of the available subcontractors. Each solution is evaluated in terms of total production plan cost and allocation time. To compute these two goals, the EMO algorithm uses the production time prediction (as shown in Fig. 1). Once the Pareto curve is optimized and for user selected trade-offs, we then compute the prediction indicators of the remaining targets (e.g., production waste), such that the user can further inspect the quality of the obtained solutions. In order to obtain a single measure per Pareto curve, we selected the Hypervolume (HV) measure, which represents the volume of the objective space when assuming a "worst" reference point [5]. The higher the HV value, the better is the Pareto curve optimization.

3 Experiments and Results

The average of the external 10-fold iteration predictive results (in terms of **MAE** and **NMAE**) are presented in Table 3. The table also presents the best ML **Model**. In general, low regression errors were achieved, with the NMAE values ranging from 3.6% to 9.2%. We particularly note that the best NMAE values were obtained for the target that is directly used by the NSGA-II MO (Prod_days produces an average NMAE error of just 3.6%). The selected ML algorithm was a stacked ensemble for three of the targets, while the Gradient Boosting Machine obtained the best results for the production waste prediction.

Table 3. AutoML predictive results for each predicted target.

Target	Model	MAE	NMAE
Ldtime	Stacked ensemble (All models)	3.31	9.20%
Prod_days	Stacked ensemble (All models)	1.60	3.63%
Waste_ratio	Gradient boosting machine	4.24	4.24%
Delay_days	Stacked ensemble (Best of each family)	3.57	5.71%

For the optimization experiments, we analyzed a production order of 10,000 units of a product that requires three textile operations (cutting, tailoring and packaging) using one raw material. Using historical data, we then selected all the subcontractors that could execute these operations along with the respective cost and production capacity to create a subcontract allocation case study to utilize in the experiments. In total, the case study includes 26 subcontractors (which corresponds to the number of searched integers by the NSGA-II algorithm): cutting - 4 candidates, tailoring – 8 candidates and packaging – 14 candidates (4+8+14=26). To compute the cost and time associated with each solution, we use four attributes from Table 1 (*Subc_cod, Capacity, Price* and *Oper_desc*) and also the predicted `Prod_days` variable (see Table 2). We assumed some reasonable assumptions (defined by the INFOS company): one subcontractor cannot execute two or more tasks simultaneously, the subcontractor is always available and there is no shortage of raw materials.

The two objective functions that need to be minimized are the Total Cost (TC) and Total Production Time (TPT). The TC function is the sum of the multiplication of the assigned quantity to a individual by price of operation for that individual operation. As for TPT, the function is the sum of the maximum days required by each sequential operation (cutting, tailoring and packaging). Since subcontractors can work simultaneously in the same operation (e.g., cutting), we consider the slowest operator time (measured in terms of number of days). Solutions that split the q_i quantities by different operators for an operation will thus contribute for a lower TPT value. The lower bound is always zero and the upper bound was set to the quantity to be produced. When needed, a repair procedure is used to convert an unfeasible solution to a feasible one, see Sect. 2.2.

The NGSA-II algorithm was configured with a check procedure that eliminates duplicates, making sure that the mating produces offspring that are different from themselves and the existing population regarding their design space values. A grid search was used to set the NSGA-II hyperparameters (e.g., the population size was ranged within {50,100,150,...,500}), assuming the HV measure as the selection criterion and a reference point of (30 days, 20,000 EUR). The best obtained values correspond to a normalized HV (when each objective is divided by the respective reference point value) of 0.71, which requires 157 s of execution time on an Intel Xeon processor. The selected NSGA-II setup includes: population size of 100, two-point crossover with 90%, polynomial mutation probability of 20% and total of 200 generations.

The left of Fig. 2 shows the Pareto front obtained after 200 generations when considering our case study. The Pareto front contains 100 solutions, with the TPT ranging from 12 to 30 working days and TC ranging from 18,000 to 20,000 EUR. The right of Fig. 2 shows the evolution of the NSGA-II algorithm, in terms of the full HV measure (y−axis) through the executed 200 generations.

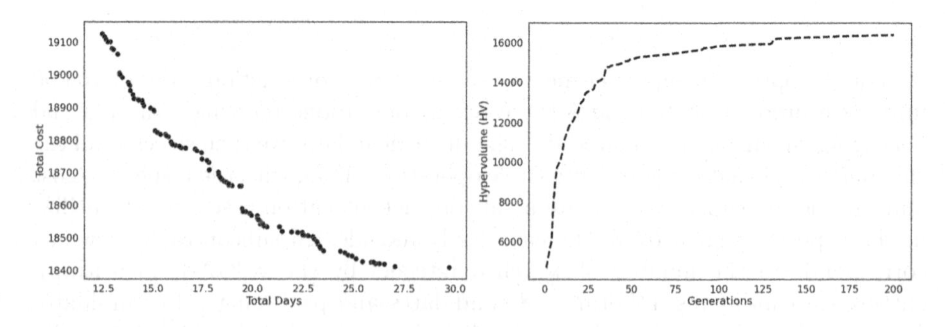

Fig. 2. Optimized Pareto front (left) and NSGA-II HV generation evolution (right).

The graph shows a substantial improvement that is obtained by NSGA-II. In effect, in the first generation the HV measure is 4,700 (normalized value of 0.2). After 200 generations, the value increased to 16,391 (normalized value of 0.71), which corresponds to an improvement of 51% points when considering the normalized HV scale. The results were shown to the INFOS company, which provided a very positive feedback. In particular, the obtained TPT and TC ranges were considered realistic. Moreover, the company signaled that the obtained Pareto front provides a more richer set of trade-off solutions, while also being faster to compute when compared with the currently adopted manual subcontractor allocation.

4 Conclusions

We propose an IDSS that creates a textile production plan to allocate subcontractors. The IDSS is based on the ABI concept that combines predictive (via ML) with prescriptive (via MO) analytics in order to provide actionable knowledge from raw data. The IDSS was designed to work with real-world data from a Portuguese software company (INFOS) that works with diverse textile clients. Firstly, an AutoML tool was adopted to automatically select the best ML model among five algorithms when targeting four relevant allocation decision context variables. Interesting results were achieved by the prediction models (error that ranges from 3.6% to 9.2%). Then, we designed a MO model that uses one of the predicted variables (production time) and historical data to automatically allocate subcontractors to execute sequential operations associated with a textile order. The MO model, based on the NSGA-II algorithm, assumes a Pareto approach and it was designed to simultaneously minimize the cost and time to execute the order. To demonstrate the MO, we selected a case study that includes four operations and 26 potential textile subcontractors.

The obtained results were shown to the INFOS company, which considered them very positive. In future work, we intend to augment the IDSS by incorporating more problem-domain constraints, such as incorporating updated data about the currenty availability of subcontractors. Furthermore, we wish to deploy the designed IDSS into the INFOS ERP system, in order to get more valuable feedback from a real environment usage.

Acknowledgments. This work was carried out within the project "Connect@Fashion" reference POCI-01-0247-FEDER-045296, co-funded by *Fundo Europeu de Desenvolvimento Regional* (FEDER), through Portugal 2020 (P2020).

References

1. Ángeles Solari, M.D.L., Ocampo, E.: Application of genetic algorithms to a manufacturing industry scheduling multi-agent system. In: Sobh, T., Elleithy, K., Mahmood, A., Karim, M. (eds.) Innovative Algorithms and Techniques in Automation, Industrial Electronics and Telecommunications, pp. 263–268. Springer, Dordrecht (2007). https://doi.org/10.1007/978-1-4020-6266-7_48
2. Arnott, D., Pervan, G.: A critical analysis of decision support systems research revisited: the rise of design science. J. Inf. Technol. **29**(4), 269–293 (2014). https://doi.org/10.1057/jit.2014.16

3. Ben-Ammar, O., Bettayeb, B., Dolgui, A.: Optimization of multi-period supply planning under stochastic lead times and a dynamic demand. Int. J. Prod. Econ. **218**, 106–117 (2019). https://doi.org/10.1016/j.ijpe.2019.05.003
4. Blank, J., Deb, K.: Pymoo: multi-objective optimization in python. IEEE Access **8**, 89497–89509 (2020). https://doi.org/10.1109/ACCESS.2020.2990567
5. Campos Ciro, G., Dugardin, F., Yalaoui, F., Kelly, R.: A nsga-ii and nsga-iii comparison for solving an open shop scheduling problem with resource constraints. IFAC-PapersOnLine **49**(12), 1272–1277 (2016). https://doi.org/10.1016/j.ifacol.2016.07.690
6. Chiandussi, G., Codegone, M., Ferrero, S., Varesio, F.: Comparison of multi-objective optimization methodologies for engineering applications. Comput. Math. Appl **63**(5), 912–942 (2012). https://doi.org/10.1016/j.camwa.2011.11.057
7. Cortez, P.: Modern Optimization with R. UR, Springer, Cham (2021). https://doi.org/10.1007/978-3-030-72819-9
8. Deb, K., Pratap, A., Agarwal, S., Meyarivan, T.: A fast and elitist multiobjective genetic algorithm: Nsga-ii. IEEE Trans. Evol. Comput. **6**(2), 182–197 (2002). https://doi.org/10.1109/4235.996017
9. Ferreira, L., Pilastri, A., Martins, C.M., Pires, P.M., Cortez, P.: A comparison of AutoML tools for machine learning, deep learning and XGBoost. In: International Joint Conference on Neural Networks, IJCNN 2021. IEEE (2021)
10. Guo, Z., Wong, W., Li, Z., Ren, P.: Modeling and pareto optimization of multi-objective order scheduling problems in production planning. Comput. Ind. Eng. **64**(4), 972–986 (2013). https://doi.org/10.1016/j.cie.2013.01.006
11. Michalewicz, Z., Schmidt, M., Michalewicz, M., Chiriac, C.: Adaptive Business Intelligence. Springer, Heidelberg (2006). https://doi.org/10.1007/978-3-540-32929-9
12. Mok, P.Y., Cheung, T.Y., Wong, W.K., Leung, S.Y., Fan, J.T.: Intelligent production planning for complex garment manufacturing. J. Intell. Manuf. **24**(1), 133–145 (2013). https://doi.org/10.1007/s10845-011-0548-y
13. Mokhtari, H., Hasani, A.: An energy-efficient multi-objective optimization for flexible job-shop scheduling problem. Comput. Chem. Eng. **104**, 339–352 (2017). https://doi.org/10.1016/j.compchemeng.2017.05.004
14. Tsai, W.H.: Green production planning and control for the textile industry by using mathematical programming and industry 4.0 techniques. Energies **11**(8), 2072 (2018). https://doi.org/10.3390/en11082072

Learning Dynamic Connectivity with Residual-Attention Network for Autism Classification in 4D fMRI Brain Images

Kyoung-Won Park[1(✉)], Seok-Jun Bu[2], and Sung-Bae Cho[1,2]

[1] Department of Artificial Intelligence, Yonsei University, Seoul 03722, Korea
{pkw408,sbcho}@yonsei.ac.kr
[2] Department of Computer Science, Yonsei University, Seoul 03722, Korea
sjbuhan@yonsei.ac.kr

Abstract. Diagnosing autism spectrum disorder (ASD) is still challenging because of its complex disorder and insufficient evidence to diagnose. A recent research in psychiatry perspective demonstrates that there are no obvious reasons for ASD. However, considering a hypothesis that abnormalities in the superior temporal sulcus (STS) connected with visual cortex regions can be a critical sign of ASD, a model is required to exploit the brain functional connectivity between STS and visual cortex to reinforce the neurobiological evidence. This paper proposes a deep learning model composed of attention and convolutional recurrent neural networks that can select and extract the time-series pattern of dynamic connectivity between the two regions within the brain based on observations. By integration of extracting autism disorder features from dynamic connectivity through attention with the structure containing interlayer connections to preserve the functional connectivity loss within a neural network, the model extracts the connectivity between STS and visual cortex, leading to the increase of generalization performance. Experiments with 800 patients' fMRI imaging data known as ABIDE (Autism Brain Imaging Data Exchange) and 10-fold cross-validation to compare its performance show that the proposed model outperforms the state-of-the-art performance by achieving a 4.90% improvement in the ASD classification. Additionally, the proposed method is analyzed by visualizing dynamic brain connectivity of the neural network layers.

Keywords: Autism spectrum disorder · Dynamic connectivity · 4D functional magnetic resonance imaging · Deep learning

1 Introduction

Autism spectrum disorder (ASD) is defined as a neurodevelopmental disorder that causes difficulties in communication and social interaction skills with others. Since autism spectrum disorder is a complex mental symptom at a wide range of levels, diagnosing autism spectrum disorder is still a challenging problem in the field of psychiatry. Over two decades, with the increase of the prevalence of ASD, many psychiatrists, researchers,

H. Yin et al. (Eds.): IDEAL 2021, LNCS 13113, pp. 387–396, 2021.
https://doi.org/10.1007/978-3-030-91608-4_38

and experts on AI in healthcare have been putting their effort into helping identify the exact factors of ASD or other developmental disabilities [1, 2]. Especially, the AI experts have been attempting to extract the decent autism features, which are hard to discover from human knowledge, with various usage of deep learning methods and numerous sources and types of data [3–7].

Among their valuable findings for the neurodevelopmental disorder, the most promising hypothesis of diagnosis ASD is caused by the abnormalities in the superior temporal sulcus (STS), which is a major role in biological motion perception as well as speech processing, connected with visual cortex. For starters, a multivariate time series approach is introduced to model 4D fMRI as high-performance deep learning for extracting structural features of the brain diagnosed with ASD. However, limitations have been pointed out while extracting dynamic connectivity between STS and visual cortex responsible for processing visual information.

Therefore, in this paper, we propose a combined architecture of an attention mechanism and a residual convolutional neural network that considers the weight of dynamic connectivity between STS and visual cortex regions with the experiments with 800 patients' fMRI imaging data known as ABIDE (Autism Brain Imaging Data Exchange) [8]. The proposed method outperforms the state-of-art method, which is the ensemble of convolutional neural networks achieving the existing highest diagnostic performance on 4D fMRI benchmark datasets, by improving a 4.90% accuracy in ASD classification [9]. In addition, we attempt to diagnose ASD by visualizing the attention weights and dynamic brain connectivity of the hidden layers in the neural network.

2 Related Works

Table 1 summarizes the recent deep learning studies over the past three years for ASD diagnosis, using the same ABIDE fMRI dataset. Ensemble methods of deep learning

Table 1. Comparison of performance over convolution networks, recurrent neural networks, and machine learning algorithms.

Year	Task	# of subject	Method	Accuracy
2018	Classification	1096	CNN-EW [10]	66.88%
2019	Classification	1110	3DCNN-LSTM [11]	53.00%
2019	Classification	1112	DF-SWSM [12]	66.05%
2019	Classification	920	DFC-CNN [13]	68.80%
2019	Prediction	1112	CNN Combination [14]	72.30%
2019	Diagnosis	279	Ensemble MVTC [15]	72.60%
2019	Classification	872	GCNN [16]	70.86%
2019	Detection	1035	ASD DiagNet [17]	82.60%
2020	Prediction	1034	GCN-CL [18]	67.52%
2020	Classification	1112	CNN Ensemble [9]	84.90%

models have been used as the main method for ASD diagnosis over the recent studies. Among those methods, an ensemble method of convolutional neural networks achieved the best performance with 0.8490 [9]. Specifically, the ensemble model is composed of four different neural networks which are DenseNet, ResNet, Xception, and Inception V3 and the convolutional operation proved to be suited for extracting spatial structural features of the brain.

Meanwhile, the graph approaches using the connection between brain regions have been studied. Among the graph approaches, A convolutional neural network with element-wise filters (CNN-EW) composed of an Edge-to-Node layer with element-wise filters, node-to-graph layer, a fully connected layer, and a softmax layer is exploited to diagnose ASD. Element-wise filter weights the edges according to structure information of the connections and it reflects on the actual task to distinguish ASD and normal group. As a result, the approach reaches a meaningful conclusion, but the limitation of its performance appears compared to other approaches [10].

The convolutional-recurrent neural networks have been attempted to simultaneously learn structural features of the brain and functional connectivity of the brain. The accuracy of the method achieved to 0.5300 and the limitation of the deep learning models for functional connectivity of the brain is addressed [11]. In this paper, we propose an integration method of three components; a recurrent neural network to learn functional connectivity, an attention mechanism to focus on dynamic connectivity of the two brain regions, and the convolutional neural network to extract spatial structural features.

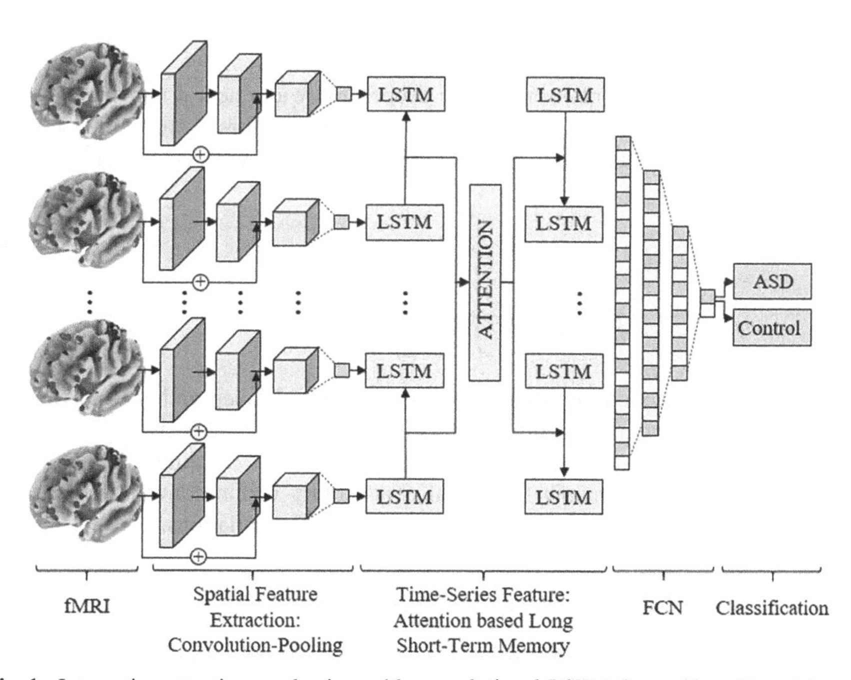

Fig. 1. Integrating attention mechanism with convolutional-LSTM (Long Short-Term Memory) for dynamic brain's functional connectivity learning.

3 The Proposed Method

3.1 Deep Learning of Dynamic Brain's Functional Connectivity

Since functional connectivity of the brain in 4D fMRI images has both various spatial variations in each patient and dynamic correlation over scan times, we consider not only minimize loss of connectivity between STS and visual cortex region but also selectively extract autism diagnostic features from the dynamic connectivity, especially inside the convolution network.

In Fig. 1, the attention mechanism and the residual convolutional-LSTM recurrent neural network is schematized. In the proposed learning phase, a neural network structure that selects and extracts characteristics of ASD by extracting patient-specific spatial variational elements and weighting connectivity between the STS and visual cortex regions via a self-diagnosis mechanism and LSTM [19, 20]. The extracted features are computed with the probability for control patients or patients with ASD between 0 and 1 via a softmax layer.

3.2 Residual Convolutional for Extraction of Brain Connectivity

Usage of convolutional neural networks based on deep learning is the common deep learning model to solve fMRI image tasks [21]. However, the convolutional neural network is exposed to the limitations for the extraction of each patients' variations. To maximize extraction of each patients' spatial variations between the two regions in the brain that are comprehensively different from patient to patient, we exploit a residual convolutional neural network, which has a deeper and wider neural network than the conventional convolutional neural networks. We define the following residual operation H(·) and the convolutional operation F(·), where the row-i, column-j, and layer-l's output is x_{ij}^{l}.

$$H(x) = F(x) + x \tag{1}$$

$$F(x) = \sum\nolimits_{a=0}^{m-1} \sum\nolimits_{b=0}^{m-1} w_{ab}^{l} x_{(i+a)(j+b)}^{l-1} \tag{2}$$

The residual connection has the gradient F'(x) + 1 over all layers and as well as figures out the vanishing gradient problem. Specifically, residual convolution is appropriate for 4D fMRI based ASD diagnosis because it fixes the gradient vanishing problem while minimizing the loss of dynamic functional connectivity within convolutional neural networks and maximizing the extraction of patient's spatial variations.

3.3 Attention Mechanism for Dynamic Brain's Functional Connectivity

Recently, attention mechanism has been used together with recurrent neural network in various time series analysis. In the field of ASD diagnosis using 4D fMRI images, we utilize attention mechanisms and LSTM to improve the modeling of dynamic time series connectivity between brain regions. LSTM based attention mechanism learns attention weights to select and focus on the major region from correlation among regions in the brain. The attention score of the attention mechanism internally computes the input value I_t of the LSTM network and the output value of i state O_i, is defined as:

$$Attention\ Score(I_t, O_i) = {I_t}^T \times O_i \tag{3}$$

The overall weight vector is defined as the sum of the attention score's product through the Dot-product of each layer and the hidden state h_i.

$$Attention\ weight_t = \sum_{i=1}^{N} (Attention\ Score_i \times h_i) \tag{4}$$

It is possible to selectively extract ASD features because detailed extraction of dynamic connectivity is learned through data and additional information with attention weights and outputs of LSTM. The final output values of the attention mechanism and the LSTM recurrent neural network are represented as the probability of either the patient with ASD or the normal patient through the last softmax layer.

4 Experiments

In this section, we perform a quantitative analysis to demonstrate the relationship between dynamics functional connectivity and diagnosis of ASD. The experiments are performed on NVIDIA DGX Station with 2,560 NVIDA Tensor Cores, Ubuntu Desktop Linux OS, 4X Tesla V100(64 GB), and 256 GB LRDIMM DDR4 Memory.

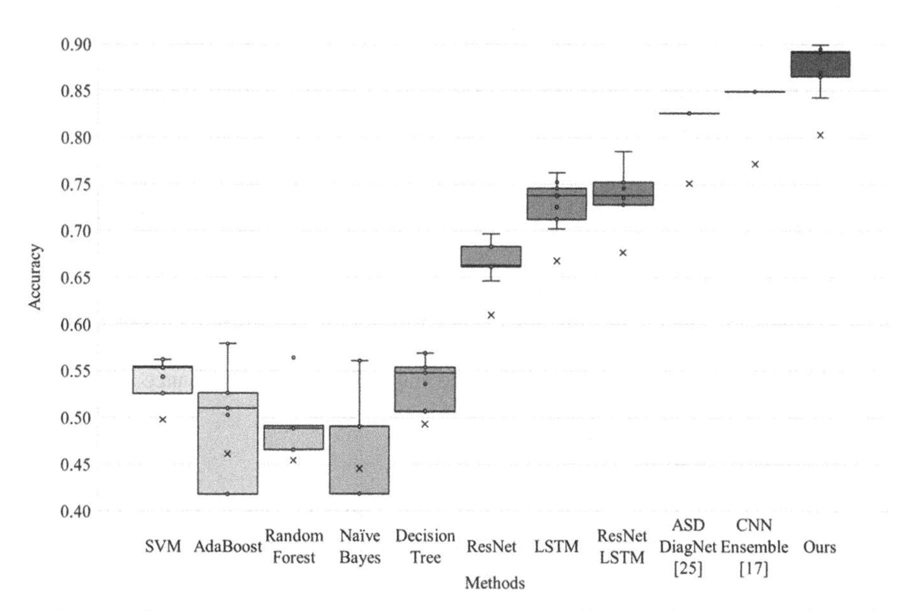

Fig. 2. Classification accuracy of autism spectrum disorder using 10-fold cross-validation on 800 patients' fMRI imaging data.

Table 2. Neural network classification performance analysis table.

Class	Precision	Recall	F1-score	Support
Autism	0.8600	0.9200	0.8900	779
Normal	0.9200	0.8500	0.8800	694
Average	0.8900	0.8900	0.8900	1473

4.1 Data Collection and Preprocessing

We utilize the preprocessed ABIDE I dataset provided in a web-based environment for validation. The total number of patients is 800 and each patient has their own 4D functional magnetic resonance imaging. The patients consist of 389 patients diagnosed with autism spectrum disorder and 411 control patients. Since each patient's fMRI image has a different size and length, we resized all images to identical $24 \times 24 \times 24$ size and used the sliding window method for the fixed length of the image. After all, we divided 14,727 images into training datasets (90%) and validation datasets (10%).

4.2 Quantitative Analysis

Table 2 shows the performance of precision, recall, and F1-score of each class via the classification performance analysis table concerning our proposed method. Especially, the meaning of the recall value of ASD class regarding our analysis is the ratio classified by our proposed classifier as ASD among the entire of ASD cases. Our proposed method achieves 0.9200 recall value in autism class as shown in the analytical table and its value represents how well our method can diagnose autism spectrum disorder and how much it is specialized to detect autism features.

In Fig. 2, 10-fold cross-validation is performed to compare its performance to the conventional convolutional networks, deep learning neural networks including residual neural network and LSTM, and various machine learning algorithm. Most machine learning algorithms shows low accuracy performance around 0.5500 or even below. Furthermore, the accuracy of the residual neural network reaches 0.6640 and the accuracy of LSTM reaches 0.7472. On the other hands, the accuracy of our proposed method reaches 0.8980 and we attain 4.90% improvement compared to the ensemble of convolutional neural networks that achieved the existing the state-of-the-art performance.

4.3 Qualitative Analysis

Class activation map allows for highlighting the activated regions of the given data. As shown in Fig. 3, (b) in terms of typical vanilla convolutional network, the activated regions of the brain are distributed out of the brain. Meanwhile, in terms of the proposed method, the activated regions of the brain image are centered around the center of the brain unlike the typical neural network. Furthermore, we confirm that the class activation image from our method generally tends to focus on the center of the brain at the beginning and end of the two-dimensional images. Thus, the proposed method is much more likely to learn the dynamic connectivity between the two regions.

We have analyzed the extracted features from the residual network of the proposed model to support the justification of learning dynamics connectivity between the regions. As shown in Fig. 4, the extracted feature vector (10, width, height, depth) keeps the dynamics connectivity without any loss while learning in the model. Moreover, the ten individual separated feature vectors maintain the dynamic connectivity as well, and it

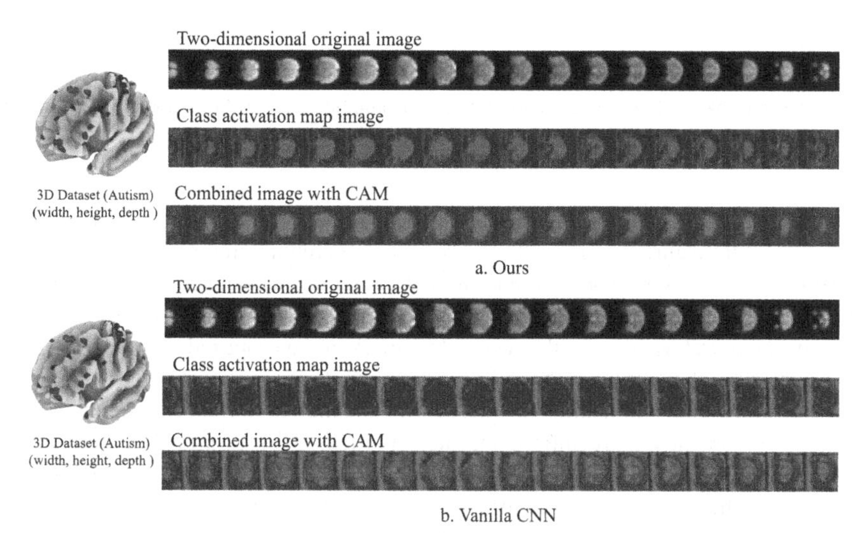

Fig. 3. Comparison of our method and the Vanilla CNN. (a) In terms of our method, the activated regions of a brain image are centered in the center of the brain. (b) Meanwhile, in terms of the vanilla CNN, the activated regions of the brain are distributed out of the brain.

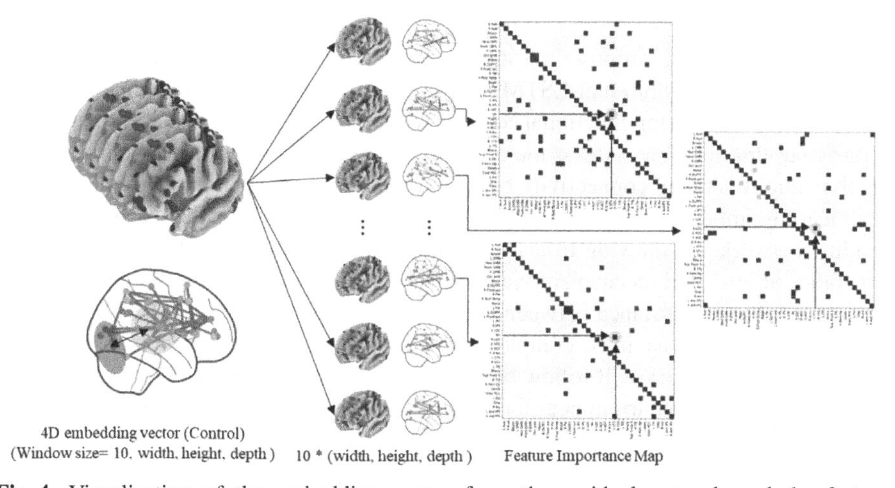

Fig. 4. Visualization of the embedding vector from the residual network and the feature importance map of the embedding vectors for the control class for the ABIDE dataset.

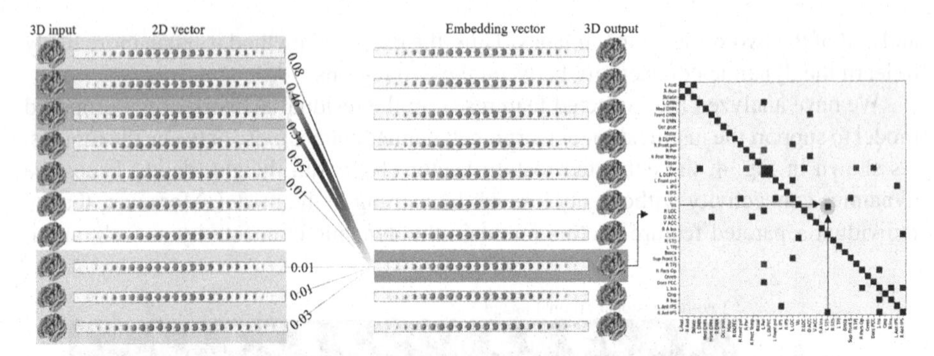

Fig. 5. An example of illustrating for weight relationship between the 8th embedding vector and the input vectors.

shows the time series patterns have existed in the feature vectors. Concerning the second, third, and eighth feature vectors in the 4D brain image, feature importance maps of each 3d feature vector contain the information of the relationship between the two regions, and such relationships form time-series patterns. This finding strongly supports utilizing the recurrent neural network to the residual features in the proposed method.

To explore further, we have observed the same input case as shown in Fig. 4. In Fig. 5, concerning the eighth embedding vector, showing the high importance between STS and visual cortex from the feature importance map, our method for learning the eighth feature vector endows the second and third input vector with 0.46 and 0.34 attention weights, respectively. Therefore, the qualitative result shows the attention mechanism leads to achieving high performance compared to other methods by giving weight attention we corresponding to the importance between the two regions while learning.

5 Conclusion

In this paper, we have proposed an integrated network with the attention mechanism and the residual convolutional-LSTM that learns the feature space learning and time-series pattern considering functional connectivity between STS and visual cortex region. It turns out that the new state-of-the-art (0.8980) is achieved by successfully learning the dynamic functional connectivity between the two regions compared to the existing deep learning models. Furthermore, the proposed method justifies the feasibility of the residual network via showing to learn dynamic connectivity without any loss and the feasibility of attention mechanism via endowing high weight attention to input vectors that contain high importance between the two regions. Meanwhile, the LSTM used for time series connectivity must complement the vanishing gradient problem over long scan times. Future work will follow regarding a new approach to resolve the vanishing gradient problem of recurrent neural networks by leveraging the attention mechanism.

Acknowledgement. This work was partially supported by an IITP grant funded by the Korean government (MSIT) (No. 2020–0- 01361, Artificial Intelligence Graduate School Program (Yonsei University)) and Electronics and Telecommunications Research Institute (ETRI) grant funded by the Korean government (21ZS1100, Core Technology Research for Self-Improving Integrated Artificial Intelligence System).

References

1. Ke, F., Choi, S., Kang, Y.H., Cheon, K.A., Lee, S.W.: Exploring the structural and strategic bases of autism spectrum disorders with deep learning. IEEE Access **8**, 153341–153352 (2016)
2. Zilbovicius, M., Meresse, I., Chabane, N., Brunelle, F., Samson, Y., Boddaert, N.: Autism, the superior temporal sulcus and social perception. Trends Neurosci. **29**(7), 359–366 (2016)
3. Aradhya, A.M., Joglekar, A., Suresh, S., Pratama, M.: Deep transformation method for discriminant analysis of multi-channel resting state fMRI. Proc. AAAI Conf. Artif. Intell. **33**(1), 2556–2563 (2019)
4. Cacha, L.A., Parida, S., Dehuri, S., Cho, S.B., Poznanski, R.R.: A fuzzy integral method based on the ensemble of neural networks to analyze fMRI data for cognitive state classification across multiple subjects. J. Integr. Neurosci. **15**(4), 1–14 (2017)
5. Li, X., et al.: 2-channel convolutional 3D deep neural network (2CC3D) for fMRI analysis: ASD classification and feature learning. In: IEEE 15th International Symposium on Biomedical Imaging, pp. 1252–1255 (2018)
6. Pandey, P., Prathosh, A.P., Kohli, M., Pritchard, J.: Guided weak supervision for action recognition with scarce data to assess skills of children with autism. Proc. AAAI Conf. Artif. Intell. **34**(1), 463–470 (2020)
7. Parida, S., Dehuri, S., Cho, S.B., Cacha, L.A., Poznanski, R.R.: A hybrid method for classifying cognitive states from fMRI data. J. Integr. Neurosci. **14**(3), 355–368 (2015)
8. Craddock, C., et al.: The neuro bureau preprocessing initiative: Open sharing of preprocessed neuroimaging data and derivatives. Neuroinformatics **7** (2013)
9. Ahmed, M.R., Zhang, Y., Liu, Y., Liao, H.: Single volume image generator and deep learning-based ASD classification. IEEE J. Biomed. Health Inform. **24**, 3044–3054 (2020)
10. Xing, X., Ji, J., Yao, Y.: Convolutional neural network with element-wise filters to extract hierarchical topological features for brain networks. In: 2018 IEEE International Conference on Bioinformatics and Biomedicine, pp. 780–783 (2018)
11. El-Gazzar, A., Quaak, M., Cerliani, L., Bloem, P., van Wingen, G., Mani Thomas, R.: A hybrid 3DCNN and 3DC-LSTM based model for 4D spatio-temporal fMRI data: an ABIDE autism classification study. In: Zhou, L. (ed.) OR 2.0/MLCN -2019. LNCS, vol. 11796, pp. 95–102. Springer, Cham (2019). https://doi.org/10.1007/978-3-030-32695-1_11
12. Li, J., Ji, J., Liang, Y., Zhang, X., Wang, Z.: Deep forest with cross-shaped window scanning mechanism to extract topological features. In: IEEE International Conference Bioinformatics and Biomedicine, pp. 688–691 (2019)
13. Chen, Z., Ji, J., Liang, Y.: Convolutional neural network with an element-wise filter to classify dynamic functional connectivity. In: IEEE International Conference on Bioinformatics and Biomedicine, pp. 643–646 (2019)
14. Khosla, M., Jamison, K., Kuceyeski, A., Sabuncu, M.R.: Ensemble learning with 3D convolutional neural networks for functional connectome-based prediction. Neuroimage **199**, 651–662 (2019)
15. Wang, J., Wang, Q., Zhang, H., Chen, J., Wang, S., Shen, D.: Sparse multiview task-centralized ensemble learning for ASD diagnosis based on age and sex-related functional connectivity patterns. IEEE Trans. Cybern. **49**, 3141–3154 (2019)

16. Anirudh,R., Thiagarajan, J.J.: Bootstrapping graph convolutional neural networks for autism spectrum disorder classification. In: IEEE International Conference on Acoustics, Speech Signal Processing, pp. 3197–3201 (2019)
17. Eslami, T., Mirjalili, V., Fong, A., Laird, A.R., Saeed, F.: ASD-DiagNet: a hybrid learning approach for detection of autism spectrum disorder using fMRI data. Front. Neuroinform. **13**, 70 (2019)
18. Chen, L., Huang, Y., Liao, B., Nie, K., Dong, S., Hu, J.: Graph learning approaches for graph with noise: application to disease prediction in population graph. In: IEEE International Conference on Bioinformatics and Biomedicine, pp. 2724–2729 (2020)
19. Kim, T.Y., Cho, S.B.: Optimizing CNN-LSTM neural networks with PSO for anomalous query access control. Neurocomputing **456**, 666–677 (2021)
20. Kim, T.Y., Cho, S.B.: Predicting residential energy consumption using CNN-LSTM neural networks. Energy **182**, 72–81 (2019)
21. Affolter, N., Egressy, B., Pascual, D., Wattenhofer, R.: Brain2Word: Improving brain decoding methods and evaluation. In: Medical Imaging Meets Neurips Workshop-34th Conference on Neural Information Processing Systems (2020)

A Profile on Twitter Shadowban: An AI Ethics Position Paper on Free-Speech

Francisco S. Marcondes[1], Adelino Gala[2], Dalila Durães[1](✉), Fernando Moreira[3], José João Almeida[1], Vania Baldi[2], and Paulo Novais[1]

[1] Algoritmi Center, University of Minho, Braga, Portugal
{francisco.marcondes,dalila.duraes}@algoritmi.uminho.pt, {jj,pjon}@di.uminho.pt

[2] Department of Communication and Art, University of Aveiro, Aveiro, Portugal
{adelino,vbaldi}@ua.pt

[3] REMIT, IJP, Portucalense University, Porto, Portugal
fmoreira@upt.pt

Abstract. Concerns have been expressed lately about content verification algorithms on social media platforms, resulting in data being presented to users or omitted from them. Twitter is one of those platforms that use this strategy, giving censorship strategies for specific content. The shadowban or practice of limiting content distribution without user acknowledgement is a current practice in online social networks, especially on Twitter. So, this paper is an AI Ethics position paper willing to expand the reflection about the impacts of programming artefacts on individual liberties in enterprise's Online social networks (OSN) to the computer science public. The conclusion to be drawn is that the limits of speech are to be imposed by the state after a properly democratic process took place [4]. The concern then turns into an international relations issue as it would be a threat for national sovereignty that one state, even inadvertently, regulates the actions of a communication enterprise on foreign states. Finally, the use of smart contracts is suggested as an alternative to be used by the base state for enforcing regulations on foreign OSNs.

Keywords: Twitter shadowban · Free-speech · Smart-contract

1 Introduction

Shadow Banning (or Stealth Banning) is an automatic online moderation approach for preventing unwanted behavior [11]. It is based on limiting the visibility and reach of profiles and posts throughout the social network **without awareness of the user**. Until version 14 of the terms of service (valid from 2020) [17],

This work has been supported by FCT – Fundação para a Ciência e Tecnologia within the R&D Units Project Scope: UIDB/00319/2020.

H. Yin et al. (Eds.): IDEAL 2021, LNCS 13113, pp. 397–405, 2021.
https://doi.org/10.1007/978-3-030-91608-4_39

Twitter declared that it does not perform any content moderation [16][1]. The problem now is not if Twitter is doing shadow banning, as experimentally [11] and declared in [17]. The issue is how is it being done and its impact.

The discussion focuses on free speech and procedures used for amplifying or reducing the reach of selected narratives or actors on Twitter. So, the following section describes the current states. Section 3 explain the methodology, and Sect. 4 presents the results based on behavior-based shadowban profile and cases of shadowban practices and impacts. Section 5 introduce an insight for using Smart-Contracts. Finally, Sect. 6 presents the conclusions and future work

2 Current Status

OSN such as Twitter and Facebook, started with the humble target of connecting people and favoring free speech. After some time such networks grew enormously throughout the world, until, finally, reach a power comparable to that of traditional media such as newspapers and television. However, unlike traditional media, OSNs had no editorial line, allowing anyone to publish whatever is wished, which is the fake-news phenomenon. However, "liberty" was never a real problem. It is not different to what happened on the pre-Internet Bulletin Board System (BBS) or on previous devices. This was also not a major problem on the several communication services exiting through the Internet, such as the Internet Relay Chat (IRC) and forums. In short, fake-news always existed in communication [13]. Then, it appears to be misleading to consider free-speech and communication causing the fake-news issue [9].

As an approximation *cf.* [10], suppose that someone interacted with "plain earth" content, the *mediation algorithm* tracking this person's activity tends to deliver more content, groups, and people related to "plain earth". However, as there are no "global earth" content, this same person will hardly interact with a content because he did not receive proper information. Yet, this person consumes more content on "plain earth", become part of groups about that subject and follows other people with this interest, the *mediation algorithm* is reinforced. Furthermore, as it tries to cluster this person's profile (currently called micro-targeting), it may happen that person to fall into a cluster that prevents scientific content in its timeline. In addition, micro-targeting is an advertising tool that can be used for marketing purposes within the OSNs (highlight that the activities on the advertisements also feeds the *mediation algorithm*). The use of OSN's advertisement systems for propaganda purposes was matter of time [8].

It was probably after Zuckerberg's hearing at the U.S. Senate in 2018 [15] that serious attempts to hinder fake-news spreading started to appear. Among them it is included the **use of fact-checking committees, diminishing the reach of some people** and **remove people from an OSN**.

[1] For quoting: "*We may not monitor or control the Content posted via the Services we cannot take responsibility for such Content*" [16] (may 2018); and "*We may also remove or refuse to distribute any Content on the Services, limit distribution or visibility of any Content on the service, suspend or terminate users, and reclaim usernames without liability to you*" [17] (jan 2020).

3 Methodology

Data collection in this paper aims two targets: 1) verify if shadowban can be associated to behavior (an attempt to reject that shadowban is based on narratives) and 2) present instances of the impact that shadowban may be producing.

The first target was performed following the steps. Initially, for providing an actual random sample, a handle set was artificially generated using onomatology structures upon fictional names composed after pooling from first-name and last-name dataset (refer to [12]). This procedure is chosen over topological crawl for avoiding groups of shadow banned profiles [11] that could bias the results. The result set is balanced, it includes both personal and professional profiles scattered through several languages. The handle set was then submitted to Twint[2] for retrieving: "joining date"; "all tweets"; "tweets_sample"; "followers"; and "following". For avoid the gathering to halt, the "tweets_sample" is limited to the last 30 days with a top of 400 tweets per profile. This same field is used for setting the "30d_volume" following a simple distribution ($\lceil n \div 30 \rceil$, note that due to sampling limitation the maximum is 14). Also, based on the "joining date" the field "new?" is set true if the account is newer than 30 days. Finally, profiles whose "30d_volume" was greater than zero is checked for shadowban[3]. Based on the number of active users in Twitter (340 million [1]), the sample size for confidence of 95% and error of 5% is of 385 individuals to be tested against shadowban. Highlight that Twitter classifies an user as active if the account logged in within six month time-span, but, for this paper, an active user is that who tweeted at least once in one month time-frame.

The second target is undertook by Internet document search. The approach is qualitative, the inclusion criteria is to be an artifact that illustrates the practice and impacts related to shadow banning in Twitter.

4 Results and Discussion

For a landscape *cf.* [1], the Twitter numbers for the first quarter of 2020 are: More than 1.3 billion accounts (including inactive), 340 million active users (users that logged in at least once in the last six months), 166 million monetizable daily active users (profiles that are able to show ads). In the United States, 80% of tweets are posted by 10% of users; 500 million tweets are sent each day; 42% of Twitter users are on the platform daily. The average U.S. adult posts twice and favorites one tweet per month, follows 89 accounts, and has 29 followers.

4.1 Behavior-Based Shadowban Profiling

This section explores the possibility that shadowban is caused by the user behavior (and not due to free-speech moderation as being issued).

[2] https://github.com/twintproject/twint.
[3] https://shadowban.eu/.

The first collection was performed on April 9th 2021 by generating 3391 handles whose data was scrapped from Twitter. It was noticed that some handles were repeated, therefore removed from the sample. Each handle was classified into non-existing or suspended (808 - 24%), inactive (2196 - 65%), active (387 profiles - 11%); only the active profiles were considered valid for this sample. When checking for shadowban on April 21th 2021, 111 profiles (29% of active handle sample) returns the error message "something unexpected went wrong" then discarded. There were then sampled 280 profiles.

A new gathering and shadowban check took place on April 24th 2021 for completing the sample. The program was refactored for avoiding repetitions and for excluding already analysed handles. After generating 1882 handles expecting gathering 15% of active profiles, a set of 227 additional active profiles were submitted to the shadowban. After discarding 49 profiles (22% of active handle sample) the evaluated profiles for the second round were of 178 profiles.

Summarizing it was generated 5273 handles for sampling 614 (12%) individuals with positive "30d_volume". The shadowban tester raised the "something went wrong" error for 156 (26%) profiles, removed from the sample. The final sample checked for shadowban is of 458 profiles, with 26 (6%) handles positive for shadowban; 4 (1%) for "Search Suggestion Ban" (hard shadowban) and 22 (5%) for "Reply Deboosted" (soft shadowban). One profile was suspended in the time gap between the sample collection and shadowban check (twelve days), and three profiles were classified as "protected".

Since the resulting data set is compatible with the Twitter numbers *cf.* [1], it is possible to undergo through a projection. Considering, 1.3bi existing profiles [1], there are ≈156.000.000 (12% × 1.3B) handles tweeting at least once a month; and ≈9.360.000 (6% × 156M) handles shadowban. None of shadowban handles in the sample were suspended until mid July 2021.

In order to verify if shadowban can be related to broad user behavior the correlation between shadowban and the other measures were computed resulting:

- shadowban × "30d_volume" 0.05681926 ($p = 0.2269$);
- shadowban × "all tweets" −0.05687502 ($p = 0.2265$);
- shadowban × "followers" −0.03090302 ($p = 0.5113$);
- shadowban × "following" −0.01769144 ($p = 0.707$).

In other words, it was not found in this sample correlation between shadowban and tweeting frequency, tweet amount, followers nor following numbers. Behaviors that were not assessed in this study includes private message sending pattern; tweeting time-lapse concentration; and liking pattern. Assess tweeting, replying and retweeting patterns separately may also provide insights on how does the shadowban is working. Nevertheless, the measurements provide enough evidence that shadowban may not be closely related to broad user behavior.

4.2 Cases of Shadowban Practice and Impacts

Since shadowban is not likely to be linked with user behavior, an alternative to consider is that it is linked to the content of posts. For guiding the following analysis it will be used a page on the Reddit[4] dedicated to understand the shadowban in Twitter posing then possible causes distilled from case reports. The presented causes are mainly related with content moderation. It worth then to understand the truth of such claim and understand what are the contents being moderated. According to that page there two types of shadowban. The hard when the profile is not retrieved by the search engine, called "Search Suggestion Ban"; or when an user's tweets are not rendered except for the origin profile and its followers, called "Ghost Ban". The soft when the tweets are "Deboosted", *i.e.* it is necessary to press something like a "show more replies" button for the tweet appear. Highlight that in addition to shadowban there is soft moderation such as using warning messages tagged to tweets considered sensitive [19].

The Reddit page suggests that hard shadowban profiles would be linked to adult oriented content and soft shadowban to offensive content. Therefore, hard shadowban is directed to profiles and soft shadowban to tweets.

Considering the four profiles with hard shadowban, `@handle1` is a profile localized in Uruguay with 194 followers that sent 4 tweets in the sampled period; it was not possible to identify nothing unusual. `@handle2` is a profile localized in Japan with 23 followers that sent 73 tweets in the sampled period; the tweets are messages from a game for earning points (an usual advertising practice). Most of tweets are tagged with soft moderation messages of sensitive content by showing drawn monsters. `@handle3` a profile localized in US with 37 followers that sent 13 tweets in the sampled period; the profile appears to be re-seller shop that uses Twitter for advertising, especially, for shoes. `@handle4`, localization not declared, with 515 followers that sent 269 tweets in the sampled period; the tweets content is often soft-porn.

Soft shadowban in turn would happen to brand new accounts (not present in the sample then not evaluated); profiles with low number of followers (not corroborated as the average number of followers on deboosted profiles is of 1347 ± 3419); replying to tweets with an image (no evidence found); tweeting about on an undesirable Twitter topic (corroborated); use negative sentiment related words on the tweets (not evaluated); and offensive language (corroborated). Going through deboosted profiles (on https://twitter.com/HANDLE/with_replies) a pattern appear to emerge suggesting that Twitter may have an editorial line for shadowbaning sensitive topics (refer to Table 1). Notice that this is also not conclusive as the sample is small and it was not verified the amount of profiles with similar timeline profile that were not shadowban. Also, may exist differences between OSN as such biases were not found in the YouTube [7].

The presented analysis, despite not conclusive, strengths the idea that shadowban in Twitter is likely to be directed by content. Therefore a potential threat to free-speech. Based on the sample it is possible to project 1.560.000 handles with hard shadowban and 7.800.000 with soft shadowban.

[4] https://www.reddit.com/r/Twitter/comments/p1ggm4/rtwitter_attempts_to_solve_twitters_shadowban.

Table 1. Deboosted profiles timeline overview profiling. For the assessment the timeline profiling was held by one author and validated by another using the eyesight, disputes were solved by joint re-evaluation and search for consensus.

Number	Location	Overall timeline profile
@handle5	US	Abortion and progressive topics
@handle6	Chile	Propagation of progressive agenda
@handle7	Brazil	Feminism and other minorities
@handle8	France	Pro-China speech
@handle9	Norway	Few pro-Arab tweets
@handle10	n/c	Propagation of progressive agenda
@handle11	n/c	Science negation and liberal speech
@handle12	Scotland	Pro-Arab speech and cryptocurrency
@handle13	US	Pattern could not be found
@handle14	Japan	–
@handle15	US	Offensive language
@handle16	US	Anti-war and peace oriented speech
@handle17	n/c	Negative exposition of brands
@handle18	Brazil	Political criticism
@handle19	n/c	Propagation of progressive agenda
@handle20	n/c	Always a same phrase
@handle21	–	–
@handle22	Vietnam	Big tech and political criticism
@handle23	n/c	Pattern could not be found
@handle24	US	Republican criticism
@handle25	US	Pro-science, pandemic-related tweets
@handle26	–	–

For a deepen illustration consider the handle `@daniel_debunker`. This handle was the public profile of an allegedly guardian[5] whose activity is based on debunking tweets that contains fake-news by replying them with common-sense. A sample of `@daniel_debunker`'s (17 following and 79 followers) activity in Twitter is presented in Fig. 1. That profile's data was gathered in May 2021; checking again in July 2021, the account is suspended by Twitter.

Notice that may exist actual reasons for the Twitter's decision such as foreign propaganda [6] but the point is the exclusion criteria is not clear, vague and ever changing. With clear statements it would be possible to undertake a proper democratic discussion about free-speech in Twitter. Therefore, how does one should read that in January 2021 (after the US's Capitol invasion) around 70.000 profiles were suspended from Twitter accused to be linked with QAnnon [2]?

[5] Guardians are people who use their free-time for spreading fact-checks and fight against misinformation [18].

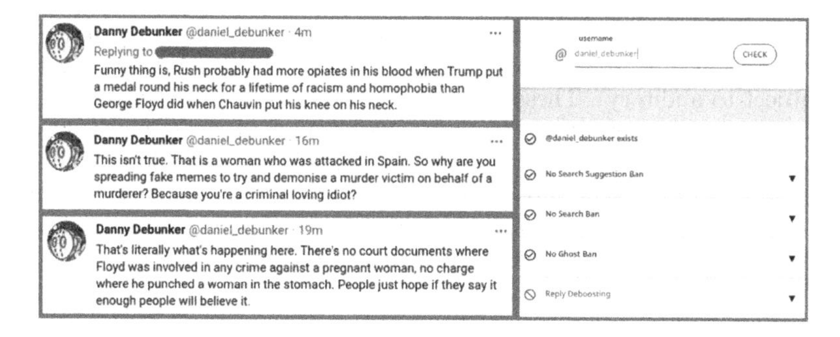

Fig. 1. Guardian shadowbaned with reply deboosting (data gathered in May, 2021)

The impacts of shadowban goes beyond free-speech and other political and social impacts and includes social concerns. OSN are programmed for triggering dopamine production leading to addiction [10], abruptly reducing the exposition for these triggers may lead people to breakdown [10].

It is not new that journalists and even presidents are being suspended from Twitter but there is another case that worth present due to its importance for the future research in this area. On August 2021, Facebook suspend the accounts of researches of the New York University alleging they are not complying with the OSN terms of service and user's privacy since they were collecting data with scrapping software and not through the API [5]. This could be a potential transparency threat that may lead to hinder OSN research.

5 An Insight for Using Smart-Contracts

The issue of democratic speech regulation on OSN is a wicked problem by its own nature. The question initially posed as "how worthy a company is to regulate people speech?" into "how worthy a **foreign** company is to regulate people speech of an **alien nation**?".

The difficulty of such matter can be illustrated by a Brazilian case. The president of that country is allegedly using a massive propaganda operation based on fake-news [14]. Since the major OSNs start to remove some of such contents, the president enforced a bill forbidding any content to be removed without previous judicial process. The alleged motives are to give clarity to the OSN ban policies and protect freedom of speech [3]. Based on the evidence presented and on the possibility of malicious legislation take place; it is possible to claim that alternatives should be explored as the debate in the political sphere advances. Considering the shadowban issue, as user abuse is being tackled by OSNs, it is necessary also to avoid OSN abuse.

Being a web-service, it is not possible to an external agent such as state governments to control the *mediation algorithm* of an OSN, it is however possible to **assess the produced side effects** and **data exchanged between client and server**. In addition, it is not due technology artifacts to questioning a country's political decisions and regulations, it is however possible to assure **fairness** (the same rules applicable for everybody) and **transparency** (everything is recorded and subject to auditory). These concerns impacts national sovereignty.

Smart contracts suits these concerns. Due to the underlying block-chain it is transparent and due to its automatism it is fair. Also, it is able to control the flow of data between client and server and detect *mediation algorithm* side effect patterns. Certainly each state smart-contract is to be build by each nation and reflect their legislation. A sketch is presented in Fig. 2.

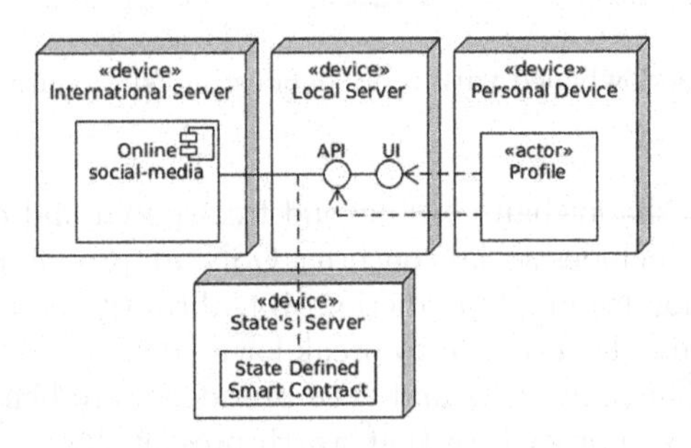

Fig. 2. UML component diagram for the smart-contract proposal.

Suppose for an instance that someone place his or hers location and that is not allowed by a nation, the contract may work as a proxy replacing the specific local such an street with a broad location such as an country. When the page retrieves, that broad location can be replaced back to the specific location. Then, since the information is held locally the alien nations would see the broad location while the base state citizen can retrieve the specific information. Control the *mediation algorithm* is not that straightforward but it is possible to assess the mediation volume and patterns for subsidize a legal process or denounced by journalists. In addition, for avoiding that a citizen keep unaware about being shadowban or such, the contract may send this information as the page passes through it. Therefore, smart-contracts is a potential alternative for preventing abuses from OSN.

6 Conclusion

The shadowban profile raised on Sect. 4.1 projected that there could be approximately 9.4 million users under shadowban and the cause, at first, is not due to their behavior. Section 4.2 suggests that the cause may be due to the tweet content, some cases are presented to support this possibility. Content moderation threatens free speech being a major concern for liberal democracies. However, there are situations when content moderation must be employed for assuring the democratic speech (Popper's paradox on intolerance). Therefore, an entangled issue to handle. From the computer science perspective there are few actions to be taken concerning the OSN and fewer concerning the political sphere. Nevertheless, it is always possible to provide transparency and a bit of control either to the user, the OSN or the government. Smart-contracts, acting a proxy managing national interests, may help on such enterprise. For future works it is necessary to increasing the data collection for better profile the shadowban phenomenon and further develop the smart contract into a proper proof-of-concept for addressing the raised issues.

References

1. Aslam, S.: Twitter by the numbers: Stats, demographics & fun facts, July 2021
2. BBC. Twitter suspends 70,000 accounts linked to QAnon, January 2021
3. Brito, R.: Brazil president Jair Bolsonaro signs decree changing social media regulations, September 2021
4. Browne, R.: Germany's Merkel hits out at Twitter over 'problematic' trump ban, January 2021
5. Hatmaker, T.: Facebook cuts off NYU researcher access, prompting rebuke from lawmakers, August 2021
6. Select Committee on Intelligence: Report of the select committee on intelligence united states senate on Russian active measures campaigns and interference in the 2016 U.S. election. Technical report, THE UNITED STATES SENATE (2019)
7. Jiang, S., Robertson, R.E., Wilson, C.: Reasoning about political bias in content moderation. In: Proceedings of the AAAI Conference on Artificial Intelligence, vol. 34, pp. 13669–13672 (2020)
8. Jungherr, A., Rivero, G., Gayo-Avello, D.: Retooling Politics: How Digital Media Are Shaping Democracy. Cambridge University Press, Cambridge (2020)
9. Kelly, J.: How 'fact-checking' can be used as censorship, February 2021
10. Lanier, J.: Ten arguments for deleting your social media accounts right now. Random House, University of California, US (2018)
11. Le Merrer, E., Morgan, B., Trédan, G.: Setting the record straighter on shadow banning. In: IEEE INFOCOM 2021-IEEE Conference on Computer Communications, pp. 1–10. IEEE (2021)
12. Marcondes, F.S., Almeida, J.J., Novais, P.: Structural onomatology for username generation: a partial account. In: Proceedings of the 9th European Starting AI Researchers' Symposium 2020 Co-located with 24th European Conference on Artificial Intelligence (ECAI 2020) (2020)
13. Sagan, C.: The Demon-haunted World: Science as a Candle in the Dark. Ballantine books, University of California, US (2011)
14. Saxena, S., Costa, F.: Bolsonaro wants to make brazil's social media safe for his fake-news factory, September 2021
15. New York Times: Mark Zuckerberg testimony: Senators question Facebook's commitment to privacy, April 2018
16. Twitter. Twitter terms of service, vol. 13, May 2018
17. Twitter. Twitter terms of service, vol. 14, January 2020
18. Vo, N., Lee, K.: The rise of guardians: fact-checking URL recommendation to combat fake news. In: The 41st International ACM SIGIR Conference on Research & Development in Information Retrieval, pp. 275–284 (2018)
19. Zannettou, S.: "I won the election!": an empirical analysis of soft moderation interventions on Twitter. arXiv:2101.07183 v1, 18 January 2021

Meta-feature Extraction Strategies for Active Anomaly Detection

Fabrizio Angiulli, Fabio Fassetti, Luca Ferragina(✉), and Prospero Papaleo

DIMES, University of Calabria, 87036 Rende, CS, Italy
{f.angiulli,f.fassetti,l.ferragina}@dimes.unical.it

Abstract. Active Learning is a machine learning scenario in which methods are trained by iteratively submitting a query to a human expert and then taking into account his feedback for the following computations. The application of such paradigm to the anomaly detection task takes the name of Active Anomaly Detection (AAD). Reinforcement Learning describes a family of algorithms that aim to teach an agent to determine a policy to deal with external factors, and are based on the maximization of a reward function. Recently some AAD methods, based on the training of a meta-policy with Deep Reinforcement Learning have been very successful because, after the training, the methods simply work on a small number of meta-features that can be directly applied to any new dataset without further tuning.

For these approaches a central question is the selection of good meta-features: actually, the most common choice is to define these meta-features in terms of the distances with the points that the expert has already labelled as either anomaly or normal. In this work we explore different strategies for selecting effective meta-features. Specifically, we take into account both direct and reverse nearest-neighbor rankings in order to build meta-features, since they are less sensitive to the specific distance distribution characterizing the training data, and experiment the combination also with related base detectors. The experiments show that there are scenarios in which our approach offers advantages over the standard technique.

Keywords: Anomaly detection · Active Learning · Meta-feature extraction · Reinforcement learning

1 Introduction

Anomaly detection is a fundamental discovery problem whose aim is to identify objects in a dataset that are suspected of not being generated by the same process as the majority of the data.

Anomalies can arise due to many reasons, like mechanical faults, fraudulent behavior, human errors, instrument error or simply through natural deviations in populations. Several statistical, data mining and machine learning approaches

H. Yin et al. (Eds.): IDEAL 2021, LNCS 13113, pp. 406–414, 2021.
https://doi.org/10.1007/978-3-030-91608-4_40

have been proposed to detect outliers, namely, statistical-based [9,13], distance-based [5–8,18], density-based [10,16], reverse nearest neighbor-based [2,4,15,21], isolation-based [20], angle-based [19], SVM-based [22,24], deep learning-based [11,14], and many others [1,12].

In this work we focus on Active Anomaly Detection (AAD), that is a special setting of anomaly detection that involves an expert advice in the models training. In the simplest AAD scenario, the model works iteratively by first selecting a query to submit to the expert that assigns a label on each object, and then using them to update the model's parameters and select the next query. In [25] is introduced an AAD model that is based on the extraction from the dataset of a fixed number of meta-features. Most of these meta-features are based on the Euclidean distance separating the selected point from the known anomalous and normal objects, but it is also considered a special meta-feature consisting in the score returned by an unsupervised anomaly detector (in the original paper it is used Isolation Forest).

We introduce a novel set of meta-features that are based on the rank of the point within the direct and reverse nearest neighbor lists of the points already labelled by the expert, rather than on their distances. We also propose to employ a different anomaly score, namely CFOF [4], which bases its decision on the reverse nearest-neighbor relationship. These features are less sensitive to the specific distance distribution characterizing the training data, and we believe that this is important in order to improve the generality of the policy and to mitigate the curse of dimensionality as the number of features of the dataset taken into account grows [3].

The contributions of the work can be summarized as follows:

- We implement and test a version of *Meta–AAD*, called *Meta–AAD–Rank*, that is neighbor-rank based rather than distance based as the original method.
- We consider CFOF as an alternative anomaly score meta-feature to insert in the meta-policy.
- We introduce a dimensionality reduction method based on Variational Autoencoders in order to deal with more complex data.

The rest of the work is organized as follows. In Sect. 2 we introduce the context and the main instruments at the basis of our contribution as well as the related work. In Sect. 3 we describe *Meta-AAD-Rank*, a novel meta-feature extraction method for Active Anomaly Detection. In Sect. 4 we show the results of the experiments. Finally, Sect. 5 concludes the work.

2 Preliminaries

A recent interesting improvement of AAD is the Active Anomaly Detection with Meta-Policy (*Meta–AAD*), which learns a meta-policy with the aim of optimizing the number of discovered anomalies keeping the same *budget*, that is to say the number of instances presented to the human for feedback.

Meta–AAD. Given a dataset $X = \{x_1, \ldots, x_n\} \subseteq \mathbb{R}^d$, at each iteration the policy selects an example x_i to submit to the expert that will assert weather x_i is actually an anomaly or not. The state of the policy is recorded in a vector $\hat{\mathbf{y}} \in \mathbf{R}^n$ such that, for each $j \in \{1, \ldots, n\}$, we have that y_j is equal to:

-1, if the item x_j has been submitted to the expert and reported as an anomaly;
1, if x_j has been submitted to the expert and reported as normal;
0, if x_j has not yet been submitted to the expert.

At the beginning of the policy we have $y_j = 0$, for each $j \in \{1, \ldots, n\}$, and then at each iteration the state of the item selected for the query is updated according to the expert feedback. The main innovations of *Meta–AAD* with respect to AAD consists in the fact that in AAD the instance presented to the human is the one obtaining the highest value of anomaly score and, due to feedbacks, the anomalous scores are adjusted to promote the anomalous instances to the top, thus the main goal is to make the top instance more likely to be anomalous so maximizing the immediate performance.

Thus, in [25], first the meta-policy is trained and any Deep Reinforcement Learning (DRL) algorithm can be used to this aim. In particular authors adopt Proximal Policy Optimization (PPO) [23], a family of methods for reinforcement learning, in which a deep neural network is trained iteratively by sampling data through interaction with the environment, and by using these data for the optimization of a state function using stochastic gradient ascent.

Then the meta-features are extracted for all the instances and the probability of the meta-policy is computed for all the instances. The instance with the highest probability is presented to the human.

A fundamental part of this process is the meta-features extraction. Formally, the aim is to define a function $g : X \times \mathbf{y} \rightarrow \mathbf{R}^{n \times l}$, where l is the number of meta-features, such that its image is as much as possible independent from the dataset. The framework allows flexible choices. Authors [25] proposes the following set of $l = 6$ meta-features, divided in 4 groups:

- *Detector Features* (**1**): score related to the unsupervised detector, in the original paper is adopted Isolation Forest (IF);
- *Binary Features* (**2**): binary indicator asserting if x_i is one of the k nearest neighbors of an instance;
- *Anomaly Features* (**3**,**4**): minimum and mean Euler distance between x_i and the set of instances currently labelled as anomalous;
- *Normality Features* (**5**,**6**): minimum and mean distance between x_i and the set of instances currently labelled as normal.

3 The *Meta–AAD–Rank* Approach

The *Meta–AAD* approach is distance-based, we argue that this fact may cause some issue in the query selection phase due to the specific distance distribution of the dataset. In order to relief this dependence, here we introduce an innovative choice of meta features that is based on nearest neighbor rankings rather than on distances. We call *Meta–AAD–Rank* the approach based on exploiting the novel features.

Let us define a rank matrix $R \in \mathbb{R}^{n \times n}$, where the element R_{ij} contains the position occupied by the point x_j inside the nearest neighbor list of the point x_i. The meta-features relative to the generic point $x_i \in X$ used in this work are listed next.

- *Detector Features* (**1**): score related to the unsupervised detector.
- *Binary Features* (**2**): binary indicator asserting if i is one of the k nearest neighbors of an instance.
- *Anomaly Features* (**3**,**4**): the position occupied by the nearest known anomaly in the nearest neighbors list of x_i, that is $\min_{j \in H_A} R_{ij}$ and the position $R_{\mu i}$ occupied by x_i in the nearest neighbor list of x_μ, where $\mu = \arg\min_{j \in H_A} R_{ij}$ and H_A is the set of the known anomaly indices;
- *Normality Features* (**5**,**6**): the position occupied by the nearest known normal example in the nearest neighbors list of x_i, that is $\min_{j \in H_N} R_{ij}$ and the position $R_{\nu i}$ occupied by x_i in the nearest neighbor list of x_ν, where $\nu = \arg\min_{j \in H_N} R_{ij}$ and H_N is the set of the known normal examples indices.

Note that now features represent direct and reverse rankings associated with point x_i with respect to known anomalous and normal points, rather than take into account inter-point distances.

Reverse Nearest-Neighbor-Based Anomalies with *Meta–AAD–Rank*$_{\mathbf{CFOF}}$. We use *Meta–AAD–Rank* (or also *Meta–AAD–Rank*$_{\mathrm{IF}}$) to refer to the method using as *Detector Feature* the anomaly score returned by the *Isolation Forest* (IF) algorithm. To investigate the effect of exploiting reverse ranking information, we consider here also the *Meta–AAD–Rank*$_{\mathrm{CFOF}}$ algorithm which exploits a *Detector Features* the CFOF anomaly score [2,4], defined as

$$CFOF(x_i) = \min_{1 \le k' \le n} \left\{ \frac{k'}{n} : \mathrm{N}_{k'}(x_i) \ge n\rho \right\},$$

where $\mathrm{N}_k(x_i) = |\{y : x_i \in \mathrm{NN}_k(y)\}|$ is the *reverse k nearest neighbor count*, that is the number of objects having x_i among their k nearest neighbors, and $\mathrm{NN}_k(y)$ is the set of the k nearest neighbor of y. The CFOF score of x_i represents the smallest neighborhood width, normalized with respect to n, for which x_i exhibits a reverse neighborhood of size at least $n\rho$.

The intuition is that isolated points will require larger values of k than inliers in order to be selected as neighbors by an equal-sized fraction of the data population. In contrast to almost all known outlier detection measures, CFOF scores do not exhibit concentration. By leveraging the closed form of the function N_k it is possible to formally see that CFOF outliers are few in number and separated from inliers even in intrinsically high-dimensional spaces, whereas the direct use of the reverse k nearest neighbor count for outlier detection is prone to false positives [3].

Semantic Neighborhood with *Meta–AAD–SemRank*. Another aspect we consider is semantic similarity. It is known that when the dimensionality of the data grows, the "geometric" neighborhood tends to lose relationship with the "semantic" neighborhood [3], and this may worse the effectiveness of meta-features based on the former kind of information.

We propose to deal with the above problem by performing semantic dimensionality reduction through *Variational AutoEncoders* (VAE). A variational autoencoder is a stochastic generative model that has the aim of outputting a reconstruction $\hat{x}$ of a given input sample x [17].

Basically, variational autoencoders encode each example as a normal distribution over the latent space and regularize the loss by maximizing similarity of these distributions with the standard normal distribution. This encoding is conducive to obtain a continuous latent space, namely a latent space for which close points will lead to close decoded representation. This also means that the geometric neighborhood in the VAE latent space is in good agreement with the notion of semantic neighborhood.

In particular, in the *Meta–AAD–SemRank* method we exploit the semantic distance

$$d_{\mathcal{L}}(x_i, x_j) := \|z_i - z_j\|_2, \tag{1}$$

where z_i (z_j, resp.) is the image of x_i (x_j, resp.) in the latent space $\mathcal{L}$ of the Variational Autoencoder after it has been trained on the dataset X.

4 Experimental Results

In this section we present experimental results obtained by the *Meta–AAD–Rank* algorithm. In order to compare the considered methods we define the measures **Precision**, that is equal to

$$\frac{a_s}{\min(s, n_a)}, \tag{2}$$

where a_s is the number of anomalies detected after s queries and n_a is the total number of anomalies in the dataset, and **Normalized AUC**, that is the area under the curve of a_s with varying s normalized by the number of the total number of anomalies in the dataset, in formula

$$\frac{2\sum_{s=1}^{B} a_s}{B'(B'+1)}, \tag{3}$$

where $B' = \min(B, n_a)$ is the minimum between the number of anomalies in the dataset and the number of query B submitted to the expert.

The aim of these measures is to mitigate the difference among the number of anomalies in the datasets that we consider, that are the same 24 datasets listed in Table 2 of [25].

We partition the 24 datasets into two groups of 12 each, train the model on each of the two groups, and test it on the other group. Table 1 reports the results

Table 1. Comparison between *Meta–AAD* and *Meta–AAD–Rank* in terms of *Normalized AUC* (Eq. (3)).

Dataset	*Meta–AAD*	*Meta–AAD–Rank*				
		Best	IF	IFstd	CFOF	CFOFstd
Annthyroid	0.76	**0.94**	0.07	**0.94**	0.85	*0.88*
Arrhythmia	0.69	**0.72**	0.57	**0.72**	*0.71*	0.69
Breastw	*0.99*	**1.00**	**1.00**	**1.00**	0.90	0.90
Cardio	**0.96**	*0.90*	0.76	*0.90*	0.78	0.81
Glass	**0.33**	*0.27*	0.04	0.16	0.16	*0.27*
Ionosphere	**0.96**	*0.95*	0.88	0.90	0.92	*0.95*
Letter	*0.41*	**0.45**	0.07	0.09	*0.41*	**0.45**
Lympho	**0.86**	**0.86**	**0.86**	**0.86**	0.52	*0.67*
Mammography	*0.86*	**0.87**	0.69	**0.87**	0.78	0.83
Mnist	*0.96*	**0.99**	**0.99**	0.90	0.92	0.91
Musk	**1.00**	**1.00**	**1.00**	**1.00**	**1.00**	**1.00**
Optdigits	**0.95**	*0.90*	0.55	0.51	*0.90*	0.44
Pendigits	0.95	**0.98**	0.88	*0.97*	**0.98**	**0.98**
Pima	0.65	**0.75**	0.68	*0.69*	0.65	**0.75**
Satellite	**1.00**	**1.00**	**1.00**	**1.00**	0.83	*0.93*
Satimage-2	**0.99**	**0.99**	*0.98*	**0.99**	0.93	**0.99**
Shuttle	**1.00**	**1.00**	*0.97*	**1.00**	0.75	0.96
Speech	0.06	**0.15**	**0.15**	0.04	*0.10*	0.06
Thyroid	0.88	**0.91**	*0.89*	**0.91**	0.82	0.86
Vertebral	0.32	**0.39**	**0.39**	0.17	*0.34*	0.33
Vowels	0.73	**0.90**	0.80	0.73	*0.88*	**0.90**
Wbc	**0.90**	*0.85*	*0.85*	*0.85*	0.58	0.81
Wine	0.27	**1.00**	0.75	0.62	**1.00**	*0.93*
Yeast	0.24	**0.50**	0.46	0.47	*0.48*	**0.50**

in terms of *Normalized AUC* (see Eq. 3) of the *Meta–AAD* and of *Meta–AAD–Rank*. In these experiments we also consider the effect of the standardization on *Meta–AAD–Rank*. Specifically, dataset features are standardized by centering values on the mean and then dividing by their standard deviation. In particular, in the table the columns "IFstd" and "CFOFstd" contain the results obtained by *Meta–AAD–Rank* when standardization is employed. We can observe that in almost all the cases the *Meta–AAD–Rank* strategies are able to improve over the standard *Meta–AAD* strategy.

Although in some dataset meta-AAD-rank does not reach the performance of meta-AAD a very important result is that on some large dimensional datasets, as *Arrhythmia*, *Mnist*, and *Speech* having at least one hundred features, the *Meta–AAD–Rank* variants are able to detect more anomalies.

To understand if *Meta–AAD* degrades its performances in the high-dimensional scenario because of the distance concentration phenomenon, we considered also three high-dimensional datasets consisting of images: *MNIST*, *Fashion-MNIST*, and *CIFAR-10*. All these datasets are multi-labelled, thus, in order to be used for anomaly detection, for each of them we consider the first class as normal and the others as anomalies. We tested the meta-policy on each of these datasets after having trained it on the other two. We also considered *Meta–AAD–SemRank* due the high dimensionality of the data. Table 2 reports the number of anomalies detected by the methods on the image datasets, while Fig. 1 reports the Normalized AUC on the same datasets. Except for *MNIST*, which is the simplest datasets of the bunch, differences among the methods are more marked on the other datasets. In particular, *Meta–AAD* performs worse than all the *Meta–AAD–Rank* strategies. Interestingly, *Meta–AAD–SemRank* achieves the best performances according to Fig. 1, but also is able to mitigate the curse of dimensionality effect.

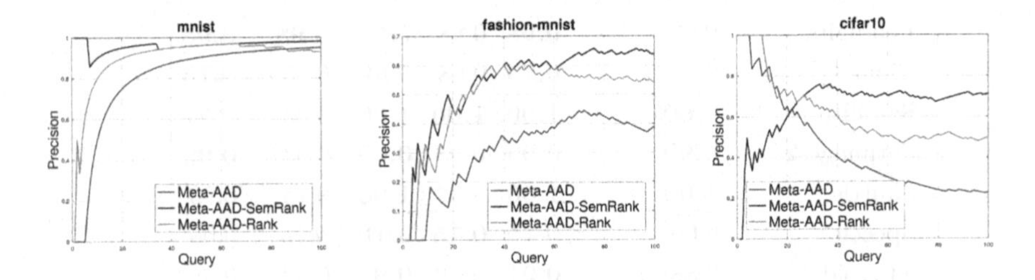

Fig. 1. Comparison on the image datasets.

Table 2. Number of anomalies detected on the image datasets.

Method	Mnist	Fashion-Mnist	Cifar10
Meta–AAD	95	39	23
Meta–AAD–Rank	94	60	54
Meta–AAD–Rank$_{\text{CFOF}}$	98	67	59
Meta–AAD–SemRank	93	54	50
Meta–AAD–SemRank$_{\text{CFOF}}$	98	63	57

5 Conclusion

In this work we presented the *Meta–AAD–Rank* approach, a meta-features extraction strategy for the task of Active Anomaly Detection after a Deep

Reinforcement Learning training. We presented both direct and reverse nearest-neighbor ranking based meta-features and also semantic meta-features exploiting the continuous latent space of variational autoencoders. We experimentally showed that our meta-features perform well in different settings and can improve performances depending on the peculiarities of the data at hand.

References

1. Aggarwal, C.C.: Outlier Analysis. Springer, New York (2013). https://doi.org/10.1007/978-1-4614-6396-2
2. Angiulli, F.: Concentration free outlier detection. In: Ceci, M., Hollmén, J., Todorovski, L., Vens, C., Džeroski, S. (eds.) ECML PKDD 2017. LNCS (LNAI), vol. 10534, pp. 3–19. Springer, Cham (2017). https://doi.org/10.1007/978-3-319-71249-9_1
3. Angiulli, F.: On the behavior of intrinsically high-dimensional spaces: distances, direct and reverse nearest neighbors, and hubness. J. Mach. Learn. Res. **18**, 170:1–170:60 (2018)
4. Angiulli, F.: CFOF: a concentration free measure for anomaly detection. ACM Trans. Knowl. Discov. Data (TKDD) **14**(1), 4:1–4:53 (2020)
5. Angiulli, F., Basta, S., Pizzuti, C.: Distance-based detection and prediction of outliers. IEEE Trans. Knowl. Data Eng. **2**(18), 145–160 (2006)
6. Angiulli, F., Fassetti, F.: DOLPHIN: an efficient algorithm for mining distance-based outliers in very large datasets. ACM Trans. Knowl. Disc. Data (TKDD) **3**(1), 1–57 (2009). Article no. 4
7. Angiulli, F., Pizzuti, C.: Fast outlier detection in large high-dimensional data sets. In: Proceedings of the International Conference on Principles of Data Mining and Knowledge Discovery (PKDD), pp. 15–26 (2002)
8. Angiulli, F., Pizzuti, C.: Outlier mining in large high-dimensional data sets. IEEE Trans. Knowl. Data Eng. **2**(17), 203–215 (2005)
9. Barnett, V., Lewis, T.: Outliers in Statistical Data. Wiley, Hoboken (1994)
10. Breunig, M.M., Kriegel, H., Ng, R., Sander, J.: LOF: identifying density-based local outliers. In: Proceedings of the International Conference on Management of Data (SIGMOD) (2000)
11. Chalapathy, R., Chawla, S.: Deep learning for anomaly detection: a survey (2019)
12. Chandola, V., Banerjee, A., Kumar, V.: Anomaly detection: a survey. ACM Comput. Surv. **41**(3), 1–58 (2009)
13. Davies, L., Gather, U.: The identification of multiple outliers. J. Am. Stat. Assoc. **88**, 782–792 (1993)
14. Goodfellow, I., Bengio, Y., Courville, A.: Deep Learning. MIT Press, Cambridge (2016)
15. Hautamäki, V., Kärkkäinen, I., Fränti, P.: Outlier detection using k-nearest neighbour graph. In: International Conference on Pattern Recognition (ICPR), Cambridge, UK, 23–26 August 2004, pp. 430–433 (2004)
16. Jin, W., Tung, A., Han, J.: Mining top-n local outliers in large databases. In: Proceedings of the ACM SIGKDD International Conference on Knowledge Discovery and Data Mining (KDD) (2001)
17. Kingma, D.P., Welling, M.: Auto-encoding variational Bayes (2013)
18. Knorr, E., Ng, R., Tucakov, V.: Distance-based outlier: algorithms and applications. VLDB J. **8**(3–4), 237–253 (2000)

19. Kriegel, H.P., Schubert, M., Zimek, A.: Angle-based outlier detection in high-dimensional data. In: Proceedings of the International Conference on Knowledge Discovery and Data Mining (KDD), pp. 444–452 (2008)
20. Liu, F., Ting, K., Zhou, Z.H.: Isolation-based anomaly detection. TKDD **6**(1), 1–39 (2012)
21. Radovanović, M., Nanopoulos, A., Ivanović, M.: Reverse nearest neighbors in unsupervised distance-based outlier detection. IEEE Trans. Knowl. Data Eng. **27**(5), 1369–1382 (2015)
22. Schölkopf, B., Platt, J.C., Shawe-Taylor, J., Smola, A.J., Williamson, R.C.: Estimating the support of a high-dimensional distribution. Neural Comput. **13**(7), 1443–1471 (2001)
23. Schulman, J., Wolski, F., Dhariwal, P., Radford, A., Klimov, O.: Proximal policy optimization algorithms. CoRR abs/1707.06347 (2017)
24. Tax, D.M.J., Duin, R.P.W.: Support vector data description. Mach. Learn. **54**(1), 45–66 (2004)
25. Zha, D., Lai, K., Wan, M., Hu, X.: Meta-AAD: active anomaly detection with deep reinforcement learning. CoRR abs/2009.07415 (2020)

An Implementation of the "Guess Who?" Game Using CLIP

Arnau Martí Sarri[1,2] and Victor Rodriguez-Fernandez[3(✉)]

[1] Valencian International University, Calle Pintor Sorolla 21, 46002 Valencia, Spain
a.marti@dimaisl.com
[2] Dimai S.L., Cam í de la Font Calda 10, 08270 Navarcles, Spain
[3] School of Computer Systems Engineering, Universidad Politécnica de Madrid, Calle de Alan Turing, 28038 Madrid, Spain
victor.rfernandez@upm.es

Abstract. CLIP (Contrastive Language-Image Pretraining) is an efficient method for learning computer vision tasks from natural language supervision that has powered a recent breakthrough in deep learning due to its zero-shot transfer capabilities. By training from image-text pairs available on the internet, the CLIP model transfers non-trivially to most tasks without the need for any data set specific training. In this work, we use CLIP to implement the engine of the popular game "Guess who?", so that the player interacts with the game using natural language prompts and CLIP automatically decides whether an image in the game board fulfills that prompt or not. We study the performance of this approach by benchmarking on different ways of prompting the questions to CLIP, and show the limitations of its zero-shot capabilites.

Keywords: CLIP · Guess who · Zero-shot learning · Language-image models

1 Introduction

The ability to learn at the same time from different data modalities (image, audio, text, tabular data...) is a trending topic in the field of machine learning in general, and deep learning in particular, with many domains of application such as self-driving cars, healthcare and the Internet of Things [5]. Among all the possibilities that multimodal deep learning provides, one of the most interesting ones is the connection of text and images in the same model.

This concept brings into play challenging tasks in the areas of computer vision and natural language processing, such as multimodal image and text classification [7], image captioning [2], and visual-language robot navigation [10]. All of these tasks have the core idea of learning visual perception from supervision contained in text, or vice-versa.

H. Yin et al. (Eds.): IDEAL 2021, LNCS 13113, pp. 415–425, 2021.
https://doi.org/10.1007/978-3-030-91608-4_41

In January 2021, the company OpenAI made a great milestone in the field of language-image models with the presentation of CLIP (Contrastive Language–Image Pre-training) [8][1]. CLIP is a deep neural network designed to perform zero-shot image classification, i.e., to generalize flawlessly to unseen image classification tasks in which the data and the labels can be different each time. The way CLIP does so is by training on a wide variety of (image, text) pairs that are abundantly available on the internet, instructing the model to predict the most relevant text snippet, given an image. Since the code and weights of CLIP were publicly released on GitHub[2], many researchers have explored its zero-shot capabilities in different areas such as art classification [1], video-text retrieval [4] or text-to-image generation [9].

In this work, we present an application of the zero-shot classification capabilities of CLIP in the popular game "Guess who?", in which the player asks yes/no questions to describe people in a game board and try to guess who the selected person is. Each time the player makes a new question, CLIP will analyse the images in the game board and decide automatically which images fulfill it. Although this could be also tackled with a multi-label image binary classification model with a fixed set of labels, the power behind using CLIP relies on the use of natural language to interact with the model, which gives freedom to the player to ask any question and tests CLIP's zero-shot generalization capabilities. We release our code in a public Github repository[3].

In summary, the contributions of this paper are:

- The implementation of the game engine based on AI, which allows for the use of any set of images instead of having a fixed board. To the best of our knowledge, there is no other version of the game "Guess who?", that uses an AI in a similar way.
- The use of CLIP as a zero-shot classifier based on textual prompts, which allows the player to interact with the game through natural language.

The rest of the paper is structured as follows: in Sect. 2 we give some background on CLIP, in Sect. 3 we present a description of the game and how CLIP is integrated in it, not as a player but as the engine of the game. Then, in Sect. 4 we show some experiments on how changing the way the game prompts CLIP about the characteristic of a person affects its classification performance, and finally, in Sect. 5 we outline the conclusions and provide future lines of research in this topic.

2 Backgrounds on CLIP

CLIP (Contrastive Language-Image Pre-training), by OpenAI, is based on a large amount of work in zero-shot transfer, natural language supervision and

[1] The paper was accompanied with a blog post publication: https://openai.com/blog/clip/.
[2] https://github.com/openai/CLIP.
[3] https://github.com/ArnauDIMAI/CLIP-GuessWho.

multi-modal learning, and shows that scaling a simple pre-training task is sufficient to achieve competitive zero-sample performance on various image classification data sets. CLIP uses a large number of available supervision sources: the text paired with images found on the Internet. This data is used to create the following agent training task for CLIP: Given an image, predict which of a set of 32,768 randomly sampled text fragments is actually paired with it in the data set. This is achieved by combining a text encoder, built as a Transformer [11], and an image encoder, built as a Vision Transformer [3], under a contrastive objective that connects them [12]. To the best of our knowledge, there is no other publicly available pretrained model with the scale of CLIP that connects text and image data.

Once pre-trained, CLIP can then be applied to nearly arbitrary visual classification tasks. For instance, if the task of a data set is classifying photos of dogs vs cats, we will check, for each image, whether CLIP predicts that the caption "a picture of a dog" is more likely to be paired with it than "a picture of a cat". In case the task does not have a fixed set of labels, one can still use CLIP for classification by specifying a text description of a target attribute and a corresponding neutral class. For example, when manipulating images of faces, the target attribute might be specified as "a picture of a blonde person", in which case the corresponding neutral prompt might be "a picture of a person". CLIP's zero-shot classifiers can be sensitive to wording or phrasing, and sometimes require trial and error "prompt engineering" to perform well [8].

3 Game Description

The aim of the game, as in the original one, is to find a specific image from a group of different images of a person's face. To discover the image, the player must ask questions that can be answered with a binary response, such as "Yes and No". After every question made by the player, the images that don't share the same answer that the winning one are discarded automatically. The answer to the player's questions, and thus, the process of discarding the images will be established by CLIP (See Fig. 1). When all the images but one have been discarded, the game is over.

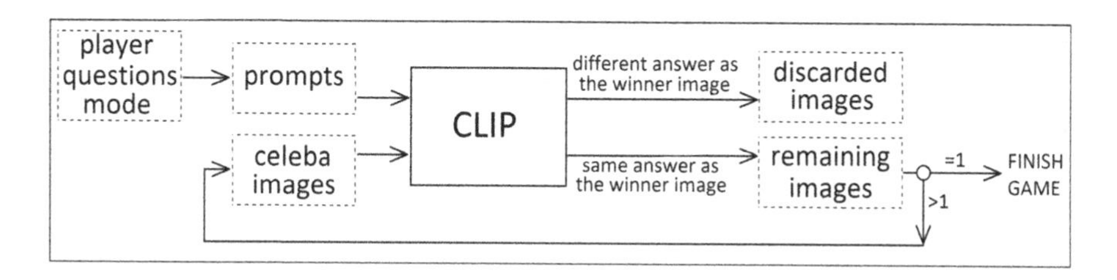

Fig. 1. Diagram of how CLIP is integrated in the game.

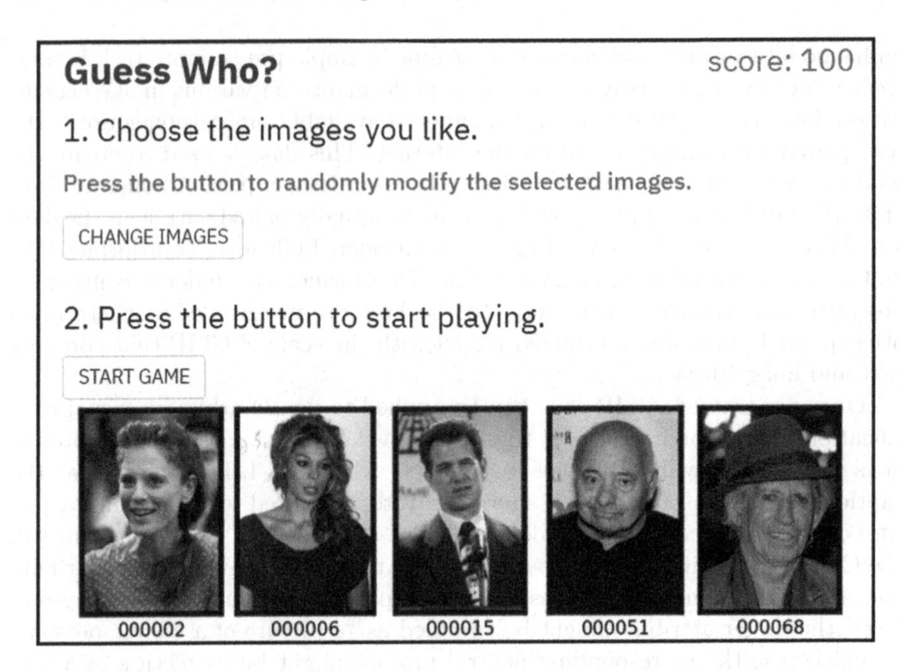

Fig. 2. Screen of the game which allows the user selecting the images to play.

The first step of the game is to select the images to play (See Fig. 2). The player can press a button to randomly change the used images, which are taken from the *CelebA* (CelebFaces Attributes) data set [6]. This data set contains 202,599 face images of the size 178–218 from 10,177 celebrities, each annotated with 40 binary labels indicating facial attributes like hair color, gender and age.

3.1 Questions

The game will allow the player to ask the questions in 4 different ways:

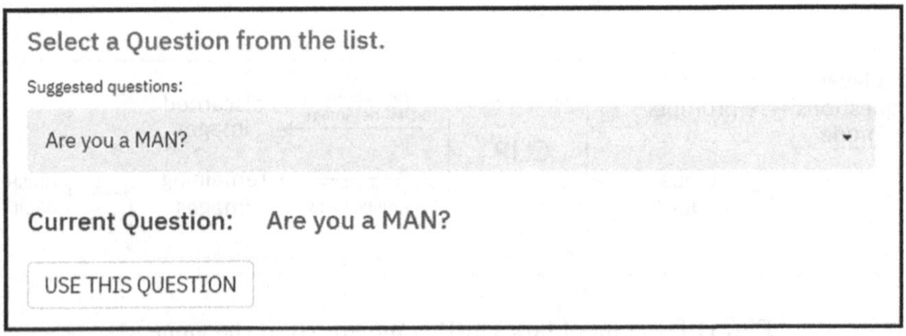

Select a Question from the list.
Suggested questions:
Are you a MAN?
Current Question: Are you a MAN?
USE THIS QUESTION

Fig. 3. Game screen that allows the user to ask a default question.

1. Asking a question from a list (See Fig. 3). A drop-down list allows the player to select the question to be asked from a group of pre-set questions, taken from the set of binary labels of the Celeba data set. Under the hood, each question is translated into textual prompts for the CLIP model to allow for the binary classification based on that question. When they are passed to CLIP along with an image, the model responds by giving a greater value to the prompt that is most related to the image.

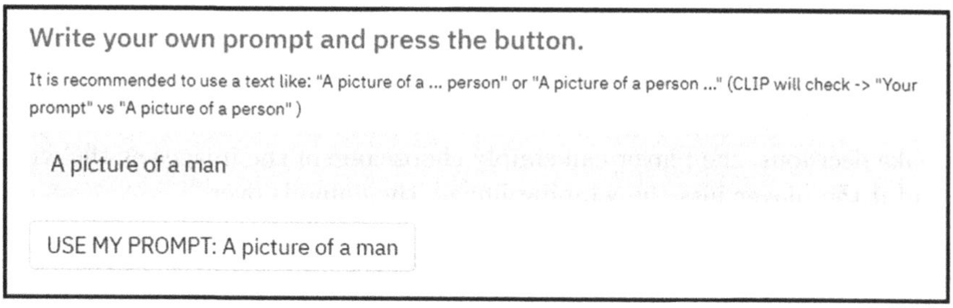

Fig. 4. Game screen that allows the user to create his own prompt using 1 text input.

2. Use one prompt (See Fig. 4). This option is used to allow the player introducing a textual prompt for CLIP with his/her own words. The player text will be then confronted with the neutral prompt, “A picture of a person”, and the pair of prompts will be passed to CLIP as in the previous case.

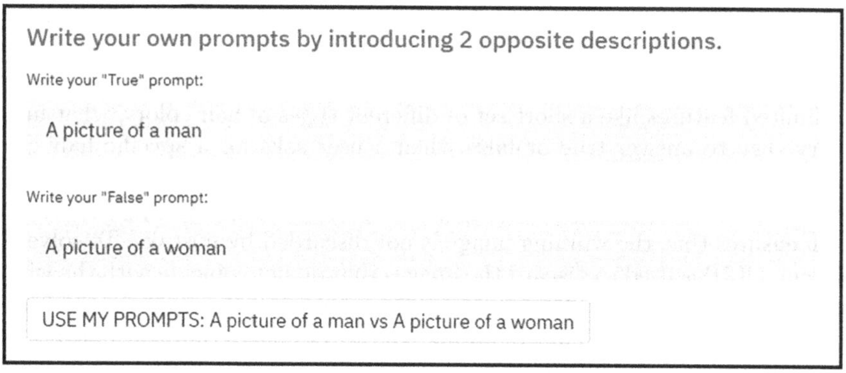

Fig. 5. Game screen that allows the user to create his own prompt using 2 text inputs.

3. Use two prompts (See Fig. 5). In this case two text input are used to allow the player write two sentences. The player must use two opposite sentences, that is, with an opposite meaning.

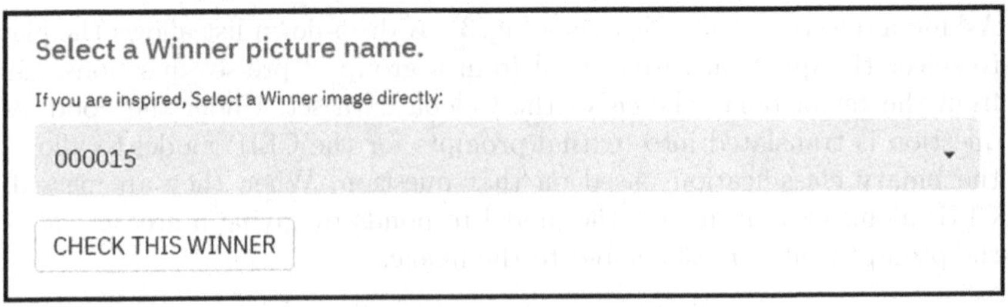

Select a Winner picture name.

If you are inspired, Select a Winner image directly:

000015

CHECK THIS WINNER

Fig. 6. Game screen that allows the user to select the winner image directly.

4. Select a winner (See Fig. 6). This option does not use the CLIP model to make decisions, the player can simply choose one of the images as the winner and if the player hits the winning image, the game is over.

3.2 Punctuation

To motivate the players in finding the winning image with the minimum number of questions, a scoring system is established so that it begins with a certain number of points (100 in the example), and decreases with each asked question. The score is decreased by subtracting the number of remaining images after each question. Furthermore, there are two extra penalties. The first is applied when the player uses the option "Select a winner". This penalty depends on the number of remaining images, so that the fewer images are left, the bigger will be the penalty. Finally, the score is also decreased by two extra points if, after the player makes a question, no image can be discarded.

The "Guess Who?" game has a handicap when it uses real images, because it is necessary to always ensure that the same criteria are applied when the images are discarded. The original game uses images with characters that present simple and limited features like a short set of different types of hair colors, what makes it very easy to answer true or false when a user asks for a specific hair color. However, with real images it is possible to doubt about if a person is blond haired or brown haired, for example, and it is necessary to apply a method which ensures that the winning image is not discarded by mistake. To solve this problem, CLIP is used to discard the images that do not coincide with the winner image after each prompt. In this way, when the user asks a question, CLIP is used to classify the images in two groups: the set of images that continue because they have the same prediction than the winning image, and the discarded set that has the opposite prediction. Figure 7 shows the screen that is prompted after calling CLIP on each image in the game board, where the discarded images are highlighted in red and the others in green.

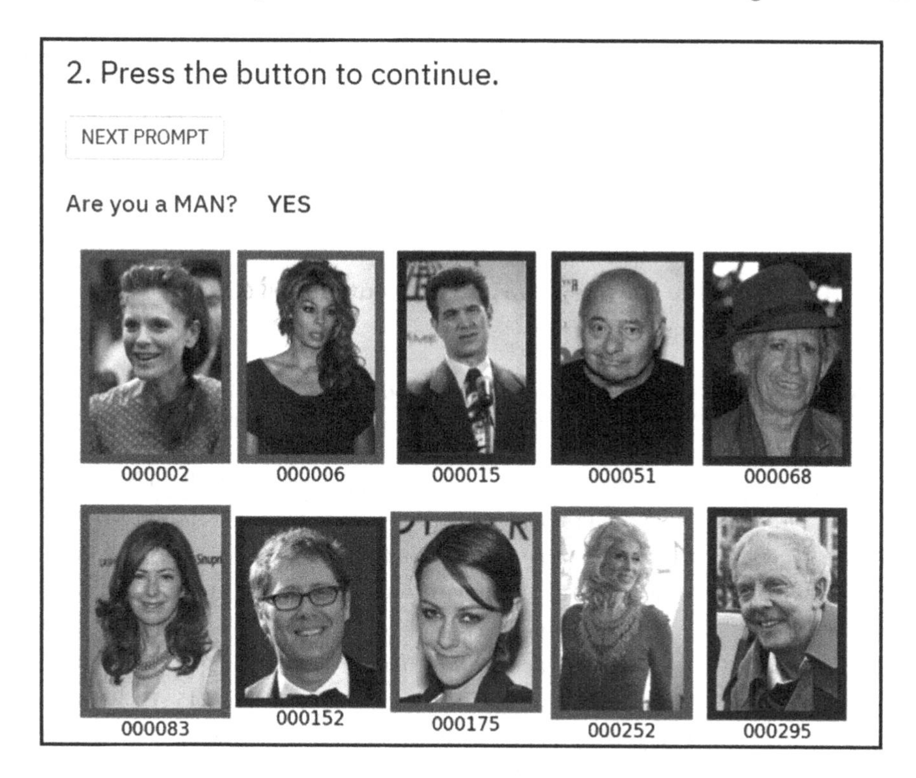

Fig. 7. Game screen showing CLIP answer after a question. (Color figure online)

4 Experiments and Prompt Analysis

To use CLIP as a zero-shot image classifier in the game, we create a pair of textual prompts for each class to address each player question as its own binary classification problem. Two basic prompting methods are proposed to create the textual descriptions:

1. Target vs neutral. This method consists in using a standard neutral prompt that fulfills all images, like "A picture of a person", and another sentence, very similar, which changes only some words and is more specific for the target class, like "A picture of a person with eyeglasses". In this way, when the additional information is added to the prompt, CLIP should return a bigger value for the second sentence than for the first for an image of a person with eyeglasses. And vice versa, i.e., when the extra information is not related to the image, CLIP should return a smaller likelihood value for this sentence. This is the method used to allow the user to introduce his own prompt.
2. Target vs contrary. This method consists in using two opposite sentences that represent opposite concepts, like "A picture of a man" and "A picture of a woman". This method is only implemented in the game for the attributes included in the set of labels of the Celeba data set.

We take advantage of the labeled images from the Celeba data set to validate the performance of the textual prompts introduced to CLIP. The True Positive Rate (TPR), True Negative Rate (TNR), and the accuracy (average of TPR and TNR) are calculated to analyze the results. We used the first 4.000 true images and the first 4.000 false images for each of the 40 binary labels of the data set to calculate these rates.

4.1 Results Table for the "Target vs Neutral" Method

Table 1. "Target vs neutral" prompting method applied on Celeba data set. True Positive Rate, True Negative Rate and Accuracy are shown in percentage.

Celeba label	Target prompt	TPR	TNR	Acc
Male	A picture of a male person	98.14	96.13	97.11
Wearing hat	A picture of a person with hat	97.29	83.67	89.34
Goatee	A picture of a person with goatee	91.6	77.05	82.78
Blond hair	A picture of a person with blond hair	74.14	97.84	82.09
Bangs	A picture of a person with bangs	88	77.71	82.05
Eyeglasses	A picture of a person with eyeglasses	87.45	77.59	81.78
Smiling	A picture of a person who is smiling	89.07	76.75	81.76
Bald	A picture of a bald person	96.63	73.86	81.58
Wearing necktie	A picture of a person with necktie	77.91	81.36	79.54
Gray hair	A picture of a person with gray hair	83.66	74.04	78.05
...	...	...	...	...
Big lips	A picture of a person with big lips	64.64	51.75	53.12
Wearing lipstick	A picture of a person with lipstick	85.34	51.53	52.94
Pointy nose	A picture of a person with pointy nose	52.74	52.16	52.41
Big nose	A picture of a person with big nose	57.63	51.09	51.91
Attractive	A picture of an attractive person	54.34	50.88	51.46
Rosy cheeks	A picture of a person with rosy cheeks	49.36	49.76	49.65
High cheekbones	A picture of a person with high cheekbones	47.33	49.23	48.8
Bags under eyes	A picture of a person with bags under eyes	47.7	49.05	48.66
Narrow eyes	A picture of a person with narrow eyes	40.67	45.36	43.8
No beard	A picture of a person with no beard	16.93	30.02	25.09

Table 1 shows the top ten and bottom ten results sorted by accuracy, as well as the labels of the Celeba data set and the CLIP textual inputs used. In this experiment, we simply use the literal Celeba labels to create the target prompt, and the neutral prompt is kept as "A picture of a person". With all this, approximately 25% of the target prompts obtained an accuracy above 70%. The 'male' and the 'wearing hat' features obtained remarkable accuracy results, 97% and 89% respectively.

In general, CLIP works sufficiently well when the label represents a physical object (e.g., hat or eyeglasses) or a common expression (e.g., smile), which will arguably be common in the data set in which CLIP has been trained on. The CLIP data set has millions of images and natural text from the Internet, so we must think about how the image descriptions in Internet often look like, in order to engineer good prompts. For example, when talking about a specific person appearing in an image that contains several people, it is common to talk about "the one who wears a hat" or "the one who has a goatee", but is unlikely to use a description like "the one who has narrow eyes" or "the one who has rosy cheeks".

Another key for CLIP performance lies in the ambiguity of elements or concepts. Some labels present no doubt about whether they are true or false, but some others are susceptible to observer interpretation. For instance, asking if a person wears a hat is a very objective concept that raises no doubt, so practically everyone would respond the same. However, asking if a person has big lips, has a pointy nose, or is attractive, are relative or subjective questions whose answer will depend on the observer. In these cases, it seems that CLIP does not work so well.

Finally, a remarkable result in this experiment is the accuracy obtained with the feature "no beard", which is related to a negation. In this case, the accuracy is 25%, but in a binary classification such as this, what CLIP really has reached is the 75 % of accuracy, if we invert its response. That means that CLIP ignored the "no" word in the prompt and classified the images of a person with beard with a 75% of accuracy. That indicates that CLIP probably is not able to deal with the concept of negation.

4.2 Results for the "Target vs Contrary" Method

Table 2. "Target vs contrary" prompting method applied on Celeba data set. True Positive Rate, True Negative Rate, Accuracy and Gain are shown in percentage.

Target prompt	Contrary prompt	TPR	TNR	Acc	Gain
A picture of a man	A picture of a woman	99.39	97.72	98.54	+1.43
A picture of a bald person	A picture of a haired person	96.08	80.43	86.65	+5.07
A picture of a person who is smiling	A picture of a person who is serious	80.05	89.88	84.28	+2.52
A picture of a person with pale skin	A picture of a person with tanned skin	68.12	75.8	71.29	+3.35
A picture of a young person	A picture of an aged person	65.17	78.2	69.72	+5.59
A picture of a person with straight hair	A picture of a person with wavy hair	60.38	67.04	62.9	+6.96
A picture of an attractive person	A picture of an unattractive person	50.44	50.13	50.2	−1.26

Table 2 shows some examples of the performance of the second proposed prompting method. The results of the first method, "target vs neutral" are also shown for easy comparison. This experiment shows how, in general, this method allows to improve the classification results by prompting "contrary words" to the target attribute, since. Even the "male" attribute, which obtained a 97 % of accuracy in the first experiment, can be improved with this method, reaching almost the 1,5% of accuracy improvement.

However, it must be remarked that it is difficult to find useful "contrary words". As an example, the label "wearing a necktie", "wearing lipstick" or "having rosy cheeks" only can be inverted using a negation, what does not work properly with CLIP. In cases where the antonym is very similar to the original word as "attractive" and "unattractive", CLIP does not seem to work properly either, but more tests must be done to ensure it.

5 Conclusions and Future Work

In this work, we present an implementation of the popular game "Guess Who?", in which the part of the game engine that decides whether an image fulfills the player question or not is made by CLIP, a language-image model that estimates how good a text caption pairs with a given image. To do that, we take each player prompt describing a person attribute and confront it with another prompt that does not represent it. In this way, we can create a binary classifier for each attribute of interest just by getting the maximum likelihood of CLIP's output to those two prompts. We have tried different prompting methods, such as confronting the target prompt with a neutral one (e.g., "A picture of a person") or using a prompt that describes the contrary of the target attribute. Experiments have been made with the labelled images of the Celeba data set to analyse the performance of these two prompting methods, showing that, as long as there is a clear contrary word to the target attribute, using the contrary prompt method normally leads to a better zero-shot classification performance.

As future work, we will continue working on "prompt engineering" due to CLIP-based zero-shot classifiers can be sensitive to wording or phrasing. Moreover, we will deploy the game as a public web application where interested users can play with their own images and give feedback about its performance.

Acknowledgements. This work has been partially supported by the company Dimai S.L, the "Convenio Plurianual" with the Universidad Politécnica de Madrid in the actuation line of "Programa de Excelencia para el Profesorado Universitario" and by next research projects: FightDIS (PID2020-117263GB-100), IBERIFIER (2020-EU-IA-0252:29374659), and the CIVIC project (BBVA Foundation Grants For Scientific Research Teams SARS-CoV-2 and COVID-19).

References

1. Conde, M.V., Turgutlu, K.: CLIP-Art: contrastive pre-training for fine-grained art classification. In: Proceedings of the IEEE/CVF Conference on Computer Vision and Pattern Recognition (CVPR) Workshops, pp. 3956–3960, June 2021

2. Cornia, M., Stefanini, M., Baraldi, L., Cucchiara, R.: Meshed-memory transformer for image captioning. In: Proceedings of the IEEE/CVF Conference on Computer Vision and Pattern Recognition, pp. 10578–10587 (2020)
3. Dosovitskiy, A., et al.: An image is worth 16x16 words: transformers for image recognition at scale. arXiv preprint arXiv:2010.11929 (2020)
4. Fang, H., Xiong, P., Xu, L., Chen, Y.: CLIP2Video: mastering video-text retrieval via image clip. arXiv:2106.11097 (2021)
5. Gao, J., Li, P., Chen, Z., Zhang, J.: A survey on deep learning for multimodal data fusion. Neural Comput. **32**, 829–864 (2020)
6. Liu, Z., Luo, P., Wang, X., Tang, X.: Deep learning face attributes in the wild. In: Proceedings of the IEEE International Conference on Computer Vision, pp. 3730–3738 (2015)
7. Nawaz, S., Calefati, A., Caraffini, M., Landro, N., Gallo, I.: Are these birds similar: learning branched networks for fine-grained representations. In: 2019 International Conference on Image and Vision Computing New Zealand (IVCNZ), pp. 1–5 (2019). https://doi.org/10.1109/IVCNZ48456.2019.8960960
8. Radford, A., et al.: Learning transferable visual models from natural language supervision. arXiv preprint arXiv:2103.00020 (2021)
9. Ramesh, A., et al.: Zero-shot text-to-image generation. arXiv preprint arXiv:2102.12092 (2021)
10. Tan, H., Yu, L., Bansal, M.: Learning to navigate unseen environments: back translation with environmental dropout. In: Proceedings of the 2019 Conference of the North American Chapter of the Association for Computational Linguistics: Human Language Technologies, Volume 1 (Long and Short Papers), pp. 2610–2621 (2019)
11. Vaswani, A., et al.: Attention is all you need. arXiv preprint arXiv:1706.03762 (2017)
12. Zhang, Y., Jiang, H., Miura, Y., Manning, C.D., Langlotz, C.P.: Contrastive learning of medical visual representations from paired images and text. arXiv preprint arXiv:2010.00747 (2020)

Directional Graph Transformer-Based Control Flow Embedding for Malware Classification

Hyung-Jun Moon[1], Seok-Jun Bu[2], and Sung-Bae Cho[2(✉)]

[1] Department of Artificial Intelligence, Yonsei University, Seoul 03722, South Korea
axtabio@yonsei.ac.kr

[2] Department of Computer Science, Yonsei University, Seoul 03722, South Korea
{sjbuhan,sbcho}@yonsei.ac.kr

Abstract. Considering the fatality of malware attacks, the data-driven approach using massive malware observations has been verified. Deep learning-based approaches to learn the unified features by exploiting the local and sequential nature of control flow graph achieved the best performance. However, only considering local and sequential information from graph-based malware representation is not enough to model the semantics, such as structural and functional nature of malware. In this paper, functional nature are combined to the control flow graph by adding opcodes, and structural nature is embedded through DeepWalk algorithm. Subsequently, we propose the transformer-based malware control flow embedding to overcome the difficulty in modeling the long-term control flow and to selectively learn the code embeddings. Extensive experiments achieved performance improvement compared to the latest deep learning-based graph embedding methods, and in a 37.50% improvement in recall was confirmed for the Simda botnet attack.

Keywords: Malware classification · Control flow graph · Graph embedding · Attention mechanism · Transformer encoder

1 Introduction

As internet and electronic devices develop, many software vaccine suppliers and security researchers have been developing the malicious software classifier for years [1]. Recent approaches usually represent the source code and the sequence of the functions from a lexical perspective to measure the malware similarities [2]. Among the prominent, the deep learning-based methods to learn the mapping function between observed malware signatures and attack types through many parameters were introduced and verified [3].

To learn the structural and functional nature of malware, research has been conducted to classify malware using Control Flow Graph (CFG) to address the limitation of signature-based approach. CFG is a data structure used to characterize the control flow of computer programs, which can be extracted from various file formats (binary files, byte codes, etc.), and can include features used in deep learning (n-gram, opcode, etc.), which can be considered as a critical features in malware classification using deep learning [4].

H. Yin et al. (Eds.): IDEAL 2021, LNCS 13113, pp. 426–436, 2021.
https://doi.org/10.1007/978-3-030-91608-4_42

However, there are two problems when using the CFG embeddings for malware classification: The first is the omission of edge information from CFG. When extracting CFG from a program, the node in the graph is represented as a function, and the edges representing the node's connection relationship are connected using opcode such as jump, call, etc. between functions. After connecting the two nodes, opcode is no longer used, but simply expresses the connection relationship between nodes. This method causes the opcode to ignore information. For example, in the case of jump opcode, there are several jmp, jo, jz, etc., and each code has a different meaning, so if express it as a simple connection, edge information disappears, which means loss of information in the program [5]. This can adversely affect the performance of the model. The second issue is that adding opcode prolongs the CFG of malicious code. If the length of the sequence is L without adding the edge information of the CFG, then the sequence after adding the opcode information will have a length of $2L - 1$. This is directly related to the limitation of LSTM, where gradient vanishes as the length increases, and makes it difficult to model the semantics of the malware attack [6]. As described above, these problems can become factors that can reduce malware's detection ability and cause major problems with cybersecurity.

In this paper, we propose the following methods to overcome the two mentioned problems. We generates a CFG including edge information to minimize omission of information and fully exploit the malware characteristics. In addition, the sequence is extracted from the CFG through the DeepWalk method, which supplements the sequence's information including the opcode information added in the previous step. To cope with the prolonged sequence, the transformer, which is the best performance in NLP, is introduced for the first time in the malware detection field to overcome the limitations of LSTM. The proposed method showed better performance on average than conventional CFG-LSTM methods and showed superior performance in diagnosing types of malicious codes that were not well matched in CFG-LSTM due to too long sequence.

2 Related Work

Research on malware analysis and malware detection is summarized in Table 1. First, there is a classic way to classify malware by extracting features from simple bytecode in a classical way. There are studies that use KNNs to classify malware with features extracted from raw features [7], and studies that perform malware classification through similarities in signature of each program [9]. Chen et al. tried detecting mobile malware including highly imbalanced network traffic by using imbalanced data gravitation-based classification [10].

However, it is known that the modeling approach of extracted features based on the bag-of-word technique is vulnerable to obfuscation involving mutation operation. Character-level obfuscation is an attack type that evades malware detectors by confusing some characters in an executable file, and a byte-image based approach to model the character-level malware features has been proposed. For the image-based malware detection, a feature extracted from the bytecode is embedded to a single vector and the association of each feature is modeled [11, 12]. Both approaches are signature-based approaches. Although it has the advantage of being widely accessible and being able

Table 1. Summary of related work on malware detection

Malware representation	Approach	Method	Challenges
Signature	Raw features	KNN [7]	o Failing to detect the polymorphic malwares o Replicating information in the huge database [8]
		Signature matching [9]	
		Imbalanced data gravitation-based classification [10]	
	Byte image	Ensemble of CNN [11]	
		GAN [12]	
Control flow graph	Embedding	GCN [13]	o Information dilution problem caused by updating node information without weight o Difficulty in improving malware detection performance due to performance degradation for many layers
		DGCNN [4]	
	Traversing	DeepWalk + LSTM [14]	o Dilution of sequence information due to LSTM's gradient vanishing problem

to identify to some extent for various malware types, it has the disadvantage of quickly identifying and replicating information from a vast database without detecting multiple malware.

On the other hand, control flow graph (CFG)-based approach has the advantage of being able to fully exploit the functionality of malware and store the sequence of function calls in graph-based structure. Furthermore, the graph neural network (GNN), which has recently been studied in deep learning, has been incorporated to model the CFG-based malware using multiple graph convolution operations and generates embeddings [4, 13]. A representative example of using CFG based on deep learning is CG-CNN, where CFG was used to localize potential bugs in code [14]. After converting code into CFG, each code feature is extracted using CNN, and sequence is extracted using DeepWalk, a type of embedding method within CFG. After embedding the extracted sequences using LSTM, the embeddings of several sequences were combined to localize them, achieving SOTA in the bug localization field. This method has the advantage of being able to embed by tokenizing information such as the name of the function by splitting the CFG into multiple sequences. Therefore, based on the above methods, we would like to introduce how to convert the program to CFG and sequence it to detect malware.

A recent study created sequence after converting a program to CFG and used LSTM to solve localization and classification problems within the program. This paper extracted and sequenced nodes from the CFG using a method called DeepWalk, which showed that the CFG converted to sequence could use RNNs and thus the connectivity between each node was learnable [14].

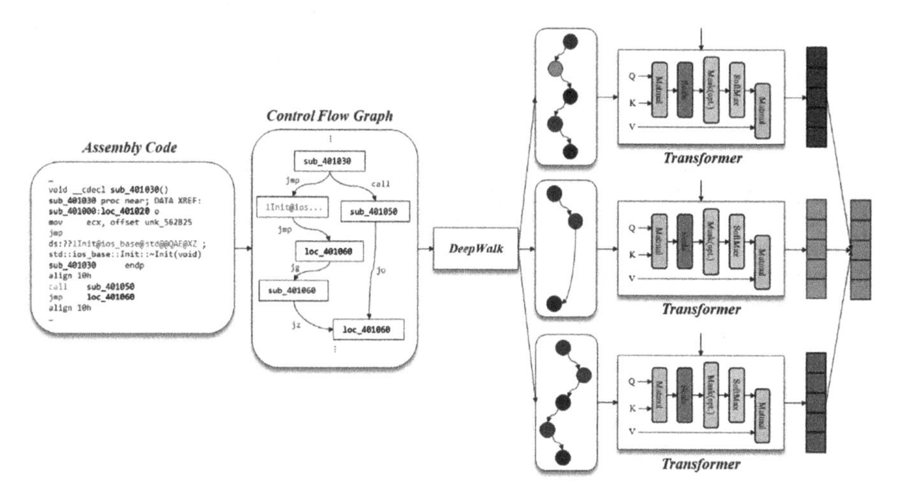

Fig. 1. Structural diagram of the proposed method

3 Proposed Method

In this section, we describe the model that we intend to propose in this paper. It will introduce how to change the assembly code to CFG including edge information, divide it into several sequences using DeepWalk method, and print embeddings by putting these sequences into the transformer.

3.1 Overview

Figure 1 is a complete structural diagram of the method proposed in this paper. First, when the assembly code enters the input, it is converted to control flow graph. To preserve the information in the opcode, we include the information in the edge. Then use the DeepWalk method to extract sequences from CFG. At this time, the opcode included in the edge is changed into a node, increasing the length of the sequence and including the edge information in the sequence. Afterwards, each sequence is embedded through transformer block and finally merged and classified.

3.2 Code Block Embedding Based on Traversing of Control Flow Graph

Algorithm 1 represents an algorithm that creates a CFG from a program. An algorithm that connects two functions if the opcode is a type of jump or call through the name of a function inside a program called Block, which converts the assembly code into CFG. These converted CFGs specifically contain edge information.

Specifically, when there is an assembly code, it reads down one line to find a function block, and if the connection between the function and the other is jump or call, it connects two functions (nodes), so the edge of the graph has direction. The presence of directionality enables the generation of sequences of interfunctional calls [15]. Figure 2 shows a brief description of where the unit of composition of CFGs occurs in the assembly code.

Algorithm 1 Code_to_CFG (P , OPCODE_LIST)

Input : Program code *P,* Predefined opcode list *OPCODE_LIST*
Output : Control Flow Graph *G*

```
For all block in program P do
    CurrBlock = getBolck( block.addr )
    If CurrBlock.start then
        Graph.add_start_node( CurrBlock )
    Else
        Graph.add_node( CurrBlock )
    End if
    Next_Block = getNextBlock( P , block )
    While Next_Block != None then
        Edge = getBetweenEdge( CurrBlock , Next_Block )
        If Edge is in OPCODE_LIST
            Graph.add_node( Next_Block )
            Graph.addEdge( Edge , CurrBlock , Next_Block )
        End if
    CurrBlock    NextBlock
End For
```

Algorithm 2 shows the overall process in which DeepWalk works [16]. Starting from the starting node of the CFG, it moves by the number of walks set, the length of the walk, and does not move by the path once passed. In addition, if no further progress is made, it fills the "None" value by the length of the remaining walk and includes empty space. The reason for adding empty space is to constantly discard the input size of the model. As mentioned earlier, since CFGs have directionality, nodes cannot return to the previous node (if there is no loop) and can be isolated without reaching the maximum length of the walk depending on the direction of DeepWalk progress.

In the list output through this algorithm, each sequence is subjected to a tokenize process and edge information additional process. First, in the process of adding edge information, there is an edge that runs between each node and these edges have information, so it is the process of converting them to nodes. In the case of tokenize, it is a process of changing the names and information of nodes set to each function to a special number, and in the case of this tokenize process, it is performed on all nodes of all graphs. Any "number" that comes through tokenize will have the same meaning as a word in NLP and can then be entered into transformer. The number of walks set for all graphs is repeated, and all walks on all graphs have the same length due to the addition of empty space. This generated walk is used as an input to the model.

3.3 Malware Subroutine Sampling and Transformer-Style Embedding

Let $X = [D_1, D_2, D_3, \ldots, D_n]$ (n means the number of walks) denote the sequence list after DeepWalk from graph G, l denote length of sequence, V denote a set of different nodes and edges in sequence. Each D is $D \epsilon R^{l \times V}$. Subsequently, Let K denote a key, Q denote a query with d_k denote dimension of key and query, and V denote a vocabulary with d_v dimensions. The results of applying the transformer for each sequence are as follows.

Algorithm 2 DeepWalk (G , num_walks , walk_length)

```
Input : Graph G, Number of Walks num_walks , length of walk walk_length
Output : Walk List W
Nodes = getnode( G )
W = [ ]
For 0 to num_walks do
    root = nodes[ 0 ]
    walk = [ root ]
    for 1 to walk_length do
        cur = walk[ k-1 ]
        valid_neighbours = ( set( G.neighbors( cur ) ) - set( walk ) )
        if getlength( valid_neighbours ) > 0:
            next_node = random_choice( valid_neighbours )
            walk.append( next_node )
        else:
            walk = walk[ :-1 ]
            walk = [ "NONE" ] * ( walk_length- getlength( walk ) ) + walk
            break
return walks_all
```

$$Attention(Q, K, V) = softmax(\frac{QK^T}{\sqrt{d_k}})V \quad (1)$$

Because this extracted attention is for one sequence, the formula with multiple attention is:

$$head_i = Attention(QW_i^Q, KW_i^K, VW_i^V), i = 1, \ldots, n \quad (2)$$

Subsequently, the formula for combining them into a single embedment is as follows:

$$Transformer(Q, K, V) = Concat(head_1, \ldots, head_h)W^O \quad (3)$$

This process produces embedding vectors of multiple sequences from a single graph, using Multi-Layer Perceptron (MLP) and softmax functions to bring them into classification problems.

$$Prediction = softmax(MLP(Concat(head_1, \ldots, head_h)W^O)) \quad (4)$$

Transformer only used attention in seq2seq architecture, solving the problem of RNNs where the longer the sentences being learned, the less information about the sentences being separated from each other [6].

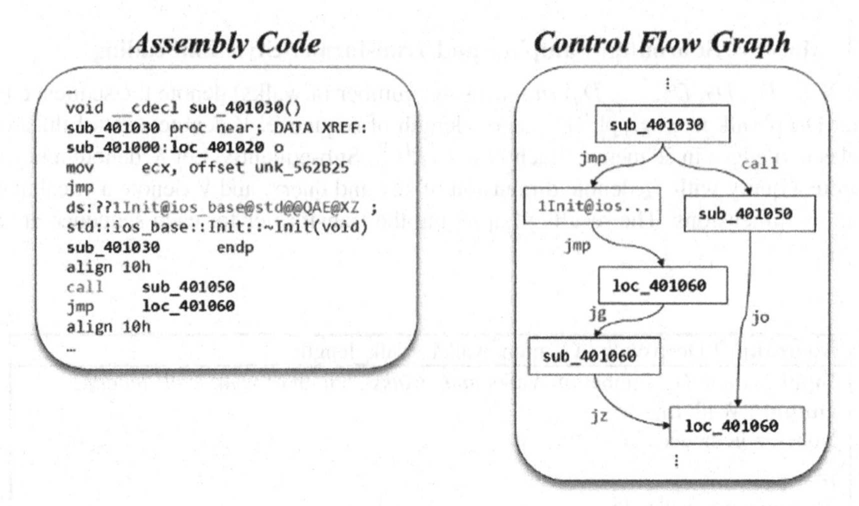

Fig. 2. Comparison of the function block and the node of the control flow graph corresponding to the assembly code.

4 Experiment Result

This section describes the datasets used in the experiment, as well as the experimental results, misclassification, and analysis.

4.1 Dataset and Implementation

We used the malware dataset from the Kaggle Microsoft Malware Classification Challenge. The malware data were written in assembly and binary codes, and we used the form of the binary code. There are a total of 10868 assembly codes, binary codes, and we only used 10813 datasets, except for 55 codes that were hard to represent as CFGs because we had only one node. This dataset has nine classification classes.

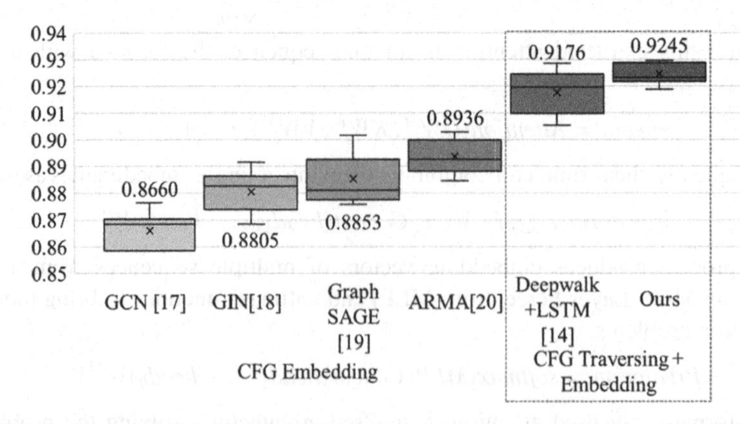

Fig. 3. 10 cross-validation results for malware detection compared to other methods

Table 2. Confusion matrix of existing methods using proposed methods and Deep-Walk + LSTM(bracket)

		Predict								
		0	1	2	3	4	5	6	7	8
Actual	0	273(223)	25(55)	3(7)	1(2)	0(0)	1(8)	4(4)	7(15)	0(0)
	1	2(3)	478(467)	2(1)	0(0)	0(0)	4(4)	2(1)	0(0)	12(24)
	2	0(0)	6(11)	543(537)	0(0)	0(0)	2(1)	3(5)	0(0)	0(0)
	3	1(1)	4(4)	0(0)	79(73)	0(0)	2(3)	1(2)	8(13)	1(0)
	4	1(2)	1(2)	0(1)	0(0)	6(3)	0(0)	0(0)	0(0)	0(0)
	5	0(0)	9(14)	3(2)	2(2)	0(0)	129(125)	3	2	0
	6	1(1)	0(0)	0(0)	0(0)	0(0)	1(2)	72(70)	0(1)	0(0)
	7	6(9)	16(24)	4(3)	2(5)	1(0)	6(9)	4(3)	211(196)	0(1)
	8	1(3)	1(4)	0(0)	1(0)	0(0)	1(6)	0(1)	0(0)	215(205)

Table 3. Classification performance measurement with respect to malware class

Malware class		0	1	2	3	4	5	6	7	8	Avg.	Acc.	F1 score
CFG-LSTM [14]	Precision	0.921	0.803	0.974	0.890	**1.000**	0.791	0.795	0.863	0.887	0.880	**0.878**	**0.842**
	Recall	0.710	0.934	0.969	0.760	**0.375**	0.844	0.945	0.784	0.936	0.806		
CFG-Transformer	Precision	0.958	0.885	0.978	0.929	**0.857**	0.884	0.809	0.925	0.943	0.908	**0.927**	**0.901**
	Recall	0.869	0.956	0.98	0.823	**0.75**	0.872	0.973	0.844	0.982	0.894		

4.2 Malware Detection

Experiments showed that the proposed method was superior, with a minimum difference of 3% compared to the graph embedding method. In addition, the proposed model achieved better performance than the existing Traversing + LSTM model.

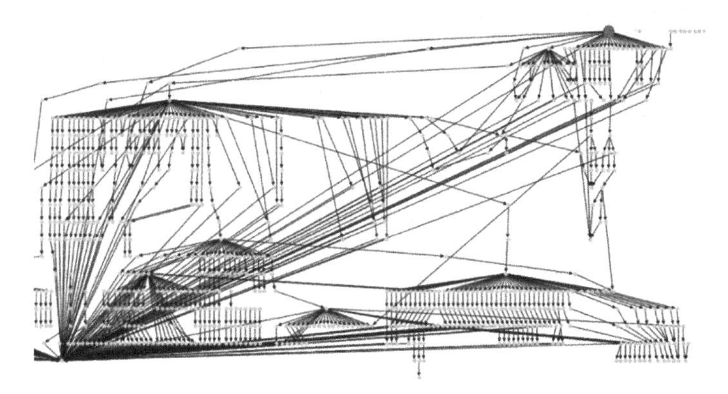

Fig. 4. The subgraph of control flow graph extracted from Simda botnet. Red dot is entry function. (Color figure online)

Figure 3 is a comparison with other graph embedding techniques using 10-cv. The comparison with graph embedding [17–20] shows good performance due to global pooling, which combines node embeddings into one and transforms them into one of the graph's embeddings, proving that the graph's information is not preserved and diluted. The comparison with CFG + LSTM shows that with the addition of edge information, resulting in higher performance.

According to Table 2, the result value of the chi-square test was calculated. As a result of the confusion matrix of the proposed model and the analysis of the existing model, the chi-square test results showed a result of 83.675 at the degree of freedom 64, so our proposed model proved statistical significance.

According to Table 3, in Simda Malware expressed as class 4, the recall of the existing method was 0.375, but the proposed method showed better performance with 0.75. The reason for this is that Simda Malware has a short lifecycle and features that are generated and distributed every few hours, and the graph shows that the depth of the graph is very deep. Therefore, since we are vulnerable to long sequences when using LSTM by sequencing using Traversing, we were able to confirm that our model using transformer performs better to improve these shortcomings. Figure 4 illustrates a situation in which sequence length can be very long in a particular situation on the Simda botnet graph.

5 Conclusion

In this paper, we propose a new method to create CFG embeddings reflecting edge information in CFG-based malware classification tasks and introduce Transformer's long sequence modeling advantages to malware classification for the first time. As suggested, it is flexible to use for various types of program files, preventing omissions of information, directly embedding functional call relationships, and ensuring performance in CFGs with long depths. In addition, the experiments showed better performance in terms of CFG than other method. For Simda botnet, which has a recall of 0.35 with existing deep learning-based graph embedding, our method has found advantages in long malware modeling by showing 0.75 recall.

On the other hand, our model explores graphs by using the DeepWalk method based on the randomness and sampling strategy between nodes, which has the disadvantage to fully exploit CFGs. In the future study, we plans to develop an integrated model in combination with byte images and signature-based methods to achieve better Malware detection performance.

Acknowledgment. This work was supported by an IITP grant funded by the Korean government (MSIT) (No. 2020-0-01361, Artificial Intelligence Graduate School Program (Yonsei University)) and Air Force Defense Research Sciences Program funded by Air Force Office of Scientific Research.

References

1. Bu, S.-J., Cho, S.-B.: Integrating deep learning with first-order logic programmed constraints for zero-day phishing attack detection. In: ICASSP 2021–2021 IEEE International Conference on Acoustics, Speech and Signal Processing (ICASSP), pp. 2685–2689. IEEE (2021)
2. Pektaş, A., Acarman, T.: Deep learning for effective Android malware detection using API call graph embeddings. Soft. Comput. **24**(2), 1027–1043 (2019). https://doi.org/10.1007/s00500-019-03940-5
3. Kim, J.-Y., Cho, S.-B.: Detecting intrusive malware with a hybrid generative deep learning model. In: Yin, H., Camacho, D., Novais, P., Tallón-Ballesteros, A.J. (eds.) IDEAL 2018. LNCS, vol. 11314, pp. 499–507. Springer, Cham (2018). https://doi.org/10.1007/978-3-030-03493-1_52
4. Yan, J., Yan, G., Jin, D.: Classifying malware represented as control flow graphs using deep graph convolutional neural network. In: 2019 49th Annual IEEE/IFIP International Conference on Dependable Systems and Networks (DSN), pp. 52–63. IEEE (2019)
5. Ding, Y., Dai, W., Yan, S., Zhang, Y.: Control flow-based opcode behavior analysis for malware detection. Comput. Secur. **44**, 65–74 (2014)
6. Moon, H.-J., Bu, S.-J., Cho, S.-B.: Learning disentangled representation of residential power demand peak via convolutional-recurrent triplet network. In: 2020 International Conference on Data Mining Workshops (ICDMW), pp. 757–761. IEEE (2020)
7. Assegie, T.A.: An optimized KNN model for signature-based malware detection. In: International Journal of Computer Engineering In Research Trends (IJCERT), ISSN 2349–7084 (2021)
8. Souri, A., Hosseini, R.: A state-of-the-art survey of malware detection approaches using data mining techniques. HCIS **8**(1), 1–22 (2018). https://doi.org/10.1186/s13673-018-0125-x
9. Sihag, V., Swami, A., Vardhan, M., Singh, P.: Signature based malicious behavior detection in android. In: International Conference on Computing Science, Communication and Security, pp. 251–262. Springer (2020)
10. Chen, Z., et al.: Machine learning based mobile malware detection using highly imbalanced network traffic. Inf. Sci. **433**, 346–364 (2018)
11. Vasan, D., Alazab, M., Wassan, S., Safaei, B., Zheng, Q.: Image-based malware classification using ensemble of CNN architectures (IMCEC). Comput Secur. **92**, 101748 (2020)
12. Kim, J.-Y., Bu, S.-J., Cho, S.-B.: Zero-day malware detection using transferred generative adversarial networks based on deep autoencoders. Inf. Sci. **460**, 83–102 (2018)
13. Yasaei, R., Yu, S.-Y., Al Faruque, M.A.: GNN4TJ: graph neural networks for hardware Trojan detection at register transfer level. In: 2021 Design, Automation and Test in Europe Conference and Exhibition (DATE), pp. 1504–1509. IEEE (2021)
14. Huo, X., Li, M., Zhou, Z.-H.: Control flow graph embedding based on multi-instance decomposition for bug localization. Proc. AAAI Conf. Artif. Intell. **34**, 4223–4230 (2020)
15. Armando, A., Costa, G., Merlo, A., Verderame, L.: Enabling BYOD through secure meta-market. In: Proceedings of the 2014 ACM Conference on Security and Privacy in Wireless and Mobile Networks, pp. 219–230 (2014)
16. Wu, Z., Pan, S., Chen, F., Long, G., Zhang, C., Philip, S.Y.: A comprehensive survey on graph neural networks. IEEE Trans. Neural Networks Learn. Syst. **32**, 4–24 (2020)
17. Kipf, T.N., Welling, M.: Semi-supervised classification with graph convolutional networks. arXiv preprint, 2016. arXiv:1609.02907
18. Xu, K., Hu, W., Leskovec, J., Jegelka, S.: How powerful are graph neural networks? arXiv preprint, 2018. arXiv:1810.00826

19. Hamilton, W.L., Ying, R., Leskovec, J.: Inductive representation learning on large graphs. In: Proceedings of the 31st International Conference on Neural Information Processing Systems, pp. 1025–1035 (2017)
20. Bianchi, F.M., Grattarola, D., Livi, L., Alippi, C.: Graph neural networks with convolutional arma filters. In: IEEE Transactions on Pattern Analysis and Machine Intelligence (2021)

In-Car Violence Detection Based on the Audio Signal

Flávio Santos[1], Dalila Durães[1(✉)], Francisco S. Marcondes[1], Niklas Hammerschmidt[2], Sascha Lange[2], José Machado[1], and Paulo Novais[1]

[1] Algorithm Center, University of Minho, Braga, Portugal
{flavio.santos,dalila.duraes,francisco.marcondes}@algoritmi.uminho.pt, {jmac,pjon}@di.uminho.pt

[2] Bosch Car Multimedia, Braga, Portugal
niklas.hammerschmidt@de.bosch.com, sascha.lange@pt.bosch.com

Abstract. When it is intended to detect violence in the car, audio, speech processing, music, and ambient sound are some of the main points of this problem since it is necessary to find the similarities and differences between these domains. The recent increase in interest in deep learning has allowed practical applications in many areas of signal processing, often surpassing traditional signal processing on a large scale. This paper presents a comparative study of state-of-the-art deep learning architectures applied for inside car violence detection based only on the audio signal. The methodology proposed for audio signal representation was Mel-spectrogram, after an in-depth review of the literature. We build an In-Car video dataset in the experiments and apply four different deep learning architectures to solve the classification problem. The results have shown that the ResNet-18 model presents the best accuracy results on the test set.

Keywords: Audio violence detection · Audio action recognition · Deep learning · Interior vehicle

1 Introduction

When it comes to detecting violence, it is mainly associated with the video. However, recent studies focus mainly on images rather than audio. Still, audio can easily be picked up by microphones. These are very powerful sensors that capture human behavior and context. Nevertheless, audio detection is highly susceptible to large fluctuations in accuracy depending on the acoustic environment in which it is inserted. Thus, a good audio representation is fundamental to perform audio-based video classification.

In recent years, there has been a different form of data modeling called deep learning [6], which resulted in new architectures and new learning algorithms that allowed entry into action recognition [13], object recognition, and automatic

H. Yin et al. (Eds.): IDEAL 2021, LNCS 13113, pp. 437–445, 2021.
https://doi.org/10.1007/978-3-030-91608-4_43

translation, among others [1,18]. Deep learning models are now the basis of study in different areas of knowledge, having been the driving force to advance in the field of the accuracy and robustness of several models in domains related to audio [2]. Due to its importance, in this work, we propose an experimental evaluation of the well-known state-of-the-art deep learning architectures to solve the In-Car violence recognition task based on the audio signal.

The paper is organized as follows: next section presents the related work with explanations of general audio classification, violence recognition on audio, and transfer learning; Sect. 3 describes our Experiments with the description of the dataset, the algorithm models, and some training details; Sect. 4, explain the results; and finally, Sect. 5 concludes this work with some future directions.

2 Related Works

In detecting violence through audio, the needs for both bandwidth, storage, and computing are much smaller than for video. Regarding the limitations that audio sensors have, we can consider minor compared to video cameras: i) a video camera has a limited angular field of view, but microphones can be omnidirectional, providing a spherical field of view; ii) audio event acquisitions are better than video acquisitions since the audio wavelength is longer and many surfaces allow acoustic wave reflections (when there are obstacles in the direct path); iii) lighting and temperature are not problems for audio processing [4], and iv) in the audio approach, the signal produced by the sound of an audio contains a large amount of information that only visual data cannot represent, as examples are: screams, explosions, abuse words and even sound passages that reveal some emotion. However, there are few applications to violence detection with audio.

Souto, Mello and Furtado [17] focus on domestic violence and the classification of acoustic scenes through machine learning. The parameters used in the extraction and processing in the short and medium term were MFCC, Energy and ZCR. The technique used for classification was the SVM. The models obtained after training were the MFCC-SVM classifier, the Energy-SVM classifier and the ZCR-SVM classifier.

Rouas, Louradour and Ambellouis [15] based on public transport vehicles, studied the detection of audio events. It created an automatic audio segmentation, which divides an audio signal into several consecutive, almost stationary zones. The developed algorithm detected activity. That is, it ignored the silence and low noise zones, focusing only on the high noise zones. An SVM classifier was used.

Crocco, Cristiani, Trucco and Murino [4] did a systematic review of audio-based surveillance. It presents several approaches, namely: i) the subtraction of the background by monomodal analysis; ii) background subtraction by multi-modal analysis; iii) the classification of the audio event; iv) the location and tracking of the source, especially the location of the audio source; v) location of audiovisual sources; and vi) audio source tracking and audiovisual source tracking.

Choi, Fazekas, Cho, and Sandler [3] comprehensively presents deep learning techniques that can be designed and used in the context of MIR. It also presents some basics of MIR in the context of deep learning.

Gaviria et al. [5] presents a device designed to accurately perform the reconnaissance task in urban areas where generally loud noise is a concern. This device is portable and inexpensive. Audios were recorded in real urban settings using an accessible microphone. The strategy was to train a classifier based on temporal and frequency data analysis and deep convolutional neural networks.

Hossaina and Muhammadb [8] proposes an audiovisual emotion recognition system using a deep net to extract resources and another deep net to merge resources. They use Big Data emotion network training and separate information based on gender. They use a two-dimensional convolutional neural network (CNN) for audio signals and a three-dimensional CNN for video signals in the proposed system.

Purwins, Virtanen, Schluter, Chang, and Sainath [14] previously used audio signal processing methods, such as Gaussian mixture models, hidden Markov models and non-negative matrix factorization, have often been outperformed by deep learning models in applications where sufficient data is available. They applied some methods like categorization, audio features, models, data, and evaluation. They made some cross-domain comparisons with speech, music, and environmentally sound. For synthesis and transformation of audio, they used source separation, speech enhancement, and audio generation.

3 Experiments

The audio recognition challenge begins with the question of how we must represent the audio to perform the classification. Some of the audio representation forms can be obtained from the following transformations: short-time Fourier transform, Mel-spectrogram, Constant-Q transform, and Chromagram. All of them have their pros and cons. Since Mel-Spectrogram is well used in the literature [3,5,8,14] we decided to use it as audio representation to our tested models. In the following subsections, we present all the experimental setup used in this work. We will describe the dataset used, the deep learning models, and all the training details employed in the training step.

3.1 In Car Dataset

In order to evaluate the deep learning models to recognize In-Car Violence through audio signals, we need a benchmark with in-car recording and with audio data. We have search in a wide range of databases in the literature, but we have not found any dataset with our specifics constraints. Thus, we built a dataset from scratch. Our dataset is composed of videos from the youtube platform and videos recording by us with non-professional actors. The full dataset has 517 videos where each one has about 20 s. From the 517 videos, 176 have violent situations while the others 341 do not have violence. Thus, our In-Car dataset is unbalanced. However, we believe that it is closer to reality because non-violence situations are more common than violence.

3.2 Models

Instead of beginning with custom architecture, we have trained well-known architectures in the literature to find the better one for our task. We have performed experiments with four architectures, they are (1) ResNet [7], (2) DenseNet [9], (3) SqueezeNet [10], and (4) Mobile [16]. Those architectures are diverse because they have different goals, such as solve the vanishing gradient problem to training deeper models and have fewer parameters to need less computational power. Next, we present a description of those models.

ResNet [7], is a deep neural network that reformulates the architecture layers as learning residual functions with reference to the layer inputs. After the ResNet had been proposed, deep models with more than 1000 layers were possible to be trained without the vanishing gradient. In this work, we have used a ResNet model composed of 18 layers.

Densely Connected Convolutional Networks (DenseNet) [9] has proposed a new CNN-based architecture after the observation that CNN models can be deeper and more accurate if they contain shorter connections between layers closer to the input and those close to the output. Thus, a DenseNet with L layers have $L(L + 1)/2$ direct connections because, for each, the feature maps of all preceding layers (and their own) are used as inputs to all next layers. According to the DenseNet authors, their model has some advantages, such as alleviating the vanishing gradient problem, reinforcing feature propagation and reuse, and briefly reducing the number of parameters.

In contrast with the huge number of parameters in ResNet and DenseNet, the SqueezeNet [10] model proposes an architecture whose goal is to be efficient in distributed training, have less overhead when used on the client-side, and be viable to embedded deployment. So, it is an architecture that presents acceptable accuracy with a low computational cost. To achieve its goal, the SqueezeNet is composed of several building blocks named Fire Module.

Following the line of the SqueezeNet, MobileNet [16] propose an efficient architecture to be used in mobile devices with computer vision applications. It is composed of depth-wise separable convolutions whose goal is to split the standard CNN layer output into two layers, one for filtering and the other for combining. The MobileNet authors claim that this factorization has is responsible for reducing the computation and model size.

3.3 Training Details

Since our dataset is unbalanced, we have used three different loss functions to decide which one is best for our problem. They are (i) Categorical Cross Entropy, (ii) Cross Entropy Weighted [12], and (iii) Focal Loss [11]. The literature widely knows the first loss function. It is a classic Cross Entropy jointly with the softmax function. Thus it considers only the example class in the loss computation. The second loss function, cross-entropy weighted, is a weighted version of the first one, it inserts the weights in the classes to down-weighted do majority class. The last loss function differs from the first two because insert the weight in the loss in a sample-by-sample fashion. Thus the model will focus on the samples that it is mostly wrong. The main idea of the focal loss is that if a sample is already classified correctly by the model, its contribution to the loss decreases. The Eq. 1 presents the focal loss.

$$FL = -\sum_{i=1}^{C=2} (1 - s_i)^{\gamma} t_i log(s_i) \tag{1}$$

The $(1 - s_i)^{\gamma}$ term in the Eq. 1 is a factor to decrease the influence of the correctly classified samples.

4 Results

This section presents all the results obtained in the experiments. It is structured based on the questions made during the experiments.

What Is the Best Loss Function? The choice of the loss function is a major step in any supervised learning setting. However, the scenario with an unbalanced dataset deserves particular attention because the samples from the minority class may need a special role in the problem. Thus, we chose three loss functions and experimented with a ResNet-18 to select the loss function that presents better results. In this analysis, for each loss function, we have trained the ResNet 10 times. Is important highlight that each training executions we use a different dataset split. The Fig. 1 present the results obtained in this analysis. We have plotted a straight line with the accuracy of 90% to improve the visualization. The results show that even though the cross-entropy weight considering the minority class more important, its results were worse than the classic cross-entropy. Although the scenario with the cross-entropy presents results close to an accuracy of 90%, one of the executions with the Focal Loss achieved better results than it and even close to 100% of accuracy. Thus, we have decided to use Focal Loss in the following experiments.

What Is the Test Accuracy Behavior of all Models? After choosing the focal loss to perform the experiments, we need to select the best model for our dataset. To answer that, we have executed 5x the training of the four architectures presented before (i.e., ResNet, DenseNet, SqueezeNet, and MobileNet). For each training execution, the models were trained during ten epochs. After the end of each training epoch, we evaluate our model with our dev set, so for each model, we obtained 50 accuracy values to the dev set. To visualize the accuracy behavior of our model, the Fig. 2 presents the box-plot of the 50 accuracies for each model. The results show that the ResNet model presents the best results,

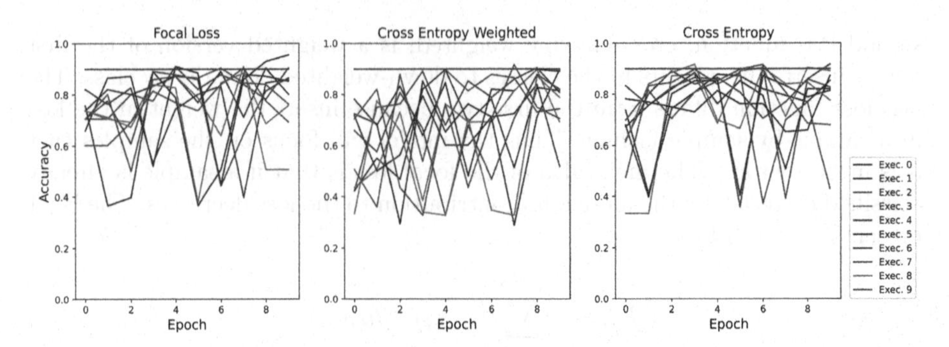

Fig. 1. Accuracy analysis.

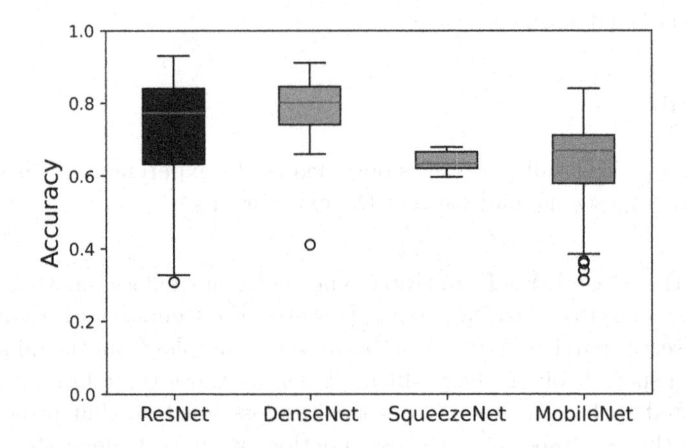

Fig. 2. Box plot of the test accuracy of every trained model.

but it has high accuracy. However, the DenseNet presents more coherent results. As the DenseNet, the SqueezeNet presents coherent results, but its results are very low compared to the DenseNet and ResNet.

What Is the Statistics of the Best Accuracy of Every Execution for Each Model? In the Table 1, for each model, we have summarized the statistics of the best accuracies obtained in each one of the 5 executions. We can see that the ResNet presents the best accuracy results in the test set, but its min accuracy was lower than the DenseNet min accuracy. Besides, the std of the ResNet also is higher than the std of the DenseNet. It shows that the ResNet can present better results, but it is not always so high, confirming the behavior presented in the boxplot.

Why the SqueezeNet has Poor Performance? The SqueezeNet presented poor results compared to others methods, as we can see from the results showed so far. In the Fig. 3 for each model, we present the loss curve in the training set.

Table 1. Statistics of the best test accuracy achieved by every model.

Model	Max.	Min.	Std.	Mean
ResNet	92.95	83.33	03.41	86.80
DenseNet	91.03	84.62	02.55	87.69
SqueezeNet	67.95	59.62	03.02	63.97
MobileNet	83.97	77.56	02.09	80.39

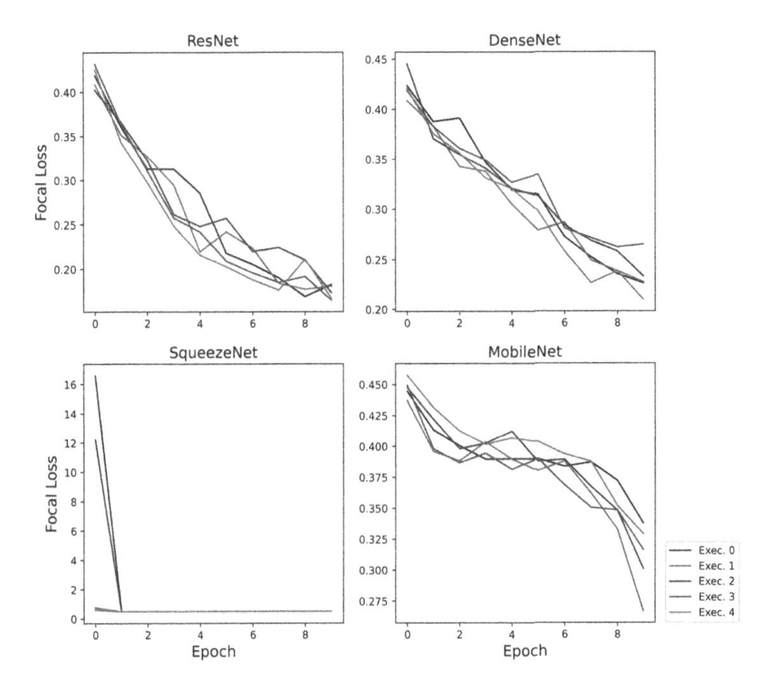

Fig. 3. Loss convergence for every model.

Its results show that the SqueezeNet starts with inferior results than the other three models and stabilizes in the ending epochs, thus being strong evidence that maybe the SqueezeNet has a slow convergence and needs more training epochs to do so present better results.

5 Conclusion

In this work, we have used deep learning to classify violence inside the car based on audio signals only. First, we have followed previous work in the literature and represented the audio signal with the Mel-Spectrogram. Next, was made an extensive experiment to choose the best classification architecture using four deep learning models, named ResNet, DenseNet, SqueezeNet, and MobileNet. The obtained results show that the ResNet model achieved the best result, but the DenseNet presents close results to ResNet, and their accuracy standard deviation was lower than ResNet.

In future work, we intend to make the fusion between audio and video signals. Applying different techniques and compare the results obtained to analyze the best fusion techniques.

Acknowledgments. This work is supported by: European Structural and Investment Funds in the FEDER component, through the Operational Competitiveness and Internationalization Programme (COMPETE 2020) [Project n° 039334; Funding Reference: POCI-01-0247-FEDER- 039334].

References

1. Arukgoda, A.S.: Improving Sinhala-Tamil translation through deep learning techniques. Ph.D. thesis (2021)
2. Cho, Y., Bianchi-Berthouze, N., Julier, S.J.: DeepBreath: deep learning of breathing patterns for automatic stress recognition using low-cost thermal imaging in unconstrained settings. In: 2017 Seventh International Conference on Affective Computing and Intelligent Interaction (ACII), pp. 456–463. IEEE (2017)
3. Choi, K., Fazekas, G., Cho, K., Sandler, M.B.: A tutorial on deep learning for music information retrieval. CoRR abs/1709.04396 (2017). http://arxiv.org/abs/1709.04396
4. Crocco, M., Cristani, M., Trucco, A., Murino, V.: Audio surveillance: a systematic review. ACM Comput. Surv. (CSUR) **48**(4), 1–46 (2016)
5. Gaviria, J.F., et al.: Deep learning-based portable device for audio distress signal recognition in urban areas. Appl. Sci. **10**(21) (2020). https://doi.org/10.3390/app10217448. https://www.mdpi.com/2076-3417/10/21/7448
6. Goodfellow, I., Bengio, Y., Courville, A.: Deep Learning. MIT Press, Cambridge (2016)
7. He, K., Zhang, X., Ren, S., Sun, J.: Deep residual learning for image recognition. In: Proceedings of the IEEE Conference on Computer Vision and Pattern Recognition, pp. 770–778 (2016)
8. Hossain, M.S., Muhammad, G.: Emotion recognition using deep learning approach from audio-visual emotional big data. Inf. Fusion **49**, 69–78 (2019)
9. Huang, G., Liu, Z., Van Der Maaten, L., Weinberger, K.Q.: Densely connected convolutional networks. In: Proceedings of the IEEE Conference on Computer Vision and Pattern Recognition, pp. 4700–4708 (2017)
10. Iandola, F.N., Han, S., Moskewicz, M.W., Ashraf, K., Dally, W.J., Keutzer, K.: SqueezeNet: AlexNet-level accuracy with 50x fewer parameters and <0.5 MB model size. arXiv preprint arXiv:1602.07360 (2016)
11. Lin, T.Y., Goyal, P., Girshick, R., He, K., Dollár, P.: Focal loss for dense object detection. In: Proceedings of the IEEE International Conference on Computer Vision, pp. 2980–2988 (2017)
12. Panchapagesan, S., et al.: Multi-task learning and weighted cross-entropy for DNN-based keyword spotting. In: Interspeech, vol. 9, pp. 760–764 (2016)
13. Peixoto, B., Lavi, B., Bestagini, P., Dias, Z., Rocha, A.: Multimodal violence detection in videos. In: ICASSP 2020–2020 IEEE International Conference on Acoustics, Speech and Signal Processing (ICASSP), pp. 2957–2961. IEEE (2020)

14. Purwins, H., Li, B., Virtanen, T., Schlüter, J., Chang, S.Y., Sainath, T.: Deep learning for audio signal processing. IEEE J. Sel. Top. Signal Process. **13**(2), 206–219 (2019)
15. Rouas, J.L., Louradour, J., Ambellouis, S.: Audio events detection in public transport vehicle. In: 2006 IEEE Intelligent Transportation Systems Conference, pp. 733–738. IEEE (2006)
16. Sandler, M., Howard, A., Zhu, M., Zhmoginov, A., Chen, L.C.: MobileNetV 2: inverted residuals and linear bottlenecks. In: Proceedings of the IEEE Conference on Computer Vision and Pattern Recognition, pp. 4510–4520 (2018)
17. Souto, H., Mello, R., Furtado, A.: An acoustic scene classification approach involving domestic violence using machine learning. In: Anais do XVI Encontro Nacional de Inteligência Artificial e Computacional, pp. 705–716. SBC (2019)
18. Uçar, A., Demir, Y., Güzeliş, C.: Object recognition and detection with deep learning for autonomous driving applications. Simulation **93**(9), 759–769 (2017)

Evaluating Unidimensional Convolutional Neural Networks to Forecast the Influent pH of Wastewater Treatment Plants

Pedro Oliveira[1](✉), Bruno Fernandes[1], Francisco Aguiar[3], Maria Alcina Pereira[2], and Paulo Novais[1]

[1] ALGORITMI Centre, University of Minho, Braga, Portugal
{pedro.jose.oliveira,bruno.fernandes}@algoritmi.uminho.pt, pjon@di.uminho.pt

[2] Centre of Biological Engineering, University of Minho, Braga, Portugal
alcina@deb.uminho.pt

[3] Águas do Norte, Guimarães, Portugal
f.aguiar@adp.pt

Abstract. One of our society's challenges today is water resources management due to its importance for human life. The monitoring of various substances present in wastewater is a crucial part of the process of Wastewater Treatment Plants (WWTPs). One of these substances is the influent's pH, which plays a fundamental role in the nitrification and nitration processes. Hence, this paper presents a study to forecast the influent pH in a WWTP for the next two days. For this purpose, several candidate models were conceived, tunned and evaluated, taking into account the one-dimensional Convolutional Neural Networks (CNNs) considering two distinct approaches in the Pooling layer: the channels' last and the channels' first. The best candidate model obtained a Mean Absolute Error (MAE) of 0.257, following the channel's last approach, compared to the channels' first that obtained a MAE of 0.272.

Keywords: Convolutional Neural Networks · Deep Learning · Influent pH · Time series · Wastewater Treatment Plants

1 Introduction

In recent years, we have seen an increase in water consumption, resulting in increased complexity in water resources management [1]. Since water is an essential element to human life, its treatment is a fundamental process today. Therefore, the Wastewater Treatment Plants (WWTPs) play a fundamental role in treating water before returning to the environment. These facilities aim to protect the aquatic environment, monitoring its effluents to ensure its quality. The water quality discharged at the WWTPs allows to promote environmental safety and reduce the pollution of water resources [2].

H. Yin et al. (Eds.): IDEAL 2021, LNCS 13113, pp. 446–457, 2021.
https://doi.org/10.1007/978-3-030-91608-4_44

The pH plays a crucial role throughout water treatment carried out in a WWTP, from chemical to biological treatment systems [3]. Monitoring and controlling the influent pH in a WWTP is essential to adjust its value by adding chemicals in the following processes. Another factor to consider when managing this substance is the crucial role that pH plays in the processes of nitrification and nitration [4]. This substance is determined by the concentration of hydrogen ions in a solution, thus determining its hydrogen ion potential. pH is presented on a scale between 1 and 14, with the pH being neutral when it has a value of 7. Furthermore, between 1 and 6, the pH is classified as acidic, and between 8 and 14, it is considered alkaline [5].

The present article aim to conceive, tune and evaluate several Deep Learning (DL) candidate models to forecast the influent pH in a WWTP for the next two days. With this in mind, we used the one-dimensional Convolutional Neural Networks (CNNs). Two types of approaches were used in the pooling layer: the channels' last and channels' first. The rest of this manuscript is structured as follows: the following section describes the literature review carried out, taking into account the pH prediction in the WWTP influent. The third section describes how the data were collected, analyzed and pre-processed, in addition to explaining the DL model used. The various experiments and the discussion of the obtained results are presented in the fourth and fifth sections, respectively. Finally, the sixth section summarizes the conclusions reached and the steps to be developed as future work.

2 State of the Art

Maleki et al. [6] carried out a study that aimed to predict a set of influent water characteristics in a WWTP for the next seven days. The authors used two distinct models to make the intended prediction, namely the Auto-Regressive Integrated Moving Average (ARIMA) and the Neural Network Auto-Regressive (NNAR). The data used in this study were based on a WWTP located in Sanandaj, Iran. The authors performed different analyzes to verify whether the features were stationary or non-stationary series. Several metrics were used to evaluate and compare the models, such as the Mean Absolute Error (MAE) and the correlation coefficient (R^2). Concerning the use of cross-validation, the authors used a particular method, the Time Series Cross-Validation. In the NNAR model, a network with only one hidden layer was always used, without any hyperparameters' optimization. On the other hand, in ARIMA models, they determined the hyperparameters according to the Maximum Likelihood method. Considering the obtained results, the authors could conclude that the NNAR-based models obtained a better performance, with an R^2 of 0.92, compared to the ARIMA-based ones, where the R^2 assumed the value of 0.41.

Another study carried out by Ansari et al. [7] aimed to predict different substances present in a WWTP influent for the next day, including the pH. With this in mind, the authors used a Fuzzy Inference System (FIS), namely the Adaptive neuro-fuzzy Inference System (ANFIS). In addition to this model, the authors used two methods to improve the prediction accuracy. One of these models combined the FIS model with Genetic Algorithms (GA), and the other integrated FIS with Particle Swarm Optimization (PSO). The data used in this study were collected from the WWTP in Kuala Lumpur, Malaysia, between 2011 and 2013. The data were analyzed and treated, namely by removing identified outliers and replacing them with the average of the previous and subsequent values in the respective feature. Regarding evaluation metrics, R^2, MAE and Root Mean Squared-Error (RMSE) were considered. The authors divided the data into 80% for training and 20% for testing. Each model used was executed with different input timesteps. With obtained results, the authors could verify that the GA-FIS and PSO-FIS models had a better performance than the ANFIS. Concerning pH prediction, the PSO-FIS model performed better than the GA-FIS, with an R^2 of 0.89 and 0.80, respectively.

Flores-Alsina et al. [8] developed a study that predicts the pH in different WWTP processes, including the influent. To perform the prediction of pH values, the authors used two simulation models, the Activated Sludge Models (ASM) and the Benchmark Simulation Model No. 2 (BSM2) in Anaerobic Digestion Model No. 1 (ADM1). This study took into account three different WWTPs. These simulation models were based on data files relating to pH in several WWTP processes. In the first two WWTPs, the authors were based on the ASM simulation models. The first WWTP used a 1 and 3 version of ASM, while the second one used a 2d version of the same model. In the case of the influent, the obtained results in the two WWTPs were quite similar, with a predicted pH value around 6.8. On the other hand, in the third WWTP, the authors use the BSM2 simulation model, with the forecasted pH value slightly increasing to 7. At the end of this study, the authors concluded that it is possible to reliably predict the pH at different stages of the WWTP using the ASM and ADM simulation models.

In conclusion, some studies do not consider some topics, such as the analysis of overfitting or underfitting of the models, which is an important aspect of the conception of DL and Machine Learning (ML) models. Another aspect to note is the non-optimization of a more extensive range of hyperparameters in the different models used, which may lead to better results. Although some studies already use a specific cross-validator to time series problems, it is still common to find studies that do not consider this.

3 Materials and Methods

This section details the materials and methods used in this study. In the following lines, data analysis and processing are detailed, as well as the DL model used and the evaluation metrics.

3.1 Data Collection

The data used throughout this study were based on a WWTP in Portugal, made available by a multi-municipal wastewater company. Furthermore, the collected data were between January 1st, 2016 and April 30th, 2017. To understand whether or not the climatological events could affect the influent pH, we collected these data for the same city of the WWTP, using the Open Weather MAP API, for the same period.

3.2 Data Exploration

The final dataset used by the several conceived candidate models throughout this study was based on two distinct datasets. The first one was related to the influent pH in a WWTP, with some observations with a bi-daily periodicity and the others daily, in a total of 1556 observations. This dataset includes two features, the *ph_value* and the corresponding timestamp of this record. On the other hand, concerning the climatological dataset, it presents 25 features, with an hourly periodicity, representing 21716 observations. Table 1 describes the two features present in the first dataset, as well as some of the climatological-related features present in the second dataset.

Table 1: Available features in the used datasets.

#	Features	Description	Unit
		Influent pH Dataset	
1	*date*	Timestamp	date & time
2	*ph_value*	Influent pH	Sorensen Scale
		Climatological Dataset	
1	*dt_iso*	Timestamp	date & time
2	*temp*	Temperature	C
3	*pressure*	Atmospheric pressure	hPa
4	*humidity*	Humidity percentage	%
5	*wind_speed*	Wind speed value	m/s
6	*rain*	Rain volume	mm

The first step in analyzing the data used in this study was to verify the missing timesteps and values. There was no absence in the set of climatological data, either of values or of timesteps. However, in the influent pH dataset, we verified 5 missing values and 25 missing timesteps.

Then, a statistical analysis of the target feature of this study, the influent pH, was performed and described in Table 2. Through this analysis, it was possible to verify that in the collected data, the pH value presents an average of 6.16 and a standard deviation of 0.57. In addition, Skewness and Kurtosis values were also analyzed. In the case of the Skewness value, it presented a negative value of -0.94, which indicates that the influent pH data are asymmetric. Concerning Kurtosis, this shows a value of 1.34, which represents that the data follow the leptokurtic distribution.

The next step in data exploration was using several box plots to analyze the pH values over the months represented in the data, illustrated in Fig. 1. It is possible to verify that some months follow a symmetric distribution, i.e., the distance between the first and second quartiles to be approximately the same as the distance between the second and third quartiles. For example, the months of

Table 2: Statistical analysis of influent pH.

Mean	Standard Deviation	Max	Min	Skewness	Kurtosis
6.16	0.57	7.65	3.77	-0.94	1.34

July and October are examples of this distribution. On the other hand, month's May and June are examples of a skewed distribution. Another analysis that can be drawn from this figure is that the pH median, which in most months, is between 6 and 6.5. Regarding the presence of outliers, it is possible to identify their existence in a few months, with the month of January being the one with the highest number, in this case, four outliers.

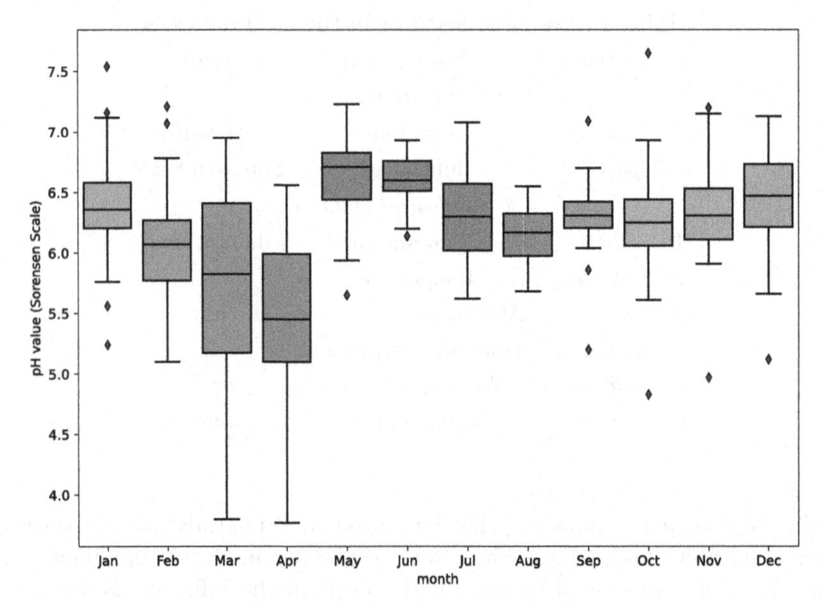

Fig. 1: Boxplot of month-wise influent pH distribution.

Finally, it was verified that the data do not follow a normal distribution, using the Kolmogorov-Smirnov test with a $p < 0.05$.

3.3 Data Preparation

Concerning data preparation, we created new features, namely the *year*, *month* and *day* from the feature *date* and *dt_iso*, in the two datasets. In the case of pH data, as these presented bi-daily periodicity in some observations, they were grouped by *day*, with the pH value assuming the corresponding mean value. In the climatological data, to have both datasets at the same frequency, we moved from hourly data to daily, grouping the different features by their average.

As mentioned above, there was a set of missing timesteps in the pH data that had to be filled. The inclusion of the missing timesteps led to more missing values, which at the end of this step totalled 29. The missing values identified were filled in, taking into account the average of the previous 7 days.

After performing the previous steps, we had two datasets with the same number of observations and periodicity, without missing values or missing timesteps. So the next step was to join the data into a single dataset. This process was carried out using the *year*, *month*, and *day* features. After the join, the dataset totalled 486 observations. With the data all together in a single dataset, it was necessary to understand the correlations of the different features with the influent pH's target feature.

Then, as the data did not follow a Gaussian distribution, the non-parametric Spearman's rank correlation coefficient test was used. With this test, it was possible to verify that no feature showed a strong correlation, either negative or positive, with the influent pH. Thus, concerning the pH value, there is no strong correlation between this value and any feature related to climatological data. With this result in mind, we removed all features except the *ph_value*, obtaining a final dataset, in ascending order by *index*, with only one feature, making this problem a uni-variate one. Table 3 presents an example of observation in the final dataset, with the *timestep* only being the index.

Table 3: Example of an observation in the final dataset.

#	**Feature**	**Observation**
0	*timestep_index*	2016-3-4
1	*ph_value*	6.51

The last step in the data processing considered the normalization of data and the transformation of the problem into supervised. Regarding normalization, this was applied between 0 and 1, taking into account the *MinMaxScaler*. As for the passage to a supervised problem, it was necessary to create a data sequence depending on the number of timesteps given as input to the model. For example, if the number of timesteps is 14, it is necessary to create an input sequence X, with 14 days and its corresponding label y, using a sliding window that goes through all the data.

3.4 Evaluation Metrics

Since we are dealing with a regression model, we used two error metrics to evaluate the various candidate models. The first metric corresponds to the MAE, which tells us the absolute value of the difference between the actual and the predicted value. As the MAE is a linear score, all individual differences are averaged.

$$\text{MAE} = \frac{1}{n}\sum_{i=1}^{n}|y_i - \hat{y}_i| \tag{1}$$

The second metric used took into account the RMSE, which represents the standard deviation of the residuals. Because errors are squared before they are calculated, the RMSE tends to penalize high errors.

$$\text{RMSE} = \sqrt{\frac{\sum_{i=1}^{n}(y_i - \hat{y}_i)^2}{n}} \tag{2}$$

3.5 CNNs

CNN's are a set of neural networks widely used in image recognition [9] problems, or the context of autonomous driving [10]. In the scope of time series problems, Recurrent Neural Networks (RNNs) perform well in this type of problem, as in the forecast of road traffic [11,12]. However, CNN's have been emerging as an option in some studies concerning time series problems [13,14], such as energy consumption forecasting [15]. In this study, two approaches were used within the pooling layer, namely in the data format hyperparameter. In this case, we used channels' last and channels' first approaches.

The channels' first approach aims to maintain the number of filters, decreasing the number of timesteps. On the contrary, the channels' last approach reduces the number of filters across the network, leaving the timesteps intact. Before arriving at the pooling layer, it is necessary to pay attention to the output of the convolutional layer, whose calculation can be verified in work performed by Oliveira et al. [15]. Note that the pooling layer used in this study was the Average Pooling. In Fig. 2 it is possible to see an example of the two approaches at the pooling layer, taking into account the output shape of the Conv1D layer. In the example, we considered 14 timesteps, 32 filters, pool size of 2 and kernel size of 4. Regarding the Input layer, taking into account the example seen in the figure, its output shape is (486, 14, 1) corresponding to (number of observations, timesteps, features).

4 Experiments

Several experiments were carried out to obtain the best model to forecast the influent pH. All models were based on a uni-variate and multi-step approach, with the candidate models fed by one feature and predicting the pH for the next two days, respectively. The experiments carried out aimed to achieve the best combination of hyperparameters for each used approach.

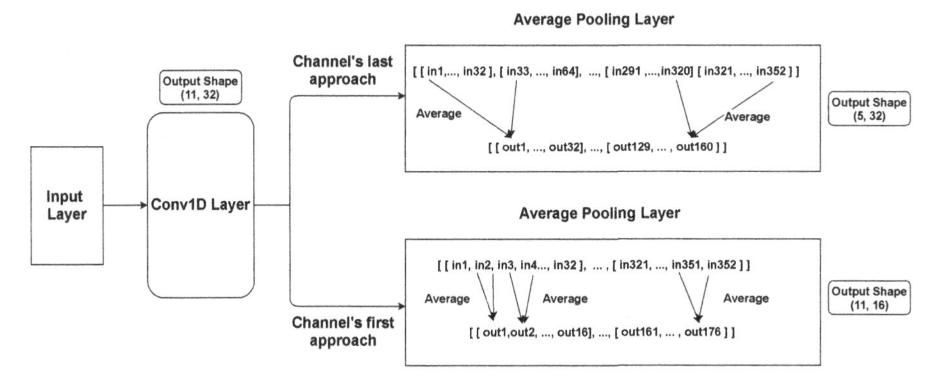

Fig. 2: Example of the two approaches in the pooling layer.

All performances achieved by the several conceived candidate models in the two approaches were evaluated in terms of the error metrics described above. Table 4 presents which hyperparameters were tuned and the range of values used in each of the approaches. It is possible to notice in the channels' last approach that the maximum value of the number of filters and kernel size is larger than the channel's first. On the other hand, there are higher maximum values regarding pool size and the number of layers in channels' first. Due to the two types of approach, certain hyperparameters are also affected in their range of values, such as the number of layers CONV1D-Average Pooling or pool size. Regarding the number of timesteps, in the case of channels' first approach, it was possible to carry out experiments with a lower value than in channels' last approach, in this case using 14 timesteps. This type of variation in some hyperparameters happens due to the nature that each approach uses, as mentioned above.

In each of the approaches used to conceive the different candidate models, we always pay attention to preventing overfitting or underfitting situations. For this, the learning curves in each of the approaches were plotted and analyzed. Bearing in mind that we face a time series problem, it is essential to use a problem-specific cross-validator. Hence, the *TimeSeriesSplit* cross-validator was used, taking into account the number of observations in the dataset, with a k value equal to 3.

Finally, Python, version 3.7, was used as a programming language to explore and pre-processing the data and conceiving the various candidate models. Several Python libraries were used, such as *Pandas* or *scikit-learn*. Concerning data exploration, the Knime platform was also used. To conceive the different models, *TensorFlow* version 2.0 was taken into account. As far as hardware is concerned, this was made available by Google's Collaboratory.

Table 4: Channels' last approach vs Channels' first approach hyperparameters' searching space.

Parameter	Channels' last	Channels' first
Timesteps	[21, 28]	[14, 21, 28]
Batch size	[10, 20, 30]	[5, 10, 20]
Conv1D-AveragePooling layer pairs	[1, 2]	[3, 4, 5]
Activation	[reLu, tanh]	[reLu, tanh]
Pool Size	[1, 2]	[2, 3]
Filters	[32, 64]	[16, 32]
Dropout	[0.0, 0.5]	[0.0, 0.5]
Kernel Size	[4, 5, 6]	[3, 4, 5]
Learning rate	Callback	Callback
Epochs	[15, 200]	[15, 150]
Multisteps	2	2
CV Splits	3	3

5 Results and Discussion

Once all the developed experiments were completed, it was necessary to analyze the obtained results for each approach. Table 5 describes the top-5 of the best-conceived models, either on the channels' last approach or on the channel's first. The results show the obtained hyperparameters' values in each candidate model, the error metrics, and each candidate model's learning time. It should be noted that the error values described in the table are already denormalized.

With the analysis of the obtained results, it is possible to verify that the best candidate model between the two approaches is based on the channels' last, with a RMSE value of 0.293 and a MAE of 0.257. In the remainder candidate models of this approach, it can be seen that there is practically a uniformity in some hyperparameters. For example, concerning the number of layers, except for one candidate model, the others need one layer. On the other hand, there is a certain heterogeneity in the kernel size and number of filters values.

Considering the channels' first approach candidate models, the best model presents a RMSE value of 0.307 and a MAE of 0.272. As with the best candidate model in channels' last approach, the best channel's first also needs to have 21 input timesteps and the same activation function or number of filters. In this approach, the five best candidate models need the same timesteps to be given to the model and almost the same values at the number of layers.

Overall, considering the execution times, it turns out that the candidate models based on the channel's last approach take significantly less time, improving their performance compared to channels' first. Other factors that can be concluded that differentiate these two approaches centre around the number of layers. All five best candidate models of the channels' last approach need a

Table 5: Channels' last approach vs Channels' first approach top-5 models. Legend: a. timesteps; b. batch size; c. nº of layers; d. filters; e. pool size; f. kernel size; g. dropout; h. activation; i. RMSE; j. MAE; k. time (in seconds).

a.	b.	c.	d.	e.	f.	g.	h.	i.	j.	k.
Channels last candidate models										
21	20	1	32	2	4	0.0	tanh	**0.293**	**0.257**	6.64
28	30	1	64	2	5	0.0	tanh	0.294	0.260	6.62
28	30	2	32	1	4	0.0	tanh	0.295	0.263	7.26
21	20	1	64	2	4	0.0	tanh	0.297	0.266	6.69
28	30	1	32	2	6	0.0	relu	0.297	0.267	6.66
Channels first candidate models										
21	10	4	32	3	3	0.0	tanh	**0.307**	**0.272**	34.58
21	20	3	32	2	3	0.0	tanh	0.308	0.274	18.67
21	20	3	32	2	4	0.0	tanh	0.313	0.279	15.75
21	10	3	16	2	3	0.0	tanh	0.321	0.287	21.41
21	20	3	16	2	4	0.0	tanh	0.324	0.29	18.37

lower value for the number of layers than channels' first. Regarding the batch size value, it is possible to verify that some of the candidate models of the channel's last approach need a higher value than the channel's first. According to the best candidate model, Fig. 3 illustrates several 2-day influent pH predictions based on the channels' last approach. This prediction was based on a total of 21 timesteps.

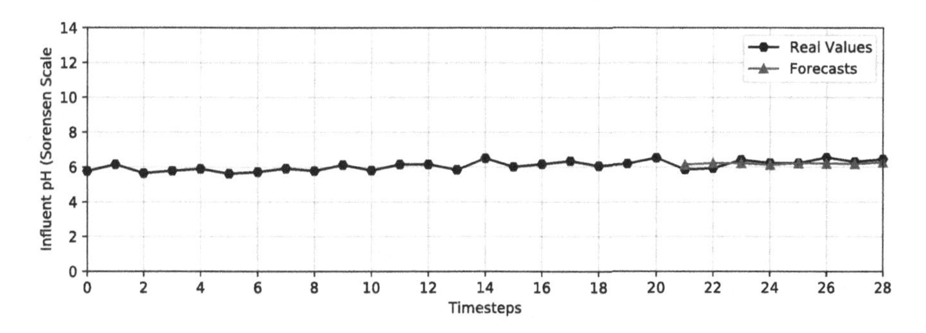

Fig. 3: Several multi-step forecast of influent pH.

6 Conclusions

pH is a very important property to monitor throughout the entire process in a WWTP. Its control concerning the influent gains a fundamental role in the remaining treatment to be carried out. Therefore, this study aimed to forecast this substance, in the influent of a WWTP, for the next two days using CNNs with two distinct approaches at the pooling layer.

To reach the outlined objective, several experiments were developed with different combinations of hyperparameters, taking into account the two defined approaches. From the analysis of the obtained results, it was possible to verify that the best candidate model was based on the channels' last approach, with a RMSE of 0.293 and a MAE of 0.257. Another conclusion was that the climatological data do not present a strong correlation with the target feature present in this study, therefore not using any feature of this dataset in the conceived models.

As future work, we aim to incorporate new features related to other substances present in water, such as total nitrogen or ammonia. Thus, our objective will be to verify if any of these substances is strongly correlated with the influent pH and if it can lead to better performances in the conceived models. Furthermore, the use of other types of networks, such as RNNs or even Multilayer Perceptron, to compare these models' performance with those developed in this study.

Acknowledgments. This work is financed by National Funds through the Portuguese funding agency, FCT - Fundação para a Ciência e a Tecnologia within project DSAIPA/AI/0099/2019.

References

1. Salgot, M., Folch, M.: Wastewater treatment and water reuse. Curr. Opin. Environ. Sci. Health **2**, 64–74 (2018). https://doi.org/10.1016/j.coesh.2018.03.005
2. Hao, X., Wang, X., Liu, R., Li, S., van Loosdrecht, M.C., Jiang, H.: Environmental impacts of resource recovery from wastewater treatment plants. Water Res. **160**, 268–277 (2019). https://doi.org/10.1016/j.watres.2019.05.068
3. Wang, X., et al.: Stepwise pH control to promote synergy of chemical and biological processes for augmenting short-chain fatty acid production from anaerobic sludge fermentation. Water Res. **155**, 193–203 (2019). https://doi.org/10.1016/j.watres.2019.02.032
4. Peng, F., et al.: Removal of high-strength ammonia nitrogen in biofilters: nitrifying bacterial community compositions and their effects on nitrogen transformation. Water **12**(3), 712 (2020). https://doi.org/10.3390/w12030712
5. Proksch, E.: pH in nature, humans and skin. J. Dermatol. **45**(9), 1044–1052 (2018). https://doi.org/10.1111/1346-8138.14489
6. Maleki, A., Nasseri, S., Aminabad, M.S., Hadi, M.: Comparison of ARIMA and NNAR models for forecasting water treatment plant's influent characteristics. KSCE J. Civ. Eng. **22**(9), 3233–3245 (2018). https://doi.org/10.1007/s12205-018-1195-z
7. Ansari, M., Othman, F., El-Shafie, A.: Optimized fuzzy inference system to enhance prediction accuracy for influent characteristics of a sewage treatment plant. Sci. Total Environ. **722**, 137878 (2020). https://doi.org/10.1016/j.scitotenv.2020.137878

8. Flores-Alsina, X., et al.: A plant-wide aqueous phase chemistry module describing pH variations and ion speciation/pairing in wastewater treatment process models. Water Res. **85**, 255–265 (2015). https://doi.org/10.1016/j.watres.2015.07.014
9. Shang, L., Yang, Q., Wang, J., Li, S., Lei, W.: Detection of rail surface defects based on CNN image recognition and classification. In: 2018 20th International Conference on Advanced Communication Technology (ICACT), pp. 45–51. IEEE, February 2018. https://doi.org/10.23919/ICACT.2018.8323642
10. Fernandes, D., et al.: Point-cloud based 3D object detection and classification methods for self-driving applications: a survey and taxonomy. Inf. Fusion **68**, 161–191 (2021). https://doi.org/10.1016/j.inffus.2020.11.002
11. Fernandes, B., Silva, F., Alaiz-Moreton, H., Novais, P., Neves, J., Analide, C.: Long short-term memory networks for traffic flow forecasting: exploring input variables, time frames and multi-step approaches. Informatica **31**(4), 723–749 (2020). https://doi.org/10.15388/20-INFOR431
12. Fernandes, B., Silva, F., Alaiz-Moretón, H., Novais, P., Analide, C., Neves, J.: Traffic flow forecasting on data-scarce environments using ARIMA and LSTM networks. In: Rocha, Á., Adeli, H., Reis, L.P., Costanzo, S. (eds.) WorldCIST'19 2019. AISC, vol. 930, pp. 273–282. Springer, Cham (2019). https://doi.org/10.1007/978-3-030-16181-1_26
13. Zhou, F., Zhou, H., Yang, Z., Gu, L.: IF2CNN: towards non-stationary time series feature extraction by integrating iterative filtering and convolutional neural networks. Expert Syst. Appl. **170**, 114527 (2021). https://doi.org/10.1016/j.eswa.2020.114527
14. Jin, X., Yu, X., Wang, X., Bai, Y., Su, T., Kong, J.: Prediction for time series with CNN and LSTM. In: Wang, R., Chen, Z., Zhang, W., Zhu, Q. (eds.) Proceedings of the 11th International Conference on Modelling, Identification and Control (ICMIC2019). LNEE, vol. 582, pp. 631–641. Springer, Singapore (2020). https://doi.org/10.1007/978-981-15-0474-7_59
15. Oliveira, P., Fernandes, B., Analide, C., Novais, P.: Forecasting energy consumption of wastewater treatment plants with a transfer learning approach for sustainable cities. Electronics **10**(10), 1149 (2021). https://doi.org/10.3390/electronics10101149

LSTM Neural Network Modeling of Wind Speed and Correlation Analysis of Wind and Waves

Carlos Serrano-Barreto[1](✉), Cristina Leonard[1], and Matilde Santos[2](✉)

[1] Computer Science Faculty, Complutense University of Madrid, Madrid, Spain
carser06@ucm.es

[2] Institute of Knowledge Technology, Complutense University of Madrid, 28040 Madrid, Spain
msantos@ucm.es

Abstract. In floating offshore wind turbines (FOWT) not only the wind but the waves have a strong impact on the dynamic of the floating device, and thus on its efficiency. So, it is important to study the relationship between wind speed and wave height. In this paper, data obtained from Casco Bay buoy (USA) in 2020 are analyzed. First, a deep learning model of the wind based on LSTM (Long Short-Term Memory) networks is developed, with inputs the mean wind speed and wind direction. To test the model, the last 876 h of the year were used. The model has been proved to be able of forecasting average hourly wind speed with good accuracy. In addition, the data of the average wave height for the same period of time and the same location are considered, and the correlation between both variables (wind and waves) is obtained. This study allows to better understand the behavior of environmental loads on floating wind turbines.

Keywords: Deep learning · Modelling · Wind · Waves · Floating offshore wind turbines · Wind energy

1 Introduction

Due to the rapid running out of fossil fuels and the increase in climate change and its various consequences, there have been many recent studies on alternative energies. Offshore wind energy emerge as an attractive alternative to fossil fuels [1].

The scientific community seeks to better understand and more accurately forecast the environmental processes related to global change, both on large and regional scales. It is well known that sea state in the ocean plays a key role in the air-sea interaction process of the coupled ocean-atmosphere system. The coexistence of wind-sea and waves significantly affects the stability of offshore structures and their design [2–4], and it is a crucial part of the performance of small craft and ship operations over harbor entrances, as well as wave forecasting [5].

In this work, the analysis of wind and wave data is intended to be a starting point for the evaluation of marine energy extraction projects as floating offshore wind turbines, in order to optimize the control strategies and improve their efficiency [6]. The study of the relationship between wind speed and wave height is proposed. In addition, a model

H. Yin et al. (Eds.): IDEAL 2021, LNCS 13113, pp. 458–466, 2021.
https://doi.org/10.1007/978-3-030-91608-4_45

to predict wind speed is developed, to use it a basis for the analysis. Deep learning techniques, that have proved efficient for dealing with these complex marine systems, have been applied [7–9]. Real data from Casco Bay [10], which is 15 km off the coast from Maine, New England, United States for year 2020.

Some previous works classify wind speed prediction into deterministic and probabilistic. Probabilistic forecasting includes three major groups: Prediction interval (PI) [11]; quantile forecasting [12], and finally methods based on probabilistic density function [13] which describes the statistical behavior of the studied phenomenon [14].

On the other hand, the short-term deterministic prediction can use physical, statistical and data-based methods [14, 15], the latter being the one applied in this work. Data-driven methods include shallow and deep structures that generate their own model of the phenomenon and can understand the nonlinear and complex characteristics associated with it. Within the data-driven methods deep learning, and specifically the Long short-term memory (LSTM) architecture, is used in this work to forecast wind speed [16].

The structure of the paper is as follows. Section 2 describes the long short term memory (LSTM) neural network model. Section 3 presents the study of correlation between wind and waves. In Sect. 4 results of the deep learning model are shown and discussed. The paper ends with conclusions and future works.

2 LSTM Neural Network Model

Data of wind and waves were collected hourly at the Casco Bay, USA, during year 2020 [10]. They include wind speed, wind direction and wave height. For wind speed, the maximum value was 18 m/s and the minimum was 0 m/s. Figure 1 shows the average values of the 12 months in year 2020 for wind speed (blue) and wave height (red).

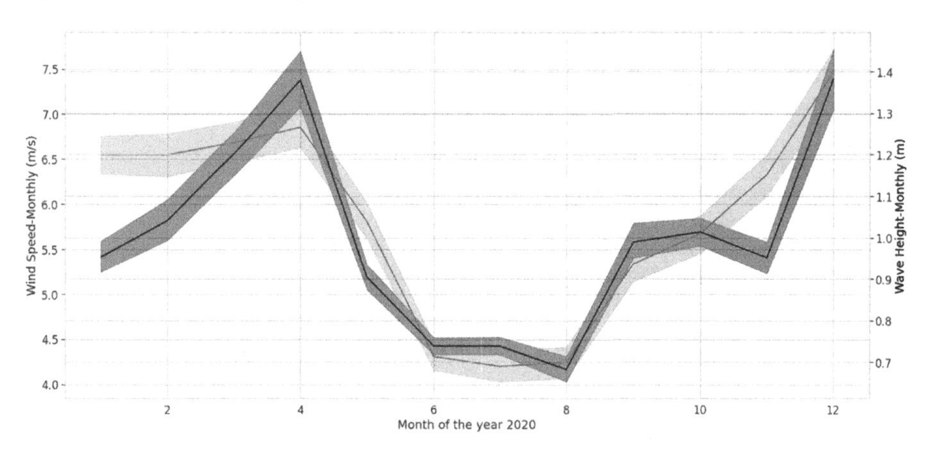

Fig. 1. Average wind speed and wave height per month in year 2020

From Fig. 1, it is possible to observe the randomness of the wind speed and, hence, the difficulty to predict it accurately. As it is possible to observe, similar peaks appear between wind and waves. That is why the correlation study of both variables will be carried out. The area around the line is the confidence interval.

In Fig. 2, the wind direction in year 2020 is represented. The legend on the right explains the color code for the wind speed.

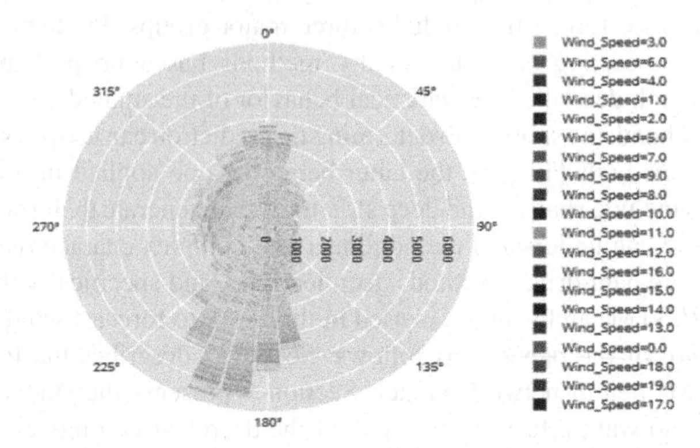

Fig. 2. Prevailing hourly wind direction in year 2020

A trend between maximum wind speed and wind direction can be observed in this figure. The wind speed and direction have a higher recurrence between 170° and 225° with respect to the buoy that takes the measurements.

2.1 Feature Description

The meteorological data studied contains the information corresponding to the hourly measure of wind and wave. The structure of the data is as shown in Fig. 3.

Time-UTC	Wind_Speed	Wave_Height	Wind_Direction	hour	day_of_month	day_of_week	month
2020-01-01 00:00:00+00:00	7.0	3.2	340.0	0	1	2	1
2020-01-01 01:00:00+00:00	5.0	2.9	330.0	1	1	2	1
2020-01-01 02:00:00+00:00	5.0	2.8	300.0	2	1	2	1
2020-01-01 03:00:00+00:00	4.0	2.4	310.0	3	1	2	1
2020-01-01 04:00:00+00:00	3.0	2.4	300.0	4	1	2	1
...	...	...	...	...	...	...	...

Fig. 3. Samples of the data

There is a total of 8347 data for each variable, of which those used to train the wind speed prediction model were wind speed and wind direction, and for the correlation study, wave height and wind speed. The wind direction is measured clockwise from the north.

2.2 Deep Leaning Model

Given these wind data, the generation of a model to forecast the wind speed using deep learning is proposed. These artificial intelligent methods are able to match inputs and outputs [17, 18]. Specifically, a recurrent neural network model known as LSTM (long time series prediction) is used, which can learn long-term dependence information and avoid the problem of gradient vanishing [19]. The LSTM networks has in the hidden layers a structure called Memory Cell to remember the past information and three gates of structures (Input, Forget, Output) [20]. Figure 4 shows the structure of a neuron of an LSTM network.

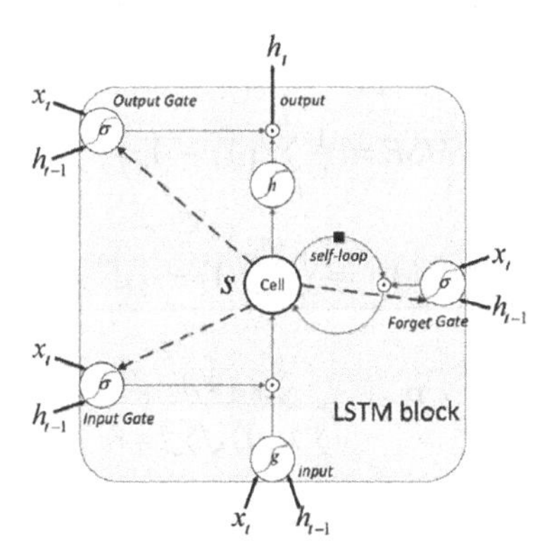

Fig. 4. LSTM block diagram [15].

Where x_t is the input vector, h_t is the output of the LSTM, h_{t-1} is the previous output of the network, h represents the hyperbolic tangent activation function, and g is the input activation function of the LSTM, In this case it is the hyperbolic tangent function.

The predictive model of the network has as inputs the historical data of wind speed and direction, with a fixed time window. In this case, 10 values have been considered, that is, the last 10 h recorded. The goal of the network is to forecast the next wind speed value. The data set was split into 80% for training, 10% for validation and 10% for testing.

The network architecture consists of 128 LSTM neurons in the hidden layer; the Adam optimizer and a dropout of 0.2 are used to avoid data overfitting. The equations of the LSTM neurons are as follows [17]:

$$i_t = sigmoid(w_{hi}h_{t-1} + w_{xi}x_t) \tag{1}$$

$$f_t = sigmoid\left(w_{hf}h_{t-1} + w_{xf}x_t\right) \tag{2}$$

$$c_t = f_t \otimes c_{t-1} + i_t \otimes tanh(w_{hc}h_{t-1} + w_{xc}x_t) \tag{3}$$

$$o_t = sigmoid(w_{ho}h_{t-1} + w_{hx}x_t + w_{co}c_t) \tag{4}$$

$$h_t = o_t \otimes tanh(c_t) \tag{5}$$

Where i_t is the input gate, f_t is the forget gate, c_t is the cell state, o_t is the output gate, the output of the LSTM is h_t; w_{ij} represents the weights of each neuron and σ is the sigmoid activation function.

3 Results of the Prediction Model

To evaluate the LSTM model, the Mean Square Error (MSE), the Mean Absolute Error (MAE) and the Cosine Proximity (CP) are taken into account. These are defined by the following equations, respectively.

$$MSE = \frac{1}{n}\sum_{i=1}^{n}\left(\widehat{Y_i} - Y_i\right)^2 \tag{6}$$

$$MAE = \frac{1}{n}\sum_{i=1}^{n}|\widehat{Y_i} - Y_i| \tag{7}$$

$$CP = \frac{\sum_{i=1}^{n}\widehat{Y_i}Y_i}{\sqrt{\sum_{i=1}^{n}\widehat{Y_i}}\sqrt{\sum_{i=1}^{n}Y_i}} \tag{8}$$

Where $\widehat{Y_i}$ it's the predicted value, Y_i is the real value and n is the number of samples. The results of the wind speed prediction are shown in Table 1:

Table 1. Prediction model results

LSTM-model	MSE	MAP	CP
Training	0.0731	0.2046	0.7364
Validation	0.0629	0.1909	0.6806
Test	0.0662	0.1983	0.6613

Figure 5 shows the results of the prediction model on the test set. These values have a one-hour resolution. As it is possible to see, the accuracy of the results is good.

Different classical regression models have been also applied in order to compare the performance. The result are show in Table 2. The best result is obtained with LSTM (smallest errors) model (bolded).

4 Correlation Study

In addition, the relationship between wind speed and wave height is studied for the available data of year 2020. In Fig. 6 the mean wind speed is represented vs. the wave height.

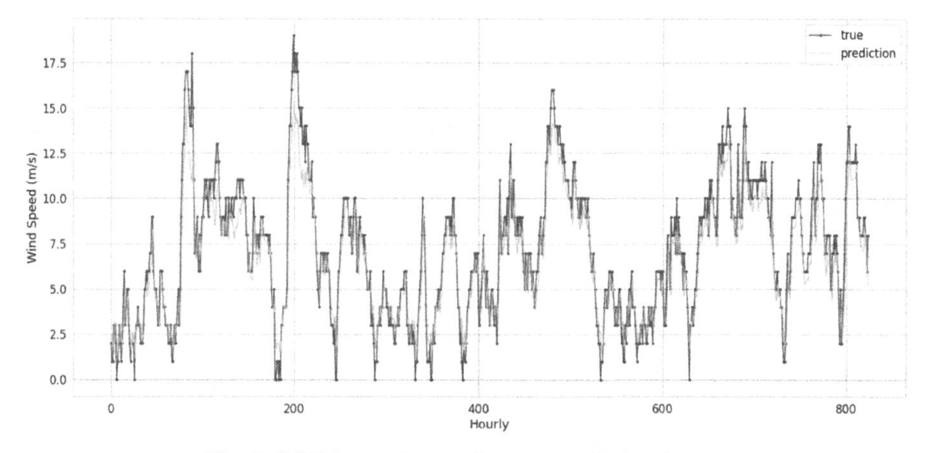

Fig. 5. LSTM neural network output applied to the test.

Table 2. LSTM-model vs other regression models

Regressions-model	MSE	MAP
LSTM	**0.0662**	**0.1983**
AdaBoost regressor	2.4737	9.2352
Lasso least angel regression	2.6150	9.7350
Gradient boosting regressor	2.6445	11.0041
Huber regressor	2.6711	11.2561
Linear regression	2.6927	11.2509

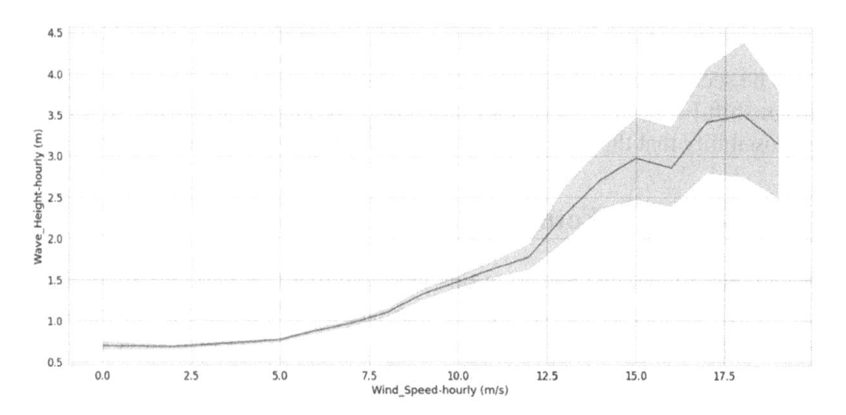

Fig. 6. Wave height against wind speed over the year.

A positive linear relationship between wave height and wind speed can be observed, with a higher standard deviation for higher speeds. Nevertheless, as expected, the stronger the wind speed, the higher the waves.

In order to compare both variables, first the linear correlation is calculated. The Pearson correlation coefficient (ρ) for the mean wave height and wind speed with a time interval of one hour was obtained using Eq. (9) [21]. It goes from -1 to 1, with 1 being the strongest possible positive correlation, -1 meaning the strongest possible negative correlation, and 0 representing no correlation.

$$\rho_{x,y} = \frac{\sum_{i=1}^{n}(x_i - \overline{x})(y_i - \overline{y})}{\sqrt{\sum_{i=1}^{n}(x_i - \overline{x})^2}\sqrt{\sum_{i=1}^{n}(y_i - \overline{y})^2}} \tag{9}$$

Where n is the number of samples, x_i an y_i are the individual sample points indexed with$i$, and $\overline{x}$, $\overline{y}$ are the mean of data. The Pearson correlation obtained was $\rho = 0.53891$. That means that there is a medium correlation between both variables (>0.5).

The cross-correlation has been also studied when the variables are out of phase [21]. Figure 7 shows the normalized cross correlation.

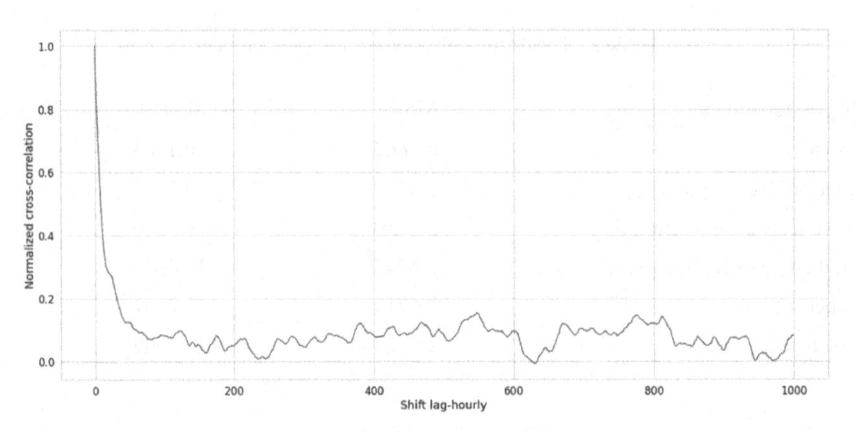

Fig. 7. Normalized cross correlation.

It can be seen that at 0, the maximum correlation is obtained. Therefore, there is no significant shift between the variables in the periods studied. However, we do not rule out the possibility that these variables could be shifted if a smaller time scale is used.

5 Conclusions and Future Works

The main conclusions obtained from this work are that it is possible to obtain a good wind speed forecasting model for one hour ahead using real data from Casco Bay. Smalls errors have been obtained. On the other hand, a small linear correlation was obtained between wind speed and wave height; nevertheless, there is no correlation if one variable is shifted with respect to the other.

The proposed deep learning model is able to predict the wind speed from the 10 previous values of wind direction and speed values. The results are given in terms of the

MSE, MAE and CP metrics, presenting improvement over classical regression models regarding the errors. This can provide important support in the operations of energy extraction from offshore wind turbines.

As future work, it is proposed to develop a deep learning model for wave prediction, in order to implement strategies to mitigate the loads on offshore structures.

Acknowledgement. This work was partially supported by the Spanish Ministry of Science, Innovation and Universities under MCI/AEI/FEDER Project number RTI2018-094902-B-C21.

References

1. Sierra-García, J.E., Santos, M.: Neural networks and reinforcement learning in wind turbine control. Rev. Iberoam. Automática e Informática Ind. **18**(4) (2021)
2. Galán-Lavado, A., Santos, M.: Analysis of the effects of the location of passive control devices on the platform of a floating wind turbine. Energies **14**(10), 2850 (2021)
3. Gomes, I.L., Melício, R., Mendes, V.M., Pousinho, H.M.: Wind power with energy storage arbitrage in day-ahead market by a stochastic MILP approach. Logic J. IGPL **28**(4), 570–582 (2020)
4. Sierra-García, J.E., Santos, M.: Performance analysis of a wind turbine pitch neurocontroller with unsupervised learning. Complexity (2020)
5. Shanas, P.R., Kumar, V.S.: Trends in surface wind speed and significant wave height as revealed by ERA-Interim wind wave hindcast in the Central Bay of Bengal. Int. J. Climatol. **35**, 2654–2663 (2015). https://doi.org/10.1002/joc.4164
6. Young, I.R., Vinoth, J., Zieger, S., Babanin, A.V.: Investigation of trends in extreme value wave height and wind speed. J. Geophys. Res. Ocean. **117**, 1–13 (2012). https://doi.org/10.1029/2011JC007753
7. Liu, D.R., Lee, S.J., Huang, Y., Chiu, C.J.: Air pollution forecasting based on attention-based LSTM neural network and ensemble learning. Expert Syst. **37**(3), e12511 (2020)
8. Sierra-Garcia, J.E.: Deep learning and fuzzy logic to implement a hybrid wind turbine pitch control. Neural Comput. Appl. 0123456789 (2021). https://doi.org/10.1007/s00521-021-06323-w
9. Paula, M., Marilaine, C., Nuno, F.J., Wallace, C.: Predicting long-term wind speed in wind farms of Northeast Brazil: a comparative analysis through machine learning models. IEEE Lat. Am. Trans. **18**(11), 2011–2018 (2020)
10. Northeastern Regional Association of Coastal Ocean Observing Systems. http://www.neracoos.org/
11. Kim, Y., Hur, J.: An ensemble forecasting model of wind power outputs based on improved statistical approaches. Energies **13**(5), 1071 (2020)
12. Tang, B., Chen, Y., Chen, Q., Su, M.: Research on short-term wind power forecasting by data mining on historical wind resource. Appl. Sci. **10**(4), 1295 (2020)
13. Khorramdel, B., et al.: A fuzzy adaptive probabilistic wind power prediction framework using diffusion kernel density estimators. IEEE Trans. Power Syst. **33**, 7109–7121 (2018)
14. Afrasiabi, M., Mohammadi, M., Rastegar, M., Afrasiabi, S.: Advanced deep learning approach for probabilistic wind speed forecasting. IEEE Trans. Ind. Informatics. **17**, 720–727 (2021). https://doi.org/10.1109/TII.2020.3004436
15. Sierra-García, J.E., Santos, M.: Switched learning adaptive neuro-control strategy. Neurocomputing **452**, 450–464 (2021)

16. Liu, H., Mi, X., Li, Y.: Smart multi-step deep learning model for wind speed forecasting based on variational mode decomposition, singular spectrum analysis, LSTM network and ELM. Energy Convers. Manage. **159**, 54–64 (2018). https://doi.org/10.1016/j.enconman.2018.01.010
17. Geng, D., Zhang, H., Wu, H.: Short-term wind speed prediction based on principal component analysis and LSTM. Appl. Sci. **10**, 4416 (2020)
18. Jove, E., Casteleiro-Roca, J., Quintián, H., Méndez-Pérez, J.A., Calvo-Rolle, J.L.: Detection of anomalies based on intelligent techniques in a plant for obtaining bicomponent material used in the manufacture of wind turbine blades. Rev. Iberoam. Automática e Informática Ind. **17**(1), 84–93 (2020)
19. Liang, T., Zhao, Q., Lv, Q., Sun, H.: A novel wind speed prediction strategy based on Bi-LSTM, MOOFADA and transfer learning for centralized control centers. Energy **230**, 120904 (2021)
20. Yao, W., Huang, P., Jia, Z.: Multidimensional LSTM networks to predict wind speed. In: 2018 37th Chinese Control Conference (CCC), pp. 7493–7497 (2018)
21. Meucci, A., Young, I.R., Aarnes, O.J., Breivik, Ø.: Comparison of wind speed and wave height trends from twentieth-century models and satellite altimeters. J. Clim. **33**, 611–624 (2020)

Finding Local Explanations Through Masking Models

Fabrizio Angiulli, Fabio Fassetti, and Simona Nisticò(✉)

DIMES Department, University of Calabria, Rende, Italy
{fabrizio.angiulli,fabio.fassetti,simona.nistico}@dimes.unical.it

Abstract. Among the XAI (eXplainable Artificial Intelligence) techniques, local explanations are witnessing increasing interest due to the user need to trust specific black-box decisions. In this work we explore a novel local explanation approach appliable to any kind of classifier based on generating *masking models*. The idea underlying the method is to learn a transformation of the input leading to a novel instance able to confuse the black-box and simultaneously minimizing dissimilarity with the instance to explain. The transformed instance then highlights the parts of the input that need to be (de-)emphasized and acts as an explanation for the local decision. We clarify differences with existing local explanation methods and experiment our approach on different image classification scenarios, pointing out advantages and peculiarities of the proposal.

Keywords: eXplainable Artificial Intelligence · Local explanations for machine learning · Deep learning

1 Introduction

With the widespread use of complex machine learning models due to the breakthrough they brought in a lot of different fields, a problem that gained more and more importance in the last years is to explain that black-box models. The efforts to respond to this need have produced different methods, also referred to XAI (eXplainable Artificial Intelligence) techniques.

There are at least two opposing families of approaches that can be adopted to solve the above problem [12]. The first family tries to explain the black-box model globally [1,7,8,13,14], that is to say to understand its behaviour when the whole input domain is considered. The second family, instead, aims to explain why the model gives a specific result when the input instance is fixed [2–4,11].

Thus, the latter kind of approach does not require considering the whole input domain. Rather, since they concentrate their analysis on the locality of the sample that needs to be explained, they are referred to as Local Explanation approaches.

Among methods that give local explanations, there is the *Occlusion Analysis* [16], that is a particular type of Perturbation Analysis and in which the result

H. Yin et al. (Eds.): IDEAL 2021, LNCS 13113, pp. 467–475, 2021.
https://doi.org/10.1007/978-3-030-91608-4_46

obtained from the black-box model is explained studying how the occlusion of some patches or some individual features affect the output of the model under consideration.

Two well-known local explanation techniques are LIME [10] and SHAP [6]. LIME requires user intervention to define user interpretable features and also to approximate the local decision by learning a white box model in the neighbourhood of the input instance. Since a linear model is adopted as a white box, it has difficulties capturing the behaviour of models which decision function is not locally near-linear. SHAP approximates features contribution using the Shapley Values of a conditional expectation function of the original model, thus measuring feature importance in terms of contribution in classification result. Both approaches generate neighbour samples by silencing some features of the image. This leads to the possibility that considered samples could be out-of-distribution and that the local model has a behaviour that does respect the black-box to explain.

In this work, we propose a novel local explanation technique, called *Mask Image to Local Explain* (MILE), whose aim is to address the above mentioned limitations. The underlying idea is to learn a transformation of the input that produces a novel instance able to confuse the black-box model by maximizing the probability to belong to a class that differs from the one assigned to the original input instance. Since we exploit the novel instance as an explanation of the black-box decision, we simultaneously require that it minimizes the dissimilarity from the instance to explain.

Further contributions are listed next: to avoid out-of-distribution samples, we directly use neighbourhood training data instead of silencing input instance features; we use a *masking model* instead of a white box model to produce an explanation, so we use a richer model to capture the behaviour of the black-box model; in presence of more than two classes, instead of considering the explanation in a one vs rest fashion as is usually done, we adopt a one vs one approach to explain the outcome associated with the considered sample for one of the other possible classes.

The paper is organized in the following way: in Sect. 2 we describe MILE and the innovative ideas under this proposed technique, and in Sect. 3 we present different experiments conducted to illustrate how the proposed approach behaves.

2 Proposed Approach

In this section we introduce the algorithm MILE. It receives in input the black-box model $f : X \mapsto Y$, together with the instance $x \in X$ to explain and the additional class label $\ell' \in Y$, representing the class we want to exploit to obtain an explanation for f returning the class label $\ell = f(x)$. We always assume that ℓ' is different from ℓ and that, for any class label $\ell \in Y$, the probability $p(y = \ell|x)$ that x belongs to the class ℓ can be obtained from f; we denote such a probability also with $p_f(y = \ell|x)$, to emphasize the dependence from f of its computation.

In order to explain the instance x, our approach is based on determining a transformation $x' = g(x)$ of x leading to a novel instance $x' \in X$ which is able to *confuse* the black-box f. Specifically, our objective is that $p_f(y = \ell'|x')$ is maximized. Moreover, for x' to represent an useful explanation, the transformation g should be *parsimonious*, that is to say we require the dissimilarity between x and x' must be simultaneously minimized.

In this work we investigate transformations g of the form $g(x) = x + m$, with m representing a *mask*, since they are enough expressive to allow to detect parts that need to be added and/or removed to influence the black-box classification outcome. Without loss of generality, from now we assume that the instance domain is $X = [0, 1]^N$, with N any tuple of positive integers, and that the mask $m \in [-1, 1]^N$; the transformed instance $x' = g(x)$ is thus obtained from $x + m$ after clipping each feature in $[0, 1]$. Hence, due to the adopted loss, the considered transformation (de-)emphasizes minimal regions that mostly influence the black-box decision.

To learn the transformation g we build a neural architecture G consisting of the following components: the first component, called *Mask Generator* M, takes charge of computing the mask m; the second component, called *Mask Applier* A, applies the mask m to x to obtain x'; the last component, consists of the *black-box Model* f that receives x'. The architecture is depicted in Fig. 1.

In order to train the above architecture we must find the mask that manages to change the black-box outcome from ℓ to ℓ' when its input is x. Hence, given the original training set $T = \{(x_1, y_1), (x_2, y_2), \ldots, (x_n, y_n)\}$, we build the novel *training set* T_x composed by the n_ℓ nearest examples of the class ℓ and the $n_{\ell'}$ nearest examples of the class ℓ'. All these examples have ℓ' as their true class label in T_x. Notice that having only examples from one class in T_x is not a problem, since the black-box model f is held fixed during the training of G.

The loss employed in training process takes care of the two contrasting objectives above stated: the first term is the main goal and considers the ability of the network to confuse the black-box model, while the second term acts as a regularizes and intends to minimize the changes to the original instance. So the formulation of the loss function is the following:

$$\mathcal{L}(x, x', \ell') = H(p_f(y|x'); \ell') + \lambda \|x - x'\|_2 \tag{1}$$

where $H(p_f(y|x'); \ell')$ is the cross entropy between the distribution $p_f(y|x')$ of the class probabilities associated to x' by f and the "true" distribution assigning x' to the class ℓ', given by the one hot representation of the class, $\|x - x'\|_2$ is the Mean Squared Error obtained when x is approximated by means of x', and λ is an hyper-parameter representing the trade-off between the two loss contributions.

As already stated, the second term of the loss is important because represents a guide for the network to change the classification without totally change the original instance.

In this work, we have investigated the use as a Mask Generator module of a three-layered dense neural network with the Rectified Linear Unit (ReLU) as an activation function for the middle layers. To ensure that the produced mask

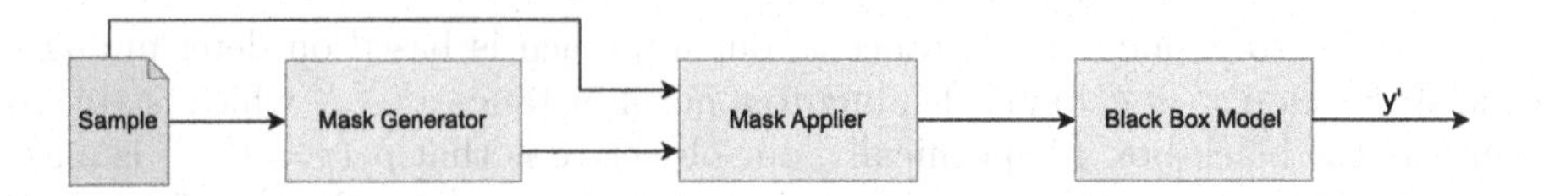

Fig. 1. Architecture of the MILE algorithm.

has each feature value ranging from -1 to 1, we use the Hyperbolic Tangent activation function for the last layer. In the Mask Applier we sum the image and the mask and then we clip values minor of 0 to 0 and values major to 1 to 1 to obtain images with values ranging from 0 to 1.

3 Experiments

This section is devoted to the experimental assessment of the proposed approach to show its effectiveness and compare it with related methods. We have considered two datasets: *MNIST* [5], composed of 60000 black and white images of handwritten digits from 0 to 9, and *Fashion-MNIST* [15], composed of 60000 black and white clothes pictures divided into 10 categories.

For the sake of intelligibility of results, only image data is considered. The rest of the section is organized as follows. First, Sect. 3.1 deals with the capability of MILE in detecting the peculiarities of data on which the base classifier is focused, subsequent Sect. 3.2 compares the explanations produced by MILE and with that provided by Occlusion Analysis.

3.1 Testing the Effectiveness of the Model

The first family of experiments is designed to show the effectiveness of MILE in explaining the behaviour of the classifier. To this aim, a customized version of MNIST is considered. In particular, we take the 12080 samples of class 3 and 9 from the original dataset and add a black rectangle in the bottom left corner to all images belonging to the class 3, as shown in Fig. 2. Thus, the added sign is fully discriminating between classes. In the following, we refer to the images with this modification as *signed.*

The experiment is then designed to test if the model focuses its attention only on the presence of the sign added or if other characteristics are used as discriminating.

The setting used is the following, the classifier is trained with the custom dataset and tested with original images for the class 3, and with signed images of class 9. In this way, the expected behaviour is that the model will not classify the sample on the basis of discriminant pattern related to different digits but exploiting the added sign.

Figure 3 reports the confusion matrix related to this experiments. What emerged from this trial is that the model misclassifies 91% of class 3 samples

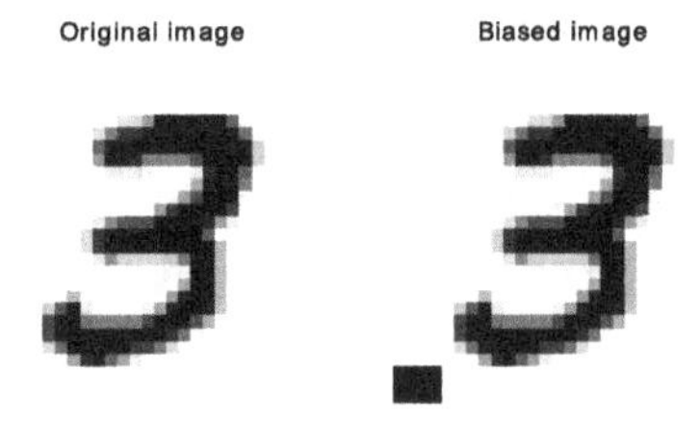

Fig. 2. On the left a original sample of class 3 is reported, on the right the modified version, obtained after adding to the image a black rectangle on the bottom left corner, is shown.

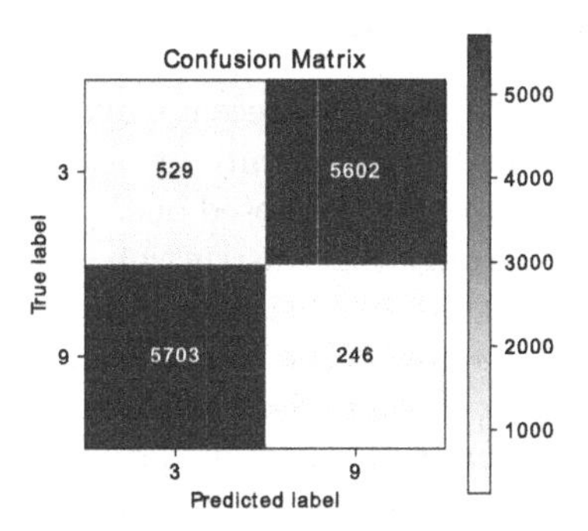

Fig. 3. Confusion matrix obtained classifying samples containing unsigned 3s and samples containing signed 9s.

and 96% of class 9 samples; this result confirms that the model behaves as we expect.

For proposed method effectiveness validation, MILE is executed to explain one of the signed 3 and one of 9 images. For this experiment, the number ns of samples of the same class equals 0 and the number no of samples of the other class equals 3, the number of epochs used for training the explanation model is 3000. The model to explain here is a one-layered dense neural network. Figure 4 reports the output of MILE. In each figure the first image is the sample, the second is the mask produced by the model, and the third one is the image obtained after patch application. In the title of the first and the third image, we report the classification of the black-box model. In the mask, blue pixels have negative values and red pixels positive ones; the intensity of the colour is as high as large is the variation.

The test highlights that the region of the image where there is the hugest variation is the one located in the bottom left corner, the change follows obliviously two different directions, in the first case the mask reduces the values of these pixels, in the second one the mask increases their value.

Thus, MILE detects that it is enough to increase or reduce the intensity of the black rectangle to change sample classification, namely that, as shown above, the classifier is focused on the sign to discriminate between classes.

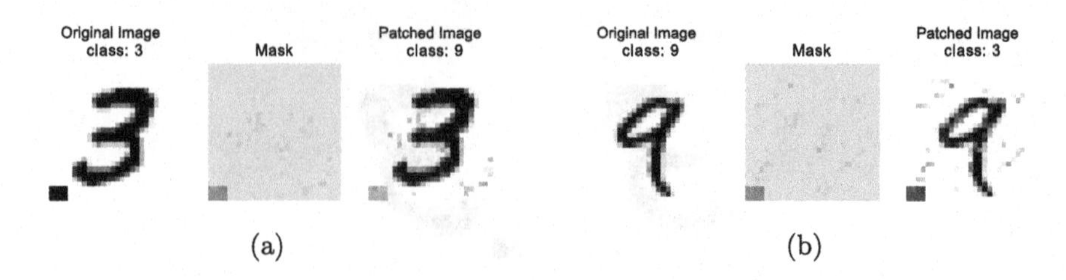

Fig. 4. Experiments about the effectiveness of the MILE algorithm to detect bias.

3.2 Comparison with Occlusion Analysis

In this Section MILE is compared with a recent explanation algorithm, Occlusion Sensitivity [9], which belongs to the family of occlusion analysis's algorithms which are the most related with the proposed one. Occlusion Sensitivity exploits the prediction of a sample through the construction of a sensitivity map, obtained occluding sequentially a patch of fixed size in an image and measuring changes in prediction confidence for a given class.

The experiment concerns a comparison between the two methods from the semantic of provided explanations point of view. To this aim, Fig. 3.2 shows some examples and related explanations obtained by executing the algorithms on MNIST and Fashion MNIST, here the model to explain is a two-layered Convolutional neural network with 128 filters for each layer. In particular, 5(a) concerns the MNIST dataset and 5(b) concerns the Fashion MNIST dataset. In both cases, the first column reports the input samples to be explained, which are a 0, a 3 and a 4 for MNIST, a "t-shirt", a "dress" and a "coat" for Fashion MNIST; the second column reports the explanations provided by the Occlusion Sensitivity; the third column reports the explanations provided by MILE using as target class a 6 for the 0, a 9 for both the 3 and the 4, a "shirt" for the "t-shirt' a "bag" for the "dress" and a "trousers" for the "coat".

The main difference between the two considered approaches is that Occlusion Sensitivity considers only one class while MILE considers the class of prediction and a second class that is the target class. Namely, given a sample s of c_0, Occlusion Sensitivity provides explanations for which s is classified as c_0 while MILE, given also a target class c_1, provides explanations for which s is in class c_0 and not in c_1.

As shown in Fig. 5(a), as far as the model for MNIST dataset classification is concerned, all three explanations pixels modified by MILE belong to areas with high sensitivity detected by Occlusion Sensitivity, but MILE, considering a couple of classes, gives more detailed information. In particular, it highlights the changes needed to fault the classifier in detecting the sample as belonging to the target class, changing, as few as possible, the input image.

For example, the image in the first row is a 0 and not a 6, because it is a closed "circle" with an empty centre or, in other words, the 0 would be classified as a 6, if it were modified to "open" on the top right part and to "close" on the bottom right part. The 3 in the second row of Fig. 5(a) belongs to that class

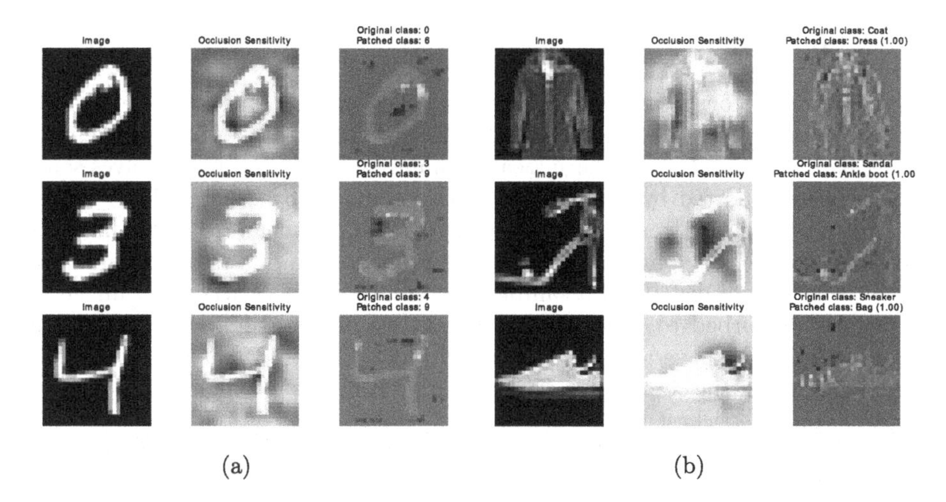

Fig. 5. Comparison between the explanations obtained from Occlusion Sensitivity and masks produced by proposed approach. In 5(a) we consider MNIST dataset, in particular showing explanations with respect to class 0, 3 and 4 for the Occlusion Analysis, and from MILE for class 0 with respect to class 6, 3 with respect to class 9 and 4 with respect to class 9. In 5(b) we consider fashion MNIST dataset considering for the Occlusion Analysis respectively classes "t-shirt", "dress" and "coat", and for MILE "t-shirt" with respect to "shirt", "dress" with respect to "bag" and "coat" with respect to "trousers".

because the upper circle of the hand-written digit is "open", or in other words, the 3 would be classified as 9 if it were modified to close the upper circle. The sample of the last column belong to 4 class because it is "open" on the upper part or, in other words, the 4 would be classified as 9, if a line were added to "close" the upper part.

Different is the situation in Fig. 5(b), at a first glance the behaviour of the model under analysis is not as good as that of the model previously considered and the two different approaches seem to bring to different conclusions.

As for the occlusion analysis, it seems to highlight that the model seems to not work badly since the model has high sensitivity in a region that can effectively reasonably change the classification and for this reason, despite the explanation given by Occlusion Sensitivity doesn't depict an excellent model, it results to be fair for the user.

Conversely, the situation highlighted by masks produced by MILE is different. Indeed, the three masks considered seem to change few peculiarities of images and to add noise. This is semantically interesting since this shows a potential weakness of the model for adversarial attacks which is not shown by explanation produced by occlusion analysis. In particular, the analysis conducted by MILE indicates that changing some pixels of the input images forces the classifier in identifying the sample as belonging to the target class even if it is not similar to other objects of the class.

4 Conclusions

In this work we considered the problem of explaining black-box classifier predictions. We presented a technique, called MILE, that approaches this problem by using *masking models* to produce local explanations. This technique solves some of the issues reported by local explanation techniques by using a richer model to capture black-box local behaviour. Moreover, when there are more than two possible classes, the approach faces up the problem in a one-vs-one fashion. To avoid using out-of-distribution samples, as usually happens when the image to explain is perturbed to generate a neighbor, we direct exploit neighbourhood training data. We have assessed the capacity of MILE to detect biased models, and, in addition, the comparison with Sensitive Analysis has highlighted how its explanations are richest and have superior quality.

As future work, we intend to explore other strategies to generate data neighbourhood avoiding out-of-distribution samples, to test other architectures for masking models and to take into account different families of transformations. Moreover, we also plan to consider further datasets and other competitors, and to investigate the application of our approach to other data types.

References

1. Altmann, A., Toloşi, L., Sander, O., Lengauer, T.: Permutation importance: a corrected feature importance measure. Bioinformatics **26**(10), 1340–1347 (2010)
2. Dabkowski, P., Gal, Y.: Real time image saliency for black box classifiers (2017). arXiv preprint arXiv:1705.07857
3. Du, M., Liu, N., Song, Q., Hu, X.: Towards explanation of dnn-based prediction with guided feature inversion. In: Proceedings of the 24th ACM SIGKDD International Conference on Knowledge Discovery & Data Mining, pp. 1358–1367 (2018)
4. Du, M., Liu, N., Yang, F., Ji, S., Hu, X.: On attribution of recurrent neural network predictions via additive decomposition. In: The World Wide Web Conference, pp. 383–393 (2019)
5. LeCun, Y., Cortes, C., Burges, C.: Mnist handwritten digit database (2010)
6. Lundberg, S., Lee, S.I.: A unified approach to interpreting model predictions (2017). arXiv preprint arXiv:1705.07874
7. Montavon, G., Binder, A., Lapuschkin, S., Samek, W., Müller, K.R.: Layer-wise relevance propagation: an overview. In: Explainable AI: Interpreting, Explaining and Visualizing Deep Learning, pp. 193–209 (2019)
8. Qi, Z., Khorram, S., Li, F.: Visualizing deep networks by optimizing with integrated gradients. In: CVPR Workshops, vol. 2 (2019)
9. Rajaraman, S., et al.: Understanding the learned behavior of customized convolutional neural networks toward malaria parasite detection in thin blood smear images. J. Med. Imaging **5**(3), 034501 (2018)
10. Ribeiro, M.T., Singh, S., Guestrin, C.: "Why should i trust you?" explaining the predictions of any classifier. In: Proceedings of the 22nd ACM SIGKDD International Conference on Knowledge Discovery and Data Mining, pp. 1135–1144 (2016)
11. Ribeiro, M.T., Singh, S., Guestrin, C.: Anchors: high-precision model-agnostic explanations. In: Proceedings of the AAAI Conference on Artificial Intelligence, vol. 32 (2018)

12. Samek, W., Montavon, G., Lapuschkin, S., Anders, C.J., Müller, K.R.: Explaining deep neural networks and beyond: a review of methods and applications. Proc. IEEE **109**(3), 247–278 (2021)
13. Simonyan, K., Vedaldi, A., Zisserman, A.: Deep inside convolutional networks: Visualising image classification models and saliency maps (2013). arXiv preprint arXiv:1312.6034
14. Smilkov, D., Thorat, N., Kim, B., Viégas, F., Wattenberg, M.: Smoothgrad: removing noise by adding noise (2017)
15. Xiao, H., Rasul, K., Vollgraf, R.: Fashion-mnist: a novel image dataset for benchmarking machine learning algorithms (2017)
16. Zeiler, Matthew D.., Fergus, Rob: Visualizing and understanding convolutional networks. In: Fleet, David, Pajdla, Tomas, Schiele, Bernt, Tuytelaars, Tinne (eds.) ECCV 2014. LNCS, vol. 8689, pp. 818–833. Springer, Cham (2014). https://doi.org/10.1007/978-3-319-10590-1_53

Wind Turbine Modelling Based on Neural Networks: A First Approach

J. Enrique Sierra-García[1](✉) and Matilde Santos[2]

[1] Department of Electromechanical Engineering, University of Burgos, 09006 Burgos, Spain
jesierra@ubu.es

[2] Institute of Knowledge Technology, Complutense University of Madrid, 28040 Madrid, Spain
msantos@ucm.es

Abstract. Wind turbine modelling is a complex task due to the rotation of the spinning rotor, the aerodynamic of the blades, and the fluid viscosity. They are normally studied by advanced flow dynamics simulators, complex aerodynamics theory, and blade element methods. There are some wind turbine modelling software such as FAST that consider all these phenomena. However, the complexity of these models makes it impossible to integrate them into program logic controllers. Neural networks can be powerful tools to help simplify these complex models and embedded them into the wind turbine controllers. In this work recurrent neural networks have been successfully used to obtain a low-complexity control-oriented computational models of a wind turbine. Graphical and numerical results prove the validity of the approach.

Keywords: Modelling · Wind turbines · Neural networks · NARX · FAST

1 Introduction

Wind turbines (WT) are one of the most widely renewable energy used. This clean energy has been proved essential to the sustainability of the current and future demand of energy worldwide. Although we could say that wind energy, mainly onshore, is a mature technology, there are still many engineering problems related to wind energy that must be addressed [1, 2].

For each turbine, some operating targets are defined in order to obtain the best possible performance. These are accomplished thanks to the control strategy defined to generate energy from the wind. The WT control system is designed to get the highest efficiency and at the same time, to ensure safe operation under all wind conditions. This may be critical for floating offshore wind turbines (FOWT) as it has been proved that the control can affect the stability of the floating device [3]. There are different actuating mechanisms, namely pitch (blade angle), yaw (yaw angle) and generator torque control systems.

To design controllers for any system is crucial to have good models of the system dynamic. This allows us to test the performance of the controller and integrate the model in the controller when a model-based control strategy is used. For instance, model

H. Yin et al. (Eds.): IDEAL 2021, LNCS 13113, pp. 476–484, 2021.
https://doi.org/10.1007/978-3-030-91608-4_47

predictive controllers (MPC) exploit the knowledge of the model to control the system in a control horizon [4].

Wind turbine modelling is a complex task because its operation is based on physical phenomena as the rotation of the spinning rotor, the aerodynamic of the blades and the fluid viscosity, and it also includes random resources such as wind. They are normally studied by advanced flow dynamics, aerodynamics theory, and blade element methods. There are some wind turbine modelling software such as FAST (by NREL) that consider all these complex phenomena and are widely used to model the behavior of wind turbines [5]. However, its complexity makes unfeasible its integration in program logic controllers (PLCs). In addition, the time required to obtain the computational solution is large and thus it is not easy to include them in real time control loops.

Neural networks can be used to developed models that can be embedded into the wind turbine controllers. Long-short term memory (LSTM) neural networks and non-linear autoregressive neural networks with exogenous inputs (NARX) are examples of architectures of recurrent neural networks that have been successfully used to model time series and processes. Indeed, LTSM are widely applied in deep learning applications to model time series and even WT, but they usually need real data [6]. However, here we focus on the generation of low-complexity control-oriented models. Thus, we have selected the NARX architecture as it has a lower computational load. As it will be shown, it is possible to obtain low complexity neural network models that fit well the output of the software FAST nonlinear and realistic models. Among all the output signals that FAST generates, the rotor angular velocity and the output power has been considered as they are the standard controlled variables in wind turbine pitch control. These signals vary with the blade angle and the wind speed; thus, they are the inputs signals of the neural networks. These lower complexity control-oriented neural network models can be easily integrated in control strategies.

The WT modelling and control have been addressed in the literature using different classical and intelligent techniques [6–8]. For example, in [9] a general methodology for the identification of reduced dynamic models of barge-type FOWTs is presented. Most of the works found in the literature use historical data [10]. Some of them model some signals, such as the wind turbine power curve for performance monitoring [11]. This work develops highly accurate non-parametric models for wind turbines based on radial basis function networks. Similarly, in [12] a deep convolutional neural network is built and is trained for defect identification method of wind turbine blades. And in [13], where the authors propose different one-class intelligent techniques to perform anomaly detection on a bicomponent mixing system used on the wind generator manufacturing. Some of these models are thought to improve the wind turbine control, such as in [14, 15].

In [16], neural networks are shown to be effective surrogate aerodynamic models of wind turbine blades. In this work, a convolutional neural network is used and it computes faster than BEM and significantly faster than CFD-based aerodynamic models. A generic power curve model dependent on main turbine characteristics is proposed in [17], where the main environmental parameters are considered. But the works where NN are used to model the wind turbine behavior are scarce.

Also, neural networks have been previously used to estimate the wind speed and the disturbances of wind turbines. In the paper by Asghar and Liu [18], an effective wind speed estimator is proposed using an adaptive neuro-fuzzy algorithm (ANFIS). In a similar paper by the same authors, [19], a hybrid intelligent learning adaptive neuro-fuzzy inference system is proposed to estimate the Weibull wind speed probability distribution. Neural networks have been also successfully used to model and control other engineering systems [20, 21].

To the best of our knowledge, there is no previous works focused on the reduction of model complexity of FAST wind turbine models using neural networks that can be later integrated into PLCs. Thus, in this work low-complexity control-oriented neural network models of the output of a wind turbine simulated by FAST have been obtained. NARX neural networks have been used as they have been proved to fit well the output signal and have low computational complexity. The output signals are generated with different amplitude steps of the wind speed and the pitch angle reference. Graphical and numeric results prove the validity of the approach.

The rest of the paper is organized as follows. Section 2 describes the wind turbine model. The modelling based on neural networks is presented in Sect. 3. Results are discussed in Sect. 4. The work ends with the conclusions and future works.

2 Description of the System

The wind power that can be harvested by a wind turbine is related to the wind speed and the area swept by the blades. This power is commonly represented by Eq. (1):

$$P_m = 0.5\rho A v^3 C_p(\theta, \lambda), \tag{1}$$

Where ρ is the air density (Kg/m^3), v is the wind speed (m/s), and A is the area swept by the blades (m^2). The term $C_p(\theta, \lambda)$ denotes the power coefficient and it represents the fraction of available power in the wind that it is effectively transformed by the wind turbine. The power coefficient has a maximum theoretical limit, the Betz's limit (59.3%).

However, this theoretical limit is never reached due to, among other reasons, the rotation of the spinning rotor, the number of blades, the aerodynamic of the blades and the fluid viscosity [12]. These physical effects are complex and nonlinear. The software FAST (by NREL) includes all these real conditions. Just to have an idea of the complexity of the models, Eq. (2) shows an approximation of the C_p coefficient [22].

$$C_p = \frac{2}{\lambda R}\int_H^R \frac{\sigma' \lambda_r^2 C_L}{\sin^2(\theta)}\left[\frac{\lambda_r^2 \sin^2(\theta)}{\cos^2(\theta) - \frac{\sigma' C_L \cos^2(\theta)}{2F} + \frac{\sigma^2 C_L^2}{16F^2}}\right]\left[\sin(\theta) - \frac{C_D \cos(\theta)}{C_L}\right] dr, \tag{2}$$

$$\lambda = \frac{w \cdot R}{v}, \quad \lambda_r = \frac{w \cdot r}{v}, \tag{3}$$

Where θ is the pitch angle (rad), w is the angular rotor velocity (rad/s), v is the wind speed (m/s), λ is the tip-speed ratio, λ_r is the local speed ratio, R is the radius of the blade (m), C_L is the lift coefficient, C_D is the drag coefficient, F is the tip loss factor, and σ' is the local solidity. A further explanation of (1–2) can be found in [22].

So it is not easy to compute and run these models in program logic controllers (PLCs). Here we proposed a low computational model based on neural.

Figures 1 and 2 show some examples of the input (left) and output (right) signals obtained with FAST that we will work with. The input signals, wind speed and pitch angle, have been designed to extract relevant information from the models. The output signals have been obtained by the software FAST with those input signals.

To identify the models, two different experiments have been carried out. In the first one the wind speed is constant and a train of pulses with different amplitude is introduced. The results are shown in Fig. 1. In the second experiment, the pitch is set to 0°, and the wind speed is a train of pulses. The results are shown in Fig. 2.

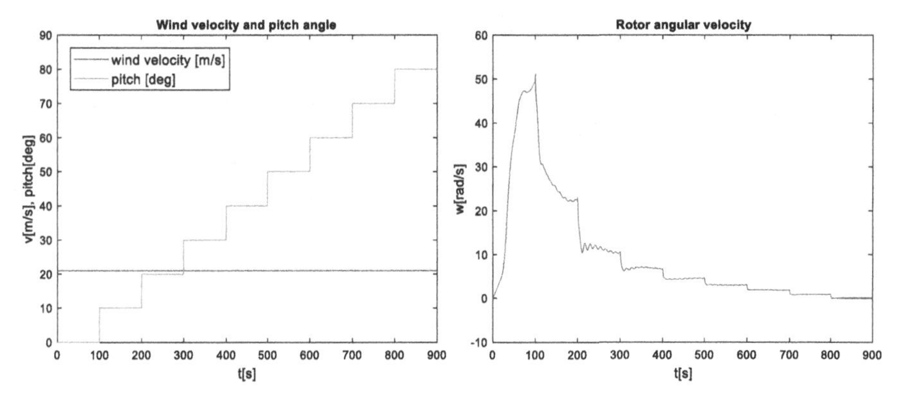

Fig. 1. Wind speed and pitch angle (left) and rotor angular velocity (right) for constant wind and a train of steps in the pitch input.

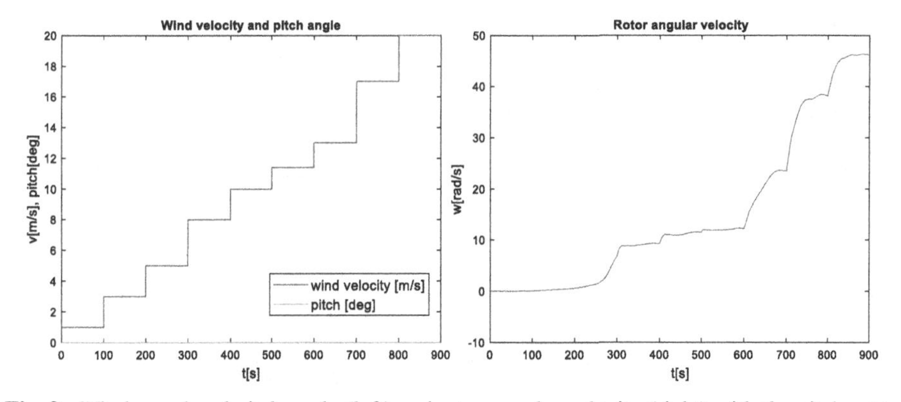

Fig. 2. Wind speed and pitch angle (left) and rotor angular velocity (right) with the pitch set to 0° and a train of steps in the wind.

In these figures it is possible to see how the rotor angular velocity decreases with the pitch and increases with the wind speed. The pitch angle regulates the area of the blade that faces the wind. When the pitch is 0° the area of this surface is maximum, hence the captured power is also maximum and the rotor angular velocity increases. When the pitch increases the blade acts as a brake and the angular velocity tends to decrease. The relation between the pitch and the decreasing rate depends on the wind turbine parameters.

As expected, the higher the wind speed the larger the rotor angular velocity. However, this growth is not linear, that is, different wind speeds produce different rates. In Fig. 2 it is also possible to see that the wind turbine needs a minimum wind speed to start to rotate due to the rotational inertia (cut-in speed).

3 Neural Networks Identification

Supervised neural networks can be successfully applied to model complex systems. In this study, recurrent multilayer feed-forward networks have been selected due to their well-known properties to approximate time series. These networks consist of several layers of neurons connected feedforward. The activation function can be different in each layer, however normally all neurons in the neural network have the same activation function. Each neuron is connected to all neurons of the next layer. The connection between the i-neuron and the j-neuron that belong to two consecutive layers is characterized by the weight coefficient w_{ij}. In addition, each neuron also has a threshold coefficient b. This way the output of the j-neuron in the k-layer is given by:

$$x_{kj} = f\left(b_j + \sum\nolimits_{i=1}^{N_{k-1}} w_{ij} x_i\right), \tag{4}$$

where f is the activation function and N_{k-1} is the number of neurons in the $(k-1)$ layer.

In this neural network the input layer has some delays. The output of these delay neurons is connected to the inputs of the first hidden layer. This way, at time t, the neurons of the first layer receive the previous value of the inputs (at time $t-1$, $t-2$, $t-3$, etc.). In addition, there is a connection between the output of the neural network and one of its inputs. This way the NN output is also a function of the previous values of the inputs and outputs signals. This feedback loop makes the neural network to be recurrent. Finally, in the output layer the outputs of the neurons of the previous layer are weighted. The architecture of this neural network configuration is shown in Fig. 3.

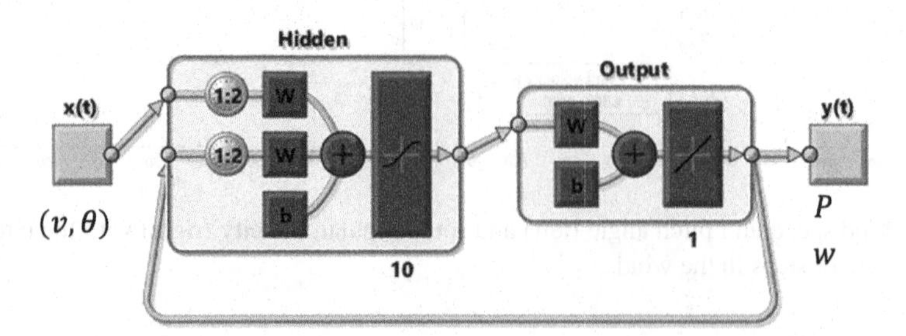

Fig. 3. Structure of the nonlinear autoregressive with exogenous inputs (NARX) neural network.

The Levenberg-Marquardt algorithm was used for training and the performance indicator was the mean squared error (MSE). The hyperparameters of the neural networks were selected by trial and error after several simulations. In almost all cases the configuration of the NN is a hidden layer of 10 neurons, and two delays in the inputs and the outputs. The only exception was the model of the output power, when the pitch angle is varied; in this case the hidden layers have 9 neurons, the inputs have 2 delays and the output 4. In our case, the two inputs have the same delay, although the architecture allows it to have a different delay in each input.

The neural networks were trained with the signals of Figs. 1 and 2. Two NARX neural networks have been implemented. One of them generates the rotor angular velocity and the other is used to model the output power. Both neural networks receive as inputs the pitch reference and the wind speed. The transfer function in the neurons of the hidden layer is tansig and in the output layer this function is linear.

During the training, the neural networks work in open loop and the expected output values are used as inputs and as outputs. Once the neural networks have been trained, the feedback loop is closed and the previous outputs generated by the neural network are used as inputs of the final neural network. This way the neural network uses past output values as input information.

4 Discussion of the Results

Simulation results of the experiments varying the wind speed and varying the pitch angle are shown and discussed. All these results have been obtained with the Deep Learning Toolbox of Matlab.

Figure 4 shows the rotor angular velocity and Fig. 5 the output power of the wind turbine. In both cases the blue line represents the output of FAST and the red line shows the output of the neural network. The figures on the left represent the output signals obtained when the pitch angle is changed and the figures on the right represent the output signals obtained when the wind speed is varied. The input signals for these experiments are the ones shown in Figs. 1 and 2.

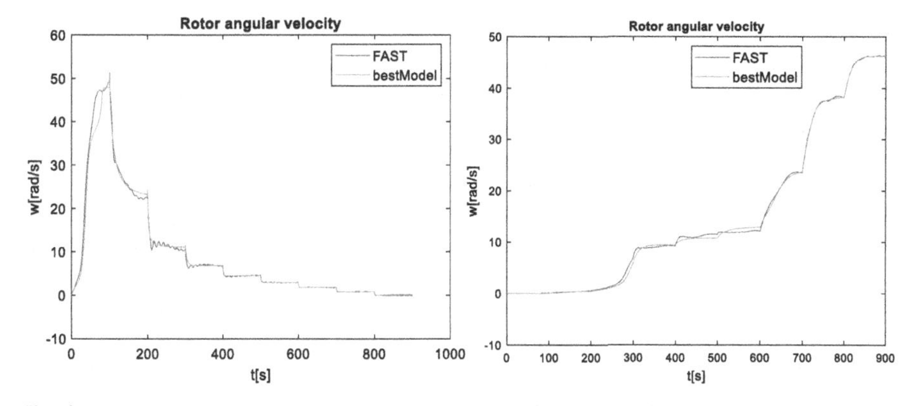

Fig. 4. Rotor angular velocity when the pitch angle (left) and the wind speed (right) are varied.

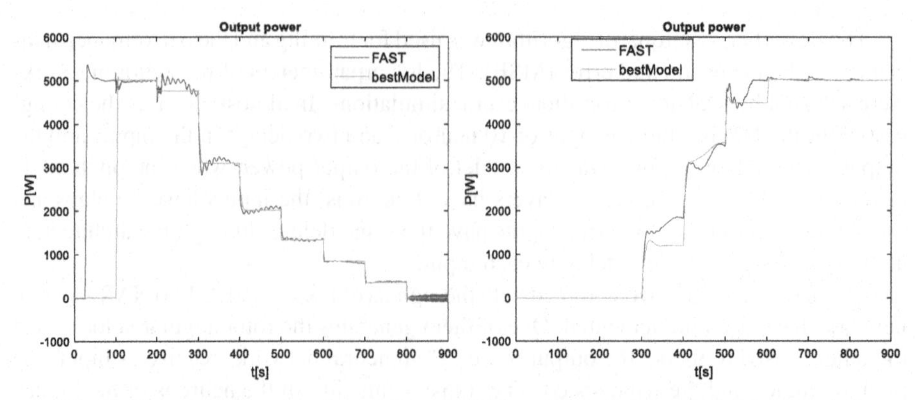

Fig. 5. Output power when the pitch angle (left) and the wind speed (right) are varied.

It is possible to observe how the model outputs fit the FAST outputs reasonably well. The signal with smallest error is the rotor angular velocity when the wind speed is varied. This can be explained because this signal is the smoothest one. The rotor angular velocity is very well fitted from 0 up to 90 s, and from 100 s to 900 s. On the other hand, the output power model when the pitch angle is varied has the biggest error. The neural model obtained has a delay to approximate this large step in the output power. This may be because the output layer of the neural network is linear.

In addition to the graphical results, numeric values are obtained to quantify the accuracy of the models. Two different metrics have been evaluated: the weighted mean absolute error (WMAE) and the root mean squared error (RMSE), as shown in Table 1. As the results vary with the initialization of the neural networks, 20 neural networks are generated in parallel, and the best model is selected. This process is repeated 5 times, the mean and median (med) values are shown in Table 1. H is the number of neurons, D the delay, and O the output signal. These results confirm what it can be observed in the

Table 1. Accuracy of the neural models

D	H	O	Pitch is varied				Wind is varied			
			Mean WMAE	Med WMAE	Mean RMSE	Med RMSE	Mean WMAE	Med WMAE	Mean RMSE	Med RMSE
2	10	*w*	0.1896	0.1873	4.1346	4.1433	0.1798	0.1901	0.0602	0.0934
2	10	*P*	0.2501	0.2732	143.40	151.3	0.0651	0.0754	237.64	257.26
2	5	*w*	0.3752	0.3423	0.8393	0.7746	0.4216	0.4749	0.4203	0.4298
2	5	*P*	0.3551	0.3552	259.74	282.66	0.2198	0.0748	931.18	1122.4
0	10	*w*	0.2315	0.2001	0.3533	0.3227	0.1533	0.1912	0.0531	0.0665
0	10	*P*	0.3115	0.2721	268.51	268.70	0.0833	0.0413	174.07	193.13
0	5	*w*	0.2645	0.2686	0.7766	0.8067	0.2681	0.2694	0.4321	0.1764
0	5	*P*	0.2682	0.2651	276.34	280.53	0.0723	0.0746	254.23	260.31

figures, the most accurate model is the rotor angular velocity when the wind is varied, and the worst one is the output power when the pitch is varied. It is a common practice to consider that a model is accurate enough when the WMAE is below or close to 0.25. In our case all the models are close or below this value, thus we can consider the neural models gives results as good as the FAST model.

Another interesting result is the fact that the RMSE of w and P can not be directly compared as the ranges of these values are very different. The WMAE gives more importance to errors in the model when the target values are low. For this reason the WMAE are similar when the rotor angular velocity is estimated but the RMSE are very different.

5 Conclusions and Future Works

In this work low-complexity control-oriented neural network models of some variables of a wind turbine have been obtained. Specifically, the output power and the rotor angular velocity of the wind turbine are modelled as they are the most relevant signals of the WT. However, any other output signals of the wind turbine could have been obtained by this approach.

Recurrent neural networks have been used to develop the low computational models of the wind system. Models with as few as 10 neurons fit well the outputs obtained with the more realistic but nonlinear and more complex FAST software models. Graphical and numerical results prove the validity of the approach. This low-complexity models can be integrated in control strategies.

As future work we can mention the integration of these models in model-based control strategies and the estimation of other output signals as the tower top displacement. In addition, low computational models of floating wind turbine could be developed.

Acknowledgment. This work was partially supported by the Spanish Ministry of Science, Innovation and Universities under Project number RTI2018-094902-B-C21.

References

1. Mikati, M., Santos, M., Armenta, C.: Electric grid dependence on the configuration of a small-scale wind and solar power hybrid system. Renew. Energy **57**, 587–593 (2013)
2. Gomes, I.L., Melício, R., Mendes, V.M., Pousinho, H.M.: Wind power with energy storage arbitrage in day-ahead market by a stochastic MILP approach. Logic J. IGPL **28**(4), 570–582 (2020)
3. Galán-Lavado, A., Santos, M.: Analysis of the effects of the location of passive control devices on the platform of a floating wind turbine. Energies **14**(10), 2850 (2021)
4. Chen, Y.L., Liu, Y.P., Sun, X.F.: The active frequency control strategy of the wind power based on model predictive control. Complexity (2021)
5. FAST 2021. https://www.nrel.gov/wind/nwtc/fast.html. Accessed 30 July 2021
6. Delgado, I., Fahim, M.: Wind turbine data analysis and LSTM-based prediction in SCADA system. Energies **14**(1), 125 (2021)

7. Tomás-Rodríguez, M., Santos, M.: Modelling and control of floating offshore wind turbines. Rev. Iberoam. Automática Inf. Ind. **16**, 381–390 (2019)
8. Sierra-Garcia, J.E., Santos, M.: Redes neuronales y aprendizaje por refuerzo en el control de turbinas eólicas. Revista Iberoamericana de Automática e Informática industrial **18**(4), (2021)
9. Villoslada, D., Santos, M., Tomás-Rodríguez, M.: General methodology for the identification of reduced dynamic models of barge-type floating wind turbines. Energies **14**(13), 3902 (2021)
10. McKinnon, C., Turnbull, A., Koukoura, S., Carroll, J., McDonald, A.: Effect of time history on normal behaviour modelling using SCADA data to predict wind turbine failures. Energies **13**(18), 4745 (2020)
11. Karamichailidou, D., Kaloutsa, V., Alexandridis, A.: Wind turbine power curve modeling using radial basis function neural networks and tabu search. Renew. Energy **163**, 2137–2152 (2021)
12. Yu, Y., Cao, H., Yan, X., Wang, T., Ge, S.S.: Defect identification of wind turbine blades based on defect semantic features with transfer feature extractor. Neurocomputing **376**, 1–9 (2020)
13. Jove, E., Casteleiro-Roca, J., Quintián, H., Méndez-Pérez, J.A., Calvo-Rolle, J.L.: Detección de anomalías basada en técnicas inteligentes de una planta de obtención de material bicomponente empleado en la fabricación de palas de aerogenerador. Revista Iberoamericana de Automática e Informática Industrial **17**(1), 84–93 (2020)
14. Hur, S.H., Reddy, Y.S.: Neural network-based cost-effective estimation of useful variables to improve wind turbine control. Appl. Sci. **11**(12), 566 (2021)
15. Sierra-García, J.E., Santos, M.: Improving wind turbine pitch control by effective wind neuro-estimators. IEEE Access **9**, 10413–10425 (2021)
16. Lalonde, E.R., Vischschraper, B., Bitsuamlak, G., Dai, K.: Comparison of neural network types and architectures for generating a surrogate aerodynamic wind turbine blade model. J. Wind Eng. Ind. Aerodyn. **216**, 104696 (2021)
17. Saint-Drenan, Y.M., et al.: A parametric model for wind turbine power curves incorporating environmental conditions. Renew. Energy **157**, 754–768 (2020)
18. Asghar, A.B., Liu, X.: Adaptive neuro-fuzzy algorithm to estimate effective wind speed and optimal rotor speed for variable-speed wind turbine. Neurocomputing **272**, 495–504 (2018)
19. Asghar, A.B., Liu, X.: Estimation of wind speed probability distribution and wind energy potential using adaptive neuro-fuzzy methodology. Neurocomputing **287**, 58–67 (2018)
20. Sierra, J.E., Santos, M.: Modelling engineering systems using analytical and neural techniques: hybridization. Neurocomputing **271**, 70–83 (2018)
21. Sierra-García, J.E., Santos, M.: Switched learning adaptive neuro-control strategy. Neurocomputing **452**, 450–464 (2021)
22. Jonkman, J.M.: Modeling of the UAE Wind Turbine for Refinement of FAST AD (No. NREL/TP-500-34755). National Renewable Energy Lab., Golden, CO (US) (2003)

Multi-Attribute Forecast of the Price in the Iberian Electricity Market

Gonçalo Peres[1], Antonio J. Tallón-Ballesteros[2(✉)], and Luís Cavique[1]

[1] Universidade Aberta, Lisbon, Portugal
1800301@estudante.uab.pt, luis.cavique@uab.pt
[2] University of Huelva, Huelva, Spain
antonio.tallon.diesia@zimbra.uhu.es

Abstract. Electricity has been acquiring a more significant presence in our lives, and it is estimated that the future will be increasingly electric. Nowadays, we have access to enormous amounts of data that do not have much-added value if they cannot support decision-making or plan systems in advance and correctly. Forecasts are vital tools to support decision-making. We believe it is possible to resort to open data available on the Internet to make electricity price forecasts that - decision-makers can use in the sector. In this work, we study the multi-attribute hourly forecast of the electricity price in MIBEL (Iberian electricity market) for the 24 h of the following day, using open data. The realization of the multi-attribute predictions fell on the TIM ('Tangent Information Modeler') tool with AutoML ('Auto Machine Learning') capabilities. The TOPSIS ('technique for order of preference by similarity to ideal solution') decision support technique was used to analyze the results.

Keywords: Economic prediction · Forecasting · Iberian electricity market (MIBEL) · Auto machine learning · Multi-attribute decision

1 Introduction

The uncertainty associated with many temporal phenomena does not allow the exact knowledge of their behavior in the future, which leads us to have to make forecasts. Forecasts play a fundamental role in the planning, decision-making, and control process in any organization.

Electricity is considered an exceptional commodity because although there have been several advances regarding batteries' performance, electricity is primarily a non-storable commodity. In this sense, the price of electricity presents unique characteristics that make forecasting a challenging task.

This work aims to develop models that will allow the multi-attribute forecasting of the electricity price in the Iberian Energy Market (MIBEL) daily market for the 24 h of the following day (multi-step), using an autoML tool.

The remaining of the paper is organized as follows. Section 2 describes the data enrichment of the MIBEL. Section 3 presents the experimental planning. In Sect. 4, the AutoML tool is detailed. In Sect. 5, the computational results are discussed. Finally, the conclusions are drawn in Sect. 6.

H. Yin et al. (Eds.): IDEAL 2021, LNCS 13113, pp. 485–492, 2021.
https://doi.org/10.1007/978-3-030-91608-4_48

2 Data Enrichment

Data enrichment enhances collected data with relevant context obtained from additional sources [3]. This section discusses some properties of the dataset to be used, presented in Table 1.

Table 1. Explanatory variables.

Variable	Explanatory	Description	Range and unit
V1	Chronological	Year	2010–2019
V2	Chronological	Month	1–12
V3	Chronological	Day	1–31
V4	Chronological	Hour	1–24
V5	Chronological	Day of the week	1–7 (1 is Monday and 7 is Sunday)
V6	Chronological	Holiday	0 or 1 (0 is not a holiday, 1 is a holiday)
V7	Prices	Hourly price	€/MWh
V8	Demand	Real demand	MW
V9	Demand	Forecast demand	MW
V10	Demand	Scheduled demand	MW
V11	Production	Wind	MW
V12	Production	Nuclear	MW
V13	Production	Coal	MW
V14	Production	Combined cycle	MW
V15	Production	Hydraulics	MW
V16	Production	International exchanges	MW
V17	Production	Solar	MW
V18	Weather	Earth temperature	K
V19	Weather	Earth or water temperature	K
V20	Weather	10 m u-component wind	ms^{-1}
V21	Weather	10 m v-component wind	ms^{-1}
V22	Weather	Solar radiation	Jm^{-2}
V23	Weather	Thermal radiation	Jm^{-2}
V24	Weather	Total cloud cover	0–1
V25	Weather	Total precipitation	mm
V26	Weather	Total sky radiation	$J\,m^{-2}$

For the development of the hourly price forecast models, we consider different types of explanatory variables, precisely:

1. Chronological data: Year, Month, Day, Hour, Day of the Week, Holiday[1] (in Spain).
2. Hourly electricity market prices (MIBEL), available from the market operator (OMIE) or the Portuguese TSO's Energy Market Information System.[2]
3. Demand data: Actual Demand, Forecast Demand, and Programmed Demand. These data come from the Spanish TSO, REE.[3]
4. Production or generation system data: Wind, Nuclear, Coal, Combined Cycle, Hydro, Solar, International Exchanges. These data were obtained from the Spanish TSO, the REE.[4] Since the production variables changed from 2015-04-30 to 2015-05-01, we consider that Solar encompasses (Solar Photovoltaic and Solar Thermal).
5. Weather or climate data: ERA-5, from the Climate Data Store[5] in Iberia. The Iberian Peninsula is considered the region delimited by the following geographic coordinates: North: 44°, West: −9.6°, South: 36°, East: 3.5°.

3 Experimental Planning

The models proposed allow performing multipoint forecasting of daily electricity prices in MIBEL. Such models may be beneficial for day-to-day market operations. To perform the forecasts, we will use TIM, an AutoML based on Information Geometry with a strong component of visual tool [2]. The experimental planning (Table 2) was inspired by Cláudio Monteiro's articles [5, 6].

Unlike in the above articles, where one-time intervals and one type of split in training-test were considered, in this paper, we consider three-time intervals and multiple splits in training-test, more specifically three: 70–30, 80–20, and 90–10. Apart from that, various metrics will be obtained: runtime (in seconds), Mean Absolute Error (MAE), Mean Absolute Percentage Error (MAPE), and Root Mean Square Error (RMSE).

Throughout this work, a total of 225 models will be developed, with variations related to the time intervals to be considered (75 models for each time interval), the percentages of the training and test samples (25 models for each train-test division), and the input variables (Table 2).

In this study, the following time intervals will be considered:

- Interval A. January 1, 2019 (00:00) to December 31, 2019 (23:00).
- Interval B. January 1, 2018 (00:00) to December 31, 2019 (23:00).
- Interval C. January 1, 2010 (00:00) to December 31, 2019 (23:00).

---

[1] The holidays in Spain were created based on this script: https://github.com/goncaloperes/TimeSeries/blob/master/IsItHoliday/Spain/SpainBusinessCalendar.py.

[2] https://www.mercado.ren.pt/PT/Electr/InfoMercado/InfOp/MercOmel/Paginas/Precos.aspx.

[3] https://demanda.ree.es/visiona/peninsula/demanda/.

[4] https://demanda.ree.es/visiona/peninsula/demanda/tablas/2.

[5] https://cds.climate.copernicus.eu/cdsapp#!/dataset/reanalysis-era5-single-levels?tab=form.

A total of 225 (25 × 3 × 3) forecast models used to predict the price of electricity in MIBEL for the 24 h following the last point in the series (multi-step forecast) were created, 75 models were considered for each time interval (25 for each training-test division).

Table 2. Models to be implemented and their input variables (M25 × V26).

Variable	M1	M2	M3	M4	M5	M6	M7	M8	M9	M10	M11	M12	M13	M14	M15	M16	M17	M18	M19	M20	M21	M22	M23	M24	M25
V1	•	•	•	•	•	•	•	•	•	•	•	•	•	•	•	•	•	•	•	•	•	•	•	•	•
V2	•	•	•	•	•	•	•	•	•	•	•	•	•	•	•	•	•	•	•	•	•	•	•	•	•
V3	•	•	•	•	•	•	•	•	•	•	•	•	•	•	•	•	•	•	•	•	•	•	•	•	•
V4	•	•	•	•	•	•	•	•	•	•	•	•	•	•	•	•	•	•	•	•	•	•	•	•	•
V5		•	•																						•
V6			•																						•
V7	•	•	•	•	•	•	•	•	•	•	•	•	•	•	•	•	•	•	•	•	•	•	•	•	•
V8				•																					•
V9					•																				•
V10						•																			•
V11							•							•											•
V12								•						•											•
V13									•					•											•
V14										•				•											•
V15											•			•											•
V16												•		•											•
V17													•	•											•
V18															•									•	•
V19																•								•	•
V20																	•							•	•
V21																		•						•	•
V22																			•					•	•
V23																				•				•	•
V24																					•			•	•
V25																						•		•	•
V26																							•	•	•

4 Auto Machine Learning

The section begins by introducing the method used, TIM, an AutoML (automatic machine learning [9]) based on Information Geometry, a method for exploring the world of information through modern geometry [1], and then the experiment planning is described. Tangent Information Modeler, or TIM, is an AutoML solution designed for time-series prediction and anomaly detection developed by Tangent Works.

In the background, TIM constantly manages a few workers that can facilitate all possible requests. Scalability is an essential characteristic of this architecture, achieved through the computational distribution to TIM workers' units.

For this research, we will be using the Client TIM-Studio, a web application that provides an intuitive interface to a Machine Learning technology for data time-series, TIM.

TIM-Studio makes the experience as intuitive as possible. One can go with the default parameters or, tuning is also possible.

TIM-Studio version 2.0 makes the predictions based on the AutoML approach. TIM's AutoML allows one to build an entirely new model each time a forecast is needed, so there is no need to establish a particular data availability scenario. TIM analyzes the availability of the data as it is in the dataset at the time of the forecast and then builds a model based on this data and applies the model in a single pass through the data, delivering a forecast using all the available data from the forecast in just seconds to minutes [4].

To perform multi-step forecasting, namely to forecast the electricity price for the next 24 h, TIM finds a separate model for each hour in the forecast horizon, F(h), as shown in Fig. 1. It is unnecessary to make contradictory optimizations because each model needs to be optimized for only one output. In this multi-step forecast approach, the outputs are not used to create other outputs, so the risk of error propagation is also eliminated.

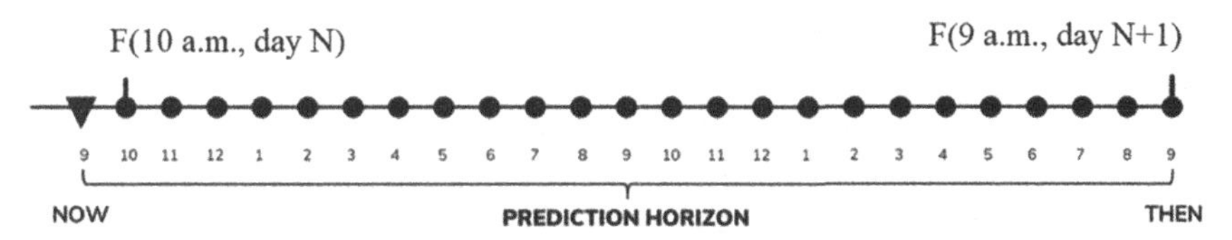

Fig. 1. Multi-step forecast, where TIM finds a separate model F(h) for each hour.

5 Computational Results

In the computational results phase, some choices must be made, such as the dataset, the algorithm, the computational environment, the performance measures, and the interpretability of the solutions.

The dataset and the experimental modeling were detailed in the previous section. AutoML TIM-Studio provides the algorithm and the computational environment in a cloud environment. Next, the performance measure and the interpretability of the solutions are presented.

5.1 Performance Measures

By analyzing the computational results of the 225 models, we observe that the fastest model does not necessarily obtain a minor error (either MAE, or MAPE, or RMSE). The one with a smaller MAPE is not the fastest. We understand that taking advantage of more metrics can be beneficial for selecting the best models in the real world. Accordingly, to facilitate the decision-making process of the best models, we applied the multi-attribute decision method TOPSIS.

We have selected TOPSIS [7] (acronym for "Technique of Order Preference Similarity to the Ideal Solution") to add as little subjectivity as possible in the decision process.

The subjectivity is associated with the weights of each attribute and with its assignment of cost or benefit.

The 225 models were ranked based on a multi-criteria TOPSIS analysis considering equal weight for each attribute in this study. TOPSIS was applied for each interval, and the solutions are shown in Table 3.

Table 3. TOPSIS computational results.

Interval	Training-test	Model	Run time (seconds)	MAE (€/MWh)	MAPE (%)	RMSE (€/MWh)
A	90–10	M6	7	5.26	62.63	7.70
B	90–10	M25	43	4.84	29.46	6.64
C	90–10	M23	170	3.96	15.52	5.42

Analyzing the results with the multi-attribute decision-making model performed at all intervals and the best models adopt the training-test 90–10.

Naturally, the best results (MAE, MAPE, RMSE) are presented for interval C, which includes ten years of data. The value of MAPE of 15.52% is competitive and reflects the hard time-series with 24 h forecast compared with similar works as in [5].

5.2 Interpretability of the Solution

TIM-Studio version 2.0, for each hour of the multi-step forecast, presents a weighted sum with an average of 20 rules. The expression for the forecast, F, at each hour, h, is given by the weighted, w, sum of *i* rules, is as follows:

$$F(h) = \sum\nolimits_{i=1}^{n} w_i.rule_i \;\; with \;\; \sum\nolimits_{i=1}^{n} w_i = 1$$

Based on the sunburst diagram of the model A-M7 at 10:00 a.m shown in Fig. 2, the weighted rules can be exemplified:

- Rule1 = PRICE(t-24) & PRICE(t-11): w1= 20.12%

Price with lag/delay/offset equal to 24 and with Price lag/delay/offset equal to 11.

- Rule2= DoW(t-7) ≤ Fri & PRICE(t-14): w2= 17.2%

Weekday (day of week) with lag/delay/offset equal to 7 and less than Friday, and Price with lag/delay/offset equal to 14.

- Rule3= DoW(t-20) ≤ Sat: w3= 12.76%

Weekday with lag/delay/offset equal to 20 and less than Saturday.

- Rule4= DoW(t-7) ≤ Sat & PRICE(t-12): w4= 10.83%

Weekday with lag/delay/offset equal to 7 and less than Saturday, and Price with lag/delay/offset equal to 12.

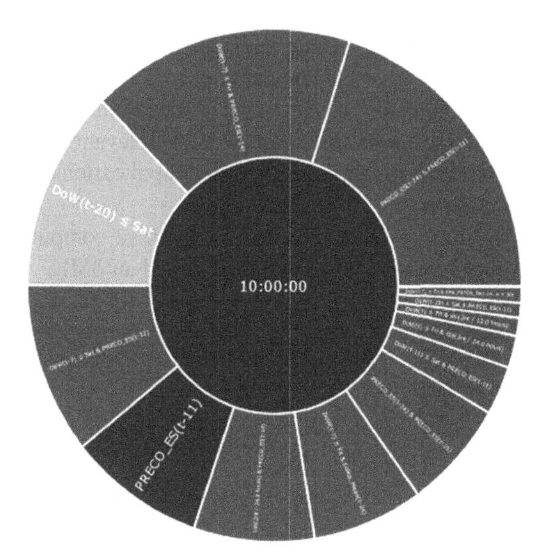

Fig. 2. A-M7 – 90–10 – sunburst diagram – 10:00

A vital issue in electricity price forecasting is the appropriate choice of explanatory variables. The typical approach has been to select predictors in an ad-hoc manner, sometimes using expert knowledge [8], as was the case in [5]. In the context of this work, a different approach was used, the result of mixing expert knowledge by selecting categories of explanatory variables and their respective variables and an automatic process of expanding variables that TIM provides. This automated process carried out by TIM provides the variables used, expansions performed, and their importance. So it is possible to forecast time-series, as is the case of MIBEL, with little specialized knowledge of how the market works. Furthermore, this functionality allows analysts working in the market to find other types of combinations of attributes that they can perform to analyze whether it produces improvements.

6 Conclusions

This work deals with the multi-attribute hourly forecast of the Electricity Price in MIBEL, namely in the Spanish part, for the 24 h of the following day, using open data. The data enrichment process adds an extensive set of input variables to the hourly price as demand forecasts, hourly power generation (wind, nuclear, coal, combined cycle, hydro, solar, international exchanges). Also, input variables include weather variables in the region and chronological data (date, year, month, day, hour, day of the week, holiday).

The accuracy of the model predictions was evaluated by the metrics MAE, MAPE, and RMSE, considering the execution time (in seconds). We believe that the evaluation's availability using several indicators becomes useful for a better comparison between papers in the area. The multi-attribute decision-making process was used, TOPSIS. The method allows ordering the models based on certain weights and criteria. The price forecasting models in this work, their performance in MIBEL, mainly in terms of the errors and the runtime, the importance of their input variables, can be helpful for electricity market agents and other players in the electricity industry.

Another relevant point to mention is the issue of model interpretability. A set of rules are generated in the presented approach, and the most relevant variables are identified. As models are being used in increasingly sensitive applications, their interpretability and reliability are key issues.

Finally, we believe that we have achieved competitive computational results for a complex time-series problem using data enrichment and autoML.

Acknowledgements. This work has been partially subsidized by these projects: TIN2017-88209-C2-2-R (Spanish Inter-Ministerial Commission of Science and Technology), FEDER funds and US-1263341 (Junta de Andalucía).

References

1. Amari, S.: Information Geometry and Its Applications. Springer, Japan (2016)
2. Cho, S.-B., Tallón-Ballesteros, A.J.: Visual tools to lecture data analytics and engineering. In: International Work-Conference on the Interplay Between Natural and Artificial Computation. Springer, Cham (2017)
3. Eric D.K., Joel, T.L.: Industrial Network Security, 2nd edn. Elsevier, Amsterdam (2015)
4. Miller, L.: Predictive Analytics for Time-series with InstantML For Dummies®, Tangent Works Special Edition, vol. 76 (2021). https://www.tangent.works/wp-content/uploads/2020/11/Predictive-Analytics-For-Dummies-E-book-by-Tangent-Works-1.pdf
5. Monteiro, C., Ramirez-Rosado, I., Fernandez-Jimenez, L., Conde, P.: Short-term price forecasting models based on artificial neural networks for intraday sessions in the iberian electricity market. Energies **9**, 721 (2016). https://doi.org/10.3390/en9090721
6. Monteiro, C., Ramirez-Rosado, I.J., Fernandez-Jimenez, L.A., Ribeiro, M.: New probabilistic price forecasting models: application to the Iberian electricity market. Int. J. Electr. Power Energy Syst. **103**, 483–496 (2018). https://doi.org/10.1016/j.ijepes.2018.06.005
7. Papathanasiou, J., Ploskas, N.: TOPSIS. In: Papathanasiou J., Ploskas N. (eds.) Multiple Criteria Decision Aid: Methods, Examples, and Python Implementations, pp. 1–30. Springer, Cham (2018). https://doi.org/10.1007/978-3-319-91648-4_1
8. Uniejewski, B., Nowotarski, J., Weron, R.: Automated variable selection and shrinkage for day-ahead electricity price forecasting. Energies **9**, 621 (2016). https://doi.org/10.3390/en9080621
9. Yao, Q., et al.: Taking Human out of Learning Applications: A Survey on Automated Machine Learning. arXiv:1810.13306 [cs, stat] (2019)

Tracking the Temporal-Evolution of Supernova Bubbles in Numerical Simulations

Marco Canducci[1](✉), Abolfazl Taghribi[2], Michele Mastropietro[3], Sven de Rijcke[3], Reynier Peletier[2], Kerstin Bunte[2], and Peter Tino[1]

[1] University of Birmingham, Edgabston B15 2TT, Birmingham, UK
{M.Canducci,p.tino}@bham.ac.uk
[2] Faculty of Science and Engineering, University of Groningen, Groningen, The Netherlands
{a.taghribi,k.bunte}@rug.nl, peletier@astro.rug.nl
[3] Department of Physics and Astronomy, Ghent University, Ghent, Belgium
{michele.mastropietro,sven.derijcke}@ugent.be

Abstract. The study of low-dimensional, noisy manifolds embedded in a higher dimensional space has been extremely useful in many applications, from the chemical analysis of multi-phase flows to simulations of galactic mergers. Building a probabilistic model of the manifolds has helped in describing their essential properties and how they vary in space. However, when the manifold is evolving through time, a joint spatio-temporal modelling is needed, in order to fully comprehend its nature. We propose a first-order Markovian process that propagates the spatial probabilistic model of a manifold at fixed time, to its adjacent temporal stages. The proposed methodology is demonstrated using a particle simulation of an interacting dwarf galaxy to describe the evolution of a cavity generated by a Supernova.

Keywords: Manifold learning · l1 and l2-regularization · Temporal generative topographic mapping · SPH simulation · Superbubble

1 Introduction

Generally, manifold learning assumes relevant information to lie on possibly noisy, low-dimensional structures embedded in higher dimensions. In order to extract this information it is essential to build a robust representation of the structure itself. Many approaches have been proposed, based on different assumptions, that address such a problem. Locally Linear Embedding (LLE) [9], Multi-Dimensional Scaling (MDS) [10] and their generalizations, aim at recovering a low-dimensional representation of the embedded manifold by covering it

Supported from the European Union's Horizon 2020 Marie Sklodowska-Curie grant agreement No. 721463 and the Alan Turing Institute Fellowship n. 96102.

H. Yin et al. (Eds.): IDEAL 2021, LNCS 13113, pp. 493–501, 2021.
https://doi.org/10.1007/978-3-030-91608-4_49

with hyper-planes (such as LLE) or building an adjacency matrix of locally linear neighbourhoods (like MDS). However, if the manifold has inherent transverse noise (e.g. measurement errors) Generative Topographic Mapping (GTM) [1], as a fully probabilistic principled model, is preferred. GTM naturally handles and preserves the information retained by noise while producing a low-dimensional latent representation of the manifold's topological structure.

In several situations the nature of the manifold is dynamic and it evolves through time, e.g. astronomical numerical simulations. In these cases, a joint spatio-temporal representation of the manifold is needed, in order to capture its underlying nature. It has been shown how GTM can be generalized to model time-series (GTM Through Time) [4], by imposing a Hidden Markov Model (HMM) structure on latent space nodes [8]. This enables one to project whole trajectories evolving around a lower-dimensional manifold onto the latent space of a single GTM. Here we tackle a different problem - we propose a methodology for tracking the evolution of a time series of low-dimensional manifolds based on the first-order Markovian assumption [5], where the spatial structure of the manifold at each time-step is modelled via GTM. The model is propagated through adjacent time-steps by minimization of a set of cost functions, aimed at preserving the position of the Gaussian Mixture's centres, while still fitting the evolved data set. The proposed regularizations are based on L_1- and L_2-norms on changes of model parameters. The methodology is applied to a particle simulation of a dwarf galaxy interacting with its host galaxy cluster. The manifold of interest is a cavity ("bubble") generated by a Supernova and detected via the Dionysus toolbox [7] (Persistent Homology, PH) at a given temporal stage. In [3] a generalization of GTM to abstract graphs has been proposed. However, knowing the topological properties of cavities, we adopt Geodesic GTM [11] as a modelling technique.

2 Methodology

Consider a sequence of $\mathcal{S} = (\mathcal{Q}^0, \ldots, \mathcal{Q}^\tau, \mathcal{Q}^{\tau+1}, \ldots, \mathcal{Q}^T)$ of point clouds $\mathcal{Q}^\tau = \{\mathbf{t}_1^\tau, \mathbf{t}_2^\tau, \ldots, \mathbf{t}_{N_\tau}^\tau\} \subset \mathbb{R}^d$, sampling a noisy, time-dependent, low-dimensional manifold $\mathcal{M}(\tau)$ of dimension ℓ throughout its evolution, $\tau = 1, 2, ..., T$. A naive approach to produce a probabilistic model of the sequence $\mathcal{S}$ is to model each temporal realization of the manifold individually, disregarding the mutual temporal dependencies of its components. However, it is sometimes useful to include the time-dependency of data sets when interested in joint space-time variations of the manifold's properties. We first describe the chosen methodology for modelling the spatial state of manifold $\mathcal{M}(\tau)$ for two topologically different cases. We then propose different measures for the propagation of such models through adjacent temporal stages.

2.1 One-Dimensional GTM

Let us assume that the manifolds $\mathcal{M}(\tau)$ are open, one-dimensional and nonlinear. As previously mentioned, each $\mathcal{M}(\tau)$ is sampled by $\mathcal{Q}^\tau$. Since we have the information about the manifold's dimensionality ($\ell = 1$), we take advantage of the Generative Topographic Mapping (GTM) formulation and define our model as a constrained mixture of Gaussians, where the latent space representation of centres is the interval $\mathrm{I} = [-1, 1]$. We sample I regularly with M centres $\mathbf{X} = \{x_1, \ldots, x_M\}$ and define a set of $K < M$ Radial Basis Functions (RBFs) $\phi(\mu_k, x_m)$ centred on uniformly spaced points $\mathbf{K} = \{\mu_1, \ldots, \mu_K\} \in \mathrm{I}$. We choose the RBFs to be spherical Gaussians:

$$\phi(\mu_k, x_m) = \exp\left[-\frac{(\mu_k - x_m)^2}{2\sigma^2}\right] , \tag{1}$$

where σ is a scale parameter chosen to be a multiple of the distance between neighboring RBFs centres. We then map the set of latent centres $\mathbf{X}$ onto the ambient space of manifold $\mathcal{M}(\tau)|_{\tau=0}$ through the non-linear mapping function $\mathbf{y}(\mathbf{X}; \mathbf{W})$ defined as:

$$\mathbf{y}(\mathbf{X}; \mathbf{W}) = \mathbf{W}\Phi(\mathbf{X}, \mathbf{K}) . \tag{2}$$

Here, Φ is the matrix with $\phi(\mu_k, x_m)$ being its $(k, m)-$th element. Each $x_m \in \mathbf{X}$ is thus mapped, through Eq. (2), to point $\mathbf{c}_m \in \mathbb{R}^d$, which acts as a centre of the Gaussian mixture. On each mapped centre $\mathbf{c}_m$ we impose a manifold-aligned noise model obtained by differentiation of the mapping function $\mathbf{y}(\mathbf{X}; \mathbf{W})$ w.r.t. the latent space coordinates, as described in [2]:

$$\mathbf{C}_m = \frac{1}{\beta}\mathbf{I} + \eta_m \sum_{p=1}^{\ell} \left[\left(\frac{\partial \mathbf{y}}{\partial x_p}\right)^{\top} \left(\frac{\partial \mathbf{y}}{\partial x_p}\right)\right]\Bigg|_{x_m} . \tag{3}$$

Here, x_p denotes the latent coordinate of dimension p. With suitable values for parameters β and η_m ($m = 1, \ldots, M$) the density of each component of the mixture reads:

$$p(\mathbf{t}|x_m, \mathbf{W}) = \left(\frac{1}{2\pi}\right)^{d/2} |\mathbf{C}_m|^{-1/2} \exp\left[-\frac{1}{2}(\mathbf{c}_m - \mathbf{t})^{\top}\mathbf{C}_m^{-1}(\mathbf{c}_m - \mathbf{t})\right] . \tag{4}$$

The overall GTM model is a flat mixture of Gaussians,

$$p(\mathbf{t}|\mathbf{W}) = \frac{1}{M}\sum_{m=1}^{M} p(\mathbf{t}|x_m, \mathbf{W}) . \tag{5}$$

whose parameters can be trained via a maximum-likelihood optimization, e.g. through the Expectation-Maximization (EM) algorithm [2].

2.2 Geodesic GTM for Cavities

In [11], we proposed a *Geodesic-GTM*, where the linear latent space of classical GTM is replaced by a spherical one with dedicated geodesic distance. Hence, every point $\boldsymbol{\xi}$ on the surface of the sphere is uniquely determined by a pair of angular coordinates: θ and λ where $(\theta, \lambda) \in \mathrm{I}_{\angle}^{\ell} = [-\pi; \pi] \times [-\pi/2; \pi/2]$ and $\mathrm{I}_{\angle}^{\ell}$

denotes the ℓ-dimensional, angular interval ($\ell = 2$). Assuming a radius $r = 1$ the geodesic distance between any pair of points $\boldsymbol{\xi}_i$ and $\boldsymbol{\xi}_j$ can be rewritten:

$$d_{\Omega}(\boldsymbol{\xi}_i, \boldsymbol{\xi}_j) = 2 \arcsin\left(\frac{\Delta\Omega}{2}\right), \qquad \Delta\Omega = \sqrt{\Delta X^2 + \Delta Y^2 + \Delta Z^2}\ , \tag{6}$$

where $\Delta X = \sin\theta_i \cos\lambda_i - \sin\theta_j \cos\lambda_j$, $\Delta Y = \sin\theta_i \sin\lambda_i - \sin\theta_j \sin\lambda_j$ and $\Delta Z = \cos\theta_i - \cos\theta_j$. As in the previous section, we define on the latent space a set $\mathbf{X} = \{\boldsymbol{x}_1, \ldots, \boldsymbol{x}_M\} \subset \mathrm{I}^{\ell}_{\angle}$ of latent points and $\mathbf{K} = \{\boldsymbol{\mu}_1, \ldots, \boldsymbol{\mu}_K\} \subset \mathrm{I}^{\ell}_{\angle}$ of centres for the RBFs $\boldsymbol{\phi} = \{\phi(\boldsymbol{\mu}_1), \ldots, \phi(\boldsymbol{\mu}_K)\}$. Using this formulation each RBF takes the form:

$$\phi(\boldsymbol{\mu}_k, \boldsymbol{\xi}_m) = \exp\left[-\frac{d_{\Omega}(\boldsymbol{\mu}_k, \boldsymbol{\xi}_m)^2}{2\sigma^2}\right]\ . \tag{7}$$

The probabilistic model is then formed analogously to the one-dimensional case, with the only exception for the estimation of the covariance matrix of the centres, that is now dependant on the geodesic distance on a sphere (Eq. (6)). The covariance matrix of each centre $\mathbf{y}(\boldsymbol{\xi}_m; \mathbf{W})$ of the Gaussian mixture model is obtained via Eq. (3), where $\ell = 2$, $x_1 = \theta$ and $x_2 = \lambda$.

2.3 Temporal Evolution

Given a data set $\mathcal{Q}^{\tau} = \left\{\mathbf{t}_1^{\tau}, \mathbf{t}_2^{\tau}, \ldots, \mathbf{t}_{N_{\tau}}^{\tau}\right\} \subset \mathbb{R}^d$ sampling the ℓ-dimensional manifold $\mathcal{M}(\tau)$ at time τ, we now wish to model its evolution in time by propagating the model created at time $\tau - 1$ through the sequence $\mathcal{S}(\tau)$. We impose the temporal smoothness of model structure by chaining the model parameters in the first-order Markov dependency structure through regularization terms. Once the probabilistic model of data set $\mathcal{Q}^{\tau-1}$ is obtained, when presented with the next data set $\mathcal{Q}^{\tau}$, we require minimization of a cost function for the recovery of the updated parameters $\mathbf{W}^{\tau}$. Multiple choices could be made for the cost function; in the following we will present and compare four options that behave differently when applied iteratively on each pair of adjacent data sets in $\mathcal{S}$:

$$\mathrm{l}_{2\mathrm{W}}(\mathbf{W}^{\tau}; \Lambda) = -\mathcal{L}(\mathcal{Q}^{\tau}|\mathbf{X}, \mathbf{W}^{\tau}) + \Lambda\|\mathbf{W}^{\tau} - \mathbf{W}^{\tau-1}\|_2^2 \tag{8}$$

$$\mathrm{l}_{2\mathrm{C}}(\mathbf{W}^{\tau}; \Lambda) = -\mathcal{L}(\mathcal{Q}^{\tau}|\mathbf{X}, \mathbf{W}^{\tau}) + \Lambda\|\mathbf{W}^{\tau}\Phi - \mathbf{W}^{\tau-1}\Phi\|_2^2 \tag{9}$$

$$\mathrm{l}_{1\mathrm{W}}(\mathbf{W}^{\tau}; \Lambda) = -\mathcal{L}(\mathcal{Q}^{\tau}|\mathbf{X}, \mathbf{W}^{\tau}) + \Lambda\|\mathbf{W}^{\tau} - \mathbf{W}^{\tau-1}\|_1 \tag{10}$$

$$\mathrm{l}_{1\mathrm{C}}(\mathbf{W}^{\tau}; \Lambda) = -\mathcal{L}(\mathcal{Q}^{\tau}|\mathbf{X}, \mathbf{W}^{\tau}) + \Lambda\|\mathbf{W}^{\tau}\Phi - \mathbf{W}^{\tau-1}\Phi\|_1\ . \tag{11}$$

Here $\mathcal{L}(\mathcal{Q}|\mathbf{X}, \mathbf{W})$ is the log likelihood of the GTM model with latent structure $\mathbf{X}$ and parameters $\mathbf{W}$, given a data set $\mathcal{Q}$. The cost functions are based on the L_1 and L_2-norms ($\|\cdot\|_1$ and $\|\cdot\|_2$) of the (vectorized) differences between weights or centres at times $\tau - 1$ and τ and are minimized following the simplex search method presented in [6]. Each time we evaluate the cost functions we also update the covariance matrices of the components according to Eq. (3), keeping fixed parameters β and η_m for $m = 1, \ldots, M$.

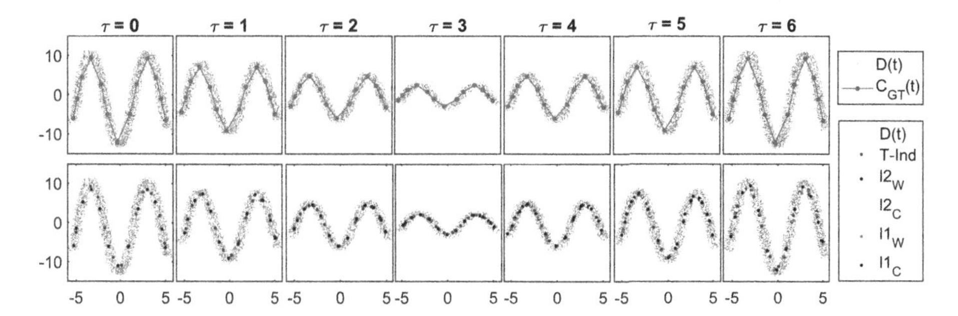

Fig. 1. Top row: 1D time-evolving data set with corresponding "Ground Truth". Bottom: centres obtained via independent (magenta line) or time-dependent modelling.

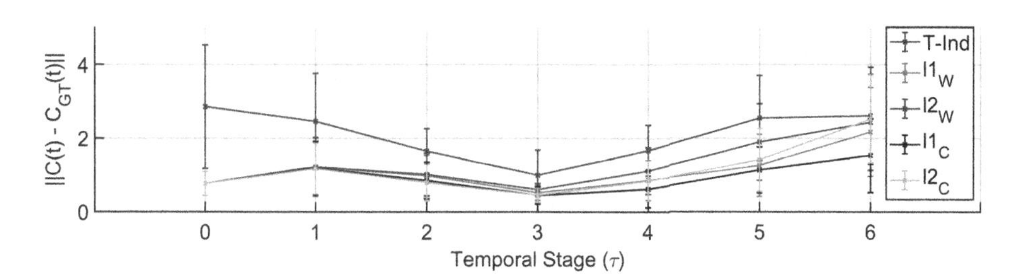

Fig. 2. Mean and standard deviation of distances between updated centres and corresponding Ground Truth at time τ.

3 Experiments on a Synthetic Data Set

In order to test the performance of the minimization procedure when presented with the four cost functions, we constructed a controlled experiment where the Ground Truth is known. A synthetic sequence $\mathcal{S}(\tau)$ is presented in Fig. 1 (top row). Each panel shows the spatial distribution of points at the corresponding time (grey points) and a pre-defined selection of centres considered optimal (Ground Truth: $C_{GT}(\tau)$) for the desription and modelling of the temporal evolotion. For example, to understand the evolution of such "wave shapes", one would need to consistently track the extremal "knot points" and the inner nodes in between them. The one-dimensional GTM presented in Sect. 2.1 is set with 13 centres and 7 RBFs. The weights of the mapping function $\mathbf{y}(\mathbf{X}, \mathbf{W})$ are initialized via linear regression imposing congruence between the mapped centres and $C_{GT}(\tau = 0)$ at stage $\tau = 0$. The Λ hyper-parameter was estimated through 5−fold cross-validation. For comparison, we also modelled each temporal stage individually with the same initialization, but additionally trained with the EM-algorithm. The centres recovered via both time-independent (T-Ind) and dependent modelling are shown in the bottom row of Fig. 1. We also compute at each temporal stage, the distance between the Ground Truth centres at that time and the centres obtained by each modelling approach, with results presented

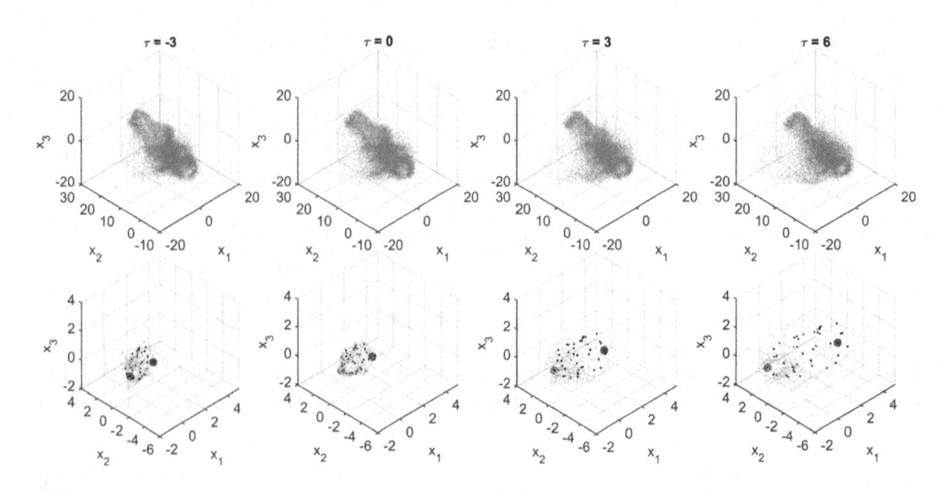

Fig. 3. Top row: Temporal stages of simulated dwarf galaxy. Grey dots are galaxy's point distribution, in cyan the recovered bubble propagated through time. Bottom row: Zoom in on the bubble with centres (black) representing opposite lobes (red and blue). (Color figure online)

in Fig. 2. We show here the mean and standard deviation of this measure at each temporal stage. The minimum distance is consistently achieved by the l1-regularization on centres (l_{1C}), followed in order by l_{1W}, l_{2C}, l_{2W} and finally time-independent modelling. Given the results on this simple synthetic experiment, we infer that $l1-$regularization over the mixture's centres is particularly convenient when identifiability of local properties of a manifold through time is required.

4 Tracking Supernovae Explosions Through Time

We now demonstrate our methodology in the context of a simulated dwarf galaxy, orbiting around the halo of its host galaxy cluster. By increasing local densities the interaction between the gas of the galaxy and that of the cluster may enforce the formation of massive stars which later approach the phase of Supernovae. When a star reaches this stage it expands outwards in a short time, sweeping away the surrounding gas. The "explosion" of the Supernova generates a cavity in the ambient gas, usually referred to as a "bubble". In [7], a methodology for detecting cavities has been proposed using PH. In [11] a sub-sampling procedure (ASAP) has been shown to regularize the detection through PH and applied to a similar astronomical simulation. We aim to track the evolution of a detected bubble through adjacent temporal-stages of the particle simulation. To preserve the position of centres in adjacent temporal stages while capturing the topological properties of the bubble, we need to adopt the probabilistic modelling technique proposed in Sect. 2.2. We adopt all four cost functions tested in the 1D case. However, since the $l1-$regularization on centres proved to be the optimal choice

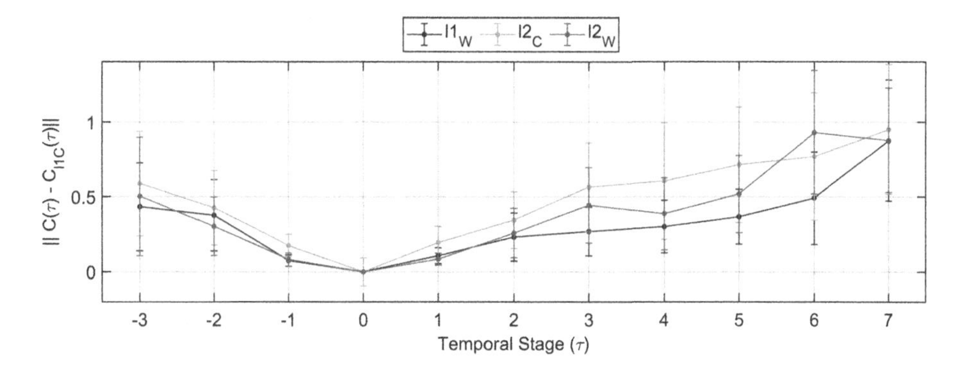

Fig. 4. As in Fig. 2 but w.r.t. the centres obtained via $l1-$regularization "on centres", for the remaining optimization formulations.

in our previous analysis, the detailed comparison will be performed w.r.t. this approach only. We regularly sample the latent space $\mathrm{I}^{\ell}_{\angle}$ with 45 centres and 25 RBFs, initializing the weights of the mapping function by assuming a spherical distribution of the embedded centres[1].

We first initialize the model at time $\tau = 0$ and run the EM-algorithm in order to obtain an accurate model at fixed time of the cavity. We then propagate the model from time $\tau = 0$ to the following and preceding stages, obtaining the spatio-temporal model of the bubble throughout 11 consecutive snapshots. A selection of snapshots equally spaced in time is shown in Fig. 3, top row. Grey dots represent the position of the simulated particles and cyan ones show the bubble's position w.r.t. the galaxy, at each temporal stage. The bubble's particles, detected at $\tau = 0$ via the Dionysus toolbox and ASAP, have been tracked through time using their unique identifiers assigned by the simulation. The bottom row shows a zoom in on the bubble's particles at the corresponding stage with the centres of the respective models in black. We additionally highlight two centres, lying on opposite lobes of the cavity. The red one is closest to the galaxy while the blue one points outwards.

The model's centres have been propagated through adjacent temporal snapshots using the four cost functions defined in Eqs. (8)–(11). Figure 4 shows the mean and standard deviation, at each temporal stage (τ), of the distances from all centres recovered by Eqs. (10) (red), (9) (cyan) and (8) (blue), to the corresponding ones obtained via Eq. (11). Except for a general agreement at temporal stage $\tau = 0$, when the models are initialized, the centres obtained via the three remaining cost functions increasingly diverge with time (both forward and backward). For each selected centre, we can now study how properties of the particles change through time nearby the two selected centres, for all adopted cost functions. In particular, we consider density ρ, temperature T and Neutral Fraction (NF). The Neutral Fraction is the ratio of the neutral Hydrogen mass to the

[1] The radius of the embedded sphere is estimated from the data similar to [11].

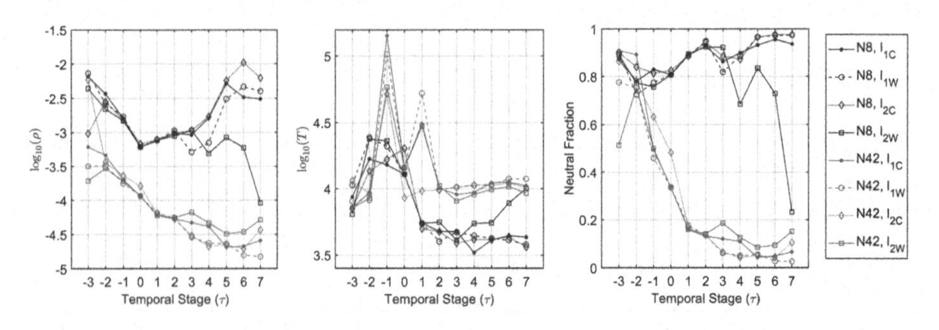

Fig. 5. Weighted mean of simulated quantities ($\log_{10}(\rho)$, T and Neutral Fraction (NF)), on centres $N8$ and $N42$ of the models propagated through time using costs (8)–(11).

total particle mass. For each temporal stage (and each propagated model), we compute the weighted mean of quantity a at centre c_m as:

$$\langle a \rangle|_{\mathbf{c}_m} = \frac{\sum_{n=1}^{N_\tau} a_n R_{mn}}{\sum_{n=1}^{N_\tau} R_{mn}} \,, \tag{12}$$

where R_{mn} is the *responsibility* of the m-th Gaussian component (around $\mathbf{c}_m$) for the n-th particle and N_τ the number of particles belonging to the bubble at time τ. When applied to properties $a \in \{\rho, T, NF\}$, at each temporal stage for both centres, we obtain the profiles shown in Fig. 5. The results confirm the intuitive expectation of $l1-$regularization to be preferable to the $l2$ one. In addition, direct $l2-$regularization of the centers' position stabilizes the results to a greater degree than $l2-$regularization on $\mathbf{W}$. Centre 8 being closer to the galaxy, has a consistently higher ρ and NF, while retaining lower T w.r.t. to centre 42 that is further away. The cluster's gas is generally hotter and less dense than the gas of the galaxy. The hot gas connected with the blue centre being pushed out can escape the galaxy, keeping a low neutral fraction and high temperature (the UV background is shining on it, keeping it hot and ionized). Gas at the red centre is inside the galaxy's interstellar medium, sheltered from the external UV and can cool and condense to higher densities, which is, indeed, favorable for enhancing star formation.

5 Conclusion

We propose a spatio-temporal probabilistic modelling of time-varying, noisy low dimensional manifolds based on the Generative Topographic Mapping (GTM). The centres of the constrained Gaussian Mixture are propagated through time by minimization of the regularized negative log-likelihood. We study the effect of four different regularizations on the model's parameters, finding that l1-regularization on centres yields the best results in a synthetic experiment. We then demonstrate our methodology to the astronomical particle simulation of a dwarf galaxy interacting with its host galaxy cluster, focusing on a Supernova bubble. We study the evolution of relevant quantities around selected centres of the Mixture and we find Supernova feedback may be enhancing Star Formation in the galaxy's interior.

References

1. Bishop, C.M., Svensén, M., Williams, C.K.I.: GTM: the generative topographic mapping. Neural Comput. **10**, 215–234 (1998)
2. Bishop, C.M., Svensén, M., Williams, C.K.I.: Developments of the generative topographic mapping. Neurocomputing **21**(1), 203–224 (1998). https://doi.org/10.1016/S0925-2312(98)00043-5
3. Canducci, M., Tiño, P., Mastropietro, M.: Probabilistic modelling of general noisy multi-manifold data sets. Artif. Intell. **302**, 103579 (2022). https://doi.org/10.1016/j.artint.2021.103579, https://www.sciencedirect.com/science/article/pii/S0004370221001302
4. C.M. Bishop, G.E. Hinton, I.S.: GTM through time. In: IET Conference Proceedings, pp. 111–116(5) (1997)
5. Gagniuc, P.A.: Markov Chains: From Theory to Implementation and Experimentation. Wiley, Hoboken (2017). https://doi.org/10.1002/9781119387596
6. Lagarias, J.C., Reeds, J.A., Wright, M.H., Wright, P.E.: Convergence properties of the nelder-mead simplex method in low dimensions. SIAM J. Optim. **9**(1), 112–147 (1998). https://doi.org/10.1137/S1052623496303470
7. Morozov, D.: Dionysus, a C++ library for computing persistent homology. https://mrzv.org/software/dionysus/
8. Rabiner, L.: A tutorial on hidden markov models and selected applications in speech recognition. Proc. IEEE **77**(2), 257–286 (1989). https://doi.org/10.1109/5.18626
9. Roweis, S.T., Saul, L.K.: Nonlinear dimensionality reduction by locally linear embedding. Science **290**, 2323–2326 (2000)
10. Shepard, R.N.: The analysis of proximities: multidimensional scaling with an unknown distance function. i. Psychometrika **27**(2), 125–140 (1962). https://doi.org/10.1007/BF02289630
11. Taghribi, A., Canducci, M., Mastropietro, M., De Rijcke, S., Bunte, K., Tino, P.: ASAP - a sub-sampling approach for preserving topological structures modeled with geodesic topographic mapping. Neurocomputing (2021). https://doi.org/10.1016/j.neucom.2021.05.108

SOMiMS - Topographic Mapping in the Model Space

Xinyue Chen[1(✉)], Yuan Shen[2], Eder Zavala[3], Krasimira Tsaneva-Atanasova[4], Thomas Upton[5], Georgina Russell[5], and Peter Tino[1]

[1] School of Computer Science, University of Birmingham, Birmingham B15 2TT, UK
{xyc588,p.tino}@cs.bham.ac.uk
[2] Nottingham Trent University, Nottingham NG1 4FQ, UK
Yuan.Shen@ntu.ac.uk
[3] Centre for Systems Modelling and Quantitative Biomedicine (SMQB), University of Birmingham, Edgbaston B15 2TT, UK
[4] University of Exeter, Exeter EX4 4PY, UK
[5] University of Bristol, Bristol BS8 1TH, UK

Abstract. Learning in the model space (LiMS) represents each observational unit (e.g. sparse and irregular time series) with a suitable model of it (point estimate), or a full posterior distribution over models. LiMS approaches take the mechanistic information of how the data is generated into account, thus enhancing the transparency and interpretability of the machine learning tools employed. In this paper we develop a novel topographic mapping in the model space and compare it with an extension of the Generative Topographic Mapping (GTM) to the model space. We demonstrate these two methods on a dataset of measurements taken on subjects in an adrenal steroid hormone deficiency study.

Keywords: LiMS · Topographic mapping · Sparse · Irregular time series

1 Introduction

Topographic visualisation techniques have been established as an important tool in data analysis and data mining, e.g. Self-Organising Map (SOM) [7,8] and its probabilistic reformulation - Generative Topographic Mapping (GTM) [3,4]. However, most of these methods were designed to operate in a vectorial data space. Also, there has been an increasing interest in formulating SOM and GTM in the model space to deal with data with more complex structures, e.g. [5,10,14]. The approach in [10] establishes the Self-Organising mixture autoregressive (SOMAR) model, in which components in the construction of the topological mixture model are AR models to model foreign exchange (FX) rates. For visualising sets of symbolic sequences, [14] attempts to extend GTM in the model space based on a constrained mixture of discrete hidden Markov models. As an extension of [14], the work in [5] formulates GTM by extending it to the space of Hidden Tree Models for tree-structured data.

H. Yin et al. (Eds.): IDEAL 2021, LNCS 13113, pp. 502–510, 2021.
https://doi.org/10.1007/978-3-030-91608-4_50

In this work we are interested in using Learning in the Model Space (LiMS) approaches to deal with potentially sparsely sampled and noisy time series. We offer a non-GTM learning method for SOM formulated directly in the model space termed *SOM in model space* (SOMiMS), which takes advantage of the probabilistic model formulation of our base inferential models. Keeping the SOM philosophy translated into the model space, we can retain control over the neighbourhood-shrinking rate and make the components move directly in the direction of gradients with respect to the model parameters of the inferential model class. We also extend the GTM to the model space. In general, GTM offers a clean formulation, but direct manipulation of dynamic neighbourhood size is not possible and prototype movements are controlled only implicitly through parameters of the embedding kernel regression function. We demonstrate the two methods on a real dataset of measurements taken on subjects in an adrenal steroid hormone deficiency study.

The rest of the paper is organised as follows: Sect. 2 briefly introduces the SOMiMS and extended GTM models. Section 3 presents the base inferential model and describes the real data set we use. Section 4 provides experimental results. We conclude the paper with a brief summary of key elements in Sect. 5.

2 Topographic Mapping of Time Series in the Model Space

Consider a set of time series $\mathcal{Y} = \{Y^{(1)}, Y^{(2)}, ..., Y^{(N)}\}$, $n = 1 : N$. The n-th time series will be denoted by $Y^{(n)} = \{Y_t^{(n)}\}_{t=1:T_n}$, where T_n is the length of n-th time series. The individual time series can be of different length, but we assume that there is a unique time grid where the observations are allowed to be taken. Generalization to time grids specific for each individual time series is relatively straightforward, but beyond the scope of this paper.

In our LiMS approach, each time series is considered as a set of partial observations of some underlying mechanistic model parametrised via $\vec{\theta} \in \mathbb{R}^d$ [12]. Mathematically, this parametric mechanistic model will be formulated as a multivariate Ordinary Differential Equation (ODE).

The topographic mapping of a vectorial data set is given by a (usually) nonlinear mapping from input vector space to a low-dimensional topographic mapping space (usually two dimensional) [8,15]. Topographic mappings we are interested in will operate in the model space rather than the original signal space. Each node of a topographic map corresponds to an instance from the inferential model class. The aim is to represent each time series as an individual projection on the topographic map.

2.1 SOMiMS

We will assume a SOM structure with $k \times k$ nodes. Each node i will be assigned an inferential model representative parameterized by a parameter vector $\vec{\theta}_i$. Considering the n-th time series $Y^{(n)}$, the log likelihood of i-th node is

$$\mathcal{L}(Y^{(n)}|\vec{\theta}_i, \Sigma) = \ln \prod_{t=1}^{T_n} p(Y_t^{(n)}|\vec{\theta}_i, \Sigma) = \sum_{t=1}^{T_n} \ln p(Y_t^{(n)}|\vec{\theta}_i, \Sigma),$$

where Σ collects parameters of the observational noise. Since each time series may have different length, we will operate with log likelihood per observation,

$$\mathcal{Q}(Y^{(n)}|\vec{\theta}_i, \Sigma) = \frac{1}{T_n}\mathcal{L}(Y^{(n)}|m_{\vec{\theta}_i}, \Sigma) = \frac{1}{T_n}\sum_{t=1}^{T_n} \ln p(Y_t^{(n)}|\vec{\theta}_i, \Sigma).$$

The "quality measure" of the i-th node, given the time series $Y^{(n)}$, is obtained by renormalization through all nodes,

$$\overline{\mathcal{Q}}(Y^{(n)}|\vec{\theta}_i, \Sigma) = \frac{\mathcal{Q}(Y^{(n)}|\vec{\theta}_i, \Sigma)}{\sum_a \mathcal{Q}(Y^{(n)}|\vec{\theta}_a, \Sigma)} = \frac{-\mathcal{Q}(Y^{(n)}|\vec{\theta}_i, \Sigma)}{\sum_a -\mathcal{Q}(Y^{(n)}|\vec{\theta}_a, \Sigma)}.$$

Note that $-\mathcal{Q}(Y^{(n)}|\vec{\theta}_i, \Sigma)$ can be thought of as the information (per observation) the node i contains about the time series $Y^{(n)}$. The quality measure $\overline{\mathcal{Q}}(Y^{(n)}|\vec{\theta}_i, \Sigma)$ is then the normalized information the node i holds on $Y^{(n)}$, renormalized in the competition across all the nodes.

Adopting a Gaussian observational noise model, we have:

$$p(Y_t^{(n)}|\vec{\theta}_i, \Sigma) = \frac{1}{(2\pi)^{\frac{D}{2}}|\Sigma|^{\frac{1}{2}}}\left\{ \exp(-\frac{1}{2}(Y_t^{(n)} - X_{i,t})^{\mathbf{T}}\Sigma^{-1}(Y_t^{(n)} - X_{i,t})\right\},$$

where $X_{i,t}$ is the (noiseless) observational vector at time t obtained from the inferential model parametrized with $\vec{\theta}_i$. Here we assume a homoscedastic process with a fixed covariance Σ. Again, generalization to time varying noise model is relatively straightforward, but out of the scope of the present paper.

During the training phase, in each iteration we randomly pick (with replacement) a time series $Y^{(n)}$ from the data set $\mathcal{Y}$. In each iteration, rather than updating the winner node (the node with maximum quality $\mathcal{Q}$) and its neighborhood as in the classic topographic mapping [8,15], we will update each node and its neighborhood according to the normalized quality measure $\overline{\mathcal{Q}}(Y^{(n)}|\vec{\theta}_i, \Sigma)$. This is needed, since especially for sparsely observed and/or noisy data, several prototypical node models can be likely and committing to a single winner node may bias the final topographic map. All nodes are considered in turn. For the i-th node, nodes c its neighbourhood are updated as

$$\vec{\theta}_c(l+1) = \vec{\theta}_c(l) \ + \ \overline{\mathcal{Q}}(Y^{(n)}|\vec{\theta}_i, \Sigma) \cdot h_{(c,i)}(l) \cdot \eta(l) \cdot \nabla_{\vec{\theta}_c} \mathcal{Q}(Y^{(n)}|\vec{\theta}_c(l)),$$

where $h_{(c,i)}(l)$ is the neighborhood function and $\eta(l)$ is the learning rate, both monotonically decreasing in algorithmic time steps l:

$$\eta(l) = \eta(0) \cdot \exp\left(-\frac{l}{\tau}\right) \tag{1}$$

$$h_{(c,i)}(l) = \exp\left\{-\frac{||c-i||^2}{2(\alpha(l))^2}\right\} \tag{2}$$

$$\alpha(l) = \alpha(0) \cdot \exp\left(-\frac{l}{\tau}\right) \tag{3}$$

Here τ is a time scale parameter and l is the current iteration index [9]. $\eta(0)$ is the initial learning rate in the power series learning rate function [13]; $\alpha(0)$ is the initial neighborhood size.

Note two crucial aspects of the SOMiMS methodology: (1) for each time series $Y^{(n)}$ we perform a double scan through the grid nodes, the outer scan through the pivotal nodes $\vec{\theta}_i$ with the inner scan through their neighbours $\vec{\theta}_c$; (2) the node updates are performed in the model space in the directions improving the node likelihoods, given $Y^{(n)}$, i.e. directions given by the gradient $\nabla_{\vec{\theta}_c} Q(Y^{(n)}|\vec{\theta}_c)$.

To visualize the time series data, we embed the $k \times k$ node grid in a square, e.g. $[-1,1]^2$, resulting in the embedded grid of points $\mathbf{g}_i \in [-1,1]^2$. The n-th time series $Y^{(n)}$ is then visualized (GTM-style) in $[-1,1]^2$ as the mean of the posterior distribution over the grid points [4,14],

$$Proj(Y^{(n)}) = \sum_{i=1}^{J} P(\mathbf{g}_i|Y^{(n)}, \vec{\theta}_i, \Sigma) \cdot \mathbf{g}_i,$$

where (imposing a uniform prior distribution over the grid),

$$P(\mathbf{g}_i|Y^{(n)}, \vec{\theta}_i, \Sigma) = \frac{p(Y^{(n)}|\vec{\theta}_i, \Sigma)}{\sum_{j=1}^{J} p(Y^{(n)}|\vec{\theta}_j, \Sigma)}.$$

2.2 Generative Topographic Mapping in the Model Space

As an alternative to SOMiMS, we also extend the Generative Topographic Mapping (GTM) [4] to the model space along the lines of [14] and [5]. Consider a 2-dimensional latent space $\mathcal{H}$, e.g. $\mathcal{H} = [-1,1]^2$. The aim is to represent each time series using this latent space through imposing a uniform prior over regular grid $\{\mathbf{g}_i\}_{i=1}^{J}$ of J points $\mathbf{g}_i \in \mathcal{H}$ covering the latent space. One imposes a function $\ell(\mathbf{g}; W)$ parametrized by W that maps the latent space into the model space:

$$\ell(\mathbf{g} : W) = W\phi(\mathbf{g}),$$

where W is a $d \times M$ matrix of parameters that governs the mapping $\ell(\mathbf{g}; W)$ and $\phi(\mathbf{g})$ contains M fixed basis functions $\phi_m(\mathbf{g}) : \mathcal{H} \to \mathbb{R}$. Note that $\ell(\mathbf{g}_i : W)$ now plays the role of the i-th prototypical model setting $\vec{\theta}_i$ in SOMiMS.

Given the n-th $Y^{(n)}$ time series in $\mathcal{Y}$ of length T_n, we can again calculate its probability, given the forward ODE model parametrized by $\ell(\mathbf{g}_i : W)$ and observational noise model parametrized by Σ:

$$p(Y^{(n)}|\mathbf{g}_i, W, \Sigma) = \prod_{t=1}^{T_n} p(Y_t^{(n)}|\ell(\mathbf{g}_i : W), \Sigma).$$

Since GTM is a flat mixture model of the latent grid, we have for the data log likelihood:

$$\mathcal{L} = \sum_{n=1}^{N} \ln \left\{ \frac{1}{J} \sum_{i=1}^{J} p(Y^{(n)} | \mathbf{g}_i, W, \Sigma) \right\}. \tag{4}$$

Expectation Maximization (EM) algorithm is used to obtain W by maximizing $\mathcal{L}$. The 'responsibilities' of grid points $\mathbf{g}_i, i = 1 : J$, for time series $Y^{(n)}$ are calculated in the E-step as

$$R_{in} = p(\mathbf{g}_i | Y^{(n)}, W, \Sigma) = \frac{p(Y^{(n)} | \mathbf{g}_i, W, \Sigma)}{\sum_j p(Y^{(n)} | \mathbf{g}_j, W, \Sigma)}.$$

The expected complete-data log likelihood then takes the form

$$\langle \mathcal{L}_{comp} \rangle = \sum_{n=1}^{N} \sum_{j=1}^{J} R_{in} \ln \{ p(Y^{(n)} | \mathbf{g}_i, W, \Sigma) \}.$$

The M-step consists of maximizing $\langle \mathcal{L}_{comp} \rangle$ with respect to W.

After training, each time series $Y^{(n)}$ can be visualized in the latent space $\mathcal{H}$ as the mean of the posterior distribution over the latent grid points [4,14],

$$Proj(Y^{(n)}) = \sum_{i=1}^{J} R_{in} \cdot \mathbf{g}_i.$$

3 Biomedical Background, Mechanistic Model, and the Data

Major adrenal steroid hormones are synthesized in different areas of the adrenal cortex[2]. We are particularly interested in the glucocorticoid and mineralocorticoid pathways. An appreciation of these pathways helps to understand the different forms of congenital adrenal hyperplasia (CAH) and isolated hypoaldosteronism characterized by defects in functionality of enzymes involved in adrenal steroid hormone synthesis[1].

The real data set we used in the experiments includes three conditions: Healthy control, Cushing's and Primary Aldosteronism. Cushing's usually results from the excessive production of Cortisol. Primary Aldosteronism is corresponding to the Aldosterone excess. The dataset contains subject-specific multivariate time series (of Corticosterone, Aldosterone, Cortisol, and Cortisone) obtained from 60 subjects covering the three conditions - Control (30), Cushing's (15) and Primary Aldosteronism (15). Each time series was sampled every 20 min within 24 h. However, there are some missing values due to certain operational problems. Thus, the length of time series may vary.

Below we introduce the mechanistic model representing the adrenal steroid hormone biosynthesis pathway that will be used to represent the observed data

Table 1. Model parameters.

Parameter	Description	Parameter	Description
k_C	Corticosterone synthesis rate	k_A	Aldosterone synthesis rate
k_F	Cortisol synthesis rate	k_E	Cortisone synthesis rate
k_b	Cortisone to Cortisol conversion rate	γ_C	Corticosterone degradation
γ_A	Aldosterone degradation rate	γ_F	Cortisol degradation rate
γ_E	Cortisone degradation rate	α_c	Amplitude of circadian drive
T_s^c	Phase shift of circadian drive	σ	Asymmetry of circadian drive
β	Offset of circadian drive	α_u	Amplitude of ultradian drive
T_s^u	Phase shift of ultradian drive	n_p	Number of ultradian pulses

in the LiMS framework. In the first instance, four hormones (Corticosterone (C), Aldosterone (A), Cortisol (F), Cortisone (E)) are modeled through a system of ODEs. The joint model reads:

$$\frac{d}{dt}C = k_C\varphi_c(t) - k_A C - \gamma_C C \qquad \frac{d}{dt}A = k_A C\varphi_u(t) - \gamma_A A$$

$$\frac{d}{dt}F = k_F\varphi_c(t) - k_E F + k_b E - \gamma_F F \qquad \frac{d}{dt}E = k_E F - k_b E - \gamma_E E,$$

where $\varphi_c(t)$ and $\varphi_u(t)$ are periodic circadian and ultradian drives specified by

$$\varphi_c(t) = \alpha_c \sin(2\pi(t + T_s^c) + \sigma * \sin(2\pi(t + T_s^c))) + \beta$$

$$\varphi_u(t) = 1 + \alpha_u \sin(2\pi(t + T_s^u)n_p).$$

All sixteen parameters used in models and their descriptions are listed in Table 1. Three drive parameters were fixed to $\alpha_c = 1$, $\alpha_u = 1$, $T_s^u = 0.5$, leaving thirteen free parameters.

4 Experiments and Results

In this section we present results of applying the SOMiMS and extended GTM methodologies on the real adrenal steroid dataset. We used a 10×10 grid and the models were initialized by applying the classic Self-Organising Map [8] in the signal space on all 60 subjects. The missing values were imputed using Gaussian Process model [11]. After training, most grid points of the classical SOM contain in their Voronoi compartments one or several time series. For grid points with no time series assigned, we used time series of their closest neighbours on the grid. Each grid point was then transformed to the model space by calculating the maximum likelihood parameter estimate obtained on the time series assigned to it. The 10×10 classic map thus became a map in the model space, each grid point corresponding to a setting of the 13-dimensional parameter vector.

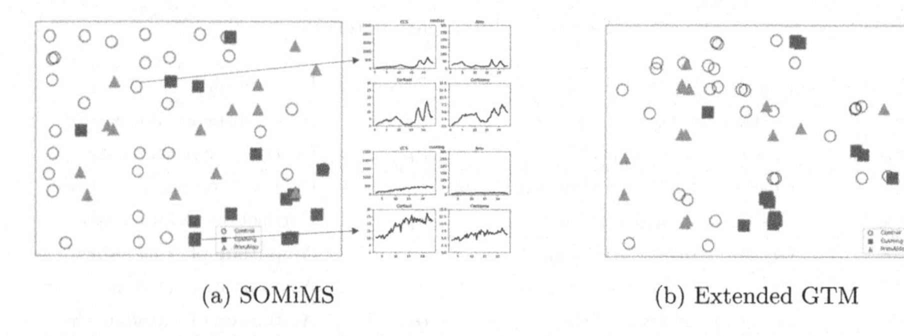

(a) SOMiMS (b) Extended GTM

Fig. 1. Topographic visualization of the data obtained by the SOMiMS (a) and Extended GTM (b) models. The Control, PrimaryAldo and Cushing's conditions are marked as blue circles, green triangles and red squares, respectively. (Color figure online)

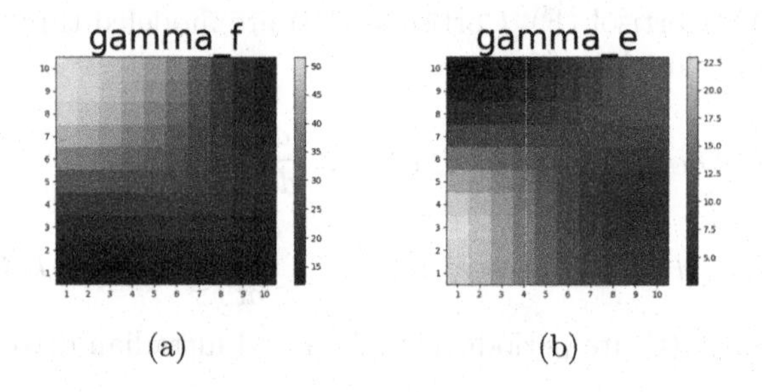

(a) (b)

Fig. 2. Parameter heat maps of γ_F (a) and γ_E (b) for the SOMiMS model.

The classic SOM (initialization stage for SOMiMS and extended GTM) was trained for 300 epochs with initial learning rate and initial neighbourhood size equal to 0.2 and 6, respectively. For SOMiMS, the initial learning rate and neighbourhood size were set to 0.1 and 2, respectively. This accounts for the fact that some very rough initial topographic organisation was already achieved in the classic SOM. SOMiMS was trained for 200 epochs. In the extended GTM, we employed $M = 4 \times 4 = 16$ basis functions ϕ_m and one additional constant basis function as the bias term. Basis functions were radial Gaussian functions with the same width $\sigma = 1$. The likelihood leveled up after 80–100 E-M cycles.

Topographic maps obatained by SOMiMS and extended GTM are shown in Fig. 1. The models were trained in a completely unsupervised manner, i.e. markers on the data projections signifying their corresponding conditions were not used during the training in any way. Overall, both topographic maps constructed in the model space show a good degree of separation of the conditions, noting that this is a noisy data set measured on real subjects. Both maps also show a tendency of sub-grouping the Cushing's cohort into at least two sub-populations. The signal plots to the right of the SOMiMS map illustrate the steroid time series corresponding to the selected projections (subjects).

Table 2. Confusion matrix

		Actual value					
		SOMiMS			Extended GTM		
Predicted value		Control	Cushing's	Primary Aldosteronism	Control	Cushing's	Primary Aldosteronism
	Control	0.80	0.10	0.10	0.68	0.27	0.05
	Cushing's	0.13	0.67	0.20	0.13	0.80	0.07
	Primary Aldosteronism	0.21	0.29	0.50	0.14	0.22	0.64

A detailed bio-medical analysis of the visualization plots is beyond the scope of this paper. We nevertheless stress that one of the primary advantages of topographic maps in the model space is the opportunity to readily interpret the topographic data organization from the mechanistic point of view of the underlying processes that generated the data. To that end, one can create parameter plots where the values learnt for each individual mechanistic model parameter across the prototypes on the grid are shown as heat maps. As an example, Fig. 2 presents parameter heat maps for Cortisol and Cortisone degradation rates, γ_F and γ_E, respectively. The two parameters have low values in the regions of the SOMiMS topographic map containing Cushing's projections. It is clear that the Cortisol excess associated with the Cushing's condition is partially caused by reduced degradation rates of Cortisol and Cortisone (which is positively coupled to Cortisol through k_b).

To quantify the amount of separation of the different conditions on the visualization plot, we also performed K-nearest neighbor (KNN) classification [6] on the map projections. Based on the cross-validated hyper-parameter tuning, we picked $K = 3$. Table 2 presents KNN confusion matrices for SOMiMS and Extended GTM projections. Thanks to the possibility of explicit control over the topographic map formation offered by SOMiMS (neighbourhood function and its shrinkage), the projections on the SOMiMS map are much more spread than those of the Extended GTM. Obviously, topographic organization does not correspond directly to the classification performance. After all, this is an unsupervised learning scenario. Such an analysis does, however, demonstrate that a full formation of a topographic map may disrupt cases of multiple projections in a very close neighbourhood of the visualisation space - a scenario that could yield good distance-based classifications, but is not preferable from the visualisation point of view.

5 Conclusion

We have presented a new learning method for SOM formulated directly in the model space, termed SOM in model space (SOMiMS), together with an extended GTM formulation in the model space for visualizing sets of sparse time series. We illustrated the methodologies on a real data set of measurements on subjects with different steroid hormone biosynthesis conditions. To that end, we formed a parameterized mechanistic inferential model in the form of coupled ordinary differential equations and demostrated how the topographic maps could be formed in the space of such inferential models, given the data.

Compared to the traditional approaches working in the signal space, SOMiMS and extended GTM are not only naturally able to deal with sparse time series, but also capable of taking the mechanistic information into account, creating scientifically interpretable readily data visualisations.

References

1. Arlt, W., et al.: Steroid metabolome analysis reveals prevalent glucocorticoid excess in primary aldosteronism. JCI Insight **2**(8), e93136 (2017)
2. Arlt, W., Stewart, P.M.: Adrenal corticosteroid biosynthesis, metabolism, and action. Endocrinol. Metab. Clin **34**(2), 293–313 (2005)
3. Bishop, C.M., Svensén, M., Williams, C.K.: Developments of the generative topographic mapping. Neurocomputing **21**(1–3), 203–224 (1998)
4. Bishop, C.M., Svensén, M., Williams, C.K.: GTM: the generative topographic mapping. Neural Comput. **10**(1), 215–234 (1998)
5. Gianniotis, N., Tino, P.: Visualization of tree-structured data through generative topographic mapping. IEEE Trans. Neural Netw. **19**(8), 1468–1493 (2008)
6. Keller, J.M., Gray, M.R., Givens, J.A.: A fuzzy k-nearest neighbor algorithm. IEEE Trans. Syst. Man Cybern. **4**, 580–585 (1985)
7. Kohonen, T.: Self-organized formation of topologically correct feature maps. Biol. Cybern. **43**(1), 59–69 (1982)
8. Kohonen, T.: Essentials of the self-organizing map. Neural Netw. **37**, 52–65 (2013)
9. Natita, W., Wiboonsak, W., Dusadee, S.: Appropriate learning rate and neighborhood function of self-organizing map (som) for specific humidity pattern classification over southern thailand. Int. J. Model. Optim. **6**(1), 61 (2016)
10. Ni, H., Yin, H.: A self-organising mixture autoregressive network for fx time series modelling and prediction. Neurocomputing **72**(16–18), 3529–3537 (2009)
11. Rasmussen, C.E., Williams, C.: Gaussian Processes for Machine Learning, vol. 32, p. 68. The Mit Press, Cambridge (2006)
12. Shen, Y., Tino, P., Tsaneva-Atanasova, K.: Classification framework for partially observed dynamical systems. Phys. Rev. E **95**(4), 043303 (2017)
13. Stefanovič, P., Kurasova, O.: Visual analysis of self-organizing maps. Nonlinear Anal. Model. Control **16**(4), 488–504 (2011)
14. Tino, P., Kabán, A., Sun, Y.: A generative probabilistic approach to visualizing sets of symbolic sequences. In: Proceedings of the Tenth ACM SIGKDD International Conference on Knowledge Discovery and Data Mining, pp. 701–706 (2004)
15. Torma, M.: Kohonen self-organizing feature map and its use in clustering. In: ISPRS Commission III Symposium: Spatial Information from Digital Photogrammetry and Computer Vision, vol. 2357, pp. 830–835. International Society for Optics and Photonics (1994)

Fair Regret Minimization Queries

Yuan Ma[1] and Jiping Zheng[1,2(✉)]

[1] College of Computer Science and Technology, Nanjing University of Aeronautics and Astronautics, Nanjing, China
{mayuancs,jzh}@nuaa.edu.cn

[2] Collaborative Innovation Center of Novel Software Technology and Industrialization, Nanjing, China

Abstract. When facing a database containing numerous tuples, users may be only interested in a small but representative subset. Unlike top-k and skyline queries, the k-regret query is a tool which does not need users to provide preferences but returns a representative subset of specified size by users with the minimum regret. However, existing regret-based approaches cannot answer the k-regret query on the dataset which is divided into groups and the result set contains fixed-size tuples in each group, which can be viewed as a metric of *fairness*. For this scenario, in this paper we generalize the k-regret query to its fair form, *i.e.,* the fair regret minimization query. Moreover, we provide an efficient algorithm named α-GREEDY which does not need to access the whole dataset at each greedy step with the help of a layer structure. We conduct experiments to verify the efficiency of the proposed algorithm on both synthetic and real datasets.

Keywords: Regret minimization · Fairness · Set cover

1 Introduction

Returning a representative subset from a large dataset[1] is an important functionality for multi-criteria decision making. Top-k, skyline and k-regret queries are three important tools to address this problem. The k-regret query proposed by Nanongkai *et al.* [21] which integrates the merits of top-k and skyline queries has attracted great attention in the last decade. For the k-regret query, the notion of the *regret ratio* is provided to quantify how regretful if a user gets the best tuple in the selected subset but not the best tuple among all tuples in the database. In vein of the k-regret query, extensions and variants are proposed to solve different problems in real applications [29]. Existing approaches to answer regret minimization queries focus on the datasets where each tuple in them is treated equally.

However, in real applications datasets might have sensitive attributes and be grouped, *e.g.,* the dataset is partitioned into several groups according to some

[1] We use the terms "database","dataset" and "tuple","point" interchangeably in the paper.

H. Yin et al. (Eds.): IDEAL 2021, LNCS 13113, pp. 511–523, 2021.
https://doi.org/10.1007/978-3-030-91608-4_51

attributes such as *gender*, *race* and *ethnicity*. For election, all candidates may be divided into two groups according to their gender or into several groups by race. Under these settings, no existing regret minimization approaches can solve the problem since previous researches only concern the overall returned k tuples and the extracted representative subset would be biased towards minor attributes or groups. For example, when users search the term "CEO" on Google Images [18], the percentage of women in top-100 results is only 11%, significantly lower than the truth of 27% (the ratio of women to men for CEO is 27/73). To solve the problem, it is essential to introduce the concept of fairness to the regret minimization query. Since fairness has different interpretations under different contexts [3,25], in this paper, we take the notion of *fairness* as the input constraint and define the *fairness* constraint as the group cardinality constraints [8,27], *i.e.*, for each group G_i, there exists a cardinality constraint k_i. That is, to be fair, we must select k_i tuples in group G_i. To speed up the process of our fair regret minimization query, we first transform our problem to the constrained set cover problem. Moreover, to avoid the inefficiency of existing greedy approach which needs traversing the whole dataset at each greedy step, we propose an approximation algorithm named α-GREEDY. With the help of a layer structure, α-GREEDY only accesses part of the dataset thus processes the fair regret minimization query efficiently. Also, the proposed α-GREEDY algorithm is with theoretical guarantee. The main contributions of this paper are summarized as follows

- We provide a formal definition of *the fair regret minimization query.*
- An algorithm named α-GREEDY is proposed transforming the fair regret minimization query to *the constrained set cover problem*, which has approximation ratio to the overall cardinality of the returned subset.
- We evaluate the performance of our proposed algorithm along with the baselines on both real and synthetic datasets.

2 Related Work

The k-regret query first proposed by Nanongkai *et al.* [21] has been investigated in the last decade [4,11,15,20–22,28,30,31,34] to avoid the drawbacks of top-k queries and skyline queries. For the k-regret query, Nanongkai *et al.* [21] propose the greedy algorithm which we call RDP-GREEDY where the tuple which contributes the most is added to the current solution in each selection. Peng *et al.* [22] interpret RDP-GREEDY from the geometric aspect and devise a geometric approach whose regret ratios are the same as RDP-GREEDY's but is more efficient. CUBE is proposed in [21] which has a known upper bound, but its empirical performance is quite poor; Xie *et al.* [30] improve the algorithm by the geometric properties of tuples.

Beside the researches of the k-regret minimization query, there are several variants in the literature. Chester *et al.* [11] propose the concept of kRMS which relaxes the maximum regret ratio to the kth maximum regret ratio. Zeighami *et al.* [34] raise the average regret minimization problem which aims at minimizing the average regret ratio instead of the maximum regret ratio. [1,31] study the problem that minimizes the output size when given a specified regret ratio ϵ, *i.e.*, they return a subset whose maximum regret ratio is at most ϵ and the size is as small as possible. A fully dynamic algorithm is proposed by Wang *et al.* [28] which can answer the k-regret query while the dataset is changing, *i.e.*, with insertions and deletions of database tuples. [20] introduces *interaction* to minimize the maximum regret ratio. All existing researches related to the regret minimization problem focus on the cardinality constraint, *i.e.*, the output size is k and do not consider the fairness constraint.

The fairness constraint is ubiquitous in our society and there is a broad literature on fairness, incorporating it as a metric [2,3,13,17,23,25,26,33]. [19,27] introduce fairness into submodular maximization problems. [12,14,32] study fairness in machine learning and classification. Also, fairness is attracting more attentions in recommendation [5,7,9,10,24], subset selection [27], and data summarization [8], etc.

3 Preliminaries

Denote D as a set of n d-dimensional points, and $p[i]$ as the value on the ith dimension for each point $p \in D$. The concepts of utility function and regret ratio are given before we define our problem.

Utility Function [21]. A utility function f is a mapping f: $\mathbb{R}_+^d \rightarrow \mathbb{R}_+$. The utility of a user with f is $f(p)$ for any point p which shows how satisfied the user is with the point. Denote $f(S)$ the highest utility among $p \in S$ with function f, *i.e.*, $f(S) = \max_{p \in S} f(p)$.

Following [4,21,30], we restrict the class of utility functions to linear utility functions, *i.e.*, for a linear function f, there exists a d-dimensional vector v, satisfies $f(p) =< v, p >= \sum_{i=1}^{d} v[i] \cdot p[i]$ for any $p \in D$ where $p[i]$ and $v[i]$ are non-negative real values.

Regret Ratio [21]. Given a dataset D, a subset $S \subseteq D$ and a set of utility functions F. The regret ratio of S with a utility function $f \in F$, represented as $rr_D(S, f)$, is defined to be

$$rr_D(S, f) = 1 - \frac{f(S)}{f(D)}$$

Moreover, we use $mrr_D(S, F) = \max_{f \in F} rr_D(S, f)$ to represent the maximum regret ratio of S under a set of utility functions F.

Fairness Constraint [24,27]. The fairness constraint is decribed as follows: assign a cardinality constraint k_i to each group G_i and ensure that $\sum_{i=1}^{l} k_i = k$. The value for each k_i can be determined by users. We can assign $k_i = \frac{|G_i|}{|D|} \cdot k$ to represent the proportion of each group in the dataset, or set $k_i = \frac{k}{l}$ to achieve a balanced representation of each group.

Problem Definition. Given a dataset D divided into l groups $G_1, \ldots, G_l$, a set of utility functions F, and a set of positive integers $k_1, \cdots, k_l$, the fair regret minimization query is to find a subset S of size k where $k = \sum_{i=1}^{l} k_i$ from D such that $mrr_D(S, F)$ is minimized and $|S \cap G_i| = k_i$ for each G_i.

4 The α-Greedy Algorithm

Inspired by [4,28,31], the equivalence between *the fair regret minimization query* and *the constrained set cover problem* is revealed. Equation 1 shows the connection between them.

$$minimize\ \epsilon \quad s.t. \quad S\ covers\ F\ \&\ |S \cap G_i| = k_i, 1 \leq i \leq l \tag{1}$$

For the constrained set cover problem, the aim is to minimize ϵ while satisfying the following two constraints: *coverage* and *fairness*. First, we define the constraint of *coverage*: if a tuple p_i and a utility function f_j satisfy $1 - \frac{f_j(p_i)}{\max_{p \in D} f_j(p)} \leq \epsilon$, *i.e.*, the regret ratio of p_i on f_j is no more than ϵ, we say f_j is covered by p_i or p_i covers f_j, and ϵ is the threshold of the coverage. For a subset $S \subseteq D$, if there exists at least one tuple $p \in S$ covering any arbitrary function f in F, then S covers F. Based on this definition, we want to find a subset S which can cover F and satisfy the fairness constraint. It means that S is a fair subset and with the mrr no more than ϵ. This reveals the equivalence between the constrained set cover problem and the fair regret minimization query under a specified threshold ϵ. A collection $\mathcal{D}$ corresponds to the dataset D and each set $P \in \mathcal{D}$ corresponds to a point $p \in D$ which is composed of utility functions covered by p. Hence, processing the regret query is equal to find a solution for the corresponding set cover problem.

By utilizing the form of Eq. 1, we can find a suitable value of ϵ by performing a binary search which is also adopted by [4,28,31]. When a solution with k tuples is returned, we can find the exact ϵ. Thus, the problem is transformed to the decision version of a set cover problem, *i.e.*, whether the returned k tuples are enough to cover all elements (utility functions). Here, k tuples can be seen as k sets for the set cover problem.

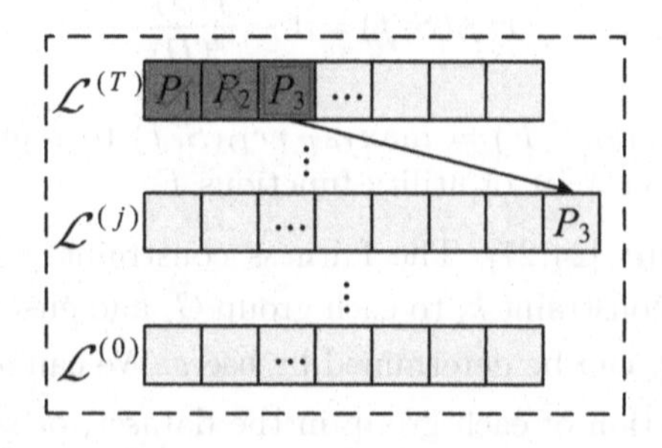

Fig. 1. The layer structure: P_1 is selected, P_2 is discarded and P_3 is moved to layer $\mathcal{L}^{(j)}$

The main idea of α-GREEDY is to adopt a layer structure shown in Fig. 1 to avoid traversing all the sets at each greedy step to find the set contributing the most to the result set. Initially, all sets $\{P_1, \ldots, P_n\}$ in $\mathcal{D}$ are put into different levels of the layer structure according to their sizes, *i.e.*, the set P is put into $\mathcal{L}^{(j)}$ if $\alpha^j \leq |P| < \alpha^{j+1}$. The layers are verified in the top-down manner. A set in jth layer would be selected if the number of new elements it covers is no smaller than α^j. At the beginning, since the covered set C is empty, we directly select the first set P_1 in the top level $\mathcal{L}^{(T)}$ ($|P_1 \backslash C| = |P_1 \backslash \emptyset| \geq \alpha^T$). The covered set $C = C \cup P_1 = P_1$ is updated after the selection. Then we check some following set P_i. P_i is discarded if one of the two cases occurs: P_i cannot cover any new element ($|P_i \backslash C| = 0$) or selecting P_i would violate the fairness constraint. Otherwise, we calculate the number of new elements P_i covers, *i.e.*, $|P_i \backslash C|$. If $|P_i \backslash C| < \alpha^T$ and $|P_i \backslash C|$ lies in the interval $[\alpha^j, \alpha^{j+1})$ ($j \geq 0$), we move P_i to the end of $\mathcal{L}^{(j)}$. In Fig. 1, we first directly select P_1 to the result set. Then P_2 is discarded for no new elements added. P_3 is moved to Layer $\mathcal{L}^{(j)}$ and appended to the end of it for $|P_3 \backslash C|$ is in the interval of this layer. Once we check the sets at the ith level ($i < T$), it means the above levels are empty since all the sets in them are selected (*e.g.*, P_1) or discarded (*e.g.*, P_2) or moved to some lower level (*e.g.*, P_3, moved to $\mathcal{L}^{(j)}$). When all elements are covered or all the layers have been accessed, the check process is terminated. In most cases, we need not access all the layers which can improve efficiency.

The pseudocode of α-GREEDY is shown in Algorithm 1. We set the upper bound and lower bound of ϵ to be 1 and 0, respectively (Line 1). Then we use the binary search to find the most suitable value of ϵ, *i.e.*, the lowest coverage threshold for the decision version of the set cover problem (Lines 2–9). We construct the set system $\Sigma = (\mathcal{D}, F)$ according the coverage along with ϵ: taking all utility functions as the ground set $F = \{f_1, \ldots, f_m\}$; for each point p_i and function f_j, if $rr_D(p_i, f_j) \leq \epsilon$, we put f_j to the set P_i, hence P_i is a set of utility functions; lastly, all P_is make up the collection $\mathcal{D} = \{P_1, \ldots, P_n\}$. The *Solver* procedure is adopted to solve the constrained set cover problem. If *Solver* returns an empty set, it implies the algorithm cannot find a feasible solution with the threshold ϵ, and the value of ϵ should be enlarged. Otherwise, we can find a feasible solution with the threshold ϵ, but we don't know whether the threshold ϵ can be lowered or not. Thus, we lower its value and try again. When the difference between ϵ_l and ϵ_h is within a parameter ξ (small enough), we use the latest feasible collection as the result, and determine its corresponding points in dataset D instead of the sets in $\mathcal{D}$. Finally, Lines 11–12 ensure the solution is fair for each group.

In the procedure of *Solver*, the solution $\mathcal{S}$ and the covered set C are initialized to be empty sets. All non-empty sets (empty sets are discarded directly) are put into appropriate layers (Line 16). We check the layers in the top-down manner and determine the operation on the set in each layer (Lines 19–27). When the loop breaks, *Solver* decides whether the solution $\mathcal{S}$ is feasible or not according to $F \backslash C$ (Line 28). $|F \backslash C| = 0$ implies that all elements are covered by $\mathcal{S}$ under the fairness constraint. Note that the α value is set to be only slightly larger than

1 to make sure that the P_is lie in different layers. When α approaches 1, there will be a large amount of layers with no sets. For this situation, when checking the P_is, we directly ignore these empty layers.

Algorithm 1: α-GREEDY

Input: A dataset of n points $D = \{p_1, p_2, \cdots, p_n\}$ divided into l groups $G_1, \ldots, G_l$, a set $F = \{f_1, \cdots, f_m\}$, fairness constraint $FC : k_1, \cdots, k_l \in \mathbb{Z}^+$, and a real-valued parameter $\alpha > 1$.

Output: A result set S of the fair regret minimization query.

1 $\epsilon_l = 0$, $\epsilon_h = 1$;
2 **while** $\epsilon_h - \epsilon_l \geq \xi$ **do**
3 $\quad \epsilon = (\epsilon_l + \epsilon_h)/2$;
4 $\quad$ Construct set system $\Sigma = (\mathcal{D}, F)$;
5 $\quad$ $temp \leftarrow Solver(\mathcal{D}, F, FC, \alpha)$;
6 $\quad$ **if** $temp == \emptyset$ **then** $\epsilon_l = \epsilon$;
7 $\quad$ **else**
8 $\quad\quad \mathcal{S} \leftarrow temp$;
9 $\quad\quad \epsilon_h = \epsilon$;
10 $S \leftarrow$ points in D corresponding to $\mathcal{S}$;
11 **while** $\exists i \in [1, l] : |S \cap G_i| < k_i$ **do**
12 $\quad$ Add points in G_i to S until $|S \cap G_i| = k_i$;
13 return S;
14 **Function** $Solver(\mathcal{D}, F, FC, \alpha)$
15 $\quad \mathcal{S}, C \leftarrow \emptyset$;
16 $\quad$ Assign each set $P_i \in \mathcal{D}$ to level $\mathcal{L}^{(t)}$ if $\alpha^t \leq |P_i| < \alpha^{t+1}$;
17 $\quad$ Let T be the largest t with non-empty $\mathcal{L}^{(t)}$;
18 $\quad$ **for** *(*$j = T; j \geq 0; j --$*)* **do**
19 $\quad\quad$ **for** *each set* P_i *in* $\mathcal{L}^{(j)}$ **do**
20 $\quad\quad\quad G_o \leftarrow \{G : P_i \in G_o, 1 \leq o \leq l\}$;
21 $\quad\quad\quad$ **if** $|\mathcal{S} \cap G_o| == k_o$ *or* $|P_i \backslash C| == 0$ **then**
22 $\quad\quad\quad\quad \mathcal{L}^{(j)} \leftarrow \mathcal{L}^{(j)} \backslash \{P_i\}$;
23 $\quad\quad\quad\quad$ **continue;**
24 $\quad\quad\quad$ **if** $|P_i \backslash C| \geq \alpha^j$ **then**
25 $\quad\quad\quad\quad \mathcal{S} \leftarrow \mathcal{S} \cup \{P_i\}, C \leftarrow C \cup P_i$;
26 $\quad\quad\quad\quad$ **if** $|F \backslash C| == 0$ **then break;**
27 $\quad\quad\quad$ **else** $P_i \leftarrow P_i \backslash C$, push P_i into $\mathcal{L}^{(j')}$: $\alpha^{j'} \leq |P_i| < \alpha^{j'+1}$;
28 $\quad$ **if** $|F \backslash C| > 0$ **then** $\mathcal{S} = \emptyset$;
29 $\quad$ return $\mathcal{S}$;

Lemma 1. *Solver is a* $1/(\alpha + 1)$*-approximation algorithm for the maximum coverage problem with the fairness constraint.*

Proof. The maximum coverage problem is to select k subsets from F such that their union has the maximum cardinality [16]. For our problem, we only prove the special case, *i.e.*, each group's constraint k_i equals 1. For $k_i > 1$, we make k_i copies of group G_i and the fairness constraint to each copy of G_i can also be 1.

Reorder the groups so that *Solver* picks a set from G_i just in the ith iteration, *i.e.*, $\mathcal{S} = \{S_1, \ldots, S_k\}$. Let O denote the optimal solution, and we reorder the sets in O to ensure that S_i and O_i are both selected in the group G_i in the ith iteration. Let $S = \bigcup_{i=1}^{k} S_i$, $O = \bigcup_{i=1}^{k} O_i$, and denote $\Delta_i = S_i - \bigcup_{h=1}^{i-1} S_h$ as the set of new elements added by *Solver* in the ith iteration. When *Solver* picks S_i, the set O_i can also be picked. Since $\frac{\alpha^i}{\alpha^{i+1}} = \frac{1}{\alpha}$, the new elements S_i covers are at least $\frac{1}{\alpha}$ of the set M_i where M_i is the set which adds the largest number of new elements in group G_i. Therefore, $|\Delta_i| \geq \frac{1}{\alpha}|M_i - \bigcup_{h=1}^{i-1} S_h| \geq \frac{1}{\alpha}|O_i - \bigcup_{h=1}^{i-1} S_h| \geq \frac{1}{\alpha}|O_i - S|$. We have

$$|S| = \sum_{i=1}^{k} |\Delta_i| \geq \sum_{i=1}^{k} \frac{1}{\alpha}|O_i - S| \overset{(*)}{\geq} \frac{1}{\alpha}(|\bigcup_{i=1}^{k} O_i| - |S|) = \frac{1}{\alpha}(|O| - |S|)$$

The derivation of the inequality $(*)$ is as follows.

$$\begin{aligned}
&\sum_{i=1}^{k} |O_i - S| = |O_1 - S| + |O_2 - S| + \sum_{i=3}^{k} |O_i - S| \\
&\geq |O_1 + O_2 - S| + \sum_{i=3}^{k} |O_i - S| \geq |O_1 + O_2 + \cdots + O_k - S| \\
&= |\bigcup_{i=1}^{k} O_i - S| \geq |\bigcup_{i=1}^{k} O_i| - |S|
\end{aligned}$$

Hence, $|S| \geq \frac{1}{\alpha+1}|O|$, the lemma is proved.

Theorem 1. *Algorithm 1 has* $1+\log_{1+\frac{1}{\alpha}} m$*-approximation with total cardinality constraint (*i.e., k*) for the fair regret minimization query.*

Proof. We assume the size of optimal cover set is k. Since the *Solver* procedure picks out k sets at each step, by Lemma 1 we have that the number of uncovered elements is not more than $(1 - \frac{1}{1+\alpha})|F| = \frac{\alpha}{1+\alpha}m$. After $1 + \log_{1+\frac{1}{\alpha}} m$ greedy steps, the number of uncovered elements is less than $(\frac{\alpha}{1+\alpha})^{1+\log_{1+\frac{1}{\alpha}} m} \cdot m = \frac{\alpha}{1+\alpha} < 1$, so there does not exist any uncovered element after these greedy steps. The α-Greedy algorithm which violates the cardinality constraint has the better performance than the optimal solution.

5 Empirical Evaluation

We first elaborate the experimental setup before we analyze the quality and efficiency of our proposed algorithm.

Datasets. We run the algorithms on both synthetic and real datasets. The synthetic dataset was created using the data generator [6], and three real-world datasets which are widely adopted in the literature [1,21,30,31]. The information of datasets is summarized as the following.

- Anti-correlated. The syntheic dataset is a 6-dimensional anti-correlated dataset of 10,000 points.
- NBA[2]. NBA is a 6-dimensional dataset with 21,961 points for each player/season combination from year 1946 to 2009. Six attributes represent the performance of each player, *e.g., total scores* and *blocks*, etc.
- Household[3]. The household dataset contains 1,048,576 family tuples with seven attributes, showing economic characteristics of each family, *e.g., annual property insurance cost*, and *selected monthly owner costs.*
- El Nino[4]. The El Nino dataset contains 178,080 tuples with 5 oceanographic attributes taken at the Pacific Ocean, such as *relative humidity, sea surface temperature*, and *air temperature.*

Following [4,21,30], we also preprocessed the datasets to only contain skyline points to answer the query. The processed anti-correlated dataset has 5,120 skyline points and there are 130, 57, 1183 skyline points for the NBA, Household and El Nino datasets, respectively. For convenience, we divide the datasets in sequence into l groups. F is consisted of 1,000 uniform sampled utility functions. All the algorithms are performed on a Core-i7 machine running Ubuntu 18.04 LTS, and implemented in C++.

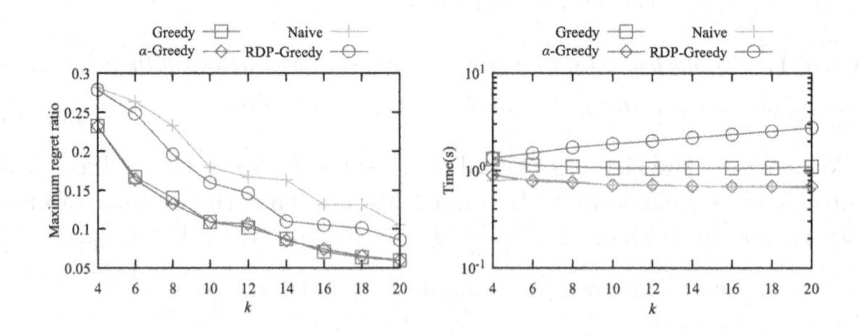

Fig. 2. Vary k on the anti-correlated dataset

[2] https://www.basketball-reference.com/.
[3] http://www.ipums.org.
[4] http://archive.ics.uci.edu/ml/datasets/El+Nino.

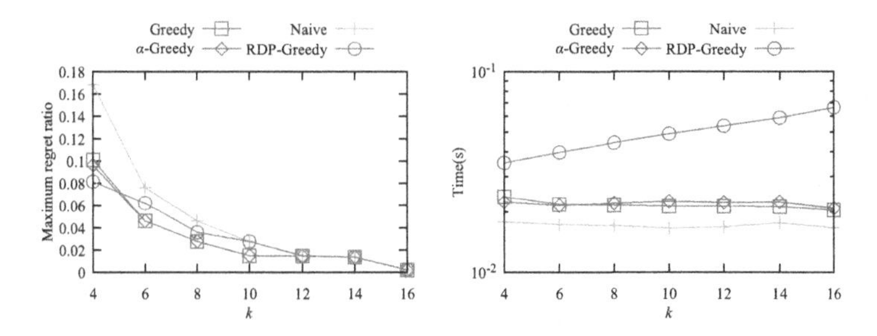

Fig. 3. Vary k on the NBA

In our experiments, we compare the following algorithms: GREEDY, α-GREEDY, NAIVE and RDP-GREEDY. GREEDY and NAIVE are mentioned in Sect. 4. RDP-GREEDY is a heuristic algorithm adapting the greedy algorithm in [21] for the fair regret minimization query. The parameter ξ is set to be 10^{-4}. For α-GREEDY, we set $\alpha = 1.001$. We compare all the algorithms from two aspects: maximum regret ratio (mrr) and running time.

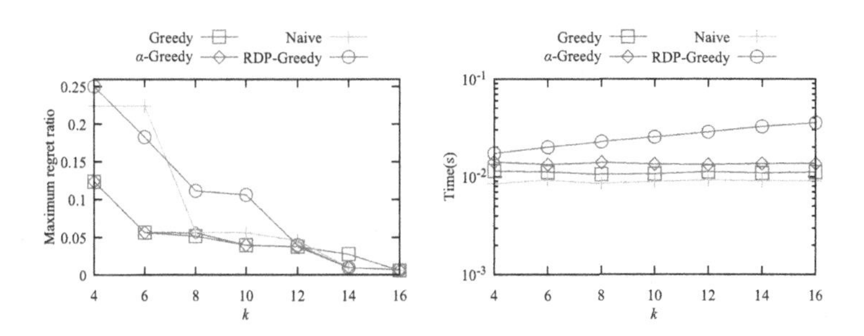

Fig. 4. Vary k on the household

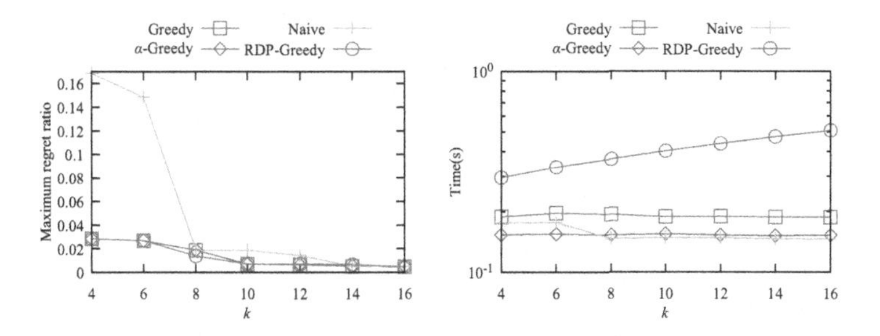

Fig. 5. Vary k on the El Nino

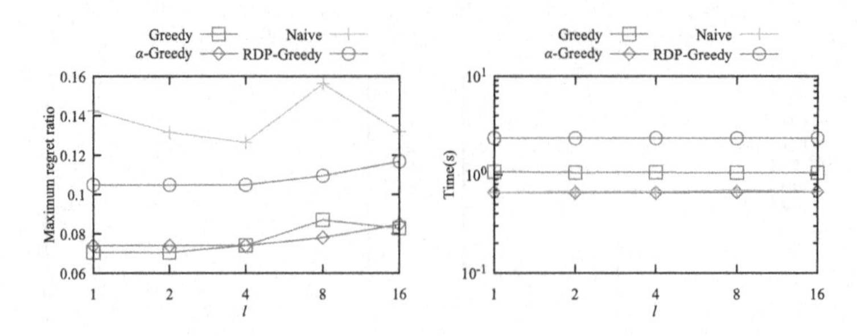

Fig. 6. Vary l on the anti-correlated dataset

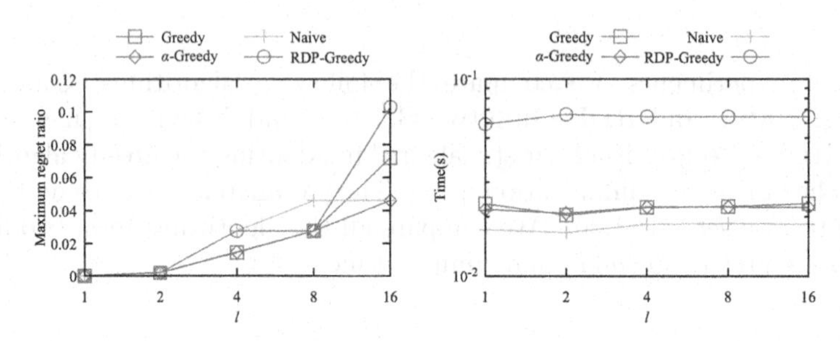

Fig. 7. Vary l on the NBA

Results. As Figs. 2, 3, 4, 5 show, Greedy and α-Greedy achieve better performance in mrr than Naive and RDP-Greedy in most cases. α-Greedy obtains the similar quality in mrr, but α-Greedy is much more efficient on the large datasets, such as the Anti-correlated dataset with 5,120 skyline points. Since the skyline sizes of the NBA and the household datasets are only 130, 57 respectively, α-Greedy does not have the advantage in running time and may be even worse. This is because α-Greedy needs time to put each set into the suitable level of the layer structure. Naive is with the best performance in running time due to its selection strategy not needing to search the whole dataset, which is similar to α-Greedy. But Naive has the worst performance in the maximum regret ratio. RDP-Greedy performs well on the NBA and El Nino datasets, but it is time-consuming to compute mrr of current solution before each selection, and the time cost increases significantly with the increase of k.

Figures 6, 7 show the results when the size of groups varies. We set the group size to be 1, 2, 4, 8, and 16, respectively. Note that, when $l = 1$ the fair regret minimization query is degraded into the k-regret query, and the lowest regret ratio is achieved due to the largest searching space for selection. As the group size l increases, the maximum regret ratio with the same output size increases. This can be viewed as the cost of fairness, *i.e.*, sacrificing some quality in mrr to ensure fairness. When the output sizes of the algorithms are set to a same value, *i.e.*, $k = 16$, the time costs of the algorithms almost remain unchanged under different group sizes.

6 Conclusion

In this paper, we propose *the fair regret minimization query* which combines the fairness constraint with the regret minimization query. To speed up the process of the fair regret minimization query, we first transform the query to the constrained set cover problem. Then, the α-GREEDY algorithm is proposed with theoretical guarantees to answer the fair regret minimization query efficiently by utilizing a layer structure. Our experiments verify that our proposed algorithm is effective and scales well on different datasets or under various constraints.

Acknowledgment. This work is partially supported by the National Natural Science Foundation of China under grant U1733112 and the Fundamental Research Funds for the Central Universities under grant NS2020068.

References

1. Agarwal, P.K., Kumar, N., Sintos, S., Suri, S.: Efficient algorithms for k-regret minimizing sets. In: Proceedings of International Symposium on Experimental Algorithms (SEA), pp. 7:1–7:23 (2017)
2. Ajtai, M., Aspnes, J., Naor, M., Rabani, Y., Schulman, L.J., Waarts, O.: Fairness in scheduling. J. Algorithms **29**(2), 306–357 (1998)
3. Asudeh, A., Jagadish, H.: Fairly evaluating and scoring items in a data set. In: Proceedings of the VLDB Endowment (VLDB), pp. 3445–3448 (2020)
4. Asudeh, A., Nazi, A., Zhang, N., Das, G.: Efficient computation of regret-ratio minimizing set: a compact maxima representative. In: Proceedings of International Conference on Management of Data (SIGMOD), pp. 821–834 (2017)
5. Beutel, A., et al.: Fairness in recommendation ranking through pairwise comparisons. In: Proceedings of International Conference on Knowledge Discovery & Data Mining (SIGKDD), pp. 2212–2220 (2019)
6. Börzsöny, S., Kossmann, D., Stocker, K.: The skyline operator. In: Proceedings of International Conference on Data Engineering (ICDE), pp. 421–430 (2001)
7. Celis, L.E., Huang, L., Vishnoi, N.K.: Multiwinner voting with fairness constraints. In: Proceedings of International Joint Conference on Artificial Intelligence (IJCAI), pp. 144–151 (2018)
8. Celis, L.E., Keswani, V., Straszak, D., Deshpande, A., Kathuria, T., Vishnoi, N.: Fair and diverse DPP-based data summarization. In: Proceedings of International Conference on Machine Learning (ICML), pp. 716–725 (2018)
9. Celis, L.E., Straszak, D., Vishnoi, N.K.: Ranking with fairness constraints. In: International Colloquium on Automata, Languages, and Programming (ICALP), vol. 107, pp. 28:1–28:15 (2018)
10. Celis, L.E., Vishnoi, N.K.: Fair personalization. arXiv preprint arXiv:1707.02260 p. 7 (2017)
11. Chester, S., Thomo, A., Venkatesh, S., Whitesides, S.: Computing k-regret minimizing sets. In: Proceedings of the VLDB Endowment (VLDB), pp. 389–400 (2014)
12. Chouldechova, A., Roth, A.: A snapshot of the frontiers of fairness in machine learning. Commun. ACM **63**(5), 82–89 (2020)
13. Dash, A., Shandilya, A., Biswas, A., Ghosh, K., Ghosh, S., Chakraborty, A.: Summarizing user-generated textual content: motivation and methods for fairness in algorithmic summaries. In: Proceedings of the ACM on Human-Computer Interaction, vol. 3(CSCW), pp. 1–28 (2019)

14. Dwork, C., Hardt, M., Pitassi, T., Reingold, O., Zemel, R.: Fairness through awareness. In: Proceedings of Innovations in Theoretical Computer Science Conference, pp. 214–226 (2012)
15. Faulkner, T.K., Brackenbury, W., Lall, A.: k-regret queries with nonlinear utilities. In: Proceedings of the VLDB Endowment (VLDB), pp. 2098–2109 (2015)
16. Feige, U.: A threshold of ln n for approximating set cover. J. ACM (JACM) **45**(4), 634–652 (1998)
17. Feige, U.: On allocations that maximize fairness. In: Proceedings of Symposium on Discrete Algorithms (SODA), vol. 8, pp. 287–293 (2008)
18. Kay, M., Matuszek, C., Munson, S.A.: Unequal representation and gender stereotypes in image search results for occupations. In: Proceedings of Conference on Human Factors in Computing Systems, pp. 3819–3828 (2015)
19. Kazemi, E., Zadimoghaddam, M., Karbasi, A.: Scalable deletion-robust submodular maximization: Data summarization with privacy and fairness constraints. In: Proceedings of International Conference on Machine Learning (ICML), pp. 2544–2553 (2018)
20. Nanongkai, D., Lall, A., Das Sarma, A., Makino, K.: Interactive regret minimization. In: Proceedings of International Conference on Management of Data (SIGMOD), pp. 109–120 (2012)
21. Nanongkai, D., Sarma, A.D., Lall, A., Lipton, R.J., Xu, J.: Regret-minimizing representative databases. In: Proceedings of the VLDB Endowment (VLDB), pp. 1114–1124 (2010)
22. Peng, P., Wong, R.C.W.: Geometry approach for k-regret query. In: Proceedings of International Conference on Data Engineering (ICDE), pp. 772–783 (2014)
23. Rabin, M.: Incorporating fairness into game theory and economics. Am. Econ. Rev. **83**(5), 1281–1302 (1993)
24. Serbos, D., Qi, S., Mamoulis, N., Pitoura, E., Tsaparas, P.: Fairness in package-to-group recommendations. In: Proceedings of International Conference on World Wide Web (WWW), pp. 371–379 (2017)
25. Singh, A., Joachims, T.: Fairness of exposure in rankings. In: Proceedings of International Conference on Knowledge Discovery & Data Mining (SIGKDD), pp. 2219–2228 (2018)
26. Stoyanovich, J., Yang, K., Jagadish, H.: Online set selection with fairness and diversity constraints. In: Proceedings of International Conference on Extending Database Technology (EDBT), pp. 241–252 (2018)
27. Wang, Y., Fabbri, F., Mathioudakis, M.: Fair and representative subset selection from data streams. In: Proceedings of The Web Conference (WWW), p. 11 (2021)
28. Wang, Y., Li, Y., Wong, R.C.W., Tan, K.L.: A fully dynamic algorithm for k-regret minimizing sets. In: Proceedings of International Conference on Data Engineering (ICDE), p. 12 (2021)
29. Xie, M., Wong, R.C.W., Lall, A.: An experimental survey of regret minimization query and variants: bridging the best worlds between top-k query and skyline query. VLDB J. **29**, 147–175 (2020)
30. Xie, M., Wong, R.C.W., Li, J., Long, C., Lall, A.: Efficient k-regret query algorithm with restriction-free bound for any dimensionality. In: Proceedings of International Conference on Management of Data (SIGMOD), pp. 959–974 (2018)
31. Xie, M., Wong, R.C.W., Peng, P., Tsotras, V.J.: Being happy with the least: achieving α-happiness with minimum number of tuples. In: Proceedings of International Conference on Data Engineering (ICDE), pp. 1009–1020 (2020)

32. Zafar, M.B., Valera, I., Rogriguez, M.G., Gummadi, K.P.: Fairness constraints: mechanisms for fair classification. In: Proceedings of International Conference on Artificial Intelligence and Statistics (AISTATS), pp. 962–970 (2017)
33. Zehlike, M., Yang, K., Stoyanovich, J.: Fairness in ranking: A survey. arXiv preprint arXiv:2103.14000 p. 58 (2021)
34. Zeighami, S., Wong, R.C.: Minimizing average regret ratio in database. In: Proceedings of International Conference on Management of Data (SIGMOD), pp. 2265–2266 (2016)

Chimera: A Hybrid Machine Learning-Driven Multi-Objective Design Space Exploration Tool for FPGA High-Level Synthesis

Mang Yu(✉), Sitao Huang, and Deming Chen

University of Illinois at Urbana-Champaign, Urbana, IL 61820, USA
{mangyu2,shuang91,dchen}@illinois.edu

Abstract. In recent years, hardware accelerators based on field programmable gate arrays (FPGA) have been widely applied and the high-level synthesis (HLS) tools were created to facilitate the design of these accelerators. However, achieving high performance with HLS is still time-consuming and requires expert knowledge. Therefore, we present Chimera, an automated design space exploration tool for applying HLS optimization directives. It utilizes a novel multi-objective exploration method that seamlessly integrates active learning, evolutionary algorithm, and Thompson sampling, which enables it to find a set of optimized designs on a Pareto curve by only evaluating a small number of design points. On the Rosetta benchmark suite, Chimera explored design points that have the same or superior performance compared to highly optimized hand-tuned designs created by expert HLS users in less than 24 h. Moreover, it explores a Pareto frontier, where the elbow point can save up to 26% of flip-flop resource with negligible performance overhead.

1 Introduction

In recent years, hardware accelerators on field-programmable gate arrays (FPGAs) has been widely adopted, since they offer performance close to customized hardware yet with the flexibility of programmable devices. However, designing accelerators for complex applications on FPGAs requires an immense amount of human effort and expert knowledge.

To address this challenge, the high-level synthesis (HLS) tools are developed to help designers describe hardware designs directly with high-level languages. However, using a software design in HLS without applying optimization directives usually leads to poor hardware performance, and as [1] pointed out, applying the directives can be difficult since they have many assumptions on the input code. Therefore, to achieve the best performance, the user needs advanced knowledge of the optimization directives that the HLS design tools provide.

This work was supported in part by the Xilinx Center of Excellence and Xilinx Adaptive Compute Clusters (XACC) program at the University of Illinois Urbana-Champaign.

H. Yin et al. (Eds.): IDEAL 2021, LNCS 13113, pp. 524–536, 2021.
https://doi.org/10.1007/978-3-030-91608-4_52

Such difficulties inspire us to develop an automated design space exploration (DSE) tool to find the optimal configuration of the optimization directives. For example, it should be able to automatically determine the directives/pragmas used to partition the arrays with the optimal partitioning factor. For such a DSE tool, there are several key challenges: the most important one is to reduce the number of invocations of the HLS synthesis, since the synthesis is usually the most time-consuming process in HLS DSE. In addition to time-saving, when designing an accelerator, the designers need multiple options in terms of latency and resource trade-offs, so that they can select the one that balances the compute performance and resource usage/power consumption. For this reason, the DSE tool needs to perform multi-objective optimization, so that it can discover Pareto optimal designs between the extreme ones. Finally, due to the complex nature of the HLS process, the performance of the synthesized hardware, as a function of the input design, is highly nonlinear and multimodal, which means an optimization method can easily fall into local optima. Therefore, it is crucial for the DSE method to have the ability to escape local optima effectively.

In this paper, we present Chimera, a machine learning-driven DSE tool that aims to solve the aforementioned challenges with the following contributions:

1. We enable multi-objective optimization with active learning, which uses the predictions from the machine learning (ML) model itself to create the training dataset. As a result, an effective performance/resource model can be built with a significantly smaller number of evaluated samples, reducing the number of HLS invocations during the exploration process.
2. We leverage the evolutionary algorithm (EA) in conjunction with ML to achieve an exceptionally high probability of discovering Pareto optimal points in each step of the exploration.
3. We utilize the Thompson sampling heuristic to create a hybrid method that adaptively switches between exploration and exploitation, which greatly helps escape local optima.

For the realistic applications in the Rosetta benchmark suite [2], in less than 24 h, Chimera can not only match or surpass the low latency design points found by human designers but also provide a Pareto curve that represents efficient trade-offs of latency and resources.

2 Backgrounds

2.1 Xilinx Vivado High-Level Synthesis

The Xilinx Vivado HLS [3] is a representative HLS tool that has been widely used for accelerator design. It supports C/C++ to register-transfer-level synthesis and provides a suite of optimization directives/pragmas. The combination of these directives/pragmas forms a large design space and the designer usually needs to explore the design space by evaluating many design points to find the optimal one. The pragmas/directives will be referred to as *directives* from here on.

Among the types of directives, the *loop directives* and *array directives* are most commonly used. As the survey by Schafer and Wang [4] pointed out, they provide the most direct and fine-grained control of the synthesized hardware. Each of the directives can have several configurations and a corresponding numerical factor. For instance, the loop directive has two configurations: *pipelining* and *unrolling.* In this work, we refer to these loops and arrays where we can apply optimization directives as *tunable knobs.* We then define a design point as the configuration of the directives for the tunable knobs. For example, for the code snippet shown on the left of Fig. 1, a design point for it can be summarized as the table on the right. Notice that a design point can be easily converted to an input sample to a machine learning model with one-hot encoding.

```
int mat[M][N];
int vec[N];
int out[M];

Outer: for (i = 0; i < M; i ++) {
    out[i] = 0;
    Inner: for (j = 0; j < N; j ++)
        out[i] += mat[i][j] * vec[j];
}
```

Tunable Knob	Configuration	Factor (s)
Array: mat	Cyclic Partitioning	1, N (for 2 dimensions)
Array: vec	Cyclic Partitioning	N
Array: out	Cyclic Partitioning	2
Loop: Outer	Pipelining	N/A
Loop: Inner	Unrolling	N

Fig. 1. Example of a design point for a code snippet

2.2 Multi-Objective Optimization

Solving an Optimization Problem for Low Latency and Low Resource Usage is the main Target of Chimera. Formally, the optimization problem we present in this paper can be described as the following: Denote the design space defined by the user for code K with $\mathbb{R}_K$. Each design point d_i within the design space $\mathbb{R}_K$ can be evaluated by the target black-box HLS tool to acquire a resource usage value $R(d_i)$ and a latency $L(d_i)$. Then we can define the objective as: Find the Pareto optimal set $\mathbb{P}_K \subseteq \mathbb{R}_K$, such that for any $p_j \in \mathbb{P}_K$, there is NO design point $d_i \in \mathbb{R}_K$ that satisfies $R(d_i) < R(p_j)$ and $L(d_i) < L(p_j)$ simultaneously.

2.3 Active Learning

Active learning is a family of machine learning methods that focuses on reducing the number of samples needed for constructing accurate models [5]. The active learning method uses the models' own predictions to determine the training samples to be evaluated and added to the training data, and construct the model by interleaving the process of evaluating new samples and re-training the models with the updated dataset. This means that it will only evaluate the most informative samples.

2.4 Thompson Sampling

Thompson sampling is a heuristic commonly used for multi-armed bandit problems [6]. For such problems, the algorithm will face multiple options that can potentially yield rewards and it needs to maximize the rewards it gains over multiple attempts. For such a problem, choosing the option that has the highest expected reward in the history of attempts can be too greedy and easily fall to local optima. The main reason is that the uncertainty of the expectation is not taken into consideration, since a smaller number of attempts means the uncertainty in the expected reward of an option is higher. On the contrary, the Thompson sampling not only considers the option that yields the highest reward but also considers the uncertainties of the expectations, such that the exploration and exploitation are well balanced. This means that, even if some options have a lower expected reward, they can still be selected for exploration purposes if there are much fewer attempts on it.

3 Related Works

There has been several recent attempts on applying analytical methods on HLS DSE [7–10]. However, unlike Chimera, they are based on single-objective optimization methods, which means they only optimize for one objective, such as low resource or low latency. On the contrary, Chimera is based on a multi-objective optimization method that can provide a wide range of design choices with different resource and latency trade-offs.

While previous works such as [11,12] also apply data-driven methods to enable multi-objective optimization, comparing to Chimera, [11] only applied random sampling instead of active learning, which means it will require more samples to construct an effective model; [12] applied selective sampling for the training dataset but does not use the predictions from machine learning model and does not integrate evolutionary algorithm. Moreover, Chimera employs Thompson sampling to combine the strength of various optimization methods and help escape local optima effectively, making it more efficient in exploring large design space. Finally, Chimera demonstrates its capability on realistic benchmarks, showing its practicality.

4 Methodology

Chimera is an automated DSE tool dedicated to optimizing the directive configurations of C/C++-based HLS designs for FPGAs. As an automated DSE tool, the user only needs to provide a comma-separated values (CSV) file containing the information of the tunable knobs.

Figure 2 demonstrates the main stages and the overall flow of Chimera. In the initial sampling stage, Chimera first randomly evaluates a small set of design points in the design space. Then, in the model training stage, it trains the machine learning models with the initial samples. These models are used for predicting the resource usage and latency for all later stages of the DSE process. Once the models are trained, in the method selection stage, it uses the Thompson sampling heuristic to pick one of the three point proposal methods to get

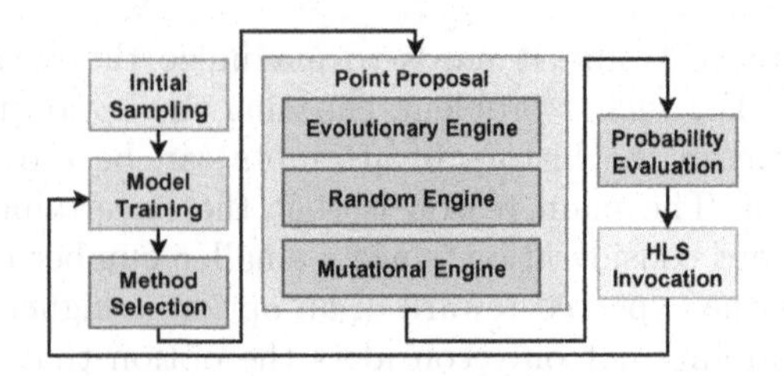

Fig. 2. Overall flow and stages of the chimera DSE process

the next design point to explore. Afterwards, it enters the probability evaluation stage, where a probability is assigned to the proposed design point based on its predicted performance and resource usage. This probability determines how likely a design point should be evaluated. Following that, if the new point is selected to be evaluated, the information of the new point will be passed to the HLS invocation stage, which will call the Vivado HLS synthesis tool. The result from HLS is then added to the training dataset. It determines whether the evaluated design point is superior to the Pareto frontier constructed so far and record it in the history of the point proposal methods. Finally, the tool returns to the model training stage, where the models will be retrained. This loop will then repeat until the stopping condition, that is, a limit on the time spent or number of points explored, is met.

4.1 Initial Sampling

In the initial sampling stage, Chimera evaluates the latencies and resource usages of a set of randomly selected design points without using the ML models. All evaluated design points will then be added to the training dataset.

4.2 Model Training

The main function of the ML models is to predict the performance, resource usages, as well as possibility of encountering an error or timeout for a design point. It also works as a shared knowledge representation that helps integrating multiple optimization methods. The Random Decision Forests (RDF) were used for all the predictive models in this work for their ability to learn non-linear relationships and high accuracy on small tabular datasets.

For resource predictions, there are four separate models, which predict the hardware resource utilization in terms of Block RAM (BRAM), Look-Up Table (LUT), Flip-Flop (FF), and DSP, respectively. The input features to the models comprise directive configurations and factors as shown in Fig. 1, and the output predictions are the proportions of the corresponding resource type consumed on the target FPGA. For latency predictions, the output is the predicted latency in μs.

In addition to predicting the resource and latency, Chimera also has a classification model dedicated to predicting the probability of encountering synthesis errors or timeout. If the synthesis for a design point finishes abnormally or exceeds the time limit, it will be recorded in the dataset. The model will then learn from these records and predict the probability of timeout/error $P_{timeout}$ when a new point is proposed. This probability will be used in the probability evaluation stage to determine the total probability of evaluating a point. That is to say, the more likely the point is predicted to timeout/error, the less likely that the design point will be evaluated. This way, we can avoid evaluating un-synthesizable design points.

4.3 Method Selection

Chimera has three methods to propose new design points to explore starting from the initial sampling set: *random*, *evolutionary*, and *mutational*. Among the three methods, the evolutionary method is especially effective in combining the beneficial directives from multiple known design points; the mutational method is a special case of evolution, in which both parents are the same point on the Pareto-frontier, making it more effective in discovering more extreme design points, such as points with lowest known latency or lowest known resource usages; the random proposal method employs random sampling, which means it is the most exploratory method and effective on introducing new information to the dataset. The three methods complement each other in terms of greediness and are tightly integrated by sharing the same set of machine learning models and sample population, making Chimera fundamentally different from a simple ensemble of exploration techniques.

Selecting the method for each iteration of exploration can be considered a beta-Bernoulli bandit problem, in which a positive reward 1 will be assigned when a method finds a new Pareto non-dominated design point that pushes the Pareto frontier, and the algorithm needs to maximize the total reward in a finite number of attempts. Notice that we refer to the points as *Pareto non-dominated* since they are the design points that are not surpassed by any other *known* design point in every objective. However, these points are not necessarily the absolute Pareto optimal points, which can only be confirmed with exhaustive searches.

For our DSE problem, we refer to finding a new Pareto non-dominated point as a success, and otherwise a failure. In this type of problems, the distribution of the possible reward for each of the options can be considered as a Bernoulli distribution; therefore, we apply the *Thompson sampling* heuristic, in which the expectation of getting a reward for each of the options obeys a separate beta distribution where β is the total number of attempts on the option and α is the number of successful attempts.

During each iteration of the exploration, a sample will be drawn from each of the three beta distributions, and the method corresponding to the sample with the highest value will be chosen. At the beginning of the exploration process, α and β will both be initialized to 1 for the distributions, giving the methods the same chance of being selected. As the exploration process goes on, the values of α and β will increase according to the result of each attempt, and the distribution will be more concentrated around the mean reward $\alpha/(\alpha+\beta)$. Since the expected value from the beta distribution of an option is the same as the mean reward of it in history [6,13], in general, the method with a higher mean reward is more likely to be selected.

4.4 Point Proposal

The point proposal stage is used to generate the next design point to be explored. The effectiveness of this stage ensures that the point explored can provide the most valuable information to the models and most likely to be Pareto non-dominant.

Random Proposal Method. The random proposal method is the simplest and the least greedy point proposal method since it generates a random directive for each of the tunable knobs. But notice that the proposed design point will still be given a probability in the probability evaluation stage and will not always be evaluated.

Evolutionary Proposal Method. The evolutionary proposal method proposes a new design point to explore using the principles of the evolutionary algorithm. Essentially, it creates a population of candidate designs that will evolve according to a fitness function, which defines the objective of the optimization process. Each individual in the population will have a certain genotype, which determines the phenotype of the individual. For our DSE problem, the genotype comprises the design parameters of the design point and the phenotype is the performance and resource usage of the design point.

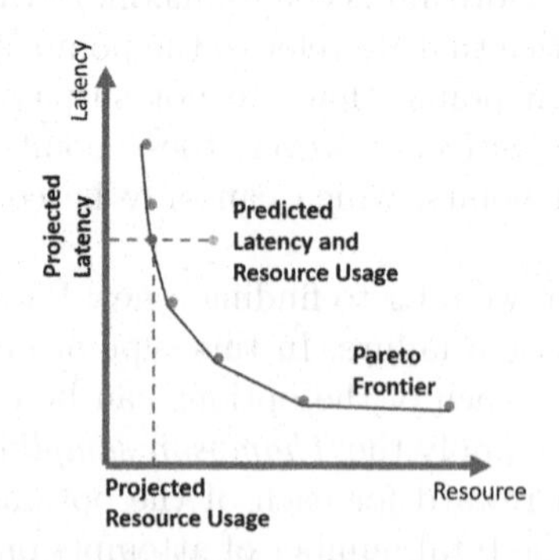

Fig. 3. Example of the latency-wise projection on the pareto frontier

The method first forms a population with known design points that are close to the Pareto frontier. This is done by comparing the resource usage of a design point with the projected point on the Pareto frontier. As shown in Fig. 3, the projection of the orange design point is latency-wise, and the resource of the projected point is found by linear interpolation. Within the population, it will randomly select several design points as the "fathers". Each of the selected points will then breed with one of the two neighboring design point on the Pareto frontier. For example, the blue points in Fig. 3 are the neighboring points for the orange design point. The crossover and mutation will happen during the breeding process, which generates a set of offspring. Then, the RDF models will be used to predict the latency and resource usage of each offspring, and a quality score will be calculated for it based on its distance to the Pareto frontier and whether it is feasible. Afterward, they will be added to a list of candidate points and their corresponding quality scores (the probability of evaluation) will also be recorded. Finally, the candidate design points will be sorted by their quality scores and the one with the highest score will be selected to pass to the next stage together with its quality score.

Comparing to proposing new design points randomly or from a simple probability distribution [14,15], EA enables the DSE tool to incorporate more information from the population of known design points so that the probability of discovering a new Pareto optimal point is higher.

Mutational Proposal Method. The mutation proposal method first randomly selects a point on the Pareto frontier constructed so far and mutates it several times to generate a set of mutants. Then, the mutants with the highest predicted probability of evaluation will be selected.

This can be seen as a greedier version of the evolutionary method since it only applies mutation to the design points on the Pareto frontier. Its greediness means that it can be more effective in discovering extreme design points, but also more likely to fall into local optima. Therefore, if the mutational method is selected for multiple iterations but fails to find any new Pareto non-dominant design point, Chimera will automatically switch to the evolutionary or random methods to help escape this potential local optima.

The Exploration-Exploitation Loop. For each exploration iteration, the point proposal method used is determined by the method selection stage. By combining the three methods with Thompson sampling, the exploration process can be seen as a conceptual loop of "Exploration-Exploitation (E-E)":

Exploration: After the first few steps of the DSE, the optimization process will likely fall into a local optimum where the more greedy non-random methods fail to escape and find new Pareto non-dominated points. During this period, the random method will have an increasingly higher chance of being selected by Thompson sampling. Since it can introduce design points that are drastically different from the known design points, it can help to discover the correlated changes that have a beneficial combined effect. For example, it might generate array and loop directives with matching partitioning and unrolling factors at once, whereas applying the two directives individually will not bring any improvement. The design points explored by the random method are not necessarily Pareto non-dominated, but they are still helpful as long as they bring in beneficial mutations to the population. Also, this does not mean that using the random method alone will be as effective because its effectiveness is dependent on the quality of the machine learning models, whose accuracy is based on the points explored by the other methods.

Exploitation: Once the new beneficial mutations are introduced to the population, they will be rapidly spread in the population by the evolutionary method. During this period, the newly introduced mutations will be refined and the frontier can be quickly pushed forward. The refining process, in turn, provides valuable information to the training dataset and helps improve the ML models that guide the next round of exploration. Without the evolutionary and mutational method, the machine learning model will have fewer training samples around the new Pareto frontier, which means the quality of the point proposed by the random method will be lower. After the exploitation period, the tool will continue the next iteration of E-E. Unlike the epsilon greedy method that only has one E-E loop with a fixed boundary, our method can infinitely continue to further improve the result of exploration.

4.5 Probability Evaluation

In this stage, we introduce the "soft-boundary" technique that assigns a probability for whether a proposed design point should be evaluated, instead of the conventional "pass-or-fail" method. This technique accounts for the inaccuracies of the models and avoids being overly greedy, making active learning less prone to poor initial sampling. We determine the probability that the given design point is to be evaluated based on the Pareto frontier, as well as predictions from the RDF models on the latency, resource usages, and timeout/error probability. The design points that are predicted to consume more resources compared to the projected point on the Pareto frontier as shown in Fig. 3, or use higher than 100% of resource will have a lower probability of being evaluated.

4.6 HLS Invocation

The HLS invocation stage is for executing the HLS synthesis tool and collecting the results from the reports generated. When a design point is passed to the HLS invocation step, it will first generate a TCL script that sets the directives accordingly. Then, it will launch the HLS synthesis using the script and extract the results from the synthesis report when the HLS tool exits. Meanwhile, it will monitor the total runtime of the HLS synthesis and report an error if the time limit is exceeded or the HLS tool exited abnormally. Once the results are acquired, it will update the datasets with the newly evaluated design point.

5 Experiments and Results

We evaluate the efficacy of the Chimera DSE tool with the Rosetta benchmark suite [2]. Specifically, it is tested on the six benchmarks in Rosetta that are dedicated to the Vivado HLS environment.

Table 1. Summary of the benchmarks in the Rosetta benchmark suite

Name	Description
3D rendering (3DF)	Renders 2D triangles into a 3D mesh model
BNN	Binarized deep neural network
Digit recognition (DR)	Recognize digits using k-NN algorithm
Face detection (FD)	Detect faces using Haar cascade classifiers
Optical flow (OF)	Compute optical flow for a set of images
Spam filtering (SF)	Use a perceptron model to filter spam emails

The information of the benchmarks is summarized in Table 1. The HLS tool used for testing is Xilinx Vivado HLS 2019.2 and the experiments are conducted on a desktop computer with an AMD 3900X CPU running at 4.1 GHz. The target device is set to the ZC706 as in [2].

The tunable knobs for each benchmark are defined by the user in the CSV file passed to Chimera, as described at the beginning of Sect. 4. In our experiments, tunable knobs that are nontrivial to tune are added to the set of tunable knobs. They not only include all loops and arrays for which the human designer inserted pragmas but also expand to other tunable knobs that have potential impacts on resource usage and latency. This enables Chimera to find relationships between the directives that the designers of Rosetta did not find, so it can explore design points that are superior to the hand-tuned design. It is worth emphasizing that, in our experiments, Chimera determines the configuration of all the directives automatically and do not use any hint from the human designer.

To acquire the Pareto curve, the resource usage value is calculated as a weighted sum of the proportion of consumed resources for the four types of resources. In the experiments, the weights are 0.4 for the BRAM and DSP, and 0.1 for LUT and FF. This is because the DSP and BRAM resources are much more scarce than LUT and FF on a typical FPGA. These weights can be adjusted by the user for different optimization targets if a lower LUT and FF usage are desirable.

Since the key target of Chimera is to be practical on real-world HLS designs, for the benchmarks in Rosetta, we aim to finish the exploration overnight on a typical workstation. Therefore, based on the preliminary profiling results on the benchmarks, we limit the total number of points to 170, such that the DSE for a benchmark can be finished within 24 h. Despite the various design space sizes and complexities of different benchmarks, we found that 170 design points is sufficient for the benchmarks to converge to a stable Pareto frontier. Such a result demonstrates the practicality of Chimera in real-world scenarios. For example, this enables the designers to run daily regression with new design changes. Among the 170 points, 20 points are the random samples evaluated in the initial sampling stage and 150 points are explored with the DSE algorithm. Notice that these numbers can be further tuned for better results for specific applications if needed. Also, due to the random nature of the algorithm, the result can vary slightly.

Table 2. Comparison between the explored and hand-tuned design points

Name	Hand-tuned		Chimera	
	Latency	Resource	Latency	Resource
3DR	5008	0.0356	5011 (+0.06%)	0.0336 (−5.6%)
BNN	409	0.0758	409 (−0.00%)	0.0757 (−0.00%)
DR	18670	0.129	18790 (+0.6%)	0.122 (−5.4%)
FD	54802	0.156	54290 (−0.93%)	0.156 (−0.00%)
OF	4496	0.155	4496 (−0.00%)	0.155 (−0.00%)
SF	11755	0.0793	11755 (−0.00%)	0.0752 (−5.17%)

The final results for the benchmarks are presented in Fig. 4. The red dot represents the hand-tuned result from the benchmark suite, the yellow curve represent the Pareto frontier explored by Chimera in a single exploration run, and the blue curve represents the Pareto frontier explored with only random proposal engine. Table 2 presents the detailed comparison between the hand-tuned design point and the lowest-latency point explored on weighted resource usage and latency. The latency number are measured in μs and the resource numbers are the weighted resource usage. The percentage numbers shows the improvement or degradation of the explored design point.

From the results, we can observe that Chimera can discover design points that are equally optimal or superior to the hand-tuned design. When optimizing for low latency, the design points found by Chimera consume fewer resources while having the same or negligibly higher latency. The diagrams also show that, in general, the introduction of EA help Chimera to discover more optimal Pareto frontiers, especially on finding low-latency or low-resource usage points.

It should be noted that, as a multi-objective optimization method, Chimera not only optimizes for lower latency but also finds a Pareto curve. In comparison, the designers of Rosetta only optimized the benchmarks for lower latency, so they miss the better designs with slightly higher latency but significantly lower resource usages. *For example, in the 3D rendering benchmark, the latency of the hand-tuned design is 5008 μs and consumes 16512 FFs, whereas the design point at the elbow of the curve (marked with a green dot in Fig. 4a) with a latency of 5016 μs only consumes 12229 FFs, which is a 26% reduction.*

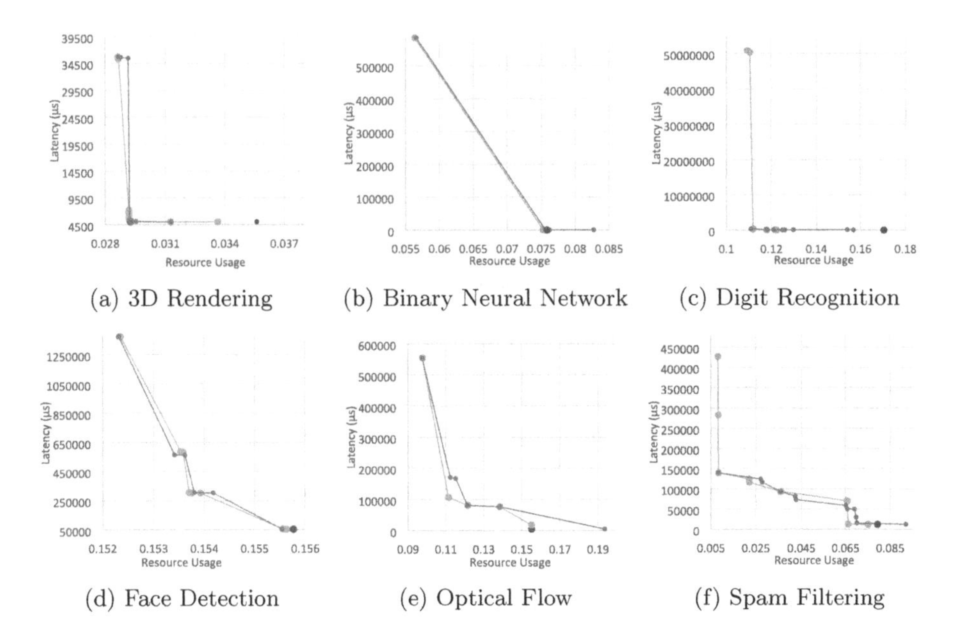

(a) 3D Rendering (b) Binary Neural Network (c) Digit Recognition

(d) Face Detection (e) Optical Flow (f) Spam Filtering

Fig. 4. Pareto frontier and the hand-tuned design point (Color figure online)

6 Conclusion

In conclusion, Chimera is a novel ML-driven software tool that facilitates the DSE process for tuning HLS directives. It combines the strengths of multiple optimization algorithms to form an hybrid efficient DSE method. The core of Chimera can also be ported to other black-box optimization problems and we will also release the source code[1] to enable future collaborations.

References

1. Rupnow, K., Liang, Y., Li, Y., Chen, D.: A study of high-level synthesis: Promises and challenges. In: 2011 9th IEEE International Conference on ASIC, pp. 1102–1105. IEEE (2011)
2. Zhou, Y., et al.: Rosetta: A realistic high-level synthesis benchmark suite for software programmable FPGAs. In: Proceedings of the 2018 ACM/SIGDA International Symposium on FPGA, pp. 269–278 (2018)
3. Xilinx, Vivado Design Suite User Guide - High-Level Synthesis (UG902). https://www.xilinx.com/support/documentation/sw_manuals/xilinx2018_2/ug902-vivado-high-level-synthesis.pdf
4. Schafer, B.C., Wang, Z.: High-level synthesis design space exploration: past, present, and future. IEEE Trans. Comput. Aided Des. Integr. Circuits Syst. **39**(10), 2628–2639 (2020)
5. Settles, B.: Active learning literature survey (2009)

[1] https://github.com/MangoY/Chimera.

6. Russo, D., Van Roy, B., Kazerouni, A., Osband, I., Wen, Z.: A tutorial on Thompson sampling, arXiv preprint arXiv:1707.02038 (2017)
7. Zhong, G., Prakash, A., Liang, Y., Mitra, T., Niar, S.: Lin-Analyzer: A high-level performance analysis tool for FPGA-based accelerators. In: 2016 53nd ACM/EDAC/IEEE DAC, pp. 1–6 (2016)
8. Cong, J., Wei, P., Yu, C.H., Zhang, P.: Automated accelerator generation and optimization with composable, parallel and pipeline architecture. In: Proceedings of the 55th Annual Design Automation Conference, ser. DAC 2018. New York, Association for Computing Machinery (2018). https://doi.org/10.1145/3195970.3195999
9. Choi, Y.-K., Cong, J.: HLS-based optimization and design space exploration for applications with variable loop bounds. ICCAD **2018**, 1–8 (2018)
10. Zhao, J., Feng, L., Sinha, S., Zhang, W., Liang, Y., He, B.: COMBA: a comprehensive model-based analysis framework for high level synthesis of real applications. In: 2017 IEEE/ACM ICCAD, pp. 430–437 (2017)
11. Schafer, B.C., Wakabayashi, K.: Machine learning predictive modelling high-level synthesis design space exploration. IET Comput. Digital Tech. **6**(3), 153–159 (2012)
12. Liu, H.-Y., Carloni, L.P.: On learning-based methods for design-space exploration with high-level synthesis. DAC **2013**, 1–7 (2013)
13. Forbes, C., Evans, M., Hastings, N., Peacock, B.: Statistical Distributions. Wiley, Hoboken (2011)
14. Nardi, L., Souza, A., Koeplinger, D., Olukotun, K.: Hypermapper: a practical design space exploration framework. In: 2019 IEEE 27th MASCOTS. Los Alamitos, CA, USA: IEEE Computer Society, pp. 425–426 (October 2019). https://doi.ieeecomputersociety.org/10.1109/MASCOTS.2019.00053
15. Nardi, L., Koeplinger, D., Olukotun, K.: Practical design space exploration, CoRR, vol. abs/1810.05236, 2018. http://arxiv.org/abs/1810.05236

Special Session on Clustering for Interpretable Machine Learning

Detecting Communities in Feature-Rich Networks with a K-Means Method

Soroosh Shalileh[1,2(✉)] and Boris Mirkin[1,3]

[1] Department of Data Analysis and Artificial Intelligence, HSE University, Pokrovsky Boulevard, 11, Moscow, Russian Federation
[2] Laboratory of Methods for Big Data Analysis (LAMBDA), HSE University, Pokrovsky Boulevard, 11, Moscow, Russian Federation
[3] Department of Computer Science and Information Systems, Birkbeck University of London, Malet Street, London WC1E 7HX, UK

Abstract. The main result of this paper is an extension of the K-means algorithm to the issue of community detection in feature-rich networks. This is based on a data-recovery criterion additively combining conventional least-squares criteria for approximation of the network link data and the feature data at network nodes. The dimension of the space at which the method operates is the sum of the number of nodes and the number of features, which may be high indeed. To tackle the so-called curse of dimensionality, we replace the innate Euclidean distance with cosine distance. We experimentally validate our proposed methods and demonstrate their efficiency by comparing them to most popular approaches using both synthetic data and real-world data.

Keywords: Node-attributed network · Feature-rich network · Cluster analysis · Community detection · Data recovery · K-means clustering

1 Introduction: Background, Motivation, and Previous Work

1.1 Background and Motivation

Community detection is widespread and applied in various applications ranging from sociology to biology to computer science. The corresponding data structure is a network, or graph, of objects, called nodes, interconnected by pair-wise links (edges). A feature-rich network adds to this a set of features associated with the nodes. We consider datasets at which the features are not necessarily categorical but may be quantitative or the combination of both, we refer to these data structures as "feature-rich" networks following [10].

A community is a group of relatively densely interconnected nodes which also are similar in the feature space. Paper [6] can be considered as a relatively recent and comprehensive survey of the approaches to detection of communities in feature-rich networks.

H. Yin et al. (Eds.): IDEAL 2021, LNCS 13113, pp. 539–547, 2021.
https://doi.org/10.1007/978-3-030-91608-4_53

In our view, they all can be classified in two directions: heuristics and data modeling. Unlike heuristic approaches, those in data modeling involve an important characteristic: evaluation of the degree of correspondence between the data and found solutions. In the data modeling direction, one may distinguish between theory-driven and data-driven approaches. A theory-driven approach involves a model for data generation leading to a probabilistic distribution, parameters of which can be recovered from the data. In contrast, data-driven approaches involve no models for data generation but rather focus on the data as is. The data in this approach is considered as an array of numbers to be recovered by decoding a model that "encodes" the data.

This paper belongs to the data-driven modeling approach. Conventionally, our data-driven model assumes a hidden partition of the nodes in non-overlapping communities, supplied with hidden parameters encoding the average link values in the network space and central community points in the feature space. According to the least-squares criterion, these are used at the decoding stage to minimize the residuals of data recovery equations.

The paper is structured as follows. Section (1.2) reviews the previous work. Section (2) describes our proposed methods. Section (3) describes the setting of our experiments for testing the algorithms. Section (4) presents results of our experiments.

1.2 Previous Work

Among heuristics approach, one can separate those modifying the classical feature-based only or network-based only clustering criteria according to the presence of two data sources [7,27] from the embedding-based approaches [5,22]. Graph-Neural Networks (GNNs) is the cornerstone of methods in [2,23,24] which can be considered as powerful heuristic.

In data modeling, we consider two approaches: theory-driven modeling and data-driven modeling.

The theory-driven approach involves both the maximum likelihood and Bayesian criteria for fitting probabilistic models for data generation, including so-called stochastic block models (SBM) [16,20]. Methods in [3,15] are based on Bayesian inferences. In [26], the authors proposed a clustering criterion to statistically model interrelation between the network structure and node attributes.

At the data-driven modeling approach, there exist several versions of non-negative matrix factorization (see [11,13]). Papers [4,25] propose combined criteria and corresponding methods for finding suboptimal solutions. The criteria are based on the least-squares approach like that by ourselves. However, these criteria involve some derived data rather than the original ones.

2 Least-Squares Criterion and Extended K-Means

2.1 Least Squares Model for Community Detection

Consider a network with features at the nodes, $A = \{P, Y\}$, over an entity set I. Here I is a set of network nodes of cardinality $|I| = N$; $P = (p_{ij})$ is an $N \times N$

matrix of mutual link weights between nodes $i, j \in I$; and $Y = (y_{iv})$ is an $N \times V$ matrix of feature values at the nodes, so that entry y_{iv} is the value of feature $v = 1, 2, ..., V$ at node $i \in I$.

Consider a partition of I in $K > 0$ crisp non-overlapping communities $S = \{S_1, S_2, ..., S_K\}$. A community S_k can be represented by an $N \times 1$ column-vector $s_k = (s_{ik})$, $i \in I$, in which $s_{ik} = 1$ if $i \in S_k$ and $s_{ik} = 0$, otherwise.

In the feature space, community S_k can be represented by a "standard" V-dimensional point $c_k = (c_{kv})$ $(v \in V, k = 1, ..., K)$ playing the role of a center of S_k. In the network space community S_k is represented by a "standard" N-dimensional point $\lambda_k = (\lambda_{kj})$ $(j \in I, k = 1, ..., K)$. According to the least-squares approach, the problem is to find a hidden membership matrix $s = (s_{ik})$, as well as community centers, $c_k = (c_{kv})$ and $\lambda_k = (\lambda_{kj})$, minimizing the sum of squared residuals:

$$F(s_{ik}, c_{kv}, \lambda_{kj}) = \rho \sum_{i,v} (y_{iv} - \sum_{k=1}^{K} c_{kv} s_{ik})^2 + \xi \sum_{i,j} (p_{ij} - \sum_{k=1}^{K} \lambda_{kj} s_{ik})^2. \quad (1)$$

The factors ρ and ξ in Eq. (1) are expert-driven constants reflecting the relative weight of each of the data sources, network and features. They are set to be equal to unity each.

2.2 The Alternating Minimization and K-Means

Optimizing criterion (1) is computationally intensive and cannot be solved exactly in a reasonable time. Consequently, various heuristic strategies have been proposed to advance to local optima. In this work, we adopt the so-called alternating minimization strategy, more specifically, the batch K-Means algorithm [21].

Our K-means-like algorithm, KEFRiN, works in iterations, each to consist of two steps: (1) given centers c_k, λ_k, find partition $\{S_k\}$ minimizing criterion (1) over S; (2) given partition $\{S_k\}$, find centers c_k, λ_k minimizing criterion (1) over all possible c_k, λ_k, $k = 1, ...K$.

In the feature space, node $i \in I$ is expressed by vector $y_i = (y_{iv})$, $v = 1, ..., V$ and in the network space, by i-th row of matrix P, $p_i = (p_{ij})$, $j = 1, ..., N$. The squared Euclidean distances between node i and standard vectors $c_k = (c_{kv})$ and $\lambda_k = (\lambda_{kj})$ are defined as

$$d_e(y_i, c_k) = \sum_{v} (y_{iv} - c_{kv})^2, \;\; d_e(p_i, \lambda_k) = \sum_{j} (p_{ij} - \lambda_{kj})^2 \quad (2)$$

Substituting these expressions in the criterion (1) and recalling that $s_{ik} = 1$ can be for unique k only, we obtain

$$F(S, c, \lambda) = \sum_{k=1}^{K} \sum_{i \in S_k} [\rho d_e(y_i, c_k) + \xi d_e(p_i, \lambda_k)] \quad (3)$$

The expression (3) leads to the following Minimum Distance rule for determining an optimal partition: for any $i \in I$ assign i to that cluster S_k for which the combined distance $d_e(i,k) = \rho d_e(y_i, c_k) + \xi d_e(p_i, \lambda_k)$ is minimum.

The standard points c_k and λ_k are to be computed as within cluster averages of the rows of matrices Y and P, as follows from the first-order optimality conditions.

This leads us to a natural extension of K-means clustering algorithm KEFRiN. Each iteration of that consists of two parts: (a) Cluster update: given K centers in the feature space and K centers in the network space, determine partition $S^{'} = \{S_k^{'}\}_{k=1}^K$ with the Minimum Distance rule; (b) Center update: Given clusters $\{S_k\}_{k=1}^K$, calculate within-cluster means in the feature space, c_k, and in the network space, λ_k.

To initialise the algorithm KEFRiN, we use a version of K-Means++ algorithm [1]. Specifically, we generate the first initial seed randomly. After a number of seeds have been generated, the next seed is the entity at which the summary distance to those already selected is the maximum.

We adopted two stopping conditions of the conventional K-means in KEFRiN. These are: 1) a pre-specified maximum number of iterations is reached; 2) the detected partitions (clustering results) are the same in two consecutive iterations. Each of them stops KEFRiN.

2.3 Cosine Distance

Given two vectors $f = (f_t)$ and $g = (g_t)$, $t = 1, ..., T$, the cosine between them is defined as

$$cos(f,g) = \frac{< f, g >}{\|f\| \|g\|} = \frac{\sum_t f_t g_t}{\sqrt{\sum_t f_t^2} \sqrt{\sum_t g_t^2}}. \quad (4)$$

One may say that the cosine of two vectors is just the inner product of them after they had been normed. The cosine gives rise to what is referred to the cosine distance between f and g, $d_c(f,g) = 1 - cos(f,g)$.

In fact, the cosine distance between f and g is proportional to the squared Euclidean distance between f and g after they had been normed. Indeed, assuming that f and g are normed $\|f\| = \|g\| = 1$, we have $d_e(f,g) = \sum_t (f_t - g_t)^2 = \sum_t f_t^2 + \sum_t g_t^2 - 2\sum_t f_t g_t = \|f\|^2 + \|g\|^2 - 2 < f, g > = 2 - 2cos(f,g) \geq d_c(f,g)$.

Therefore, under the assumption that all vectors y_i, c_k, p_i, and λ_k occurring in the Eq. (3) are normed, that equation can be rewritten:

$$F_n(S, c, \lambda) = 2 \sum_{k=1}^{K} \sum_{i \in S_k} [\rho d_c(y_i, c_k) + \xi d_c(p_i, \lambda_k)] \quad (5)$$

This leads us to propose one more version of K-means extended to feature-rich networks: that using the cosine metric d_c as the distance in KEFRiN algorithm above. We use abbreviation KEFRiNe for the version based on the squared Euclidean metric and KEFRiNc for that based on the cosine distance. A Python source code of both versions of KEFRiN method is publicly available at https://github.com/Sorooshi/KEFRiN.

Table 1. Real world datasets: Symbols N, E, and F stand for the number of nodes, the number of edges, and the number of node features, respectively.

Name	Nodes	Edges	Features	Number of communities	Ground truth	Ref.
Malaria HVR6	307	6526	6	2	Cys labels	[12]
Parliament	451	11646	108	7	Political parties	[3]
COSN	46	552	16	2	Region	[8]
Cora	2708	5276	1433	7	Computer science research area	[17]
Amazon Photo	7650	71831	745	8	Product categories	[19]

It should be mentioned that other distances can be used in this framework, such as, say, that proposed in [14]. We, however, feel that the cosine distance fits into the least-squares criterion real smoothly.

3 Setting of Experiments

Algorithms Under Comparison
We compare the performance of our proposed methods with three most popular algorithms of the model-based approach, CESNA [26], SIAN [15], and DMoN [23]. Also, we add another algorithm, SEANAC, developed within our least-squares approach [18]. Author-made codes of all these algorithms are publicly available.

Real World Datasets: Two of the algorithms under comparison, CESNA and SIAN, restrict the features to be categorical. Thus, whenever a data set contains a quantitative feature, we convert it into a categorical version. For more information on this refer to [18]. The datasets are publicly available at https://github.com/Sorooshi/KEFRiN/tree/main/data. A brief overview of the five real-world data sets under consideration can be found in Table 1.

Synthetic Datasets: We apply the proposed framework and the corresponding software in [18] for experimentally validating and comparing the algorithms under consideration at synthetic datasets generated according to what is referred to as stochastic block modeling.

Evaluation Metrics. To evaluate and compare obtained clustering results in our experiments, we use the Adjusted Rand Index (ARI) [9].

4 Experimental Results

4.1 Comparison on Real-World Datasets

Table 2 reports the computational results by all the algorithms under consideration over real-world datasets. KEFRiNc and DMoN dominate this table.

In contrast, CESNA results are inferior here. Finally, the SIAN algorithm performs poorly; however, it produces a good result at the Parliament dataset.

It ought to mention that the authors of DMoN reported somewhat different results on Amazon photo dataset, probably because they used a subset of the original dataset [23]. In this paper, we use the entire Amazon photo dataset.

Table 2. Comparison of the methods on real-world data sets: average values of ARI are presented over 10 random initializations. The best results are highlighted in bold-face; those second-best are underlined.

Dataset	CESNA	SIAN	DMoN	SEANAC	KEFRiNe	KEFRiNc
HRV6	0.20(0.00)	0.39(0.29)	0.64(0.00)	0.49(0.11)	0.34(0.02)	**0.69(0.38)**
Parliament	0.25(0.00)	**0.79(0.12)**	0.48(0.02)	0.28(0.01)	0.15(0.09)	0.41(0.05)
COSN	0.44(0.00)	0.75(0.00)	0.91(0.00)	0.72(0.02)	0.65(0.18)	**1.00(0.00)**
Cora	0.14(0.00)	0.17(0.03)	**0.37(0.04)**	0.00(0.00)	0.00(0.00)	0.21(0.01)
Amazon Photo	0.19(0.000)	N.A	**0.44(0.04)**	N.A	0.06(0.01)	0.43(0.06)

Table 3. Comparison of the methods on small-size synthetic data sets with categorical attributes: the average ARI index and its standard deviation over ten generated datasets. The best results are highlighted in bold-face; second-best ones are underlined.

Dataset	CESNA	SIAN	DMoN	SEANAC	KEFRiNe	KEFRiNc
0.9, 0.3, 0.9	**1.00(0.00)**	0.554(0.285)	0.709(0.101)	0.994(0.008)	0.886(0.116)	0.922(0.119)
0.9, 0.3, 0.7	0.948(0.105)	0.479(0.289)	0.380(0.107)	**0.974(0.024)**	0.835(0.138)	0.819(0.142)
0.9, 0.6, 0.9	0.934(0.075)	0.320(0.255)	0.412(0.109)	**0.965(0.013)**	0.963(0.072)	0.726(0.097)
0.9, 0.6, 0.7	**0.902(0.063)**	0.110(0.138)	0.213(0.051)	0.750(0.117)	0.694(0.096)	0.711(0.145)
0.7, 0.3, 0.9	0.965(0.078)	0.553(0.157)	0.566(0.105)	**0.975(0.018)**	0.788(0.117)	0.877(0.130)
0.7, 0.3, 0.7	**0.890(0.138)**	0.508(0.211)	0.292(0.077)	0.870(0.067)	0.836(0.115)	0.795(0.117)
0.7, 0.6, 0.9	0.506(0.101)	0.047(0.087)	0.345(0.064)	**0.896(0.067)**	0.762(0.169)	0.834(0.132)
0.7, 0.6, 0.7	0.202(0.081)	0.030(0.040)	0.115(0.058)	**0.605(0.091)**	0.574(0.142)	0.540(0.107)

4.2 Comparison on Synthetic Datasets with Categorical Features

Table 3 and Table 4 compare the performance of the algorithms under consideration at synthetic networks with categorical features, small-sized and medium-sized, respectively.

Two methods, SEANAC and CESNA, dominate the Table 3 at small-size networks, although the superiority of SEANAC is clearly seen at the "difficult" settings for (p, q, ϵ) in the last two rows. SEANAC remains the only winner at medium-sized networks, whereas CESNA's performance decisively declines at the settings of the last two rows.

KEFRiN methods perform relatively well getting the second-best position at many settings. One should also notice their relatively low computational cost. DMoN and SIAN performances are relatively poor, especially at the medium-sized datasets at which the performance of them are undermined by the convergence issues.

Table 4. Comparison of the methods on medium-size synthetic data sets with categorical attributes: the average ARI index and its standard deviation over ten generated datasets. The best results are highlighted in bold-face and second ones are under-lined.

Dataset	CESNA	SIAN	DMoN	SEANAC	KEFRiNe	KEFRiNc
0.9, 0.3, 0.9	0.894(0.053)	0.000(0.000)	0.512(0.137)	**1.000(0.000)**	0.508(0.205)	0.724(0.097)
0.9, 0.3, 0.7	0.849(0.076)	0.000(0.000)	0.272(0.073)	**0.996(0.005)**	0.777(0.129)	0.742(0.182)
0.9, 0.6, 0.9	0.632(0.058)	0.000(0.000)	0.370(0.063)	**0.998(0.002)**	0.279(0.204)	0.652(0.110)
0.9, 0.6, 0.7	0.474(0.089)	0.000(0.000)	0.168(0.030)	**0.959(0.032)**	0.766(0.180)	0.733(0.083)
0.7, 0.3, 0.9	0.764(0.068)	0.026(0.077)	0.446(0.099)	**1.000(0.001)**	0.364(0.247)	0.641(0.111)
0.7, 0.3, 0.7	0.715(0.128)	0.000(0.000)	0.228(0.077)	**0.993(0.002)**	0.829(0.085)	0.797(0.088)
0.7, 0.6, 0.9	0.060(0.024)	0.000(0.000)	0.332(0.051)	**0.998(0.001)**	0.426(0.246)	0.591(0.094)
0.7, 0.6, 0.7	0.016(0.008)	0.000(0.000)	0.133(0.016)	**0.909(0.035)**	0.671(0.196)	0.773(0.070)

5 Conclusion

The least-squares criterion leads to computationally hard problems, which are customarily addressed with various heuristics. In our paper [18], we applied a greedy one-by-one strategy. In this work, we follow a different – alternating optimization – strategy. This leads us to propose here KEFRiN method, an extension of the classical K-Means clustering algorithm developed for feature space only data [21], to the case of feature-rich networks. Since the dimension of our combined space can be rather high indeed, we apply both squared Euclidean distance and cosine distance, after establishing a strong mathematical relation between them.

The proposed KEFRiN method works quite well on the real-world datasets, especially in comparison with the popular state-of-the-art methods (see Table 2). To be able to compare KEFRiN with these methods on synthetic data, we use here only those with categorical features. At such a setting the advantage of KEFRiN – applicability to larger datasets – does not work, so that our previous method, SEANAC [18], appears more effective. However, at synthetic data combining both quantitative and categorical data, the KEFRiNc method is frequently the best (not shown because of length constraints).

The future work should address some shortcomings of our approach, such as lack of advice regarding the balancing factors ρ and ξ, and include real-world applications.

References

1. Arthur, D., Vassilvitskii, S.: k-means++: The advantages of careful seeding. In: Proceedings of the Eighteenth Annual ACM-SIAM Symposium on Discrete Algorithms, pp. 1027–1035 (2006)
2. Bianchi, F., Grattarola, D., Alippi, C.: Spectral clustering with graph neural networks for graph pooling. In: International Conference on Machine Learning (PMLR), pp. 874–883 (November 2020)

3. Bojchevski, A., Günnemanz., S.: Bayesian robust attributed graph clustering: joint learning of partial anomalies and group structure. In: Thirty-Second AAAI Conference on Artificial Intelligence (2018)
4. Cao, J., Wanga, H., Jin, D., Dang., J.: Combination of links and node contents for community discovery using a graph regularization approach. Future Gener. Comput. Syst. **91**, 361–370 (2019)
5. Cavallari, S., Zheng, V.W., Cai, H., Chang, K.C., Cambria., E.: Learning community embedding with community detection and node embedding on graphs. In: Proceedings of the 2017 ACM Conference on Information and Knowledge Management, pp. 377–386. ACM (2017)
6. Chunaev., P.: Community detection in node-attributed social networks: a survey. Comput. Sci. Rev. **37**, 100286 (2020)
7. Combe, D., Largeron, C., Géry, M., Egyed-Zsigmond., E.: I-louvain: an attributed graph clustering method. In: E. Fromont, T. De Bie, M. van Leeuwen (eds.), Advances in Intelligent Data Analysis XIV, pp. 181–192 (2015)
8. Cross, R., Parker., A.: The Hidden Power of Social Networks: Understanding How Work Really Gets Done in Organizations. Harvard Business Press, Boston (2004)
9. Hubert, L., Arabie, P.: Comparing partitions. J. Classif. **2**(1), 193–218 (1985)
10. Interdonato, R., Atzmueller, M., Gaito, S., Kanawati, R., Largeron, C., Sala, A.: Feature-rich networks: going beyond complex network topologies. Appl. Netw. Sci. **4** (2019). https://doi.org/10.1007/s41109-019-0111-x
11. Jin, D., He, J., Chai, B., He, D.: Semi-supervised community detection on attributed networks using non-negative matrix tri-factorization with node popularity. Front. Comput. Sci. **15**(4), 1–11 (2021)
12. Larremore, D., Clauset, A., A, C.B.: network approach to analyzing highly recombinant malaria parasite genes. PLoS Comput. Biol. **9**(10), e1003268 (2013)
13. Luo, X., Liu, Z., Shang, M., Zhou, M.: Highly-accurate community detection via pointwise mutual information-incorporated symmetric non-negative matrix factorization. IEEE Trans. Netw. Sci. Eng. **8**(1), 463–476 (2020)
14. Naranjo, R., andand L. Garmendia, M.S.: A convolution-based distance measure for fuzzy singletons and its application in a pattern recognition problem. Integr. Comput. Aided Eng. **28**(1), 51–63 (2021)
15. Newman, M., Clauset, A.: Structure and inference in annotated networks. Nat Commun. **7**, 11863 (2016)
16. Peel, L., Larremore, D., Clauset, A.: The ground truth about metadata and community detection in networks. Sci. Adv. **3**(5), e1602548 (2017)
17. Sen, P., Namata, G., Bilgic, M., Getoor, L., Galligher, B., Eliassi-Rad, T.: Collective classification in network data. AI Mag. **29**(3), 93–106 (2008)
18. Shalileh, S., Mirkin, B.: A one-by-one method for community detection in attributed networks. In: Analide, C., Novais, P., Camacho, D., Yin, H. (eds.) IDEAL 2020. LNCS, vol. 12490, pp. 413–422. Springer, Cham (2020). https://doi.org/10.1007/978-3-030-62365-4_39
19. Shchur, O., Mumme, M., Bojchevski, A., Günnemann, S.: Pitfalls of graph neural network evaluation. arXiv preprint arXiv:1811.05868 (2018)
20. Stanley, N., Bonacci, T., Kwitt, R., Niethammer, M., Mucha, P.: Stochastic block models with multiple continuous attributes. Appl. Netw. Sci. **4**(1), 1–22 (2019)
21. Steinley, D.: K-means clustering: a half-century synthesis. Br. J. Math. Stat. Psychol. **59**(1), 1–34 (2006)
22. Sun, H., et al.: Network embedding for community detection in attributed networks. ACM Trans. Knowl. Discov. Data (TKDD) **14**(3), 1–25 (2020)

23. Tsitsulin, A., Palowitch, J., Perozzi, B., Müller, E.: Graph clustering with graph neural networks. arXiv preprint arXiv:2006.16904 (2020)
24. Wang, C., Pan, S., Hu, R., Long, G., Jiang, J., Zhang, C.: Attributed graph clustering: A deep attentional embedding approach. arXiv preprint arXiv:1906.06532. (2019)
25. Wang, X., Jin, D., Cao, X., Yang, L., Zhang, W.: Semantic community identification in large attribute networks. In: Proceedings of the Thirtieth AAAI Conference on Artificial Intelligence, AAAI 2016, pp. 265–271 (2016)
26. Yang, J., McAuley, J., Leskovec, J.: Community detection in networks with node attributes. In: IEEE 13th International Conference on Data Mining, pp. 1151–1156 (2013)
27. Ye, W., Zhou, L., Sun, X., Plant, C., Böhm., C.: Attributed graph clustering with unimodal normalized cut. In: Ceci, M., Hollmén, J., Todorovski, L., Vens, C., Džeroski, S. (eds.) Machine Learning and Knowledge Discovery in Databases, pp. 601–616 (2017)

Meta-Learning Based Feature Selection for Clustering

Oleg Taratukhin(✉) and Sergey Muravyov

ITMO University, St. Petersburg, Russia
smuravyov@itmo.ru

Abstract. Clustering is a highly demanded task nowadays. However, it requires the attention of human experts and it might certainly benefit from automation. This paper presents a method to perform simultaneous automatic algorithm selection and hyperparameter tuning with feature selection for clustering. The algorithm also features a meta-model to predict promising algorithm configurations based on dataset properties. Experimental results are provided for a set of benchmark datasets, it is shown that the proposed method outperforms known alternatives.

Keywords: Machine learning · Clustering · Feature selection · Meta-learning

1 Introduction

1.1 Overview

Clustering is a classical and fundamental task in machine learning, it is applied in different life domains. Importance of unsupervised learning and semi-supervised learning techniques are steadily growing as datasets are getting bigger and more complex. Most of the collected data is not labeled and labeling requires additional resources such as crowd sourcing or other use of human experts. The clustering results are than examined and interpreted by data analysts. As there can be hundreds and more features in the data interpreting the clusters might be difficult. Feature selection can be used to reduce the number of used features, simplify the model, make it more robust and increase quality of clustering. It might be business critical to derive key insights from best partition, but in order to get it experts have to spend much time to get the best solution. This explains the relevance of automating process of arriving to best clustering solution using methodological and algorithmic approach, for instance by developing tools for automated machine learning. Some parts of the data processing pipelines are notoriously difficult, as practitioners have to collect the data manually, store and process it, derive features from raw data and deploy it to production with monitoring. While the goal of fully automated machine learning systems is still not feasible we can automate certain parts of it.

H. Yin et al. (Eds.): IDEAL 2021, LNCS 13113, pp. 548–559, 2021.
https://doi.org/10.1007/978-3-030-91608-4_54

There is a significant research of AutoML for supervised learning problems, i.e. classification and regression. Comparatively, there is not much progress in unsupervised applications of AutoML, including clustering. So far proposed algorithms for AutoML in clustering focus on algorithm selection and hyperparameter tuning, it is possible to include feature engineering parts of the data science pipeline and potentially other parts as well.

There exists many different clustering and feature selection algorithms each with their hyperparameters. This results in huge number of combinations of two algorithms with their hyperparameters. It takes a lot of time to evaluate all of them for each particular dataset by brute force even if done automatically, each algorithm pair has limited computation resources in order to explore configuration space and arrive at optimal results. In order to select only the most promising configurations experienced practitioners use their knowledge of which algorithms worked best for similar datasets. One of the options to formalize this idea is to use meta-learning.

It is possible to reframe the original problem of selecting algorithms to supervised learning task of classification, each algorithm is treated as a class and a meta-model is trained to predict top class based on the meta-features of the dataset. This method can be used for selection of a set of potentially best algorithm combinations and only tune their hyperparameters instead of going through all algorithms, this allows to allocate more resources for exploring configuration space of the algorithms that showed better results of the training set of datasets.

1.2 Clustering

Mathematically correct and commonly agreed definition of clustering does not exist in the literature now [1]. There exists a number of ways to asses clustering results. Some measures use external information about the task and called external measures, internal measures on the other hand use solely partition structure [2]. Many clustering algorithms and their variation exists and are widely available, such as K-Means, DBSCAN, Gaussian Mixture, Spectral clustering and many others [3]. However, it is known that no single algorithm is able to achieve the best quality metrics over all datasets [4]. This explains why we need to choose the most appropriate algorithm for each given task.

1.3 Feature Selection

The dimensionality of available data in many common applications is increasing, while the intrinsic dimentionality is often much lower than the original feature space. Feature selection algorithms can be used to discard redundant and less informative features, thus the models get smaller and easier to interpret, faster as less data should be processed and more robust as some of the noise gets discarded.

The other way to reduce dimentionality is to use feature extraction methods, these methods construct new feature space from the original features using some transformation, but semantics of the original feature space is generally not preserved in the process. This results in hard explainability of both feature space and produced clustering. While it might not matter in some applications this has crucial importance on applications where interpretability is a hard requirement. Some of the well known and widely used algorithms from this family are PCA [5], Isomap [6] and t-SNE [7].

Feature selection is known to be NP-hard problem [8], thus any algorithm that is not brute force is guaranteed to fail to find optimal feature subspace. From this many different feature selection algorithms have emerged, each algorithm may behave differently in its goal to find "good" feature subset and may yield different results.

Feature selection algorithms for clustering, similarly to feature selection methods for supervised learning, can be classified into filter-method, wrapper-methods and embedded methods.

Filter methods for feature selection tend to be the least computationally demanding and unbiased towards any clustering algorithm, that makes them good candidates for algorithms to consider in unsupervised AutoML pipelines [9].

Wrapper algorithms use a clustering algorithm to evaluate the quality of selected features, first it finds a candidate subset, than it evaluates the clustering quality using selected feature subset, finally the steps are repeated until stopping criteria is not met. Evaluating all possible subsets is not feasible for real world high-dimensional data, therefore different algorithms adopt heuristics to reduce the search space. Wrapper models are very computationally intensive, which is undesirable property for resource and time-bound scenarios of AutoML. Because of this in some cases it might be better to use filter methods as it allows more exploration of the clustering algorithm configuration space in the same time.

Unsupervised feature selection algorithms has been studied for text-based data and general data [9,10].

1.4 Combined Algorithm Selection and Hyperparameter Optimization

Rice et al. has formulated [11] general framework to work general algorithm selection. Essentially, training database of datasets gets trained and evaluated using certain evaluation criterion, finally evaluation information is feed back to algorithm selection phase to aid optimization process. This framework can be reformulated to work with meta-learning [12].

There are two similar problems - algorithm selection [11] and hyperparameter tuning [13]. These problems has been formulated as a single CASH problem [14], Combined Algorithm Selection and Hyperparameter optimization, in the following way:

Let $\mathcal{A} = \{A^1, A^2, A^3, ..., A^R\}$ be set of algorithms, and let the hyperparameters of each algorithm A^j have domain Λ^j. Further, let D_{train} be training set that is split into K cross-validation folds: $\{D^1_{train}, D^2_{train}, D^3_{train}, ..., D^K_{train}\}$ and $\{D^1_{valid}, D^2_{valid}, D^3_{valid}, ..., D^K_{valid}\}$. Finally, let $\mathcal{L}(A^j, \lambda, D^i_{train}, D^i_{valid})$ denote that cost function that the algorithms achieves on validation subset when trained on training subset with given hyperparameters. Then CASH problem is to find pair of algorithm and hyperparameters that minimize the loss:

$$A^*, \lambda^* = argmin_{A^j \in \mathcal{A}, \lambda \in \Lambda^j} \frac{1}{K} \mathcal{L}(A^j, \lambda, D^i_{train}, D^i_{valid}) \quad (1)$$

In order to solve this problem a number of approaches has been proposed. The problem can be viewed naively without any assumptions of the search space, assuming each individual configuration is completely independent, than possible approaches are grid and random search in the configuration space. Both these approaches are easy to implement, however random search is superior as it is able to take advantage of hyperparameter hierarchy. More sophisticated approaches use Bayesian optimization and other techniques, most notably SMAC - sequential model-based algorithm configuration [15]. SMAC being state-of-the-art algorithm configuration optimizer has different achievements for both supervised and unsupervised tasks.

There exist a number of approaches to solve this problem for supervised case, such as Auto-WEKA [16], hyperopt [17], Auto-sklearn [18], TPOT [19] and Auto-Keras [20]. These frameworks perform a combined optimization across different preprocessors, algorithms and hyperparameters. For unsupervised learning no well known and publicly available frameworks exists yet. In clustering, feedback information is expensive as we have to compute measure value for the whole clustering result. So approaches that work so well in supervised setting might not work well in clustering.

1.5 Meta-Learning

There are many different options how to select and tune algorithms, meta-learning approach is one of them and it is effective in both supervised and unsupervised cases. The approach is essentially to reformulate original problem of algorithm selection to multi-class classification problem, where each class is an algorithm. A model is then trained to predict top performing algorithm or rank them. In contrast to SMAC, this approach allows to precompute meta-knowledge in a form of the classifier, that is used in inference. Compared to approaches of TPOT, SMAC and other algorithms it is relatively cheap for run-time performance for the practitioner, and can be used on top, so drop some candidates prior to SMAC or others.

In the literature on Meta-learning meta-features are measures that are used to characterize datasets and their relationship with algorithm bias [21]. In our implementation we used *pyMFE* library [22] in unsupervised way that has many features and feature groups implemented including but not limited to general information (number of instances, attributes), standard statistical measures, information-theoretic features, and different variations of landmarking [23].

1.6 Related Works

A number of achievements in different domains attributed as foundation for my research – clustering algorithms, unsupervised feature selection method, meta-learning and optimization techniques for algorithm configuration.

Ferrari et al. proposed use of meta-learning to select clustering algorithm [24], the paper proposed new approach to characterize clustering problems based on the similarity among objects and new methods to combine internal indices for ranking algorithms based on their performance.

Muravov et al. proposed method to perform simultaneous algorithm selection and hyperparameter optimization using reinforcement learning-based algorithm [25].

Another alternative have been proposed by Blumenberg et al. [26]. This paper presents a library to solve clustering algorithm selection problem in Python. It uses arbitrary measure to evaluate results, the examples assume that the library is used with external measures as optimization target. However, external measures are not always applicable as there might be no additional information about the problem.

Another library is implemented by Wong et al. is called *autocluster* [27]. This library performs clustering algorithm selection, is based on SMAC and has meta-learning. Meta-learning is used for warmstarting SMAC optimizer with similar configurations, discovered on training phase.

2 Proposed Method

The method is based on framing the original problem as optimization problem and solve it using the SMAC algorithm to select and tune algorithms.

In order to choose both feature selection and clustering algorithm algorithms algorithm space is constructed as Cartesian product of feature selection algorithm and clustering algorithm spaces. Configuration spaces of individual algorithms are also combined in a similar fashion. More formally, let $A = A_{fs} \times A_{cl}$ be combined algorithm space of A_{fs} (feature selection algorithm space) and A_{cl} (clustering algorithm space). Hyperparameters space is constructed as follows:

$$\begin{aligned} A &= A_{fs} \times A_{cl} \\ \Lambda &= \cup_{A^k \in A_{cl}} \cup_{A^j \in A_{fs}} \Lambda^j_{fs} \times \Lambda^k_{cl} \end{aligned} \tag{2}$$

In implementation all algorithm pairs are expressed as categorical variable and each attribute is prefixed with algorithm name, this allows to distinguish between same-named hyperparameters for different algorithms and tune them separately.

In order to build meta-knowledge about the task a dataset of clustering dataset has been collected and annotated. Annotation was required because real world datasets have different structure, numerical and categorical features should be handled differently. This allows us not to limit our meta-dataset to numerical-only datasets and utilize all available features. For instance, titanic benchmark dataset is annotated as follows:

Listing 1.1. An example dataset annotation for Titanic

```
{
    ``path'': ``titanic.csv",
    ``target": ``Survived",
    ``categorical": [``Sex", ``Embarked"],
    ``ignore": [``PassengerId", ``Name", ``Ticket", ``Cabin"],
    ``numerical": [``Age", ``SibSp", ``Parch", ``Fare"],
    ``ordinal": {
        ``Pclass": [``1", ``2", ``3"]
    }
}
```

Meta-dataset has been augmented by synthetic data to increase dataset size, diverse generated datasets has been added. Data augmentation is needed to collect more data for the meta-model. Depending on the exact number of algorithm pairs considered by a practitioner there might not be enough only real world selected annotated (as in the example above) datasets to train a classifier. Additional synthetic data helps to produce more samples for the classifier to learn from. The training pipeline is illustrated in Fig. 1.

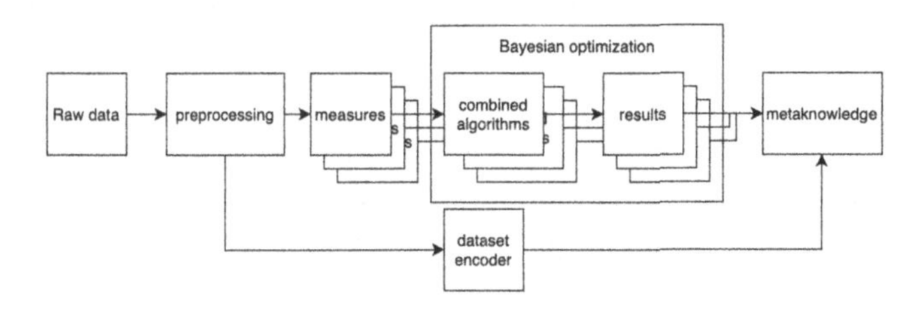

Fig. 1. Pretraining pipeline

As all results from datasets has been collected as meta-knowledge a meta-model is trained to predict top performing configurations on meta-feature representation of the dataset. This allows to test less configurations after pretraining has been performed. The learned meta-model is then used in the inference pipeline as illustrated in Fig. 2.

3 Results

3.1 Experiment Setup

Total of 35 real datasets and 200 synthetic datasets has been collected or generated. Real datasets were collected from different domains and different formats as the most used in clustering benchmarking, datasets were retrieved from

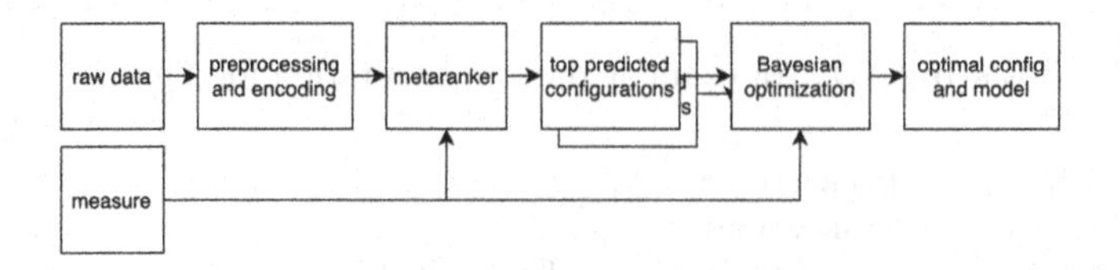

Fig. 2. Pretraining pipeline

OpenML [28]. Generated datasets were produced by several benchmark generators with different parameters that are present is standard packages (i.e. sklearn).

There exists many different CVIs, following 4 measures are selected as it is shown [29] that these measures correlate the most with human assessor scores:

1. Calinski-Harabasz (CH) measure – measure that has distance from elements to each cluster centroid as linkage, the difference between clusters is based on distance t
2. Silhouette (Sil) measure – normalized summation measure, linkage is based on distance within cluster. Difference is based on distance to closest element from cluster.
3. OS index – SV index variation, which is one of the most recent proposed measures. However, the dissimilarity is calculated in a more complicated way.
4. gD41 – a variation of generalized Dunn measure. This particular variation uses distance between cluster centroids as dissimilarity and cluster diameter as similarity.

We considered the following clustering algorithms: DBSCAN, KMeans, MiniBatchKMeans, AffinityPropagation, MeanShift, SpectralClustering, OPTICS, Birch and Gaussian Mixture models. And the following unsupervised feature selection methods RFE, Lasso, SPEC, LFBSS, MFSC, WKmeans, NormalizedCut, filter methods are considered as in review [9]. Additionally a dummy feature selection model has been added, it returns the original feature set and is used to emulate pipeline without feature selection.

Parallel variation of SMAC algorithm has been used to train and optimize algorithms. On each dataset top k ($k = 5$ in the provided results) algorithms with their respective configurations and results has been saved. SMAC has been configured to output top k configuration per run on each dataset. The meta-model is trained to perform multi-label classification: 1 if algorithm pair has been selected among top k algorithms and 0 otherwise.

3.2 Algorithm Performance

Different algorithms has seen different results on dataset from the training data. Some algorithms were often selected as best. Also, algorithm pairs with NullModel (no feature selection) were better on small subset of training datasets. The order and specific numbers is different for different measures. Tables 1, 2, 3 and 4 show top 10 number of times each algorithm configuration was selected among top 5 configuration. Note that the tables only show numbers for top 10 configuration for each measure, however, the values were obtained for every configuration.

Table 1. Top 10 algorithm results for Silhouette measure

	Configuration	#Best
0	FixedSPEC_MiniBatchKMeans	180
1	FixedSPEC_AffinityPropagation	112
2	MCFS_AffinityPropagation	104
3	NormalizedCut_SpectralClustering	75
4	Lasso_MiniBatchKMeans	34
5	MCFS_Birch	30
6	GenericSPEC_SpectralClustering	29
7	FixedSPEC_KMeans	24
8	MCFS_KMeans	22
9	Lasso_DBSCAN	18
10	NormalizedCut_MiniBatchKMeans	17

Table 2. Top 10 algorithm results for Calinksi-Harabasz measure

	Configuration	#Best
0	FixedSPEC_MiniBatchKMeans	150
1	FixedSPEC_AffinityPropagation	98
2	MCFS_AffinityPropagation	95
3	NormalizedCut_SpectralClustering	40
4	MCFS_Birch	39
5	Lasso_MiniBatchKMeans	37
6	GenericSPEC_SpectralClustering	30
7	Lasso_AffinityPropagation	25
8	FixedSPEC_KMeans	24
9	NormalizedCut_MiniBatchKMeans	22
10	MCFS_KMeans	21

3.3 Meta-Algorithm Results

Meta model has been learned to predict top 5 configurations from vectorized dataset description. In order to derive meta-feature description we used [22]. Dataset is composed from meta-feature representation of the dataset and vectorized encoding of top algorithms. This reframes the problem to supervised

Table 3. Top 10 algorithm results for OS measure

	Configuration	#Best
0	FixedSPEC_MiniBatchKMeans	64
1	MCFS_AffinityPropagation	54
2	FixedSPEC_AffinityPropagation	49
3	NormalizedCut_SpectralClustering	42
4	NullModel_GaussianMixture	37
5	NormalizedCut_Birch	32
6	GenericSPEC_SpectralClustering	28
7	MCFS_Birch	25
8	Lasso_MiniBatchKMeans	20
9	GenericSPEC_Birch	20
10	FixedSPEC_GaussianMixture	19

Table 4. Top 10 algorithm results for gD41 measure

	Configuration	#Best
0	FixedSPEC_MiniBatchKMeans	103
1	MCFS_AffinityPropagation	94
2	FixedSPEC_AffinityPropagation	77
3	NormalizedCut_SpectralClustering	67
4	MCFS_Birch	35
5	Lasso_MiniBatchKMeans	28
6	Lasso_DBSCAN	22
7	NullModel_GaussianMixture	18
8	NormalizedCut_GaussianMixture	15
9	Lasso_GaussianMixture	15
10	FixedSPEC_SpectralClustering	15

learning. For each individual algorithm a model is fit to predict whether algorithm will be selected among top algorithms. For each algorithm we now have a binary classification problem. We get different classification accuracy for different algorithms. Individual scores are represented in the Table 5.

3.4 Comparison with Autocluster

Experiments were conducted to compare the proposed approach with autocluster package. Autocluster doesn't have feature selection, but has other dimensionality reduction techniques. The set of clustering algorithms is common for the experiments, however in both packages the user is able to configure desired set of algorithms to use. Autocluster package required multiple modification to make it work with all 4 measures mentioned above and also fix some stability issues. In order to evaluate the obtained scores we used Wilcoxon test with alternative that autocluster produces better measure values: $H_1 : score_{afsfc} > score_{autocluster}$.

The following results in Table 6 show that for 3 out of 4 measures our method outperforms autocluster, on Calinksi-Harabasz measure the difference is not significant. This result in certain is expected as in the worst case NullModel gets selected for feature selection and the results would be similar to the ones of

Table 5. Individual results of classification per algorithm

	Configuration	F1		Configuration	F1
0	NullModel_MeanShift	0.890698	26	NormalizedCut_KMeans	0.735981
1	Lasso_SpectralClustering	0.832558	27	MCFS_GaussianMixture	0.613797
2	Lasso_Birch	0.867991	28	FixedSPEC_MiniBatchKMeans	0.765803
3	NullModel_SpectralClustering	0.625581	29	FixedSPEC_GaussianMixture	0.805213
4	NullModel_GaussianMixture	0.830065	30	GenericSPEC_AffinityPropagation	0.582944
5	FixedSPEC_KMeans	0.781357	31	WKMeans_DBSCAN	0.969767
6	NullModel_MiniBatchKMeans	0.672573	32	MCFS_MiniBatchKMeans	0.901587
7	MCFS_DBSCAN	0.801402	33	GenericSPEC_MiniBatchKMeans	0.759921
8	GenericSPEC_SpectralClustering	0.848597	34	WKMeans_KMeans	0.703791
9	Lasso_MiniBatchKMeans	0.870579	35	WKMeans_MiniBatchKMeans	0.824334
10	NullModel_AffinityPropagation	0.527123	36	NormalizedCut_SpectralClustering	0.898488
11	Lasso_AffinityPropagation	0.832039	37	GenericSPEC_DBSCAN	0.967442
12	NullModel_Birch	0.731722	38	FixedSPEC_Birch	0.836286
13	GenericSPEC_GaussianMixture	0.630340	39	NullModel_KMeans	0.792891
14	WKMeans_AffinityPropagation	0.679365	40	WKMeans_GaussianMixture	0.629384
15	Lasso_DBSCAN	0.908860	41	WKMeans_SpectralClustering	0.761203
16	MCFS_Birch	0.734233	42	MCFS_SpectralClustering	0.913915
17	NormalizedCut_Birch	0.792645	43	NullModel_DBSCAN	0.995349
18	NormalizedCut_MeanShift	0.865116	44	NormalizedCut_AffinityPropagation	0.706368
19	GenericSPEC_KMeans	0.707792	45	MCFS_KMeans	0.779820
20	NormalizedCut_DBSCAN	0.974419	46	FixedSPEC_AffinityPropagation	0.692161
21	MCFS_AffinityPropagation	0.761336	47	WKMeans_Birch	0.952830
22	FixedSPEC_SpectralClustering	0.616270	48	GenericSPEC_Birch	0.718149
23	Lasso_GaussianMixture	0.790873	49	Lasso_KMeans	0.657429
25	NormalizedCut_MiniBatchKMeans	0.683370	50	NormalizedCut_GaussianMixture	0.606971

Table 6. Wilcoxon test results

Clustering measure	p-value
Calinksi-Harabazs	0.13
Silhouette	0.034
OS	0.0015
gD41	0.0083

autocluster, in case feature selection improved the scores, than the results would apriori be better.

Autocluster is able to leverage past experiments and perform warm-starting of the algorithm, it also has dimensionality reduction in a form of UMAP, TSNE and other algorithms, this additional steps and improvements are considered for future work.

4 Conclusion

In this paper we present and examine a solution for automatic feature selection for clustering and a meta-model to narrow down search for optimal algorithm configuration. The proposed approach uses sequential model based Bayesian optimization and meta-learning. It appears that the proposed approach to add feature selection increases clustering quality measure values and brings benefits of feature selection. The suggested method can be improved by fine tuning on more specific dataset distributions, additional dimensionality reduction via other methods and by using warm-starting. Proposed method for selecting potentially top algorithm configuration is flexible in the number of desired solutions. More algorithm configuration means more robust top scores, but increases computational resources.

References

1. Hennig, C.: What are the true clusters? (2015)
2. Arbelaitz, O., Gurrutxaga, I., Muguerza, J., Pérez, J., Perona, I.: An extensive comparative study of cluster validity indices. Pattern Recogn. **46**, 243–256 (2013)
3. Pedregosa, F., et al.: Scikit-learn: machine learning in Python. J. Mach. Learn. Res. **12**, 2825–2830 (2011)
4. Wolpert, D., Macready, W.: No free lunch theorems for optimization. IEEE Trans. Evol. Comput **1**(1), 67–82 (1997)
5. Pearson, K.: Liii. on lines and planes of closest fit to systems of points in space. Lond. Edinburgh Dublin Phil. Mag. J. Sci. **2**(11), 559–572 (1901)
6. Zhang, Y., Zhang, Z., Qin, J., Zhang, L., Li, B., Li, F.: Semi-supervised local multi-manifold isomap by linear embedding for feature extraction. Pattern Recogn. **76**, 662–678 (2018)
7. van der Maaten, L., Hinton, G.: Visualizing data using t-sne. J. Mach. Learn. Res. **9**, 2579–2605 (2008)
8. Chen, B., Hong, J., Wang, Y.: The minimum feature subset selection problem. J. Comput. Sci. Technol **12**, 145–153 (2008)
9. Alelyani, S., Tang, J., Liu, H.: Feature selection for clustering: a review (2014)
10. Jiang, S., Wang, L.: An unsupervised feature selection framework based on clustering. In: Cao, L., Huang, J.Z., Bailey, J., Koh, Y.S., Luo, J. (eds.) PAKDD 2011. LNCS (LNAI), vol. 7104, pp. 339–350. Springer, Heidelberg (2012). https://doi.org/10.1007/978-3-642-28320-8_29
11. Rice, J.R.: The algorithm selection problem**this work was partially supported by the national science foundation through grant gp-32940x. This chapter was presented as the george e. forsythe memorial lecture at the computer science conference, 19 february 1975, washington, d. c." vol. 15 of Advances in Computers, pp. 65–118. Elsevier (1976)

12. Muravyov, S.: System for automatic selection and evaluation of clustering algorithms and their parameters (2019)
13. Yang, L., Shami, A.: On hyperparameter optimization of machine learning algorithms: theory and practice. Neurocomputing **415**, 295–316 (2020)
14. Feurer, M., Klein, A., Eggensperger, K., Springenberg, J.T., Blum, M., Hutter, F.: Auto-sklearn: efficient and robust automated machine learning. In: Hutter, F., Kotthoff, L., Vanschoren, J. (eds.) Automated Machine Learning. TSSCML, pp. 113–134. Springer, Cham (2019). https://doi.org/10.1007/978-3-030-05318-5_6
15. Hutter, F., Hoos, H.H., Leyton-Brown, K.: Sequential model-based optimization for general algorithm configuration. In: Coello, C.A.C. (ed.) LION 2011. LNCS, vol. 6683, pp. 507–523. Springer, Heidelberg (2011). https://doi.org/10.1007/978-3-642-25566-3_40
16. Thornton, C., Hutter, F., Hoos, H., Leyton-Brown, K.: Auto-weka: combined selection and hyperparameter optimization of classification algorithms. In: KDD (2012)
17. Komer, B., Bergstra, J., Eliasmith, C.: Hyperopt-sklearn: automatic hyperparameter configuration for scikit-learn, pp. 32–37 (2014)
18. Feurer, M., Klein, A., Eggensperger, K., Springenberg, J., Blum, M., Hutter, F.: Efficient and robust automated machine learning. In: Cortes, C., Lawrence, N.D., Lee, D.D., Sugiyama, M., Garnett, R. (eds.) Advances in Neural Information Processing Systems, vol. 28, pp. 2962–2970, Curran Associates Inc. (2015)
19. Olson, R., Bartley, N., Urbanowicz, R., Moore, J.: Evaluation of a tree-based pipeline optimization tool for automating data science, pp. 485–492 (2016)
20. Jin, H., Song, Q., Hu, X.: Auto-keras: an efficient neural architecture search system, pp. 1946–1956 (2019)
21. Pinto, F., Soares, C., Mendes-Moreira, J.: Towards automatic generation of metafeatures. In: Bailey, J., Khan, L., Washio, T., Dobbie, G., Huang, J.Z., Wang, R. (eds.) PAKDD 2016. LNCS (LNAI), vol. 9651, pp. 215–226. Springer, Cham (2016). https://doi.org/10.1007/978-3-319-31753-3_18
22. Alcobaça, E., Siqueira, F., Rivolli, A., Garcia, L.P.F., Oliva, J.T., de Carvalho, A.C.P.L.F.: MFE: towards reproducible meta-feature extraction. J. Mach. Learn. Res. **21**(111), 1–5 (2020)
23. Rivolli, A., Garcia, L.P.F., Soares, C., Vanschoren, J., de Carvalho, A.C.P.L.F.: "Characterizing classification datasets: a study of meta-features for meta-learning (2019)
24. Ferrari, D., De Castro, L.: Clustering algorithm selection by meta-learning systems: a new distance-based problem characterization and ranking combination methods. Inf. Sci. **301**, 181–194 (2015)
25. Muravyov, S., Efimova, V., Shalamov, V., Filchenkov, A., Smetannikov, I.: Automatic hyperparameter optimization for clustering algorithms with reinforcement learning. Sci. Tech. J. Inf. Technol. Mech. Opt. **19**, 508–515 (2019)
26. Blumenberg, L., Ruggles, K.: Hypercluster: a flexible tool for parallelized unsupervised clustering optimization. BMC Bioinf. **21**, 428 (2020)
27. Autocluster - github repository. https://github.com/wywongbd/autocluster, Accessed 14 June 2021
28. Vanschoren, J., Van Rijn, J.N., Bischl, B., Torgo, L.: Openml: networked science in machine learning. ACM SIGKDD Explor. Newsl **15**(2), 49–60 (2014)
29. Filchenkov, A., Muravyov, S., Parfenov, V.: Towards cluster validity index evaluation and selection. In: 2016 IEEE Artificial Intelligence and Natural Language Conference (AINL), pp. 1–8 (2016)

ACHC: Associative Classifier Based on Hierarchical Clustering

Jamolbek Mattiev[1,2] and Branko Kavšek[2,3](✉)

[1] Urgench State University, Khamid Olimjan 14, 220100 Urgench, Uzbekistan
mattiev.jamolbek@urdu.uz, jamolbek.mattiev@famnit.upr.si
[2] University of Primorska, Glagoljaška 8, 6000 Koper, Slovenia
branko.kavsek@upr.si
[3] Jožef Stefan Institute, Jamova cesta 39, 1000 Ljubljana, Slovenia

Abstract. The size of collected data is increasing and the number of rules generated on those datasets is getting bigger. Producing compact and accurate models is being the most important task of data mining.

In this research work, we develop a new associative classifier – ACHC, that utilizes agglomerative hierarchical clustering as a post-processing step to reduce the number of rules and a new method is proposed in the rule-selection step to increase classification accuracy.

Experimental evaluations show that the ACHC method achieves significantly better results than classical rule learning algorithms in terms of rules on bigger datasets while maintaining classification accuracy on those datasets. More precisely, ACHC achieved the highest (43) result on the average number of rules and the third-highest (84.8%) result in terms of average classification accuracy among 10 classification algorithms.

Keywords: Frequent itemsets · Class association rules · Associative classification · Agglomerative hierarchical clustering · Cluster center

1 Introduction

Since information technologies are developing very rapidly and the amount of collected data is growing, analyzing such big data is an important task of data mining.

The number of rules generated from "real-life" datasets can easily grow very large, which may cause a combinatorial explosion. Therefore, mining association rules from these data and reducing their number to produce compact models for end-users is becoming a crucial data mining task [17].

To overcome this problem and achieve more compact as well as understandable models, rules have to be pruned and/or clustered.

Association [1] and classification rule mining [2, 7, 26] are two important fields of data mining. Classification rule mining aims at building accurate models to forecast the class value of a future object by selecting small subsets of rules, while association rule mining algorithms find all existing rules in a dataset based on some user-specified constraints by exploring the entire search space.

H. Yin et al. (Eds.): IDEAL 2021, LNCS 13113, pp. 560–571, 2021.
https://doi.org/10.1007/978-3-030-91608-4_55

Associative Classification (AC) [5, 12, 14, 18, 21, 22] is another data mining technique that combines classification and association rule mining. The main goal of AC methods is to produce compact, accurate and descriptive models based on association rules. Thus, the performance of AC methods can sometimes be better than some of the traditional classification methods on accuracy, in spite of worse efficiency, because AC methods are sensitive to user-defined parameters such as minimum support and confidence.

Another area in data mining is clustering [13, 25, 29]. Clustering methods can be usually grouped in two groups: partitional clustering [24, 30] that aims at grouping similar objects together by using partitioning techniques, and hierarchical clustering [28], which is a nested sequence of partitions.

In this research work, we present a novel cluster-based associative classification method. Firstly, we describe a new normalized combined distance metric to find the similarity of two class association rules (CARs). Secondly, we cluster the CARs by using a bottom-up approach of hierarchical agglomerative clustering. In this step, we automatically identify the optimal number of clusters. Thirdly, after CARs are clustered, we present a novel method of extracting the "representative" CAR for each cluster. Finally, we develop a compact and accurate associative classification model by including just the representative CARs from each cluster.

The performance of our method (ACHC) is evaluated on 12 selected datasets taken from the UCI ML Repository [6] and compared with 9 popular (associative) classification approaches, such as Decision Table and Naïve Bayes (DTNB) [10], Decision Table (DT) [15], PART (PT) [7], C4.5 [26], CBA [18], Simple Associative Classifier (SA) [20], FURIA (FR) [11], Ripple Down rules (RDR) [27], and J&B [19].

2 Related Work

The first related approach [3] is clustering of association rules based on the k-means clustering algorithm. Similarly to our approach, this approach utilizes the "APRIORI" algorithm [1] to generate the association rules. The key differences between ACHC and this method are (1) it uses a different algorithm in the clustering step, that is, it clusters the rules based on interestingness measures by using the k-means clustering method and (2) a novel distance metric is used to cluster the rules in our method.

The fuzzy clustering algorithm on association rules (FCAR) [4] is proposed based on rule simulation. In this method, researchers aimed to develop a fuzzy clustering method which partitions n objects into k subsets. FCAR uses the "APRIORI" algorithm in the rule-generation part as in our method, but they use a different technique (partitional) in the clustering phase. Another key difference is that they cluster the association rules while we intend to cluster class association rules.

Another related approach [8] is distance-based clustering of association rules. In this method, an "indirect" distance metric (based on CARs support and coverage) is used to group the association rules, while we develop a novel "combined" distance metric (based on direct and indirect measures) and an agglomerative clustering method to cluster the rules.

In [16], authors mine the clusters with association rules. The APRIORI algorithm is used to find strong CARs in the first step and then, it clusters these CARs by using a hierarchical clustering method as in our method. However, they used a different distance metric based on an "indirect" distance metric (based on coverage probabilities) in the clustering step.

The algorithm, described in this paper extends our previous work from [21] and [22]. In [21], the CMAC algorithm used a direct distance metric in the clustering phase of CARs and the cluster centroid approach to select the representative CAR for each cluster, while in [22] CMAC is compared to two similar algorithms, one using the direct distance metric and covering approach in the representative CAR selection phase and the other using combined (direct and indirect) distance metric with the same covering approach to select the representative CAR. This paper presents the only remaining combination, i.e., using a combined distance metric in the CAR clustering phase and the cluster centroid approach to select the representative CAR for each cluster.

3 Our Proposed Method – ACHC

It is assumed that we have given a relational table consisting of S examples (transactions). Each example is defined by A different attributes and classified into one of the C known classes.

Our main goals and contributions within this research are listed below:

- Developing a novel distance metric to measure the similarity of class association rules;
- Clustering of class association rules by utilizing the hierarchical clustering algorithm and finding the optimal number of clusters for each class value;
- Defining a new method of selecting a representative class association rule for each cluster to represent the compact and accurate classifier;
- Performing an experimental evaluation to show the advantages and disadvantages of the developed algorithm.

3.1 Class Association Rules

This section describes how to produce the strong class association rules. We first need to discover the frequent itemsets and we then generate the class association rules from these frequent itemsets. To generate the frequent itemsets, we apply the minimum support threshold, that is, frequent itemsets satisfied by minimum support constraint are generated in the first step. After generating the frequent itemsets, it is a straightforward approach to discover the class association rules from frequent itemsets. In this step, we apply the minimum confidence threshold to produce strong CARs.

We utilize the APRIORI algorithm to generate the frequent itemsets, because APRIORI is a well-known and frequently used algorithm for association rule mining.

Once we generate the frequent itemsets, our next goal is to produce the strong class association rules which satisfy the minimum confidence constraint. We check the confidence of each rule, if it satisfies the required threshold, then we generate that rule. Confidence of the *rule:* $A \rightarrow B$ is computed as follows:

$$confidence(rule) = \frac{support\_count(A \cup B)}{support\_count(A)} \tag{1}$$

Equation (1) describes the confidence of a rule based on the support count of a frequent itemset, where A represents the antecedent (left-hand side of the rule) and B the consequence (class value in the case of CARs) of a rule. Moreover, *support_count* $(A \cup B)$ is the number of examples in the dataset that match the itemset $A \cup B$, and *support_count* (A) is the number of examples that match the itemset A. We generate the strong class association rules that satisfy the minimum confidence constraint based on Eq. (1), as follows:

1. All nonempty subsets S of frequent itemset L belonging to class C are generated;
2. For every nonempty subset S of L, output the strong rule R in the form of "$S \rightarrow C$" if, $confidence(R) \geq min\_conf$, where min_conf is the minimum confidence threshold.

3.2 The New "combined" Distance Metric

In our previous research [21], we proposed a novel "direct" distance metric and defined its advantages and disadvantages. In this research, we use the previously developed "direct" distance metric and also focus on the "indirect" distance metric based on the Conditional Market-basket Probability (CMBP) [8]. Using a probability estimate for distance computation has many advantages. Probabilities are well understood, are intuitive, and a good measure for further processing, and it is appropriate for rules only with the same consequent. The distance d^{CMBP} between two rules *rule1* and *rule2* is the (estimated) probability that one rule does not hold for a basket, given at least one rule holds for the same baskets. This distance is defined as follows:

$$d^{CMBP}_{rule1,rule2} = 1 - \frac{|m(BS_{rule1} \cup BS_{rule2})|}{|m(BS_{rule1})| + |m(BS_{rule2})| - |m(BS_{rule1} \cup BS_{rule2})|}, \tag{2}$$

where BS is both sides of the rule, that is, the itemset for the association rule. $m(BS) and\ m(BS_{rule1} \cup BS_{rule2})$ denotes the set of transactions (baskets) matched by BS and $(BS_{rule1} \cup BS_{rule2})$ respectively. |m(BS)| and $|m(BS_{rule1} \cup BS_{rule2})|$ is the number of such transactions.

With this metric, rules having no common market baskets are at a distance of 1, and rules valid for an identical set of baskets are at a distance of 0.

In this research, we combine the "direct" and "indirect" distance metric to produce a new Weighted and Combined Distance Metric (WCDM). WCDM combines direct measure (rule items) and indirect measure (rule coverage). The weighted distance d^{WCDM} between two rules, *rule1* and *rule2* is defined in Eq. (3).

$$d^{WCDM}_{rule1,rule2} = \alpha \times d^{direct}_{rule1,rule2} + (1 - \alpha) \times d^{indirect}_{rule1,rule2}, \tag{3}$$

where α is a weigh balancing parametr in (3). We set $\alpha = 0.5$ parameter in the distance metric developing part, that is, the contribution of direct and indirect is considered equal to make a weighted and balanced distance metric. The resulting distance metric is defined in Eq. (4).

$$d^{WCB}_{rule1,rule2} = 0.5 \times d^{direct}_{rule1,rule2} + 0.5 \times d^{indirect}_{rule1,rule2}. \tag{4}$$

3.3 Clustering

Clustering algorithms are usually split into two groups: partitional and agglomerative hierarchical clustering. In our research, we apply the complete linkage method of agglomerative hierarchical clustering (in the bottom-up fashion) because it is frequently used and more consistent than other algorithms.

In the algorithms based on the bottom-up fashion, every example is considered as a unique cluster at the beginning and the two the nearest clusters are merged in each iteration until all clusters have been merged into a unique cluster.

In the complete linkage (farthest neighbor) method of agglomerative hierarchical clustering, the similarity between two clusters is computed based on the similarity of their most dissimilar examples, that is, the farthest groups are taken as an intra-cluster distance.

Algorithm 1: Computing the natural number of clusters

Input: Cluster heights.
Output: Optimal number of clusters.
1: Set Max to 0
2: Set O pt_number_of_cls to 1
3: Set N to length(cls_heights)
4: **FOR each** index in *N* **DO**
5:     **IF** cls_heights[index]-csl_heights[index-1] > Max **THEN**
6:         Set Max to cls_heights[index]-cls_heights[index-1]
7:         Set Opt_number_of_cls to N-index
8:     **END IF**
9: **END FOR**

To cluster the class association rules, we first identify the optimal number of clusters. In this step, we utilize the most-common technique (described in Algorithm 1), where the dendrogram is cut from the point which represents the maximum difference of two consecutive cluster heights.

Algorithm 1 gets cluster heights (distance between two clusters) which are computed during the dendrogram construction process as an input, and it outputs the optimal number of clusters.

N stores the total number of clusters in line 3. For each two consecutive cluster heights (lines 4–9), we compute the differences and store the maximum difference to Opt_number_of_cls parameter.

3.4 Selecting the "Representative" CAR

After clustering the CARs, we define an algorithm (presented in Algorithm 2) to select a "representative" CAR for each cluster. Only representative CARs are included in the final classifier.

Algorithm 2: Selecting the representative class association rule

Input: Class association rules in the cluster;
1. **Computation:** For each CAR in the cluster, we sum up all distances from the selected CAR to all other CARs in cluster;
2. **Update:** check if the distance of the selected CAR is lower than the representative CAR's distance, then update the selected CAR as a representative CAR.
3. **Output:** after checking all the CARs in the cluster, return the representative CAR which has the minimum distance.

Algorithm 2 describes the method of extracting the representative CAR. Firstly, the distances between the selected CAR and all other CARs are calculated for each CAR. Secondly, we find the CAR which obtains the minimum distance and returns that class association rules as a representative.

Algorithm 3 summarizes all the above-mentioned steps and describes the final ACHC classifier. It gets the training dataset, minimum support as well as confidence thresholds, as input parameters and outputs the compact and accurate associative classification model – ACHC.

Algorithm 3: ACHC associative classifier

Input: A training dataset *D*, *minimum support*, *minimum confidence.*
Output: ACHC associative classifier.

1: *Data*:={all examples in the dataset}, *s*:=minimum support, *c*:=minimum confidence;
2: *F*:= {find the frequent itemsets from *Data* which support $\geq s$}
3: C:= {Generate the CARs from frequent itemsets *F* which confidence $\geq c$ and sort them in confidence and support descending order};
4: G:= {Group the CARs *C* based on class label}
5: **FOR** each group in G **DO**
6:     Build distance_matrix of CARs in the same group;
7:     Compute cluster_heights by using hierarchical_clustering;
8:     Compute optimal_number_of_clusters N by using cluster_heights;
9:     Cluster CARs into N clusters by using hierarchical_clustering;
10:     **FOR** each cluster in N **DO**
11:         Extract representative_CAR;
12:         Add representative_CAR to Associative_Clssifier;
13:     **END FOR**
14: **END FOR**
15: **RETURN** Associative_Classifier

The first three lines generate the strong CARs defined in Sect. 3.1 and sort them by confidence and support descending order due to the following criterion:

If $Rule_1$ and $Rule_2$ represent two CARs, $Rule_1$ is said to have higher rank than $Rule_2$ ($Rule_1 > Rule_2$):

- if and only if, $conf(Rule_1) > conf(Rule_2)$ or
- if $conf(Rule_1) = conf(Rule_2)$, but $supp(Rule_1) > supp(Rule_2)$ or
- if $conf(Rule_1) = conf(Rule_2)$ and $supp(Rule_1) = supp(Rule_2)$, but $Rule_1$ has fewer items in its left-hand side than $Rule_2$.

Once the CARs are sorted, we group them according to class value (line 4) and then we build the distance matrix (line 6) for each group of CARs (defined in Sect. 3.2) to apply the hierarchical clustering algorithm with complete linkage (For example: if we have 3 class value, we build the model for three group of CARS and merge them at the end). Line 7 finds the cluster heights (distances between clusters) and the natural number of clusters is identified by using the cluster heights (line 8). After determining the optimal number of clusters, the hierarchical clustering algorithm is again utilized to find the cluster of CARs (*Cluster* array stores the list of clustered CARs) in line 9. In the last step (lines 10–12), we select the representative class association rule for every cluster (described in Subsect. 3.4) to produce the final compact and accurate model.

4 Experimental Setting and Results

Our developed algorithm is compared against 9 rule learners in terms of classification accuracy and rules. All associative classifiers were run with default parameters set up by WEKA [9] software. Some parameters (minimum support and confidence) were modified on "imbalanced" datasets to achieve the intended number of rules (at least 10 rules for each class value) for AC methods.

Statistical significance testing is performed based on the paired t-test (significance difference threshold was set to 95%) method. The description of the datasets and input parameters are shown in Table 1.

Moreover, a 10-fold cross-validation assessment technique was employed to represent all experimental results. Table 2 shows the experimental results for classification accuracies (with standard deviations).

Table 2 shows that our proposed associative classifier (ACHC) achieved comparable average accuracies (84.8%) to other classification models on selected datasets. More precisely, ACHC gained the third-highest average accuracy among 10 classification and association rule mining algorithms.

Our developed classifier obtained the best accuracy on "Balance" (except DTNB), "Breast.Can", "Spect.H", "Hayes.R" and "Connect4" datasets among all algorithms, while on "Car.Evn" and "Nursery" datasets, ACHC is beaten by all "classical" classification models.

Unexpectedly, standard deviation of all rule-learners was high (this situation happens mainly with imbalanced datasets which affect the rule-generation part of "APRIORI" algorithm) on the "Hayes.R", "Lymp" and "Connect4" datasets.

It can be seen from Fig. 1 that all classification algorithms achieved almost similar result on average accuracy except DTNB, DT and SA.

Table 3 represents the statistical significance testing (wins/losses counts) on accuracy between ACHC and other classification models. **W**: winning count (our approach was significantly better than the compared algorithm); **L**: losing count (our approach was significantly worse than the selected algorithm); **N**: no statistically significant difference has been detected in the comparison.

Table.1 Description of datasets and AC algorithm parameters

Dataset	# of attributes	# of classes	# of records	*Min support*	*Min confidence*	# of analyzed rules
Breast.Can	10	2	286	1%	60%	1000
Balance	5	3	625	1%	50%	218
Car.Evn	7	4	1728	1%	50%	1000
Vote	17	2	435	1%	60%	500
Tic-Tac	10	2	958	1%	60%	3000
Nursery	9	5	12960	0.5%	50%	3000
Hayes.R	6	3	160	0.1%	50%	1000
Lymp	19	4	148	1%	60%	1500
Spect.H	23	2	267	0.5%	50%	3000
Adult	15	2	45221	0.5%	60%	5000
Chess	37	2	3196	0.5%	60%	3000
Connect4	43	3	67557	1%	60%	5000

Table.2 Overall accuracies with standard deviations:

Dataset	DTNB	DT	C4.5	PT	FR	RDR	CBA	SA	J&B	ACHC
Breast.Can	70.4 ±4.1	69.2 ±6.7	75.0 ±6.9	74.0 ±4.0	75.1 ±5.3	71.8 ±5.7	71.9 ±9.8	79.3 ±4.4	80.5 ±4.7	**80.6** ±5.1
Balance	**81.4** ±8.1	66.7 ±5.0	64.4±4.3	76.2±5.6	76.3±7.6	68.5±4.3	73.2±3.8	74.0±4.1	74.1±2.6	76.5 ± 2.1
Car.Evn	**95.4**±0.8	91.3±1.7	92.1±1.7	94.3±1.0	91.8±1.1	91.0±1.8	91.2±3.9	86.2±2.1	89.4±1.4	86.9 ± 1.9
Vote	94.7±3.4	94.9±3.7	94.7±4.4	94.8±4.2	94.4±2.8	**95.6**±4.1	94.4±2.6	94.7±2.3	94.1±1.8	91.8 ± 1.9
Tic-Tac	69.9±2.7	74.4±4.4	85.2±2.7	94.3±3.3	94.1±3.1	94.3±2.9	**100.0**±0.0	91.7±1.5	95.8±2.0	91.0±1.4
Nursery	94.0±1.5	93.6±1.2	95.4±1.4	**96.7**±1.7	91.0±1.4	92.5±1.5	92.1±2.4	91.6±1.2	89.6±1.1	89.7 ± 0.7
Hayes.R	75.0±7.2	53.4±8.3	78.7±8.4	73.1±9.7	77.7±8.7	74.3±7.1	75.6±10.9	73.1±6.0	79.3±5.9	**80.7 ± 6.0**
Lymp	72.9±9.0	72.2±8.3	76.2±8.7	**81.7**±9.0	80.0±8.2	78.3±7.3	79.0±9.7	73.7±5.1	80.6±5.7	81.5 ± 7.1
Spect.H	79.3±2.7	79.3±1.6	80.0±9.0	80.4±5.6	80.4±2.2	80.4±2.2	79.0±1.6	79.1±2.1	79.7±3.1	**80.6 ± 1.1**
Adult	73.0±4.1	82.0±2.3	**82.4**±4.7	82.1±4.7	75.2±3.2	80.8±2.7	81.8±3.4	80.8±2.6	80.8±2.9	81.3 ± 2.4
Chess	93.7±3.0	97.3±3.1	98.9±3.6	98.9±3.1	96.4±2.1	95.8±3.3	95.4±2.9	92.2±3.8	94.6±2.7	97.0 ± 1.6
Connect4	78.8±5.9	76.7±7.7	80.0±6.8	81.0±7.9	80.6±7.1	80.0±6.4	80.9±8.1	78.7±6.0	81.0±5.2	81.0 ± 6.9
Avg(%):	81.5 ± 4.4	79.3 ± 4.5	83.6±5.2	85.6±4.9	84.4±4.4	83.6 ± 4.1	84.5±4.9	82.3±3.4	84.9±3.3	84.8 ± 3.2

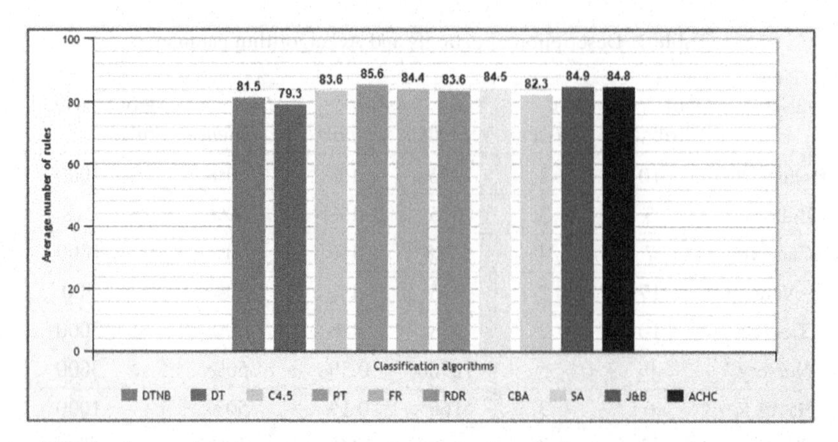

Fig. 1. Comparison between our method and other methods on average accuracy

Table.3 Statistically significant wins/loss counts of ACHC method on accuracy

	DTNB	DT	C4.5	PT	FR	RDR	CBA	SA	J&B
W	7	6	5	2	3	4	4	5	2
L	4	3	4	5	2	4	4	1	3
N	1	3	3	5	7	4	4	6	7

Table 3 illustrates that the performance of the ACHC method on accuracy is better than DTNB, DT and SA methods while this performance is similar or the same to C4.5, FR, J&B, RDR and CBA according to win/losses counts. Table 4 illustrates the size of all classification methods.

Experimental evaluations on the number of rules (Table 4) show that ACHC produced the best result in terms of the average number of rules among 10 rule-learners. Our approach produced a statistically smaller classifier on "Car.Evn" and "Nursery" datasets, although not achieving the best classification accuracies on those datasets.

On "Hayes-root" and "Balance" datasets, ACHC obtained an unexpectedly larger number of rules (due to imbalanced datasets) but it produced accurate classifiers for those datasets.

Figure 2 illustrates that associative classifiers achieved better result than "classical" classification models on the average number of rules. The main reason is that "classical" classification models are sensitive to the size of the dataset.

Table 5 shows that ACHC produced smaller classifiers than DTNB, DT, C4.5, PT, FR, SA, and J&B algorithms on more than 8 datasets out of 12 (by win/losses count).

5 Conclusion and Future Work

Overall, our developed method achieved the highest result on the average number of rules by exhaustively searching the entire example space using constraints and clustering while maintaining a classification accuracy that was comparable to state-of-the-art rule-learning classification algorithms. Experimental evaluations showed that ACHC

Table.4 Number of CARs

Dataset	DTNB	DT	C4.5	PT	FR	RDR	CBA	SA	J&B	ACHC
Breast.Can	122	22	10	20	13	13	63	20	47	9
Balance	31	35	35	27	44	22	77	45	79	79
Car.Evn	144	432	123	62	100	119	72	160	41	32
Vote	270	24	11	8	17	7	22	30	13	6
Tic-Tac	258	121	88	37	21	13	23	60	14	17
Nursery	1240	804	301	172	288	141	141	175	109	80
Hayes.R	5	8	22	14	11	10	34	45	34	80
Lymp	129	19	20	10	17	11	23	60	29	7
Spect.H	145	2	9	13	17	12	4	50	11	5
Adult	737	1571	279	571	150	175	126	130	97	88
Chess	507	101	31	29	29	30	12	120	24	17
Connect4	3826	4952	3973	3973	403	341	349	600	273	102
Avg(%):	618	674	409	411	93	75	79	125	64	43

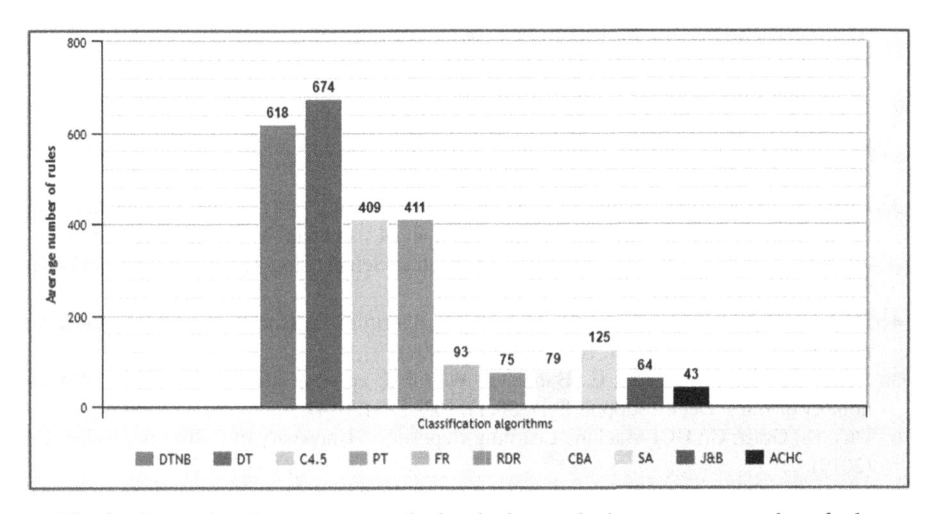

Fig. 2. Comparison between our method and other methods on average number of rules

Table.5 Statistically significant wins/loss counts of ACHC method on rules

	DTNB	DT	C4.5	PT	FR	RDR	CBA	SA	J&B
W	10	9	9	8	10	6	7	10	9
L	2	3	2	2	2	4	2	2	2
N	0	0	1	2	0	2	3	0	1

produced an accurate and compact classifier that was able to reduce the number of classification rules in the classifier by 2–4 times on average compared to the other "classical" rule-learners, while this ratio is even bigger on datasets with a higher number of examples.

The main drawback of our proposed method (ACHC) is its time efficiency. In future work we plan to optimize ACHC to bring its time complexity at least a bit closer to state-of-the-art "divide-and-conquer" rule-learning algorithms.

We plan to develop new methods by setting up different values for α parameter and perform experiments to show the advantages and disadvantages of those methods.

Acknowledgement. The authors gratefully acknowledge the European Commission for funding the InnoRenew CoE project (Grant Agreement #739574) under the Horizon2020 Widespread-Teaming programme and the Republic of Slovenia (Investment funding of the Republic of Slovenia and the European Union of the European Regional Development Fund). They also acknowledge the Slovenian Research Agency ARRS for funding the project J2-2504. Jamolbek Mattiev is also funded for his Ph.D. by the "El-Yurt-Umidi" foundation under the Cabinet of Ministers of the Republic of Uzbekistan.

References

1. Agrawal, R., Srikant, R.: Fast algorithms for mining association rules. In: Bocca, J.B., Jarke, M., Zaniolo, C. (eds.) VLDB 1994 Proceedings of the 20th International Conference on Very Large Data Bases, pp. 487–499. Chile (1994)
2. Cohen, W.W.: Fast Effective rule induction. In: Prieditis, A., Russel, S.J. (eds.) ICML 1995 Proceedings of the Twelfth International Conference on Machine Learning, pp. 115–123. California (1995)
3. Dahbi, A., Mouhir, M., Balouki, Y., Gadi, T.: Classification of association rules based on K-means algorithm. In: Mohajir, M.E., Chahhou, M., Achhab, M.A., Mohajir, B.E. (eds.) 4th IEEE International Colloquium on Information Science and Technology, pp. 300–305. Tangier, Morocco (2016)
4. Dechang, P., Xiaolin, Q.: A new fuzzy clustering algorithm on association rules for knowledge management. Inf. Technol. J. **7**(1), 119–124 (2008)
5. Deng, H., Runger, G., Tuv, E., Bannister, W.: CBC: an associative classifier with a small number of rules. Decis. Support Syst. **50**(1), 163–170 (2014)
6. Dua, D., Graff, C.: UCI Machine Learning Repository. University of California, Irvine, CA (2019)
7. Frank, E., Witten, I.: Generating accurate rule sets without global optimization. In: Shavlik, J.W. (eds) Fifteenth International Conference on Machine Learning, pp. 144–151. USA (1998)
8. Gupta, K.G., Strehl, A., Ghosh, J.: Distance based clustering of association rules. In: Proceedings of Artificial Neural Networks in Engineering Conference, pp. 759–764. USA (1999)
9. Hall, M., Frank, E., Holmes, G., Pfahringer, B., Reutemann, P., Witten, I.H.: The WEKA data mining software: an update. SIGKDD Explor. 11(1), 10–18 (2009)
10. Hall, M., Frank, E.: Combining Naive Bayes and Decision Tables. In: Wilson, D.L, Chad, H. (eds.) Proceedings of Twenty-First International Florida Artificial Intelligence Research Society Conference, pp. 318–319, Florida, USA (2008)
11. Hühn, J., Hüllermeier, E.: FURIA: an algorithm for unordered fuzzy rule induction. Data Min. Knowl. Disc. **19**(1), 293–319 (2019). https://doi.org/10.1007/s10618-009-0131-8

12. Hu, L.Y., Hu, Y.H., Tsai, C.F., Wang, J.S., Huang, M.W.: Building an associative classifier with multiple minimum supports. SpringerPlus **5**, 528 (2016). https://doi.org/10.1186/s40064-016-2153-1
13. Kaufman, L., Rousseeuw, P.J.: Finding Groups in Data: An Introduction to Cluster Analysis. Wiley, USA (1990).
14. Khairan, D.R.: New associative classification method based on rule pruning for classification of datasets. IEEE Access **7**, 157783–157795 (2019)
15. Kohavi, R.: The power of decision tables. In: Lavrač, N., Wrobel, S. (eds) 8th European Conference on Machine Learning, pp. 174–189. Crete, Greece (1995)
16. Kosters, W.A., Marchiori, E., Oerlemans, A.A.J.: Mining clusters with association rules. In: Hand, D.J., Kok, J.N., Berthold, M.R. (eds.) IDA 1999. LNCS, vol. 1642, pp. 39–50. Springer, Heidelberg (1999). https://doi.org/10.1007/3-540-48412-4_4
17. Lent, B., Swami, A., Widom, J.: Clustering association rules. In: Gray, A., Larson, P. (eds.) Proceedings of the Thirteenth International Conference on Data Engineering, pp. 220–231. England (1997)
18. Liu, B., Hsu, W., Ma, Y.: Integrating classification and association rule mining. In: Agrawal, R., Stolorz, P. (eds.) Proceedings of the 4th International Conference on Knowledge Discovery and Data Mining, pp. 80–86. New York, USA (1998)
19. Mattiev, J., Kavšek, B.: A compact and understandable associative classifier based on overall coverage.In: Procedia Computer Science, vol. 170, pp. 1161–1167. Warsaw, Poland (2020).
20. Mattiev, J., Kavšek, B.: Simple and accurate classification method based on class association rules performs well on well-known datasets. In: Nicosia, G., Pardalos, P., Umeton, R., Giuffrida, G., Sciacca, V. (eds.) LOD 2019. LNCS, vol. 11943, pp. 192–204. Springer, Cham (2019). https://doi.org/10.1007/978-3-030-37599-7_17
21. Mattiev, J., Kavšek, B.: CMAC: clustering class association rules to form a compact and meaningful associative classifier. In: Nicosia, G., et al. (eds.) LOD 2020. LNCS, vol. 12565, pp. 372–384. Springer, Cham (2020). https://doi.org/10.1007/978-3-030-64583-0_34
22. Mattiev, J., Kavšek, B.: Distance-based clustering of class association rules to build a compact, accurate and descriptive classifier. Comput. Sci. Inf. Syst. **18**(3), 791–811 (2021). https://doi.org/10.2298/CSIS200430037M
23. Mattiev, J., Kavsek, B.: Coverage-based classification using association rule mining. Appl. Sci. **10**, 7013 (2020). https://doi.org/10.3390/app10207013
24. Ng, T.R., Han, J.: Efficient and effective clustering methods for spatial data mining. In: Bocca, J., B., Jarke, M., Zaniolo, C. (eds.) Proceedings of the 20th Conference on Very Large Data Bases (VLDB), pp. 144–155, Santiago, Chile (1994)
25. Phipps, A., Lawrence, J.H.: An overview of combinatorial data analysis. clustering and classification, pp. 5–63, World Scientific, New Jersey (1996)
26. Quinlan, J.: C4.5: programs for machine learning. Mach. Learn. **16**(3), 235–240 (1993)
27. Richards, D.: Ripple down rules: a technique for acquiring knowledge. Decision-making support systems: achievements, trends and challenges for, pp. 207–226. IGI Global, USA (2002)
28. Theodoridis, S., Koutroumbas, K.: Hierarchical algorithms. Pattern Recogn. **4**(13), 653–700 (2009)
29. Zait, M., Messatfa, H.: A comparative study of clustering methods. Futur. Gener. Comput. Syst. **13**(2–3), 149–159 (1997)
30. Zhang, T., Ramakrishnan, R., Livny, M.: BIRCH: an efficient data clustering method for very large databases. In: Widom, J. (ed) Proceedings of the 1996 ACM-SIGMOD International Conference on Management of Data, pp. 103–114. Montreal, Canada (1996)

Special Session on Machine Learning towards Smarter Multimodal Systems

Multimodal Semi-supervised Bipolar Disorder Classification

Niloufar AbaeiKoupaei(✉) and Hussein Al Osman

School of Electrical and Computer Engineering, University of Ottawa, Ottawa, Canada
{nabae040,hussein.alosman}@uottawa.ca

Abstract. The objective of this study is to classify the states of individuals with bipolar disorder. We employ a dataset that uses the Young Mania Recall Scale to distinguish the manic states of patients as: Mania, Hypo-Mania, and Remission. The dataset comprises audio-visual recordings of bipolar disorder patients undergoing a structured interview. Having a small dataset and confidential test labels have motivated us to train a classifier using a semi-supervised ladder network, which benefits from unlabeled data during training. The key advantage of developing a semi-supervised model is removing the manual annotation training data, which is an expensive and time-consuming. We collect informative audio, visual, and textual features from the recordings to realize a multi-model classifier of the manic states. The proposed model achieved a 53.7% UAR and 60.0% UAR on the test and development sets, respectively.

Keywords: Bipolar disorder · Automated detection · Semi-supervide learning

1 Introduction

Bipolar Disorder (BD), originally known as *manic-depressive* disorder or *manic depression*, is a chronic and episodic mental disorder. Around sixty million individuals suffer from this illnesses globally [1]. BD usually causes extreme and unstable changes in mood, activity, and energy. These irregular episodes are often separated by normal mood (remission) periods. Manic episodes that involve elevated mood, high emotions and increased activity levels, are usually preceded and/or followed by hypomanic episodes characterized by relatively lower energy and emotion level. Such mood swings can affect sleep, energy, judgment, and behavior of individuals with BD. Therefore, BD can continuously impair an individual's wellbeing and ability to work.

Clinical diagnosis of BD often depends on the psychiatrist experience and knowledge, and may in some cases lack objectivity [2]. Moreover, difficulty in accessing relevant clinical services in some contexts may hinder early detection [3]. Hence, there is a need for additional strategies to address these challenges.

H. Yin et al. (Eds.): IDEAL 2021, LNCS 13113, pp. 575–586, 2021.
https://doi.org/10.1007/978-3-030-91608-4_56

Automated detection systems may improve early detection and in some cases reduce treatment resistance that stem from societal stigma [4].

In the last few years, we have been witnessing improvements in automated detection methods of some mental health disorders which highlight the feasibility of such techniques [5]. Hence, in this paper, we propose a machine learning method that analyzes several input modalities to assess BD states. Each modality may provide meaningful insight to the biological markers of BD state recognition. The goal of automated systems is not to replace health professionals. Such systems can be considered as an additional tool that assist health professionals in accurately assessing patients.

Many of the recent improvements in deep learning methods are associated with supervised learning. However, the effective application of such methods requires a large labeled dataset. This can present a challenge when the dataset size is limited. Hence, we are motivated to design a semi-supervised model, which can address the issue of small labeled datasets. Semi-supervised learning enables the label prediction of a large number of unlabeled samples by training with a small number of labeled samples. Such models use a mix of labeled and unlabeled data during training. The superior of a semi-supervised model over supervised one is that it doesn't require time-consuming, expensive, and labor-intensive manual annotation task for training data. In addition, semi-supervised approach alleviates the need for domain experts to label data, which is a challenging task particularly in healthcare domain. Semi-supervised learning has shown promising performance in many pattern recognition applications, such as emotion recognition [6], depression detection [7], and facial expression recognition [8].

Ladder network is a novel approach for semi-supervised learning, which delivered impressive results on the MNIST handwritten digit classification problem with only 100 labeled training examples [9]. This network extends the *denoising autoencoders* [10], which solely rely on unsupervised learning, by complementing them with a supervised component. Therefore, the ladder network is deep feedforwad network that combines supervised and unsupervised learning.

In this study, we propose an multimodal BD assessment framework that classifies three manic states associated with BD: mania, hypomania, and remission. We rely on three modalities of information for the classification: video, audio, and text. We propose a semi-supervised ladder network to classify obtained features. To the best of our knowledge, this is the first study that implements a ladder network for BD classification. We introduce this method due to the difficulty we experienced in training supervised models given the limited size of the available dataset.

The remainder of this paper is organized as follows. In Sect. 2 we review state-of-the-art studies on automated BD assessment. In Sect. 3 we provide details about the extracted features we use for classification. In Sect. 4, we describe the proposed classification model. We discuss the experimental setting, describe the BD dataset, and detail the experimental results in Sect. 5. Finally, we present our conclusions in Sect. 6.

2 Related Works

Access to a rich dataset is one of the main challenges in this field. Recent studies relied on datasets depicting face-to-face interviews. Therefore, in this work, we adopt a an interview-based dataset that was collected by a Turkish research group [11] and used for the AVEC2018 challenge [12] which focused on automated BD assessment using multimodal approach. The baseline approach proposed for the AVEC2018 challenge used audio and visual features with late fusion of the best feature sets. The extracted features included Mel-Frequency Cepstral Coefficients (MFCCs), extended Geneva Minimalistic Acoustic Parameter Set (eGeMAPs), Bags-of-audio-words (BoAW) as audio features and Facial Action Units (FAUs), and Bag-of-video-words (BoVW) as visual descriptors. A linear SVM classifier modeled the best combination of audio and visual features to an Unweighted Average Rate (UAR) of 50.0% on the test dataset.

Mania and hypo-mania refer to higher energetic states compared to remission [13]. Thus, higher arousal levels characterize the emotional state of patients in the mania or hypo-mania states. Given that audio features tend to reflect arousal [14], Du et al. proposed a multi-scale temporal audio features empowered IncepLSTM model for BD assessment [15]. The IncepLSTM model, which integrates an Inception and Long Short-Term Memory (LSTM) networks, was trained with a triplet loss to obtain discriminative representations of BD states. Video features can also reflective of BD states. Hence, in [16], a hybrid model consisting of an LSTM classifier predicted the BD severity utilizing extracted visual features from a fine-tuned Deep Neural Network (DNN).

Due to the elevated energy that subjects in manic or hypomanic states often display, rapid changes in audio and visual features may reflect the patient's condition. To this end, in [17], the authors employ turbulence features and fisher vector encoding of some audio features. They propose a Greedy Ensembles of Weighted Extreme Learning Machines (GEWELMs) classifier to model encoded ComParE low level descriptors (LLDs). [18] introduced a Histogram of Displacement (HDR) based upper body posture feature and histogram-based arousal features. Both the audio and visual cues were modeled by a DNNs and Random Forest algorithms.

DNN networks used for feature extraction or classification require large datasets for training. To address the limited amount of data in [11], the authors of [19] divided speech recording into smaller weakly labelled segments. They fed each segment into a DNN to extract features and employed a multi-instance ensemble classifier to predict the segement's label. Similarly, Amiriparian et al. [20] captured the hierarchical relationships of extracted spectrogram of audio files and classified them using Capsule Neural Network (CapsNet).

Although, audio and visual signals can reveal important insights about the individuals' mental state, adding other informative modalities may increase the capacity of the recognition system. For the turkish corpus [11], textual features can be extracted from interview transcripts. Therefore, [21] established a novel hierarchically recall model using a Gradient Boosted Decision Tree (GBDT) on MFCCs, FAUs, and textual features. Moreover, Zhang et al. [22] proposed a

Multimodal Deep Denoising Autoencoder (multi- DDAE) followed by a Fisher Vector encoder to produce audio-visual descriptors. In addition, they used a Paragraph Vector (PV) representation of the text to produce textual features. A multitask DNN classifier predicted the final label using the early fusion of extracted feature sets. Abaeikoupaei and Al Osman extracted a combination of textual and audio features classifying with a stacked ensemble classifier [23]. A reinforcement learning algorithm was utilized for parameter tuning of stacked model.

3 Bipolar Disorder Extracted Features

In this section, we present the features that we employ to classify the BD states in the proposed approach.

3.1 Visual Features

The patient's facial expressions and body movements can reveal valuable information about their BD state [21]. In fact, researchers have expanded a significant effort in exploring the inherent relation between emotions and mood disorders. For instance, Dalili et al. [24] proposed a depression recognition model that relies on emotional information extracted from facial expressions.

In this work, we extract emotion features from facial expressions to inform our BD assessment model. We detect the subject's face in a each video frame and using the Face++ toolkit [25], we classify the depicted emotion into one of seven discrete emotions: fear, happiness, anger, disgust, surprise, neutral, and sadness. We map the obtained emotion into the valence-arousal emotion dimensional space to facilitate the examination of changes in arousal and valence over time. Thus, we add 13 statistical valence and arousal features to the feature vector. Furthermore, we add 24 statistical functions of the Euclidean distance between emotion points of successive frames in the valence-arousal coordinate space to the feature vector. In total, we generate 61 emotion related features. Moreover, we capture facial expression using FAUs. The AVEC2018 [12] provided the intensities of 17 key FAUs along with their confidence measures for each frame. We also apply 16 statical functions to each FAU, which results in 272 features.

Individuals in manic, or even hypomanic, state suffer from lack of sleep, which can result in a change to eye movement patterns [26]. Thus, we calculate mean, variance, and covariance for the right and left eye movements. Head movements can also provide cues about the BD state. Patients in aroused states during manic or hypomanic episodes, tend to move their head more frequently compared to the remission state. Therefore, we calculate pixel variation between facial frames to estimate the head movement of each subject.

3.2 Audio Features

Recent studies have proven the efficiency of using acoustic features for several affective computing applications such as emotion recognition [14] and depression detection [27]. Choosing the most suitable audio features highly depends on the problem domain.

Spectral, cepstral, prosodic, and voice quality information can be captured through LLDs of speech signals. Summarising LLDs with a set of statistical measures computed over time is commonly utilized for different mental health recognition systems [27]. The AVEC2018 [12] employed LLDs of each subject's voice for BD assessment.

Using the open-source openSMILE toolkit [28], we extract the MFCCs and eGEMAPs acoustic features. All eGeMAPs descriptors at the speaker turn levelwere botained using an LSTM-RNN netowkr. The 23 extracted eGeMAPs features consist of 3 energy/amplitude related parameters, 6 frequency related parameters, and 14 spectral parameters. However, we segment the audio record into small parts and calculate the MFCCs over each segment to account for the non-stationary nature of speech signals. We measure all MFCCs, including 13 Mel-frequency cepstral coefficients and 26 dynamic coefficients (delta and double-delta) over a segment length of 60 ms and shift of 10 ms. Moreover, we calculate the maximum, minimum, average, and standard deviation for each of the audio features.

3.3 Textual Features

A group of researchers conducted a study to distinguish between depression and anxiety from verbal behaviour during psychotherapy [29]. The study stated that the patients' words selection were noticeably affected by their condition. Andreasen and Pfohl [30] used word choice analysis to distinguish between manic and depressive episodes. They pointed out that manic and depressed subjects differed in how they expressed their emotions. The depressed patients mostly used vague, qualified, and personalized sentences, whereas manic patients utilized more colorful and concrete sentences. Hence, bipolar disorder mental states can impact an individual's language content.

Following the suggestions of [21], we capture both linguistic and sentiment features using the Suite of Automatic Linguistic Analysis Tools (SALAT) [31]. SALAT is an open-source toolkit that has been widely used for emotion and depression recognition systems [32]. This toolkit consists of different tools including: Simple Natural Language Processing (SiNLP) [33] and Sentiment Analysis and Cognition Engine (SEANCE) [34] for linguistic and sentiment analysis, respectively.

The SiNLP tool [33] analyses high-level linguistic representations such as word frequency, number of sentences, and grammar structure. To obtain mentioned linguistic features, the toolkit permits researchers to select custom dictionaries relating to the text content, which enables us to specify our own words lists to analyze texts specific to our topic. It outputs 14 linguistic features.

Sentiment analysis reflects human opinions and attitudes. Using the sentiment analysis feature of the SALAT toolkit [34], we derive four notable sentiment indices. These indices measure affective norms for arousal-valence, token words related to each emotion, and a list of eight sentiment groups such as affection, enlightenment, power, rectitude, respect, skill, wealth, and well-being.

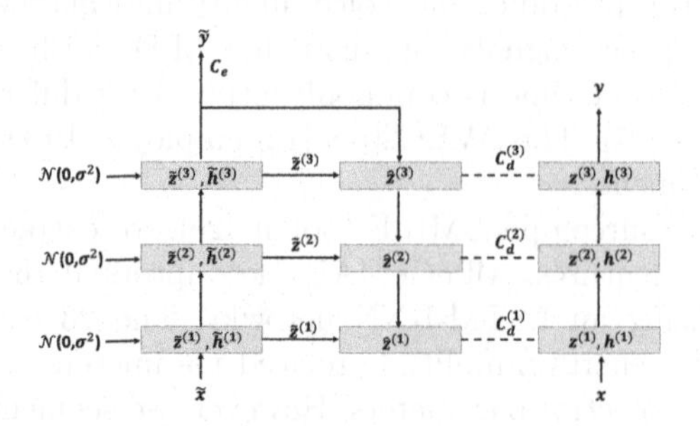

Fig. 1. An illustration of the ladder network with two encoders on the right side (noiseless) and the left side (noisy) and one decoder in the middle. Encoders : at each layer, $\tilde{\mathbf{z}}^{(K)}$ and $\mathbf{z}^{(K)}$ are captured by applying some linear transformation on $\tilde{\mathbf{h}}^{(k-1)}$ and $\mathbf{h}^{(k-1)}$, respectively (Eqs. 1 to 3). Decoder : at each level it receives two sets of information, the lateral connection $\tilde{\mathbf{z}}^{(K)}$ and $\boldsymbol{\nu}^{(k+1)}$ to reconstruct $\hat{\mathbf{z}}^{(K)}$ (Eqs. 5 to 7). The final objective function is a weighted sum of all cross entropy (C_e) and the unsupervised reconstruction (C_d) cost.

4 Ladder Network Classifier

In this section, we describe the proposed ladder network classifier. We refer readers to [9,10] for a detailed description of ladder networks. The dataset consists of M labeled data points, $\{(\mathbf{x}_i, \mathbf{y}_i) \mid 1 \leq i \leq M\}$, and N unlabeled data points, $\{(\mathbf{x}_j) \mid M+1 \leq j \leq N+(M+1)\}$. As the ladder network is a *denoising autoencoder network*, we inject noise in each layer of the network to force the autoencoder to learn how to denoise the corrupted input. Thus, we have two different encoder paths with shared parameters, where one produces noisy data and the other provides noiseless data. For the encoder that produces noisy data, we combine of the input vector with a noise vector, $\tilde{\mathbf{x}} = \mathbf{x} + \mathbf{noise}$, and transform the resulting vector into a latent representation, $\tilde{\mathbf{z}}^{(k)} \mid 1 \leq k \leq K$ where K is the number of network layers. The noiseless encoder is similar to the noisy one, except that it does not add a noise vector to the input. Hence, we denote the noisy and noiseless encoders as $\tilde{\mathbf{x}}, \tilde{\mathbf{z}}^{(1)}, ..., \tilde{\mathbf{z}}^{(K)}$ =Encoder$_{noisy}(x)$ and $\mathbf{x}, \mathbf{z}^{(1)}, ..., \mathbf{z}^{(K)}$ =Encoder$_{noiseless}(x)$, respectively.

Each layer of the noisy encoder is connected through lateral connections to its corresponding layer in the decoder. This enables the higher layer features to focus on more abstract and task-specific features. Therefore, the decoder combines the two outputs, one from the layer above and one from the corresponding layer in the noisy encoder, to yield the reconstructed observation at each layer, $\hat{\mathbf{z}}^{(k)} \mid 1 \leq k \leq K$ where K is the number of network layers. Hence, the decoder is denoted as $\hat{\mathbf{x}}, \hat{\mathbf{z}}^{(1)}, ..., \hat{\mathbf{z}}^{(K)}$ =Decoder($\tilde{\mathbf{z}}^{(1)}, ..., \tilde{\mathbf{z}}^{(K)}$). Figure 1 illustrates the ladder network.

Inspired by [35], where a learnable Multi Layer Perceptron (MLP) improved the performance of the ladder network, we train the ladder network with an MLP. The ladder network is trained to minimize the combination of weighted sum of supervised cross entropy and unsupervised reconstruction cost function from the encoder and decoder paths, respectively. We formulate these cost functions in this section.

Encoder: Each layer of the encoder is modeled by a linear transformation as shown below:

$$\tilde{\mathbf{z}}_{pre}^{(k)} = \mathbf{W}^{(k)} \tilde{\mathbf{h}}^{(k-1)} \tag{1}$$

Where $\mathbf{W}^{(k)}$ is the weight matrix between layer $(k-1)$ and layer k and $\tilde{\mathbf{h}}^{(k-1)}$ is the activation at layer $k-1$ for $1 \leq k \leq K$.

Then, we follow [36] to apply batch normalization to each layer:

$$\tilde{\mathbf{z}}^{(k)} = \mathbf{N_B}(\tilde{\mathbf{z}}_{pre}^{(k)}) + \mathbf{noise} = \frac{\tilde{\mathbf{z}}_{pre}^{(k)} - \boldsymbol{\mu}^{(k)}}{\boldsymbol{\sigma}^{(k)}} + \mathcal{N}(\mathbf{0}, \boldsymbol{\sigma}^2) \tag{2}$$

The batch normalized form of $\tilde{\mathbf{z}}_{pre}^{(k)}$ is calculated by the mean and standard variance from min-batch, $\boldsymbol{\mu}^{(k)}$ and $\boldsymbol{\sigma}^{(k)}$, respectively. Then, we acquire $\tilde{\mathbf{z}}^{(k)}$ adding a Standard Gaussian noise with mean $\mathbf{0}$ and variance $\boldsymbol{\sigma}^2$. We calculate the activation function of layer k:

$$\tilde{\mathbf{h}}^{(k)} = \phi(\boldsymbol{\gamma}^{(k)}(\tilde{\mathbf{z}}^{(k)} + \boldsymbol{\beta}^{(k)})) \tag{3}$$

$\boldsymbol{\gamma}^{(k)}$ and $\boldsymbol{\beta}^{(k)}$ denote the trainable scaling and biasing parameters of the nonlinear activation function $\phi(.)$. All mentioned equations formulate the noisy encoder. However, one can find all equations for the noiseless encoder by inserting the noiseless components instead of the noisy ones($\tilde{\mathbf{z}}_{pre}^{(k)} \rightarrow \mathbf{z}_{pre}^{(k)}$, $\tilde{\mathbf{z}}^{(k)} \rightarrow \mathbf{z}^{(k)}$, and $\tilde{\mathbf{h}}^{(k)} \rightarrow \mathbf{h}^{(k)}$) into the equations and removing the added Standard Gaussian noise from Eq. 2.

The objective of the encoder is to minimize the weighted sum of the supervised cross entropy function. Therefore, for the given $\boldsymbol{x}(m)$ inputs, the cost of matching the noisy output $\tilde{\boldsymbol{y}}(m)$ to the true target vector $\tilde{\boldsymbol{y}}_{true}(m)$, is formulated below:

$$C_e = -\frac{1}{M} \sum_{m=1}^{M} logP(\tilde{\boldsymbol{y}}(m) = \tilde{\boldsymbol{y}}_{true}(m) | \boldsymbol{x}(m)) \tag{4}$$

Decoder: Each layer of the decoder at layer k is responsible for combining the output of the preceding layer, $\hat{\mathbf{z}}^{(k+1)}$, and the corresponding output from the noisy encoder, $\tilde{\mathbf{z}}^{(k)}$. Then, the reconstructed signal $\hat{\mathbf{z}}^{(k)}$ is obtained based on the following equations:

$$\boldsymbol{\nu}_{pre}^{(k+1)} = \boldsymbol{\Gamma}^{(k)} \hat{\mathbf{z}}^{(k+1)} \tag{5}$$

$$\boldsymbol{\nu}^{(k)} = \mathbf{N_B}(\boldsymbol{\nu}_{pre}^{(k+1)}) = \frac{\boldsymbol{\nu}_{pre}^{(k+1)} - \boldsymbol{\mu}^{(k+1)}}{\boldsymbol{\sigma}^{(k+1)}} \tag{6}$$

$$\hat{\mathbf{z}}^{(k)} = g(\tilde{\mathbf{z}}^{(k)}, \boldsymbol{\nu}^{(k)}) \tag{7}$$

Where $\boldsymbol{\Gamma}^{(k)}$, with the same dimension of $\mathbf{W}^{(k)}$ on the encoder side, is the weight matrix between layer $(k+1)$ and layer k. $\boldsymbol{\nu}^{(k+1)}$ denotes the batch normalized version of the projection vector $\boldsymbol{\nu}_{pre}^{(k+1)}$. Lastly, a mapping function, $g(.,.)$, is applied on $\tilde{\mathbf{z}}^{(k)}$ and $\boldsymbol{\nu}^{(k)}$ to reconstruct the observation.

The decoder's objective is mitigating the unsupervised reconstruction cost, which is formulated as:

$$C_d = \sum_{k=1}^{K} \lambda_k C_d^{(k)} \tag{8}$$

Where λ_k is a denoising cost multiplier and

$$C_d^{(k)} = \sum_{m=M+1}^{N} \parallel \frac{\hat{\mathbf{z}}^{(k)} - \boldsymbol{\mu}^{(k)}}{\boldsymbol{\sigma}^{(k)}} - \mathbf{z}^{(k)} \parallel^2 \tag{9}$$

where $\boldsymbol{z}^{(k)}$ is an observation at layer k from the noiseless encoder path.

5 Experimental Results

We model a ternary bipolar disorder classification task using a semi-supervised model. A defined ladder based classifier predicts the individuals' mental states using audio, visual, and textual features. We train and test our model on the Turkish Audio/Visual Bipolar Disorder Corpus [11]. In this section, we briefly provide information about the dataset and present our experimental results.

5.1 BD Dataset

The Turkish corpus [11] consists of 218 video recordings collected from 46 patients. The recordings depict structured interviews with the patients and were annotated with the patient's state of BD coupled with their Young Mania Rating Scale (YMRS) [37] score as indicated by the psychiatrists. The original training, validation, and test sets include 104, 64, and 54 recordings, respectively. Although the training and development labels are available for researchers, the labels for the testing set are not published. Hence, the classification results on the testing set is validated by the data collectors. Reader may find more information in [11].

5.2 Experimental Settings

The BD dataset [11] is highly imbalanced within each of the training and development partition. Therefore, to achieve better classification results, we apply an oversampling approach to balance the dataset. Accordingly, minority classes are randomly up-sampled to equate the numbers of majority class/classes.

We consider the performance of the model for both the development and test sets. For each set we need to divide data to two main partitions including unlabeled and labeled data then we will apply the proposed model for each set. The data providers considered three original divisions including training, development, and test partition. Therefore, we use the term "original training/development/test data" to refer to their defined data partitions. The setting for each assessment is as follows:

- *Assessment of performance on the development set*: in this scenario we consider the original development set as our unlabeled set ($N = 60$). Then, we use the original training samples as our labeled data ($M = 104$) for training the model.
- *Assessment of performance on the test set*: this time we have the original test set as our unlabeled data ($N = 54$) and we train the classifier with the original training samples ($M = 104$). The original development set, with 60 samples, is used for evaluating the model structure and hyperparameters tuning.

We select the MLP model layer size of 1000-500-250-250-250-3. We need to assign a denoising cost value, λ, for each layer. Thus, we set the vector of denoising cost to 10, 0.1, 0.1, 0.1, 0.1, 0.1, where $\lambda_0 = 10$. We set the value for the standard deviation of the Gaussian corruption noise, which is added in Eq. 2, to 0.001. We train the network with the Adam optimizer with a learning rate of 10^{-3} for 500 epochs. The batch size for both levels is set to 64. We report the performance of the model using the Unweighted Average Recall (UAR) metric as it is the common metric between all existing studies on this dataset.

6 Discussion and Conclusion

As Table 1 demonstrates, we compare our proposed model with all existing studies on the same dataset using the UAR performance indicator. As labels of the test set are confidential and researchers have to send their predicted labels to the data provider to get assessed, comparing the performance of the models on the test samples is more reliable. Although the highest UAR on the test data reported by existing works is 59.3%, the proposed model in this paper achieves a comparable result with UAR of 53.7%. Considering the development and test UAR, it is obvious that the models with a sizable gap between the development and test UAR, such as [21] and [18], suffer from an overfitting issue. However, our results show that the model has overcome this problem.

Table 1. Comparison of the ternary classification task of bipolar disorder

24emModel	Dev.	Test
	UAR(%)	UAR(%)
Proposed model	**60.0**	**53.7**
SVMs [12]	55.82	50.0
ELMs [11]	47.3	–
Hierarchical recall model [21]	86.7	57.4
GEWELMs [17]	55.0	48.2
DNNs and random forest [18]	78.3	40.7
IncepLSTM [15]	65.1	–
Multi-instance learning [19]	61.6	57.4
CapsNet [20]	46.2	45.5
LSTM [16]	60.67	57.4
Stacked ensemble [23]	64.1	59.3

This work not only produces comparable performance to state-of-the-art, but also establishes the possibility of employing semi-supervised methods, such as the ladder network, for BD classification, especially for datasets with a limited labelled set. Therefore, this study can inform data collectors on how to maximize the usefulness of their published datasets. The presented semi-supervised model's result motivates investigating this model in more depth. As annotation is a time and cost consuming process, such model could provide more potentials to take advantage of unlabeled data. Despite the lower UAR of the proposed model, this study focuses on an important aspect that has been ignored by the past studies on bipolar disorder detection. A combination of labelled and unlabelled data can alleviate the need to exclusively collect a large labelled set, given the added complexity of cost associated with this exercise, especially for mental health assessment tasks. As future work, we plan to investigate the proposed framework with more details and improve the performance of classification task. As well as this, it is also interesting to evaluate the efficacy of the proposed model on different mental disorder corpora and show the effectiveness of considering unlabeled data along with labeled data.

References

1. World Health Organization: Mental disorders affect one in four people. http://www.who.int/whr/2001/media_centre/press_release/en/ (2001)
2. Mitchell, P.B., Goodwin, G.M., Johnson, G.F., Hirschfeld, R.M.: Diagnostic guidelines for bipolar depression: a probabilistic approach. Bipolar Disord. **10**(1p2), 144–152 (2008)
3. Kazdin, A.E., Blase, S.L.: Rebooting psychotherapy research and practice to reduce the burden of mental illness. Perspect. Psychol. Sci. **6**(1), 21–37 (2011)

4. Bauer, I.E., Soares, J.C., Selek, S., Meyer, T.D.: The link between refractoriness and neuroprogression in treatment-resistant bipolar disorder. Neuroprogression Psychiatr. Disord. **31**, 10–26 (2017)
5. Ma, X., Yang, H., Chen, Q., Huang, D., Wang, Y.: Depaudionet: an efficient deep model for audio based depression classification. In: Proceedings of the 6th International Workshop on Audio/Visual Emotion Challenge, pp. 35–42 (2016)
6. Zhang, Z., Ringeval, F., Dong, B., Coutinho, E., Marchi, E., Schüller, B.: Enhanced semi-supervised learning for multimodal emotion recognition. In: 2016 IEEE International Conference on Acoustics, Speech and Signal Processing (ICASSP), pp. 5185–5189. IEEE (2016)
7. Yazdavar, A.H., et al.: Semi-supervised approach to monitoring clinical depressive symptoms in social media. In: Proceedings of the 2017 IEEE/ACM International Conference on Advances in Social Networks Analysis and Mining 2017, pp. 1191–1198 (2017)
8. Cohen, I., Sebe, N., Cozman, F.G., Huang, T.S.: Semi-supervised learning for facial expression recognition. In: Proceedings of the 5th ACM SIGMM International Workshop on Multimedia Information Retrieval, pp. 17–22 (2003)
9. Rasmus, A., Berglund, M., Honkala, M., Valpola, H., Raiko, T.: Semi-supervised learning with ladder networks. In: Advances in Neural Information Processing Systems, pp. 3546–3554 (2015)
10. Vincent, P., Larochelle, H., Lajoie, I., Bengio, Y., Manzagol, P.A., Bottou, L.: Stacked denoising autoencoders: learning useful representations in a deep network with a local denoising criterion. J. Mach. Learn. Res. **11**(12), 3371–3408 (2010)
11. Çiftçi, E., Kaya, H., Güleç, H., Salah, A.A.: The turkish audio-visual bipolar disorder corpus. In: 2018 First Asian Conference on Affective Computing and Intelligent Interaction (ACII Asia), pp. 1–6. IEEE (2018)
12. Ringeval, F., et al.: Avec 2018 workshop and challenge: Bipolar disorder and cross-cultural affect recognition. In: Proceedings of the 2018 on Audio/Visual Emotion Challenge and Workshop, pp. 3–13 (2018)
13. Grunze, H.: Bipolar disorder. In: Neurobiology of Brain Disorders. Elsevier, pp. 655–673 (2015)
14. El Ayadi, M., Kamel, M.S., Karray, F.: Survey on speech emotion recognition: features, classification schemes, and databases. Pattern Recogn. **44**(3), 572–587 (2011)
15. Du, Z., Li, W., Huang, D., Wang, Y.: Bipolar disorder recognition via multi-scale discriminative audio temporal representation. In: Proceedings of the 2018 on Audio/Visual Emotion Challenge and Workshop, pp. 23–30 (2018)
16. Abaei, N., Al Osman, H.: A hybrid model for bipolar disorder classification from visual information. In: ICASSP 2020–2020 IEEE International Conference on Acoustics, Speech and Signal Processing (ICASSP), pp. 4107–4111. IEEE (2020)
17. Syed, Z.S., Sidorov, K., Marshall, D.: Automated screening for bipolar disorder from audio/visual modalities. In: Proceedings of the 2018 on Audio/Visual Emotion Challenge and Workshop, pp. 39–45 (2018)
18. Yang, L., Li, Y., Chen, H., Jiang, D., Oveneke, M.C., Sahli, H.: Bipolar disorder recognition with histogram features of arousal and body gestures. In: Proceedings of the 2018 on Audio/Visual Emotion Challenge and Workshop, pp. 15–21 (2018)
19. Ren, Z., Han, J., Cummins, N., Kong, Q., Plumbley, M.D., Schuller, B.W.: Multi-instance learning for bipolar disorder diagnosis using weakly labelled speech data. In: Proceedings of the 9th International Conference on Digital Public Health, pp. 79–83 (2019)

20. Amiriparian, S., et al.: Audio-based recognition of bipolar disorder utilising capsule networks. In: 2019 International Joint Conference on Neural Networks (IJCNN), pp. 1–7. IEEE (2019)
21. Xing, X., Cai, B., Zhao, Y., Li, S., He, Z., Fan, W.: Multi-modality hierarchical recall based on gbdts for bipolar disorder classification. In: Proceedings of the 2018 on Audio/Visual Emotion Challenge and Workshop, pp. 31–37 (2018)
22. Zhang, Z., Lin, W., Liu, M., Mahmoud, M.: Multimodal deep learning framework for mental disorder recognition. In: 2020 15th IEEE International Conference on Automatic Face & Gesture Recognition (FG 2020), IEEE (2020)
23. Abaeikoupaei, N., Al Osman, H.: A multi-modal stacked ensemble model for bipolar disorder classification. IEEE Trans. Affect. Comput. (2020). https://doi.org/10.1109/TAFFC.2020.3047582
24. Dalili, M., Penton-Voak, I., Harmer, C., Munafò, M.: Meta-analysis of emotion recognition deficits in major depressive disorder. Psychol. Med. **45**(6), 1135–1144 (2015)
25. L. M. T. Co., "Face++". https://www.faceplusplus.com.cn/ (2018)
26. Carvalho, N., et al.: Eye movement in unipolar and bipolar depression: a systematic review of the literature. Front. Psychol. **6**, 1809 (2015)
27. Valstar, M., et al.: Avec 2014: 3d dimensional affect and depression recognition challenge. In: Proceedings of the 4th International Workshop on Audio/Visual Emotion Challenge, pp. 3–10 (2014)
28. Eyben, F., Weninger, F., Gross, F., Schuller, B.: Recent developments in opensmile, the munich open-source multimedia feature extractor. In: Proceedings of the 21st ACM International Conference on Multimedia, pp. 835–838 (2013)
29. Sonnenschein, A.R., Hofmann, S.G., Ziegelmayer, T., Lutz, W.: Linguistic analysis of patients with mood and anxiety disorders during cognitive behavioral therapy. Cogn. Behav. Ther. **47**(4), 315–327 (2018)
30. Andreasen, N.J., Pfohl, B.: Linguistic analysis of speech in affective disorders. Arch. Gen. Psychiatry **33**(11), 1361–1367 (1976)
31. Kyle, K.: The suite of linguistic analysis tools (salat). http://www.kristopherkyle.com (2017)
32. Dang, T., et al.: Investigating word affect features and fusion of probabilistic predictions incorporating uncertainty in avec 2017. In: Proceedings of the 7th Annual Workshop on Audio/Visual Emotion Challenge, pp. 27–35 (2017)
33. Crossley, S.A., Allen, L.K., Kyle, K., McNamara, D.S.: Analyzing discourse processing using a simple natural language processing tool. Discourse Process. **51**(5–6), 511–534 (2014)
34. Crossley, S.A., Kyle, K., McNamara, D.S.: Sentiment analysis and social cognition engine (seance): an automatic tool for sentiment, social cognition, and social-order analysis. Behav. Res. Methods **49**(3), 803–821 (2017)
35. Pezeshki, M., Fan, L., Brakel, P., Courville, A., Bengio, Y.: Deconstructing the ladder network architecture. In: International Conference on Machine Learning, pp. 2368–2376 (2016)
36. Ioffe, S., Szegedy, C.: Batch normalization: Accelerating deep network training by reducing internal covariate shift, arXiv preprint arXiv:1502.03167 (2015)
37. Young, R.C., Biggs, J.T., Ziegler, V.E., Meyer, D.A.: A rating scale for mania: reliability, validity and sensitivity. Br. J. Psychiatry **133**(5), 429–435 (1978)

Developments on Support Vector Machines for Multiple-Expert Learning

Ana C. Umaquinga-Criollo[1,2](✉), Juan D. Tamayo-Quintero[3,4], María N. Moreno-García[1], Yahya Aalaila[5,6,7], and Diego H. Peluffo-Ordóñez[6,7]

[1] Universidad de Salamanca, Salamanca, Spain
acumaquinga@usal.es
[2] Universidad Técnica del Norte, Ibarra, Ecuador
[3] Universidad Nacional de Colombia, Manizales, Colombia
[4] Tecnológico de Antoquia, Medellín, Colombia
[5] Cadi Ayyad University, Marrakesh, Morocco
[6] Mohammed VI Polytechnic University, Ben Guerir, Morocco
[7] SDAS Research Group, Ben Guerir, Morocco
https://sdas-group.com

Abstract. In supervised learning scenarios, some applications require solve a classification problem wherein labels are not given as a single ground truth. Instead, the criteria of a set of experts is used to provide labels aimed at compensating for the erroneous influence with respect to a single labeler as well as the error bias (excellent or lousy) due to the level of perception and experience of each expert. This paper aims to briefly outline mathematical developments on support vector machines (SVM), and overview SVM-based approaches for multiple expert learning (MEL). Such MEL approaches are posed by modifying the formulation of a least-squares SVM, which enables to obtain a set of reliable, objective labels while penalizing the evaluation quality of each expert. Particularly, this work studies both two-class (binary) MEL classifier (BMLC) and its extension to multiclass through one-against all (OaA-MLC) including penalization of each expert's influence. Formal mathematical developments are stated, as well as remarkable discussion on key aspects about the least-squares SVM formulation and penalty factors are provided.

Keywords: Multiple expert learning · Supervised learning · Support vector machines

1 Introduction

Nowadays, the computer-aided systems for decision-making on multivariate data are ubiquitous, as the multidisciplinary field of machine learning (ML) is greatly versatile and offers a wide range of suitable solutions and research domains for industry, education, and medical settings, among others [1]. Among the main ML tasks is the automatic classification, which can be single-labeling or multiple labeler or expert (here referred

Supported by SDAS Research Group (www.sdas-group.com).

H. Yin et al. (Eds.): IDEAL 2021, LNCS 13113, pp. 587–598, 2021.
https://doi.org/10.1007/978-3-030-91608-4_57

as multiple expert learning - MEL). Remarkable applications requiring MEL are found in the diagnosis of a patient under the criterion of a group of physicians [2,8,11], evaluation of academic performance of a student by a panel of teachers [4,14], and high-risk decisions in companies with criteria from different parties [3], among others.

In scientific literature, we can find few related benchmark works worth mentioning such as: Gaussian Process-Based classification with Positive Label frecuency Threshold (GPC-PLAT) [21], Multiple Annotators Distinguish Good from Random Labelers (MA-DGRL) [15], Multiple Annotators Modeling Annotator Expertise (MA-MAE) [20], Multiple Annotators using Learning From Crowds (MA-LFC) [13], Kernel Alignment-based Annotator Relevance analysis (KAAR) [5], and Gaussian-Processes-based Classification (GPC) [16]. Also, some works have devoted to the design of consistent, diverse set of simulated experts (Bag of experts) [19].

In this context, we unfold the intuition and interpretability value of extending Support Vector Machines(SVM)-based approaches for MEL problems. By design, we solely focus on developing the theoretical tools and formulations providing a solid mathematical foundation to deal with MEL problems. To this end, as a general framework, the formulation of a supervised least-squares SVM is considered, which is extended to deal with a set of experts through both generating objective final labels and penalizing the quality of each expert's labeling. We first outline a two-class (binary) MEL classifier (BMLC). Then, through a one-against-all approach, a natural extension to multiclass settings (OaA-MLC) is devised. All approaches are developed in such a manner that each expert's influence is taken into consideration. As a remarkable aspect, formal and unified mathematical developments are stated, as well as remarkable discussion on key aspects about the least-squares SVM formulation and penalty factors are provided.

The remaining of this paper is organized as follows. Section 2 outlines the concept of Multiple-Expert Learning (MEL) through an example, as well as presents a walk-through of some basics of binary SVMs. Section 3 explains the so-named Binary Multiple Expert Learning Classifier (BMLC). Section 4 outlines a natural extension to a multi-class formulation for BMLC. Finally, Sect. 5 draws some final remarks.

2 Basics and Notation

2.1 Notation

For further formal statements, let us define the ordered pair $\{\mathbf{x}_i, \bar{y}_i\}$ to denote the i-th sample or data point, where $\mathbf{x}_i = (x_i^{(1)}, \ldots, x_i^{(d)})^\top$ is its d-dimensional feature vector and $\bar{y}_i \in \{1, -1\}$ is its binary class label. Assuming that m data points are available, all feature vectors can be gathered into a $m \times d$ data matrix $\mathbf{X}$ such that $\mathbf{X} = (\mathbf{x}_1, \ldots, \mathbf{x}_m)^\top$, whereas labels into a labeling vector $\bar{\mathbf{y}} \in \mathbb{R}^m$. Also, let us define a parametric, generalization or classification function $\mathbf{e} = f_{\mathbf{w}}(\mathbf{x})$, whose parameters $\mathbf{w}$ are to be defined.

2.2 A Few Initial Words on Multiple Expert Learning (MEL)

To depict the relevance and need for MEL, we, respectfully, invite the readers to consider the following explanatory example being simple (somewhat comical) but worth going over -to the authors' criterion. Similarly to the example referenced in the preface of [17], wherein a shepherd dog in charge of herding sheep becomes able to classify them between white and black sheep, let us now consider the same shepherd dog attempting to classify sheep from French poodles, as depicted in Fig. 1. We may refer to the shepherd dog as $\mathbf{y}^{(1)}$. Given its experience, $\mathbf{y}^{(1)}$ is expected to recognize sheep with complete accuracy, however it might misclassify some poodle as a sheep. Now, let us also consider a second shepherd dog ($\mathbf{y}^{(2)}$) being expert in recognizing poodles. Similarly, we would expect $\mathbf{y}^{(2)}$ to correctly classify all poodles and it might make mistakes in classifying sheep. Also, let us consider a third, amateur shepherd dog ($\mathbf{y}^{(2)}$) with little experience.

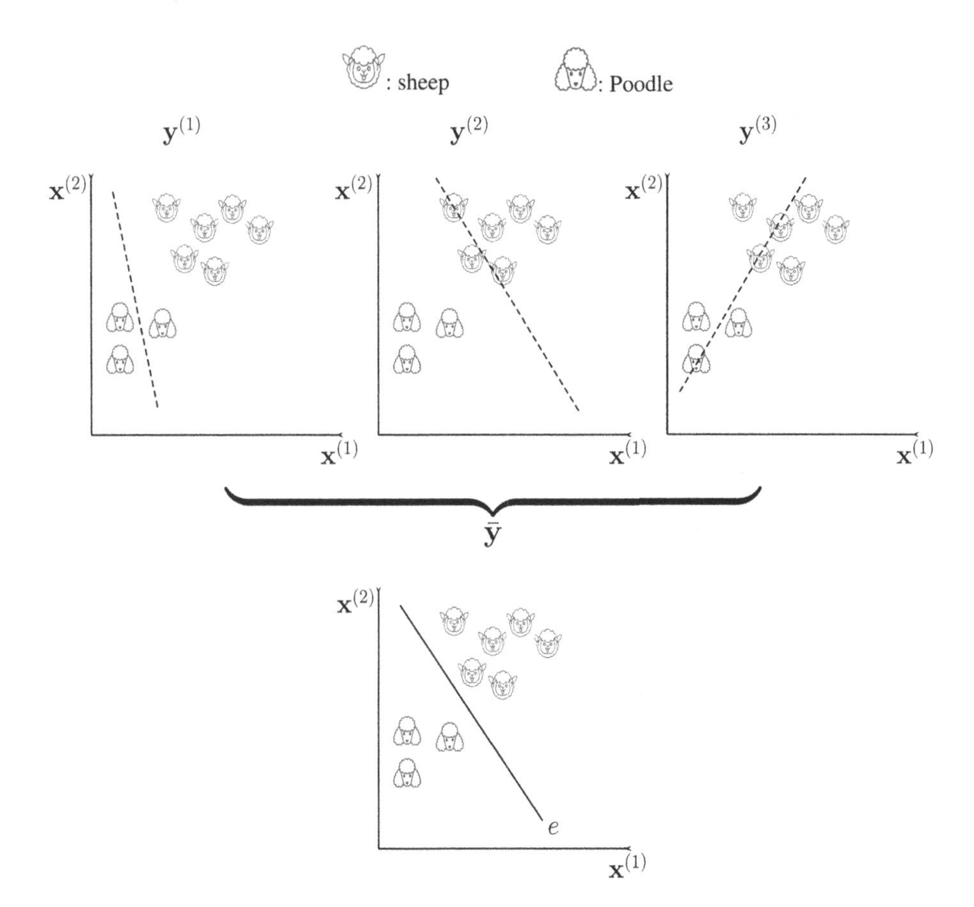

Fig. 1. A cartoon graphically explaining the concept of Multiple Expert Learning (MEL) through the example of three shepherd dogs $\{\mathbf{y}^{(1)}, \mathbf{y}^{(2)}, \mathbf{y}^{(3)}\}$ classifying a mixed herd between sheep and poodles. A ground truth $\bar{\mathbf{y}}$ is built which enables a more accurate classification e.

By assuming that the shepherd dogs linearly classify any member i of the entire mixed herd with respect to two attributes $\mathbf{x}_i = (x_i^{(1)}, x_i^{(2)})^\top$, the above situation can be depicted with the three scatter plots at the top of Fig. 1. Therefore, no single shepherd dog can correctly classify the entire mixed herd. Instead, by simultaneously exploiting the ability of $\mathbf{y}^{(1)}$ to classify sheep and the ability of $\mathbf{y}^{(2)}$ to classify poodles, a better final classification can be reached. Even the poor knowledge of $\mathbf{y}^{(3)}$ can also be of usefulness and should be taken into account. In other words, with the set of shepherd dogs $\{\mathbf{y}^{(1)}, \mathbf{y}^{(2)}, \mathbf{y}^{(3)}\}$, a ground truth $(\bar{\mathbf{y}})$ can be generated and thus a better generalization criterion e can be constructed. Now, let us assume that the mixed herd has become more diverse by including alpacas. According to the previous intuition, the shepherd dogs are able to perform the classification between sheep and poodle (a binary classification task). Such an intuition can be generalized to more than two classes by assembling a set of binary classifiers (one per class) in such a manner that each will classify between class of interest and rest of the herd members as shown in Fig. 2. That said, instead of a single e, a collection of generalization criteria $\{e^{(1)}, e^{(2)}, e^{(3)}\}$ is then generated. As the classification task has been carried out by three shepherd dogs, three linear decisions per class can be built (as seen in left side of Fig. 2). Therefore, the goal for this classification task is to determine the set of generalization criteria trained with the most suitable ground truth as depicted in right side of Fig. 2.

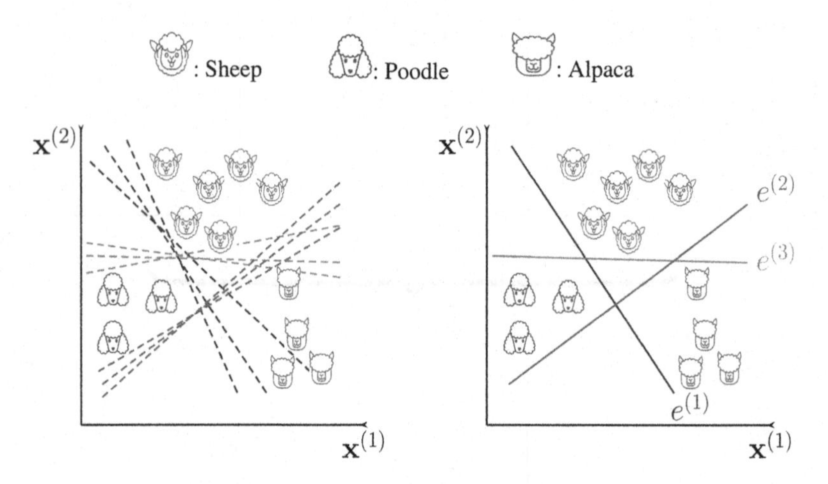

Fig. 2. A multiclass mulitple-expert classification example where the final generalization criteria are obtained one per class $\{e^{(1)}, e^{(2)}, e^{(3)}\}$. The three shepherd dogs from the example are now classifying the mixed herd among sheep, poodles and alpacas.

In general terms, a standard supervised classification problem assumes a set of examples with their correct labels. The underlining assumption however, is the existence of absolute standard that dictates the correct label of each instance, also known as ground truth or gold standard. Yet, such ground truth is mostly unavailable in real life problems, because more often than not acquiring it is expensive, unattainable or practically impossible. Two of the famous applications where this issue manifest itself is computer-aided diagnosis and automatic assessment of voice quality. In such problems, the labels of the instances are given by multiple experts, which naturally leads to various complications. Several experts may have different subjective opinion about the label of the same example case, which depends on their expertise level, bias factors, or mere knowledge. Therefore, a substantial amount of disagreement among experts is to be expected. To the light of all above mentioned, where the labels are dictated by multiple experts, an interesting questions may arise: In the absence of an absolute truth, how one can estimate the influence and reliability of the experts, and how to reduce and/or compensate for the negative effect of wrongly labeled samples? This ML scenario is here referred as MEL (multiple expert learning) which is the topic of interest exploited in this work.

2.3 Bi-Class SVM Classifier

To pose the classifier's objective function, it is assumed a latent variable model in the form

$$
\begin{aligned}
e_i = f_{\mathbf{w}}(\mathbf{x}_i) &= \underbrace{(w_1, \dots, w_d)}_{\mathbf{w}} \cdot \underbrace{\begin{pmatrix} x_i^{(1)} \\ \vdots \\ x_i^{(d)} \end{pmatrix}}_{\mathbf{x}_i} + b = \sum_{j=1}^{d} w_j x_i^{(j)} + b \\
&= \mathbf{w}^\top \mathbf{x}_i + b = \mathbf{w} \cdot \mathbf{x}_i + b = \langle \mathbf{x}_i, \mathbf{w} \rangle + b, \qquad (1)
\end{aligned}
$$

where $i \in \{1, \dots, m\}$, $\mathbf{w}$ is a d-dimensional vector, b is a bias term, and notations $\cdot$ and $\langle \cdot, \cdot \rangle$ stands for dot product and Euclidean inner product, respectively. As can be readily noted, vector $\mathbf{e} = (e_1, \dots, e_m)$ results from a linear mapping of elements of $\mathbf{X}$, which, from a geometrical point of view, is a hyperplane and can thus be seen as a projection vector. By design, if assuming $\mathbf{w} \in \mathbb{R}^d$ as an orthogonal vector to the hyperplane, the projection vector $\mathbf{e}$ can be used to encode the class assignment by a decision function in the form $\text{sign}(e_i)$, as illustrated in Fig. 3.

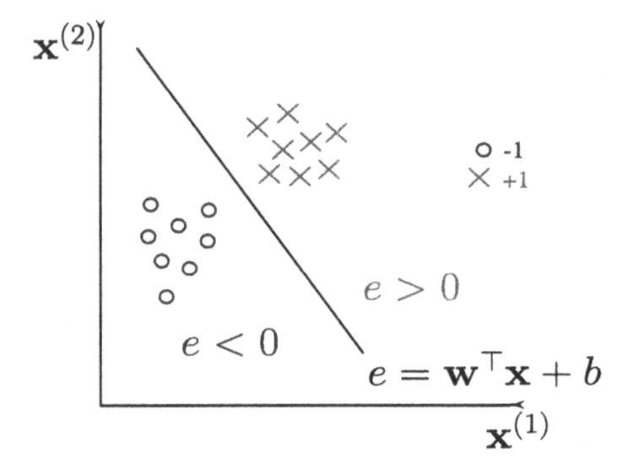

Fig. 3. Binary SVM classifier. The decision function is formed through the optimal hyperplane $e = \mathbf{w}^\top \mathbf{x} + b$.

Alternatively, projection vector can be expressed in matrix terms as $\mathbf{e} = \mathbf{X}\mathbf{w} + b\mathbf{1}_m$, being $\mathbf{1}_m$ an m-dimensional all ones vector. Moreover, in order to avoid that data points lie in an ambiguity region for making the decision as well as sample complexity complications [18], the distance between the hyperplane and any data point can be constrained to be at least 1 by fulfilling the condition: $\bar{y}_i e_i \geq 1, \forall i$.

The distance between a data point $\mathbf{x}_i$ and the hyperplane $\mathbf{e}$ is given by $d(\mathbf{e}, \mathbf{x}_i) = |\bar{y}_i e_i|/||\mathbf{w}||$, where $||\cdot||$ denotes Euclidean norm. The task is to maximize the classifier's objective function $\min_i d(\mathbf{e}, \mathbf{x}_i)$. This leads, after simple algebraic manipulation, to maximizing $1/||\mathbf{w}||$. For accounts of optimization convenience, we can write the problem as:

$$\min_{\mathbf{w}} \frac{1}{2}||\mathbf{w}||^2 \quad \text{s. t.} \quad \bar{y}_i e_i \geq 1, \; \forall i. \tag{2}$$

Notice that previous formulation is attained under the *hard* assumption that $\bar{y}_i = e_i$ ($\bar{y}_i = \text{sign}(e_i)$), and can then be named as hard-margin SVM. By relaxing (2), and by adding slack terms, a soft-margin SVM (SM-SVM) can be written as:

$$\min_{\mathbf{w},\xi} f(\mathbf{w}, \xi|\lambda) = \min_{\mathbf{w},\xi} \frac{\lambda}{2}||\mathbf{w}||^2 + \frac{1}{m}\sum_{i=1}^{m} \xi_i^2 \quad \text{s. t.} \quad \xi_i \geq 1 - \bar{y}_i e_i, \tag{3}$$

where λ is a regularization parameter and ξ_i is a slack term associated to data point i. A key aspect of this formulation is the fact that rather than linear ones, squared slack terms are used. In other words, instead of averaging, a mean square value is considered. This variation is useful to yield a robust quadratic problem, which is easy to extend to multiple-expert settings as will be seen in statements below.

3 Binary Multiple Expert Learning Classifier – BMLC

In previous works [9,10], a binary approach studied here for dealing with multiple expert settings, termed Binary Multiple Expert Learning Classifier (BMLC), is developed, which is based on least-squares SVMs framework. In the present paper, some remarkable aspects are mentioned and extended. BMLC can be basically outlined as follows: From the simplest formulation for a bi-class or binary SVM-based classifier, a generalized version incorporating a set of different labeling vectors is posed. Moreover, this formulation is also intended to estimate certain weights associated with the corresponding labeling vectors or labelers (experts). Such weighting factors are aimed at depicting or giving an indication of the quality of each labeler on the class assignment, and then they can be seen as penalty factors. Henceforth, since it is assumed that each single labeling vector is provided by a different labeler, there will make no further distinction between the terms labeling vectors and labelers (or experts). Both of them indistinctly refer to specific labels.

Let us consider a set of k labelers or experts who singly provide their corresponding labeling vectors as $\{\mathbf{y}^{(1)}, \ldots, \mathbf{y}^{(k)}\}$ as presented in Fig. 4.

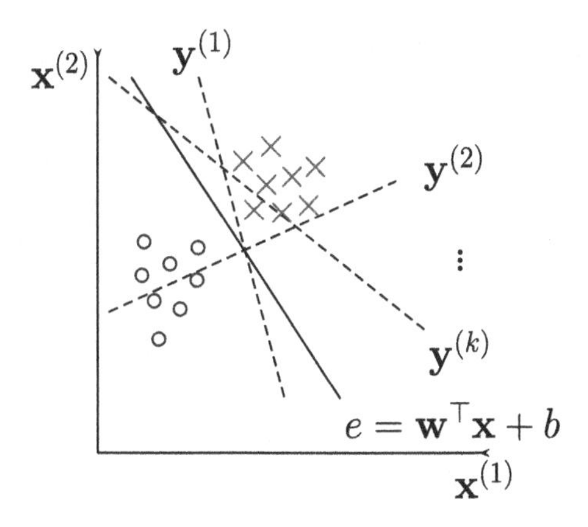

Fig. 4. An MEL problem within a binary SVM framework. The effect of a set of k diverse experts $\{\mathbf{y}^{(1)}, \ldots, \mathbf{y}^{(k)}\}$ is depicted.

Aimed at designing a multiple-expert classifier, a penalty factor θ_t is included within the functional stated in Eq. (3). This factor is intended to make f increases when adding wrong labels otherwise f should not or insignificantly decrease. Then, the t-th expert's quality is quantified by the penalty factor θ_t. Accordingly, by incorporating the penalty factors $\boldsymbol{\theta} = (\theta_1, \ldots, \theta_k)$, a new binary classification problem is introduced as follows:

$$\min_{\mathbf{w},\xi} \hat{f}(\xi, \mathbf{w}, \theta | \mathbf{C}, \lambda) = \min_{\mathbf{w},\xi} \frac{\lambda}{2}||w||^2 + \frac{1}{2m}\sum_{i=1}^{m}\left(\xi_i + \frac{1}{k}\sum_{t=1}^{k} C_{it}\theta_t\right)^2 \tag{4}$$

$$\text{s. t.} \quad \xi_i \geq 1 - \bar{y}_i e_i - \frac{1}{k}\sum_{t=1}^{k} C_{it}\theta_t,$$

where C_{it} is a weighting coefficient relating the labeler t and sample i.

It is noteworthy that the $\bar{\mathbf{y}}$ is a single vector representing a set of k labeling vectors $\{\mathbf{y}^{(1)}, \ldots, \mathbf{y}^{(k)}\}$, since our task is multi-expert-type. Then, $\bar{\mathbf{y}}$ is a reference vector to be determined. To set it, we can intuitively opt by an expectation operator over the input labeling vectors, i.e. the simple average as done in [4]. Nevertheless, another mixture alternatives can be considered as linear combinations, mode, median as well as non-linear mixtures. As seen, to fulfill their aim, penalty factors are incorporated into the cost function by intuitively adding them in the functional. It is important to highlight that not directly the weighting factors but a linear combination of them is used. Thereby, coefficients C_{it} are introduced, which aim to take into account the coordinates of data sample $(\mathbf{x}_i, \bar{y}_i)$ in the estimation of θ_t. Then, C_{it} is related to the accuracy of expert t for assigning a class label to data point $\mathbf{x}_i$. For simplicity, our approach takes advantages of the mismatch between every labeling vector and the reference one estimated by simple relative frequencies. Also, an indication of the distance between data points and boundary decision is considered. Concretely, coefficients are given by $C_{it} = (m - \mathrm{d}_H(\mathbf{y}^{(t)}, \bar{\mathbf{y}}))|e_i|/m$ denoting d_H the Hamming distance.

The idea of adding penalty factors in the SVM functional was inspired by [4]. Such work is addressed to reduce the effect of low-confidence labellers (experts) on the learning process without using any prior knowledge and without considering repeated labels. This is done under the principle that SVM determines which training examples are support vectors and which are not, so that non-support vectors can be removed from the training set without altering the learned classifier. Therefore, this approach can quantify the influence of each expert regarding the cumulative effect of some examples. BMLC can be seen as a variation as it introduces a mixture of weights in the SVM formulation. BMLC approach differs from that presented in [4] on the solution of the problem since it resorts to a primal-dual formulation, prompting then the use of quadratic programming problem instead of a heuristic search. As well, as BMLC is not designed to deal with missing values, every expert should provide a label for the whole training set, as usually happens in conventional settings.

To solve the problem (4), define an auxiliary variable $\hat{\boldsymbol{\xi}}$ as $\hat{\xi}_i = \xi_i + (1/k)\sum_{t=1}^{k} C_{it}\theta_t$, and by design write the modified SM-SVM in matrix terms as

$$\min_{\mathbf{w},\boldsymbol{\xi}} \hat{f}(\hat{\boldsymbol{\xi}}, \mathbf{w}|\mathbf{C}, \lambda) = \min_{\mathbf{w},\hat{\xi}} \frac{\lambda}{2}\mathbf{w}^{\top}\mathbf{w} + \frac{1}{2m}\hat{\boldsymbol{\xi}}^{\top}\hat{\boldsymbol{\xi}} \quad \text{s. t.} \quad \hat{\boldsymbol{\xi}} \geq \mathbf{1}_m - \mathbf{e} \circ \bar{\mathbf{y}}, \tag{5}$$

where $\circ$ stands for Hadamard product and matrix $\mathbf{C} \in \mathbb{R}^{m \times k}$ gathers the coefficients for the mixture of penalty factors, such as $\mathbf{C} = [C_{it}]$. The corresponding Lagrangian is then

$$\mathcal{L}(\boldsymbol{\alpha}, \mathbf{e}, \mathbf{w}, \hat{\boldsymbol{\xi}}|\lambda) = \frac{\lambda}{2}\mathbf{w}^{\top}\mathbf{w} + \frac{1}{2m}\hat{\boldsymbol{\xi}}^{\top}\hat{\boldsymbol{\xi}} + (\boldsymbol{\xi} - \mathbf{1}_m + \mathbf{e} \circ \bar{\mathbf{y}})^T\boldsymbol{\alpha}, \tag{6}$$

where $\boldsymbol{\alpha}$ is the vector of lagrange multipliers, which becomes the dual variable. Once solved the Karush-Kuhn-Tucker (KKT) conditions for the Lagrangian and eliminated primal variable by expressing them depending only on the dual variable, constants, and known variables, a dual problem is reached as detailed in a previous work [10]. Such a dual formulation is a quadratic problem with linear constraints easy to be solved by a quadratic programming scheme. Following from this, vector $\mathbf{w}$ is found, and then recalling Eq. 5 penalty factors can be directly calculated by

$$\theta = \mathbf{C}^{\dagger}(\mathbf{1}_m - (\mathbf{y} \circ \mathbf{e}) - \xi), \tag{7}$$

where $\mathbf{C}^{\dagger}$ is the pseudo-inverse matrix of $\mathbf{C}$.

As an important characteristic of BMLC, it is worth to mention that despite being, in principle, designed to calculate the influence of each labeller, it also outputs a suitable classifier modeled by $\mathbf{e} = \mathbf{X}\mathbf{w} + b\mathbf{1}_m$. Likewise, BMLC provides an estimated labeling vector $\bar{\mathbf{y}}$ to match labels and samples regarding the reliable subset of labellers, and redressing the influence of those having low-quality by using the hyperplane. To do so, it suffices setting $\hat{y}_i = \text{sign}(e_i)$. This is widely discussed in [9]. Finally, to achieve the full model $\mathbf{e}$, it remains to determine the bias parameter b. For easiness, data can be centered to force b to be 0. Otherwise, it can be easily estimated by solving (2) via a simple first order condition over its Lagrangian since $\mathbf{w}$ is already known. The classifier model can be used for determining the class of test data points $\mathbf{x}_{\text{test}_i}$ by means of $\hat{y}_{\text{test}_i} = \text{sign}(\mathbf{w}^\top \mathbf{x}_{\text{test}_i} + b)$. The steps of BMLC are gathered in the pseudo-code of Algorithm 1.

Algorithm 1. Binary multiple expert learning classifier (BMLC)

$[\bar{\mathbf{y}}_{\text{train}}, \bar{\mathbf{y}}_{\text{test}}, \boldsymbol{\theta}] = \text{BMLC}(\mathbf{X}, \mathbf{X}_{\text{test}}, \{\mathbf{y}^{(1)}, \ldots, \mathbf{y}^{(\ell)}\})$

1: **Input**: $\mathbf{X}, \mathbf{X}_{\text{test}}, \{\mathbf{y}^{(1)}, \ldots, \mathbf{y}^{(k)}\}$
2: Calculate the labeling reference vector $\bar{\mathbf{y}}$
3: Solve the problem (5) to determine $\mathbf{w}$ and b
4: Form the model $\mathbf{e} = \mathbf{X}\mathbf{w} + b$
5: Assign training labels $\bar{\mathbf{y}}_{\text{train}} = \text{sign}(\mathbf{e})$
6: Determine testing labels $\bar{\mathbf{y}}_{\text{test}}$ by $\bar{y}_{\text{test}_i} = \text{sign}(\mathbf{w}^\top \mathbf{x}_{\text{test}_i} + b)$
7: Form matrix $\mathbf{C}$ as $C_{it} = (m - \text{d}_H)(\mathbf{y}^{(t)}, \bar{\mathbf{y}})|e_i|/m$
8: Calculate penalty factors $\theta = \mathbf{C}^\dagger(\mathbf{1}_m - (\mathbf{y} \circ \mathbf{e}) - \xi)$
9: **Output:** $\bar{\mathbf{y}}_{\text{train}}, \bar{\mathbf{y}}_{\text{test}}, \boldsymbol{\theta}$

4 Multi-class Extension for BMLC

So far we have discussed Binary classification problems where the labels at hand can only take two values $+1$ and -1. This section is devoted to tackle multi-class problems. To do so, formulations above described can be naturally extended through conventional approaches such as: one against one, directed acyclic graph and one against all (OaA). As reported by [6,7], all of them work nearly the same. Here, the OaA approach is studied, and therefore the only step missing is its adaptation to BMLC.

One Against All extension This approach consists of fitting c bi-class SVMs as detailed in Sect. 3. Each time fixating a class ℓ ($\ell \in \{1, \ldots, c\}$) as the positive label, and comparing it with the remaining $c-1$ classes combined as the negative label [6]. Concretely, the labeling reference vector $\bar{\mathbf{y}}^{(\ell)}$ associated to a class ℓ is assumed in the form:

$$\bar{y}_i^{(\ell)} = \begin{cases} 1 & \text{if } \mathbf{x}_i \text{ belongs to class } \ell \\ -1 & \text{otherwise} \end{cases}. \tag{8}$$

In this sense, the proposed approach stated in (4) can be generalized as:

$$\begin{aligned} &\min_{\mathbf{w}^{(\ell)}, \xi^{(\ell)}} \hat{f}(\mathbf{w}^{(\ell)}, \xi^{(\ell)}, \mathbf{C}^{(\ell)}|\lambda_\ell) = \\ &\min_{\mathbf{w}^{(\ell)}, \xi^{(\ell)}} \frac{\lambda_\ell}{2}\|\mathbf{w}^{(\ell)}\|^2 + \frac{1}{2m}\sum_{i=1}^{m}\left(\xi_i^{(\ell)} + \frac{1}{k}\sum_{t=1}^{k} C_{it}^{(\ell)}\theta_t^{(\ell)}\right)^2 \\ &\text{s. t.} \quad \xi_i^{(\ell)} \geq 1 - \bar{y}_i^{(\ell)} e_i^{(\ell)} - \frac{1}{k}\sum_{t=1}^{k} C_{it}^{(\ell)}\theta_t^{(\ell)}. \end{aligned} \tag{9}$$

Consequently, the decision hyperplanes are given by $\{e_i^{(1)}, \ldots, e_i^{(c)}\}$, where $\mathbf{e}^{(\ell)} = \mathbf{X}\mathbf{w}^{(\ell)} + b^{(\ell)}\mathbf{1}_m$. Figure 5 depicts the multi-class extension effect over a three-class example.

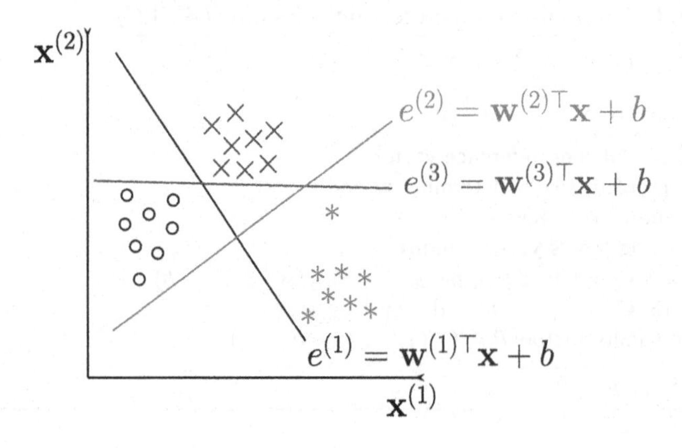

Fig. 5. Three-class example of One-against-all extension of BMLC.

Once vectors $\mathbf{w}^{(\ell)}$ are calculated by solving BMLC, corresponding penalty factors $\theta^{(\ell)}$ are calculated with (7). Finally, global penalty factors $\bar{\theta} \in \mathbb{R}^k$ can be estimated by averaging those obtained per each class

$$\bar{\theta} = \frac{1}{c}\sum_{\ell=1}^{c} \theta^{(\ell)}. \tag{10}$$

As BMLC, the set of vectors $\mathbf{w}^{(\ell)}$ also provides a reliable labeling vector $\bar{\mathbf{y}} \in \{1, \ldots, c\}^m$ given by:

$$\hat{y}_i = \arg \max_{\ell \in \{1,\ldots,c\}} e_i^{(\ell)}. \tag{11}$$

Similarly as explained in Sect. 3, OaA for BMLC can be evaluated on testing data points as well. This multi-class muliple-expert classifier is widely explained in [12].

5 Final Remarks

In supervised classification, when ground truth is not unique but it is obtained from the judgment of a group of experts, a multiple expert learning (MEL) strategy is needed to compensate for the erroneous influence of a single labeler as well as the level of perception and experience of each expert. This paper presents relevant, recent mathematical developments on support-vector-machines(SVM)-based approaches to address the MEL problem. Concretely, a modified least-squares SVM formulation is used to derive the binary MEL classifier (BMLC) and its one-against-all multi-class extension (OaA-MLC), which incorporate penalty factors for the sake of impartial decision-making.

Since OaA-MLC outcomes are given per each class, a global result estimation entails applying a statistical estimator, e.g. conventional average. Thus, the effect of each class regarding the output parameters can be biased or hidden. Therefore, more efforts need to be done to tackle multiclass MEL problems while considering impartially all the classes within the same formulation. As well, more sophisticated SVM data-driven approaches are to be explored, such as kernel-based formulations.

As a complement to the theoretical study explored in this paper, the experimental aspects of the algorithms are to be analysed as well. As a future work, a computational study must be conducted to further investigate the intricacies of the proposed algorithm. Also, a comparative experiment with the state-of-the-art approaches to MEL problems is expected.

Acknowledgments. This work is supported by the project PN223LH010-005 *"Desarrollo de nuevos modelos y métodos matemáticos para la toma de decisiones"*. As well, authors acknowledge the valuable support given by the SDAS Research Group (www.sdas-group.com).

References

1. Alzubi, J., Nayyar, A., Kumar, A.: Machine learning from theory to algorithms: an overview. In: Journal of Physics: Conference Series, vol. 1142, p. 012012. IOP Publishing (2018)
2. Chang, V., et al.: Generation of a HER2 breast cancer gold-standard using supervised learning from multiple experts. In: Stoyanov, D. (ed.) LABELS/CVII/STENT -2018. LNCS, vol. 11043, pp. 45–54. Springer, Cham (2018). https://doi.org/10.1007/978-3-030-01364-6_6
3. Danenas, P., Garsva, G., Simutis, R.: Development of discriminant analysis and majority-voting based credit risk assessment classifier. In: Proceedings of the 2011 International Conference on Artificial Intelligence, ICAI 2011, vol. 1, pp. 204–209 (2011)
4. Dekel, O., Shamir, O.: Good learners for evil teachers. In: Proceedings of the 26th Annual International Conference on Machine Learning, ICML 2009, pp. 233–240. Association for Computing Machinery, New York (2009). https://doi.org/10.1145/1553374.1553404
5. Gil-Gonzalez, J., Alvarez-Meza, A., Orozco-Gutierrez, A.: Learning from multiple annotators using kernel alignment. Pattern Recogn. Lett. **116**, 150–156 (2018). https://doi.org/10.1016/j.patrec.2018.10.005
6. Hsu, C.W., Lin, C.J.: A comparison of methods for multiclass support vector machines. Neural Netw. IEEE Trans. **13**(2), 415–425 (2002)
7. Liu, Y., Zheng, Y.F.: One-against-all multi-class SVM classification using reliability measures. In: IEEE International Joint Conference on Neural Networks, vol. 2, pp. 849–854. IEEE (2005)
8. Mahapatra, D.: Combining multiple expert annotations using semi-supervised learning and graph cuts for medical image segmentation. Comput. Vis. Image Underst. **151**, 114–123 (2016). https://doi.org/10.1016/j.cviu.2016.01.006
9. Murillo, S., Peluffo, D.H., Castellanos, G.: Support vector machine-based approach for multi-labelers problems. In: European Symposium on Artificial Neural Networks, Computational Inteligence and Machine Learning (2013)
10. Murillo-Rendón, S., Peluffo-Ordóñez, D., Arias-Londoño, J.D., Castellanos-Domínguez, C.G.: Multi-labeler analysis for bi-class problems based on soft-margin support vector machines. In: Ferrández Vicente, J.M., Álvarez Sánchez, J.R., de la Paz López, F., Toledo Moreo, F.J. (eds.) IWINAC 2013. LNCS, vol. 7930, pp. 274–282. Springer, Heidelberg (2013). https://doi.org/10.1007/978-3-642-38637-4_28

11. Nir, G., et al.: Automatic grading of prostate cancer in digitized histopathology images learning from multiple experts. Med. Image Anal. **50**, 167–180 (2018)
12. Peluffo-Ordóñez, D.H., Rendón, S.M., Arias-Londoño, J.D., Castellanos-Domínguez, G.: A multi-class extension for multi-labeler support vector machines. In: European Symposium on Artificial Neural Networks, Computational Inteligence and Machine Learning (2014)
13. Raykar, V., et al.: Learning from crowds. J. Mach. Learn. Res. **11**, 1297–1322 (2010). http://jmlr.org/papers/v11/raykar10a.html
14. Raykar, V.C., et al.: Supervised learning from multiple experts : whom to trust when everyone lies a bit. In: ACM International Conference Proceeding Series. vol. 382, pp. 1–8. ACM Press, New York (2009). https://doi.org/10.1145/1553374.1553488, http://portal.acm.org/citation.cfm?doid=1553374.1553488
15. Rodrigues, F., Pereira, F., Ribeiro, B.: Learning from multiple annotators: distinguishing good from random labelers. Pattern Recogn. Lett. **34**(12), 1428–1436 (2013). https://doi.org/10.1016/j.patrec.2013.05.012
16. Rodrigues, F., Pereira, F., Ribeiro, B.: Gaussian process classification and active learning with multiple annotators. In: Xing, E.P., Jebara, T. (eds.) Proceedings of the 31st International Conference on Machine Learning. Proceedings of Machine Learning Research, vol. 32, pp. 433–441. PMLR, Bejing, China (22–24 Jun 2014). http://proceedings.mlr.press/v32/rodrigues14.html
17. Schölkopf, B., Smola, A.J.: Learning with Kernels: Support Vector Machines, Regularization, Optimization, and Beyond. MIT Press, Cambridge (2001)
18. Shalev-Shwartz, S., Ben-David, S.: Understanding Machine Learning: From Theory to Algorithms. Cambridge University Press, Cambridge (2014)
19. Umaquinga-Criollo, A.C., Tamayo-Quintero, J.D., Moreno-García, M.N., Riascos, J.A., Peluffo-Ordóñez, D.H.: Multi-expert methods evaluation on financial and economic data: introducing bag of experts. In: de la Cal, E.A., Villar Flecha, J.R., Quintián, H., Corchado, E. (eds.) HAIS 2020. LNCS (LNAI), vol. 12344, pp. 437–449. Springer, Cham (2020). https://doi.org/10.1007/978-3-030-61705-9_36
20. Yan, Y., Rosales, R., Fung, G., Subramanian, R., Dy, J.: Learning from multiple annotators with varying expertise. Mach. Learn. **95**(3), 291–327 (2014). https://doi.org/10.1007/s10994-013-5412-1
21. Zhang, J., Wu, X., Sheng, V.S.: Imbalanced multiple noisy labeling. IEEE Trans. Knowl. Data Eng. **27**(2), 489–503 (2015)

WalkingStreet: Understanding Human Mobility Phenomena Through a Mobile Application

Luís Rosa[1](✉), Fábio Silva[2](✉), and Cesar Analide[1](✉)

[1] Department of Informatics, University of Minho, ALGORITMI Center, Braga, Portugal
id8123@alunos.uminho.pt, analide@di.uminho.pt
[2] CIICESI, ESTG, Politécnico do Porto, Felgueiras, Portugal
fas@estg.ipp.pt

Abstract. Understanding human mobility patterns requires access to timely and reliable data for an adequate policy response. This data can come from several sources, such as mobile devices. Additionally, the wide availability of communications networks enables applications (mobile apps) to generate data anytime and anywhere thanks to their general adoption by individuals. Although data is generated from personal devices, if a relevant set of metrics is applied to it, it can become useful for the authorities and the community as a whole. This paper explores new methods for gathering and analyzing location-based data using a mobile application called WalkingStreet. The article also illustrates the great potential of human mobility metrics for moving spatial measures beyond census units, key measures of individual, collective mobility and a mix of the two, investigating a range of important social phenomena, the heterogeneity of activity spaces and the dynamic nature of spatial segregation.

Keywords: Geo-location data · Human mobility metrics · Mobile application · Smart cities

1 Introduction

Over the past decade smart cities have been enhanced with the introduction of new technological devices, increasing the data capture points generated from interaction with citizens. Along with this ubiquity of devices, there is growing concern about the impact human mobility phenomena have on the sustainability of cities [12]. Additionally, both academia and industry have introduced a term and definition related to human mobility in cities by Information and Communication Technologies (ICTs) which is smart human mobility [15]. Therefore, it is not surprising that the study of human mobility has contributed to solve the problems of smart cities.

H. Yin et al. (Eds.): IDEAL 2021, LNCS 13113, pp. 599–610, 2021.
https://doi.org/10.1007/978-3-030-91608-4_58

Human mobility prototypes are proposed in the literature to measure the moving spatial beyond census units [11], investigate the heterogeneity of activity spaces [1], to predict movement patterns based on personality or spatio-temporal routines [16], to provide real time information of contrasting social and non-social sources of predictability in human mobility [2], to calculate the travel time of any trip [17], and to monitor human mobility during big events [18]. These pilot studies illustrate the great potential of mobile phone methodology and the role of mobile-based innovation in phenomena related to people's movements.

To promote more human mobility analysis, current strategies have used metrics to measure and discover daily mobility patterns taking advantage of mobile devices data such as smartphones. [14] uses a non-parametric Bayesian modeling method such as Infinite Gaussian Mixture Model (IGMM) to estimate the probability density and Kullback-Leibler (KL) divergence as the metric to measure the similarity of different probability distributions of daily mobility. In [20], the authors re-examined human mobility patterns via cell-phone position data recorded and considered four metrics to quantify the trajectories of individuals. However, this work revealed that Markov process can quantitatively reproduce the observed travel patterns at both individual and population levels at all relevant time-scales. Furthermore, [9,10,19] analyze data-driven human mobility metrics and their correlations with the transmissibility of COVID-19 using mobility data collected from mobile phone users. Thus, thanks to the availability of timely data and choosing appropriate metrics, mobile platforms can be developed for near-real-time mapped and predicted human mobility to help policy makers and the public (e.g., companies, communities and cities).

Inspired by existing metrics and mobile solutions, we aim to build a mobile application called WalkingStreet App. This novel human mobility app enables a visual interpretation of the phenomena related to people's movements. It incorporates (1) different strategies for human mobility analysis that use a potentiality wide range of the Google Maps Android API like clustering—handles the display of a large number of points, heat maps—displays a large number of points as a heat map, poly encoding and decoding—compact encoding for paths and spherical geometry (i.e., distance, heading and area); (2) several mobility metrics and patterns from data, both at the individual and collective level (i.e., length of displacements and typical distance); (3) synthetic individual trajectories using standard mathematical models such as random walk models; (4) assessing the privacy risk associated with a mobility datasets, and (5) different geographical formats like geoshape (polygon), or geotrace (polyline), that can be associated with other types of information. However, our mobile application does not consider prediction algorithms for now, since that it requires access to a huge amount of human mobility data. Moreover, this work is a prototype mobile device application designed to use georeferenced datasets with different formats such as Keyhole Markup Language (KML) and GeoJSON data supported by an API infrastructure in background based on sensor data.

With the WalkingStreet App, we aim to investigate the geographical regularity and variability of human mobility in local public amenities; to quantify the extent of spatial and temporal regularity and variability in these places; to explore geometry mapping features for georeferencing human mobility trajectories; to measure the availability and accessibility of local urban amenities or Point-Of-Interests (POIs); and, to assess the correlation between human mobility and socio-demographic characteristics of Barcelos population, a city located in the North of Portugal. Normally, internet-based travel diary instruments or traditional survey approaches are used to solve these tasks. However, we draw a prototype to estimate human mobility based on GPS traces of two simple datasets (i.e., KML and GeoJSON data). This app also is a simpler deployment and scalable alternative. Moreover, it provides a suite for offline mapping and visualization of geo-referenced data from a soft data mining process and support for a basic infrastructure.These may be the main differences between this and other existing projects [3,6].

The rest of this paper presents the WalkingStreet prototype and is organized as follows. Section 2 describes the geo-referenced datasets and human mobility metrics employed to analyze the relationship between local amenities and mobility flows. We also discuss our prototype. Section 3 presents the details of the graphical view interface incorporated in the prototype. In Sect. 4, we present and discuss the results and their implications for the prototype. Lastly, we summarize the key insights from our analysis and discuss potential avenues for future work in Sect. 5.

2 WalkingStreet Prototype

Based on the potential of georeferenced data, we define Barcelos City as our area of study, assigning a coordinate system on two datasets. We have also chosen a set of metrics to estimate human mobility. Finally, we present the architecture of a novel open mobile mapping tool that allow views, queries, and analyzes proposed datasets.

2.1 Geo-Referenced Datasets

A pair of datasets is used to verify and validate the human mobility data and include geo-referenced information, i.e. placename and geospatial referencing. This data is collected based on innovative capture methods such as MyGeodata Converter, a converter for geospatial data [7]. Although this makes it difficult to analyze results obtained, since the idea is to propose a scope/prototype, for the time being without commercial purposes and they are easily delivered on the Internet and viewed in a free application, we believe that the innovative data capture model comparing with another modern sources is more adequate to achieve the objectives of this article.

In this work we use a KML dataset, an official Open Geospatial Consortium (OGC) standard. It is an XML-based format for storing geographic, associated content of several POIs of Barcelos City such as Tourist Spots, or user-specific places such Commercial Points. KML is a common format for sharing geographic data with non-Geographic Information System (GIS) users. Regarding KML elements, this dataset is composed of feature and raster elements including points, lines, polygons, and pictures. Whereas this kind of dataset is typically seen as separate and homogeneous elements (for example, point feature classes can only contain points, rasters can only contain cells or pixels and not features), a single KML file can contain features of different types.

The second dataset includes Geodata data package, providing GeoJSON polygons of the Barcelos City. It defines several types of JSON objects and combines them to represent data about geographic features, their properties, and their spatial extents [8]. Finally, GeoJSON consists of the different parts (1) Geometry object is the location information; (2) Feature object is a geometry object which associates random ad hoq data and does not take into consideration what data is associated with the location information; and (3) FeatureCollection is a list of feature objects. Therefore, GeoJSON dataset typically consists of a FeatureCollection containing a list of data. In this work we use GeoJSON as a means to collect and share location data acquired from mobile devices. The struture of poligons and data collected will further developed in future work to account for data privacy as the objective is to determine flows of movement and not every single person's movement.

2.2 Human Mobility Metrics

We leverage a set of metrics to analyze the patterns of human mobility in the two datasets. One of them is trajectory preprocessing method. As any analytical process, mobility data analysis requires validation metrics. Thus, the trajectory preprocessing step allows the user to perform two metrics: stop detection and stop clustering. In these metrics, some points in a trajectory are called Stay Points or Stop Points. To detect them we apply spatial clustering algorithms to cluster trajectory points by looking at their spatial proximity. In other words, based on heatmaps of WalkingStreet app we find these visited points by a moving object. For example, we can identify the stops where the object spent at least minutes within a distance spatial radius, from a given point. In this work, our mobile application merges all the points that are closer than 0.5 km from each other.

Another metric is origin-destination matrix. It represents the flow of objects between two locations based in specific column names and data types such as Origin, Destination and Flow. In human mobility tasks, the territory is covered by the bi-dimensional space using a countable number of geometric shapes (e.g., squares, hexagons) with no overlaps and no gaps. For instance, for the analysis of human mobility flows, we aggregate flows of people moving among locations.

Other metrics are useful to capture the patterns of human mobility at individual and collective levels. Individual measures summarize the mobility patterns of a single moving object, while collective measures summarize mobility patterns of a population as a whole. The WalkingStreet app provides individual synthetic trajectories corresponding to a single moving object, assuming that an object is independent of the others. It provides the most data volume observed during each time. With collective trajectories we can estimate spatial flows between a set of discrete locations and compute thematic maps of the territory. For example, we identify trips between neighborhoods, migration flows between municipalities or freight shipments between states. Therefore, through these metrics, we can obtain several important human mobility patterns [4,5].

2.3 Design and Architecture

In this paper the WalkingStreet architecture is inspired in Layered Architecture. This approach combines the ideas of several other architectural approaches where (1) to be tested and audited; (2) not depend on UI; and (3) not depend on the database, external frameworks and libraries [13]. In addition, each layer is also independent while still being able to transmit information and data. Thus, our architecture (as illustrated in Fig. 1) is composed of three components: Presentation Layer, Domain Layer and Data Layer.

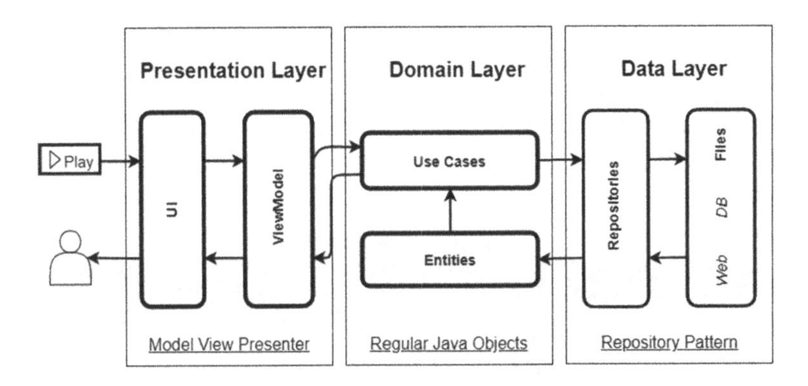

Fig. 1. WalkingStreet application architecture.

The user event goes to the Presenter. In this layer, the user selects a metric and, then, it passes to the Use Case. Use Case of Domain layer makes a request to the Repository, located in the Data layer. Repository gets the data stored in GeoJSON and KML datasets, creates an Entity, passes it to Use Case. The Use Case gets all the Entity needs, being the place for all the domain processes and operations. Then, applying them and his logic, he gets the result, and passes it back to Presenter. And that, in turn, displays the result in the User Interface (UI). Finally, the user visualizes the map of selected metric with several points.

As we mentioned, since we are presenting a prototype, we do not consider it relevant to propose a concrete data source. In other words, in this phase of our project the data source level is not important and do not use any database located on a machine or the API of a service available on the internet, all available data is stored by GeoJSON and KML files.

3 Implementation and Graphical View

In this section, we present methods for the exploration of human movement patterns using the proposed tracking formats (i.e., KML and GeoJSON). But, first we define the user flow of mobile app interface. This flow is the path that the user follows to access the several available maps, performed by selected metric (e.g., the user can see the map associated with the clustering metric). Then, we present some metrics that can be analysed based on mobility flow maps in the WalkingStreet prototype. These maps are used to extract flows from proposed datasets, rendering trajectory lines.

3.1 Mobility Presentation

The WalkingStreet app design has a focus on a user-friendly flow, promoting easy browsing through its contents. The process of designing the user flow also helped to prioritize content requirements, in terms of human mobility metrics that can be accomplished and put in the right place in order to achieve this in the most efficient way possible. Additionally, we avoid potential barriers in the navigation flow and implement quick routes to complete the intended actions.

Figure 2 explains the process of a user getting the map that applies spatial clustering algorithms to cluster trajectory points by looking at their spatial proximity. At each step in the flow is shown the wireframe available to users.

Fig. 2. Flow of the mobile application interface.

In this example, when a user selects an available metric in left menu the wireframe corresponds to the same app page, rather than representing different app pages. Each step clearly indicates the hotspots that connect to the next step in the task flow. An arrow is used to indicate the specific UI component where the user takes action such as a tap on a button and selecting a metric, and points to another wireframe image of what happens as a result of the interaction, i.e., a list of several map options associated with the selected human mobility metric.

The second "node" of that interaction shows the same page with the result of that interaction, e.g., display a large number of points as a heat map. Thus, the arrows clearly indicates the clickable "hotspots" that lead to the next step in the flow, in order to decrease the ambiguity in the wireflow.

3.2 Analytic Metrics

Understanding human mobility and how it manifests across temporal and spatial scales can be conceptually possible by implementing a set of metrics using GoogleMaps API. This section shows how our prototype drives important metrics to analyze the patterns of human mobility.

Trajectory Extraction. This flow extraction is based on the georeference data and determine flow of the human mobility in specific time. In Fig. 3, the resulting flows can be visualized, for example, to explore the popularity of different paths of movement:

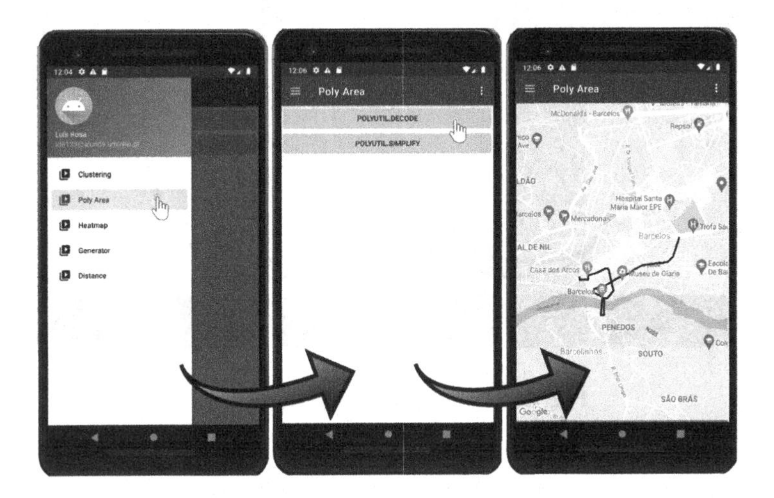

Fig. 3. Flow of extracting trajectory.

After the prototype has been computed, the flow algorithm computes transitions between points. An individual moving from point A to point B triggers an update of the corresponding flow. Additionally, the raw human mobility data records need to be converted into trajectories. Afterwards, each trajectory is processed independently ensuring (1) the distance to the match is below the distance threshold; (2) ensuring (2) the flow is retrieved or created in the prototype; and (3) and flow direction matches the record's direction.

This approach scales two datasets where the flow results and the trajectory have to be kept in memory for each iteration. However, this algorithm does not allow for continuous updates. Flows would have to be recomputed (at least locally) whenever datasets changed. Therefore, the algorithm does not support the exploration of continuous data streams.

Generating Trajectories. This method explores origin-destination relationships in several trajectories. The GeoJSON and KML databases identify individual trips with their start/end locations and trajectories between them. Moreover, trip trajectories were generated by consecutive records into continuous tracks and then splitting them at stops. For example, in our paper, we extracted human mobility paths which meant that we also had to account for observation gaps when individuals had no access to wi-fi, no longer had wireless network access due to lack of mobile data or weak network. Figure 4 explores trajectories for two users.

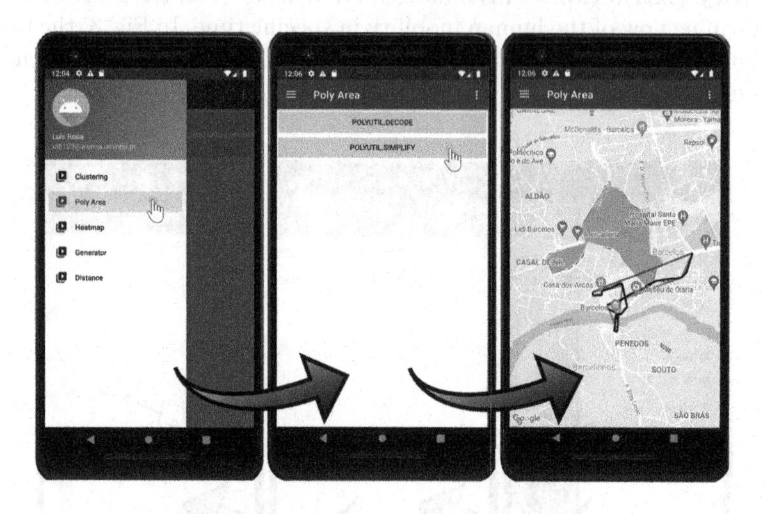

Fig. 4. Flow of generating trajectory.

Like extracting trajectory metric, the trajectory aggregation approach uses human trajectory data and the fact that operations only produce correct results if applied to a complete and chronologically sorted set of location records. This means that an aggregator needs to collect and sort the entire track on the map. Although the volume of datasets used in this prototype is reduced and we still don't have to worry about out-of-memory errors potentially being frequently encountered, it will be a challenge when the WalkingStreet app deals with large datasets.

Human Mobility Data Aggregation. Visualizations of point density maps called heat maps provide data exploration capabilities for reduced datasets. Although be limited in volume of data comparing with existing aggregation approaches to large datasets, these aggregation approaches can reveal movement patterns.

Using gespatial tools such as HeatmapTileProvider, it is quite straightforward to query timestamped location records. Thanks to the integration of Heatmap-TileProvider with Google Maps for Android it is also possible to redraw the tiles with the available options such as Radius, Gradient and Opacity which can be equally visualized and explored. Although these setups get point maps and point density maps using an aggregation method, other important movement characteristics like speed and direction are not included.

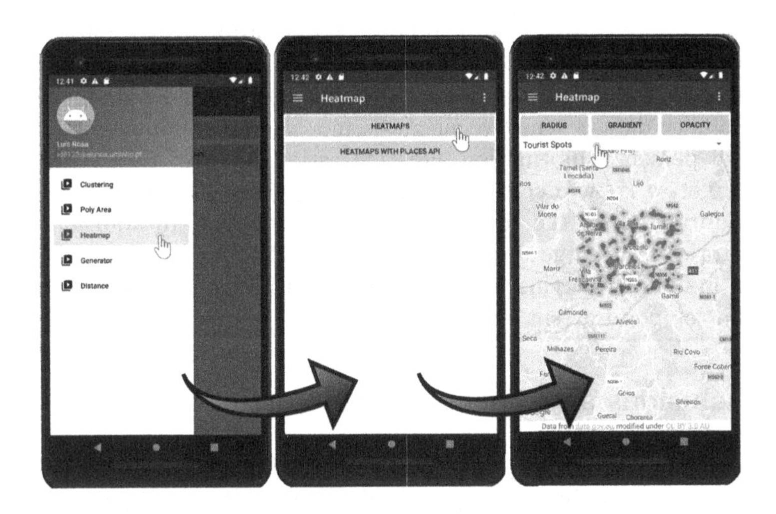

Fig. 5. Flow of the aggregating human mobility data.

Figure 5 shows how our application explores human movement data. It gives a first impression of the spatial distribution of records. The real value becomes clearer when we zoom in and start exploring local patterns. Then we can discover more details about movements. Even though the points we use are rather distributed, the densities are clearly formed in some locations. We can see exactly where these densities are and human mobility there, without having to increase the grid resolution to impractical values. The marker size shows the concentration of records on the location and thus helps distinguish heavily traveled zones from minor ones.

4 Results and Discussion

As shown in this paper, to explore human mobility datasets we needed to aggregate the data. These aggregations helped to discover patterns and visual exploration of human movements. But, applied in large datasets can be computationally expensive and therefore slow to generate. On the other hand, density maps are readily available and quick to compute but they provide only very limited insight. However, these metrics are a starting point for a new approach to exploring human mobility. Using raw location records, different forms of aggregation can be useful to learn more about GeoJSON and KML datasets:

1. aggregating raw location records to summarize human movements;
2. connecting consecutive records into continuous tracks to generate trajectories;
3. representing trajectory clusters by extracting flows.

Besides clever aggregation approaches, the human mobility datasets used in this article also require appropriate computing resources. To ensure that we can efficiently explore these datasets, we have implemented the aggregation steps in the WalkingStreet app. This enables us to run the computations on general-purpose computing clusters that can be scaled according to the dataset size.

However, during the development of WalkingStreet application problems were identified and some common statements were considered. In particular, human mobility data is sensitive since the movements of individuals can reveal confidential personal information, creating serious privacy risks. Or malicious actors can get to know a certain number of locations visited by an individual, but they do not know the temporal order of the visits. In addition, extracting trajectories from large datasets can also be challenging, particularly if the records of individual moving objects don't fit into memory anymore and if the spatial and temporal extent varies widely.

5 Conclusions and Future Work

This project has showcased how a human mobility app works/could work a human mobility application, using Google Maps API, GeoJSON, KML, and API in Python. The app helps to review the information about a zone and calculate a trajectory between points on maps from georeference previously defined in proposed datasets. As these datasets contain multiple trajectories from few users, the preprocessing methods automatically apply to the single trajectory and, when necessary, to the collective moving. In order to promote the interaction of the user with maps, different metric types can be selected, a few standard Google Maps controls are added, such as zoom controls; map scale control; and tooltips (e.g. place name). Additionally, the applicability of a set of metrics to facilitate their understanding until decision-making by the authorities should be as narrow as possible. Finally, choosing Clean Architecture is a good solution to this problem because it is not tied to a specific framework or database. In addition, each layer is also independent while still being able to transmit information and data.

In future work, we plan to introduce more users/individuals records in our research, investigating urban community patterns. Our aim is to avoid traditional methods to collect data and use self-reported data method. Self-reported data can provide additional value compared to traditional data since this data might be more spatially accurate, not outdated and with a frequent sampling time to make comparisons. Thus, in the next release, we will provide a way for users to send geo-referenced data via the proposed application. Predicting human mobility is another crucial component of urban planning and management. We will predict the individual or collective behaviors over time based on the person's past trajectory and the geographical features of the area. The effectiveness of the prediction process (e.g., in terms of points of interest and trip distance) will be possible using a massive mobile phone location dataset along with an improvement in our API features in Python.

Acknowledgments. This work has been supported by FCT - Fundacao para a Ciencia e Tecnologia within the R&D Units Project Scope: UIDB/00319/2020. It has also been supported by national funds through FCT – Fundação para a Ciência e Tecnologia through project UIDB/04728/2020.

References

1. Bu, J., Yin, J., Yu, Y., Zhan, Y.: Identifying the daily activity spaces of older adults living in a high-density urban area: a study using the smartphone-based global positioning system trajectory in shanghai. Sustainability (Switzerland) **13**(9), 5003 (2021). https://doi.org/10.3390/su13095003
2. Chen, Z., Kelty, S., Welles, B.F., Bagrow, J.P., Menezes, R., Ghoshal, G.: Contrasting social and non-social sources of predictability in human mobility, April 2021. http://arxiv.org/abs/2104.13282
3. Clouse, K., Phillips, T.K., Mogoba, P., Ndlovu, L., Bassett, J., Myer, L.: Attitudes toward a proposed GPS-based location tracking smartphone app for improving engagement in HIV care among pregnant and postpartum women in South Africa: focus group and interview study. JMIR Form. Res **5**(2), e19243 (2021). https://doi.org/10.2196/19243
4. van Duynhoven, A., Dragićević, S.: Analyzing the effects of temporal resolution and classification confidence for modeling land cover change with long short-term memory networks. Remote Sens. **11**(23) (2019). https://doi.org/10.3390/rs11232784
5. Feng, X., Li, J.: Evaluation of the spatial pattern of the resolution-enhanced thermal data for urban area. J. Sens. **2020** (2020). https://doi.org/10.1155/2020/3427321
6. Frith, J., Saker, M.: It is all about location: smartphones and tracking the spread of COVID-19. Soc. Media Soc. **6**(3) (2020). https://doi.org/10.1177/2056305120948257
7. GeoCzech, I.: MyGeodata Cloud - GIS Data Warehouse, Converter, Maps (2021). https://mygeodata.cloud/
8. Internet Engineering Task Force (IETF): The GeoJSON Format (2016). http://www.rfc-editor.org/info/rfc7946
9. Kraemer, M.U., et al.: Inferences about spatiotemporal variation in dengue virus transmission are sensitive to assumptions about human mobility: a case study using geolocated tweets from Lahore, Pakistan. EPJ Data Sci. **7**(1) (2018). https://doi.org/10.1140/epjds/s13688-018-0144-x
10. Kraemer, M.U., et al.: The effect of human mobility and control measures on the COVID-19 epidemic in China. Science **368**(6490), 493–497 (2020). https://doi.org/10.1126/science.abb4218
11. Liu, Y., et al.: Associations between changes in population mobility in response to the COVID-19 pandemic and socioeconomic factors at the city level in China and country level worldwide: a retrospective, observational study. Lancet Digit. Health **3**(6), e349–e359 (2021). https://doi.org/10.1016/s2589-7500(21)00059-5

12. Louro, A., da Costa, N.M., da Costa, E.M.: Sustainable urban mobility policies as a path to healthy cities-the case study of LMA, Portugal. Sustainability (Switzerland) **11**(10) (2019). https://doi.org/10.3390/su11102929
13. Martin, R.C.: Clean Coder (2014). https://blog.cleancoder.com/uncle-bob/2012/08/13/the-clean-architecture.html
14. Qian, W., Lauri, F., Gechter, F., et al.: A probabilistic approach for discovering daily human mobility patterns with mobile data. In: Lesot, M.-J. (ed.) IPMU 2020. CCIS, vol. 1237, pp. 457–470. Springer, Cham (2020). https://doi.org/10.1007/978-3-030-50146-4_34
15. Rosa, L., Silva, F., Analide, C.: Mobile networks and Internet of Things: contributions to smart human mobility. In: Dong, Y., Herrera-Viedma, E., Matsui, K., Omatsu, S., González Briones, A., Rodríguez González, S. (eds.) DCAI 2020. AISC, vol. 1237, pp. 168–178. Springer, Cham (2021). https://doi.org/10.1007/978-3-030-53036-5_18
16. Stachl, C., et al.: Predicting personality from patterns of behavior collected with smartphones. Proc. Natl. Acad. Sci. U.S.A. **117**(30), 17680–17687 (2020). https://doi.org/10.1073/pnas.1920484117
17. Thomas, T., Geurs, K.T., Koolwaaij, J., Bijlsma, M.: Automatic trip detection with the Dutch mobile mobility panel: towards reliable multiple-week trip registration for large samples. J. Urban Technol. **25**(2), 143–161 (2018). https://doi.org/10.1080/10630732.2018.1471874. https://www.tandfonline.com/action/journalInformation?journalCode=cjut20
18. Watson, J.R., Gelbaum, Z., Titus, M., Zoch, G., Wrathall, D.: Identifying multiscale spatio-temporal patterns in human mobility using manifold learning. PeerJ Comput. Sci. **6**, e276 (2020). https://doi.org/10.7717/peerj-cs.276
19. Yabe, T., Tsubouchi, K., Fujiwara, N., Wada, T., Sekimoto, Y., Ukkusuri, S.V.: Non-compulsory measures sufficiently reduced human mobility in Tokyo during the COVID-19 epidemic. Sci. Rep. **10**(1), 18053 (2020). https://doi.org/10.1038/s41598-020-75033-5
20. Zhao, C., Zeng, A., Yeung, C.H.: Characteristics of human mobility patterns revealed by high-frequency cell-phone position data. EPJ Data Sci. **10**(1), 1–14 (2021). https://doi.org/10.1140/epjds/s13688-021-00261-2

Indoor Positioning System for Ubiquitous Computing Environments

Pedro Albuquerque Santos[1,2](✉), Rui Porfírio[1](✉), Rui Neves Madeira[1,2](✉), and Nuno Correia[1](✉)

[1] NOVA LINCS, NOVA University of Lisbon, Lisbon, Portugal
{pe.santos,r.porfirio}@campus.fct.unl.pt, nmc@fct.unl.pt
[2] Sustain.RD, ESTSetúbal, Polytechnique Institute of Setúbal, Setúbal, Portugal
rui.madeira@estsetubal.ips.pt

Abstract. We developed an Indoor Positioning System (IPS) as part of the effort of creating Ubicomp applications with user interfaces distributed across different co-located devices. It relies on a Client that runs on the devices that we intend to locate and a Server that determines their positions. It currently supports three positioning methods: fingerprinting, trilateration and proximity. Bluetooth Low Energy and Wi-Fi are used as the underlying technologies for the positioning methods. We tested multiple machine learning algorithms during the development of the system to choose the ones providing satisfactory results. A Mean Absolute Error around or below 1 m and 95th percentile errors in the 2 m range were considered acceptable according to the type of target applications. We were also able to integrate the system into our framework and built a cross-device application that took advantage of it.

Keywords: Indoor positioning system · Ubiquitous computing · Proxemics · Machine learning · Wi-Fi · Bluetooth low energy

1 Introduction

We are living in a ubiquitous world as we face widespread ubiquitous computing (UbiComp) due to the billions of devices shaping the very fabric of an active world [18]. Nowadays, computing devices can be found everywhere around people, who are increasingly using a large number of them in their daily lives [13]. Therefore, we took on the challenge and opportunity of leveraging the presence of these co-located devices to build applications that can distribute their user interface (UI) across them (cross-device applications).

We soon identified that we would need a way to continuously pinpoint the location of the multiple devices that are part of a pervasive environment in order to automatically determine which devices are in range of use of each other to automatically distribute UI elements. The Global Positioning System (GPS)

H. Yin et al. (Eds.): IDEAL 2021, LNCS 13113, pp. 611–622, 2021.
https://doi.org/10.1007/978-3-030-91608-4_59

and commercial indoor location-based system work well outdoors and in large indoor spaces (e.g., airports and shopping malls). However, there is no readily available solution for indoor positioning that could be used throughout the rooms in a house or in offices to determine the location of the devices present in the surroundings of the potential users of cross-device applications.

Given the lack of options, we set out to create an indoor positioning system (IPS) that could be easily used by the framework that we implemented for supporting cross-device applications development. We were also concerned with ease of access, cost effectiveness and adaptability to the environment. Therefore, we set out to use commonly available technologies as starting point for our system, e.g., Bluetooth and Wi-Fi. We decided to support multiple positioning techniques, so we created a decision system to select the most appropriate one. Moreover, we avoided having to directly deal with the intricacies of radio wave propagation by employing machine learning (ML) models that were trained according to their behavior. We tested multiple ML algorithms during the development of the system, including: k-Nearest Neighbors, Support Vector Machines, Random Forest, Multilayer Perceptron, Linear Regression, and the BFGS optimization algorithm. The resulting IPS was developed in a way that it can be reused by any other UbiComp applications and systems that may require its services, thus constituting a useful contribution for the UbiComp community.

In the remaining of the paper, we introduce previous research related to our work (Sect. 2). We present the architecture of the IPS and explain how it works (Sect. 3). Section 4 presents some performance metrics for each of the supported positioning techniques, including a discussion regarding the machine learning algorithms and other parameters that enabled us to choose the best default values for our system. We finish by presenting conclusions, based on the experience we had using IPS with application prototypes, and future work.

2 Related Work

The research on indoor positioning systems has been extensive throughout the years resulting in efficient solutions based on a comprehensive set of technologies and techniques [7]. Radio frequency covers a substantial amount of studies that apply technologies like *Bluetooth*, *Wi-Fi*, and *Ultra-Wideband* to estimate the indoor position of a device. Due to its potential to achieve ubiquitous, low-power consumption, and low-cost solutions, *Wi-Fi* and *Bluetooth* established a solid reputation and popularity in the realm of indoor positioning systems [7,16].

Considering *Wi-Fi* approaches, the *RADAR* system was a pioneering work on *Wi-Fi* fingerprinting application for indoor positioning, reaching an accuracy of 2.94 m [4]. It is also worth mentioning Xia et al. work, which provides an in-depth overview of the application of *Wi-Fi*-based fingerprinting to indoor positioning settings [2]. The authors provide an analysis of the benefits (low-cost and high precision) and disadvantages (heavy labor cost to maintain the radio map) of fingerprinting and its influence factors, such as signal attenuation from people's presence in the environment.

Regarding Bluetooth-based research, Faragher and Harle research the impact of Bluetooth Low Energy (*BLE*) devices in advertising mode on fingerprint-based indoor positioning schemes and also propose a comparison between *Wi-Fi* and *BLE* fingerprinting [10]. Feldmann et al. introduce a *Bluetooth*-based system that applied trilateration and the least square estimation for location finding [11].

The design of UbiComp systems with an intrinsic proxemic-based context-awareness has been the core of a significant amount of research throughout the years [1,5]. In particular, the concept of proxemics interactions (i.e., interactions in which the devices have a fine-grained knowledge of nearby people and other devices) has further enhanced the use of proxemics in Human-Computer Interaction (*HCI*) [12].

Furthermore, the research on indoor positioning solutions has been quite fond of the adoption of ML techniques. The interest arises from the benefits ML offers in fields as pattern recognition, which can easily be transposed to indoor positioning environments, such as the online matching process in fingerprinting approaches [2]. A common approach is using a deterministic positioning algorithm like k-Nearest Neighbors (*k-NN*) regression algorithm. *k-NN* compares a device's *RSSI* with previously scanned *RSSI* in distinct reference points called fingerprints. It applies the matching procedure by finding the k most similar fingerprints to the real-time signal strength captured, using a distance metric as Euclidean distance. The k fingerprints will hold the most likely positions of a user. A final step usually applied to increase accuracy is to average the k positions and consider the mean value as the user's final location.

Among the studies that apply ML in indoor positioning solutions, the research of Blasio et al. leverages a modified version of *kNN* (Weighted K-Nearest Neighbor) to employ *Wi-Fi* and *Bluetooth* fingerprinting for harsh environments by applying the Euclidean distance and the weighted *k-NN* as a matching algorithm between fingerprints and real-time scans [6]. To aid in increasing the pervasiveness of an indoor positioning solution, there have been several studies that propose a fuzzy logic support system that exploits the uncertainty of a user's location. Orujov et al. present a fuzzy logic scheme, applying the Mamdani method, to aid in the decision support system aiming to select the most fitting positioning technique, depending upon three crisp inputs: the size of the room, *RSSI* from *BLE* beacons, and the number of available beacons [17].

3 Indoor Positioning System

We now present how the Indoor Positioning System (*IPS*) that we developed to be used as part of our efforts to build more engaging ubiquitous computing applications was developed. It enables applications to react proxemic dimensions of the relationships between the computing devices present in the environment, namely distance, orientation and identity [12].

3.1 Architecture

A general overview of the architecture of the system is provided in Fig. 1.

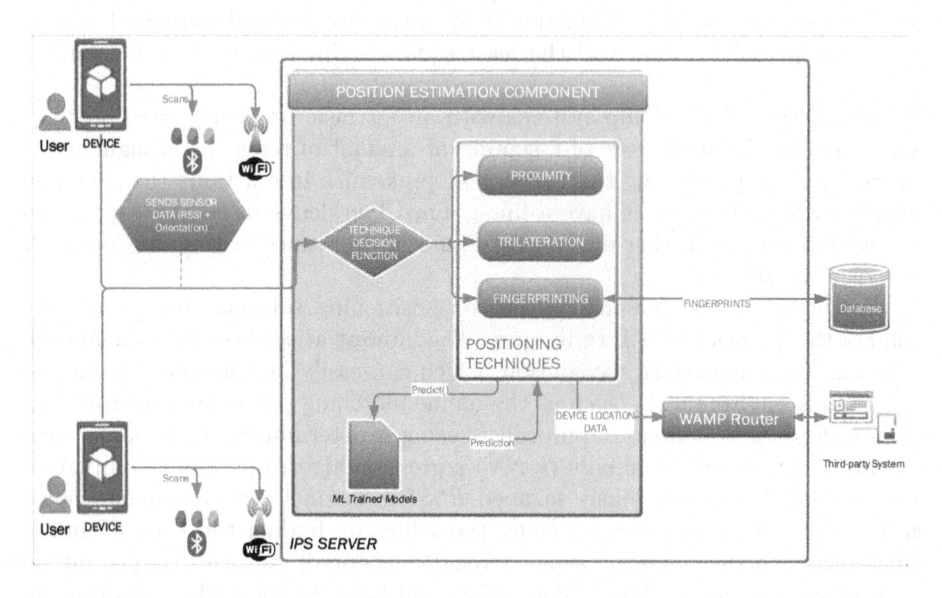

Fig. 1. Architecture of the *IPS* system

The indoor positioning system was implemented by following a *client-server model*. The *Client* runs on each device the system intends to locate. It scans for Bluetooth Low Energy (*BLE*) beacons, *Wi-Fi* access points, and collects the orientation of the device. It then continuously sends that information to the *Server*. The *Client* component has two implementations: one supporting mobile devices running *Android 6.0* or higher; and another for *Linux PCs*.

The *Server* was implemented using the *Python* programming language because it gives us access to useful data analysis and ML libraries. The *Server* is responsible for determining the position of the devices from that information using the supported positioning techniques. It is also capable of communicating with the *Clients* and third-party systems that are interested in the position of the devices. The communication between *Client* and *Server* relies on a *REST API*. However, the communication between the *Server* and third-party systems requires the ability to continuously push positioning events containing the current position of client devices. Therefore, we used the Web Application Messaging Protocol (*WAMP*). It provides the implementation of a *publish-subscribe* pattern over *WebSocket* connections [20].

3.2 Scanning Procedure

The *Client* applications hold the responsibility of continuously scanning for *BLE* beacons, *Wi-Fi* access points, and for collecting the orientation of the devices.

The data saved in each sample includes the Received Signal Strength Indicator (*RSSI*) values for the *BLE* beacons and *Wi-Fi* access points, and the *azimuth* angle in *radians*, i.e., rotation about the *-z-axis* [3]. An Universally Unique Identifier (*UUID*) is also included in order to identify the devices in a particular environment.

Each *Wi-Fi* scan takes a couple of seconds and reports all the access points it finds in a single batch [9]. Conversely, the *BLE* scan continuously reports any beacon as soon it is found. Moreover, previous research highlights a more severe fast fading multipath interference problem with *BLE* in indoor positioning settings than with *Wi-Fi* mainly due to a shorter channel length [10]. From an usability perspective, according to Kim et al. the response time of a proximity detection positioning system should range between 1 to 3 s [14]. Therefore, to alleviate such problem and accommodate the different scanning behaviors, the scanning rate is set to 3 s and in the case of the *BLE* beacons a moving average is applied to help smoothing out any errors.

3.3 Position Estimation

We chose to use *BLE* and *Wi-Fi* because they are widely supported technoligies in modern computing devices. Moreover, *Wi-Fi* access points are widely deployed, *BLE* beacons are relatively inexpensive, and many devices can be configured to emit their own beacon signals. There was also an extensive body of research in indoor positioning systems using *Bluetooth* and *Wi-Fi* technologies with satisfactory results for our use case from which we could draw inspiration.

The *Server* is responsible for estimating the position of the client devices through a set of positioning techniques: *fingerprinting*, *proximity*, and *trilateration*. It uses *fuzzy logic* to choose the most appropriate positioning technique for the current environment. The Mamdani system takes four variable features of the environment into consideration [15]: number of scanned *BLE* beacons; percentage of scanned *Wi-Fi* access points and number of scanned *BLE* beacons that have a substantial impact on position prediction (high *feature importance*); and number of scanned *BLE* beacons with known coordinates.

The *fingerprinting* technique consists of an *offline phase* responsible for data acquisition, and an *online phase* in which position predictions are made. The *offline phase* is performed by the mobile *Client* application which collects fingerprint samples scanned in a particular environment and sends them to the *Server*. After being signaled that an area has been completely scanned, the *Server* retrieves the samples from the database and converts them to a Comma-separated Values *CSV* file representing a radio map. The *Server* can then loads the radio map and train a ML model. During the *online phase*, the samples scanned in real-time by a *Client* application are sent to the server. If *fingerprinting* is selected as the positioning technique to be used by the decision system, the server will have to select the best matching radio map based on the detected access points and beacons. It will then used the corresponding ML model that was previously trained to produce a prediction of the absolute position of the device based on the received samples.

The **proximity** technique focuses on the notion of Proximity-based Services, which is the notion of relative location or context-based position (e.g., user U is near beacon B). Therefore, it only predicts the distance between a device and a *BLE* beacon, i.e., it only returns information about the relative position of a device. It is also comprised of an *offline phase* and an *online phase* since we use a ML approach instead of calculating distances from *RSSI* values based signal on path loss. An *offline phase* is required to collect samples to train the algorithm. The mobile *Client* application is responsible for data collection, sending the scanned samples at specific distances from a reference beacon to the *Server*. Although the system could rely on *Wi-Fi* access points, it is best designed to work with *BLE* beacon signals since their deployment in the environment is more flexible and best suited for real-life scenarios. This radio map collected during the *offline phase* is then used to train the ML model. During the *online phase*, if the decision system chose *proximity* as the positioning technique to use, the *BLE* beacon samples collected by a client device will be used by the trained model to make a distance prediction.

Trilateration does not require a dedicated *offline phase*. However, it relies on the *proximity* method to compute the distance between a client device and a set of beacons present in the surrounding environment. It also requires that the *Server* is configured with the coordinates of at least three beacons or access points. Similarly to the *proximity* technique, *trilateration* is designed to work with both radio technologies, but *BLE* beacons are recommended. When *trilateration* is chosen as the most appropriate positioning technique by the decision system, the *Server* will retrieve the locations of the beacons captured by the *Client*, compute the distance between the client device and the beacons using the *proximity* technique and it will estimate the absolute position of the device using least-squares optimization to minimize the sum of squared errors. This optimization process is required because it is very unlikely that the circles defined by the calculated distances intersect each other on a single point which leads to a set of equations without an analytical solution.

For all of the positioning techniques, once the prediction for the position of a client device is made, the *Server* publishes the prediction to the *onLocationUpdate* topic on the *WAMP* router. This pattern allows any interested third-party service to subscribe to this topic to continuously receive updates about the positions of client devices.

4 Evaluation

Each indoor positioning technique has multiple variables that can be adjusted. They also have their own advantages and disadvantages when it comes to accuracy and precision. Therefore, we designed a series of experiments to determine the default configuration for each technique and also to determine which technique performs better in certain situations in order to configure the decision system based on *fuzzy logic*. In this section, we present a summary of the results of these experiments and the decisions that were taken.

4.1 Fingerprinting

The dataset used to predict the absolute position of a device in a cartesian coordinate system using *fingerprinting* was a radio map scanned in 4 m × 4 m room with 25 reference points and 30 fingerprints scanned per reference point (750 samples).

The online matching phase in *fingerprinting* assesses the similarity between samples scanned in real-time by a device and the fingerprints previously saved in a radio map. Since it is assumed that the fingerprints already have labels associated, supervised learning algorithms are commonly used as part of the matching stage. Therefore, a set of five experiments were developed to test the following ML algorithms: k-Nearest Neighbors (*k-NN*), Random Forest (*RF*), Support Vector Machine (*SVM*), and Multilayer Perceptron (*MLP*).

The experiment encompassed tuning the hyperparameters of each algorithm using *Stratified K-Fold* cross-validation. The Leave One Group Out (*LOGO*) strategy was also used to simulate the worst-case scenario, i.e., the algorithm is trained with samples from all reference points except the one currently being tested. Tables 1 show the comparison between the regression performance results of the multiple ML algorithms we considered for fingerprinting.

Table 1. Performance Evaluation (best results in green, selected algorithms in bold)

Algorithm	**MAE**	**RMSE**	r^2	P_{95}
	Fingerprinting			
k-NN	0.292113	0.602620	0.909212	1.489585
SVM	0.548867	0.705709	0.875494	1.496627
RF	**0.260042**	0.485201	0.941145	1.187140
MLP	0.879875	1.028073	0.735767	1.875221
k-NN (LOGO)	1.462987	1.603255	0.357393	2.597401
SVM (LOGO)	1.390924	1.547562	0.401263	2.631749
RF (LOGO)	1.394118	1.508732	0.430932	2.520780
MLP (LOGO)	**1.390476**	1.515245	0.401356	3.060226
	Proximity			
k-NN	**0.483617**	0.723037	0.746707	1.150000
SVM	0.471081	0.806262	0.685040	1.353657
LR	0.784444	0.960112	0.553372	1.547085
MLP	0.579432	0.793403	0.694958	1.253388
k-NN (LOGO)	**0.945925**	1.295717	0.372113	1.816667
SVM (LOGO)	1.144035	1.402022	0.047464	1.726999
LR (LOGO)	1.078450	1.325616	0.148456	1.722772
MLP (LOGO)	1.190065	1.456349	0.027785	1.985990
	Trilateration			
Localization Package	1.041392	1.140057	-	1.895565
SciPy Minimize	**1.041344**	1.140028	-	1.89558
Brute-force	0.948854	1.131371	-	1.882843

RF is not always the best algorithm, but it achieved good performance across all of the experiments and it was selected as the one to be used by default. Another aspect that had to be considered is that collected fingerprints often have missing data for certain access points and beacons because they may appear in a certain place and during a certain scan, and yet be missing from other scans. Therefore, the missing data must be replaced somehow. We experimented with multiple replacement strategies, including replacing the missing values with the *maximum*, *minimum*, *median* or *mean* value of the dataset. We also experimented with calculating these globally (i.e., across the whole dataset) or for each access point or beacon.

The best results were obtained with the *mean* or *median* strategies when applied to each access point or beacon. However, the *minimum* replacement strategy also provided good results and it has a more sound reasoning behind it. Since lower *RSSI* values should correspond to access points or beacons that are further away, it seems logical that those that are not found during a scanning cycle are represented as being as further away as possible. Therefore, this was the strategy adopted for the remaining experiments and the final solution.

4.2 Proximity

Similarly to *fingerprinting*, the *proximity* technique also applies a ML algorithm to match the samples collected during the *online phase* with the data gathered during the *offline phase*. The following algorithms were compared to select the one which performed better: *k-NN*, *SVM*, *MLP*, and Linear Regression (*LR*). The dataset used to determine which ML algorithm to use was a radio map with 10 reference points placed between 0 and 4.5 m away from the beacon (24834 samples in total). Table 1 depicts the performance results of the various regression algorithms that were tested for the proximity technique.

Another experiment compared if collecting more data at each reference point improved results and in general the results showed that it did have an appreciable effect on results. Therefore, we recommend collecting as much data as possible when configuring our system. We were also interested in assessing if the model trained to estimate the distance from one beacon could be used interchangeably for other beacons. Similarly, we were interested in determining if a model trained from a dataset collected in an environment could be used on a different one (e.g., on another room). In both cases, there was a loss in performance but they were small enough that we believe that the trade-off between the additional work of having to collect data for each specific beacon/environment and the performance loss is generally worth it. Therefore, the system makes predictions based on the same radio map, regardless of the target beacon.

4.3 Trilateration

Trilateration leverages the *proximity* technique to predict the distance between a device and reference beacons in the environment, meaning there is a dependency between the performance of *trilateration* and *proximity*. The goal was to study the best approach to predict the position of a device based on the distance estimates provided by the *proximity* technique.

The analytical approach of determining the intersection of multiple circles from their equations cannot be used because the inaccuracy of the distance estimates often leads to the definition of circles that do not intercept in a single point. Therefore, the *trilateration* experiments focused on solving the problem from an optimization point of view. The optimization process tries to find the point X that provides the best approximation to the device position P. Specifically, X is the point whose distance to the known beacons best fits the distance between a target device's position and the beacons in a particular environment. A common approach is to find the point X that minimizes the sum of squared errors, i.e., a least-squares optimization problem. Consequently, the experiments studied how the following optimization approaches perform:

- **Brute Force Approach**: Finds the point X out of a list of possible P that best fits the estimated distances to each beacon.
- **Optimization Algorithm**: Given the positions of the beacons, use an optimization algorithm that iteratively finds the point X that best fits the estimated distances to each beacon.

It is assumed that the device is at one of the 25 points with saved *BLE RSSI* samples and will analyze the prediction errors at each point according to the Mean Absolute Error (*MAE*) metric. The brute force approach finds the point X that minimizes the mean squared distance error between X and the set of known beacons, with X being one of a fixed set of available points. In order to analyze its performance, the experiment considered three possible datasets with varying granularity: a $4\,\text{m} \times 4\,\text{m}$ room divided into a grid of 25 points spaced 1 m apart; 100 points spaced 0.5 m apart; or 2500 points spaced 0.1 m apart. This results in a *MAE* of 1.04747, 1.044542 and 0.948854, respectively. As expected, the increased granularity improves results at the cost of increased computational complexity. Moreover, the brute force approach is a pretty native algorithm and it requires a dataset to be generated beforehand, which includes information about all possible candidate points that form the grid pattern.

Optimization algorithms can be used as an alternative that only requires the positions of the beacons in the environment. Therefore, we experimented with the following libraries that implement least-squares optimization:

- **SciPy Minimize Function**: It was used by applying a bound-constrained optimization algorithm (L-BFGS-B [8,21]) with the sum of the squared error as the target error function to be minimized [19]. The initial prediction of the function is the target device's nearest beacon.
- **Localization package**[1]: It applies *trilateration* by trying to find a point that minimizes the sum of squared distance errors to the position of the devices. The system only requires the distance of the target device to known beacons and their respective positions. The underlying optimization algorithms are also provided by SciPy [19].

[1] https://github.com/kamalshadi/Localization.

Table 1 also outlines the performance results for trilateration, with the best brute force result as a baseline. The two tested implementations performed identically. The *SciPy Minimize Function* was selected since it is part of a well established library that offers more options to further improve the results.

4.4 Discussion

Since *proximity* only provides relative positions, it cannot be directly compared with *fingerprinting* and *trilateration*, which provide the absolute location of a device. Absolute positions offer more information and can also be used to infer the relative positions in respect to other devices and *PoIs*. Moreover, although *fingerprinting* present good overall results, its performance results in the worst-case scenario (*LOGO*) degrade considerably. Therefore, *trilateration* presents itself as the positioning technique with the good performance and more consistent results. It also has the added advantage of requiring considerably less data collection during the *offline phase*.

The decision system's rules support the performance results by establishing a hierarchy among the available positioning techniques according to their overall performance. The rules outline a hierarchy of the positioning techniques with *trilateration* being the default technique that is used when all conditions are *good enough*. Therefore, the hierarchy is *trilateration* > *fingerprinting* > *proximity*.

Nonetheless, the prioritization of *fingerprinting* above *proximity* is not so much about performance results but about the information provided. *Fingerprinting* provides the absolute position of a device allowing to compute the relative position in relation to another absolute position. However, *proximity* only displays the relative position in relation to a beacon in the environment, a beacon attached to a device, or a beacon signal emitted by a device.

5 Conclusions and Future Work

The current version of the IPS has been able to meet most of our requirements and expectations. The errors that we presented in Sect. 4 were acceptable given our use case. Mean absolute errors (*MAE*) around or below 1 m and a 95th percentile errors in the 2 m range seem reasonable for the type of applications that we envisioned. Being capable of determining if two devices are in the same area within a room is enough for most cross-device applications and may be also true for many other potential UbiComp applications.

We have integrated successfully IPS with our framework, allowing to create applications that take into consideration the proxemic relationships established between the devices running them. We used one of those applications in a user study to evaluate the concept of cross-device applications. Most users agreed that the application detected quickly the presence of the other devices during the experiment, also meaning the underlying IPS responded positively to the users' expectations. During the user study, we used the *proximity* technique to get estimates of the distances between devices. This decision was taken to simplify the deployment of our test environment, because building a radio map for *fingerprinting*, or carefully placing *BLE* beacons for *trilateration*, was not required. Besides, we only needed the relative distance and the orientation of the devices for the scenarios that were being tested. Moreover, the estimates returned by *proximity* in previous experiments were more stable and accurate than relative distances calculated from absolute positions provided by the *fingerprinting* or *trilateration* methods.

These issues are probably due to the fact that we are dealing with two position estimates, each one already with some associated error, which gets magnified when the euclidean distance between two points is calculated. Furthermore, there is currently no direct relationship between the position of a device at moment t and at $t + 1$ because the computing of each position is stateless, which means that subsequent position estimates may diverge from the previous one by an unrealistic amount (e.g., distance between positions is larger than it would be possible at walking speed). This effect may be attenuated if it incorporates information about previous estimates into the current one, i.e., by using LSTM (Long short-term memory) neural networks or Hidden Markov Models (HMM).

We also believe that we should still be able to improve the accuracy and response times of the IPS. It is still possible to explore better ways to process data during the *online* and *offline phases*, and to improve the usage of the available ML algorithms through better parameterization or by customizing them to better fit this application domain. Fine grained centimeter level accuracy may be desirable to build more complex and richer interaction scenarios. There is the possibility of extending the system with other technologies such as Wi-Fi Round Trip Time (RTT) and Ultra-wideband (UWB), which are becoming more easily available and have the potential of reaching higher levels of accuracy.

References

1. Mueller, F., et al.: Proxemics play: understanding proxemics for designing digital play experiences. In: Proceedings of the Conference on Designing Interactive Systems: Processes, Practices, Methods, and Techniques, DIS, pp. 533–542. Association for Computing Machinery (2014). https://doi.org/10.1145/2598510.2598532
2. Indoor fingerprint positioning based on Wi-Fi: An overview (May 2017). https://doi.org/10.3390/ijgi6050135
3. SensorManager - Android Developers (2021). https://developer.android.com/reference/android/hardware/SensorManager
4. Bahl, P., Padmanabhan, V.N.: RADAR: An in-building RF-based user location and tracking system. Technical Report (2000). https://doi.org/10.1109/infcom.2000.832252
5. Ballendat, T., Marquardt, N., Greenberg, S.: Proxemic interaction: Designing for a proximity and orientation-aware environment. In: ACM International Conference on Interactive Tabletops and Surfaces, ITS 2010, pp. 121–130 (2010). https://doi.org/10.1145/1936652.1936676
6. de Blasio, G., Quesada-Arencibia, A., García, C.R., Molina-Gil, J.M., Caballero-Gil, C.: Study on an indoor positioning system for harsh environments based on Wi-Fi and bluetooth low energy. Sensors (Switzerland) **17**(6), 1299 (2017). https://doi.org/10.3390/s17061299, http://www.mdpi.com/1424-8220/17/6/1299

7. Brena, R.F., García-Vázquez, J.P., Galván-Tejada, C.E., Muñoz-Rodriguez, D., Vargas-Rosales, C., Fangmeyer, J.: Evolution of Indoor Positioning Technologies: a survey (2017). https://doi.org/10.1155/2017/2630413
8. Byrd, R.H., Lu, P., Nocedal, J., Zhu, C.: A limited memory algorithm for bound constrained optimization. SIAM J. Sci. Comput. **16**(5), 1190–1208 (1995). https://doi.org/10.1137/0916069
9. Faragher, R., Harle, R.: An analysis of the accuracy of bluetooth low energy for indoor positioning applications. In: Proceedings of the 27th International Technical Meeting of The Satellite Division of the Institute of Navigation (ION GNSS+ 2014), pp. 201–210 (2014). http://www.ion.org/publications/abstract.cfm?jp=p&articleID=12411
10. Faragher, R., Harle, R.: Location fingerprinting with bluetooth low energy beacons. IEEE J. Sel. Areas Commun. **33**(11), 2418–2428 (2015). https://doi.org/10.1109/JSAC.2015.2430281, http://ieeexplore.ieee.org/document/7103024/
11. Feldmann, S., Kyamakya, K., Zapater, A., Lue, Z.: An indoor Bluetooth-based positioning system: concept, implementation and experimental evaluation. In: Proceedings of the International Conference on Wireless Networks, pp. 109–113 (2003)
12. Greenberg, S., Marquardt, N., Ballendat, T., Diaz-Marino, R., Wang, M.: Proxemic interactions: the new ubicomp? interactions **18**(1), 42 (2011). https://doi.org/10.1145/1897239.1897250, http://portal.acm.org/citation.cfm?doid=1897239.1897250
13. Kantar TNS Germany: The Connected Consumer (2019). https://www.google.com/publicdata/explore?ds=dg8d1eetcqsb1_
14. Kim, D.Y., Kim, S.H., Choi, D., Jin, S.H.: Accurate indoor proximity zone detection based on time window and frequency with bluetooth low energy. Procedia Comput. Sci. **56**(1), 88–95 (2015). https://doi.org/10.1016/j.procs.2015.07.199
15. Mamdani, E.H., Assilian, S.: An experiment in linguistic synthesis with a fuzzy logic controller. Int. J. Man. Mach. Stud. **7**(1), 1–13 (1975). https://doi.org/10.1016/S0020-7373(75)80002-2, https://linkinghub.elsevier.com/retrieve/pii/S0020737375800022
16. Mautz, R.: Indoor Positioning Technologies. Ph.D. thesis (2012). https://doi.org/10.3929/ethz-a-007313554, http://e-collection.library.ethz.ch/eserv/eth:5659/eth-5659-01.pdf
17. Orujov, F., Maskeliūnas, R., Damaševičius, R., Wei, W., Li, Y.: Smartphone based intelligent indoor positioning using fuzzy logic. Future Gener. Comput. Syst. **89**, 335–348 (2018). https://doi.org/10.1016/j.future.2018.06.030
18. Rodden, T.: Living in a ubiquitous world. Philos. Trans. Royal Soc. Math. Phys. Eng. Sci. **366**(1881), 3837–3838 (2008). https://doi.org/10.1098/rsta.2008.0146
19. Virtanen, P., et al.: SciPy 1.0: fundamental algorithms for scientific computing in Python. Nat. Methods **17**(3), 261–272 (2020). https://doi.org/10.1038/s41592-019-0686-2, http://www.nature.com/articles/s41592-019-0686-2
20. WAMP: The Web Application Messaging Protocol - Web Application Messaging Protocol version 2 documentation (2021). https://wamp-proto.org/
21. Zhu, C., Byrd, R.H., Lu, P., Nocedal, J.: Algorithm 778: L-BFGS-B. ACM Trans. Math. Softw. **23**(4), 550–560 (1997). https://doi.org/10.1145/279232.279236

Special Session on Computational Intelligence for Computer Vision and Image Processing

T Line and C Line Detection and Ratio Reading of the Ovulation Test Strip Based on Deep Learning

Libing Wang[1], Li He[2], Junyu Li[1], Deane Zeng[3], and Yimin Wen[1](✉)

[1] Guangxi Key Laboratory of Image and Graphic Intelligent Processing, Guilin University of Electronic Technology, Guilin 541004, China
19032201034@mails.guet.edu.cn, ymwen2004@aliyun.com, lijunyu@visiongo.ai, 248539087@qq.com

[2] Guangxi Police College, Nanning 530028, China
heli0228@aliyun.com

[3] Easy Healthcare Corporation, Guangzhou 510000, China
deane@healthcare-manager.com

Abstract. The ovulation test strip is a tool for ovulation detection. At present, many APPs have been developed to analyze the photos taken from ovulation test strips to read the T/C ratio for detecting the level of luteinizing hormone (LH) in human urine. However, in the process of detecting the T and C lines, the background colors of the area near the T line or C line may be red, the T and C lines are sometimes fuzzy, and the colors of the T and C lines are sometimes distributed unevenly. In these cases, these APPs will be confronting difficulty to accurately detect the T and C lines and further read their ratio. To solve these problems, we proposed a method that consists of two steps. The first step is to use Mask R-CNN to locate the T and C lines, and the second step is to use a trained Pseudo-Siamese ratio reading network (PSRRNet) to read the ratio value based on the output results of the first step. The proposed PSRRNet consists of a Pseudo-Siamese network for simultaneously extracting the features of the T and C lines and a fully connected network to use the extracted features to predict the T/C ratio value. The experimental results illustrated that the proposed method suits to handle ovulation test strip reading.

Keywords: Ovulation test strip · Mask R-CNN · Deep learning

1 Introduction

The ovulation test strip is a tool used to qualitatively detect the level of luteinizing hormone (LH) in human urine to determine the time of ovulation and the "safe period" in women's menstrual cycle, and then to select the best time for pregnancy, or to select the "safe period" for contraception. The ovulation test strip has a "MAX" sign on one side, which must not be exceeded this mark

H. Yin et al. (Eds.): IDEAL 2021, LNCS 13113, pp. 625–636, 2021.
https://doi.org/10.1007/978-3-030-91608-4_60

when it enters the urine. When the test strip reacts with urine, the test line (T line) which is near the "MAX" sign and the control line (C line) which is farther away from the "MAX" sign than T line will display the color. The two ovulation test strip pictures are shown in Fig. 1. Figure 1(a) is an example of the image taken by the user after using the ovulation test strip. It can be seen that two red lines appear in the chemical reaction area of the ovulation test strip, among which the left line is "T line" and the right line is "C line". Figure 1(b) shows the user's unused ovulation test strip. It can be seen that the chemical reaction area of the strip is white.

By comparing the color depth of the T line and C line, the user can determine the level of LH. If a sequence of LH is collected, the user can reason the time of her ovulation. (We use the "T/C ratio" to indirectly express LH levels. The change in the T/C ratio indicates the increase or decrease of LH levels in urine.) However, the user should carefully and accurately record each value of LH during her observation. Sometimes, it is a boring work for a would-be mother, and further, it is difficult for a user to accurately compare the color of the T line and C line.

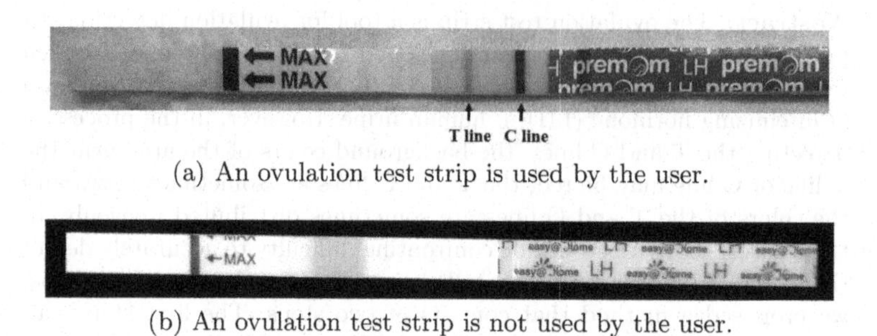

(a) An ovulation test strip is used by the user.

(b) An ovulation test strip is not used by the user.

Fig. 1. Photos of the ovulation test strip.

Nowadays, computer vision technology is widely used in object detection, regression, and classification. However, in the process of the T line and C line detection, the background color of the area near the T line and C line may show red, as shown in Fig. 2(a), the T line and C line are sometimes fuzzy, as shown in Fig. 2(b), the color of the T line and C line are distributed unevenly, as shown in Fig. 2(c), and there are also some slanted test strips, as shown in Fig. 2(d). In these cases, it is difficult to accurately detect the T line and C line and read the T/C ratio. In addition, in the process of taking photos, photos are inevitably susceptible to the illumination and the camera's shake, blur and defocus.

With the development of deep learning technology, many image-related tasks can be solved by deep learning methods. A series of object detection algorithms, such as the one-stage approach, including [1–5], and the two-stage approach, including [6–10], and some segmentation algorithms, such as [11,12], make it possible to detect and segment the reaction region of various test strip. [13] developed a novel image segmentation method for quantitative analysis of the

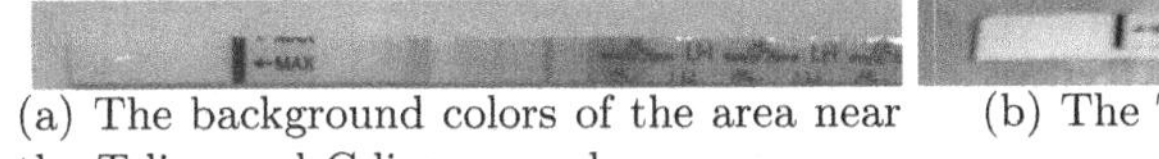

(a) The background colors of the area near the T line and C line are red.

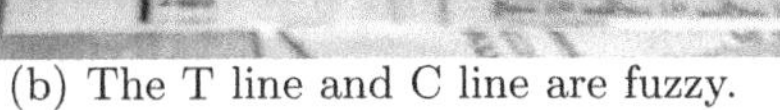

(b) The T line and C line are fuzzy.

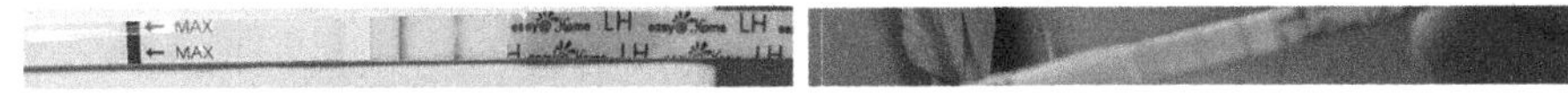

(c) The colors of the T line and C line are uneven.

(d) The ovulation test strip is slanted.

Fig. 2. Some test strips that are difficult to detect and read ratio value. (Color figure online)

gold immunochromatographic strip based on deep reinforcement learning. In [14], CNN is applied to image segmentation of the gold immunochromatographic strip. The grayscale features of the pre-processed images are learned by the CNN network, and then, the control lines and test lines are accurately extracted and further quantitative analysis is performed.

Being inspired by the above works, we propose a method to read the T/C ratio from a photo of an ovulation test strip. The proposed method includes two steps in which the first is to apply Mask R-CNN [6] to locate the T line and C line and the second is to apply a trained Pseudo-Siamese ratio reading network (PSRRNet) to read the ratio value based on the output of the first step. Our main contributions in this paper are as follows: (1) We constructed an annotated ovulation test strip dataset. (2) We proposed a novel network of PSRRNet, which consists of a Pseudo-Siamese network [15] to simultaneously extract the T line and C line features and a fully connected network to utilize the extracted features to predict the T/C ratio value.

The remainder of this paper is organized as follows. Section 2 introduces our methods to detect the T line and C line and read ratio value from a photo of an ovulation test strip. Section 3 presents our experimental details and results. Finally, the conclusion is given in Sect. 4.

2 Methodology

2.1 Pipeline

The detection and ratio reading process of the T line and C line from a photo of an ovulation test strip is shown in Fig. 3. First, the image of the ovulation test strip taken by the user is sent to Mask R-CNN. Then, the T line and C line are segmented from the test strip image by post-processing according to the result of Mask R-CNN. Next, the T line and C line are resized to 90×30 by linear interpolation, because in the subsequent ratio reading step, the features of the T line and the C line need to be fused with the same shape, and then they are input into the trained PSRRNet at the same time. The T/C ratio is obtained at last.

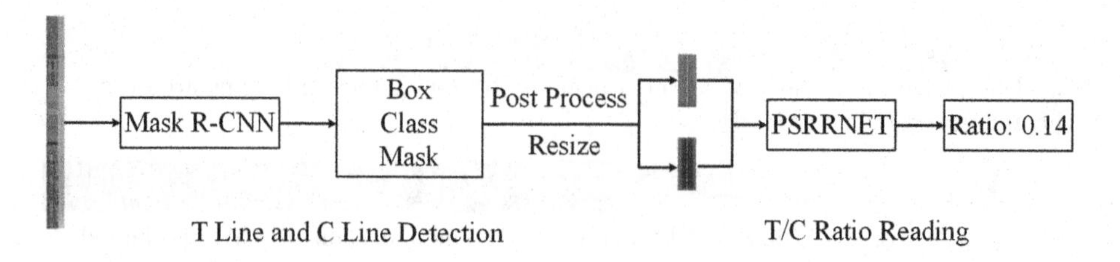

Fig. 3. The flow chart of the T line and C line detection and ratio reading from a photo of an ovulation test strip.

2.2 T Line and C Line Detection

What Is Mask R-CNN? Mask R-CNN is a two-stage instance segmentation algorithm proposed by He et al. The first stage is to use RPN [8] to generate region proposal, and the second stage is to predict the class, bounding box offset and binary mask for each RoI. During training, Mask R-CNN define a multi-task loss on each sampled RoI as follows:

$$L = L_{cls} + L_{box} + L_{mask} \tag{1}$$

Why is Mask R-CNN Chosen? Mask R-CNN is chosen because it can accurately segment the T line and C line from the background with the obtained mask, while some other object detection algorithms only get the information of their boundary box, which contains part of the background information when it is cut off from the tilted test strip, affecting the subsequent reading of the ratio. On the other hand, Mask R-CNN not only has high detection accuracy but also has strong scalability.

Detection Process. First, Mask R-CNN extracts the deep convolution feature map of the test strip image using ResNet [16] combined with FPN [17], and then sends the feature map to RPN to obtain high-quality region proposals (find out the possible position of the T line and C line in the image in advance and correct the boundary box). Then, through further selecting, the region of interest (RoI) is obtained and each RoI is mapped to the convolutional feature map using the RoI Align [6] to achieve accurate alignment between the output feature map and the input RoI, so the target location information is accurate. Then, the high-quality binary segmentation mask is output through FCN [18], and the prediction box and class are output through the full connection layer. Finally, the T line and C line are cut out from the test strip image by post-processing.

Post Processing. Considering the subsequent T/C ratio reading process, we need to cut out the T line and C line from the test strip image. We use the findContours function and minAreaRect function in OpenCV to get the precise position information (x, y, w, h, θ) of the T line and C line in the binary segmentation mask of the detection result, where, (x, y) represents the coordinates of the center of the object, (w, h) represents the width and height of the object respectively, and θ represents the rotation angle of the object. Using this position information, we can accurately cut out the T line and C line from the test strip image.

2.3 Ratio Reading

The architecture of PSRRNet is shown in Fig. 4. There are two parts of the network, the first part is to extract the features of T line and C line, and the second part is to use the extracted features to predict the T/C ratio value.

The input of PSRRNet is a pair of images of T line and C line each with the size of 90 × 30. Because we consider that there is a certain correlation between the T line and C line, such as they are taken in the same lighting environment and collected from the same test urine, etc. In the feature extraction network, we adopt Pseudo-Siamese network structure, that is, two branches compute simultaneously and parameters do not share. Each feature extraction branch is composed of three depthwise separable convolutional layers [19] with the number of channels of 32, 64 and 128, respectively. Each depthwise separable convolutional layer is followed by BatchNorm and ReLU. This can further reduce the computation and over-fitting, and improve efficiency. Then, the features of the extracted T line and C line are concatenated, which will further maintain the correlation between them.

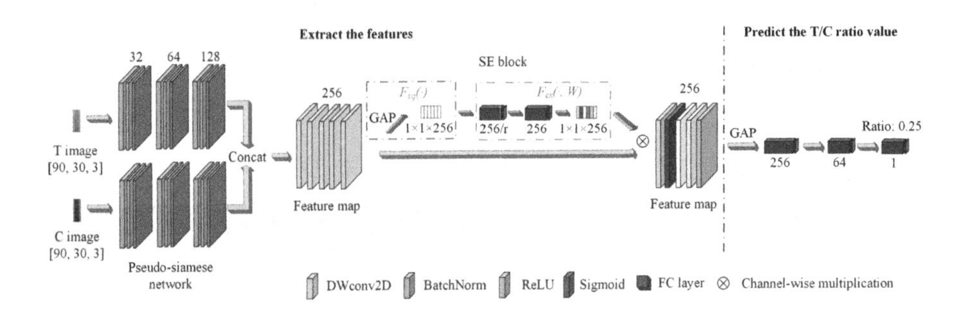

Fig. 4. The architecture of PSRRNet.

In addition, we apply an attention mechanism called Squeeze-and-Excitation block (SE block) [20] to the concatenated features to further enhance the expressive ability of the network. SE block adaptively recalibrates channel-wise feature responses by explicitly modeling interdependencies between channels. A SE block consists of two steps: squeeze (F_{sq}) and exception (F_{ex}). F_{sq} obtains the global compressed feature quantity of the current feature map by performing global average pooling (GAP) [21] on the feature map. F_{ex} gets the weight of each channel in the feature map through the two-layer fully connected bottleneck structure, and multiplies the weight with the feature map as the input of the next layer network.

In the ratio value prediction part, we use a design combining GAP and three fully connected layers with the number of channels being 256, 64 and 1. The last fully-connected layer obtains the T/C ratio value after the ReLU activation function, because the T/C ratio value is greater than or equal to 0.

3 Experiments

In this section, we will introduce the T line and C line detection and ratio reading experiments respectively, and also introduce the combined experiment of detection and ratio reading. We will provide experimental details for the datasets, implementation details, and experimental evaluation.

3.1 Experiment of T Line and C Line Detection

Datasets. We collect 3500 ovulation test strip images taken by real users, including 2000 test strip images in which the T line and C line can be easily located and 1500 test strip images in which the T line and C line cannot be easily located.

Label Generation. The ground truth of test strip instance is exemplified in Fig. 5. First, we use the quadrilateral box to label each line, and then convert it into a global binary mask image, that is, we treat the pixels inside the quadrilateral box as a target, the pixel value is 255, and the pixels outside the quadrilateral box as non-target, the pixel value is 0. The coordinate information of the quadrilateral box and the global binary mask are the ultimate ground truths.

(a) Quadrilateral box. (b) Global binary mask.

Fig. 5. Ground truth.

It should be noted that we also think of another labeling scheme, which is to label the T line and C line and the middle area between the two as a target. We will compare these two annotations in subsequent experiments.

Implementation Details. For the model, we first pre-train it with the COCO dataset. Then we fine-tune the model on our ovulation test strip dataset for 1000 epochs. When fine-tuning, we first freeze the parameters of ResNet, FPN, and RPN, train the classification branch and mask branch, then freeze the classification branch and mask branch, train ResNet, FPN, and RPN, and finally perform global training. The training batch size is set to 8. The SGD with a momentum of 0.9 is used as an optimizer and the learning rate is set to 0.001.

We do a five-fold cross-validation experiment on our dataset. The 3500 data are evenly divided into five parts. In each fold experiment, the number of the training set was 2800, and the number of test set is 700, then we randomly extract 560 data from the 2800 training set as the validation set. We use the validation set to adjust hyperparameters and monitor whether the model is over-fitting during training.

During training, we use the online augmentation method, that is, use the generator to transform the data on the fly during training [22]. The data augmentation for the training data includes (1) Random rotation with an angle range of (−30°, 30°); (2) Motion blur.

Evaluation and Analysis. We respectively use two labeling schemes and whether to use online data augmentation to do related comparative experiments. The experimental results are shown in Table 1. "Labeling scheme 1" represents our label generation method, and "Labeling scheme 2" represents the method of labeling the T line and the C line and the area between the two as a target. We evaluate the performance of the model by calculating Precision, Recall, F-Measure, mAP@0.5 on test set. Figure 6 shows the detection results using Mask R-CNN.

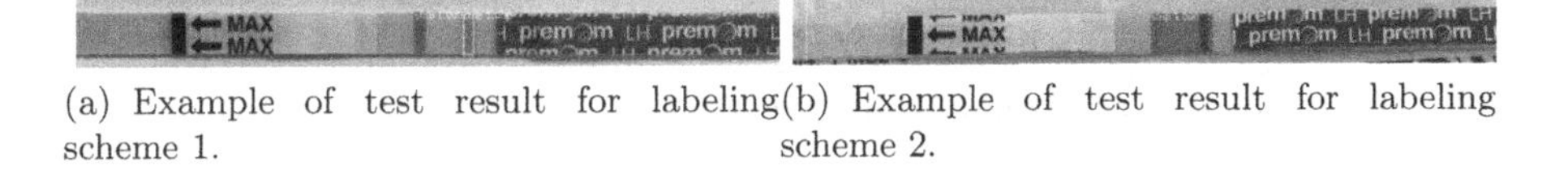

(a) Example of test result for labeling scheme 1.

(b) Example of test result for labeling scheme 2.

Fig. 6. Examples of Mask R-CNN detection results.

It can be seen from Table 1 that when online data augmentation is not used, the mAP of the two labeling methods reached 0.984 and 0.975, respectively. At the same time, Precision, Recall, and F-Measure all have good performance. The mAP of the label generation method we used (Labeling scheme 1) is higher than that of labeling scheme 2. In addition, the use of online augmentation has further improved the performance of our model, and some metrics have been slightly improved. However, we can see from Table 1 that, after using online augmentation, the result of labeling scheme 1 is slightly worse than that of labeling scheme 2. I think the reason is that the advantage of using online augmentation to detect T and C lines as a large target is brought into play. Although the result of labeling scheme 2 is better, it is difficult to use the detection result of labeling scheme 2 to crop the T line and C line from the original image. As shown in Fig. 6(b), the color of the T line is lighter, at this time, it is difficult to accurately cut out the T line, which will have a huge impact on the subsequent reading of the T/C ratio. So, we finally choose labeling scheme 1 to generate labels.

Table 1. The five-fold cross-validation experiment results of the two labeling schemes.

Methods	Metrics	Number of cross-validations					Avg
		1	2	3	4	5	
Labeling scheme 1	Precision	0.916	0.916	0.915	0.941	0.926	0.924
	Recall	0.981	0.988	0.991	0.986	0.986	0.988
	F-Measure	0.948	0.950	0.951	0.963	0.955	9.954
	mAP@0.5	0.977	0.983	0.991	0.986	0.984	0.984
Labeling scheme 2	Precision	0.890	0.951	0.901	0.959	0.954	0.931
	Recall	0.991	0.996	0.996	0.999	0.997	0.995
	F-Measure	0.938	0.973	0.946	0.978	0.975	0.962
	mAP@0.5	0.902	0.995	0.989	0.997	0.994	0.975
Labeling scheme 1 + Online aug	Precision	0.937	0.905	0.935	0.913	0.960	0.930
	Recall	0.997	0.984	0.990	0.988	0.983	0.984
	F-Measure	0.956	0.943	0.962	0.949	0.971	9.956
	mAP@0.5	0.983	0.988	0.991	0.987	0.989	0.988
Labeling scheme 2 + Online aug	Precision	0.927	0.939	0.955	0.967	0.973	0.952
	Recall	0.990	0.997	0.99	0.994	0.997	0.994
	F-Measure	0.958	0.967	0.972	0.981	0.985	0.973
	mAP@0.5	0.991	0.998	0.996	0.998	0.998	0.996

3.2 Experiment of Ratio Reading

Datasets. In this experiment, we don't use all 3500 data, because we can not get the corresponding ratio of some data. We only use 2425 data, each of which consists of a pair of images of T line and C line, color depth value of T line and C line and their corresponding T/C ratio value. The ratio is greater than or equal to 0. In particular, when the T line is as dark as the C line, the ratio is 1. It should be noted that instead of inputting the entire test strip image into the network, we use the ground truth information mentioned in Sect. 3.1 to crop the T line and C line from the test strip image as the input, as shown in the Fig. 7.

Fig. 7. An example of cropping.

Implementation Details. The PSRRNet is trained on the single Nvidia Tesla V100 GPU using the Keras framework. We do a five-fold cross-validation experiment on our dataset. The 2425 data are evenly divided into five parts. In each fold experiment, the number of training set is 1940 and the number of test set is 485, then we randomly extract 388 data from the 1940 training set as the validation set. We use the validation set to adjust hyperparameters and monitor whether the model is over-fitting during training.

We train the model on the training set for 2000 epochs. The training batch size is set to 16. The SGD with a momentum of 0.9 is used as an optimizer and the learning rate is set to 0.0001. The loss function during the training is the mean square error (MSE) between the true ratio value and the predicted ratio value. MSE can be calculated using the following formula:

$$MSE = \frac{1}{m}\sum_{i=1}^{m}\left(Y_i - \hat{Y}_i\right)^2 \tag{2}$$

Metrics. We evaluate the performance of the model by calculating root mean square error (RMSE). RMSE can be calculate using the following formula:

$$RMSE = \sqrt{\frac{1}{m}\sum_{i=1}^{m}\left(Y_i - \hat{Y}_i\right)^2} \tag{3}$$

Ablation Study. We design several ablation experiments to analyze whether it is necessary to input the T line and C line at the same time to train a model (use the correlation between the T line and the C line) and whether the SE block is effective. The results are shown in Table 2. In Table 2, "One model" represents that T line and C line are input to train a model at the same time, and "Two models" means to train a model with T line and C line respectively, and then the final T/C ratio is obtained by dividing the color depth values output by the two models.

Table 2. Ablation study.

Methods	Metrics	Number of cross-validations					Avg
		1	2	3	4	5	
One model	RMSE	0.15	0.10	0.10	0.08	0.08	0.102
One model + SE block (PSRRNet)		0.12	0.10	0.09	0.09	0.07	0.094
Two models		0.10	0.12	0.12	0.13	0.10	0.114
Two models + SE block		0.11	0.10	0.13	0.12	0.08	0.108

Table 2 shows that our model (PSRRNet) achieves a RMSE of 0.094. The RMSE value of using SE block (PSRRNet) is 0.008 lower than that of unused, which seems to be very small, but it is still meaningful in this high accuracy task. When we use T line and C line to train respectively a model, the RMSE value (use SE block) is 0.014 higher than PSRRNet, and the RMSE value (no SE block) is 0.012 higher than "One model". It can be seen that using the correlation between the T line and the C line and SE block helps to accurately read the T/C ratio value. Qualitative results (such as in Fig. 8) shown that our proposed solution can predict the T/C ratio value of a pair of T line and C line.

T, C lines							
Real ratio	0.69	0.57	0.37	0.17	0.08	1.00	0.66
Predicted ratio	0.67	0.57	0.41	0.18	0.08	0.96	0.66

Fig. 8. Examples of T line and C line with ratio prediction by PSRRNet.

3.3 Combination Experiment of Detection and Ratio Reading

Considering that the number of datasets of the above-mentioned T line and C line detection and T/C ratio reading experiments are inconsistent, and our idea is that the T line and C line input by PSRRNet are the results detected by Mask R-CNN (The T line and C line of the input of the above ratio reading experiment are cropped according to the ground truth mentioned in Sect. 3.1.), we do this combination experiment.

Implementation Details. The dataset used in the experiment is mentioned in Sect. 3.2, with a total of 2425 data, and each test strip image corresponds of a T/C ratio value.

The hyperparameter settings in this experiment are consistent with the settings of the related experiments described earlier. We also do a five-fold cross-validation experiment. In each fold experiment, we first train the Mask R-CNN stage, and after the Mask R-CNN training converges, use the detection results obtained to train PSRRNet.

Table 3. Experimental results of five-fold cross-validation.

	Number of Cross-validations					Avg
	1	2	3	4	5	
RMSE	0.12	0.12	0.10	0.10	0.05	0.098
Number of failed detections	5	5	2	2	3	4

Evaluation and Results. We use the test set (485 data) to test our model, and the experimental results are shown in Table 3. We can see that by combining the detection and ratio reading process, our method is still valid, and the RMSE value is only 0.098. But inevitably, each fold has several test data whose ratio value cannot be obtained because the T line and C line are not detected.

4 Conclusion

In this paper, we apply deep learning to an actual application scenario. We propose a framework that uses Mask R-CNN to detect T line and C line and uses our proposed PSRRNet to predict T/C ratio value, which effectively solves the problem that it is difficult to accurately detect T and C lines and read the T/C ratio value when the image quality of T line and C line is poor. Experimental results have shown its effectiveness and merit.

Acknowledgment. This work was partially supported by the Innovation Project of GUET Graduate Education (2020YCXS057), Natural Science Foundation of Guangxi District (2018GXNSFDA138006), the National Natural Science Foundation of China (61866007), and Image Intelligent Processing Project of Key Laboratory Fund (GIIP201505).

References

1. Redmon, J., Divvala, S., Girshick, R., Farhadi, A.: You only look once: unified, real-time object detection. In: Proceedings of the IEEE Conference on Computer Vision and Pattern Recognition, pp. 779–788 (2016)
2. Liu, W., et al.: SSD: single shot multibox detector. In: Leibe, B., Matas, J., Sebe, N., Welling, M. (eds.) ECCV 2016. LNCS, vol. 9905, pp. 21–37. Springer, Cham (2016). https://doi.org/10.1007/978-3-319-46448-0_2
3. Lin, T., Goyal, P., Girshick, R., He, K.M., Dollár, P.: Focal loss for dense object detection. In: Proceedings of the IEEE International Conference on Computer Vision, pp. 2980–2988 (2017)
4. Jeong, J., Park, H., Kwak, N.: Enhancement of SSD by concatenating feature maps for object detection. ArXiv, arXiv:1705.09587 (2017)
5. Zhang, S.F., Wen, L.Y., Bian, X., Lei, Z., Li, S.: Single-shot refinement neural network for object detection. In: Proceedings of the IEEE Conference on Computer Vision and Pattern Recognition, pp. 4203–4212 (2018)
6. He, K.M., Gkioxari, G., Dollár, P., Girshick, R.: Mask R-CNN. In: Proceedings of the IEEE International Conference on Computer Vision, pp. 2980–2988 (2017)
7. Girshick, R.: Fast R-CNN. In: Proceedings of the IEEE International Conference on Computer Vision, pp. 1440–1448 (2015)
8. Ren, S.Q., He, K.M., Girshick, R., Sun, J.: Faster R-CNN: towards real-time object detection with region proposal networks. In: IEEE Transactions on Pattern Analysis and Machine Intelligence, pp. 1137–1149 (2017)
9. Dai, J.F., Li, Y., He, K.M., Sun, J.: R-FCN: Object detection via region-based fully convolutional networks. ArXiv, arXiv:1605.06409 (2016)
10. Cai, Z.W., Vasconcelos, N.: Cascade R-CNN: delving into high quality object detection. In: Proceedings of the IEEE Conference on Computer Vision and Pattern Recognition, pp. 6154–6162 (2018)
11. Badrinarayanan, V., Kendall, A., Cipolla, R.: SegNet: a deep convolutional encoder-decoder architecture for image segmentation. In: IEEE Transactions on Pattern Analysis and Machine Intelligence, pp. 2481–2495 (2017)
12. Bolya, D., Zhou, C., Xiao, F.Y., Lee, Y.J.: YOLACT: real-time instance segmentation. In: Proceedings of the IEEE International Conference on Computer Vision, pp. 9157–9166 (2019)

13. Zeng, N.Y., Li, H., Wang, Z.D., Liu, W.B., Liu, X.H.: Deep-reinforcement-learning-based images segmentation for quantitative analysis of gold immunochromatographic strip. Neurocomputing **425**, 173–180 (2021)
14. Zeng, N.Y., Li, H., Li, Y.R., Luo, X.: Quantitative analysis of immunochromatographic strip based on convolutional neural network. IEEE Access **7**, 16257–16263 (2019)
15. Gao, J.Y., Xiao, C., Glass, L., Sun, J.M.: COMPOSE: cross-modal pseudo-siamese network for patient trial matching. In: Proceedings of the 26th ACM SIGKDD International Conference on Knowledge Discovery & Data Mining, pp. 803–812. Association for Computing Machinery, New York (2020)
16. He K.M., Zhang X.Y., Ren S.Q., Sun J.: Deep residual learning for image recognition. In: Proceedings of the IEEE Conference on Computer Vision and Pattern Recognition, pp. 770–778 (2016)
17. Lin T.Y., Dollár P., Girshick R., He K., Hariharan B., Belongie S.: Feature pyramid networks for object detection. In: Proceedings of the IEEE Conference on Computer Vision and Pattern Recognition, pp. 2117–2125 (2017)
18. Long, J., Shelhamer, E., Darrell, T.: Fully convolutional networks for semantic segmentation. In: Proceedings of the IEEE Conference on Computer Vision and Pattern Recognition, pp. 640–651 (2015)
19. Howard, A.G., et al.: MobileNets: Efficient convolutional neural networks for mobile vision applications. ArXiv, arXiv:1704.04861 (2017)
20. Hu, J., Shen, L., Sun, G., Wu E.H.: Squeeze-and-excitation networks. In: Proceedings of the IEEE Conference on Computer Vision and Pattern Recognition, pp. 2011–2023 (2018)
21. Lin, M., Chen, Q., Yan, S.C.: Network in network. CoRR, arXiv:1312.4400 (2014)
22. Shorten, C., Khoshgoftaar, T.M.: A survey on image data augmentation for deep learning. J. Big Data **6**(1), 60 (2019)

Radar Echo Image Prediction Algorithm Based on Multi-scale Encoding-Decoding Network

Xingang Mou(✉), Chang Liu, and Xiao Zhou

Mechanical and Electrical Engineering, Wuhan University of Technology, Wuhan, China
sunnymou@whut.edu.cn

Abstract. The traditional technology of radar echo image extrapolation for rainfall nowcasting faces such problems as insufficiently high accuracy, the incomplete analysis of the data on radar echo images, and the image blurring from the stacked LSTM (Long Short-Term Memory). In order to more accurately and clearly predict the radar echo image at a future moment, an adversarial prediction network based on multi-scale U-shaped encoder-decoder is proposed. To overcome the problem of insufficient details of the predicted image, the generator of the network adopts a U-shaped encoder-decoder structure with jump-layer connection. At the same time, in order to capture the echo movement at different scales, multi-scale convolution kernels is introduced to the encoder-decoder units. Then the conventional discriminator structure is improved and stacked ConvLSTM(Convolutional Long Short-Term Memory) layers were proposed to classify sequence. Based on the prediction of next ten frames from the given ten frames of images, this paper tests the network on the SRAD(Standardized Radar Dataset), and compares the prediction results of different networks. The test results show that the proposed model reduces image blurring, enhances the prediction accuracy while retaining sufficient prediction details .

Keywords: Encoder-decoder · Multi-scale · Prediction · Rainfall nowcasting

1 Introduction

Meteorological disasters such as typhoon, rainstorm, and hail often inflict huge damage on the development of human society. For example, short-time heavy rainfall will dramatically increase rainfall in a short period of time and cause concurrent disasters such as mountain torrents and mudslides, resulting in extremely great hazard. Therefore, more accurate and timely rainfall forecast is an important task in the field of meteorology. The weather forecast is divided into five types: long-range forecast, extended forecast, short-range forecast, short-time forecast and nowcasting [1]. This study focuses on the nowcasting, which refers to the forecast within 0–2 h. Nowcasting is mainly realized by the extrapolation of radar echo images, which means, based on a given sequence of historical radar echo image, the radar echo image at a future moment is generated through algorithmic analysis and processing.

H. Yin et al. (Eds.): IDEAL 2021, LNCS 13113, pp. 637–646, 2021.
https://doi.org/10.1007/978-3-030-91608-4_61

At present, the traditional methods of radar echo image prediction mainly include centroid method, cross-correlation method, and optical flow method [2]. The centroid method fits the centroid of a cloud cluster cell, predicts the future position of the centroid according to the movement path of the centroid, and thereby obtain the future position of the cloud cluster. This method can forecast the movement trend of the cell, but fails to predict the changes of the echo [3, 4]. The cross-correlation method divides the radar echo image into multiple rectangular area, calculates the correlation coefficient between rectangular areas of the previous and the next frames to match the echo area and then calculate the echo movement vector [5]. The optical flow method calculates the optical flow field between two consecutive frames, then uses the optical flow field for semi-Lagrangian [6] interpolation to extrapolate the echo image [7]. The problem of this method is that the used data are not sufficient, and the calculation of the optical flow field and echo extrapolation are conducted separately, which easily produces multi-step cumulative errors [8]. Besides, the optical flow method can only calculate the translational transformation of the echo, but fails to model the internal evolution.

Considering that the convolutional neural networks can extract the features of the image, and the recurrent neural networks can predict the sequence of the features, the combination of these two networks provides a new idea for the prediction of radar echo. With rainfall forecast based on radar echo as the research goal, the existing network structures were improved, and a radar echo prediction network based on multi-scale U-shaped encoder-decoder with the generative adversarial network was proposed in this paper. First, the goal of radar echo prediction was introduced in section I. The section II summarized the existing work results on radar echo prediction tasks in recent years. Section III described the overall framework of the proposed network, and the design details. Finally, the proposed network was tested for radar echo image prediction based on SRAD dataset, and compared with other exiting networks.

2 Relater Work

Srivastava et al. [9] proposed a FC-LSTM encoder-decoder network structure, by which the input of a series of video frames can achieve learning and the prediction of future video frames. Later, based on the FC-LSTM network, Xingjian et al. [10] developed the ConvLSTM network structure, in which the FC-LSTM fully-connected calculation is replaced by convolution calculation in order to fully learn the changes in the spatial and temporal characteristics of the radar echo image sequence. The conversion between input and state and between states is completed by convolution. The radar echo images are predicted by the stacked ConvLSTM encoder-decoder structure, and the prediction results are better than the FC-LSTM and traditional methods. However, the predicted images generated only by the ConvLSTM network are blurry. The longer the prediction time, the worse the effect. Thus, Singh et al. [11] combined the ConvLSTM with the generative adversarial network, using ConvLSTM as the prediction generator, and convolutional neural network as the discriminator. A predicted frame, the corresponding real value, and the first three frames from the historical sequence and were input into the discriminator simultaneously, So, the extrapolated image generated in the process of adversarial training between the generator and the discriminator had a better effect, and the final effect was superior to a pure ConvLSTM structure network. Based upon Unet and SegNet networks, Georgy et al. 8 presented the RainNet network, which has 20 convolutional layers, 4 maximum pooling layers, 4 upsampling layers, 2 dropout layers, and 4 residual connection layers. The network generates a predicted image by inputting multiple frames of radar images. It uses the method of recursive call to input the combination of the input and output into the network to predict the image for a longer time.

3 Improved Radar Echo Prediction Network

The radar echo image sequence refers to a sequence of continuous gray-scale images. The goal of this research is to predict 10 frames of radar echo images in the future through the given 10 frames of historical radar echo images. That is, the sequence P_1, P_2, $P_3 \ldots P_n$ can predict the subsequent frame sequence P_{n+1}, P_{n+2}, $P_{n+3} \ldots P_{n+m}$ by the established model, where n is the length of the original sequence and m is the length of the predicted sequence.

Since the rainfall system is a chaotic system, there are more uncertainties in the prediction for radar echo sequence than ordinary video sequence [13]. For example, the sudden emergence and disappearance of echoes in the future, and different motion speeds of different parts of echoes are difficult to predict. The problem of image blurring exists in the radar echo prediction by various network models that have been proposed [14]. This is because as the prediction time increases, the uncertainty will accumulate at each time step, making the extrapolated image increasingly blurring. Thus, the extrapolated image in a longer time also loses its reference value [15]. To solve the above problem in radar echo prediction, a Unet structured encoder-decoder echo prediction network with a multi-scale convolution kernel based on the generative adversarial network (UGAN-MSCLSTM) was proposed in this paper. The network captures echo motions at different scales by learning the evolution law of radar echo image sequence in space and time series, and achieves clearer and more accurate prediction of radar echo images with the discriminator and multi-weight loss function.

The biggest advantage of the generative adversarial network [16] is that it can generate more real images during the mutual adversarial training of the two networks. This advantage makes it extensively applied in various fields, and adopted by the network in this paper. The UGAN-MSCLSTM network proposed in this paper includes a generator and a discriminator. The generator is designed to predict the next ten frames through the input of ten frames of echo images. The discriminator is made to recognize the real image sequence as true and the sequence generated by the generator as false. Finally, the quality of the generator's predicted image is gradually improved in the process of adversarial training.

3.1 U-Shaped Encoding-Decoding Cenerator with Multi-Scale Convolution Kernels

With the existing methods of prediction of radar echo images, it was often found that the predicted image is not clear and the details are insufficient. The Unet [17–19] structure has a bilaterally symmetrical encoder-decoder structure, and each coding layer combines the features with the corresponding decoding layer through jump layer connection, which effectively uses the shallow and deep features in the image. Therefore, in this paper, the structure of the conventional generator was improved to a Unet structure to solve the above problem. The overall framework of the generator network is displayed in Fig. 1. It contains three encoders, three down-sampling layers, three decoders, three up-sampling layers, and the 3D convolutional layer. In order to allow the time sequence space memory to be continuously transferred in the encoders and decoders, each encoding layer copies the state of ConvLSTM to the initial state of the decoder at the corresponding layer. The image will change the receptive field of the convolution kernel and extract different levels of features during the down-sampling and up-sampling of the image. But it still loses a lot of detailed information of the image. Thus, the concat jump layer connection is used in to combine the shallow features with the deep features in the image sequence to supplement the detailed information of image, so that the network can finally predict the image sequence with more details.

In the image sequence prediction task, the common encoding and decoding units are combined by a single recurrent neural network layer with up-sampling and down-sampling. The encoding and decoding units in the network were improved in this paper. The larger the convolution kernel in the network layer, and the receptive field, the faster movement can be captured. The small convolution kernel can capture relatively small movement. The radar echo is often unstable and the movement speed of each part is

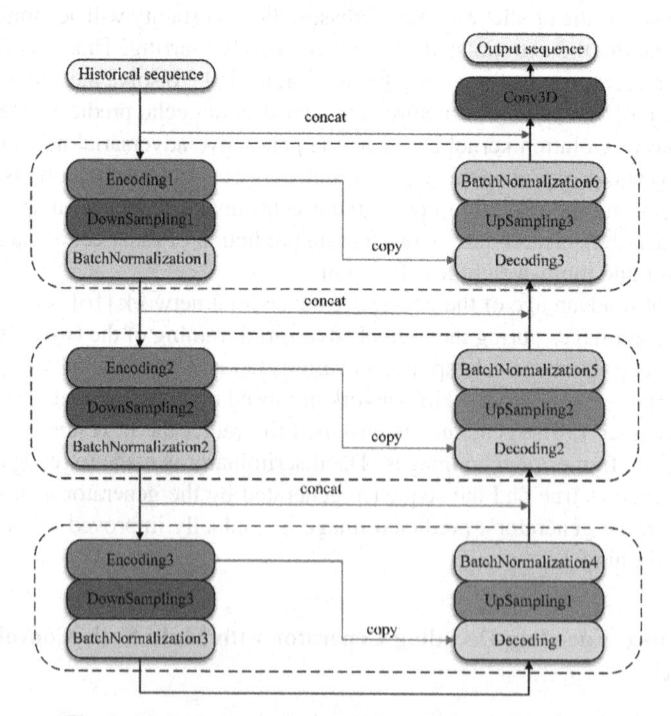

Fig. 1. U-shaped encoding-decoding generator framework

also different. In order to simultaneously consider the movement of the radar echo at various scales between two adjacent frames, the ConvLSTM structure with multi-scale convolution kernels was proposed as encoder and decoder. Based upon multiple convolution kernels, the network can capture the movement of radar echo at various scales as much as possible. In the encoding and the decoding units, the inputs are passed to the ConvLSTM layer with the convolution kernel size of 1, 3, and 5 respectively. Then the extracted sequence features are connected to perform uniform batch normalization and then generate output. The down-sampling part of the encoding unit sets the step size of ConvLSTM as 2 to replace the pooling layer. The up-sampling part of the decoding unit uses a 3D deconvolution layer. This paper tried Sigmoid, ReLU and tanh activation functions respectively to compare the realization effect and finally used tanh as the final activation function of each layer.

3.2 Image Sequence Discriminator

In this paper, the conventional discriminator structure was modified. The convolutional network in the discriminator was changed to a ConvLSTM network, with a purpose to extract features in time series and space and distinguish the predicted sequence from the real sequence of the next ten frames. The discriminator classifies the sequence predicted by the generator as 0 and the real sequence of the next ten frames as 1. The structure is shown in Fig. 2. The discriminator contains three ConvLSTM layers, a global average pooling layer and a fully-connected layer. The purpose of using the stacked ConvLSTM layer is to extract the spatial-temporal features of the image sequence, and to classify the image sequence at the level of time series and space. The global average pooling was adopted not only to reduce the dimensionality, but also to effectively decrease the number of training parameters and occupy fewer resources. The features after dimensionality

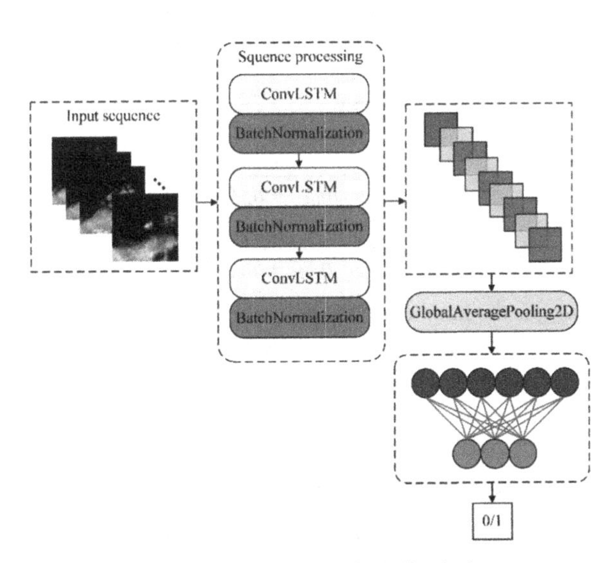

Fig. 2. Internal structure of the discriminator.

reduction were finally classified through the fully connected layer to restrict the training of the generator.

4 Experiments

The data set used in this paper comes from the radar echo image data set (SRAD) of the Guangdong-Hong Kong-Macao Greater Bay Area provided by the Meteorological Bureau of Shenzhen Municipality. The dataset has undergone quality processing on the original data, so that the range of the echo value is 0–80 dBz(z stands for radar reflectivity, dBz is defined as 10 times the logarithm of the reflectivity). SRAD contains a total of 20,000 samples, each of which has 4 h of coverage, and the time interval between frames is 6 min. The ratio of training set to test set is 10:1 in this data. The horizontal resolution of all images in the same sequence is 0.01° (about 1 km). The length and width of the image are both 256 (that is, approximately covering an area of 255 km × 255 km).

4.1 Evaluation Indicators

Since this study focuses on the rainfall prediction based on radar echo, the evaluation method for rainfall nowcasting was adopted in this test. The evaluation includes three indicators: CSI (critical success index), FAR (false alarm rate), and POD (probability of detection) [20], to assess the accuracy of the prediction. First, a rainfall threshold (30 dBz in this paper) was set. The grid points of predicted image and the real image within the threshold were treated as rain points (marked as 1), and the grid points higher than the threshold as no-rain points (marked as 0). Then the predicted images and the real images were traversed on pixel level. For each point, if the predicted value is 1, and the real value is 1, it is denoted as a hit. If the predicted value is 1 and the real value is 0, it is denoted as an alarm. If the predicted value is 0 and the real value is 1, it is denoted as a miss. The following equations calculate the three indicators:

$$CSI = \frac{n_{hits}}{n_{hits} + n_{alarms} + n_{misses}} \tag{1}$$

$$FAR = \frac{n_{alarms}}{n_{hits} + n_{alarms}} \tag{2}$$

$$POD = \frac{n_{hits}}{n_{hits} + n_{misses}} \tag{3}$$

In (1) (2) (3), n_{hits} is the number of hits points, n_{alarms} is the number of alarms points, and n_{misses} is the number of misses points. The larger the value of *CSI* and *POD*, the more accurate the prediction. The smaller the value of *FAR*, the lower the false alarm rate.

4.2 Analysis of the Experiment Results

In order to evaluate the accuracy of the proposed network model in radar echo image prediction, a collection of 550 testing samples was selected and comparative analysis was performed on the stacked ConvLSTM (CLSTM-ST), encoding-decoding

ConvLSTM(CLSTM-ED), PredRNN and UGAN-MSCLSTM proposed in this paper. In the test, the echo threshold was set to 30 dBz. Based on this threshold, the ten-frame image sequence predicted by the four networks and the actual ten-frame image sequence were compared and analyzed in terms of the four indicators- CSI, POD, FAR and MSE. Table 1 shows the mean values of the four indicators calculated by the four models on all 550 samples from testing collection and Table 2 shows the average time required for the four models to obtain output on the test set.

Table 1. Average index of different network models on all test set samples

Model	CSI	FAR	POD	MSE
CLSTM-ST	0.5262	0.1815	0.7108	72.2433
CLSTM-ED	0.5308	0.3035	0.7597	63.5084
PredRNN	0.6085	0.1948	0.7115	69.1082
UGAN-MSCLSTM	0.6513	0.2212	0.7663	58.3020

Table 2. Average index of different network models on all test set samples

Model	Time(s)
CLSTM-ST	0.1047
CLSTM-ED	0.1250
PredRNN	0.1821
UGAN-MSCLSTM	0.1576

It can be seen from Table 1 that the UGAN-MSCLSTM network is superior to the other three models in terms of the mean values of CSI, POD and MSE, and better than CLSTM-ED in terms of FAR. This shows that the proposed model has a higher prediction accuracy than the other three models. And Table 2 shows that the time consumed by UGAN-MSCLSTM for one calculation is 0.1576 s, which is slightly longer than CLSTM-ST and CLSTM-ED and better than PredRNN in time consumption.

Table 3. Average index of different network models on all test set samples

Model	CSI	FAR	POD	MSE
CLSTM-ST	0.7336	0.1285	0.8210	36.3102
CLSTM-ED	0.7130	0.1827	0.8494	52.5729
PredRNN	0.7719	0.0539	0.8069	27.8047
UGAN-MSCLSTM	0.7877	0.1731	0.9033	24.7057

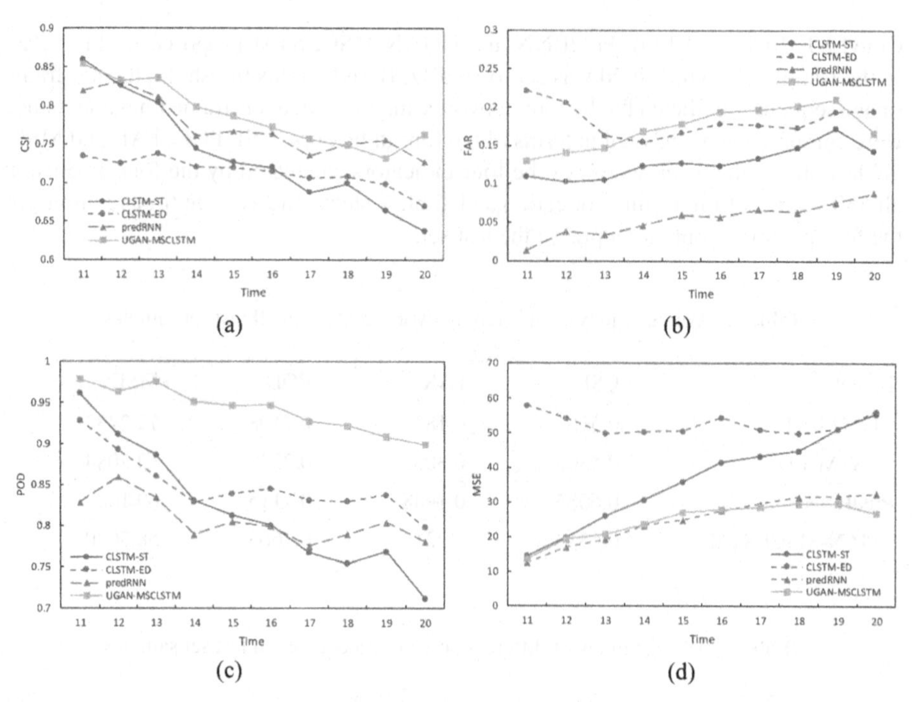

Fig. 3. Evaluation index graph. (a) CSI value change curve. (b) FAR value change curve. (c) POD value change curve. (d) MSE value change curve.

In order to conduct a more detailed analysis on individual cases, the radar echo image sequence of one sample from the testing collection was analyzed and displayed. Table 3 shows the mean values of the four indicators to predict subsequent images calculated by the four models.

Based on the above Table 3, the variation curves of the predicted 10 frames of image in terms of the values of CSI, FAR, POD and MSE were drawn, as shown in Fig. 3. It can be seen that the CSI and POD values of the 11th to the 20th frames predicted by UGAN-MSCLSTM network are higher than those predicted by other three models, and the FAR value of the 14th to the 19th frame is better than that of the ConvLSTM encoding-decoding model. The MSE value is stronger than that of CLSTM-ST and the CLSTM-ED network and is also stronger than PredRNN in the later stage of prediction, indicating that the network has strong rainfall prediction capability.

In order to show the effect of image prediction more clearly, the first, fifth, and tenth frames were extracted from the prediction results. It can be seen from Fig. 4 that the (b) (d) (e) algorithms obtain a lot of details in the prediction for the first frame. This shows that the prediction effect is not much different in the short-term. Figure 5 shows that when the fifth frame is predicted, (b), (c), (d), and (e) show blurring to varying degrees. The UGAN-MSCLSTM network has the highest definition, and its prediction for the movement of small echoes is also close to the real value. It can be found from Fig. 6 that when the 10th frame is predicted, (b), (c), and (d) present more serious blurring, but the

UGAN-MSCLSTM network still obtains a lot of details, and can precisely predict the movement trends of large and small echoes.

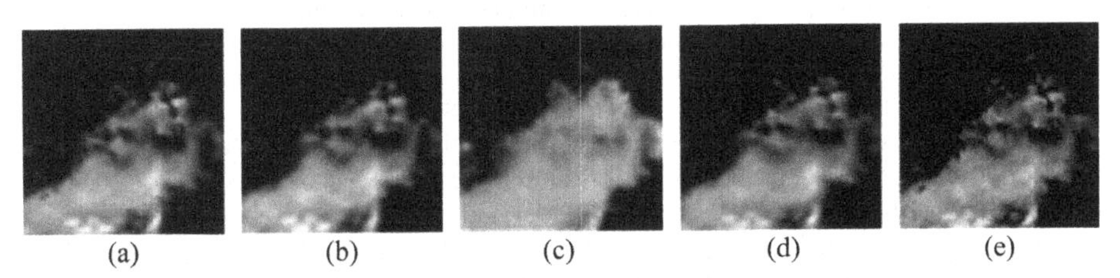

Fig. 4. Comparison of the first frame of prediction results. (a) Ground truth. (b) CLSTM-ST. (c) CLSTM-ED. (d) PredRNN. (e) UGAN-MSCLSTM.

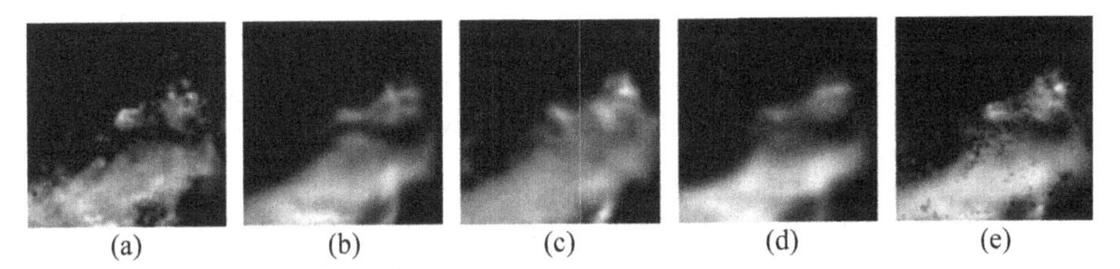

Fig. 5. Comparison of the fifth frame of prediction results. (a) Ground truth. (b) CLSTM-ST. (c) CLSTM-ED. (d) PredRNN. (e) UGAN-MSCLSTM.

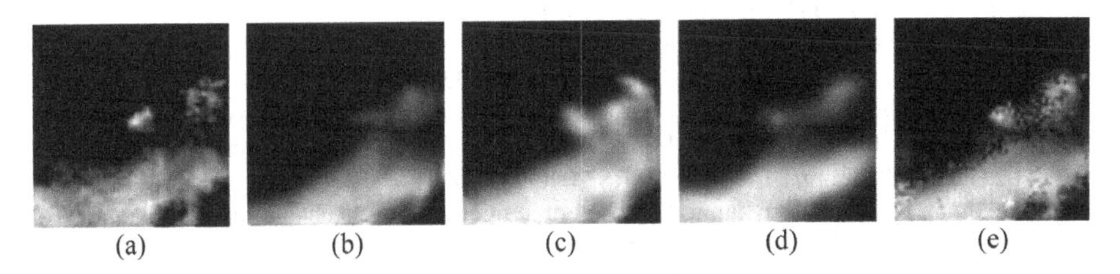

Fig. 6. Comparison of the tenth frame of prediction results. (a) Ground truth. (b) CLSTM-ST. (c) CLSTM-ED. (d) PredRNN. (e) UGAN-MSCLSTM.

5 Conclusion

Based on the generative adversarial network in deep learning and the Unet network model, a UGAN-MSCLSTM network model for rainfall prediction through radar echo was proposed in this paper. This algorithm uses the SRAD dataset as the training set and the test set, with an aim to accurately predict the future 10 frames by the given 10 frames. Experimental comparison and analysis revealed that the proposed network achieves higher accuracy of rainfall prediction than other network models, and obtains better results in the prediction of small echoes. In addition, it also has greatly solved the problem of image blurring. Compared with other models, the proposed model has made a huge improvement in the three indicators and human vision.

References

1. Kaoji, X.: Doppler weather radar echo images clutter suppression and storm clouds segmentation and tracking. M.S. Thesis, Tianjin University, Tianjin, China (2012)
2. Zhang, L., Huang, Z., Liu, W., Guo, Z., Zhang, Z.: Weather radar echo prediction method based on convolution neural network and long short-term memory networks for sustainable e-agriculture. J. Clean. Prod. **298**, 126776 (2021). ISSN 0959-6526
3. Dixon, M., Wiener, G.: TITAN: thunderstorm identification, tracking, analysis, and nowcasting—a radar-based methodology. J. Atmosph. Oceanic Technol. https://doi.org/10.1175/1520-0426(1993)0102.0.CO
4. Jacobs, I.S., Bean, C.P.: Fine particles, thin films and exchange anisotropy. In: Rado, G.T., Suhl, H. (eds.) Magnetism, vol. III, pp. 271–350. Academic, New York (1963)
5. Yuanyi, X.: Research on tracking and extrapolation of rain clouds based on Doppler radar images. M.S. Thesis, Wuhan University of Technology, Wuhan, China (2012)
6. Srivastava, N., Mansimov, E., Salakhutdinov, R.: Unsupervised learning of video representations using LSTMs. In: Proceedings of International Conference on Machine Learning, pp. 843–852 (2015)
7. Xingjian, S., Chen, Z., Wang, H.: Convolutional LSTM network: a machine learning approach for precipitation nowcasting. In: Proceedings of Advances Neural Information Processing System, pp. 802–810 (2015)
8. Singh, S., Sarkar, S., Mitra, P.: A deep learning based approach with adversarial regularization for Doppler weather radar ECHO prediction. In: Proceedings of the 2017 IEEE Geoscience and Remote Sensing Symposium (IGARSS), pp. 5205–5208 (2017)
9. Georgy, A., Tobias, S., Maik, H.: RainNet v1.0: a convolutional neural network for radar-based precipitation nowcasting. Geoentific Model Dev. https://doi.org/10.5194/gmd-13-2631-2020
10. Staniforth, A., Ct, J.: Semi-lagrangian integration schemes for atmospheric models a review. Mon. Weather Rev. **119**(9), 2206–2223 (1991)
11. Woo, W.C., Wong, W.K.: Operational application of optical flow techniques to radar-based rainfall nowcasting. Atmosphere. https://doi.org/10.3390/atmos8030048
12. Zhang, C., Zhou, X., Zhuge, X., Xu, M.: Learnable optical flow network for radar echo extrapolation. IEEE J. Selected Topics Appl. Earth Observ. Remote Sens. **14**, 1260–1266 (2021). https://doi.org/10.1109/JSTARS.2020.3031244
13. Sivakumar, B.: Rainfall dynamics at different temporal scales: a chaotic perspective. Hydrol. Earth Syst. Sci. **5**(4), 645–652 (2001)
14. Li, Y., Wang, Z., Dai, G., Wu, S., Yu, S., Xie, Y.: Evaluation of realistic blurring image quality by using a shallow convolutional neural network. IEEE Int. Conf. Inf. Autom. (ICIA) **2017**, 853–857 (2017). https://doi.org/10.1109/ICInfA.2017.8079022
15. Bingcong, L.: Research on doppler radar map estimation and forecasting based on variational autoencoder. M.S. Thesis, Guangdong University, Guangdong, China (2019)
16. Arjovsky, M., Chintala, S., Bottou, L.: Wasserstein generative adversarial networks. In: Proceedings of ICML, Sydney, NSW, Australia, pp. 214–223 (2017)
17. Ronneberger, O., Fischer, P., Brox, T.: U-Net: convolutional networks for biomedical image segmentation. In: Navab, N., Hornegger, J., Wells, W.M., Frangi, Alejandro F. (eds.) MICCAI 2015. LNCS, vol. 9351, pp. 234–241. Springer, Cham (2015). https://doi.org/10.1007/978-3-319-24574-4_28
18. Piao, S., Liu, J.: Accuracy improvement of UNet based on dilated convolution. J. Phys.: Conf. Ser. **1345**, 5. IOP Publishing (2019)
19. Sun, S., Mu, L., Wang, L., Liu, P.: L-UNet: an LSTM network for remote sensing image change detection. In: IEEE Geoscience and Remote Sensing Letters, pp. 1–5 (Early Access). https://doi.org/10.1109/LGRS.2020.3041530
20. Hualian, F.U., et al.: Cloud detection method of FY-2G satellite images based on random forest. Bull. Surveying and Mapp. **3**, 61–66 (2019). https://doi.org/10.13474/j.cnki.11-2246.2019.0079

Author Index